Photoelectron Spectroscopy

Advanced Texts in Physics

This program of advanced texts covers a broad spectrum of topics which are of current and emerging interest in physics. Each book provides a comprehensive and yet accessible introduction to a field at the forefront of modern research. As such, these texts are intended for senior undergraduate and graduate students at the MS and PhD level; however, research scientists seeking an introduction to particular areas of physics will also benefit from the titles in this collection.

Springer-Verlag Berlin Heidelberg GmbH

Stefan Hüfner

Photoelectron Spectroscopy

Principles and Applications

Third Revised and Enlarged Edition
With 461 Figures and 28 Tables

Springer

Professor Dr. Stefan Hüfner
Universität des Saarlandes
Fakultät für Physik und Elektrotechnik
Fachrichtung 7.2 - Experimentalphysik
Postfach 151150
66041 Saarbrücken
Germany

ISSN 1439-2674

ISBN 978-3-642-07520-9

Library of Congress Cataloging-in-Publication Data.
Hüfner, Stefan, 1935- Photoelectron spectroscopy : principles and applications / Stefan Hüfner.
– 3rd rev. and enlarged ed. p. cm. –(Advanced texts in physics) Includes bibliographical references.
ISBN 978-3-642-07520-9 ISBN 978-3-662-09280-4 (eBook)
DOI 10.1007/978-3-662-09280-4
1. Photoelectron spectroscopy. I. Title. II. Series. QC454.P48 H84 2003
543'.0858–dc21 2002191151

http://www.springer.de

Originally published by Springer-Verlag Berlin Heidelberg New York in 2003
Softcover reprint of the hardcover 3rd edition 2003

Typesetting: Data prepared by the author using a Springer TEX macro package
Cover design: *design & production* GmbH, Heidelberg

Printed on acid-free paper SPIN 11016083 57/3111/di 5 4 3 2 1

Preface

Since the completion of the manuscript for the first edition of Photoelectron Spectroscopy, the field has undergone a steady growth.

Firstly, the theory has been refined and condensed into a manageable form. Secondly two important experimental developments have occurred. The resolution that can be obtained is now of the order of 3 meV, which corresponds approximately to an energy of 30 k_BK. This means that photoelectron spectroscopy can now obtain data with an accuracy similar to that achieved in standard thermodynamic experiments (such as specific heat experiments), thus facilitating a direct comparison of data from the two different types of experiment. The second important experimental advance is that one can now readily measure electron energy distributions over a solid angle of almost 2π. This yields valuable information whenever these electron energy distributions have anisotropies.

It was decided, in view of these developments, to rework and expand the volume so as to do justice to the full potential of today's photoelectron spectroscopy. I have benefitted very much from the help of my group namely R. de Masi, D. Ehm, B. Eltner, F. Müller, G. Nicolay, F. Reinert, D. Reinicke and in particular S. Schmidt. Without the dedicated effort of these collaborators the present edition could not have been produced. I am grateful to S. Neumann who typed the complete text with great skill. Thanks are due to the Springer Verlag for their expert help and patience.

Saarbrücken, *Stefan Hüfner*
February 2003

Preface to the Second Edition

This new edition has provided me with the possibility to correct, with the assistance of R. Zimmermann (Saarbrücken), some errors that appeared in the previous edition and to prepare a somewhat more detailed subject index which was done with the help of Th. Engel (Saarbrücken). In addition, some references have been added from publications that appeared in 1994/1995 to give the reader the chance to find in some areas the most recent literature.

A word on the nomenclature should be added. The field treated with in this monograph is called photoelectron spectroscopy (if one wants to name it by the particle that is being detected) or photoemission spectroscopy (if one wants to name it by the primary process that takes place). Both names are and have been used in the literature on an equal footing, and in this book this practice has been adopted.

In preparing this second edition I have enjoyed the expert and friendly help of Dr. H. Lotsch from Springer Verlag.

Saarbrücken, *Stefan Hüfner*
November 1995

Preface to the First Edition

Molecules and solids can be characterized by two main types of qualities, namely their vibrational (elastic) properties and their electronic properties, which are of course intimately connected with each other. The study of vibrations in molecules and solids is mostly performed by means of optical spectroscopy. This spectroscopy can also determine the electronic excitations of molecules and solids. In solids, compared to molecules, the phonon and electron excitations depend on an additional quantum number, which originates from the periodicity of the crystal solid, namely the wave vector $\boldsymbol{k}$. In order to perform wave-vector-dependent measurements one has to work with exciting particles which can transmit or absorb wave vectors of the same magnitude as those present in a solid. Therefore the optical technique is no longer sufficient to scan the phonon or electron distributions over the whole Brillouin zone (except with the difficult technique of two-photon spectroscopy).

With respect to the elastic properties of solids, the neutron diffraction technique has provided much information on the phonon dispersion curves of a great number of systems. Today we have a fair understanding of these phonon dispersion curves. With respect to the electron dispersion curves the situation was different up to about 1980, when the first electron dispersion curves were measured by photoemission spectroscopy. In the meantime photoemission spectroscopy has been developed further and is now the method of choice to study the electron dispersion curves of solids. Of course such dispersion curves can be, and also have been, measured for electronic surface states.

This volume deals with some, although by no means all, aspects of photoemission spectroscopy. This technique has been developed in the last 25 years and, with the extensive use of synchrotron radiation, can now be employed for such diverse fields as the investigation of the chemical properties of specially treated surfaces of semiconductors or high polymers, for the study of the electronic structure of molecules absorbed on surfaces, and for the measurement of dispersion curves of bulk and surface electronic states. We have tried to write this volume at an elementary level such that the newcomer to the field can find some basic information that will then allow him to study recent reviews and the original literature.

After an introductory chapter, core levels, which are mostly used for chemical investigations, are treated in Chap 2. In Chaps. 3 and 4 the different final states that can arise in the photoemission process and the relation to the initial ground state are discussed. Chapters 5, 6 and 7 deal with valence bands in molecules and in particular solids, where we try to present in some detail the methods by which electron dispersion relations can be obtained by this kind of spectroscopy. Finally, in the last four chapters we discuss specific fields of photoemission spectroscopy, namely the study of surface effects and then three particular modes of this technique, namely inverse photoemission spectroscopy, spin polarized photoemission spectroscopy and photoelectron diffraction.

Saarbrücken,
February 1994

Stefan Hüfner

Acknowledgements

This book has profited tremendously from long standing cooperation and many discussions with colleagues in the field. My early interest in photoemission spectroscopy was stimulated by G.K. Wertheim of AT&T Laboratories more than twenty years ago. I enjoyed the collaboration with him over many years. Later, in Saarbrücken, R. Courths, A. Goldmann, H. Höchst, F. Reinert, and, in particular, P. Steiner, have worked with me and much of the material presented in this book derives from that collaboration.

A very successful collaboration that I have enjoyed during the last years was that with the group of L. Schlapbach (Zürich, now Fribourg), and with many of his colleagues, notably J. Osterwalder, T. Greber and A. Stuck. From that collaboration I have learned everything that I know about photoelectron diffraction.

Over the years I have profited from discussions with many people in the field, notably Y. Baer (Neuchâtel), A. Bradshaw (Berlin), P. Echenique (San Sebastian), P. Fulde (Stuttgart), A. Fujimori and A. Kotani (Tokyo), O. Gunnarsson (Stuttgart), K. Schönhammer (Göttingen), G. Kaindl (Berlin), F. Meier (Zürich), W.D. Schneider (Lausanne), G.A. Sawatzky (Groningen) and H.C. Siegmann (Zürich).

A. Goldmann (Kassel), M. Cardona (Stuttgart) and R. Claessen (Saarbrücken) have made various extremely useful comments and suggestions about the manuscript, for which I thank them.

K. Fauth, A. Jungmann, M. Weirich and, in particular, R. Zimmermann (all from Saarbrücken) have helped me in the proof reading procedure, which I acknowledge gratefully.

The manuscript originated from a series of lectures given at the ETH Zürich in 1983 and repeated at the University of Fribourg and Lausanne in 1989. The first draft of the manuscript was written during the tenure of an Akademie Stipendium granted by the Volkswagen Foundation during the academic year 1986/87, which I spent at the Cavendish Laboratory in Cambridge. I thank the Volkswagen Foundation for the financial support and the mentioned institutions for their hospitality.

I thank the Deutsche Forschungsgemeinschaft, the Bundesministerium für Forschung und Technologie and the Volkswagen Stiftung for the financial support that kept my laboratory running and made the work reported in this volume possible.

I have to thank my secretary H. Waack for typing with great skill the various (many) drafts of the manuscript and for helping me in reading the proofs.

This book would not have been completed without the expert and friendly cooperation and help from Dr. A. Lahee and in particular Dr. H. Lotsch from Springer Verlag. I appreciate in particular that H. Lotsch worked intensively on the manuscript even under the most difficult personal conditions.

Contents

1. Introduction and Basic Principles

1.1 Historical Development

PhotoElectron Spectroscopy (PES) is an old experimental technique still in extensive use. Frequently, this method is also called PhotoEmission Spectroscopy (PES) if instead of the emitted particle, the process that leads to the emission is utilized for its characterization. The phenomenon of photoemission was detected by Hertz [1.1] in 1887 with the experimental arrangement shown in Fig. 1.1. It consists of two arcs *d* and *f*, which are driven by the induction coils *a* and *c*, respectively. A battery *b* provided the necessary power

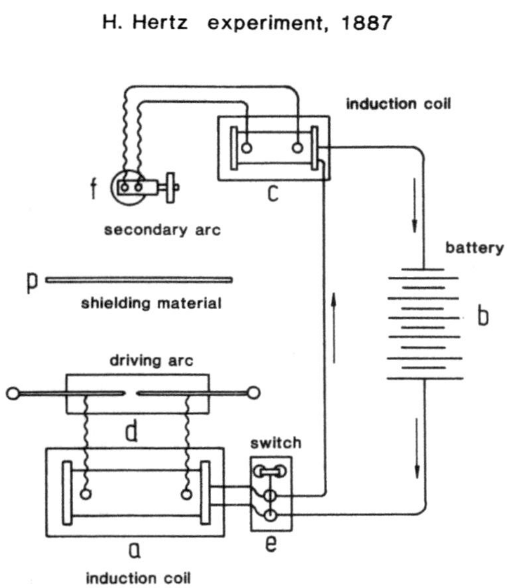

Fig. 1.1. Experimental arrangement used by Hertz [1.1] for his "discovery" of the photoelectric effect. An arc *d* emits light that can stimulate a second arc *f*; *a* and *c* are coils that produce the voltages for the arcs. A switch *e* is used to activate the coils from the battery *b*, *p* indicates shielding material that was placed between arcs *d* and *f*

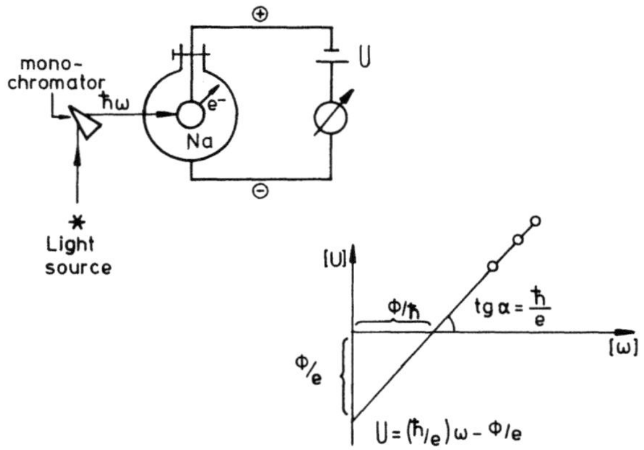

Fig. 1.2. Schematic drawing of an early "photoemission" experiment. Light is sent through a prism monochromator and impinges on a film of an alkali metal (Na, K). The energy of the photoemitted electrons is measured by applying a retarding voltage. A plot of the retarding voltage U needed to make the current disappear as a function of the frequency ω of the light is a straight line

and the coils could be activated by the switch *e*. Hertz studied the behavior of arc *f* under the radiation produced in arc *d* if different materials *p* were placed between the two arcs *d* and *f*. By varying the materials in a systematic way he found out that the ultraviolet radiation emitted by arc *d*, triggered arc *f*. We now know that the electrons photoemitted from the electrodes of arc *f* by the ultraviolet radiation of arc *d* were actually responsible for triggering arc *f*. In the following years these experiments were refined [1.2, 1.3] and Einstein in 1905 was able to explain their systematics by invoking the quantum nature of light [1.4].

The experiments at that time were performed, as depicted in Fig. 1.2. Light from a continuous source was monochromatized by a prism spectrograph. This light was then focused onto a surface of potassium or sodium in a vacuum tube. The kinetic energy of the liberated electrons was determined by measuring the voltage required to suppress the current across the vacuum vessel. This means, in essence, that the potential energy supplied by the externally applied voltage equals the maximum kinetic energy of the photoemitted electrons.

By performing such experiments as a function of the frequency of the light ($\omega/2\pi$), one was able to determine the ratio of Planck's constant to the electronic charge and also the *work function* ϕ of the metal under study.[1] The

[1] The experiment, as illustrated by Fig. 1.2, actually measures the difference of the work functions, $\Delta\phi$, of the anode and the cathode. The work function of a particular material is obtained by measuring this difference for various combinations of materials.

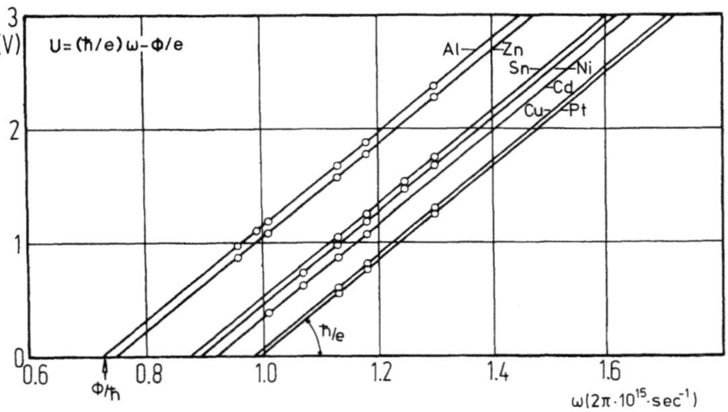

Fig. 1.3. Retarding voltage U as a function of light frequency ω in the experiment as sketched in Fig. 1.2 for a number of metals. The intercept with the abscissa gives the work function. All the lines are parallel because their slope is given by $\hbar/e$ [1.5]

schematic plot of the experimental results is given as an insert in Fig. 1.2. The basic equation governing the experiment is [1.4]:

$$eU = E_{\text{kin, max}} = \hbar\omega - \phi\ , \tag{1.1}$$

where U is the retarding voltage, and $E_{\text{kin, max}}$ the maximum electron kinetic energy.

Systematic data on $U(\omega)$ for some metals are given in Fig. 1.3 [1.5]. As expected, all lines have the same slope which, according to (1.1), is equal to $\hbar/e$. The data for different metals are, however, displaced with respect to each other due to their different work functions.

In principle, PhotoEmission (PE) is performed today in the same way as it was almost a hundred years ago. Figure 1.4 exhibits the ingredients of a modern PE experiment. The light source is either a gas-discharge lamp, an X-ray tube, or a synchrotron-radiation source. The light (vector potential

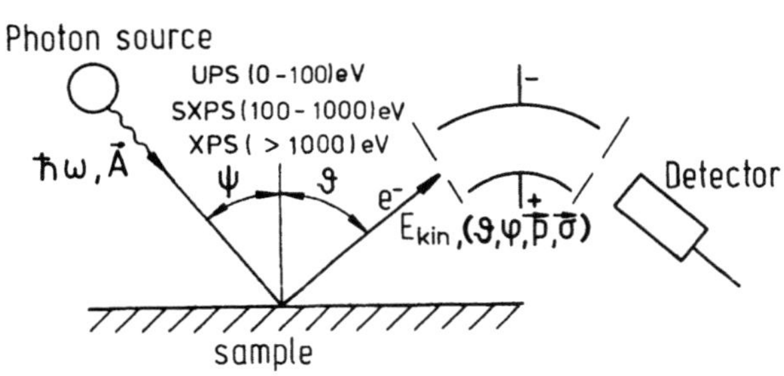

Fig. 1.4. Sketch of a modern PES experiment. The light source is a UV discharge lamp, an X-ray tube or a storage ring. The electrons are detected (in most cases) by an electrostatic analyzer

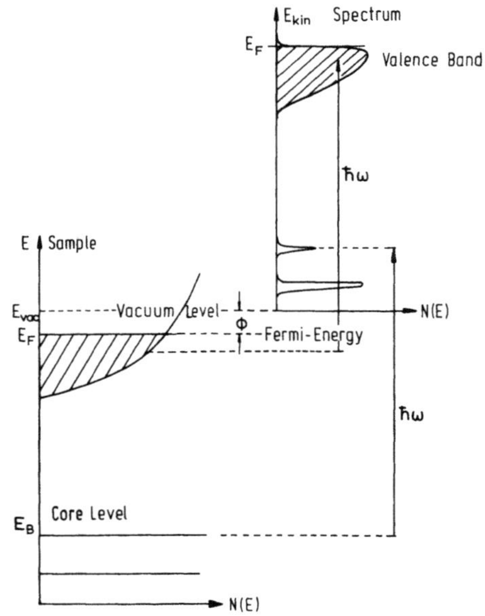

Fig. 1.5. Relation between the energy levels in a solid and the electron energy distribution produced by photons of energy $\hbar\omega$. The natural abscissa for the photoelectrons is the kinetic energy with its zero at the vacuum level of the sample ($E_{\mathrm{kin}} = \hbar\omega - |E_{\mathrm{B}}| - \phi$). Experimenters generally prefer to use E_{B} as the abscissa. E_{B} is the so-called binding energy of the electrons, which in solids is generally referred to the Fermi level and in free atoms or molecules to the vacuum level

$\boldsymbol{A}$) impinges on the sample, which is a gas or the surface of a solid, and the electrons excited by the photoelectric effect are then analyzed with respect to their kinetic energy E_{kin} and their momentum $\boldsymbol{p}$ (wave vector $\boldsymbol{p}/\hbar$) in an electrostatic analyzer. The polarization of the light is a useful property in an angle-resolved PE experiment. The important parameters to be measured are then the kinetic energy E_{kin} of the photoemitted electron, and its angle with respect to the impinging light ($\psi + \vartheta$) and the surface (ϑ, φ). Knowing the energy of the light and the work function, one can determine the binding energy [2] E_{B} of the electrons in the sample from the following equation:

$$E_{\mathrm{kin}} = \hbar\omega - \phi - |E_{\mathrm{B}}| \; . \tag{1.2}$$

[2] The binding energy E_{B} of an energy level measured by PES in solids is generally referred to the Fermi level, and in free atoms or molecules to the vacuum level. Most researchers give it as a positive number, a practice that we shall also adopt for this text. This produces the difficulty that energies measured by IPES (Inverse PhotoEmission Spectroscopy), which, e.g., in a solid are situated above the Fermi energy, come out as negative numbers, which is somewhat counterintuitive. Therefore, a second way of giving the energies is in use, too. Here the term initial-state energy E_{i} is employed. It has the same reference point as the binding energy, but the opposite sign. In this text the binding energy E_{B} will generally be used, except in cases where it is completely contrary to the accepted usage in the literature, as for example in IPES. Note that (1.1)does not contain the term E_{B} in contrast to (1.2). The former equation describes the photocurrent for the electrons with the maximum kinetic energy which are the electrons coming from the Fermi energy, meaning $E_{\mathrm{B}} = 0$ in the definitions chosen here.

The momentum p of the outgoing electron is determined from its kinetic energy by

$$\begin{aligned} E_{\mathrm{kin}} &= \frac{p^2}{2m} , \\ p &= \sqrt{2mE_{\mathrm{kin}}} . \end{aligned} \tag{1.3}$$

The direction of $\boldsymbol{p}/\hbar$ is obtained from ϑ and φ which are the polar and azimuth angles under which the electrons leave the surface.

Figure 1.5 shows schematically how the energy-level diagram and the energy distribution of photoemitted electrons relate to each other. The solid sample has core levels and a valence band. In the present case of a metal, the Fermi energy E_{F} is at the top of the valence band and has a separation ϕ from the vacuum level E_{vac}. If photoabsorption takes place in a core level with binding energy E_{B} ($E_{\mathrm{B}} = 0$ at E_{F}) the photoelectrons can be detected with kinetic energy $E_{\mathrm{kin}} = \hbar\omega - \phi - |E_{\mathrm{B}}|$ in the vacuum. (Thus, while in a solid the "physical" reference is E_{F}, the kinetic energy in the vacuum is, of course, measured with respect to E_{vac}). If the energy distribution of the emitted electrons is plotted as in Fig. 1.5, their number per energy interval often gives a replica of the electron-energy distribution in the solid. This is an attractive feature of PES; it is able to provide us information on the electron energy distribution in a material. In actual data accumulation, the reference point for molecules and atoms is taken to be E_{vac}, while in solids E_{F} is taken as the "natural" zero. However, in general, one plots not the kinetic energy but E_{B}, according to (1.2).

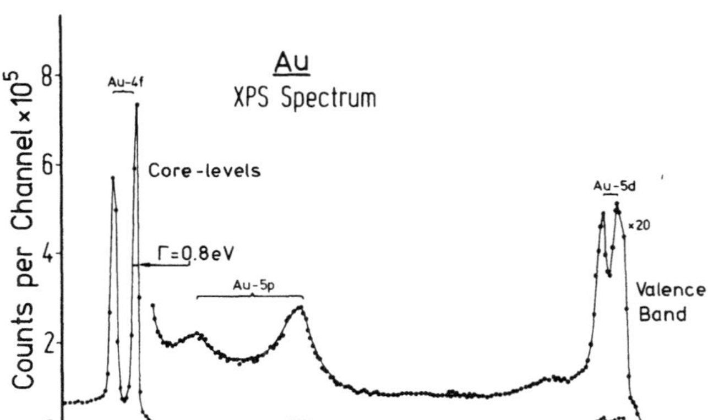

Fig. 1.6. XPS spectrum ($\hbar\omega = 1487\,\mathrm{eV}$) of polycrystalline Au for $0 \leq E_{\mathrm{B}} \leq 100$ eV. The 5d valence band and the $4\mathrm{f}_{7/2,5/2}$ doublet (often used for calibration: $E_{\mathrm{B}}(4\mathrm{f}_{7/2}) = 84.0\,\mathrm{eV}$) are clearly seen

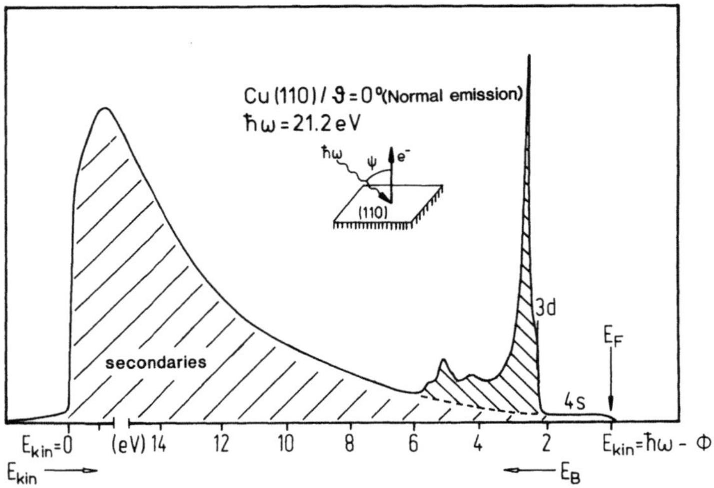

Fig. 1.7. UPS (He I, $\hbar\omega = 21.2\,\mathrm{eV}$) spectrum from a (110) face of Cu (normal emission, $\vartheta = 0$, ϑ being the polar angle with respect to the surface normal). The flat 4s band and the structured 3d band are seen. The cutoff marks the point where $E_{\mathrm{kin}} = 0$ and via (1.2) the work function can then be derived

Figure 1.6 shows a spectrum taken with $\hbar\omega = 1487\,\mathrm{eV}(\mathrm{Al\,K}_\alpha)$ from a surface of polycrystalline gold. The is clearly seen near E_{F} and at larger binding energies the 5p and 4f core levels are visible. The Au 4f levels are often used as a reference to calibrate an instrument or to calibrate the position of the Fermi energy in a particular material. In the latter case, a thin film of Au is evaporated onto the sample. The positions of the Au 4f levels are at $E_{\mathrm{B}} = 84.0 \pm 0.1\,\mathrm{eV}(4\mathrm{f}_{7/2})$ and $87.7 \pm 0.1\,\mathrm{eV}(4\mathrm{f}_{5/2})$ below E_{F}.[3]

The structure in the valence band is best investigated with lower photon energies because of the superior energy and momentum resolution then available. An angle-resolved spectrum taken with $\hbar\omega = 21.2\,\mathrm{eV}$ (He I) from a Cu(110) surface for normal electron emission is presented in Fig. 1.7. In the first 2 eV below the onset of the photoemission signal at E_{F} it shows the flat 4s valence band, followed (between 2 and 6 eV binding energy) by the structured 3d band. The third feature visible is a sloping background created by secondary electrons with a sharp cutoff at $E_{\mathrm{kin}} = 0$. This cutoff energy, measured with respect to the Fermi energy, can be used to determine the *work function* ϕ using (1.2): the width of the spectral distribution is given by $\hbar\omega - \phi$.

Finally in this section, Fig. 1.8 gives a PE spectrum for the energy region around the Fermi energy (polycrystalline silver, $\hbar\omega = 21.2\,\mathrm{eV}$). The energy

[3] It is not uncommon to use the Au $4\mathrm{f}_{7/2}$ energy of 84.00 eV as an internal standard, in which case no error has to be attached to this number.

resolution of the instrument was $\simeq 25\,\mathrm{meV}$. A fit of the measured spectrum with a Fermi function for $T = 300\,\mathrm{K}$ demonstrates good agreement, as expected. The width of the Fermi function is approximately $4k_{\mathrm{B}}T \simeq 100\,\mathrm{meV}$, which is large compared to the resolution, and determines the shape of the measured curve.

The "bare" resolution function can be measured at low temperatures (Fig. 1.9) [1.6]. This figure depicts the spectrum measured for Ag by Patthey et al. [1.6] and also gives an analysis of the data by a Fermi function convoluted with a Gaussian slit function (width $\sim 18\,\mathrm{meV}$), which describes the measured data reasonably well.

The most recent data [1.7] on the resolution that can be achieved in UPS experiments on a solid, are reproduced in Fig. 1.10, again using the Fermi edge of Ag as a test material. In that figure a convolution of the Fermi function of the measuring temperature ($T = 8\,\mathrm{K}$) and a 3-meV Gaussian resolution function fits the data very well.

The series of data in Figs. 1.8–1.10 indicates nicely the improved performance of PES instrumentation over the years.

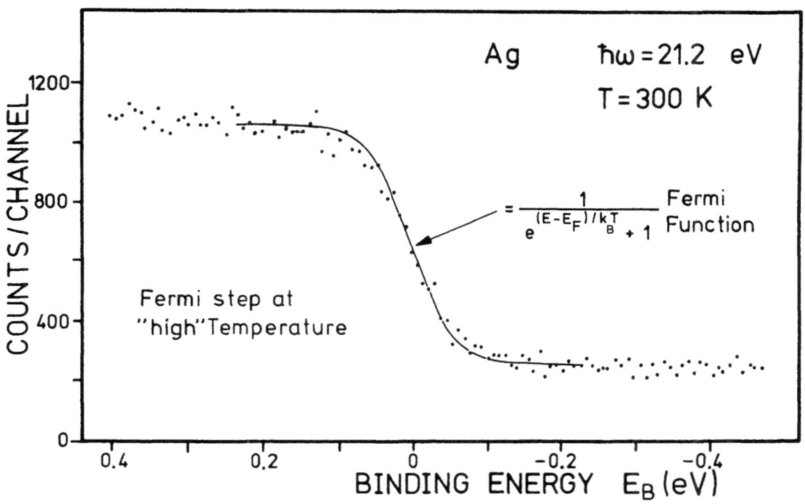

Fig. 1.8. Energy distribution around E_{F} in an UPS spectrum ($\hbar\omega = 21.2\,\mathrm{eV}$) from a polycrystalline Ag sample. The solid line is the Fermi function at room temperature. The resolution of the electron spectrometer was $\sim 0.025\,\mathrm{eV}$ and is hardly detectable in the data. The background intensity above E_{F} is produced by weak photon lines with $\hbar\omega > 21.2\,\mathrm{eV}$

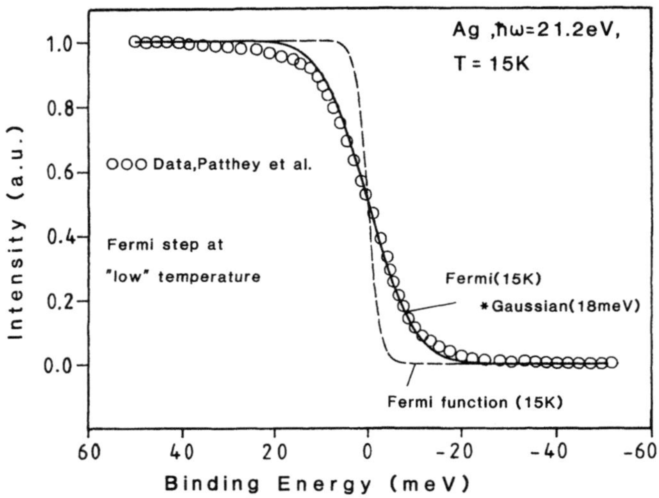

Fig. 1.9. Energy distribution around E_F in an UPS spectrum ($\hbar\omega = 21.2\,\mathrm{eV}$) from a polycrystalline Ag sample at 15 K [1.6]. The resolution ΔE is obtained by convoluting a Fermi function (T = 15 K, *dashed line*) with a Gaussian function of width $\Delta E = 18\,\mathrm{meV}$ (FWHM)

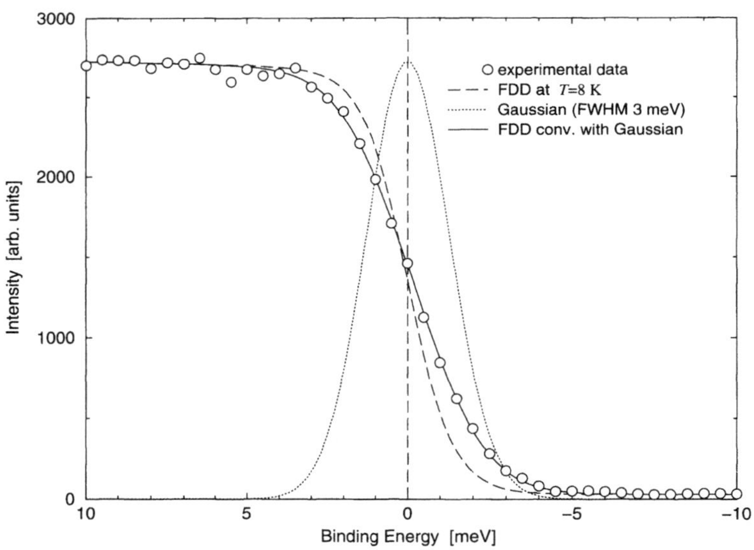

Fig. 1.10. Energy distribution around E_F in an UPS spectrum ($\hbar\omega = 21.2\,\mathrm{eV}$) from a polycrystalline Ag sample at 8 K (*open circles*). The resolution ΔE is obtained by convoluting a Fermi function ($T = 8\,\mathrm{K}$, *dashed line*) with a Gaussian function of width $\Delta E = 3\,\mathrm{meV}$ (FWHM). The resolution function is also shown by the points. The convoluted curve is represented by a full line

1.2 The Electron Mean Free Path

Interest in electron-spectroscopic methods has grown significantly since about 1970. One reason for the revival of electron spectroscopy, at least as far as solids are concerned, stems from the fact that routine methods have been developed to obtain Ultra-High Vacuum (UHV) conditions. The necessity for such UHV conditions for the study of solid surfaces can be judged from Fig. 1.11, which gives the "universal" *electron mean free path*electron mean free path λ in Å as a function of the electron kinetic energy for a few selected metals. One sees that in the kinetic energy range of interest, between about 10 and 2000 eV, the mean free path is only of the order of a few Å. This means that any spectroscopy of a solid surface involving clectrons samples only electrons from a very thin layer of the sample. Thus, if one wishes to learn about the bulk properties of the solid, one has to work with atomically clean surfaces. But also the investigation of surface states or adsorbed molecules requires UHV conditions to prevent interference from adsorbed contaminants.

One can further quantify these considerations. With a sticking coefficient $S = 1$, which means that every molecule or atom impinging on a surface also sticks there, one needs an exposure of approximately 2.5 Langmuirs ($1\text{L} = 10^{-6}\,\text{torr}\cdot\text{s}$) to obtain a coverage of 1 monolayer on a surface. At a pressure of 10^{-9} torr it therefore takes about 1000 seconds, or less than an hour, until a coverage of one monolayer is achieved. Sticking coefficients

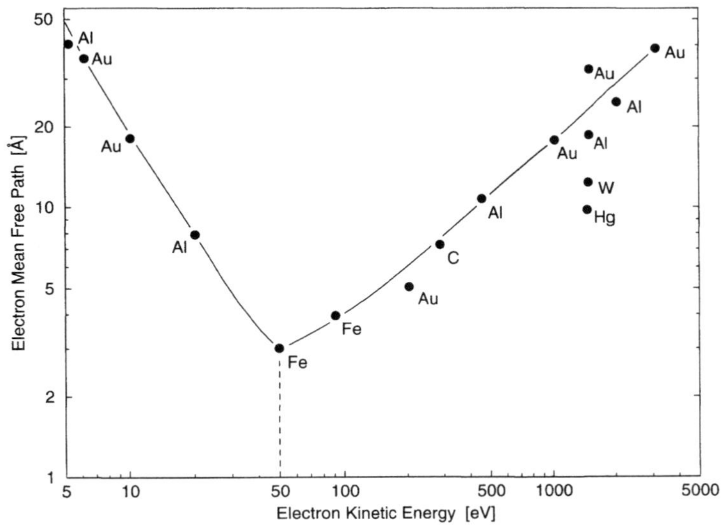

Fig. 1.11. Electron mean free path as a function of their kinetic energy for various metals. The data indicate a universal curve with a minimum of 2–5 Å for kinetic energies of 50–100 eV. The scatter of the data is evident from the values obtained at $E_{\text{kin}} = 1480\,\text{eV}$

of one or nearly one are typical for very reactive materials like alkali or rare earth metals. Thus in order to get spectra from clean surfaces of such materials, vacua of even better than 10^{-10} torr are mandatory. Fortunately, for many materials the sticking coefficients are much smaller than one and then a vacuum of 10^{-10} torr is sufficient to do meaningful experiments over an extended period of time. The standard pumping equipment consists of turbomolecular pumps, diffusion pumps, cryo pumps and sublimation pumps in various combinations. These pumping units are commercially available and will not be dealt with here.

At this point, it is perhaps useful to specify explicitly the advantages of electrons for surface spectroscopy, as compared to photons, atoms or ions, since electrons have found the widest application in this field:

(1) The escape depth of electrons is only a few Å. Electrons are therefore suited to probe electronic states in the surface region of a sample (e.g., surface states).
(2) Electrons are easily focused and their energy is easily tunable by electric fields.
(3) Electrons are easy to detect and to count.
(4) Using electrostatic fields it is easy to analyze the energy and the angle of an electron (thus determining the momentum). These instruments can be built small and handy, and can therefore easily be moved (e.g., in angle-resolved studies).
(5) Electrons vanish after they have been detected. This is an appreciable advantage compared, for instance, to ions or atoms [1.8] which are used in surface physics, too.

Of course, working with electrons also involves some disadvantages. The problem of obtaining an excellent vacuum necessary for the experiments has already been mentioned. Secondly, the small escape depth of the electrons sometimes makes it difficult to distinguish between surface and bulk properties of the sample.

The mean free path of the electrons is determined by electron–electron and electron–phonon collisions. Except in special cases, electron–phonon scattering plays a role only at very low energies. The mean free path at the energies of interest here is determined largely by the electron–electron interaction. The cross-section σ for electron–electron scattering is given by [1.9]

$$\frac{\mathrm{d}^2\sigma}{\mathrm{d}\Omega\mathrm{d}\omega} = \frac{\hbar^2}{(\pi e a_0)^2}\frac{1}{q^2}\mathrm{Im}\left\{-\frac{1}{\varepsilon(\boldsymbol{q},\omega)}\right\}, \tag{1.4}$$

where $\hbar\boldsymbol{q}$ is the momentum transfer and ω the energy transfer in the scattering process, $a_0 = 0.529$ Å (which is the Bohr radius), and Ω is the solid angle into which the electrons are scattered.

From (1.4) the inverse of the average mean free path λ^{-1} is obtained by an integration over all energy transfers (ω) and momentum transfers ($\hbar\boldsymbol{q}$). We see that λ^{-1} is essentially determined by the dielectric function $\varepsilon(\boldsymbol{q},\omega)$.

Since the dielectric function is specific to the material under study, one sees that, in principle, the mean free path is a characteristic property of each material. Notice that for insulators and semiconductors the loss function $\mathrm{Im}\{-1/\varepsilon\}$ is large only for energy transfers in excess of the gap energy. In turn, at smaller energies the loss function decreases rapidly and subsequently the mean free path becomes very large (because $\lambda^{-1} \sim \mathrm{Im}\{-1/\varepsilon\}$), as was indeed observed [1.9].

Why then does the mean free path, as a function of energy, yield a roughly "universal" curve for all materials (Fig. 1.11) although the dielectric function differs from material to material? For the energies of interest here (except for the very smallest), the electrons in the solid can be approximately described by a free-electron gas, because the bonding properties are no longer very important.[4] In this case the plasma frequency which is a function of only the electron density (or the mean electron–electron distance r_s) determines the loss function. The inverse mean free path λ^{-1} is then described by the mean electron–electron distance r_s which is roughly equal for all materials, and one obtains [1.9]

$$\lambda^{-1} \simeq \sqrt{3}\frac{a_0 R}{E_\mathrm{kin}} r_\mathrm{s}^{3/2} \ln\left[\left(\frac{4}{9\pi}\right)^{2/3} \frac{E_\mathrm{kin}}{R} r_\mathrm{s}^2\right] , \tag{1.5}$$

where $a_0 = 0.529\,\text{Å}$, $R = 13.6\,\mathrm{eV}$, and r_s is measured in units of the Bohr radius a_0. Therefore, as is observed experimentally, almost all materials show a similar energy dependence of the mean free path.

The measurement of the mean free path is not a trivial problem. It is usually done by the "overlayer method" [1.10]. However, it is not easy to produce a homogeneous island-free overlayer and thus results from various groups can differ. On inspecting the actual data, one concludes that the variation in the numbers obtained for λ by different groups for the same material is about as large as the variation between different materials.

During recent years the problem of the mean free path (or more correctly the inelastic mean free path) has been investigated intensively in order to establish electron spectroscopies, in particular PES as a quantitative analytical tool [1.13–1.16]. The reason for this desire is obvious from Fig. 1.11, where at the kinetic energy of 1400 eV the scatter in the data for different elements is between 10 Å and 30 Å, which is large, and only does not appear as such, because the inelastic mean free path [1.16] is plotted on a logarithmic scale.

[4] The binding energy of a valence electron in a solid is of the order of 10 eV. By the argument just given this is then an approximate measure of the energy up to which specific properties of the material under investigation will be felt in the mean free path. Thus, while for $E_\mathrm{kin} \gg 10\,\mathrm{eV}$ the mean free path will follow approximately a universal curve, for $E_\mathrm{kin} < 10\,\mathrm{eV}$ material effects will be observed.

We add that a detailed expression for the mean free path contains the plasmon energy and the gap energy [1.10–1.12]. The gap energy influences, in particular, the mean free path for small kinetic energies.

Even the most recent plots of inelastic mean free paths indicate that for a particular kinetic energy, the numbers for different materials can differ by factors between 3 and 5, which qualifies the term "universal" mean free path curve. Yet, for general information the curve in Fig. 1.11 is still quite useful and the above-mentioned most recent plot of the available data comes to a result similar to that given in Fig. 1.11 [1.13, 1.14].

In order to achieve an accuracy in elemental composition of 30 % or better in the chemical analysis of a sample, a more sophisticated analysis is necessary, and it is beyond the scope of this volume to treat the difficult problems in any detail, but the reader can find excellent reviews on this topic. Here only a few remarks will be added to outline the problem, and also to introduce the nomenclature used in this field [1.14, 1.15].

In a PES experiment the photons impinging on the sample produce a photocurrent, which, in order to indicate that it has a depth distribution (relative to the surface), is generally called the emission depth distribution function. Neglecting elastic scattering events, the electron inelastic mean free path (IMFP) λ can be defined as the distance over which the probability of an electron escaping without significant energy loss due to inelastic processes drops to e^{-1} of its original value. Therefore the electron current $\mathrm{d}I$, originating from a layer of thickness $\mathrm{d}z$ at depth z detected at an angle ϑ with respect to the surface normal is given by

$$\mathrm{d}I \propto \exp(-z/\lambda\cos\vartheta)\mathrm{d}z \ ; \tag{1.6}$$

here ϑ is the angle at which the electrons are detected with respect to the surface normal. In order to evaluate the escape depth Δ one has for $\Delta = z$

$$\exp(-\Delta/\lambda\cos\vartheta) = \mathrm{e}^{-1} \tag{1.7}$$

and

$$\Delta = \lambda\cos\vartheta \tag{1.8}$$

and therefore for normal detection ($\vartheta = 0$) one has $\Delta = \lambda = \mathrm{IMFP}$.

The actual determination of λ is often performed by what is known as the overlayer method, in which the intensity of the photoemission (or Auger) signal of a substrate is monitored as a function of the thickness of the overlayer ℓ; likewise, the change of the signal from the overlayer can be used. This leads to the so-called effective attenuation length (EAL) obtained from the electron current from the overlayer signal (I_ℓ):

$$I_\ell = I_\ell^\infty\left[1 - \exp\left(-t/\lambda_\ell\cos\vartheta\right)\right] \tag{1.9}$$

or that from the substrate (s) signal (I_s):

$$I_\mathrm{s} = I_\mathrm{s}^0\exp(-t/\lambda_\mathrm{s}\cos\vartheta) \ , \tag{1.10}$$

where α is again the electron detection angle measured with respect to the surface normal and I_ℓ^∞ is the photocurrent from the thick overlayer and

I_s^0 is the signal from the substrate without the overlayer; t is the overlayer thickness.

Plots of $\ln(1 - I_\ell/I_\ell^\infty)$ and $\ln(I_s/I_s^0)$ as a function of t yield $\lambda_\ell \cos\vartheta$ and $\lambda_s \cos\vartheta$ respectively.

Since in these experiments the instrumental parameters often influence the electron current and furthermore, since the overlayers are hardly ever homogeneous, one often calls $\lambda_\ell \cos\vartheta$ and $\lambda_s \cos\vartheta$ determined by this type of experiment the effective attenuation length (EAL). Under "ideal" conditions they are identical to the mean escape depth Δ, meaning that the λ determined in this way corresponds to the inelastic mean free path (IMFP).

The main effect leading to deviation from the exponential functions is elastic scattering and its effects have been treated by Monte Carlo simulations and also by transport equations [1.14, 1.15]. In effect, elastic scattering, leading to non-exponential decay curves, makes the determination of the inelastic mean free path with any accuracy a very difficult problem. We note, however, that for a rough estimate of inelastic mean free path of electrons (or small escape depths) the curve given in Fig. 1.11 is of considerable use. This does not mean that one can use it for chemical analysis with an accuracy in the percent range. If this is desired a careful analysis along the lines given, e.g., in [1.14] and [1.15] is required.

A convenient form for the inelastic electron mean free path λ has been given by Tanuma et al. [1.15], which has approximately the form of the full curve in Fig. 1.11 (in the literature often called the TPP2 relation):

$$\lambda = E/\left\{E_p^2\left[\beta\ln(\gamma E) - (C/E) + (D/E^2)\right]\right\} ,$$

where

λ	=	IMFP in Å
E	=	electron kinetic energy in eV
E_p	=	$28.8(N_v\varrho/M)^{1/2}$ is the free electron plasmon energy
ϱ	=	density in $g \cdot cm^{-3}$
N_v	=	number of valence electrons per atom (for elements) or molecule (for compounds)
M	=	atomic or molecular weight
β, γ, C, D		parameters.

Analyzing data for a number of elements the authors found the following empirical relations for the parameters:

β	=	$-0.0216 + 0.944/(E_p^2 + E_g^2)^{1/2} + 7.39 \cdot 10^{-4}\varrho$
γ	=	$0.191\varrho^{-0.50}$
C	=	$1.97\text{–}0.91\ U$
D	=	$53.4\text{–}20.8\ U$
U	=	$N_v\varrho/M = E_p^2/829.4$

where E_g is the band gap energy in eV for insulators.

1.3 Photoelectron Spectroscopy and Inverse Photoelectron Spectroscopy

The principles of PhotoElectron Spectroscopy (PES) have already been illustrated in Fig. 1.4. A photon impinges on a sample and, via the photoeffect (or the photoelectric effect as it is sometimes called), an electron is liberated, which then escapes into the vacuum. The energy of the incoming photon can be in the ultraviolet regime (5 to 100 eV, UPS), in the soft X-ray regime (100 to 1000 eV, SXPS) or in the X-ray regime ($>$ 1000 eV, XPS). In this simple picture, PES measures electron states of the sample under consideration. We shall see in what follows that in a solid, it is predominantly $\boldsymbol{k}$-conserving or "direct" transitions that are excited. From the energy and momentum distribution of the electrons, and with some reasonable assumptions, one can determine the electronic dispersion curves $E(\boldsymbol{k})$ in the solid (see below). With PES using an X-ray source (XPS) one can, in addition, observe photoionization of core levels. Their energies depend on the chemical state of the sample. Therefore the spectroscopy of core levels can often be used for chemical analysis, and the name ESCA (Electron Spectroscopy for Chemical Analysis) has been invented for this technique [1.17]. Since the escape depth of the electron is small, PES samples predominantly electrons from near the surface. Thus PES, like AES (Auger Electron Spectroscopy), can be used for the quantitative analysis of chemisorbed or physisorbed species.

For the interpretation of PE experiments the so-called *three-step model* has proven extremely useful. The basis for this model is shown in Fig. 1.12. In the three-step model the PE experiment is broken up into three distinct and independent processes. In the first step photoionization takes place; locally the photon is absorbed and an electron is excited. In the second step this electron travels through the sample to the surface. Finally, in the third step the electron escapes through the surface into the vacuum where it is detected. This separation of stages is somewhat artificial and, in principle, the whole process of photoexcitation, travel to and escape through the surface should be treated as one step. It will be seen later, however, that the division of the process into distinct steps can be justified reasonably well and that a one-step theory actually gives results that are not very different from the conceptually much simpler three-step model.

At this point, it is useful to summarize some aspects of PES:

(1) Because of the small escape depth of the electrons, one always has to ask to what extent the measured spectrum is representative of bulk or of surface properties.
(2) The final state in a photoemission experiment has a positive hole and can thus be distinctly different from the initial state. It is not always a trivial problem to infer properties of the initial (ground)-state from the final-state spectra.

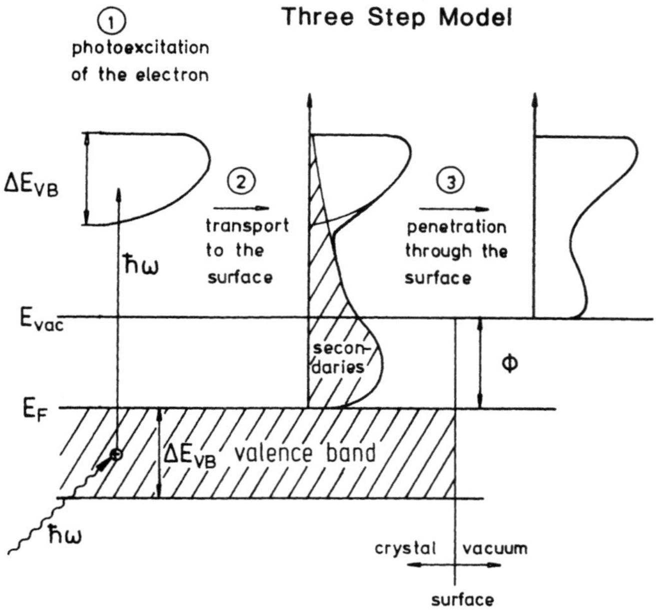

Fig. 1.12. PES as a three-step process: (1) photoexcitation of electrons; (2) travel to the surface with concomitant production of secondaries (*shaded*); (3) penetration through the surface (*barrier*) and escape into the vacuum

(3) The simplest formulation of the momentum-conservation rule yields that the final-state wave vector $\boldsymbol{k}_{\mathrm{f}}$ equals the initial state wave vector $\boldsymbol{k}_{\mathrm{i}}$ plus a reciprocal lattice vector $\boldsymbol{G}$, the exact magnitude of which is determined by the experimental conditions, the crystal structure, etc. It is not quite clear yet how rigorously momentum conservation is obeyed in actual experiments in this simple form. The occurrence of a "Fermi step" in all experiments in metals shows that additional sources of momentum (e.g., the surface) modify the above rule.[5]

[5] The reasoning behind this statement is as follows. In the band structure of a material (the energy vs. wave-vector diagram) the bands intersect the Fermi energy only at distinct points in $\boldsymbol{k}$-space. These are incidentally the points which are, e.g., determined by de Haas–van Alphen experiments. They are the points in $\boldsymbol{k}$-space at which occupied states exist at the Fermi energy for the material. If a PE experiment on a single crystal is performed in a way, that the initial state is such a state at the Fermi energy, the intensity at the Fermi energy in the PE experiment should be and is observed. If, however, conditions are chosen such that the initial state of the PE experiment is some electron Volts below the Fermi energy, no intensity at the Fermi energy should be observed because no transitions take place from this energy. A case in point is the [111] direction (ΓL direction) in the noble metals, for which there is no band that intersects the Fermi energy ("neck" direction). Therefore, in the PE experiments from

(4) The selection rules for the optical transitions between the initial and the final state have been worked out, even for the relativistic case. For an accurate interpretation of the intensities in the PE spectra, however, the photoemission matrix element has to be evaluated, which is not always possible exactly.
(5) The electromagnetic field impinging on the sample becomes modified when it penetrates into the sample. This effect can be described by the Fresnel equations. These equations assume, however, a sharp boundary between the vacuum and the solid, whereas a real surface has a certain width, and its effect on the photon field is not exactly known.

We finally want to point out that Inverse PhotoEmission Spectroscopy (IPES) has gained in importance in recent years. It is, however, used much less than PES due to its limited resolution (0.3 eV at best) and the low cross section of the inverse photoemission process. In this spectroscopy which is the inverse of PES, an electron with energy E impinges on the sample (Fig. 1.13 and by being de-accelerated emits) Bremsstrahlung which is then detected.

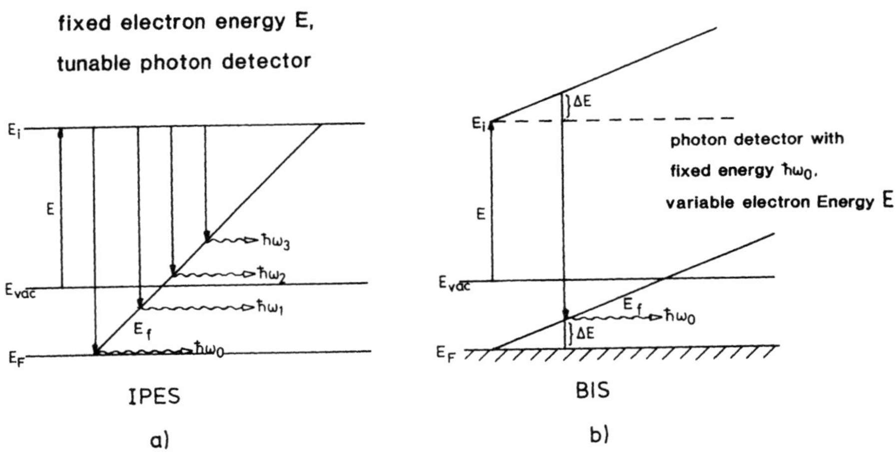

Fig. 1.13. (**a**) Energy diagram for Inverse PhotoEmission Spectroscopy (IPES). An electron is excited to an energy E_i and decays radiatively to energies E_f above the Fermi energy. The photon spectrometer is swept to record photons with different energies. (**b**) Energy diagram for Bremsstrahlung Isochromate Spectroscopy (BIS). Electrons with variable energy E impinge on a sample, thereby producing Bremsstrahlung. Its intensity is monitored with a spectrograph at a fixed ("isochromate") energy $\hbar\omega_0$

---

such surfaces no Fermi-energy emission should be visible, which is, however, in contrast to experimental experience.

The momentum of an electron with the wave vector $\boldsymbol{k}$ is $\hbar\boldsymbol{k}$. Sometimes $\boldsymbol{k}$ is, in a short-hand notation, called the momentum.

This can be viewed as a process in which the electron makes a transition from an initial state with energy E_i (the kinetic energy of the incoming electron) to an empty final state E_f, a process in which a photon with the energy difference $E_\mathrm{i} - E_\mathrm{f}$ is emitted and detected yielding the energy E_f (making the assumption that E_i is known from the conditions of the experiment).

If one works with a fixed electron energy and varies the energy of the detected photons the technique is usually called IPES. In contrast, one talks of BIS (Bremsstrahlung Isochromate Spectroscopy) if a detector for photons of a fixed energy $\hbar\omega_0$ (Fig. 1.13) is used and the energy of the incident electrons is varied.

With IPES one is able to obtain information about the density of unoccupied states. For a long time this technique was used only in the energy regime of soft X-rays. Due to the poor efficiency of the monochromators it was necessary to work with high electron currents to get a reasonable signal. This produced high temperatures and samples often got heavily damaged, which in turn led to spectra of low resolution. In recent years it has been shown that XPS spectrometers can be modified to take IPES spectra – so far exclusively as BIS spectra. This was done in the soft X-ray regime for energies corresponding to aluminium K_α-radiation [1.18]. The group of Dose [1.19] managed to show that IPES (in the BIS mode – or, more properly, UV isochromate mode) can also be performed in the UV range. This made it possible to study the unoccupied states of band structures in the same way as the occupied states (if the radiative transition takes place between two band states E_i and E_f). In addition, unoccupied surface states, which are important for the understanding of chemisorption problems, were made accessible by this technique.

With the aid of Fig. 1.14, another very useful property of PES with respect to extended valence orbitals is illustrated, namely, the dependence of transition matrix elements on the photon energy. For simplicity, the wave functions of a harmonic oscillator are used although the $1/r$ potential is shown. The principle of the argument will be explained for final-state electron wave functions reached from 3d and 3p valence orbitals by He I and He II radiation. The dipole transition matrix element is proportional to an overlap integral between the initial-state wave function and the wave function of the outgoing electron:

$$M \propto \int \psi_\mathrm{f} r \psi_\mathrm{i} \mathrm{d}\tau \quad (\mathrm{d}\tau = \mathrm{d}x\mathrm{d}y\mathrm{d}z) .$$

(This is a very simplified description leaving all factors out that are irrelevant to the following argument).

At low kinetic energy of the outgoing electron one has a "large" cross-section for states with low angular momentum. For both the low-kinetic-energy final-state electron and the low-angular-momentum initial-state electron, the variation of the wave function is small over the atomic volume, giving a large dipole matrix element, as indicated in Fig. 1.14. The oppo-

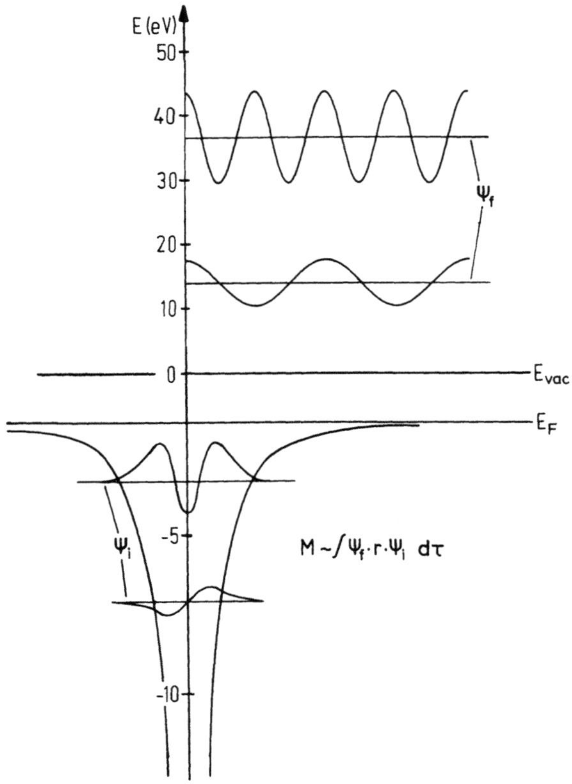

Fig. 1.14. Schematic representation of the photoexcitation process using photons of two different frequencies (20 and 40 eV). For extended valence orbitals with low angular momentum (not, e.g., 4f!) the transition matrix element is larger for low energy photons than for high energy photons [1.20]. With increasing final state energy one expects an increase of the overlap of the final state wave function with those of higher angular momentum initial states

site holds for high-energy final states and high-angular-momentum initial states.

For a high final-state energy one expects a better overlap for states with high angular momentum. This trend is generally verified by the experimental findings. Figure 1.15 presents a comparison of spectra taken with He I (21.2 eV) and He II (40.8 eV) radiation for the valence band of CuCl. One sees that the strength of the signal from the copper 3d electrons increases relative to the signal strength from the chlorine 3p electrons in going from He I to He II radiation [1.21]. This is a very useful property and can be employed to analyze complicated spectra.

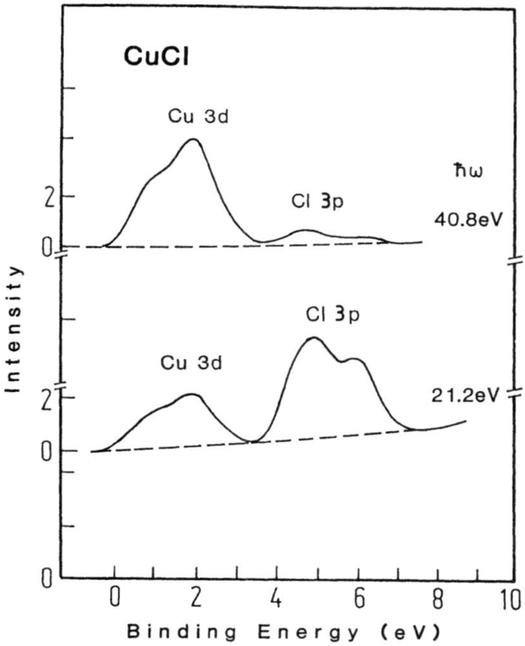

Fig. 1.15. Comparison of PE spectra of solid CuCl taken with He I (21.2 eV) and He II (40.8 eV) radiation. The intensity of the higher momentum (*less extended*) 3d electrons increases with increasing photon energy relative to the lower momentum (*more extended*) 3p orbitals (Fig. 1.14) [1.21]

Finally, a brief comparison of the advantages and disadvantages of UPS and XPS is given. UPS has relatively high sensitivity, high resolution and is advantageously used to study valence-band features. Only low-binding-energy ("shallow") core levels can be excited in this type of spectroscopy. One should also keep in mind that in solids direct (i.e., $\boldsymbol{k}$-conserving) transitions influence the spectra heavily. One may miss information in UPS, because of transitions forbidden by the $\boldsymbol{k}$-conservation selection rule. On the other hand, XPS can be used in the valence band as well as in the core-level regime. However, at present, its resolution is about one order of magnitude worse than that of UPS.[6] XPS is hardly restricted in solids by direct transition selection rules because the relaxation of $\boldsymbol{k}$-conservation at high energy due to the finite acceptance angles of the electrons makes transitions in the whole Brillouin zone possible. Therefore, with this technique, one measures essentially a density of valence-band states. Although some angular momentum matrix elements can still be operative, experience shows that they do not severely influence the spectra. The greatest disadvantage of this technique so far has been its rather

[6] With the new high-intensity synchrotron-radiation sources, this disadvantage of XPS with respect to UPS no longer exists [1.22-30] [1.22–1.30].

low resolution, which itself is determined by the strength of present-day X-ray sources. High-intensity synchrotron sources have improved this situation. On the other hand, an advantage of XPS is the fact that a very large number of core levels are accessible, thus making the technique very useful in many diverse fields including chemistry, materials research and materials testing.

High-resolution energy loss spectrometers have been reviewed by Ibach in greater detail [1.31]. Results on the newly evolving field of spatially resolved PES can be found in the article by Westhof et al. [1.32] and the references therein.

1.4 Experimental Aspects

Basic equipment for PES experiments is commercially available and so questions of instrumentation will not be dealt with in much detail. The reasons for working under sufficiently good vacuum conditions have already been outlined. Ultra-high vacuum is achieved with turbomolecular pumps, ion pumps, sublimation pumps and diffusion pumps.

The arrangement of a photoemission apparatus is shown in Fig. 1.16. In a laboratory setup usually one has two light sources (a UV source and an X-ray source[7]), an electron gun with electron optics to focus the electrons in order to execute Auger and electron-energy-loss experiments and an energy analyzer followed by an electron detector. A LEED and an Auger system are desirable for characterizing the sample. In most cases electrons are digitally counted.

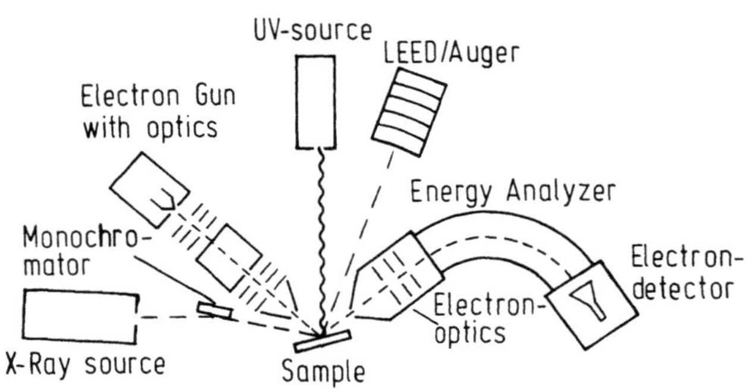

Fig. 1.16. Diagram of a typical modern setup used for PES experiments involving UV source, monochromatic X-ray source, LEED–Auger facility, and electron-energy analyzer usable for PES, Auger analysis and EELS (electron energy-loss spectroscopy)

[7] Since the X-ray spectrum contains satellites, in addition to the characteristic lines, a large background due to Bremsstrahlung, and moreover, since the X-ray lines are relatively broad, monochromatization of the X-rays is desirable.

The laboratory photon sources can, of course, be replaced by a *synchrotron-radiation source*, e.g., a "dedicated" storage ring [1.22–1.30].

The different types of electron energy analyzers will now be mentioned briefly. In principle, there are four methods to analyze the energy of a charged particle [1.9]:

(1) The use of resonances in a scattering process;
(2) the time-of-flight method;
(3) the deceleration of the particles by a retarding electric field;
(4) the change of the orbit of a particle by an electric or magnetic field.

Of these principles, the last two are best suited for PES. The time-of-flight method (2) is used for special experiments at synchrotron-radiation sources, operating in a "single bunch" mode.

Electric analyzers [1.33] (employing principles 3 or 4 from above) have now replaced magnetic analyzers (using principle 4) practically everywhere. Likewise, the retarding-field analyzer has become virtually obsolete in PES, although it is still the instrument of choice for lower resolution AES. Thus for PES almost exclusively electric analyzers are used. Technically one distinguishes between the Cylindrical Deflection Analyzer (CDA), the Spherical Deflection Analyzer (SDA), the Plane Mirror Analyzer (PMA), and the Cylindrical Mirror Analyzer (CMA). The Magnetic Deflection Analyzer (MDA) and the Wien Filter Analyzer (WFA) are omitted here [1.33, 1.34].

The simplest electric analyzer [1.33, 1.35, 1.36] is the Plane Mirror Analyzer (PMA) depicted in Fig. 1.17.[8] It is a condenser made from two parallel plates a distance d apart. The two slits are in one of the plates and have a separation L_0. This geometry, which is illustrated by Fig. 1.17, applies to the case where the entrance and exit angles of the electrons are 45°. With an angle of 30° the focusing is better (namely of second order), but one has to sacrifice transmission, because a second pair of slits needs to be installed below the position plate. Since the potential difference V between the two plates is constant, one obtains parabolic trajectories. The condition for transmission is $V = 2E_0 d/eL_0$, where E_0 is the electron's kinetic energy in electron volts and e the electron charge.

The PMA is simple but has the disadvantage of a relatively low transmission. This disadvantage can be overcome by using the 2π geometry of the Cylindrical Mirror Analyzer (CMA) [1.35, 1.36], as shown in Fig. 1.18 [1.37, 1.38]. It consists of two cylinders at a potential difference of V, with the entrance and exit slits on the inner cylinder. Focusing to second order can be achieved if the electrons enter the CMA at an angle $\vartheta = 42.3°$ with respect to the axis of the cylinders. Geometry then requires that $V = 1.3\, E_0 \ln(R_{\text{out}}/R_{\text{in}}$ and $L_0 = 6.1\, R_{\text{in}}$, where L_0 is explained in Fig. 1.18, E_0 is measured in Volts, and R_{in} and R_{out} are the radii of the inner and outer cylinders, respectively.

[8] This analyzer is sometimes called the parallel-plate analyzer.

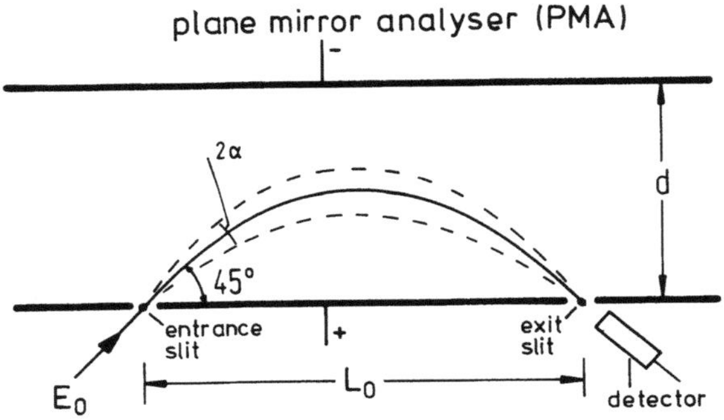

Fig. 1.17. Parallel-plate (plane) mirror analyzer [1.35]

The CMA is the best analyzer for applications requiring high sensitivity at moderate resolution [1.41–1.43].

An analyzer with high resolution but not very good transmission is the Cylindrical Deflection Analyzer (CDA) [1.42–1.54] depicted in Fig. 1.19. It focuses if the cylinders span an angle of $\pi/\sqrt{2} \approx 127°$, hence it is frequently called a 127° analyzer. In the CDA the potential difference between the inner and the outer cylinder has to be $2V = E_0(R_{\text{out}}/R_{\text{in}})$, where E_0 is the energy of those incoming electrons (in electron volts) that are focused.

An analyzer similar to the CDA is the spherical deflection analyzer (SDA) [1.41–1.64], which consists of two concentric hemispheres (Fig. 1.20). For transmission of electrons with initial energy E_0 along a path with $R_0 = (R_{\text{in}} + R_{\text{out}})/2$ the potential of the outer hemisphere has to be $V_{\text{out}} = E_0 \times [3 - 2(R_0/R_{\text{out}})]$, and that of the inner hemisphere $V_{\text{in}} = E_0[3 - 2(R_0/R_{\text{in}})]$.

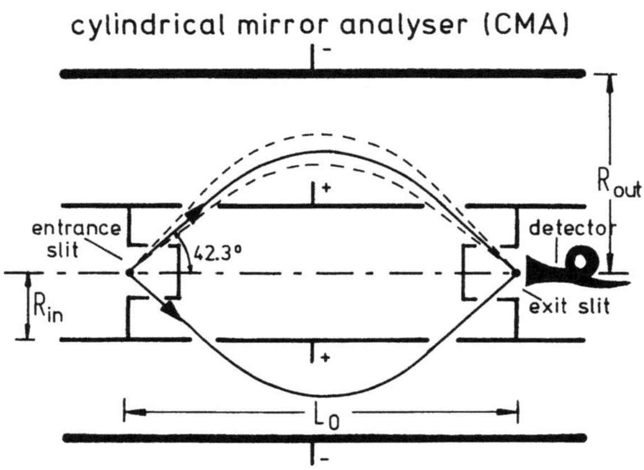

Fig. 1.18. Cylindrical mirror analyzer [1.35]

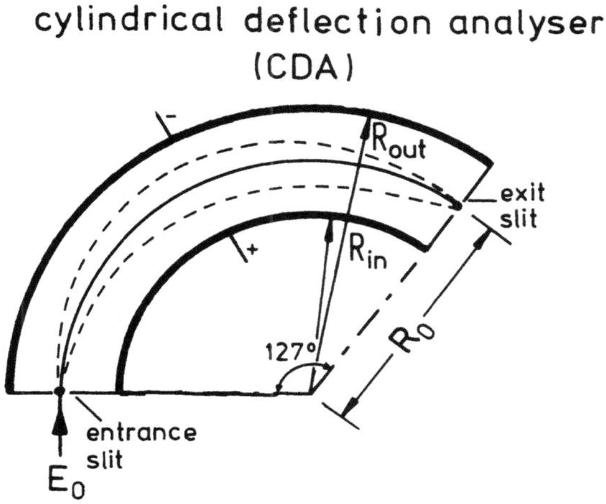

Fig. 1.19. Cylindrical deflection analyzer [1.35]

Most instruments use an additional preretardation stage prior to the energy analysis. The energy at which the electrons enter the analyzer is usually called the pass energy. This mode of operation is used because one can decelerate (or accelerate) electrons (almost) without changing their absolute energy spread. The advantage of preretardation is the following. Assume an electron leaves the sample with E_{kin}; it is then retarded to an energy E_0, before entering an analyzer of intrinsic resolving power $\Delta E_{1/2}(E_0) = KE_0$ (to first order). For the unretarded electrons, the resolution would be $\Delta E_{1/2}(E_{\mathrm{kin}}) = KE_{\mathrm{kin}}$, meaning that by retardation the effective resolution has been enhanced by $[\Delta E_{1/2}(E_{\mathrm{kin}})]/[\Delta E_{1/2}(E_0)] = E_{\mathrm{kin}}/E_0$, which in XPS can be a factor of ≈ 100 for valence-band electrons.

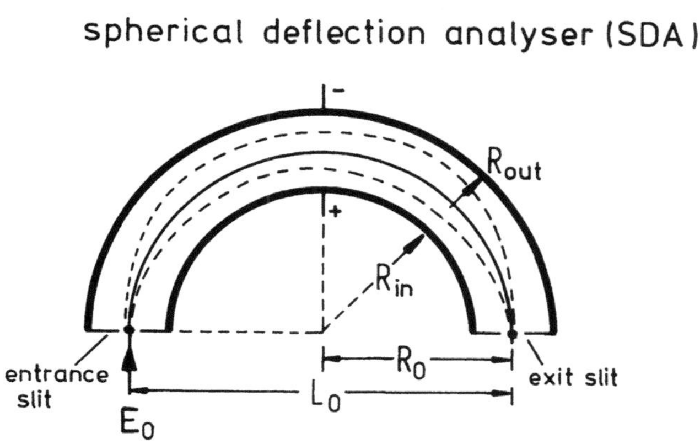

Fig. 1.20. Spherical deflection analyzer [1.35]

Photon sources for a PES experiment can be X-ray tubes, *gas-discharge lamps* or synchrotron radiation sources. The principles of X-ray tubes are well known, and do not have to be dealt with here. A typical gas-discharge lamp is sketched in Fig. 1.21. Operating such a lamp on, e.g., the He I (21.2 eV) resonance line presents no problem. However, if one wants to achieve sizable count rates with He II radiation (40.8 eV), special precautions with respect to the cleanliness of the gas-handling system are necessary.

The most versatile light source is undoubtedly the synchrotron radiation source, a storage ring for electrons or positrons (note that any charged particle on a non-linear trajectory will emit electromagnetic radiation).

The angular distributions of the radiation emitted from an electron on a circular orbit for $v \ll c$ (c: velocity of light) and for $v \simeq c$ are indicated in Fig. 1.22 [1.29,1.30,1.66,1.67], which indicates the strong forward pattern of the synchrotron radiation of the relativistic electrons in a storage ring. The synchrotron radiation is also highly polarized, with the polarization vector lying in the plane of the circulating electrons. The properties of the synchrotron radiation can be calculated by relativistic electrodynamics and can be found in handbooks on this topic [1.22–1.30]. Therefore one can be brief here.

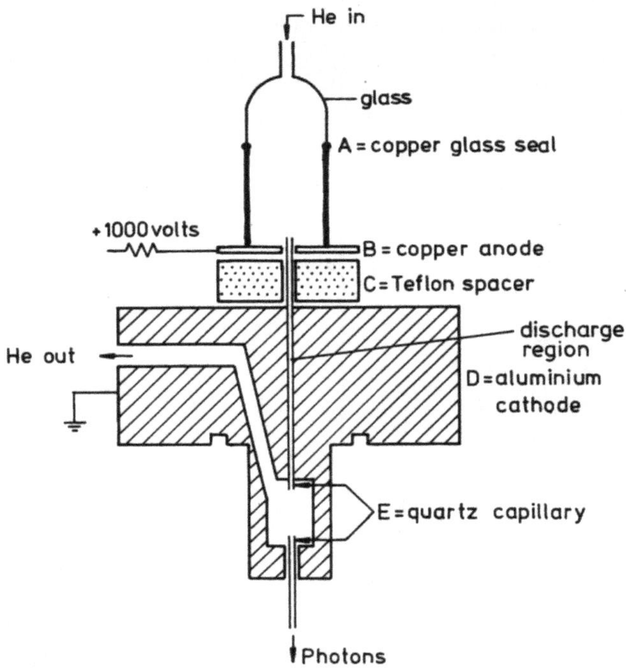

Fig. 1.21. Gas-discharge lamp [1.65]

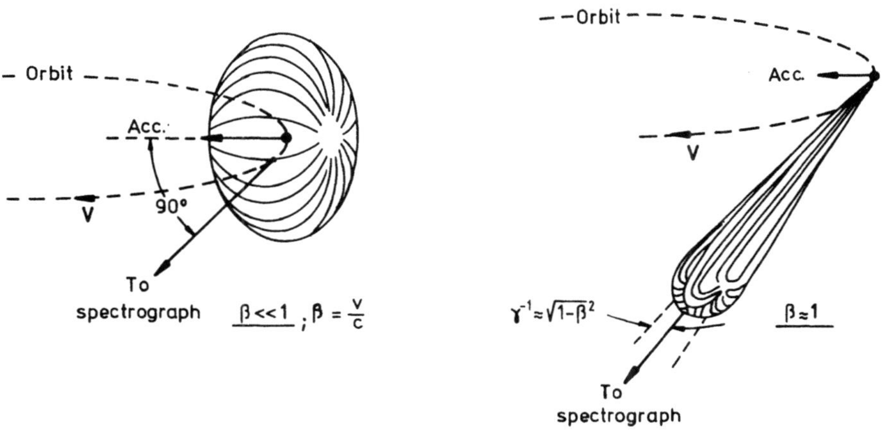

Fig. 1.22. Radiation characteristic of an electron moving in a circular orbit at small velocity ($v/c = \beta \ll 1$) and at high velocity ($v/c \simeq 1$). At high velocities the angular distribution of the radiation peaks strongly in the forward direction, whereas at low velocities it is approximately that of a classical dipole [1.66]

The total power (in kilowatt $\widehat{=}$ kW) radiated by a synchrotron (storage ring) is given by [1.68]

$$P = 88.5\, E^4 I/R \,(\mathrm{kW})\,, \tag{1.11}$$

where E is the electron energy in GeV, I the current in the storage ring in mA and R the radius of the storage ring in m.

A convenient parameter for the characterization of the radiation output of a synchrotron is the critical wavelength λ_c, which is the wavelength that divides the power spectrum into two equal parts:

$$\lambda_c = 5.8 \frac{R}{E^3}\,, \tag{1.12}$$

where λ_c is given in Å, R in m and E in GeV. Important for an experimentalist is the normalized number of photons, which is given by

$$N(\lambda) = 2.5 \times 10^{14}\, E \cdot G_1 \frac{\lambda}{\lambda_c} \tag{1.13}$$

with

$$G_1 = \frac{\lambda_c}{\lambda} G_0 \frac{\lambda_c}{\lambda}$$

$$G_0\left(\frac{\lambda_c}{\lambda}\right) = \int_{\lambda_c/\lambda}^{\infty} K_{5/3}(\eta)\mathrm{d}\eta \tag{1.14}$$

where $K_{5/3}(\eta)$ is a modified Bessel function.

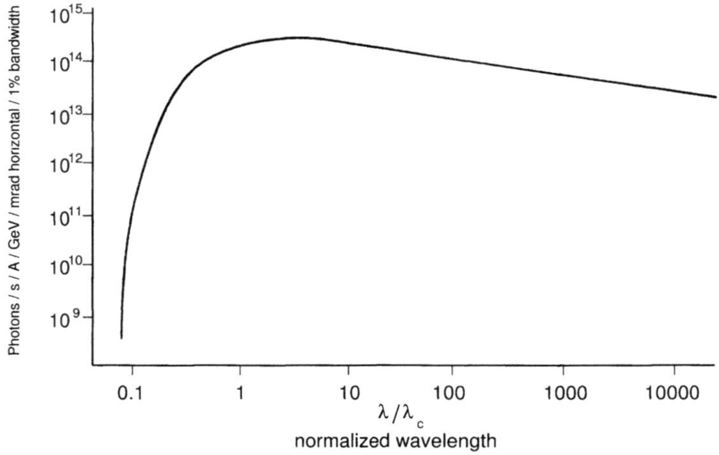

Fig. 1.23. Normalized photon flux from a synchrotron as a function of the wavelength normalized to the critical wavelength [1.68]

This normalized photon flux distribution function is shown in Fig. 1.23, where 1 % bandwidth means that at λ_1 the bandwidth is $0.01\,\lambda_1$.

The equations just given were for a "theoretical" synchrotron, where it is assumed that there is one magnet, with the field perpendicular to the electron orbit. In reality there are a number of bending magnets with straight sections between them that allow the installation of correction magnets and microwave cavities in order to supply the power that is lost by the emitted radiation.

The synchrotron radiation is emitted along the curved orbit in a bending magnet. One can in principle increase the radiation emission by placing combinations of magnets (insertions devices) into straight sections of the storage ring. These bend the electron beam back and forth, each bend producing radiation but with the overall direction of the beam remaining unchanged.

There are three different types of insertion devices. The wavelength shifter is a three pole arrangement, where the central magnet operates at very high magnetic field and therefore reduces λ_c by as much as a factor of five. This device is used to produce, even in low energy synchrotrons, a beam of high energy radiation.

The second type of insertion device is the multipole wiggler, which consists of a series of magnets with changing polarity along the length of the device. This device will act as N independent bending magnets and therefore increase the radiation strength N-fold.

Finally the undulator is a multipole wiggler with many (up to 100) magnets, which work in a regime (low field) that confines the radiation of the total device to a small angular range ($\sim 1/\gamma$ in the vertical and horizontal directions) (see Fig. 1.22). This results in intense, relatively monochromatic radiation, which, for a narrow wavelength range, yields up to 10^4 times the intensity available from a single bending magnet.

At the end of this chapter, a very simplified argument is presented enabling one to estimate the contamination of a surface as a function of pressure and time (numbers were given earlier without proof). One would like to know how many particles Δn impinge on a certain area ΔF during a certain time. This is given by the kinetic theory of gases as

$$\frac{\Delta n}{\Delta t \Delta F} = \frac{1}{4} n \overline{v} \,, \tag{1.15}$$

The average velocity $\overline{v}$ of the gas molecules can be written as

$$\overline{v} = \sqrt{8RT/\pi M} \,,$$

where R is the gas constant, T the temperature in Kelvin, M the molecular weight of the gas, and n the density of gas particles in cm^{-3}.

Substituting for $\overline{v}$ in (1.15) yields

$$\frac{\Delta n}{\Delta t \Delta F} = n \left(\frac{RT}{2\pi M} \right)^{1/2} \simeq \frac{p}{\sqrt{MT}} \cdot 3 \times 10^{22} \quad \left[\frac{1}{\mathrm{cm}^2 \cdot \mathrm{s}} \right] \tag{1.16}$$

where p is the pressure in torr.

One monolayer corresponds to a coverage of roughly 6×10^{14} particles per cm^2. Therefore, for $T = 300\,\mathrm{K}$ and $M = 28(\mathrm{CO})$, one obtains

$$\frac{\Delta n}{\Delta t \Delta F} \simeq p \cdot 0.6 \times 10^6 \quad [\mathrm{monolayers/s}] \tag{1.17}$$

This is the number of particles sticking on the area ΔF if we assume a sticking coefficient $S = 1$. Then the time t in which one monolayer is deposited on the surface area ΔF, with the assumption of a sticking coefficient $0 \leq S \leq 1$, is given by

$$t = \frac{1.7 \times 10^{-6}}{0.6 \cdot p \cdot S} \quad [\mathrm{s}] \,. \tag{1.18}$$

For a sticking coefficient of 1 and a pressure of 10^{-9} mbar it takes about 1500 seconds to produce one monolayer, i.e., of the order of half an hour. Thus for low-sticking-coefficient materials there is sufficient time to do PES experiments – even at 10^{-9} mbar. However, a pressure of 10^{-11} mbar is clearly more desirable to comfortably perform experiments on materials with large sticking coefficients. Fortunately, the sticking coefficient is usually less than one which increases the time available for an experiment.

1.5 Very High Resolution

The standard resolution in photoemission experiments is about 30 meV in UPS and 500 meV in XPS experiments, where the angular resolution is usually on the order of 1° to 2°. Since about 1990 many groups have tried to increase the resolution and one can now obtain in UPS experiments at 20 eV kinetic energy a resolution of 3 meV and in XPS experiments at 500 eV kinetic energy a resolution of 50 meV (or slightly better) with an angular resolution approaching 0.1°. These improved performances lead to a new quality in results. A resolution of 3 meV corresponds roughly to a thermal scale of 30 K such that the results obtained with photoemission spectroscopy now have the same sort of accuracy as those obtained from thermodynamic measurements. This is a new degree of sophistication and makes photoemission spectroscopy a much more general tool in solid state and surface physics than it has been previously. Of course it has to be emphasized that this kind of performance has a price not only with respect to the complexity of the experimental setup but also in the running of the experiments.

We now briefly review a few experimental results obtained with this very high resolution. These show the kind of information that is available in this field when working at the best possible performance. Of course this does not apply to most experiments reported in this book; however, we think it is useful to demonstrate the kind of detailed knowledge that can now be obtained by PES.

The first example is the measurement of the linewidth of surface states (see Chap. 8) in a metal. Briefly, a surface state is an electronic state confined to the surface of a material and it is produced by the termination of the three dimensionally infinite crystal by the two-dimensional surface. The first surface state for which data are presented is a so-called Tamm state, which is split off from the Cu 3d band in the surface layer. It arises because of the missing neighbor for the topmost ions on the Cu(100) surface. Figure 1.24 shows the result of the experiment with He I radiation at $T = 10$ K and a resolution of 5 meV [1.69]. The solid line is a fit to the data with a Lorentzian (to simulate the lifetime) convoluted with a Gaussian of 5 meV full width at half maximum to take into account the instrumental resolution. The width of the Lorentzian extracted from the data is 7 meV, which is one of the narrowest lines so far measured directly in a photoemission experiment on a solid.

Most interesting are the data in Fig. 1.25, which shows the spectrum of Fig. 1.24 measured in equal time intervals between 0 and 10 hours after surface preparation with constant illumination of the sample during the time of the experiment. We see that the signal increases in width and falls in intensity with time. This shows that the technique is in principle "destructive". The authors report that already at 70 K no such deterioration of the signal with time is observed. At this point it is not clear whether this effect is due to radiation damage or to the adsorption of residual gases, although we favor the latter possibility.

Figure 1.26 shows results for another surface state, namely the L-gap Shockley surface state (see Chap. 8 for details) on an Ag(111) surface, as it appears in measurements over more than two decades in time starting in 1977 up to the year 2000 (note that the topmost spectra have been obtained at

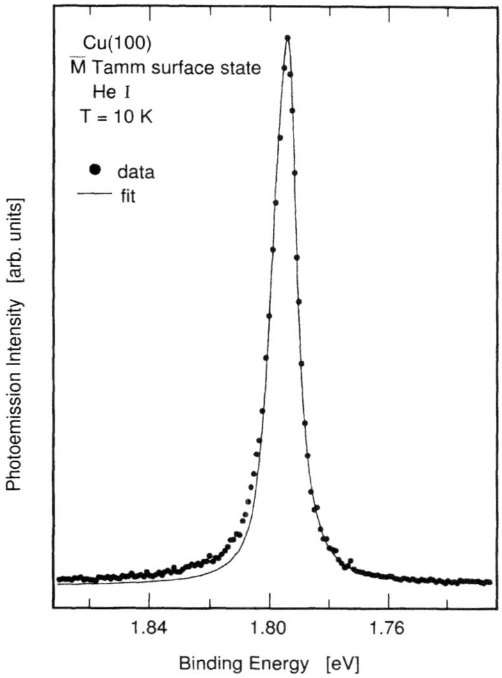

Fig. 1.24. Cu(100)$\overline{\mathrm{M}}$ Tamm surface state measured with He I radiation at $T = 10\,\mathrm{K}$ [1.69]. Fit is with a Gaussian of 5 meV. The fit to the data results in a Lorentzian peak with a width of 7 meV FWHM

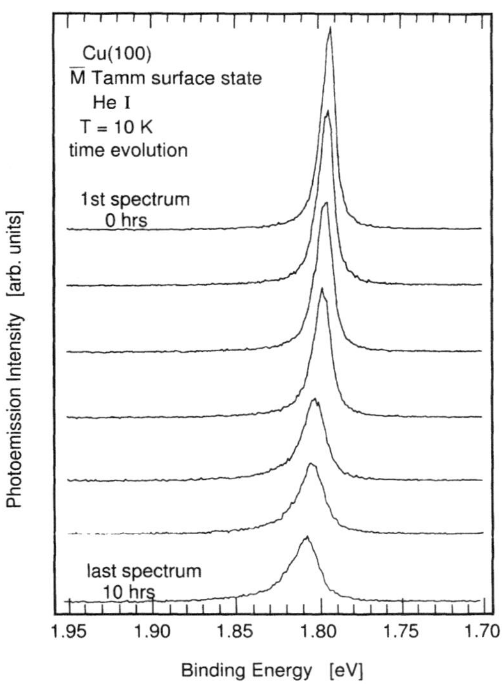

Fig. 1.25. Surface state as given in Fig. 1.24, measured as a function of time. The He I radiation was impinging on the sample continuously during the measuring time of 10 hours. An attenuation of the signal with time is clearly visible

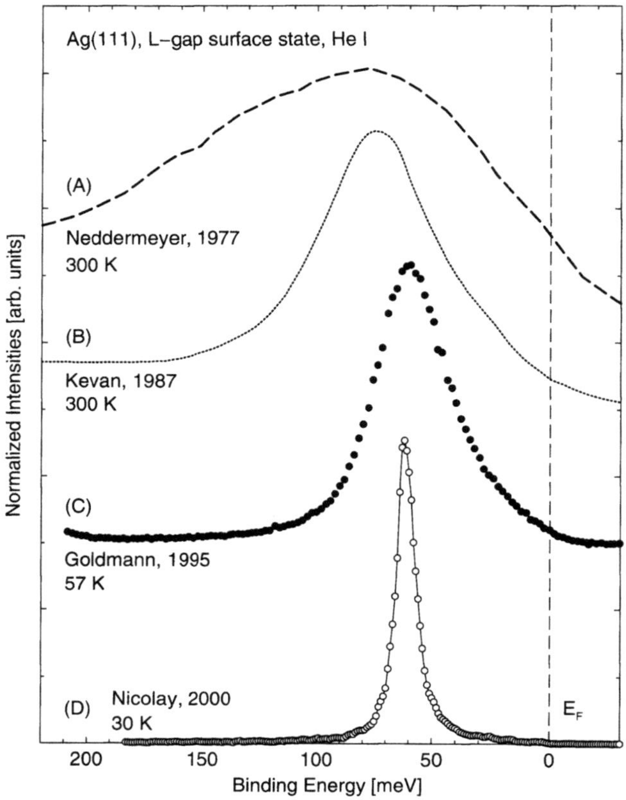

Fig. 1.26. The Ag(111) L-gap surface state, measured with He I radiation in normal emission, shown as a function of the time of the experiment. This diagram indicates the improvement in photoemission spectroscopy with time. Some of the broadening in the 300 K spectra is due to the higher temperature used to take these data [1.70]

room temperature, the two lowest ones below 100 K). It is obvious again that the improved resolution allows more detailed information to be obtained. The data in trace D which are analyzed in Fig. 1.27 yield an internal linewidth of this surface state (full line in Fig. 1.27) of 6.2 ± 0.5 meV, which is close to the theoretical value of 7.2 meV [1.70], and identical to the results of a measurement made by tunnel spectroscopy (dashed curve in Fig. 1.27).

The L-gap surface state has now been investigated with very high resolution in all three noble metals. The results of the spectra, measured at the bottom of the surface-state band are shown for all three noble metals in Fig. 1.28. The widths indicated are the lifetime widths corrected for instrumental broadening.

A detailed investigation of the surface state on a Mo(110) surface [1.71] is reported in Fig. 1.29, which shows the temperature and energy dependence of

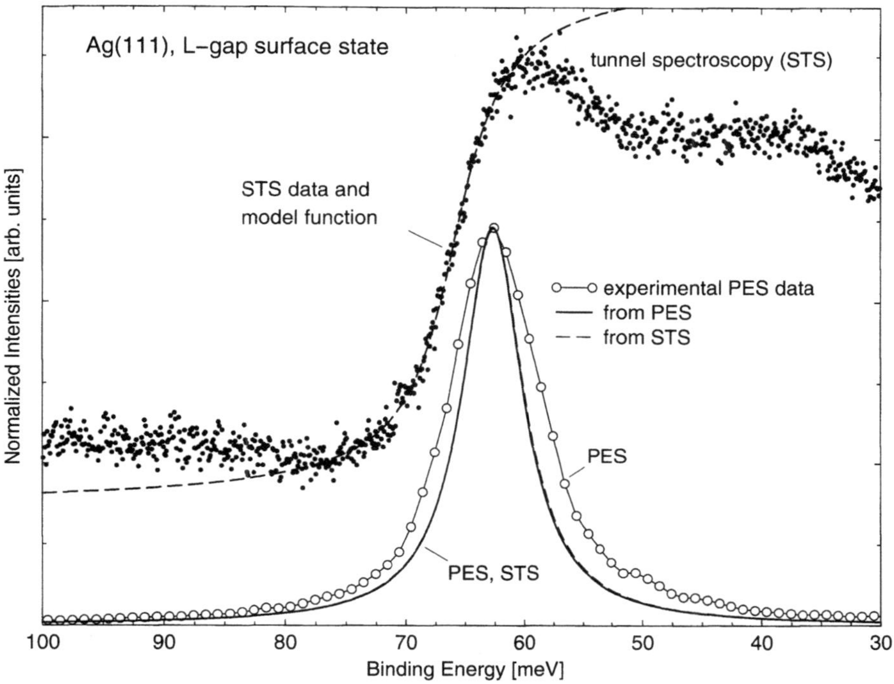

Fig. 1.27. Comparison of the width of the Ag(111) L-gap surface state measured in normal emission with scanning tunneling spectroscopy (STS) and photoelectron spectroscopy. The full line shows the photoemission data with an internal width of 6.2 meV as obtained by deconvoluting the experimental data (*points*). The tunnel data can be converted to a surface state as given by the dashed curve. The comparison shows that PES and STS give identical results and are able to resolve lifetimes corresponding to 10 meV and longer; in this case the width corresponds to a lifetime of 120 fs

this state. Figure 1.29a presents spectra at a fixed temperature as a function of the distance from the Fermi energy, indicating an increase in the width, and Fig. 1.29b gives the data for $E = E_{\mathrm{F}}$ as a function of temperature. Note that in all cases the data have been normalized to the Fermi function, which explains their slightly unusual shape. The energy dependence of the width as deduced from Fig. 1.29a is displayed in Fig. 1.29c and shows two regimes, one between the Fermi energy and 50 meV and one below 50 meV. The data can be fitted by assuming that at small energies the increase in width is produced by electron–phonon coupling, while at high energies it is due to the electron–electron coupling. The numbers deduced from this fit are in reasonable agreement with expectations.

While the atoms in the surface of a solid produce surface states in the valence band, they give rise to different binding energies in the core levels, because the core-level energies are also influenced by the valence band con-

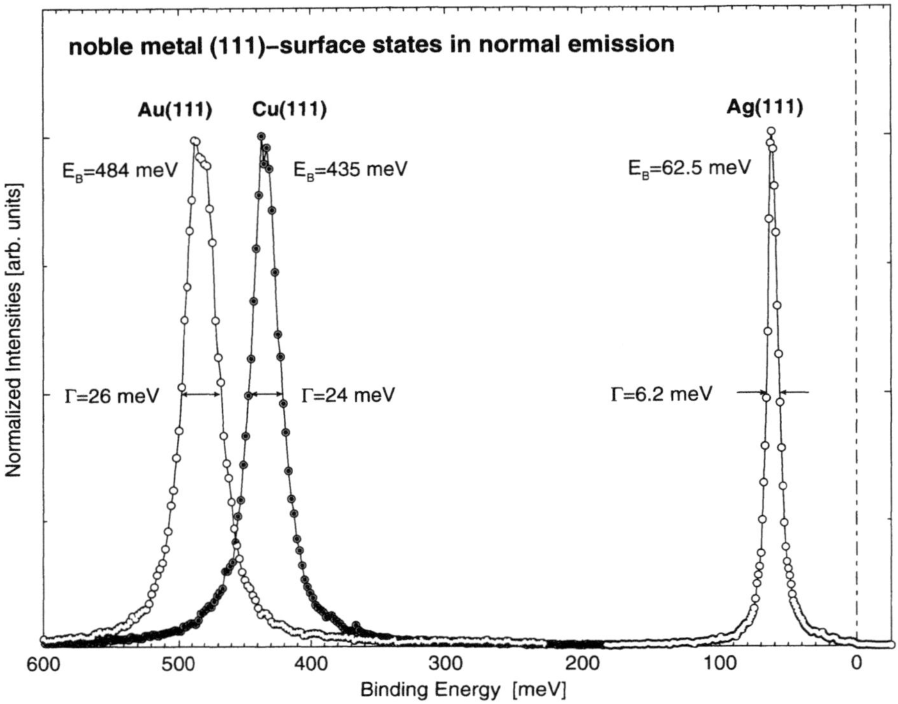

Fig. 1.28. L-gap surface states from the (111) surfaces of the three noble metals as measured at low temperature with a resolution of 3 meV. The indicated widths are the lifetime widths corrected for the instrumental broadening

figuration, which is different for atoms in the bulk and in various surface configurations. An illustration of this effect is displayed in Fig. 1.30a, which shows the 2p core level of Si measured with high resolution at a Si(111) surface at 55 K, where in addition to the bulk (B) level, five core-level contributions C_1 to C_5 with shifts relative to the bulk energy of $253, -706, 553, -188$ and 961 meV can also be observed [1.72]. Note that every component is in principle a doublet, because the 2p level splits into a $p_{3/2}$ and $p_{1/2}$ level by the spin–orbit interaction, which explains the many lines.

The final example is an investigation of the O 1s and C 1s core level of CO on Ni(100) measured with high resolution; the results are shown in Fig. 1.30b. Here it could be shown for the first time that each core level is accompanied by a number of vibrational satellites, which by comparison with other experimental data could be ascribed to the CO stretching frequency [1.73]. The excitation of vibrations is due to the fact that the potential energy surface for the atomic motion is different before and after the ionization. The difference in vibrational spacing for C 1s (217 meV) and O 1s (173 meV) is due to the fact that, in the O 1s photoionization, the elongation of the CO molecule is

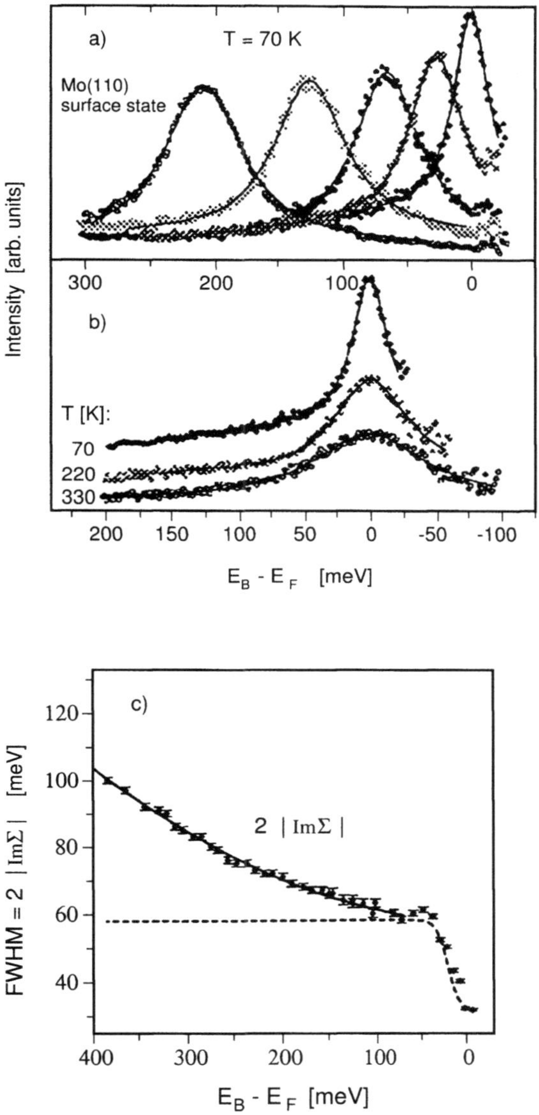

Fig. 1.29. High resolution photoemission data of the Mo(110) surface state [1.71]. (**a**) Surface state as a function of binding energy at $T = 70\,\mathrm{K}$. The data have been divided by the Fermi function, in order to get a clearer picture of the data near the Fermi energy. (**b**) Surface state of Mo(110) as a function of temperature at the Fermi energy. (**c**) Lifetime width of the surface state of Mo(110), deduced from data in (**a**), plotted as a function of the binding energy. Two regimes are visible, which are discussed in the text. The steep increase in linewidth near the Fermi energy is due to electron–phonon interactions, whereas at higher binding energies the increase in width is produced by electron–electron interactions

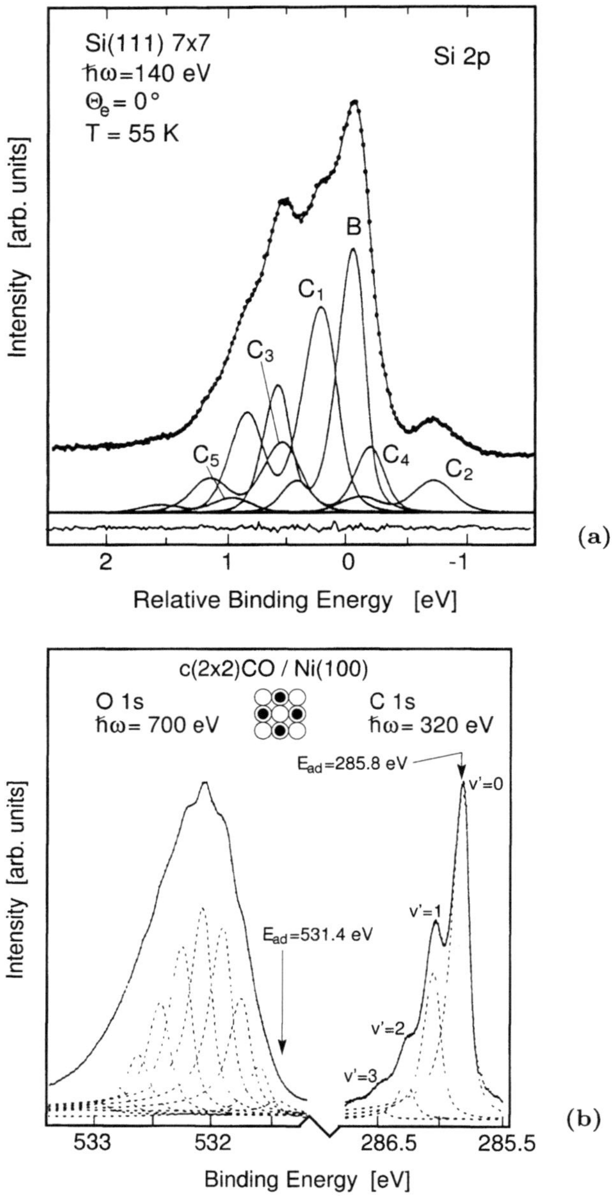

Fig. 1.30. (**a**) High resolution Si 2p photoemission spectra from a Si(111)7×7 surface, measured at $T = 55$K in normal emission. The data have been analyzed by one bulk component (B) and five surface components C1–C5 [1.72]; there is a spin–orbit splitting of each component. (**b**) High resolution O 1s and C 1s photoemission spectra of the C(2×2) CO-Ni(100) structure in normal emission. The core lines show a structure which can be analyzed by a superposition of the CO stretching frequency [1.73]. The adiabatic ionization energy is defined in Fig. 5.2

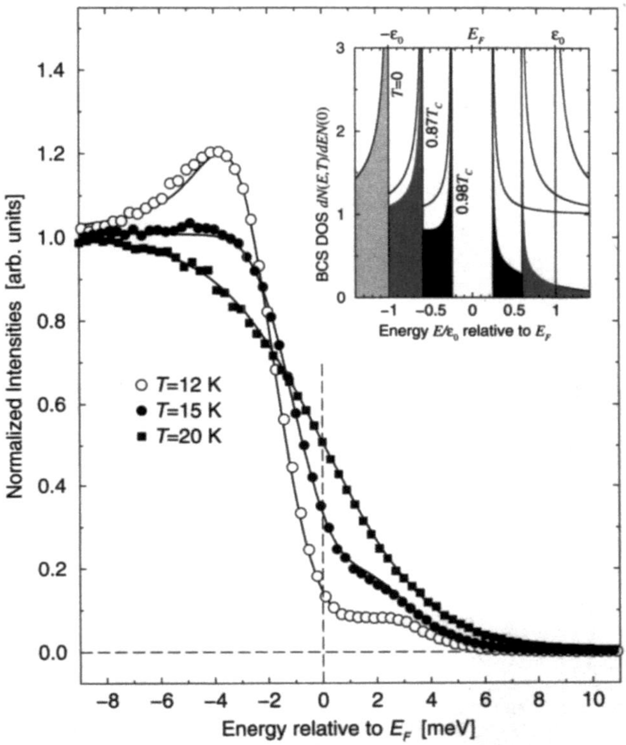

Fig. 1.31. Valence band photoemission in the energy regime around the Fermi energy for V_3Si, at temperatures around the superconducting transition temperature ($T_c = 17\,K$). The inset shows the theoretical BCS density of states

larger (weaker bonding and hence smaller frequency) than in the C 1s ionization (stronger bonding and hence higher vibrational frequency). This fact is consistent with the known bonding of CO on Ni by the C atom (Chap. 5).

Next we present an example in which the high resolution now possible in PES is employed to measure an interesting property in bulk solid state physics, namely the superconducting gap in the conventional superconductor V_3Si ($T_C = 17\,K$). Figure 1.31 (inset) shows the density of states for different temperatures below T_C expected in the BCS theory for a conventional superconductor. It consists of two singular parts at the lower (occupied) and the upper (unoccupied) part of the superconducting gap. For finite temperatures, the Fermi distribution function reduces the occupation in the lower part of the density of states and weakly populates the unoccupied (at $T = 0K$) part of the density of states. In addition, an increasing temperature reduces the gap which is $2\varepsilon_0$ at $T = 0K$. In the measured spectrum this results in a hump in the lower part of the density of states and a small signal in the upper part of the density of states (note that actual data will be more smeared out, because

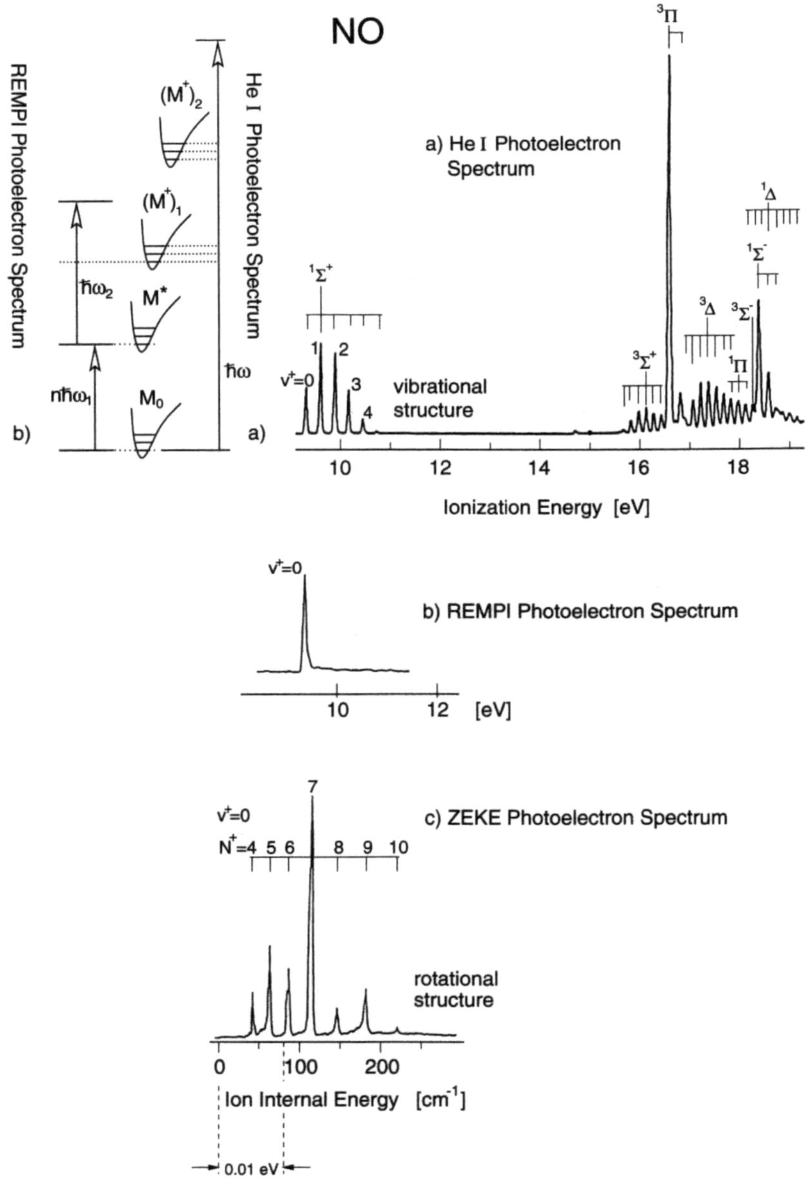

Fig. 1.32. Photoelectron spectroscopy on NO. The diagrams on the left side show the direct photoemission process and the Resonantly Enhanced Molecular PhotoIonization (REMPI) process. The spectra are: (**a**) standard NO photoemission spectrum showing the electronic transitions and their vibrational structure; (**b**) REMPI spectrum displaying only one level ($\nu^+ = 0$) of the $^1\sum^+$ final state; (**c**) ZEKE spectrum of the level in (**a**) displaying the rotational structure [1.74]

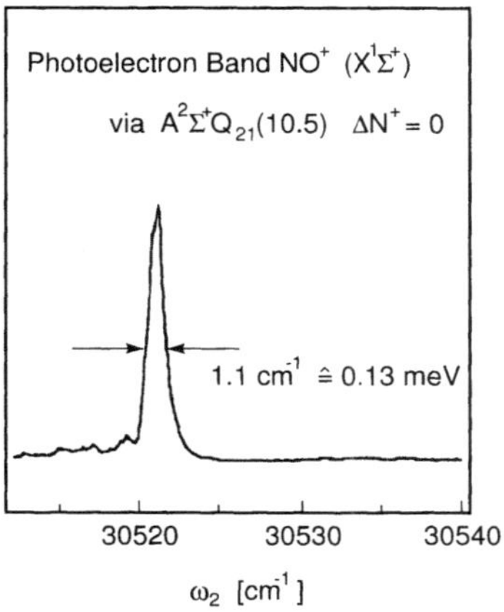

Fig. 1.33. High resolution spectrum with ZEKE of one particular level in the spectrum of NO from Fig. 1.32c showing that this method has an inherent resolution in the order of 0.1 meV [1.74]

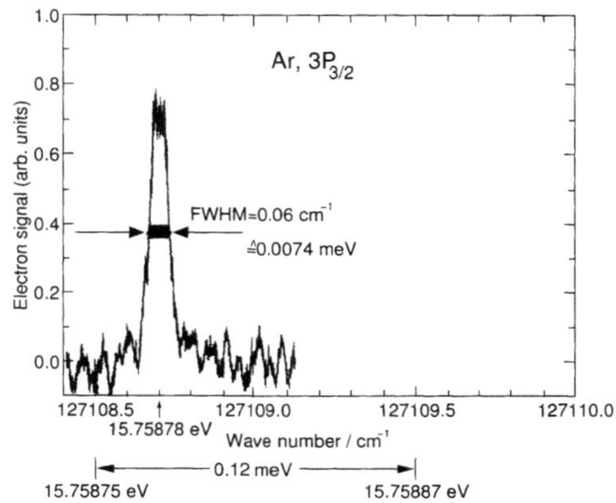

Fig. 1.34. High resolution ZEKE photoemission spectrum from gaseous Argon. Only the $3p_{3/2}$ line is shown with an inherent width of 0.0074 meV [1.93]

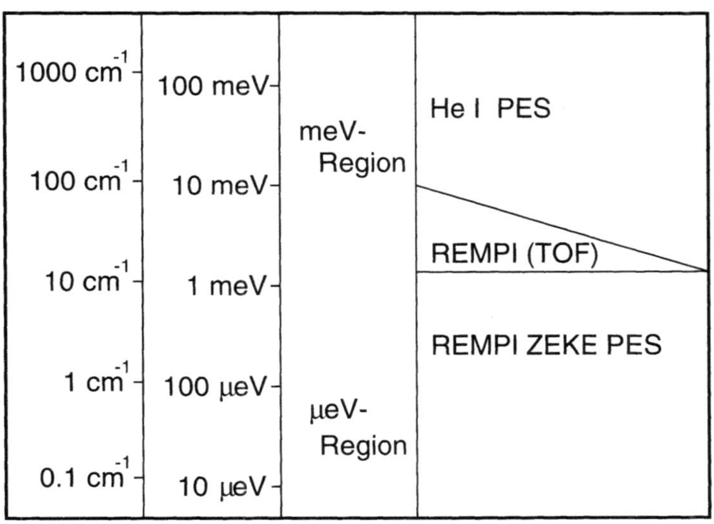

Fig. 1.35. Diagram showing the resolution that can be achieved by ordinary photoemission spectroscopy and special forms of photoelectron spectroscopy, which so far, however, have only been applied to molecules [1.74]; ZEKE stands for Zero Kinetic Energy; REMPI for Resonantly Enhanced Multiphoton Ionization

of a 3 meV resolution width). The spectra presented in the figure are in full agreement with the theoretical expectations. This example demonstrates convincingly that it is now possible to obtain data by PES with a resolution that is compatible with that achieved in other techniques like tunnel spectroscopy or thermodynamic experiments, e. g. specific heat measurements. The data in Figs. 1.24 –1.31 demonstrate the power of photoemission spectroscopy when used under optimal experimental conditions.

Better resolution than the 3 meV just mentioned for "conventional" PES can be obtained by using PES with two or more Laser-photon excitations to study molecular energy levels [1.74]. This technique will not feature in the remainder of this volume, but at this point a brief demonstration of its potential will be presented. For further details the reader is referred to the literature [1.74]. In principle the method consists of exciting an electron in a molecule with a first laser pulse (using single- or multi-photon absorption) into an excited intermediate state, from which a subsequent Laser pulse (consisting generally, although not always of one laser photon) liberates the electron into the detector where it is recorded with high energy resolution. The two analyzing techniques used in this field are the time of flight (TOF) method, or the zero kinetic energy (ZEKE) method, yielding a resolution in the sub meV range.

As an example of the potential of the method some data for the photoemission spectroscopy of the NO molecule are presented. This molecule has in the valence band ($2s^2 2p^3$ for N and $2s^2 2p^4$ for O) the following electronic structure: $(\sigma_g\,2s)^2(\sigma_u\,2s)^2(\sigma_g\,2p)^2(\pi_u\,2p)^4(\pi_g\,2p)^1\ {}^2\Pi$. The highest occupied orbital is the singly occupied $(\pi_g\,2p)^1$ orbital and its photoionization by conventional PES leads to the ${}^1\Sigma^+$ final state at 10 eV (Fig. 1.32a). The fine structure of the spectrum is due to the excitation of vibrational states (see Chap. 5). The energy levels with an ionization energy of 16 eV and more in the conventional PE spectrum are due to the photoionization out of the σ_g and π_u states. The width of the lines is of the order of 30 meV, which corresponds to about 300 cm^{-1} in the spectroscopic notation ($1\,cm^{-1} = 1.24 \times 10^{-4}$ eV).

A ZEKE photoelectron spectrum is also given in Fig. 1.32c, where in Fig. 1.33 the scan through a single rotational level is presented. Here, ω_2 is the energy of the second photon which is scanned through the rotational level, yielding a linewidth of the order of 1 cm^{-1} or 100 μeV, considerably below the 3 meV that can be achieved these days by conventional PES. The nomenclature describing molecular energy levels can be found in the work of Herzberg [1.75].

The highest resolution reported in the literature so far is slightly below 0.01 meV and a typical example is shown in Fig. 1.34 [1.93]. This figure shows the ZEKE photoemission spectrum of the $3p_{3/2}$ final state from gaseous argon at a binding energy of 15.75878 eV. This spectrum should be compared to the $3p_{3/2}$ line in Fig. 5.1, recorded with a conventional photoelectron spectrometer, where the width is 28 meV. The comparison of these two figures shows that practically all widths observed in Fig. 5.1 are due to the combined resolution of the lamp and the photoelectron spectrometer.

Finally in Fig. 1.35 we give a summary of the resolution obtainable these days with standard PES but also with the techniques that so far have only been applied to molecules.

1.6 The Theory of Photoemission

Photoemission Spectroscopy is a technique in which a photon of energy $\hbar\omega$ liberates an electron from a system. Except for the hydrogen atom (and hydrogen-like systems such as He^+) the systems under study always contain more than one electron, and therefore the system remaining after the photoelectron has left the sample still contains electrons (and protons). The final state can therefore be viewed as one where, in contrast to the initial state, one electron has been removed, or a positive potential has been added. This can make the final state substantially different to the initial state. In systems with many degrees of freedom (many electrons) the change can be quite complex. Therefore the theory describing PES is a complicated many-body theory. It is not the object of this chapter or this book to describe this theory in any detail; instead the reader is referred to the corresponding literature [1.76–1.83].

On the other hand, the understanding of many experimental results is difficult, if not impossible, without some theoretical insight into the underlying principles of PES. Thus this section attempts to sketch the theoretical approaches used in the literature to explain the complicated many-body process of photoemission. In some of the following chapters the theory presented in this section will be elaborated upon. This section may seem difficult and lengthy: However, it is our belief that a working knowledge of the basic concepts of PES will help readers to understand the literature.

Since PES is used in various fields, different degrees of theoretical sophistication are necessary for the analysis of the data. If PES is employed as an analytical tool, its many-electron aspects are generally unimportant. If on the other hand the spectral function of a metal is investigated, a detailed theory of the response of the conduction-electron system to the photohole must be known. In order to cope with this problem we shall present here a brief introduction to the theory of PES and give additional material in those chapters where this is necessary.

One can, without any calculation, give some qualitative arguments about the nature of a PE spectrum. First there will be a difference between valence and core spectra. The energy of a core level does not depend on the wave vector, because of the localization of its wave function. Therefore a core-level spectrum will, in the simplest view, be a Lorentzian (where the measured width is given by a convolution of the instrumental width and the lifetime width of the core-hole state). In the next step of sophistication, one

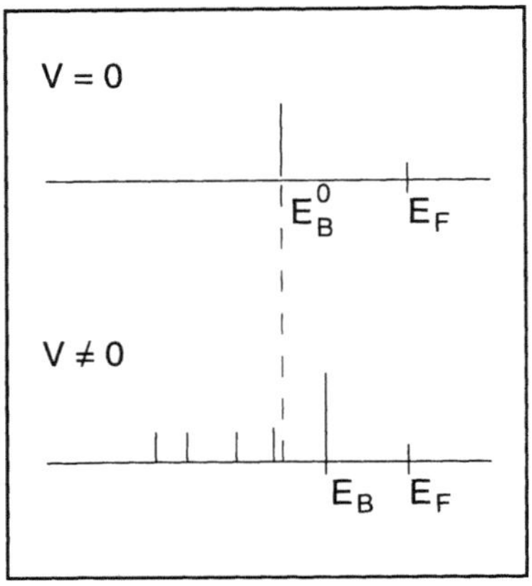

Fig. 1.36. Schematic diagram of a photoemission spectrum with no electron–electron interaction leading to a single line (*above*) and with electron–electron interaction (*below*) leading to a line accompanied by so-called satellites, which reflect internal excitations of the system from which photoionization has taken place. The center of gravity of the two spectra is identical

has to take into account the interaction of the remaining photoionized system with the potential created by the core hole. This interaction will most strongly affect the charge contained in the valence band. Assuming the valence band consists of discrete energy levels, these levels can be excited by the core-hole potential and show up in the core-level spectrum as so-called satellites at lower kinetic energies (higher binding energies). Figure 1.36 shows schematic spectra for both a system of non-interacting electrons (V = 0; V = electron–electron interaction) and for interacting electrons (V ≠ 0). In both spectra the center of gravity is the same, namely at the so-called Koopmans' binding energy $E_B^0 = -\varepsilon_k$ (see later), which is the binding energy in the non-interacting case. In the interacting case additional lines (satellites) show up in the spectrum together with the main line (now with binding energy E_B), however, the center of gravity of the spectral weight remains unchanged. If the material under investigation is a metal, with an (almost) infinite number of excitation possibilities in the valence electron sea, these will show up in the core-level spectra. In a metal these excitations are either electron–hole pairs, whose creation probability obeys an inverse power law of their energy, thus diverging towards infinity for $\Delta E_{eh} \to 0$ (ΔE_{eh} being the energy of an electron–hole pair) and giving rise to an asymmetric line shape, or collective oscillations of the system of electrons relative to the system of positive lattice ions. The latter excitations are known as the plasmons, which also show up in the core-level spectra as (intrinsic) sidebands (satellites) (Chap. 4). One can thus anticipate that the bare core line (neglecting any interactions in the system under investigation) will in this situation get convoluted with the interactions by the response of the system to the core-hole potential. In principle the situation is the same for the valence band photoemission. However, additional complications arise here because the hole is created in the responding system (i.e. the sea of valence electrons) and, in addition, valence states generally have a wave-vector dependence due to the delocalization of the valence electrons.

At the heart of all theories of PES is the so-called sudden approximation, which means that the response of the system to the creation of the photohole is assumed instantaneous and that there is no interaction between the escaping photoelectron and the remaining system (but see [1.83]). This approximation is designed for the high-energy photoelectron limit.

In what follows the photoemission intensity will be calculated for various cases, which are intimately connected with each other. Note, however, that different communities have developed slightly different nomenclatures in the literature.

The photocurrent produced in a PES experiment results from the excitation of electrons from the initial states i with wave function Ψ_i to the final states f with wave function Ψ_f by the photon field having the vector potential $\boldsymbol{A}$.

Assuming a small perturbation Δ the transition probability w per unit time between the N-electron states Ψ_i (initial) and Ψ_f (final) is calculated by Fermi's Golden Rule, which yields

$$w \propto \frac{2\pi}{\hbar}|\langle\Psi_f|\Delta|\Psi_i\rangle|^2\delta(E_f - E_i - \hbar\omega) . \tag{1.19}$$

In the most general form one has

$$\Delta = \frac{e}{2mc}(\boldsymbol{A}\cdot\boldsymbol{p} + \boldsymbol{p}\cdot\boldsymbol{A}) - e\phi + \frac{e^2}{2mc^2}\boldsymbol{A}\cdot\boldsymbol{A} , \tag{1.20}$$

where $\boldsymbol{A}$ and ϕ are the vector and scalar potentials and $\boldsymbol{p}$ the momentum operator $\boldsymbol{p} = \mathrm{i}\hbar\boldsymbol{\nabla}$ (ϕ should not be confused with the work function). The commutation relation leads to the relation $\boldsymbol{A}\cdot\boldsymbol{p}+\boldsymbol{p}\cdot\boldsymbol{A} = 2\boldsymbol{A}\cdot\boldsymbol{p}+i\hbar(\boldsymbol{\nabla}\cdot\boldsymbol{A})$. If one uses the gauge $\phi = 0$, neglects the term $\boldsymbol{A}\cdot\boldsymbol{A}$, which represents the two photon processes, and assumes that div $\boldsymbol{A} = \boldsymbol{\nabla}\cdot\boldsymbol{A} = 0$ because of the translational invariance in the solid, one retains

$$\Delta = \frac{e}{mc}\boldsymbol{A}\cdot\boldsymbol{p} . \tag{1.21}$$

For $\hbar\omega = 10\,\mathrm{eV}$ one has $\lambda = 10^3$ Å and therefore one can assume, at least for a wide range of experiments, that the wavelength is large compared to the atomic distances and therefore $\boldsymbol{A}$ can be taken as constant $\boldsymbol{A} = \boldsymbol{A}_0$, where the zero will be dropped in what follows. The commutation relations lead to the following equivalence:

$$\langle\Psi_f|\boldsymbol{A}\cdot\boldsymbol{p}|\Psi_i\rangle \propto \langle\Psi_f|\boldsymbol{A}\boldsymbol{\nabla}\cdot V|\Psi_i\rangle \propto \langle\Psi_f|\boldsymbol{A}\cdot\boldsymbol{r}|\Psi_i\rangle . \tag{1.22}$$

if the Hamiltonian of the electron (e.g., in the solid in the absence of the electromagnetic field) is

$$H = (\boldsymbol{p}^2/2m) + V(\boldsymbol{r}) . \tag{1.23}$$

One realizes that for a free-electron gas ($V(\boldsymbol{r}) =$ const leading to $\nabla\cdot V = 0$) no photoemission is possible, a problem, which will be elaborated upon later.

1.6.1 Core-Level Photoemission

For core-level photoemission, the commonly used form of the transition probability can therefore be written as ($\boldsymbol{A} = \text{const}$)

$$w \propto \frac{2\pi}{\hbar} |\langle \Psi_f | \Delta | \Psi_i \rangle|^2 \delta(E_f - E_i - \hbar\omega) . \qquad (1.24)$$

Using the equivalence in (1.22) and realizing that $\boldsymbol{A}$ is constant over the interaction volume, the matrix element in (1.24) is often written with $\boldsymbol{r}$ as the operator instead of Δ.

In order to arrive at a first view of PES, it is assumed that the system under investigation has N electrons and it is implicitly understood that one is dealing with core-level PES, although this is not necessary.

For a discussion of the transition matrix element we have to make certain assumptions about the wave functions contained in it. In the simplest approximation one can take a one-electron view for the initial- and final-state wave function. In the final state one has, in addition, a free electron with kinetic energy E_{kin}. The initial-state wave function is then written as a product of the orbital ϕ_k from which the electron is excited and the wave function of the remaining electrons $\Psi^k_{i,\text{R}}(N-1)$, assuming that the system under consideration has N electrons, where the index k indicates that the electron with quantum number k has been left out (because it was photoexcited):

$$\Psi_i(N) = C\phi_{i,k}\Psi^k_{i,\text{R}}(N-1) \quad \text{(R stands for remaining)} \qquad (1.25)$$

(C is the operator that antisymmetrizes the wave function properly).

In the same spirit, the final state is written as a product of the wave function of the photoemitted electron $\phi_{f,E_{\text{kin}}}$ and that of the remaining $(N-1)$ electrons $\Psi^k_{f,\text{R}}(N-1)$

$$\Psi_f(N) = C\phi_{f,E_{\text{kin}}}\Psi^k_{f,\text{R}}(N-1) , \qquad (1.26)$$

so that the transition matrix element in (1.24) is obtained as

$$\langle \Psi_f | \boldsymbol{r} | \Psi_i \rangle = \langle \phi_{f,E_{\text{kin}}} | \boldsymbol{r} | \phi_{i,k} \rangle \langle \Psi^k_{f,\text{R}}(N-1) | \Psi^k_{i,\text{R}}(N-1) \rangle . \qquad (1.27)$$

The matrix element is thus a product consisting of a one-electron matrix element and an $(N-1)$-electron overlap integral. In the first step of evaluating the overlap integral, one can assume that the remaining orbitals (often called the *passive orbitals*) are the same in the final state as they were in the initial state (*frozen-orbital approximation*), meaning that $\Psi^k_{f,\text{R}}(N-1) = \Psi^k_{i,\text{R}}(N-1)$, which renders the overlap integral unity, and the transition matrix element is just the one-electron matrix element. Under this assumption, the PES experiment measures the negative Hartree–Fock orbital energy of the orbital k, i.e.

$$E_{\text{B},k} \simeq -\varepsilon_k \qquad (1.28)$$

which is sometimes called Koopmans' binding energy.

Of course, this cannot be a very good approximation: One realizes intuitively that the system, after ejection of the electron from orbital k will try to readjust its remaining $N-1$ charges in such a way as to minimize its energy (relaxation).

We now assume that the final state with $N-1$ electrons has s excited states with the wave function $\Psi^k_{f,s}(N-1)$ and energy $E_s(N-1)$. We shall use, for convenience, s as the number of states and the running index.

The transition matrix element must then be calculated by summing over all possible excited final states yielding:

$$\langle \Psi_f | \boldsymbol{r} | \Psi_i \rangle = \langle \phi_{f,E_{\mathrm{kin}}} | \boldsymbol{r} | \phi_{i,k} \rangle \sum_s c_s \; , \tag{1.29}$$

with

$$c_s = \left\langle \Psi^k_{f,s}(N-1) | \Psi^k_{i,\mathrm{R}}(N-1) \right\rangle \; . \tag{1.30}$$

Here $|c_s|^2$ is the probability that the removal of an electron from orbital ϕ_k of the N-electron ground state leaves the system in the excited state s of the $N-1$-electron system. For strongly correlated systems many of the c_s are non-zero. In terms of the PE spectrum, this means that for $s=k$ one has the so-called *main line* and for the other non-zero c_s additional *satellite lines* occur. On the other hand, if correlations are weak one has

$$\Psi^k_{f,s}(N-1) \simeq \Psi^k_{i,\mathrm{R}}(N-1) \; , \quad \text{for} \quad s=k \tag{1.31}$$

meaning that $|c_s|^2 \simeq 1$ for $s=k$ and $|c_s|^2 \simeq 0$ for $s \neq k$, i.e. one has only one peak.

This now allows us to express the photocurrent I detected in a PE experiment as

$$\begin{aligned} I \propto & \sum_{f,i,k} \left| \langle \phi_{f,E_{\mathrm{kin}}} | \boldsymbol{r} | \phi_{i,k} \rangle \right|^2 \sum_s |c_s|^2 \\ & \times \delta(E_{f,\mathrm{kin}} + E_s(N-1) - E_0(N) - \hbar\omega) \; , \end{aligned} \tag{1.32}$$

where $E_0(N)$ is the groundstate energy of the N-electron system. The photocurrent thus consists of the "lines" created by photoionizing the various orbitals k, where each line can be accompanied by satellites according to the number of excited states s created in the photoexcitation of that particular orbital k (Fig. 1.36).

One often writes:

$$A = \sum_s |c_s|^2 \tag{1.33}$$

and calls A the spectral function (not to be confused with the vector potential of the incoming radiation).

Note that (1.32) contains the main elements of a core PE spectrum as discussed at the beginning of this chapter (Fig. 1.36) namely a main line accompanied by satellites which reflect the excitation spectrum of the system excited by the core hole.

For core-level PES in metals, with (almost) infinitely many degrees of freedom, a slightly different formulation for the photoemission spectrum is necessary [1.81]. One writes (note that c_s was an overlap matrix element) [1.81],

$$A(E) = \sum_s |\langle i|f_s\rangle| \,. \tag{1.34}$$

The time variation of the outer electron system due to its interaction with the core hole $|f(t)\rangle$ is:

$$|f(t)\rangle = \exp(-(\mathrm{i}/\hbar)(H - E_0)t)|i\rangle \,.$$

Here $H = H_0 + H_{\mathrm{int}}$ is the final state Hamiltonian with $H_0|i\rangle = E_0$ and $H|f\rangle = E_f|f\rangle$.

From this, the dynamical response of the outer electron system to the core hole is obtained as $g(t) = \langle i|f(t)\rangle$.

The spectral function is calculated by a Fourier transform of $g(t)$ as:

$$A(E) = \frac{1}{\pi}\mathrm{Re}\int_0^{+\infty} \mathrm{d}t \exp\left[\mathrm{i}\left(\frac{E - E_{\mathrm{B}}}{\hbar}\right)t\right] g(t) \,, \tag{1.35}$$

where the many-body effects are contained in $g(t)$, the dynamical response of the valence electron system to the core hole. If there is no interaction between the core hole and the valence electrons ($H_{\mathrm{int}} = 0$), one has $g(t) = 1$ and

$$A(E) = \frac{1}{\pi}\delta(E - E_{\mathrm{B}}) \,. \tag{1.36}$$

If the conduction electrons screen the core hole, then

$$g(t) = 1/t^{\alpha} \tag{1.37}$$

and

$$A(E) = \frac{1}{(E - E_{\mathrm{B}})^{1-\alpha}} \tag{1.38}$$

with $\alpha = 2\sum_{\ell}(2\ell + 2)(\delta_{\ell}/\pi)^2$. Here ℓ is the angular momentum of the conduction electron and δ_{ℓ} the phase shift of a conduction electron with angular momentum ℓ. This is the asymmetric line with a divergence at the binding energy mentioned before. It has its origin in the creation of the electron–hole pairs. This problem will be dealt with extensively in Chap. 4.

1.6.2 Valence-State Photoemission

We shall now deal with the photoemission in valence states because this is an area where the technique has produced much new information. The new aspects that enter the treatment are the wave vector dependence of the electronic states (because of the delocalization of the electrons) and the fact that the response is now (mostly) in the system from which the electron is photoexcited.

In working out the matrix element in (1.19) use is made of the periodicity of the Bloch function with respect to $\boldsymbol{G}$, the reciprocal lattice vector, yielding

$$\mathrm{I} \propto \left|\tilde{M}_{if}^{1}\right|^{2} \delta\left(\boldsymbol{k}_{i||} - \boldsymbol{k}_{f||} + \boldsymbol{G}_{||}\right) \delta\left(\boldsymbol{k}_{i} - \boldsymbol{k}_{f} + \boldsymbol{G}\right) \times \delta\left(E_{f} - E_{i} - \hbar\omega\right) \delta\left(E - E_{f} + \phi\right) . \quad (1.39)$$

(The matrix element is written as $\tilde{M}^{1}$ in order to indicate that the momentum conservation has been written out explicitly and that it is taken between Bloch states. The matrix element containing the momentum conservation is written as M^{1}, and if the matrix element is taken between an initial Bloch state and a final inverse LEED state (see later) it will be written as M or $\tilde{M}$.)

Because of the Fresnel equations, the momentum conservation parallel to the surface shows up separately. We have added an additional δ-function $\delta(E - E_f + \phi)$ in order to indicate the fact that only photoemitted electrons with an energy above the vacuum level ϕ can be detected in the experiment.

Next, it has to be realized that the photoexcited electrons have a relatively short inelastic mean free path within the solid. This can be taken into account by assuming a complex wave vector perpendicular to the surface of the solid leading to:

$$k_{\perp} = k_{\perp}^{(1)} + \mathrm{i}k_{\perp}^{(2)} . \quad (1.40)$$

This leads to a smearing of the momentum conservation resulting in the following relation for the photocurrent:

$$\mathrm{I} \propto \frac{\left|\tilde{M}_{if}^{1}\right|^{2}}{\left(k_{i\perp}^{(1)} - k_{f\perp}^{(1)}\right)^{2} + \left(k_{f\perp}^{(2)}\right)^{2}} \delta\left(\boldsymbol{k}_{i||} - \boldsymbol{k}_{f||} + \boldsymbol{G}_{||}\right) \delta(\boldsymbol{k}_{i} - \boldsymbol{k}_{f} + \boldsymbol{G}) \times \delta\left(E_{f} - E_{i} + \hbar\omega\right) \delta\left(E - E_{f} + \phi\right) . \quad (1.41)$$

One then has to take into account that the electrons in a solid interact with one another. This means that, generally, the final state will be one, in which not only is a hole created by the photoemission process (whereby the one hole ground state is obtained), but, via the electron–electron interacting, the final hole state of the system can be found in any one of the possible s final states $|\boldsymbol{k}_f, N-1, s\rangle$ of the many electron system. To analyze (1.19) with electron–electron interactions included explicitly, one can start out in the same way as in the discussion of the core-level photoemission. In the spirit of (1.25)–(1.30) one can write

$$\Psi_f(N) = \phi_{f,E_{\text{kin}}}(\boldsymbol{k}_f) \cdot \sum_s \Psi_{f,s}(N-1) \tag{1.42}$$

and the initial state $\phi_i(\boldsymbol{k}_i)$ is the wave function of the state from which the electron is photoemitted):

$$\Psi_i(N) = \phi_i(\boldsymbol{k}_i)\Psi_i(N-1) , \tag{1.43}$$

where the state $\Psi_i(N)$ has the energy $E_0(N)$ and the final states, s, the final state energies $E_s(N-1)$.

This leads to the following form of the photocurrent:

$$\begin{aligned} \mathrm{I} \propto \sum_{s,i} & \left|\left\langle \tilde{\phi}_{f,E_{\text{kin}}} |\boldsymbol{r}| \tilde{\phi}_i \right\rangle\right|^2 \left|\left\langle \tilde{\Psi}_{f,s}(N-1) | \tilde{\Psi}_i(N-1) \right\rangle\right|^2 \\ & \times \frac{1}{\left(k_{i\perp}^{(1)} - k_{f\perp}^{(1)}\right)^2 + \left(k_{f\perp}^{(2)}\right)^2} \delta\left(\boldsymbol{k}_{i||} - \boldsymbol{k}_{f||} + \boldsymbol{G}_{||}\right) \delta\left(\boldsymbol{k}_i - \boldsymbol{k}_f + \boldsymbol{G}\right) \\ & \times \delta(E_{\text{kin}} + E_s(N-1) - E_0(N) - \hbar\omega)\delta(E - E_f + \phi) . \end{aligned} \tag{1.44}$$

The $\tilde{\phi}_{f,E_{\text{kin}}}$ differs from $\phi_{f,E_{\text{kin}}}$ in that the momentum conservation has been worked out; the tilde in the other wave functions has the same meaning.

The term $|\langle \tilde{\Psi}_{f,s}(N-1)|\tilde{\Psi}_i(N-1)\rangle|^2$ has a simple interpretation, which becomes evident if we start with the independent particle approximation. If there are no electron–electron interactions present, one has:

$$\tilde{\Psi}_{f,s}(N-1) = \tilde{\Psi}_i(N-1) , \tag{1.45}$$

and therefore the overlap integral yields unity.

In a system with electron–electron interactions the $(N-1)$ state is generally not an eigenstate of the system. We observe intensity in the spectrum only for those final states $\tilde{\Psi}_{f,s}(N-1)$, that have a finite overlap with the $(N-1)$ groundstate $\tilde{\Psi}_i(N-1)$. One can therefore write

$$A(\boldsymbol{k}, E) = \sum_s \left|\left\langle \tilde{\Psi}_{f,s}(N-1) | \tilde{\Psi}_i(N-1) \right\rangle\right|^2 , \tag{1.46}$$

where $A(\boldsymbol{k}, E)$ is called the spectral function of an electron with energy E and momentum $\boldsymbol{k}$. In the nomenclature of second quantization

$$A(\boldsymbol{k}, E) = \sum_s |\langle N-1, s|c_{\boldsymbol{k}}|N\rangle|^2 , \tag{1.47}$$

where $c_{\boldsymbol{k}}$ is an electron annihilation operator. A($\boldsymbol{k}$,E) describes the probability with which an electron can be removed from an electron system in its ground state. It thus has the same meaning as the overlap integral used before.

The spectral function is intimately connected to a fundamental property of an N-electron system, namely its one-particle Green's function $G(r_1, r_2, t)$, which describes the probability that an electron at r_1 at time $t = 0$ will be found at r_2 at a later time t. Transforming the Green's function into reciprocal space $\boldsymbol{k}$ and into the energy domain (E) one obtains the representation $G(\boldsymbol{k}_1, \boldsymbol{k}_2, E)$. In a shorthand notation we can say that the one-electron Green's function describes the probability that an electron in a state $\boldsymbol{k}_1$ is found in a state $\boldsymbol{k}_2$ if a scattering process with energy transfer E has occurred. Neglecting the non-diagonal parts of the Green's function one obtains $G(\boldsymbol{k}, E)$. Its imaginary part is equal to the spectral function $A(\boldsymbol{k}, E)$.

To make the connection with (1.10) apparent, the one-electron Green's function is written as (the following equations obviously contain some redundancies):

$$G(\boldsymbol{k}, E) = \sum_s |\rangle \Psi_s(N-1)|\Psi_i(N-1)\langle|^2 \frac{1}{E - E(\boldsymbol{k}) + \mathrm{i}\delta}$$
$$= \left\langle \Psi_i(N) \left| c_{\boldsymbol{k}} \frac{1}{E + H - \mathrm{i}\delta} c_{\boldsymbol{k}}^{+} \right| \Psi_i(N) \right\rangle , \qquad (1.48)$$

where $\Psi_i(N)$ is the N-particle initial state ground state wave function, $\Psi_i(N-1)$ is the $(N-1)$-particle initial state wave function, $c_{\boldsymbol{k}}^{+}$ is the creation operator of an electron with wave vector $\boldsymbol{k}$, $c_{\boldsymbol{k}}$ the annihilation operator, H the interaction Hamiltonian, and δ a positive infinitesimal number.

The one-electron Green's function *G(k,E)* and the one-electron spectral function *A(k,E)* are related by the equation

$$G(\boldsymbol{k}, E) = \int_{-\infty}^{+\infty} \mathrm{d}E' \frac{A(\boldsymbol{k}, E')}{E - E' - \mathrm{i}\delta} . \qquad (1.49)$$

Using the Dirac identity

$$\frac{1}{E - E' - \mathrm{i}\delta} = P\left(\frac{1}{E - E'}\right) + \mathrm{i}\pi\delta(E - E') , \qquad (1.50)$$

one obtains

$$\mathrm{Im}G(\boldsymbol{k}, E) = \pi \int_{-\infty}^{+\infty} \mathrm{d}E' A(\boldsymbol{k}, E')\delta(E - E') , \qquad (1.51)$$

or

$$A(\boldsymbol{k}, E) = \frac{1}{\pi}|\mathrm{Im}G(\boldsymbol{k}, E)| , \qquad (1.52)$$

If one is dealing with a non-interacting electron system, the Green's function is given by

$$G^0(\boldsymbol{k}, E) = \frac{1}{E - E^0(\boldsymbol{k}) - \mathrm{i}\delta} , \qquad (1.53)$$

where again δ is an infinitesimal small number and $E^0(\boldsymbol{k}) = \hbar^2\boldsymbol{k}^2/(2m)$ is the energy of a free electron with momentum $\hbar\boldsymbol{k}$.

With (1.52) this yields

$$A^0(\boldsymbol{k},E) = \frac{1}{\pi}\delta(E - E^0(\boldsymbol{k})) , \tag{1.54}$$

namely a δ-function at $E^0(\boldsymbol{k})$. This is equivalent to the Koopmans' binding energy. If we switch on the electron–electron interaction, this can be taken into account in the energy by adding to the single-particle electron energy $E^0(\boldsymbol{k})$, a so-called self-energy $\Sigma(\boldsymbol{k},E)$ (provided perturbation theory is valid):

$$\Sigma(\boldsymbol{k},E) = \mathrm{Re}\Sigma + \mathrm{iIm}\Sigma , \tag{1.55}$$

yielding for the Green's function

$$G(\boldsymbol{k},E) = \frac{1}{E - E^0(\boldsymbol{k}) - \Sigma(\boldsymbol{k},E)} , \tag{1.56}$$

and for the spectral function

$$A(\boldsymbol{k},E) = \frac{1}{\pi}\frac{\mathrm{Im}\Sigma}{(E - E^0(\boldsymbol{k}) - \mathrm{Re}\Sigma)^2 + (\mathrm{Im}\Sigma)^2} , \tag{1.57}$$

Since for vanishing electron–electron interaction the result from the non-interacting electron system must be recovered, we look for the poles in $G(\boldsymbol{k},E)$ to obtain the spectroscopic energies in the interacting system, in the same way as the poles in $G^0(\boldsymbol{k},E)$ for the non-interacting system yielded the one-electron energies. To first order the poles are given by (assuming that $\mathrm{Re}\Sigma \gg \mathrm{Im}\Sigma$!):

$$E^1(\boldsymbol{k}) - E^0(\boldsymbol{k}) - \Sigma(\boldsymbol{k},E^1(\boldsymbol{k})) = 0 , \tag{1.58}$$

(for $\Sigma = 0$ the pole is at $E^1(\boldsymbol{k}) = E^0(\boldsymbol{k})$ as it should be), where frequently the pole energy is written as

$$E^1(\boldsymbol{k}) = \mathrm{Re}(E^1(\boldsymbol{k})) + \mathrm{iIm}(E^1(\boldsymbol{k})) , \tag{1.59}$$

meaning that the pole occurs at an energy $\mathrm{Re}(E^1(\boldsymbol{k}))$ with a width $\mathrm{Im}(E^1(\boldsymbol{k}))$.

If the self-energy is small, it is convenient to decompose the Green's function $G(\boldsymbol{k},E)$ into a pole part (which yields $G^0(\boldsymbol{k},E)$ for vanishing Σ) and a term that contains the "rest" ($G_{\mathrm{incoherent}}$ or G_{inc})

$$G(\boldsymbol{k},E) = \frac{Z_{\boldsymbol{k}}}{E - (\mathrm{Re}E^1(\boldsymbol{k}) + \mathrm{iIm}E^1(\boldsymbol{k}))} + (1 - Z_{\boldsymbol{k}})G_{\mathrm{inc}} , \tag{1.60}$$

or

$$A(\boldsymbol{k},E) = \frac{1}{\pi}\frac{Z_{\boldsymbol{k}}\,\mathrm{Im}\left\{E^1(\boldsymbol{k})\right\}}{\left[E - (\mathrm{Re}\left\{E^1(\boldsymbol{k})\right\}\right]^2 + \left[\mathrm{Im}\left\{E^1(\boldsymbol{k})\right\}\right]^2} + (1 - Z_{\boldsymbol{k}})A_{\mathrm{inc}} , \tag{1.61}$$

where $Z_{\boldsymbol{k}}$ is a normalization constant, namely the weight of the quasi-particle pole and $E^1(\boldsymbol{k}) = E^0(\boldsymbol{k}) + \Sigma(\boldsymbol{k},E^1(\boldsymbol{k}))$.

Equation (1.61)can be used to justify the quasi-particle picture of the electrons in an interacting electron system. The pole part of G yields a spectrum of the electron excitations with a renormalized energy (mass) and a finite width, in just the same way as G^0 yields a spectrum with the single-electron energy $E^0(\boldsymbol{k})$(free electron mass) and infinite lifetime. The electrons in the interacting system are "dressed" by virtual excitations that move with them coherently. These "particles", made up of the electrons with their cloud of excitations, are called the quasi-particles. The pole part of the Green's function is often called its coherent part. The remaining part cannot be defined rigorously (this applies to the experimental level; mathematically one can distinguish the coherent and the incoherent part rigorously) because the coherent part already contains low-energy excitations (e.g., electron–hole pairs). This distinguishes (1.61) from the case in which only discrete excitations were possible (1.44). In that former case (1.44) a separation into the main line (~coherent part of the spectrum) and satellite lines (~incoherent part of the spectrum) can (in principle) be performed rigorously.

The complete equation for the photocurrent from a crystalline solid is now given (using (1.57) by:

$$I(E,\hbar\omega) \propto \sum_{i,f} \frac{\operatorname{Im}\sum(\boldsymbol{k}_i)}{(E-E^0(\boldsymbol{k}_i)-\operatorname{Re}\Sigma(\boldsymbol{k}_i))^2+(\operatorname{Im}\Sigma(\boldsymbol{k}_i))^2}$$

$$\times \frac{\left|\tilde{M}_{i,f}\right|^2}{\left(k_{i\perp}^{(1)}-k_{f\perp}^{(1)}\right)^2+\left(k_{f\perp}^{(2)}\right)^2}\delta\left(\boldsymbol{k}_{i||}-\boldsymbol{k}_{f||}+\boldsymbol{G}_{||}\right)\delta(\boldsymbol{k}_i-\boldsymbol{k}_f+\boldsymbol{G})$$

$$\times\delta(E^1(\boldsymbol{k}_f)-E^1(\boldsymbol{k}_i)-\hbar\omega)\delta(E-E^1(\boldsymbol{k}_f)+\phi)\cdot f(E,T)\ ,\tag{1.62}$$

where *f(E,T)* is the Fermi distribution function. This formulation of the spectrum (photocurrent) follows from the total spectral function as given in (1.57), and contains the coherent and incoherent parts of the spectral function.

For the analysis of linewidths from photoemission data on solids, some kinematic considerations are important [1.84]. It is obvious that the measured line width, Γ_{exp}, is a combination of the width of the hole state Γ_{h} (which one is generally interested in, because it can be related to the imaginary part of the Green's function) and the width of the electron state Γ_{e}. The width of the electron state is, in the approximation used so far, related to the electron damping or, formally, to the imaginary part of the electron wave vector perpendicular to the surface. In full generality the relation for the measured linewidth is

$$\Gamma_{\mathrm{exp}}=\frac{(\Gamma_{\mathrm{h}}/|v_{\mathrm{h}\perp}|)+(\Gamma_{\mathrm{e}}/|v_{\mathrm{e}\perp}|)}{\left|(1/v_{\mathrm{h}\perp})\left[1-(mv_{\mathrm{h}||}\sin^2\vartheta/\hbar k_{||})\right]-(1/v_{\mathrm{e}\perp})\left[1-(mv_{\mathrm{e}||}\sin^2\vartheta/\hbar k_{||})\right]\right|}\ ,\tag{1.63}$$

where m is the free-electron mass, $v_{\mathrm{h}||}, v_{\mathrm{h}\perp}, v_{\mathrm{e}||}, v_{\mathrm{e}\perp}$ are the group velocities defined by $v_{\mathrm{h}||}=\hbar^{-1}(\partial E_{\mathrm{h}}/\partial k_{||})$, etc., and ϑ is the angle between the electron-detection direction and the surface normal. In the case of a normal emission experiment one has

$$\Gamma_{\mathrm{exp}}(\vartheta=0)=\frac{\Gamma_{\mathrm{h}}+\Gamma_{\mathrm{e}}|v_{\mathrm{h}\perp}/v_{\mathrm{e}\perp}|}{1-|v_{\mathrm{h}\perp}/v_{\mathrm{e}\perp}|}\ .\tag{1.64}$$

When the dispersion of the initial (hole) state is small compared to that of the final (electron) state, the equation can be further simplified:

$$\Gamma_{\exp}(\vartheta = 0) = \Gamma_{\mathrm{h}} + \Gamma_{\mathrm{e}}|v_{\mathrm{h}\perp}/v_{\mathrm{e}\perp}| \, . \tag{1.65}$$

Finally, in the case where $v_{\mathrm{h}\perp}$ vanishes, i.e., where the dispersion of the band perpendicular to the surface is perfectly flat (which holds for surface states, or two- and one-dimensional electronic states) one can identify $\Gamma_{\exp}(\vartheta = 0)$ with Γ_{h}, which is the imaginary part of the self energy.

These considerations demonstrate the importance of two-dimensional (and one-dimensional) systems for the investigation of hole state lineshapes. As a matter of fact only with these low dimensional systems does one have the possibility to measure the width of these states directly.

For a Fermi liquid, the self-energy assumes a particularly simple form. A Fermi liquid describes a system of interacting electrons whose properties can be mapped in a one-to-one fashion onto those of a system of non-interacting electrons. As before, in the Fermi liquid the electrons are called quasi-particles. These can be viewed as electrons dressed by virtual excitations that move with them coherently. The mass of the quasi-particles is renormalized and their self-energy is given by [1.79]

$$\Sigma(\boldsymbol{k}, E) = aE + \mathrm{i}bE^2 \, . \tag{1.66}$$

This yields for the coherent part of the spectral function, cf.(1.61);

$$A(\boldsymbol{k}, E) = \frac{1}{\pi} \frac{b' E^2}{(E - E^1_{\boldsymbol{k}})^2 + b'^2 E^4} \tag{1.67}$$

with

$$E^1_{\boldsymbol{k}} = Z_{\boldsymbol{k}_{\mathrm{F}}} E^0_{\boldsymbol{k}'}, \quad b' = Z_{\boldsymbol{k}_{\mathrm{F}}} b$$

and

$$Z_{\boldsymbol{k}_{\mathrm{F}}} = \frac{1}{1-a} \, , \tag{1.68}$$

where $Z_{\boldsymbol{k}_{\mathrm{F}}}$ is the renormalization constant at the Fermi wave vector $\boldsymbol{k}_{\mathrm{F}}$ and generally $(Z_{\boldsymbol{k}})^{-1} = 1 - (\partial\Sigma/\partial\Sigma)_{E=E_{\boldsymbol{k}}}$. The spectral function of a two-dimensional Fermi liquid has been investigated by Claessen et al. [1.85, 1.86].

1.6.3 Three-Step and One-Step Considerations

Without mentioning it explicitly, in deriving from (1.19) the formula for the photocurrent in a solid (1.62) we have used what is known in the literature the three-step model. This very successful model breaks up the photoemission process into three steps (see Fig. 1.12):

(1) Photoexcitation of an electron in the solid. This means that the matrix element in (1.39) is evaluated between two Bloch states in the solid.

Therefore it is written as $M^1_{i,f}$ in order to distinguish it from the "correct" matrix element $M_{i,f}$, which is evaluated by using a Bloch state for the initial state and a free electron state (inverse LEED state) in the final state. The additional tilde in $\tilde{M}^1_{i,f}$ indicates that momentum and energy conservation have been taken out of the matrix element.

(2) Propagation of the photoexcited electron to the surface. This results in a damping of the electron wave described by a complex wave vector (1.40). Alternatively one can simulate this effect by using a complex (optical) potential for the crystal potential.

(3) Escape of the electron from the solid into the vacuum, which demands conservation of $\boldsymbol{k}_{||}$.

While the three-step model makes the photoemission process quite comprehensible and (at times) easy to discuss, it is no more than a useful approximation. In order to treat the process correctly the true final state has to be introduced into (1.19) in addition to the initial Bloch state. For this final state the so-called inverse LEED wave function is used. This means the following: In LEED an incoming monochromatic beam of electrons is scattered from the ions in the crystal and the scattered waves sum up to yield the LEED diffraction pattern. Now, if one reverses the direction of the incoming beam, one obtains a monochromatic wave of electrons, which originates from the ions of the crystal, similar to the electron wave produced by the photoemission process. The use of the inverse LEED wave functions as the final state of the photoemission wave function has been very successful.

Here, we shall not deal with these procedures in greater detail (see [1.80]) but will return to this matter in Chap. 6.

At the end of this section it should be emphasized that the most complete discussion of the present state of the theory of photoemission (and other high energy spectroscopies) is given by Hedin [1.82, 1.83].

1.7 Deviations from the Simple Theory of Photoemission

The description of photoemission presented above has been very successful, although there is an evident shortcoming. Although in the three-step model the surface is explicitly mentioned its possible effect on the photoemission spectrum is barely taken into account. There are a number of obvious ways in which the surface will have an effect on the photoemission spectrum. First of all, and perhaps most importantly, the electronic structure in the topmost layers will always be different from that in the bulk, because the atomic configurations are different (for the surface atoms neighbors are missing). In other words, the surface layer will have a smaller electron density compared to the bulk because of a leakage of electronic charge into the vacuum. The different electronic structure at the surface will then lead to different energies

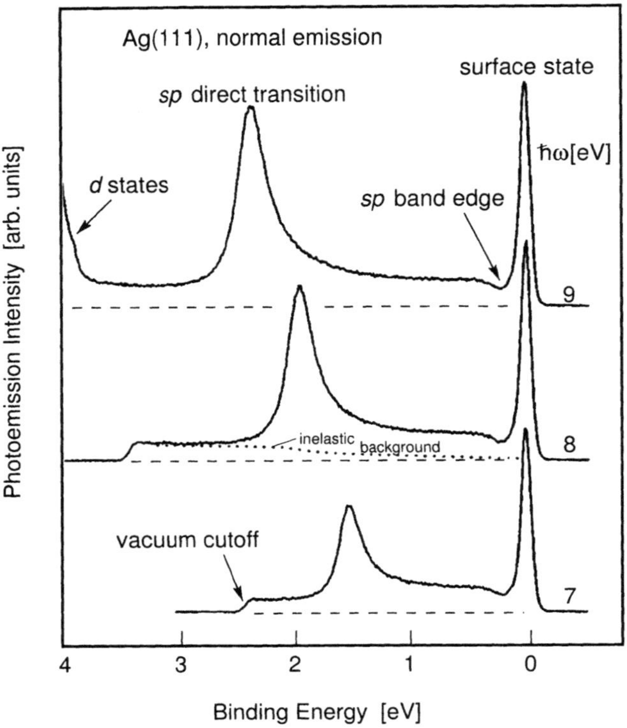

Fig. 1.37. High-resolution normal emission photoemission spectra from the Ag(111) surface for different photon energies. The surface state and the sp direct transition are visible. Also visible is a tailing out of the sp direct transition to smaller binding energies [1.87]

of the surface valence and core spectra. These have indeed been observed, and a separate chapter (Chap. 8) is devoted to this topic.

The next problem is revealed when one inspects the two parts of the sum in the operator of the transition matrix elements (1.20). While $\boldsymbol{A} \cdot \boldsymbol{p}$ describes the bulk photoemission effect, the second term, namely div $\boldsymbol{A}$, is generally set to zero. This is certainly a very valid approximation in the bulk; however, at the surface, where the dielectric constant changes rapidly in going from the surface to the vacuum, the gradient of the vector potential will also change, and therefore this term can no longer be neglected. For a long time this effect was considered to be small, but recent experiments [1.87], have shown that this is not necessarily the case. Here we will give two examples to demonstrate the importance of this effect, often called the surface photoelectric effect. Figure 1.37 shows valence band spectra of Ag(111) in normal emission (see Fig. 1.38 for the corresponding energy level diagram). These spectra, taken for different photon energies, show the direct sp transition and the well known surface state, which will be discussed in Chap. 8 (see also Sect. 1.5). The direct transition shows an unexpected tailing out to smaller binding energies (note that the creation of electron–hole pairs, which will be discussed in Chap. 4, produces the tail to larger binding energies). Figure 1.39 gives an analysis of the data in terms of a model that takes into account the difference in $\boldsymbol{A} \cdot \boldsymbol{p}$ and $\boldsymbol{p} \cdot \boldsymbol{A}$ in the transition matrix elements. In order to show the difference from the simple interpretation of photoemission, the figure shows in the lowest trace the density of states calculated for the sp band, which has a singularity at the Brillouin zone boundary, and the simple one-electron bulk band transition, which is a symmetric Lorentzian curve. If the surface photoemission effect is taken into account, the full curve, superimposed on the data in the top trace, is obtained and it can be seen that it fits the data well.

A similar surface effect, but of a different origin, was found in measurements of the band structure of the metal lithium [1.88]. The results of this experiment are depicted in Fig. 1.40, where the inset gives the initial state energies obtained from the experimental data. While the spectra show a dispersive state for photon energies between 8 eV and 6 eV, for the photon energies between 5.5 eV and 4.5 eV only a non-dispersive state near the Fermi energy is observed. The band structure obtained from the spectra, given in the inset,

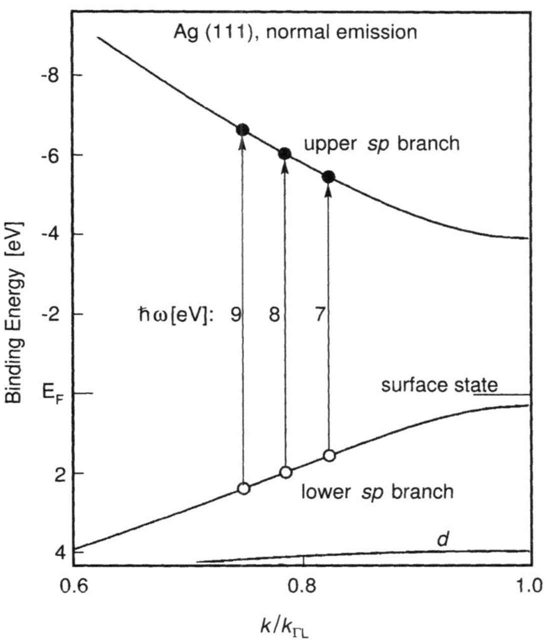

Fig. 1.38. Band structure diagram corresponding to the spectra shown in Fig. 1.37 [1.87]; shown are three representative direct transitions, the surface state and the top of the d band

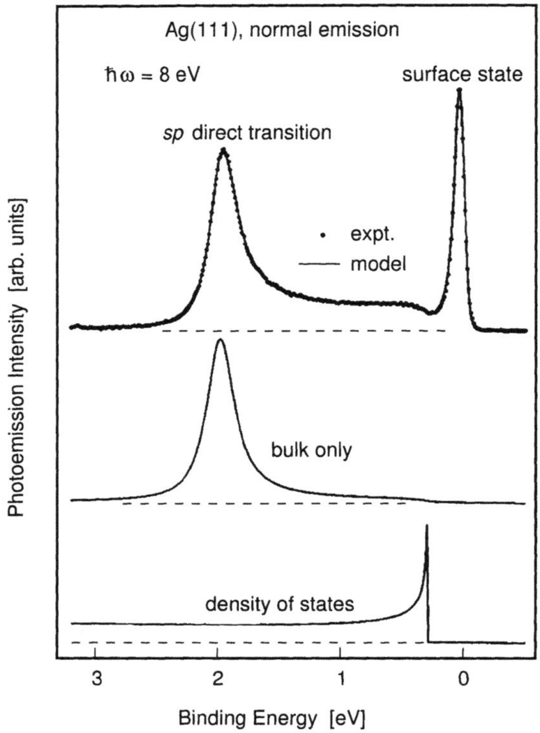

Fig. 1.39. Model calculations for photoemission from the Ag(111) surface (Fig. 1.37). Lowest trace: density of states. Middle trace: bulk direct transition. Top trace: photoemission taking into account the $\boldsymbol{A}\cdot\,\boldsymbol{p}$ term, which is usually the only one employed to analyze the spectra, and also the div $\boldsymbol{A}$ term, which takes care of the surface photoemission. This procedure leads to a calculated spectrum in excellent agreement with the measured spectra of Fig. 1.37

thus shows an apparent gap of almost 1 eV, which is incompatible with the known band structure of Li. The data can again be explained by taking into account the surface photoelectric effect via the operator div$\boldsymbol{A}$. In this case it is found that a gradient of the vector potential couples to the surface plasmon of Li, which has an energy of 5.5 eV and, over a limited energy interval of 2 eV, leads to a peak in the photoemission energy distribution curve. The theory constructed along these lines is found to reproduce the observed data very well.

Finally we give an example, where the electron–phonon coupling plays a role in high-resolution photoemission spectra. Generally this is not a strong effect on the binding energy (although it contributes to the linewidth of course). However, in the surface the phonons can become "softer" and thereby couple strongly to the electrons [1.89]. An experiment examining this effect is reported in Fig. 1.41. Photoemission out of the surface state of the Be(0001) surface is shown, and it can be observed that, very near to the Fermi energy, namely in spectra Nos. 5–7, a strong narrow peak is visible on top of the broad surface peak. The width of the phonon density of states in this surface is roughly 70 meV, and once the surface peak energy reaches this energy, a strong interaction is possible. This yields two states one being the narrow electron–phonon state near the surface as seen in Fig. 1.41, and the other the relatively broad surface state.

With the high resolution now available, especially in valence-band photoemission experiments, one can anticipate that structures due to surface effects will increasingly show up in the data. These will require careful analysis, but on the other hand, will contribute to the power of photoemission as a tool for investigating solids.

The lifetime of electronic states in solids can now be measured directly by time-resolved two-photon photoemission or by very high-resolution photoe-

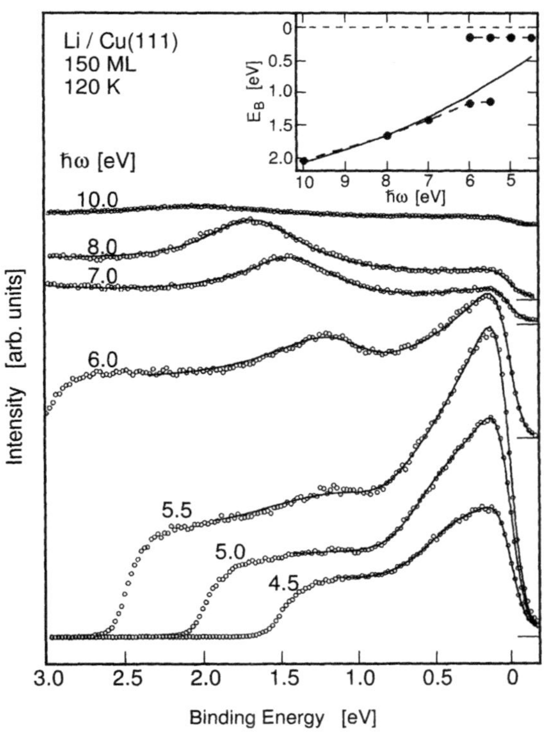

Fig. 1.40. Normal emission photoemission spectra with different photon energies from 150 ML Li evaporated on Cu(111). The inset shows the measured dispersion, which seems to indicate a gap between the Fermi energy and 1.2 eV. Again the data can be analyzed by taking into account the div $\boldsymbol{A}$ term (surface photoemission). In this case the deviation from the expected bulk behavior arises from the coupling to a surface plasmon [1.88]

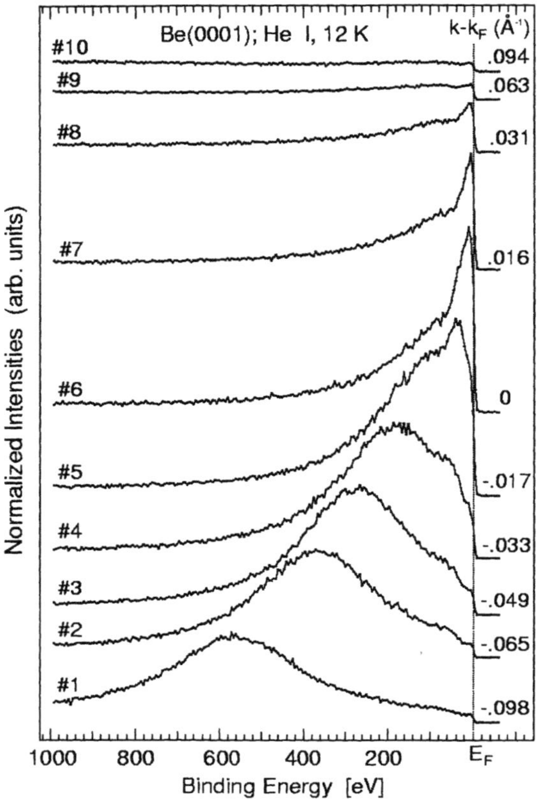

Fig. 1.41. Photoemission from the Be(0001) surface, displaying the well known surface state. If the dispersion is measured by changing the electron detection angle with respect to the surface normal, one observes that near zero binding energy a second peak with smaller width. The analysis of these data shows that in this energy regime there is a coupling to a surface phonon leading in principle to two branches, of which both are indeed observed in the data [1.89]

mission spectroscopy. These experimental results have stimulated new calculations for electronic lifetime in solids and these calculations have provided new insight into the mechanisms determining the lifetimes. A review of this area has recently been provided by Echenique et al. [1.90], and it shows that the agreement between experiment and theory in that field is surprisingly good.

Finally it is worth noting that additional experiments on conventional superconductors have appeared [1.91, 1.92], where in one case even a detail such as the anisotropy of the superconducting gap could be observed.

References[9]

1.1 H. Hertz: Über einen Einfluss des ultravioletten Lichtes auf die elektrische Entladung (On the influence of ultraviolet light on the electric discharge). Ann. Physik **31**, 983 (1887); for a very good review on the history of photoemission see: H.P. Bartel, Ch. Kleint: Prog. Surf. Sci. **49**, 107 (1995)
1.2 J.J. Thompson: Phil. Mag. **48**, 547 (1899)
1.3 P. Lenard: Ann. Physik **2**, 359 (1900); ibid **8**, 149 (1902)
1.4 A. Einstein: Ann. Physik **17**, 132 (1905)
1.5 P. Lukirsky, S. Prilezaev: Z. Physik **49**, 236 (1928)
1.6 F. Patthey, J.-M. Imer, W.D. Schneider, Y. Baer, B. Delley: Phys. Rev. B **42**, 8864 (1990); M. Grioni: Private communication
In standard photoemission instruments the energy resolution in the XPS regime is ≈(0.3–0.8) eV and in the UPS regime ≈(15–30) meV, the best angular resolutions being of the order of 2°; lateral resolutions of ≈ 5 µm can be achieved. There is also an instrument available that combines the energy resolution of 0.3 eV for XPS and 3.0 meV for UPS with an angular resolution of 0.2°.
G. Beamson, D. Briggs: *High Resolution XPS of Organic Polymers* (Wiley, Chichester 1992)
1.7 F. Reinert, G. Nicolay, S. Hüfner, U. Probst, E. Bucher: J. Electron. Spectrosc. Relat. Phenom. **114–116**(2001)
1.8 E. Hulpke (ed.): *Helium Atom Scattering from Surfaces*, Springer Ser. Surf. Sci., Vol. 27 (Springer, Berlin, Heidelberg 1992)
1.9 H. Ibach: In *Electron Spectroscopy for Surface Analysis*, ed. by H. Ibach, Topics Curr. Phys., Vol. 4 (Springer, Berlin, Heidelberg 1977)
1.10 M.P. Seah: Surf. Interface Anal. **I**, 86 (1979)
1.11 C.J. Powell: J. Electron Spectrosc. Relat. Phenom. **47**, 197 (1988)
1.12 S. Tanuma, C.J. Powell, D.R. Penn: J. Vac. Sci. Techn. A **8**, 2213 (1990)
1.13 J. Electron Spectrosc. Relat. Phenom. Vol. 100 (1999)
1.14 V.I. Nefedov in Ref. [1.13]
1.15 A. Jablonski, C.J. Powell in Ref. [1.13]; see also: S. Tanuma, C.J. Powell, O.R. Penn: Surface and Interface Analysis **21**, 165 (1993)
1.16 D. Briggs and M.P. Seah (eds.): *Practical Surface Analysis*, John Wiley, Chichester (1990); M.P. Seah in Ref. [1.13]
1.17 K. Siegbahn, C. Nordling, R. Fahlman, R. Nordberg, K. Hamrin, J. Hedman, G. Johansson, T. Bergmark, S.-E. Karlsson, I. Lindgren, B. Lindberg: *ESCA, Atomic, Molecular and Solid State Structure Studies by Means of Electron Spectroscopy*, Nova Acta Regiae Soc. Sci., Upsaliensis, Ser. IV, Vol. 20 (1967)
1.18 Y. Baer, G. Busch: Phys. Rev. Lett. **30**, 280 (1973)
1.19 V. Dose: Appl. Phys. **14**, 117 (1977)
1.20 W.C. Price: *In Electron Spectroscopy, Theory, Techniques and Applications*, ed. by C.R. Brundle, A.D. Baker (Academic, New York 1977) Vol. 1
1.21 A. Goldmann, J. Tejeda, N.J. Shevchik, M. Cardona: Phys. Rev. B **10**, 4388 (1974)
1.22 E.E. Koch (ed.): *Handbook on Synchrotron Radiation* (North-Holland, Amsterdam 1983) Vol. 1a
1.23 E.E. Koch (ed.): Handbook on *Synchrotron Rudiation* (North-Holland, Amsterdam 1983) Vol. 1b

[9] The references listed for each chapter are not intended to cover comprehensively the literature which exists on the topics of the particular chapter. They have been used by the author and may serve to provide the reader with additional information and further sources.

1.24 E.V. Marr (ed.): *Handbook on Synchrotron Radiation* (North-Holland, Amsterdam 1987) Vol. 2
1.25 G.S. Brown, D.E. Moncton (eds.): *Handbook on Synchrotron Radiation* (North-Holland, Amsterdam 1991) Vol. 3
1.26 S. Ebashi, M. Koch, E. Rubenstein (eds.): *Handbook on Synchrotron Radiation* (North-Holland, Amsterdam 1991) Vol. 4
1.27 R.Z. Bachrach (ed.): *Synchrotron Radiation Research* (Plenum, New York 1992) Vol. 1
1.28 R.Z. Bachrach (ed.): *Synchrotron Radiation Research* (Plenum, New York 1992) Vol. 2
1.29 C. Kunz (ed.): *Synchrotron Radiation*, Topics Curr. Phys., Vol. 10 (Springer, Berlin, Heidelberg 1979)
1.30 D. Attwood: *Soft X-rays and Extreme Ultraviolett Radiation*, Cambridge University Press (1999)
1.31 H. Ibach: *Electron Energy Loss Spectrometers*, Springer Ser. Opt. Sci., Vol. 63 (Springer, Berlin, Heidelberg 1991)
1.32 J. Westhof, G. Meister, F. Lodders, R. Matzdorf, R. Henning, E. Janssen, A. Goldmann: Appl. Phys. A **53**, 410 (1991)
Another good review of that field can be found in: B.P. Tonner, D. Dunham, T. Droubay, J. Kikuma, J. Denlinger, E. Rotenberg, A. Warwick: J. Electr. Spectrosc. **75**, 309 (1995)
1.33 D. Roy, J.D. Carette: In *Electron Spectroscopy for Surface Analysis*, ed. by H. Ibach, Topics Curr. Phys., Vol. 4 (Springer, Berlin, Heidelberg 1977) Chap.2
1.34 A.F. Orchard: In *Handbook of X-Ray and Ultraviolett Photoelectron Spectroscopy*, ed. by D. Briggs (Heyden, London 1977)
1.35 A. Barrie: *In Handbook of X-Ray and Ultraviolet Photoelectron Spectroscopy*, ed. by D. Briggs (Heyden, London 1977)
1.36 G.A. Harrower: Rev. Sci. Instrum. **26**, 850 (1955)
1.37 V.V. Zashkvara, M.J. Korsunskii, O.S. Kosmachev: Sov. Phys. Techn. Phys. **11**, 96 (1966)
1.38 H.Z. Sar-El: Rev. Sci. Instrum. **38**, 1210 (1967)
1.39 E. Blauth: Z. Phys. **147**, 228 (1957)
1.40 W. Mehlhorn: Z. Physik **160**, 247 (1960)
1.41 K.L. Wang: J. Phys. E **5**, 1193 (1972)
1.42 D. Roy, J.-D. Carette: Can. J. Phys. **49**, 2138 (1971)
1.43 D. Roy, J.-D. Carette: J. Appl. Phys. **42**, 3601 (1971)
1.44 A.L. Hughes, V. Rojansky: Phys. Rev. **34**, 284 (1929)
1.45 A.L. Hughes, J.H. McMillan: Phys. Rev. **34**, 291 (1929)
1.46 E.M. Clarke: Can. J. Phys. **32**, 764 (1954)
1.47 R. Francois, M. Barat: C.R. Acad. Sc. (Paris) B **266**, 1306 (1968)
1.48 J.J. Leventhal, G.R. North: Rev. Sci. Instr. **42**, 120 (1971)
1.49 D. Roy, M. De Celles, J.D. Carette: Rev. Phys. Appl. **6**, 51 (1971)
1.50 H. Ibach: J. Vac. Sci. Technol. **9**, 713 (1972)
1.51 P. Marmet, L. Kerwin: Can. J. Phys. **38**, 787 (1960)
1.52 Y. Ballu: Rev. Phys. Appl. **3**, 46 (1968)
1.53 H. Froitzheim, H. Ibach: Z. Physik 269, **17** (1974)
1.54 H. Froitzheim, H. Ibach, S. Lehwald: Rev. Sci. Instr. **46**, 1325 (1975)
1.55 E.M. Purcell: Phys. Rev. **54**, 818 (1938)
1.56 W. Steckelmacher: J. Phys. E **6**, 1061 (1973)
1.57 H.Y. Sar-El: Rev. Sci. Instr. **41**, 561 (1970)
1.58 F.H. Read, J. Comer, R.E. Imhof, J.N.H. Brunt, E. Harting: J. Electr. Spectros. **4**, 293 (1974)
1.59 D.W.O. Heddle: J. Phys. E **4**, 589 (1971)

1.60 W. Schmitz, W. Mehlhorn: J. Phys. E **5**, 64 (1972)
1.61 J.A. Simpson: Rev. Sci. Instrum. **35**, 1698 (1964)
1.62 C.E. Kuyatt, J.A. Simpson: Rev. Sci. Instrum. **38**, 103 (1967)
1.63 J.A. Simpson: In *Methods of Experimental Physics*, ed. by V.W. Hughes, H.L. Schultz (Academic, New York 1967) Vol. 4a
1.64 C.E. Kuyatt: In *Methods of Experimental Physics*, ed. by B. Bederson, W.L. Fite (Academic, New York 1968) Vol. 7a
1.65 C.R. Brundle, A.D. Baker (eds.): *Electron Spectroscopy, Theory, Techniques and Applications*, Vol. I (Academic, New York 1977)
1.66 D.H. Tomboulian, P.L. Hartmann: Phys. Rev. **102**, 1423 (1956)
1.67 W. Gudat: In *Elektronenspektroskopische Methoden an Festkörpern und Oberflächen I* and *II* Kernforschungsanlage Jülich, 1980)
1.68 I.H. Munro, G.V. Marr in Ref. [1.24]
1.69 D. Purdie, M. Hengsberger, M. Garnier, Y. Baer: Surface Science **407**, L671 (1998)
1.70 G. Nicolay, F. Reinert, D. Ehm, T. Finteis, P. Steiner, S. Hüfner: Phys. Rev. B, **62** 1631 (2000); ibid B, **63**, 115415 (2001)
1.71 T. Valla, A.V. Fedorov, P.D. Johnson, S.L. Hulbert: Phys. Rev. Lett. **83**, 2085 (1999)
1.72 R.I.G. Uhrberg, T. Kaurila, Y.C. Chao: Phys. Rev. B **58**, 1730 (1998)
1.73 A. Fröhlisch, N. Warsdahl, J. Hasselströhm, O. Karis, D. Menzel, N. Mårtensson, A. Nilsson: Phys. Rev. Lett. **81**, 1730 (1998)
1.74 K. Kimura in Ref. [1.13] and references therein
1.75 G. Herzberg: *Molecular Spectra and Molecular Structure*, van Nostrand, Princeton (1964)
1.76 C.S. Fadley: In *Electron Spectroscopy: Theory, Techniques and applications II*, ed. by C.R. Brundle, A.D. Baker (Academic, New York 1978) and references therein
1.77 W. Bardyszewski, L. Hedin: Phys. Scripta 32, 439 (1985) and references therein
1.78 C.O. Almbladh, L. Hedin: In Ref. [1.23]
1.79 L. Hedin, S. Lundquist: *Solid State Physics* **23**, 1 (Academic, New York 1969) The present author has profited from discussions with R. Claessen and L. Hedin
1.80 J. Braun: Rep. Prog. Phys. **59**, 1267 (1996)
1.81 A. Kotani, T. Jo, C.J. Parlebas: Adv. Phys. **37**, 37 (1988)
1.82 L. Hedin: J. Phys. Condens. Matter **11** **R** 489 (1999)
1.83 L. Hedin, J. Michiels, J. Inglesfield: Phys. Rev. B **58**, 15565 (1998)
1.84 N.V. Smith, P. Thiry, Y. Petroff: Phys. Rev. B **47**, 15476 (1993)
1.85 R. Claessen, R.O. Anderson, J.W. Allen, C.G. Olson, C. Janowitz, W.P. Ellis, S. Harm, M. Kalning, R. Manzke, M. Skibowski: Phys. Rev. Lett. **69**, 808 (1992).
The spectral function of a Fermi liquid is reported in this communication
1.86 The spectral function observed in a PES experiment is modified in the following way if the experiment is performed in an interacting electron gas instead of in a non-interacting one: (see Fig. 5.88):
(1) The energy (pole position) is renormalized [$E^0_{\boldsymbol{k}}$ is changed to $E^0_{\boldsymbol{k}} + \mathrm{Re}\{\Sigma(\boldsymbol{k}, E)\}$].
(2) The coherent signal has the finite width $\mathrm{Im}\{\Sigma(\boldsymbol{k}, E)\}$.
(3) An incoherent background shows up.
(4) PES intensity develops outside the Fermi surface, namely for $\boldsymbol{k} > \boldsymbol{k}_F$
1.87 T. Miller, W.E. Mc Mahon, T.C. Chiang: Phys. Rev. Lett. **77**, 1167 (1996)

1.88 D. Claesson, S.Å. Lundgren, L. Wallden, T.C. Chiang: Phys. Rev. Lett. **82**, 1740 (1999)
1.89 M. Hengsberger, D. Purdie, P. Segovia, M. Garnier, Y. Baer: Phys. Rev. Lett. **83**, 592 (1999)
1.90 P.M. Echenique, J.M. Chulkov, A. Rubio: Chem. Phys. **251**, 1 (2000)
1.91 A. Chainani, T. Yokoya, T. Kiss, S. Shin: Phys. Rev. Lett. **85**, 1966 (2000)
1.92 T. Yokoya, T. Kiss, T. Watanabe, C. Shin, M. Nohara, H. Takagi, T. Oguchi: Phys. Rev. Lett. 85, 4952 (2000)
1.93 U. Hollenstein, R. Seiler, H. Schmutz, M. Andrist, F. Merkt: J. Chem. Phys. **115**, 5461 (2001)

2. Core Levels and Final States

Photoemission produces a final state that is lacking one electron with respect to the initial state. Therefore, PES always measures final-state energies which can be related to initial-state energies only after some theoretical considerations. The problem is illustrated in Fig. 2.1. On the left-hand side are shown the one-electron orbital energies $-\varepsilon$ in the ground state of the neon atom. On the right-hand side are the final state configurations, as measured by PES for the same atom.[1] These are, of course, hole-state energies. The relation between these two sets of energy levels is far from trivial. However, one should add that the orbital energies (left side) cannot be measured. In general, one can only measure the energy levels given on the right-hand side, which are (many-body) hole-state energies [2.1].

These statements have to be modified for wide valence bands in solids. The PES final state in the valence band of Cu metal agrees well with the result of a *one-electron band structure* calculation (Chap. 7). In this case the screening of the photohole is almost perfect and therefore the final state energy equals the initial state *one-electron energy.*

Thus, except for the case of valence states in wide-band solids, the most elementary "final-state effect" in PES is observed in the binding energy. If we consider, for example, a molecule with N electrons, the kinetic energy (with respect to the vacuum level E_{V}) of a photoelectron is given by

$$E_{\mathrm{kin}} = \hbar\omega - \left(E_f^{N-1} - E_i^N\right) . \tag{2.1}$$

For the sake of clarity we have to add a short word concerning energy-reference points. Although it may seem somewhat inconsistent, the energies of free atoms and molecules will be given with respect to the vacuum level (E_{vac}) and those of solids with respect to the Fermi level (E_{F}). The logical way of referencing would be to refer the energies of solids also to the vacuum level. This, however, would be contrary to general usage. As far as (2.1) is concerned, this means that for a solid the work function ϕ has to be subtracted on the right-hand side.[2]

---

[1] These energies are sometimes just called the orbital energies. The term orbital here does not mean that the spin gives no contribution to the energy.

[2] Here, ϕ is the smallest energy required to take an electron from inside the solid to the vacuum level.

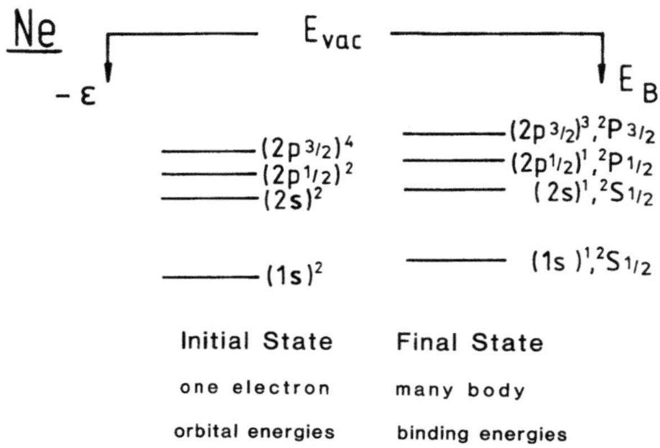

Fig. 2.1. Orbital energies and final (hole) state energies for neon. On the left-hand side one sees the orbital energies. The right-hand side shows the energy level obtained after ejection of an electron by a photon. The resulting electronic configurations (they are all one-hole states) are also indicated. The left-hand diagram is only of theoretical interest since the orbital energies themselves cannot be measured. The right-hand energy level include the rearrangement of the charge due to the created hole (relaxation)

Thus PES measures the energy difference between the *total energies* of a state with N electrons E_i^N and one with $N-1$ electrons E_f^{N-1}. In the most rigorous formulation, the binding energy of an electron measured by PES is therefore given by

$$E_{\mathrm{B}} = E_f^{N-1} - E_i^N \,. \tag{2.2}$$

This definition is exact. It has, however, the disadvantage that total energies are not very easy to calculate, and one has to rely on approximations to obtain them. The best known one is Koopmans' approximation, see (1.10). It states that the binding energy equals the negative energy of the orbital from which the photoelectron is emitted. This is only approximately true because it assumes that the PE process does not produce any change in the other orbitals. Stated more exactly, Koopmans' approximation neglects relaxation energies. In order to relate Koopmans' energies to the binding energies actually measured, one has to correct for the relaxation energy. This is the amount of energy necessary for the adjustment of the total system to the hole left in a specific orbital by the photoeffect. In a solid this relaxation energy consists of two basic contributions: One results from the relaxation of the orbitals on the same atom (*intra-atomic relaxation*). The other one is due to a charge flow from the crystal onto the ion that carries the hole (*extra-atomic relaxation*).

A particular final-state effect is the exchange splitting of certain levels. This can be demonstrated for the lithium atom, where PE from the 1s orbital can be expressed by the following reaction:

$$\mathrm{Li}(1\mathrm{s}^2 2\mathrm{s};\, {}^2\mathrm{S}) + \hbar\omega \rightarrow \begin{cases} \mathrm{Li}^+\,(1\mathrm{s}(\downarrow)\,2\mathrm{s}(\uparrow);\, {}^1\mathrm{S}) + e^- \\ \mathrm{Li}^+\,(1\mathrm{s}(\uparrow)\,2\mathrm{s}(\uparrow);\, {}^3\mathrm{S}) + e^-. \end{cases}$$

Lithium has a $1s^2$ inner shell and one additional electron in the 2s shell. Photoionization of the $1s^2$ shell can lead to two sets of final states: In one of them the spin of the remaining 1s electron and that of the 2s electron are parallel to each other leading to a ^{3}S triplet of final states, while in the other one, in which the spin of the remaining 1s electron and that of the 2s electron are antiparallel, we obtain a ^{1}S singlet final state. The singlet and the triplet states have different energies because the Coulomb integrals depend on the spin, and this difference shows up in the PE spectrum as an exchange splitting of the 1s line.

We have separated out the exchange splitting in the final state of the photoemission line from an s level, because it is in some cases an effect easily recognized in the measured spectra. In principle, this exchange splitting is part of a many-body final state and can only be separated out in an approximate way. In the following sections the binding energies in atoms, molecules and solids will be dealt with separately, since, in each case, different considerations are necessary.

2.1 Core-Level Binding Energies in Atoms and Molecules

To obtain an impression of the information that can be obtained from core-level XPS spectra, Fig. 2.2 shows the 1s spectra of the first row elements from lithium to fluorine. These spectra have been taken from solid samples and therefore cannot be considered as spectra from atomic species. However, in those cases where comparison is possible, the spectrum of the "true" atomic sample looks identical [2.2, 2.3].

2.1.1 The Equivalent-Core Approximation

The most frequently used approximation for the analysis and interpretation of core-level binding energies is the so-called *equivalent-core approximation*. In this model one assumes that the spatial extension of the core electron is small with respect to that of the valence electrons. The effect of photoionization of a core level can then be approximated with good precision by the hypothetical addition of a proton to the nucleus. This model is thus often called the *(Z+1)-approximation* [2.4, 2.5].

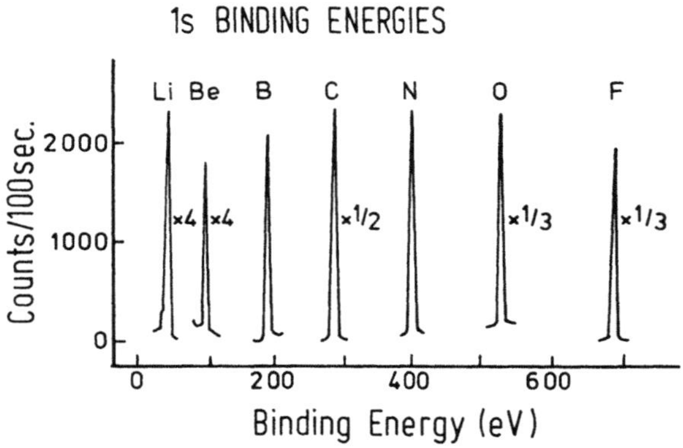

Fig. 2.2. XPS spectra of the 1s core levels of Li, Be, B, C, N, O, F [2.2, 2.3]

The usefulness of this method is demonstrated by employing it to estimate the binding energy of the 1s level of lithium [2.5]. The reaction steps that are discussed below are shown in graphical form in Fig. 2.3 [2.4, 2.6, 2.7]. To produce singly ionized lithium one has to provide the first ionization energy of lithium which is 5.39 eV. The second ionization potential of lithium is 75.62 eV and produces Li^{2+}. Thus to obtain Li^{2+} we need 81.01 eV. Next we

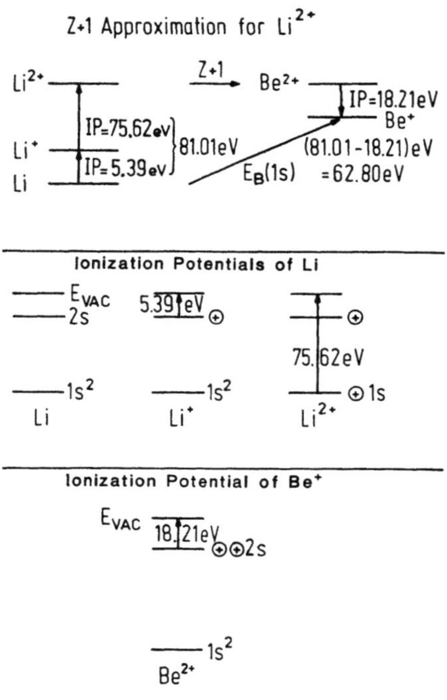

Fig. 2.3. Use of the $(Z+1)$-approximation to calculate the 1s binding energy of Li [2.4, 2.6, 2.7]

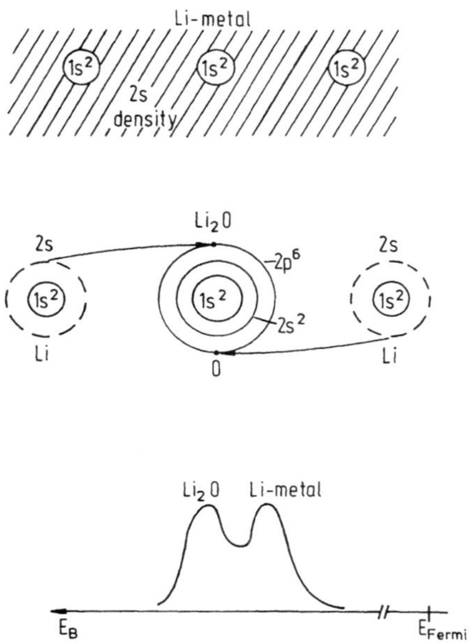

Fig. 2.4. Schematic drawing of the electron configuration of Li metal and Li_2O, and the corresponding (*schematic*) Li 1s PES spectra. In Li_2O the 1s level is more tightly bound than in Li metal because in the metal there is screening by the 2s conduction electrons

need the energy gained if an electron is placed into the 2s shell of Li with a 1s hole. To that purpose, via the $(Z+1)$-approximation, Li^{2+} is approximated by Be^{2+}. This means that the energy gained in putting one electron back into the 2s shell of Li^{2+} is approximated by the negative first ionization energy of Be^+, which is 18.21 eV. This number has to be subtracted from 81.01 eV to obtain the binding energy of the 1s electron of Li. Thus the $(Z+1)$-approximation yields 62.8 eV for the binding energy of the 1s level of Li. This value is to be compared with the measured binding energy of 64.8 eV [2.4]. The difference between the two numbers gives an estimate of the accuracy attainable with this approximation.

2.1.2 Chemical Shifts

In the investigation of molecules and solids, one is not usually interested in the absolute binding energy of a particular core level, but in the change in binding energy between two different chemical forms of the same atom. This energy difference is called the *chemical shift.*

Using a very simplified model we will discuss how this chemical shift arises. As an example we choose the lithium 1s levels in lithium metal and lithium oxide. Figure 2.4 gives a simplified representation of the electronic structure of these two systems. In lithium metal the lithium 2s electrons form a band, and their wave function is therefore only partly at the site of a particular lithium atom. However, in lithium oxide each lithium atom donates its 2s electron totally into the 2p shell of oxygen such that a closed $2p^6$ configuration is obtained. Thus in lithium oxide the Li 2s electron has no part of its wave function near the lithium atom. The 1s electrons in lithium oxide therefore feel a somewhat stronger Coulomb interaction than in Li metal, where the lithium nucleus is screened from the $1s^2$ shell through the 2s valence electron. Therefore the binding energy of the Li 1s level is larger in Li_2O than in Li metal and a "chemical shift" between the two compounds is observed.

This is a very typical example. The chemical shift between a metal and its oxide(s) very often serves to monitor the surface cleanliness. For example, the absence of oxygen is proven by the absence of the chemically shifted (to larger binding energies) metal oxide line.

Chemical shifts have been investigated in a large number of molecules. To demonstrate the power of the method, three typical examples are given in Fig. 2.5 [2.3]. The observed chemical shifts are most easily interpreted for sodium azide NaN_3 for which the structural formula is depicted in Fig. 2.5c. There are two equivalent negatively charged nitrogen ions N^- and one positively charged nitrogen ion N^+. The negative charges screen the nuclear charge of the nitrogen atom and therefore the binding energy of N^- is smaller than that of N^+, for which the screening is absent. The intensity ratio of the two N 1s lines further substantiates this interpretation.

The equivalent-core approximation and chemical-reaction energies can be used to estimate chemical shifts for core levels of molecules. We want to demonstrate this by estimating the chemical shift ΔE_B between molecular nitrogen N_2 and gaseous nitrogen dioxide [2.4]. One can describe ΔE_B by the following chemical reaction:

$$NO_2 + N_2^{+^*} \rightarrow N^{+^*}O_2 + N_2 + \Delta E \; ; \quad \Delta E = \Delta E_B \; , \tag{2.3}$$

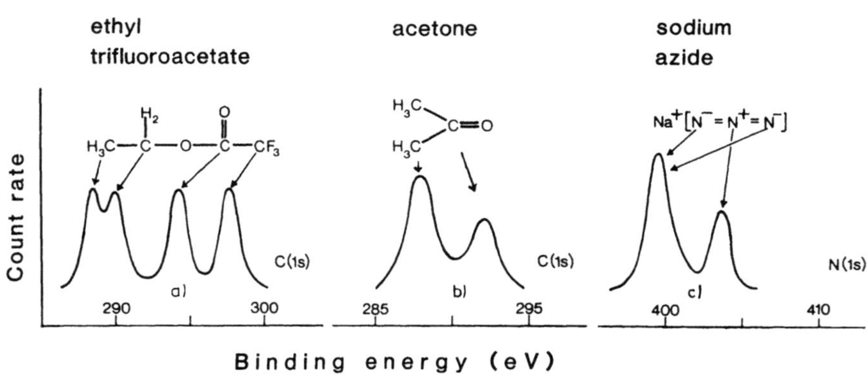

Fig. 2.5. Chemical shifts for the C 1s levels in ethyl trifluoroacetate (**a**), and acetone (**b**), and the N 1s levels in sodium azide (**c**). Chemical shifts can be crudely related to electronegativity differences: The known [2.8] electronegativity differences (Δx) are C – H : $\Delta x = 0.4$, C – O : $\Delta x = 1.0$; C – F : $\Delta x = 1.5$, which rationalize the chemical shifts in ethyl trifluoroacetate [2.3]

where ΔE is the difference between the energy needed to form a 1s core hole in N_2 and that for NO_2. This is also the chemical-shift energy ΔE_B (the asterisk indicates the 1s hole). One now has to convert this equation into one that contains only ground-state chemical species. The equivalent-core approximation is used to derive the following reaction which takes place without any energy change:

$$N^{+^*}O_2 + NO^+ \rightarrow O_3^+ + N_2^{+^*} , \quad \Delta E = 0 \quad (N^{+^*} \simeq O^+) . \tag{2.4}$$

Thus one replaces a nitrogen atom ionized in the 1s shell by an oxygen ion with a vacancy in the valence shell, and vice versa. By adding the two equations one obtains:

$$NO_2 + NO^+ \rightarrow O_3^+ + N_2 + \Delta E ; \quad \Delta E = \Delta E_B . \tag{2.5}$$

This is a chemical reaction. Its energy can be found from thermodynamic data as $\Delta E = 3.3$ eV [2.4] and is, of course, the desired chemical shift energy. It agrees well with the experimentally measured shift of $\Delta E = 3.0$ eV [2.9].

Figure 2.6 presents a summary of experimental chemical shifts ΔE_B measured by XPS and plotted with respect to the chemical shifts ΔE_B estimated from thermodynamic data [2.4]. All energies are referenced to that of molecular nitrogen. There is a reasonable correlation indicating the usefulness of the

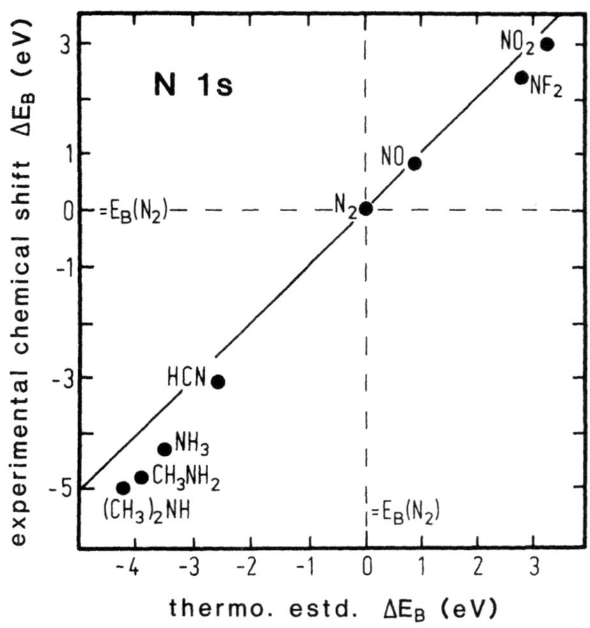

Fig. 2.6. Nitrogen 1s binding energies relative to that of N_2 as measured by XPS plotted versus values calculated from thermodynamic data [2.4]

method. (The straight line is the 45° line giving the locus of exact correlation). The scatter in the data points, on the other hand, indicates remaining inadequacies of the model. One can, of course, reverse the procedure and use chemical shifts from XPS to estimate thermodynamic reaction energies. For solids, this can be done with great success (Sect. 2.2.3).

In chemistry, binding energies are frequently correlated with the charge on an atom. With increasing positive charge on an ion in a molecule the binding energy also increases and with increasing negative charge the binding energy decreases. This is a consequence of the screening discussed above. Thus, in a first approximation, one can write for the binding energy of a particular core level in a molecule:

$$E_{\mathrm{B}} = K \cdot Q + L \; . \tag{2.6}$$

Here K and L are empirical constants, and Q is the charge on the ion in the particular molecule relative to that of the neutral atom (we do not use q, as is frequently done in the literature because this symbol will later designate the total charge in an orbital). Thus L is the binding energy of the electron in question in the free neutral atom.

In a molecule, however, an electron on a particular atom experiences not only the charges of that atom, but also the fields produced by charges on the other atoms, leading to an additional potential energy. In an accurate determination of E_{B}, this potential energy V, produced by all the other constituents of the molecule, has to be added

$$E_{\mathrm{B}} = K \cdot Q + V + L \; . \tag{2.7}$$

Finally, the *relaxation energy* E_{R} has to be taken into account. This is the energy gained when the electrons of the molecule respond to the photohole produced by the photoemission process. Thus one has the following equation for the binding energy

$$E_{\mathrm{B}} = K \cdot Q + V + L - E_{\mathrm{R}} \; . \tag{2.8}$$

(The relaxation decreases the positive binding energy; therefore a negative sign has been used for the relaxation energy). Relaxation energies are difficult to estimate from first principles. The simplest and most popular approach to describe them semi-quantitatively is the so-called *transition-state concept* (see Sect. 2.2.2 for further details). In this, one tries to model the relaxation energy by ascribing to Q and to V values which are between the corresponding values in the initial and the final state [2.10–2.14]

$$E_{\mathrm{B}} = K \cdot Q + K\Delta Q + V + \Delta V + L \; . \tag{2.9}$$

Here ΔQ and ΔV are the changes in Q and V which occur when one goes from the initial state to the transition state. In the most frequently used approximation one assumes that the charge and the potential in the transition state have values which are halfway between those in the initial and the final states. For the relaxation energy this gives

$$-E_{\mathrm{R}} = K\Delta Q + \Delta V = \frac{1}{2}K(Q_f - Q - 1) + \frac{1}{2}(V_f - V) . \tag{2.10}$$

(The index f refers to the final state. The photoelectron has especially been indicated in this formulation; one could also assume that it is already contained in Q_f). Although not exact, this approach has been used with considerable success in analyzing chemical-shift data from free molecules.

2.2 Core-Level Binding Energies in Solids

In solids, as in molecules, one is primarily interested in understanding and using the chemical shifts ΔE_{B} of binding energies rather than their absolute values E_{B}. The core-level binding-energy difference between two compounds A and B is given by, making use of (2.7),

$$\Delta E_{\mathrm{B}}(A, B) = K(Q^A - Q^B) + (V^A - V^B) . \tag{2.11}$$

(Note that this equation also holds for the binding energy difference of a particular core level between two molecules A and B). The first term describes the difference in the electron–electron interaction between the core orbital and the valence charges. The coupling constant K is, in essence, the Coulomb interaction between the valence and the core electrons. The second term represents the interaction of the atom to be photoionized with the rest of the crystal. For insulators this is, in the most naive approximation, a Madelung-type energy. The simple equation above describes only initial state energies and neglects relaxation. This inevitably leads to inaccuracies, sometimes even in the qualitative interpretation.

2.2.1 The Born–Haber Cycle in Insulators

The binding energy in ionic insulators will be discussed in this section in more detail. For this discussion, a Born–Haber cycle is used to calculate E_{B} for a particular orbital k in an atom with nuclear charge Z [2.15, 2.16]. The intention of this procedure is to break up the binding energy which one is interested in, into a sum of energies which are known from other sources. The principle of this method is depicted in Fig. 2.7. If the ion with nuclear charge Z is taken out of compound A its energy changes by the Madelung energy E_{M}^Z. The result is a free ion. This ion is then photoionized, a process which requires the binding energy of the orbital under consideration in the free ion E_{B} (ion). In the "equivalent core approximation", the photoionized ion is approximated by a $(Z{+}1)$-ion. The ion is then put back into the lattice whereby the Madelung energy for the $(Z{+}1)$-ion is gained. Thus the net gain of Madelung energy is that of one unit charge. The sum of the energies of the cycle just gives the binding energy E_{B} measured by PES for the chosen orbital k in the solid. It is the sum of the binding energy of that orbital in

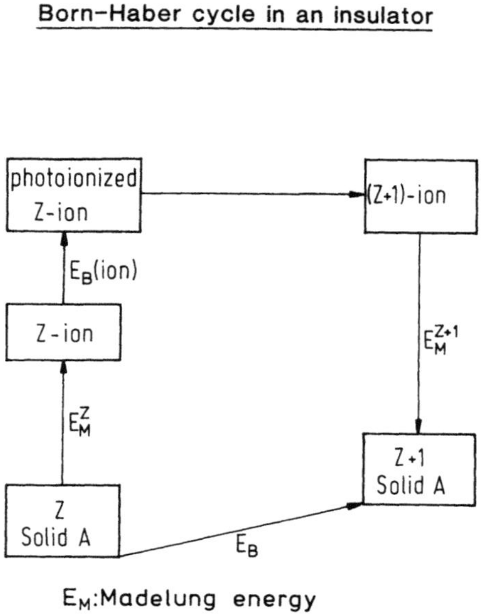

Fig. 2.7. Born–Haber cycle for the calculation of the binding energy in an ionic solid. E_{M} is the Madelung energy, and the $(Z+1)$-approximation is used to simulate the core-ionized ion [2.16]

the free ion and the Madelung term for unit charge. One can finally break up the binding energy of the free ion into that of the corresponding atom plus the interaction energy of the orbital with the charge required to produce the ion. Therefore, for the binding energy of orbital k in the compound we obtain the following equation:

$$\begin{aligned} E_{\mathrm{B}}(k, \text{solid}) &= E_{\mathrm{B}}(k, \text{ion}) + V\ , \\ E_{\mathrm{B}}(k, \text{ion}) &\simeq E_{\mathrm{B}}(k, \text{atom}) + KQ\ , \\ E_{\mathrm{B}}(k, \text{solid}) &= E_{\mathrm{B}}(k, \text{atom}) + KQ + V\ , \\ E_{\mathrm{B}}(k, \text{solid}) &= E_{\mathrm{B}}(k, \text{atom}) + Q/\bar{r}_{\mathrm{v}} + Q\alpha_{\mathrm{M}}/d\ , \end{aligned} \tag{2.12}$$

where $Q/\bar{r}_{\mathrm{v}} = KQ$; $\bar{r}_{\mathrm{v}}$ is the average radius of the valence shell. $Q\alpha_{\mathrm{M}}/d = V$ is the Madelung energy with d being the lattice constant and α_{M} the Madelung constant.[3]

From this equation the chemical shift of a core level, (2.11), between compounds A and B can be derived. On taking the difference, the free-atom or free-ion binding energy drops out. We see that the potential energy term in the chemical shift equation (2.11) is equal to the Madelung term and that

[3] The procedure just outlined produces out of a solid consisting of ions A with nuclear charge Z a solid in which for one ion the nuclear charge has been augmented by one unit. Within the limitations of the $(Z+1)$-approximation – replacement of a hole in the orbital k by an additional nuclear charge – the impurity state is the same as that created by the PE process. Therefore, the Born–Haber cycle can be used for the determination of binding energies in core levels of solids.

Table 2.1. Binding energies [eV] of the Na 1s and the Cl 2s levels in NaCl [2.18]

	E_B (atom)	$Q/\bar{r}_v$	$Q\alpha_M/d$	E_{rep}	E_{pol}	E_B (calc)	E_B (exp)
Na 1s	1079.0	+9.2	−8.9	−0.8	−2.5	1076.0	1077.4
Cl 2s	281.0	−12.1	+8.9	−1.2	−1.5	275.1	275.0

the expression containing the charge of the ion can be written as the charge divided by the average valence orbit radius.

In the literature one finds two corrections to (2.12). The first is usually called the polarization term [2.17]. It describes final state effects, and takes into account the fact that the photohole can polarize its environment, thereby enabling the whole system to lower its energy. This is why the *polarization energy* is a negative contribution to the binding energy. The polarization is different for anions and cations, e.g., for Na^+ and Cl^- in NaCl. There is also a second correction called the dielectric or repulsion term. This represents an initial-state effect and accounts for the fact that an ion repels its nearest neighbors in the initial state, whereby again the energy is lowered. This term, however, is somewhat uncertain and is often not taken into account in the calculation.

The above considerations are substantiated by numbers for NaCl [2.18]. Table 2.1 shows the contributions to the 1s binding energy of Na and the 2s binding energy for Cl. The atomic binding energies and the ionization potential of Na are taken from [2.6]. There is surprisingly good agreement between theory and experiment if we consider that the various contributions to the binding energy are of almost the same magnitude [2.19].

It is evident that the polarization and repulsion terms are rather small, and that the Madelung term represents the main solid state effect on the binding energies measured by PES.

This discussion of the core-level binding energy in an ionic solid assumes that relaxation by charge transfer onto the photoionized ion does not take place. The good agreement between the measured and calculated binding energies (Table 2.1) suggests that this approximation is valid for the alkali halides. However, it will be seen in Chap. 3 that this approximation breaks down for insulators containing transition metal ions (e.g., NiF_2). In such compounds a charge transfer from the ligand ion (F) onto the metal ion (Ni) can take place yielding large contributions to the binding energy.

2.2.2 Theory of Binding Energies

Binding energies measured by PES will now be considered from a more basic point of view [2.20–2.22]. Rigorously, the binding energy of an orbital k is given by the difference between the *total energies* of the final state $(N-1)$

and of the initial (ground) state N. For a system containing N electrons, we may write:

$$E_{\text{tot}}(N) + \hbar\omega = E_{\text{tot}}(N-1) + E_{\text{kin}} + \phi \,. \tag{2.13}$$

For convenience, the binding energies are measured with respect to the Fermi energy because most applications deal with solids. This does not restrict our considerations, however, and for molecules we simply set $\phi = 0$, thereby obtaining energies relative to the vacuum level.[4] One can rearrange (2.13) to give

$$\hbar\omega - E_{\text{kin}} - \phi = E_{\text{tot}}(N-1) + E_{\text{tot}}(N) \,. \tag{2.14}$$

The left-hand side is, see (1.2), the binding energy and therefore

$$E_{\text{B}} = E_{\text{tot}}(N-1) - E_{\text{tot}}(N) \,. \tag{2.15a}$$

To be more specific we assume that PE has occurred from an orbital k with charge $q_k = n$, where n is the number of electrons in orbital k. In keeping with the literature we have used Q for the excess charge on an atom compared to the neutral free atomic state, while q is the total charge of an orbital.

In terms of the orbital occupation, we can explicitly rewrite (2.15a) as

$$E_{\text{B}}(k) = E_{\text{tot}}(q_k = n-1) - E_{\text{tot}}(q_k = n) \,. \tag{2.15b}$$

At this point a word about sign convention is necessary: The binding energy $E_{\text{B}}(k) = \hbar\omega - E_{\text{kin}} - \phi$ is always a positive number, and this is used widely in the literature. However, with the advent of Inverse Photoemission Spectroscopy (IPES, see Sect. 1.3 for details), which measures energies above the Fermi energy, the situation has become unsatisfactory in some researchers' view, since the definition just given makes the binding energies above E_{F} negative. Others have thus adopted the habit of calling the binding energies in valence- and conduction-bands initial-state energies E_i and giving them a negative sign below the Fermi energy and a positive one above the Fermi energy. Although this situation is, admittedly, slightly confusing, we have decided in dealing with IPES data to follow this procedure in a slightly modified way in order to conform to the use of signs in the literature. Thus, we shall use positive binding energies in PES and negative binding energies in IPES, written as $(E_{\text{F}} - E_{\text{B}})$ which results in positive numbers.

If one wants to calculate a chemical shift with the help of (2.15) four total energies are required and the procedure is tedious. To overcome this, approximations have been developed to determine binding energies more readily. One starts from ground state energies and attempts to correct them by relaxation energies to obtain the energy of the final $(N-1)$ electron state. The most commonly used ground state approximation for the determination of binding energies is Koopmans' approximation [2.23].

---

[4] The work function on the right-hand side of (2.13) is necessary because the kinetic energy is always specified with reference to the vacuum level.

Koopmans assumed that the binding energy difference can be calculated from Hartree–Fock wave functions for the state with N electrons and the state with $(N-1)$ electrons. The binding energy then is the negative one-electron energy of that orbital from which the electron has been expelled by the PE process (1.10), yielding $E_{\mathrm{B}}(k) \simeq -\varepsilon_k$.

The Koopmans' theory binding energy leaves out the fact that after the ejection of an electron from a particular orbital k of an ion, the other orbitals of that ion will readjust to the new situation (relaxation). Therefore the correct binding energy is a difference of the Koopmans' approximation binding energy $-\varepsilon_k$ and a (positive) relaxation energy E_{R}, i.e.,

$$E_{\mathrm{B}}(k) = -\varepsilon_k - E_{\mathrm{R}} \ . \tag{2.16}$$

The separation of the binding energy into the Koopmans' binding energy $-\varepsilon_k$ and the relaxation energy E_{R} is convenient. However, this partition makes a comparison to the exact binding energy (1.16) difficult. Koopmans' energy is the negative Hartree–Fock energy of the orbital k. In a system with no correlations we have, on the other hand, $A = \delta(E - E_{\boldsymbol{k}}^0)$, where $E_{\boldsymbol{k}}^0$ is the one-electron energy, which may be identified with Koopmans' binding energy (where $\boldsymbol{k}$ now is the wave vector of a Bloch wave). In a system with electron–electron interactions, the single particle energy $E_{\boldsymbol{k}}^0$ is changed by the self-energy $\Sigma(\boldsymbol{k}, E_{\boldsymbol{k}}^1)$ and in an approximate way we can identify $\Sigma(\boldsymbol{k}, E_{\boldsymbol{k}}^1)$ with the relaxation energy E_{R}.

The binding energy calculated in the Koopmans' approximation "produces" an excited state of the $N-1$ electron system. Its ground state, which has a lower energy, is reached by relaxation. Since $E_{\mathrm{B}}(k) = E_{\mathrm{tot}}(q_n = n-1)$ $-E_{\mathrm{tot}}(q_n = n)$ we see that the binding energy of the unrelaxed state is larger than that of the relaxed state meaning that E_{R} is positive with the sign conventions used here.

Many procedures to correct for the lack of relaxation energies in Koopmans' energy employ the "transition-state concept" [2.14, 2.20, 2.24]. In its formulation one assumes that the total energy is a function of a variable x, which is defined by

$$q_k = n - x \ . \tag{2.17}$$

One can now expand the total energy $E(q_k)$ as a power series in x

$$E(q_k = n - x) = E(q_k = n) + ax + bx^2 + \dots \ . \tag{2.18}$$

This gives for the binding energy of an electron in orbital k, (2.15b) with $x = 1$:

$$\begin{aligned} E_{\mathrm{B}}(k) &= -E(q_k = n) + E(q_k = n-1) \\ &= -E(q_k = n) + E(q_k = n) + a + b + \dots \simeq a + b \ . \end{aligned} \tag{2.19}$$

On the other hand, one has $(x = n - q_k)$

$$\left.\frac{\partial E(q_k = n - x)}{\partial q_k}\right|_{q_k = n - \frac{1}{2}} = \left.\frac{\partial\{E(q_k = n) + a(n - q_k) + b(n - q_k)^2\}}{\partial q_k}\right|_{q_k = n - \frac{1}{2}}$$
$$= [-a - 2b(n - q_k)]|_{q_k = n - \frac{1}{2}}$$
$$= -a - b \equiv -E_{\mathrm{B}}(k)\ . \tag{2.20}$$

This last equation is the derivative of the total energy with respect to the charge, taken at a charge halfway between the ground state ($q_k = n$) and the state left after photoemission ($q_k = n - 1$). A comparison of (2.19) and (2.20) reveals that (apart from a sign which depends on the convention used to define the binding energy) (2.20) can be employed to calculate binding energies, within the approximation defined by (2.19). Therefore, in the way just outlined, any method to determine total energies can be used to determine binding energies including some approximate relaxation corrections, where the approximation comes from that used for the specific calculation of the binding energy and from that of the transition-state procedure, i.e. (2.19).

Details of the calculations of relaxation energies will not be presented here. Rather, examples will be given to demonstrate orders of magnitude and how well computational procedures can account for them.

Hedin and Johansson [2.10] have calculated E_{R} for atomic potassium with (2.16) and their result is given in Table 2.2. One sees that for orbitals whose principal quantum number n is smaller than that of the orbital from which photoemission takes place, the relaxation energy is small. This is easily explained qualitatively: if we consider the atom as a system of concentric charged spheres and if the charge outside a particular charge sphere is changed, the potential inside that sphere does not change. For orbitals where

Table 2.2. Relaxation energies [eV] for the energy levels of atomic potassium calculated by Hedin and Johansson [2.8]. The top line gives the orbital from which the electron has been photoionized

Relaxation energies in the orbitals (Coulomb plus exchange)	Photoionized electron						
	1s		2s		2p	3s	3p
1s	1.4		0		0	0	0
2s	3.2	}	0.3		0.5	0	0
2p	15.3	}	1.7		2.2	0	0
3s	2.3	} 96 %	1.4	}	1.6	0.2	0.2
3p	9.4	}	6.3	} 82 %	6.7	1.1	0.9
4s	1.3	}	1.1	}	1.1	0.9 ≡ 41 %	0.9
E_{R}	32.9		10.8		12.1	2.2	2.0

Table 2.3. Calculated atomic relaxation energies [eV] for some light atoms [2.25, 2.26]

atom	Photoionized orbital						
	1s	2s	2p	3s	3p	3d	4s
He	1.5	0	0				
Li	3.8	0	0				
Be	7.0	0.7	0				
B	10.6	1.6	0.7				
C	13.7	2.4	1.6				
N	16.6	3.0	2.4				
O	19.3	3.6	3.2				
F	22.1	4.1	3.9				
Ne	24.8	4.8	4.7				
Si	27.1	7.0	8.0	1.2	0.4		
Ti	35.4	13.0	14.4	3.6	3.4	2.0	0.3
Cu	48.2	23.7	25.7	7.7	7.2	5.3	0.3

the principal quantum number n equals that of the orbital from which photoionization takes place, the relaxation energy is not very large either. This is because, to a first approximation, all electrons with equal n have comparable radial distributions, and so a redistribution of the electrons gives only a moderate relaxation energy. Finally, the electrons with n larger than that of the photoionized orbital will give the main part of the relaxation energy, since these electrons experience an additional positive charge. This is also the reason why the $(Z+1)$-approximation works so well.

A compilation of relaxation energies calculated by Gelius [2.25] with an optimized Hartree–Fock–Slater method is given in Table 2.3. These data supply a good order-of-magnitude estimate of the various contributions, and trends can easily be recognized in this table.

The contributions mentioned above are usually termed *intra-atomic relaxation*. However, the orbitals on all other ions in the crystal or molecule also contribute to the relaxation energy. This is called the *extra-atomic relaxation* [2.26].

Relaxation energies have been calculated by a large number of methods. The results indicate a surprising agreement between "exact" calculations and those based on the transition state concept [2.1].

2.2.3 Determination of Binding Energies and Chemical Shifts from Thermodynamic Data

The methods to understand and interpret binding energies and chemical shifts discussed so far rely on the quality of the atomic calculations. One should keep in mind that such calculations are cumbersome and need a lot of computational know-how. In addition, they can only be performed by specialists and for simple experimentalists these calculations are not exactly self-evident. Therefore, it is interesting to explore possible ways of obtaining binding energies and chemical shifts from other independent sources, which lead to an understanding of measured energies and chemical shifts on a more physical and also more intuitive basis.

A procedure which has long been known and has successfully been applied to the analysis of optical spectra employs a Born–Haber cycle. The underlying concept was already discussed in the context of binding energies in insulators. It is particularly powerful in applications to the determination of binding energies in metals and chemical shifts in alloys and metallic compounds [2.27–2.33].

A discussion of this method is presented with reference to Fig. 2.8 where the Born–Haber cycle is given for the determination of binding energies in a metal. One starts with an element (nuclear charge Z) in its metallic form. The cohesion energy E^Z_{coh} is required to promote the atom from the metallic state

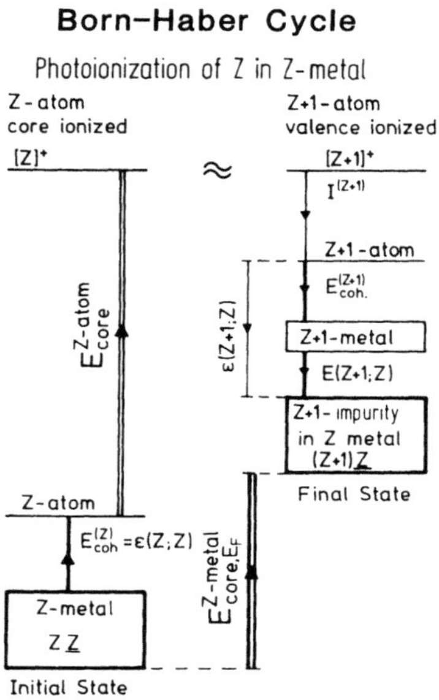

$$E^{Z\text{-metal}}_{\mathrm{core},E_F} = E^{Z\text{-atom}}_{\mathrm{core}} - I^{(Z+1)} + E^{(Z)}_{\mathrm{coh}} - E^{(Z+1)}_{\mathrm{coh}} - E(Z+1,Z)$$
$$= E^{Z\text{-atom}}_{\mathrm{core}} - I^{(Z+1)} + \varepsilon(Z,Z) - \varepsilon(Z+1;Z)$$

Fig. 2.8. Born–Haber cycle for the calculation of the binding energy in a metal. The $(Z+1)$-approximation is used to simulate the core-ionized ion [2.28]

to a free atom. Then the photoionization ($E_{\mathrm{core}}^{Z-\mathrm{atom}}$) leads to the production of an atomic core-ionized state. Using the (Z+1)-approximation this core-ionized state of the atom with nuclear charge Z is replaced by an atom which has nuclear charge (Z+1) and is lacking one electron in the valence shell. If this singly charged (Z+1) atom is neutralized one gains the ionization energy I^{Z+1}. This atom with nuclear charge (Z+1) is then transformed into its metallic state whereby the cohesion energy of the (Z+1) metal (E_{coh}^{Z+1}) is gained. Finally, this metallic atom of nuclear charge (Z+1) is dissolved in the original metal with charge Z, whereby the solution energy $E(Z+1, Z)$ is gained. Thus a final state is produced, consisting of a (Z+1) impurity in a metal with nuclear charge Z. This final state is exactly the one produced by photoemission from an atom in a metal with nuclear charge Z, provided that two assumptions hold: First, the Z+1)-approximation must be valid and, second, complete screening of the photoionized atom occurs in the metal. This means that the photoionized atom attracts one electronic charge which serves to completely screen the photohole. Under these conditions one obtains for the core-electron binding energy (element Z, metallic state, referred to E_{F})[5]

$$E_{\mathrm{c},E_{\mathrm{F}}}^{Z,\mathrm{metal}} = E_{\mathrm{c}}^{Z,\mathrm{atom}} - I^{Z+1} + E_{\mathrm{coh}}^{Z} - E_{\mathrm{coh}}^{Z+1} - E(Z+1, Z) \, , \tag{2.21}$$

bearing in mind that the difference $E_{\mathrm{c},E_{\mathrm{F}}}^{Z,\mathrm{metal}} - E_{\mathrm{c}}^{Z,\mathrm{atom}}$ is the same for all core levels of a given atom. All the quantities appearing in this equation are clearly accessible from experiments: atomic core-electron binding energy, ionization potential, cohesion energy and solution energy. Johansson and Mårtensson [2.28] have used this approach together with values for the parameters in (2.21) available from the literature to calculate the binding energies for many metals.

Figure 2.9 shows the difference in core-level binding energy in the 4d transition metal series between the free atom and the same atom in the metallic state $\Delta E_{\mathrm{c}} = E_{\mathrm{c}}^{Z,\mathrm{atom}} - E_{\mathrm{c},E_{\mathrm{F}}}^{Z,\mathrm{metal}}$. One observes quite good agreement between the theoretical curve and the experimental points. In Fig. 2.10 a compilation of selected binding energy differences in the metallic state is presented where the difference, calculated with (2.21) with respect to the atomic state is plotted against the corresponding value derived from experimental binding energies. Again one finds a very good correlation between theory and experiment. Note that in Fig. 2.9 the impurity term $E(Z+1, Z)$ from (2.21) is also indicated, and it is seen to be small or almost zero. Therefore, only a small error is made by neglecting it, which is frequently done.

The good agreement between the calculated and measured binding energies in Fig. 2.9 indicates that the assumptions underlying the calculations are valid. These are, as mentioned, the (Z+1)-approximation and the concept of a fully screened final state. This latter point indicates that the screening picture, although it is only a qualitative representation of the PE final state, describes it quite well. Of course the exact description of the final state is by

[5] The subscript "c" stands for "core" from here on.

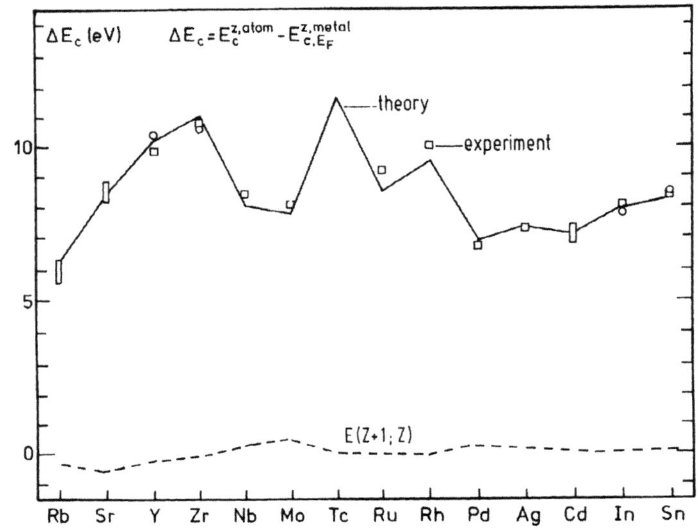

Fig. 2.9. Calculated and measured core-electron binding energy differences between the atomic and the metallic state ($\Delta E_c = E_c^{Z,\text{atom}} - E_{c,E_F}^{Z,\text{metal}}$); also shown is the solution energy $E(Z+1, Z)$ for the element $Z+1$ in a host Z, for the 4d elements [2.28]

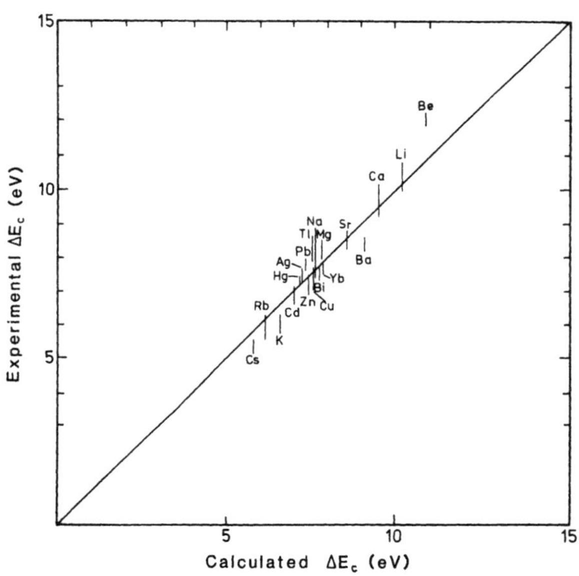

Fig. 2.10. Difference ΔE_c of core-level binding energy between the atomic and the metallic state as measured, plotted against that same quantity as calculated by the thermodynamic approach [2.28]

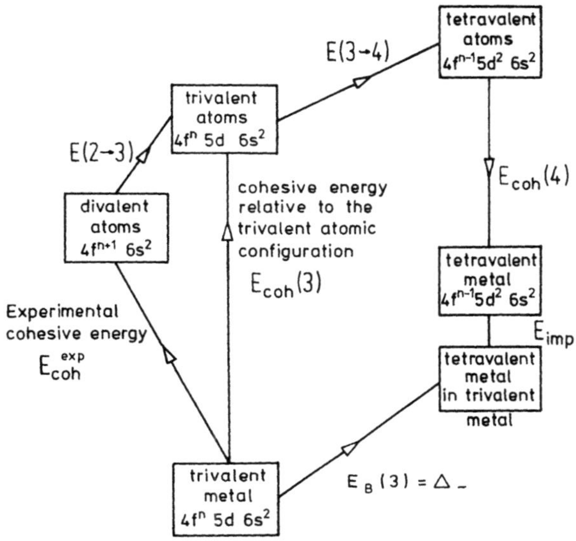

Fig. 2.11. Born–Haber cycle used to calculate the 4f binding energies in trivalent rare earth metals, where the term trivalent means that there are 3 electrons in the conduction band ($5d6s^2$) (only Eu and Yb metal are divalent). Only La, Ce, Gd and Lu are trivalent as free atoms and therefore their cohesive energy is measured with respect to that trivalent state. For the divalent rare earth ions the cohesive energy is defined with respect to the divalent free atoms. To obtain the value for the free trivalent atom, the energy $E(2 \to 3) = E(4f^{n+1}6s^2) \to E(4f^n5d6s^2)$ has to be added [2.27, 2.29]

the spectral function (1.16). However, for many semiquantitative estimates the simple screening picture is a very convenient tool.

The method employing the Born–Haber cycle (often called the thermodynamic approach because of its use of thermodynamic data such as cohesion and solution energies) has also been used with success to calculate the positions of the occupied and empty 4f levels in the rare-earth (RE) metals [2.29]. The main reason for this success stems from the fact that, as far as we know, even in a conducting solid the 4f electrons can be described very well by their atomic wave functions (with the possible exception of Ce) and can therefore be regarded, to a good approximation, as core electrons. An assumption that enters here (as before) is that of complete screening of the impurity produced in the final state by the photoemission process.

The Born–Haber cycle by which the 4f binding energy in a metal can be calculated is depicted in Fig. 2.11. Let us consider a trivalent metal, such as Gd. The term trivalent is slightly misleading at this point and has the following meaning: In an ionic solid such as GdF_3, the Gd ion occurs as Gd^{3+} with an electronic configuration $4f^7$. In the metal, Gd also has a $4f^7$ configuration (as determined, e.g., from susceptibility measurements), but now conduction electrons are also present such that the electronic configuration of Gd as a metal is $4f^7 5d6s^2$. The statement that Gd is trivalent in Gd metal therefore means that it has a 4f configuration which is equal to the one in its trivalent free ion state, namely $4f^7$ (plus the conduction electrons $5d6s^2$). When, in what follows, the additional terms divalent and tetravalent are used, they mean, e.g., for Gd, that it is in a $4f^8 6s^2$ ("divalent") or $4f^6 5d^2 6s^2$ ("tetravalent") configuration, respectively.

In the Born–Haber cycle of Fig. 2.11 the RE ion is first taken out of the trivalent metal to form a free trivalent atom ($4f^n 5d^1 6s^2$). This requires the cohesive energy $E_{\mathrm{coh}}(3)$. The atom is then converted to its tetravalent configuration ($4f^{n-1} 5d^2 6s^2$) for which the excitation energy $E(3 \to 4)$ has to be supplied. Now the tetravalent atom is put back into its metallic state whereby the cohesive energy with respect to the free tetravalent state, $E_{\mathrm{coh}}(4)$, is gained. Finally, the metallic tetravalent atom is dissolved in the trivalent metal and the solution energy E_{imp} is gained. The binding energy of a 4f electron in a trivalent 4f metal is therefore given by

$$E_{\mathrm{B}}(3) = E_{\mathrm{coh}}(3) + E(3 \to 4) - E_{\mathrm{coh}}(4) - E_{\mathrm{imp}} \,. \tag{2.22}$$

(In the literature one often writes $E_{\mathrm{B}}(3) = \Delta_-$). The term E_{imp} is small and frequently neglected.

In carrying out the cycle one encounters a difficulty: Except in La, Ce, Gd and Lu the free RE atoms occur in the divalent configuration ($4f^{n+1} 6s^2$) and not in the trivalent one ($4f^n 5d6s^2$). The experimental cohesive energies $E_{\mathrm{coh}}^{\mathrm{exp}}$ are therefore measured only for the divalent atoms. In order to reach the trivalent atomic state, one has to add the atomic excitation energy $E(2 \to 3)$ which changes the $4f^{n+1} 6s^2$ configuration into the $4f^n 5d6s^2$ configuration, whereby one obtains the hypothetical cohesive energy $E_{\mathrm{coh}}(3)$ with respect to the trivalent atomic configuration. Therefore, as can be seen from Fig. 2.11, the binding energy for the 4f electrons in a trivalent metal can contain five different terms.

One can, however, make use of the systematics of cohesive energies [2.29] in order to considerably facilitate the procedure. To that purpose Fig. 2.12 gives a compilation of cohesive energies relevant to the present problem [2.29]. It shows that E_{coh} is roughly constant throughout the series of divalent, trivalent and tetravalent metals. Thus, by neglecting the impurity term and taking $E_{\mathrm{coh}}(3) \simeq 100\,\mathrm{kcal/mole}$ and $E_{\mathrm{coh}}(4) \simeq 145\,\mathrm{kcal/mole}$ one finds from (2.22), noting that $1\,\mathrm{eV/atom} \mathrel{\hat{=}} 23\,\mathrm{kcal/mole}$,

$$E_{\mathrm{B}}(3) = E(3 \to 4) - 45\,\mathrm{kcal/mole} \simeq E(3 \to 4) - 2\,\mathrm{eV} \,. \tag{2.23}$$

This means that the 4f electron binding energy for trivalent metals is just the excitation energy $E(3 \to 4)$ minus $2\,\mathrm{eV}$.

The excitation energies $E(3 \to 4)$) can be obtained from atomic data (for some elements one has to perform interpolations, which, however, can

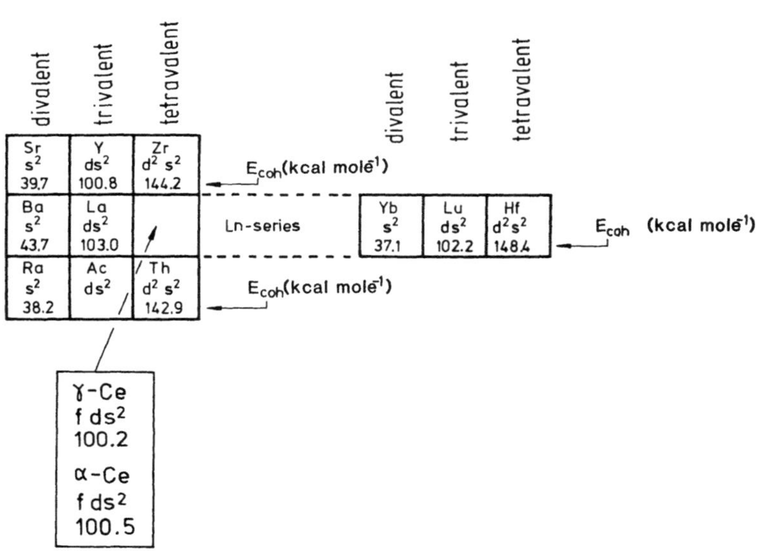

Fig. 2.12. Cohesive energies (in kcal/mole) for divalent, trivalent and tetravalent ions of the rare earth metals and their neighbors (23 kcal/mole = 1 eV/atom) [2.29]

be done reliably) [2.34, 2.35]. Experimental 4f binding energies are plotted in Fig. 2.13, and compared to calculated values.

Using similar thermodynamic arguments the electron addition energies for the trivalent RE metals (called Δ_+ or $E_B(2)$ and measured by BIS or IPES) can be calculated (Fig. 2.14). These energies are, of course, situated above the Fermi energy. Two rare earth metals (Eu and Yb) are divalent in the ground state. For them photoexcitation leads to a trivalent state and since $E_B(2)$ is below E_F, it can be obtained by reversing the arrows in the upper diagram of Fig. 2.14.

In the trivalent RE metals with the atomic divalent configuration, the binding energy of the divalent metallic 4f state is just the experimental cohesive energy E_{coh}^{exp} minus the divalent cohesive energy $E_{coh}(2) \approx 40$ kcal/mole. For those metals where the atomic state is trivalent (La, Ce, Gd, Lu), the contribution of an additional atomic excitation step, $E(4f^n 5d6s^2 \to 4f^{n+1} 6s^2)$, has to be added, as shown in Fig. 2.14. Figure 2.13 displays very good agreement between the results obtained by the simple thermodynamic approach [2.16, 2.29], a renormalized atom calculation [2.36–2.38] and the experimental data [2.39].

The discussion so far has assumed that the screening after photoionization takes place in the conduction band, meaning essentially that an additional 5d electron is put onto the photoionized site, which is generally the case. However, as will be pointed out in Sect. 3.2.2, in Ce metal, besides conduction

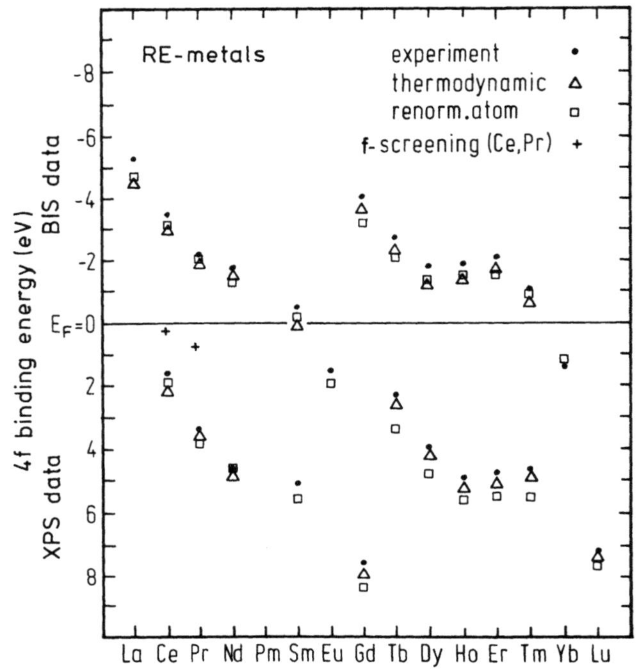

Fig. 2.13. Binding energies of the 4f electrons (above and below E_F) as measured [2.39], calculated by the thermodynamic approach [2.27, 2.29], and from a renormalized atom calculation [2.36–2.38]. Since it was decided in this text to give the binding energies below the Fermi energy a positive sign, those above E_F have a negative sign

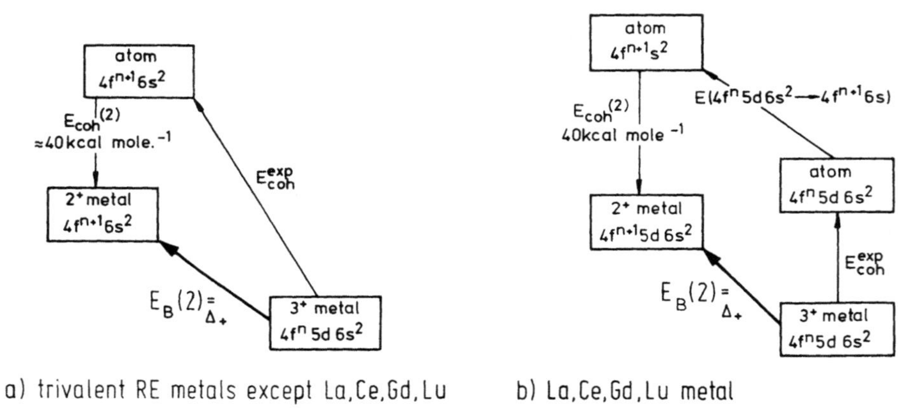

Fig. 2.14. Born–Haber cycles used to obtain the position of the unoccupied 4f levels; (**a**) for the trivalent rare earth metals (except La Ce Gd Lu) and (**b**) for La, Ce, Gd and Lu which appear as trivalent atoms [2.29]

electron screening by a 5d electron one also has the possibility that a 4f electron provides the necessary screening [2.40–2.42]. Consequently, a two-peaked structure for the 4f emission of Ce metal is expected and indeed observed. We mention that this view is not completely compatible with that held by other workers [2.43, 2.44] who prefer a many-body interpretation over the simple screening arguments (Sect. 3.4).

The final state produced by 4f screening is not much different from the initial state. The experiment locates the f-screened level (and thus also the 4f level) close to the Fermi energy. This position agrees with that deduced from most of the thermodynamic data. In contrast to this, the second state observed at about 2 eV below the Fermi energy was often interpreted as "the 4f state (level, band, etc.) in Ce metal" (the same holds, of course, for a Ce compound). A position of about 2 eV below E_F, however, was hard to reconcile with the then generally accepted knowledge [2.27] about the electronic structure of Ce which required a position of the 4f level near the Fermi energy. One sees from (2.23) that the position of the 4f hole screened by a 5d conduction electron is [2.27] $E(3 \to 4) - 2\,\mathrm{eV} = (4.5 - 2)\,\mathrm{eV} = 2.5\,\mathrm{eV}$ in agreement with the measured value of 2 eV. This is a 4f *hole* state and *not* the 4f state as often claimed. This is an example of a case where the uncritical use of PE data, without due regard of the complicated nature of the final state, produced a confusion which could have been avoided.

The 4f screened peak has also been observed in Pr metal and in a number of Pr and Nd intermetallic compounds [2.45–2.47]. Its interpretation, however, is not yet agreed upon. Nevertheless, we prefer the possibility that the peak near E_F is produced by 4f screening, whereas the peak further below E_F is due to screening by a ($5d6s^2$) conduction electron [2.48]. Other researchers prefer an interpretation in terms of a many-body calculation [2.49].

2.3 Core Polarization

Core-level PE always produces a final state with spin and angular momentum. If the system under investigation has an open valence-state configuration ℓ^n, with corresponding spin and orbital momenta, the coupling with the core-hole orbital and angular momenta leads to a number of different final states. A particularly simple situation arises if the core level is an s^2 state, resulting in a net spin of $s = 1/2$ for the photohole. This spin can couple parallel or antiparallel to the spin S of the valence-shell configuration. This then leads to an exchange splitting ΔE_s [2.22, 2.50]

$$\Delta E_s = \left(\frac{2S+1}{2\ell+1} \right) G^\ell(s,\ell) , \tag{2.24}$$

where $G^\ell(s,\ell)$ is the exchange integral [2.50].

The first observation of *core polarization* in a PE experiment was made for molecular oxygen and the results are shown in Fig. 2.15 [2.2]. This figure

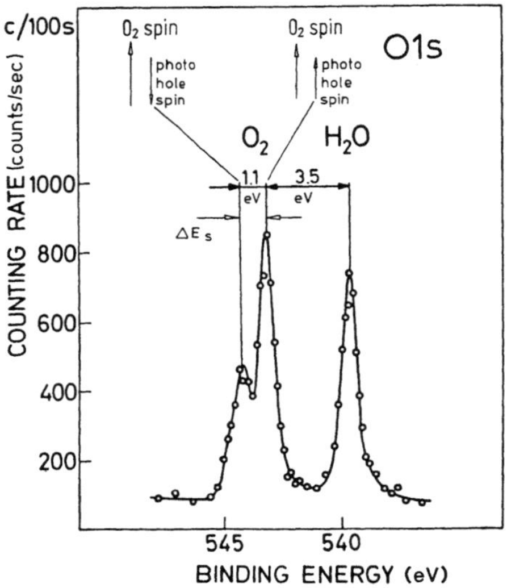

Fig. 2.15. Oxygen 1s core level (XPS) spectrum of O_2 and H_2O [2.2]. In paramagnetic O_2 the spin of the photohole ($s = 1/2$) can be parallel or antiparallel to the spin of the valence orbital ($S = 1$) and the two configurations show an energy difference of 1.1 eV due to core polarization; the intensity ratio reflects the multiplicity ratio of the two lines (3:1)

shows the O 1s lines for a mixture of gaseous O_2 and H_2O. A chemical shift of 3.5 eV is observed between the two species. However, while the line from H_2O is a singlet, that from O_2 exhibits a splitting which can only be explained by invoking the mechanism of core polarization. Molecular oxygen in its ground state has a total valence-electron spin of $S = 1$. After photoionization in the O 1s shell the spin $s = 1/2$ of the photohole can couple to the valence spin producing two final states, with total spins $J = 1/2$ or $J = 3/2$. These two states are separated by 1.1 eV, as depicted in Fig. 2.15. As might be expected, the intensities reflect the $(2J+1)$-degeneracy of the total angular momentum of the states. Water has a valence-band spin of $S = 0$ and therefore no core polarization splitting is observed.

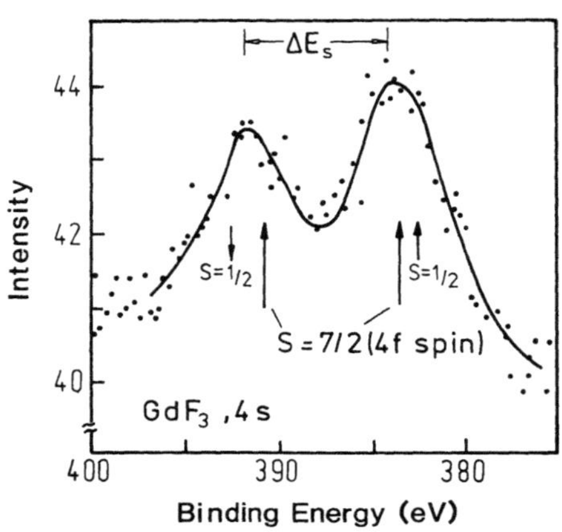

Fig. 2.16. 4s core level XPS spectrum of Gd^{3+} in GdF_3 [2.53]. The spin of the 4s photohole can be oriented parallel or antiparallel to the spin ($S = 7/2$) of the $4f(^8S_{7/2})$ configuration of Gd^{3+} in GdF_3 . Note the suppressed zero of the ordinate scale

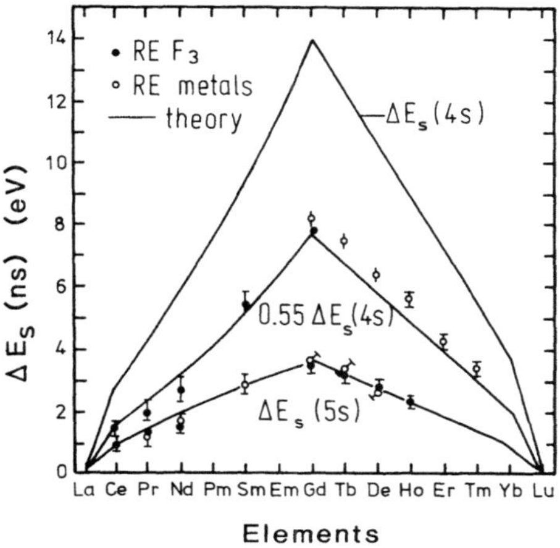

Fig. 2.17. Measured and calculated 4s and 5s core-line splittings in the rare earths [2.53–2.55]. The results agree for the 5s line but, due to configuration interaction, the calculation for the 4s line overestimates ΔE_s by almost a factor of two

Core polarization has been studied extensively for compounds of 3d transition elements [2.51,2.52] and for those containing rare-earth elements [2.53]. A typical example of core polarization in a solid is depicted in Fig. 2.16, where a photoemission spectrum of the 4s core level in GdF_3 is reproduced [2.53]. The exchange splitting is clearly resolved. A compilation of such splittings is seen in Fig. 2.17 [2.53–2.55], together with results calculated by the Hartree–Fock method. Theory and experiment agree for the 5s shell but not for the 4s shell. This is an effect of electron–electron correlations: The matrix elements for configuration interaction between the 4f and the 4s shells are large, reducing the splitting calculated in the Hartree–Fock approximation. Between the 4f and the 5s shells, however, these matrix elements are small.

In order to evaluate the situation with respect to the analysis of core polarization for 3d transition metal–ion compounds, a deeper understanding of core-level PES in systems with open shell ions (e.g., the 4f or 3d ions) is necessary. This is however a complicated topic and will be dealt with in a separate chapter (Chap. 3). Here only those facts will be presented which are necessary to the analysis of the core polarization in 3d transition metal ion compounds.

Let the ground state configuration of an open shell ion be $d^n L$, where d stands for the angular momentum of the open shell ion and L for the ligand configuration. A core hole in the s shell will be designated by s^{-1}. One can then observe two final states after the photoemission process namely $s^{-1}d^n L$ and $s^{-1}d^{n+1}L^{-1}$. The latter one is produced by a ligand-to-metal charge transfer. These two final states are also observed in RE compounds (f^n). However, there they do not interfere with the analysis of the core polarization splittings.

This is different for 3d transition metal ion compounds. For them the spectra consist of a mixture of different lines resulting from different screening channels and from the additional exchange splitting. Because of this compli-

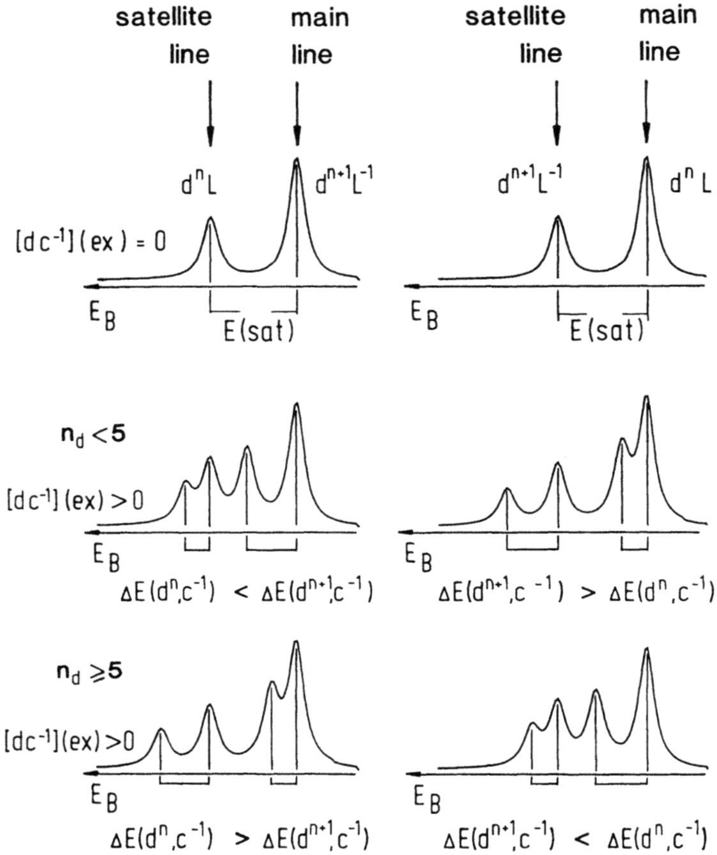

Fig. 2.18. Schematic picture of the s-core level splitting in a transition metal compound. The left column gives the situation, when the main line has a valence-band configuration of $d^{n+1}L^{-1}$ and the satellite has d^nL. In the right column the labeling is reversed. In the top row the exchange interaction between the core level (c^{-1}) and the valence d electrons $[dc^{-1}](ex)$ is assumed to be zero, while it is non-zero in the next two rows. This describes the situation for $n_d < 5$ *and* $n_d \geq 5$, respectively

cated situation the question of the interpretation of the core-level spectra in 3d transition metal compounds is not yet settled [2.56–2.59]. The discussion centers on whether in the core-level spectra it is the state $s^{-1}d^{n+1}L^{-1}$ or the state $s^{-1}d^nL$ which has the lowest energy. We mention that often in the spectra of 3d transition metal compounds the line corresponding to the final state with the lowest energy is called the *main line*, while the line (or the lines) corresponding to the excited final state (or states) is called the *satellite* (or satellites).

The problem involved in the analysis of the core polarization for core lines of 3d transition metal ions can be demonstrated with the help of Fig. 2.18. The right column indicates the situation when the lowest-energy final state (main line) has the same d^n configuration as the d^n initial ground state. The satellite line ($d^{n+1}L^{-1}$), separated from the main line by the energy $E(\mathrm{sat})$, has a larger binding energy, meaning that it is not the final ground state. One realizes now that the exchange splitting in the main line is smaller than that in the satellite for $n_d < 5$ and that the exchange splitting $\Delta E(d^n c^{-1})$ is larger in the main line than in the satellite line for $n_d \geq 5$. In the case shown in the left column of Fig. 2.18, namely that the main line has the configuration $d^{n+1}L^{-1}$, the situation with respect to the exchange splitting in the main line and satellite line is reversed.

The situation just described is the simplest one. It has been shown [2.58] that for $n_d < 5$ the possibility exists that the main line is of the type d^nL while the satellite line corresponds to a valence state configuration of the type d^nL^* (L^* stands for a ligand excitation, the satellite is then called an *exciton satellite*). A systematic investigation of 3s splittings in 3d transition metal di- and trifluorides was performed by Veal and Paulikas [2.56, 2.57]. They concluded from their systematic study that for the whole series of 3d ions a situation exists that corresponds to the left column of Fig. 2.18, meaning that the valence ground state configuration after core-level photoionization is always of the type $d^{n+1}L^{-1}$. This would mean that the d count for the main line is always increased by one relative to that in the initial state.

There is evidence indicating that this simple interpretation is not completely correct. Figure 2.19 presents the 3s core-level spectrum of CuO ($3d^9$ initial-state configuration). It shows that the satellite line has a splitting containing core polarization and therefore must correspond to a valence-band spin that is non-zero.[6] This points to the d^9L configuration for the valence band in the 3s satellite line. The main line has no splitting and is therefore ascribed to a $3d^{10}L^{-1}$ valence-band configuration. The reverse situation is observed by an analysis of the 3s line splitting in CoF_2 (Fig. 2.20) indicating that the final state configurations have changed with respect to the situation in CuO, i.e., that for CoF_2 the main line is d^7L (which is also the initial-state configuration) and the satellite has the configuration d^8L^{-1}. An analysis of the 3s core-line exchange splittings in a number of different compounds leads to the following conclusions. For $n_d = 5$ and 6, depending on the particular compound the change in the main line occurs from a $d^{n+1}L^{-1}$ final state to a d^n final state, which in fluorides and oxides is the one actually observed. Two pieces of data can help to substantiate this statement. Figure 2.21 gives the 3s splittings of FeF_2, MnF_2 and CrF_2 with $n_d = 6$, 5 and 4 in the initial ground state, respectively. If the d occupation for the 3s line

[6] The splitting is larger than expected from systematics; it probably contains contributions, like crystal–field interaction, valence-band structure, and multiplet interactions (spin–orbit and Coulomb interaction) in addition to core polarization [2.60–2.62].

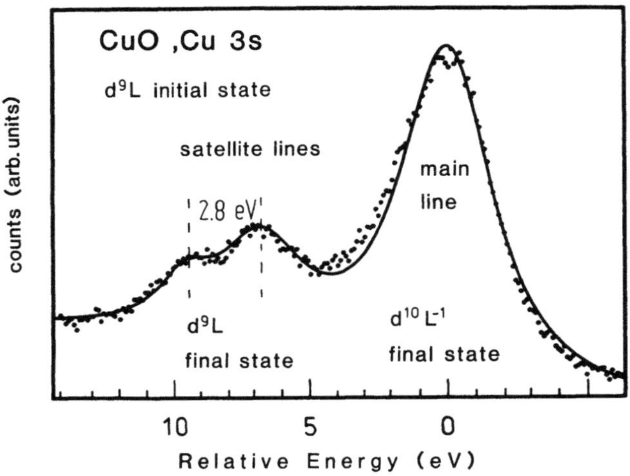

Fig. 2.19. 3s core line in CuO. In this compound the main line is $d^{10}L^{-1}$ (no exchange interaction possible) and the satellite line is d^9L and therefore exhibits an exchange interaction. CuO is a case that corresponds to the left column of Fig. 2.18 and to the bottom row. The splitting in the satellite is larger than expected from the systematics[6]

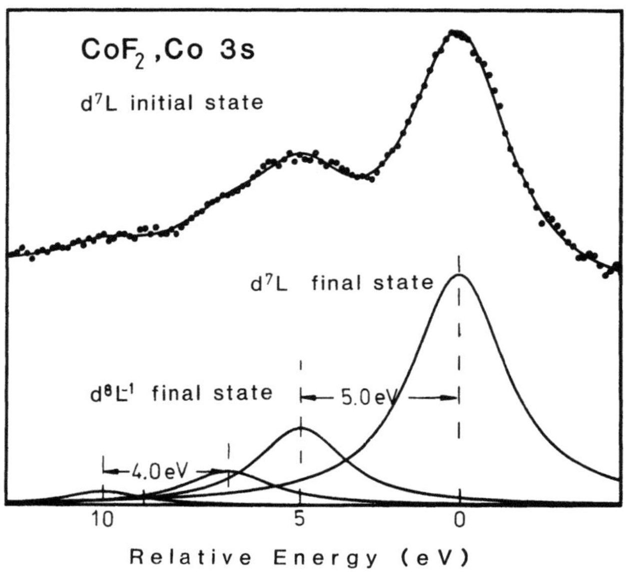

Fig. 2.20. 3s core line in CoF_2. The analysis, indicated at the bottom, gives a larger exchange splitting for the main line than for the satellite. This indicates that the main line is d^7L and the satellite is d^8L^{-1}. This situation corresponds to the right column of the diagram in Fig. 2.18 and to its bottom row

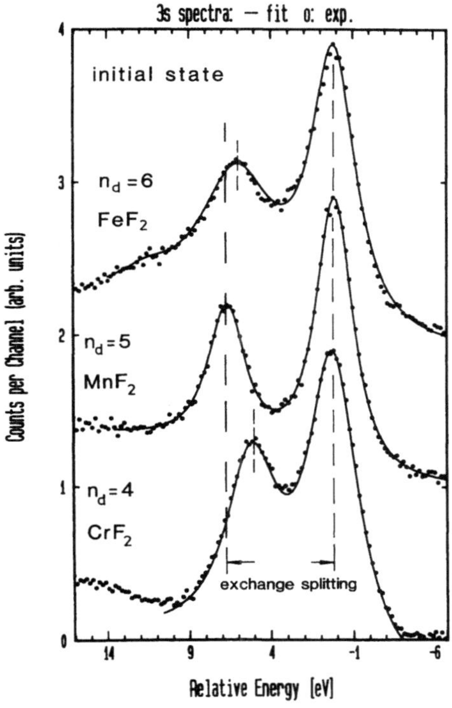

Fig. 2.21. 3s core lines for CrF_2, MnF_2 and FeF_2. The satellites are weak in these compounds and therefore the spectra give practically only the main lines. MnF_2 (ground state $n_d = 5$) has the largest core polarization splitting indicating that for these compounds the situation is that of the right column in Fig. 2.18

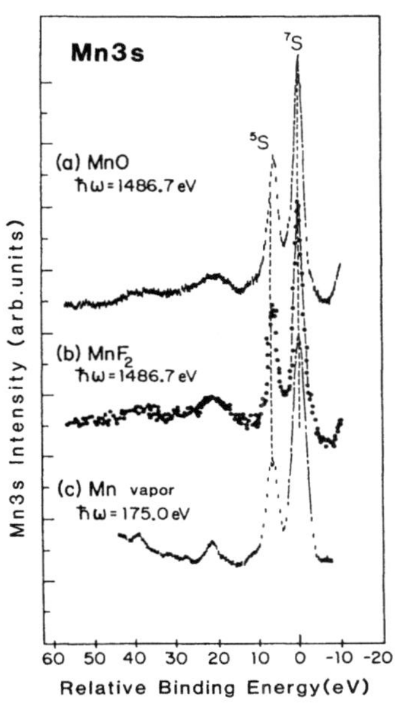

Fig. 2.22. 3s core line for MnO, MnF_2 and atomic Mn [2.62]. The exchange splittings are very similar in all three states of Mn indicating that one is dealing with the situation depicted in the right column of Fig. 2.18, because atomic Mn has a final state valence electron configuration of d^5, which must also apply to the MnO and MnF_2

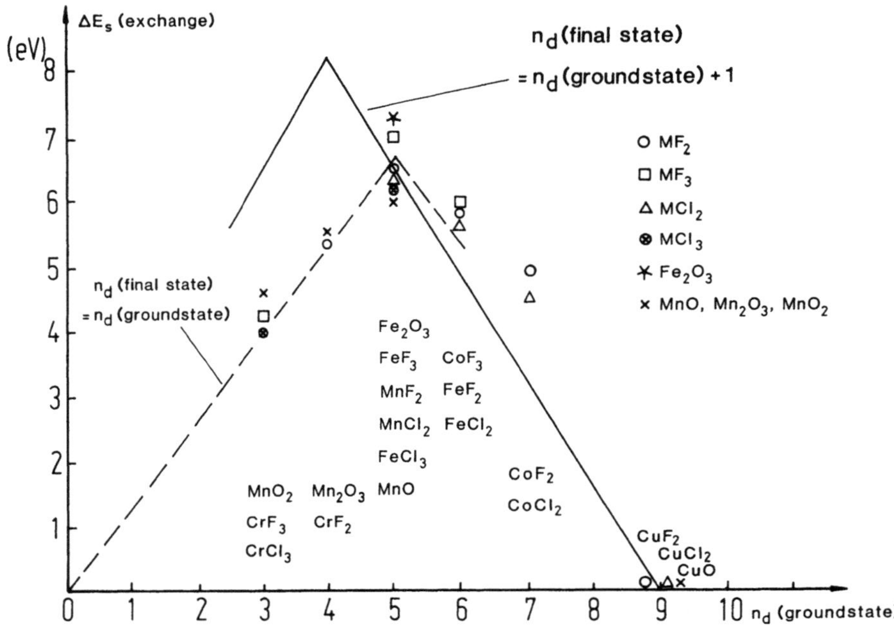

Fig. 2.23. Summary of 3s core-level exchange splittings (main line!) as a function of the d occupancy (n_d) in the initial ground state. The solid line is from Veal and Paulikas [2.56, 2.57] and probably describes the data for $n_d = 9$ and 8; the dashed line presents the data for $n_d \leq 9$; for $n_d = 6$ and 7 each compound has to be analyzed separately to find out to which systematics it belongs

(the figure shows only the main line, the satellite is barely visible for these compounds) is the same as in the initial state, the 3s splitting should be the largest in MnF_2 and smaller in FeF_2 and CrF_2 as is actually observed. If the model of Veal and Paulikas [2.56, 2.57] were correct, the final states would be d^5L^{-1}, d^6L^{-1}, and d^7L^{-1} for CrF_2, MnF_2 and FeF_2, respectively, and the 3s splittings should decrease in that order. Secondly we give in Fig. 2.22 the 3s splittings in MnO, MnF_2 and for atomic Mn (in the gas phase) [2.63]. It is known that atomic manganese has a d^5 configuration which is not changed by the core-level photoemission process. Therefore the 3s splitting reflects the core polarization of the 3s-core hole with that d^5 configuration. MnO and MnF_2 have the identical 3s core-level splitting as atomic Mn, which demonstrates that they have the same d^5 final-state ground-state configuration and not a d^6L^{-1} configuration, as suggested by Veal and Paulikas [2.56, 2.57].

Finally we present in Fig. 2.23 a compilation of the main line 3s splitting for a number of 3d transition metal compounds. Also shown is the result of Veal and Paulikas [2.56, 2.57] (full line) which assumes that the final-state d-occupation is increased by one relative to that of the initial ground state and (as a dashed line) that based on our assumption that the final state d

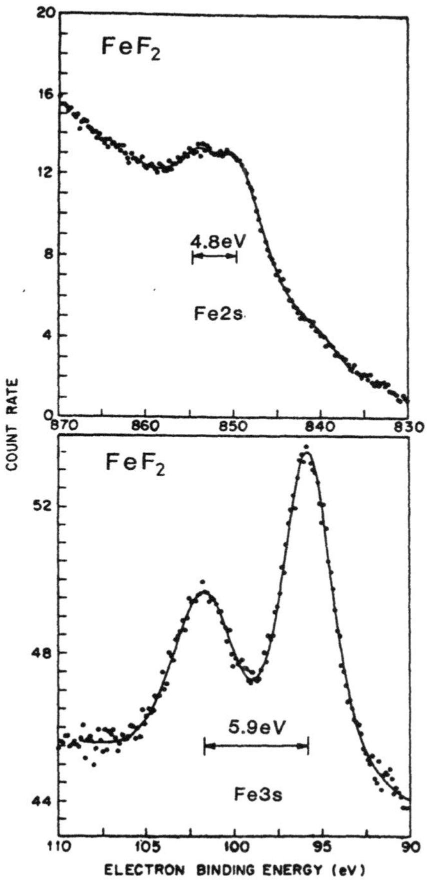

Fig. 2.24. 2s and 3s core line splittings in FeF_2 [2.52]

count equals the initial state d count. The data as they stand suggest that for divalent nickel and copper compounds ($n_d = 8$ and $n_d = 9$ in the initial state) the photoemission final state has a d-count which is increased by one relative to that of the ground state, that for $n_d \leq 6$ the d-count in the photoemission final state equals that of the initial state and that for $n_d = 7$ (divalent cobalt) the situation depends on the degree of covalency in the metal-ligand band.

The exchange interaction due to core polarization can be observed in the 3d transition metal compounds for the 3s and the 2s-core levels. Figure 2.24 compares spectra of the 2s and 3s line [2.52] obtained for FeF_2. The data for the 2s line have a much higher background (because of the larger binding energy) and a smaller exchange splitting. Therefore they are rarely used.

It should be emphasized that the interaction of the angular momentum of an inner shell with that of the valence shell is not restricted to inner s levels. However, this is the only case for which the situation is simple. As an

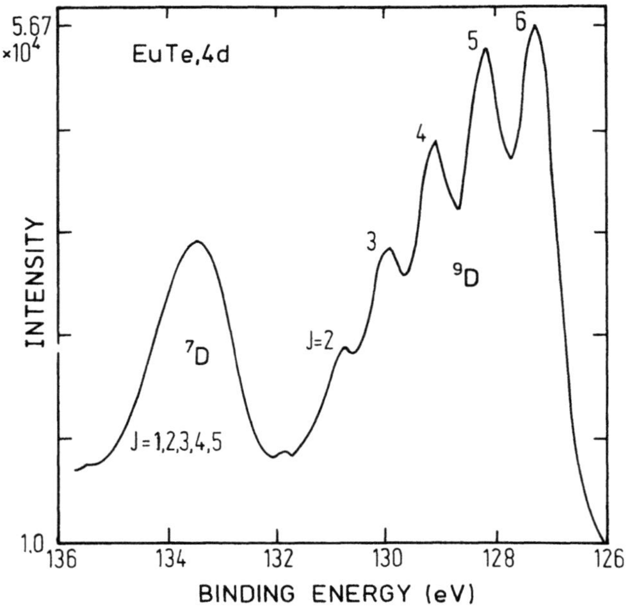

Fig. 2.25. 4d core line of Eu in EuTe (Eu^{2+}). The $^2d_{3/2}$ and $^2d_{5/2}$ final hole states couple to the $^8S_{7/2}$ state of the 4f electrons yielding a $^7D_{1...5}$ multiplet and a $^9D_{2...6}$ multiplet [2.54]

example for a more complicated situation Fig. 2.25 shows the 4d spectrum of Eu metal [2.54]. Here one has a coupling between a d-hole (in either a $^2D_{3/2}$ or a $^2D_{5/2}$ state) and the ground-state configuration ($^8S_{7/2}$) of Eu metal. This coupling can lead to a $^7D_{1...5}$ state or a $^9D_{2...6}$ state. The ^7D state splitting is not resolved, while in the case of the ^9D state the $J = 2$ to $J = 6$ components are easily recognized in the spectrum. This example serves to illustrate that final hole states can have a very complicated structure due to multiplet interactions.

2.4 Final-State Multiplets in Rare-Earth Valence Bands

Final-state coupling is especially important for an understanding of the valence-band PE spectra of RE metallic systems. They lead to significant differences compared to the photoelectron spectra of simple or transition metals. In the case of simple (sp electron) metals or d electron metals, it is generally assumed (with the exception of Ni metal) that a valence-band PE spectrum reflects a final state, screened by a charge similar to that photoemitted. Thus, for these metals, one may say that a PE signal is observed which is not too different from that expected if a one-electron description

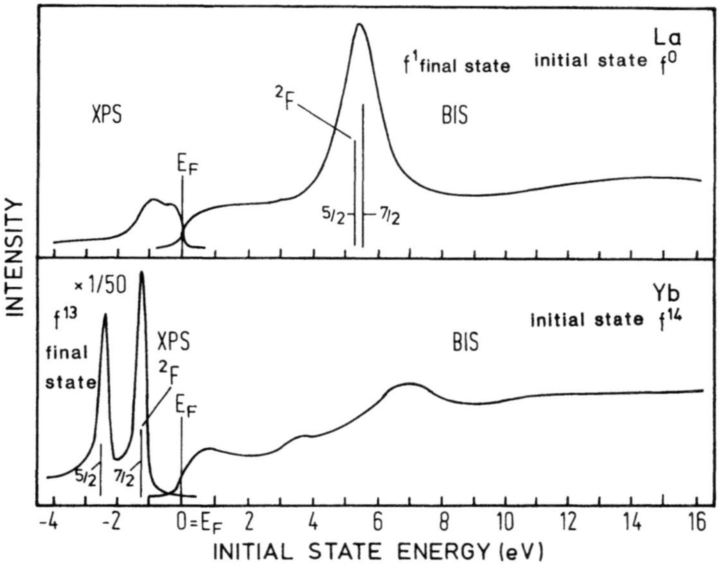

Fig. 2.26. Combined XPS and BIS spectra of La metal ($4f^0$ initial state) and Yb metal ($4f^{14}$ initial state). The XPS spectrum of La metal shows only the $(5d6s)^3$ conduction band and the BIS spectrum the $4f^1$ final state (the spin–orbit interaction is not resolved). The XPS spectrum of Yb metal ($4f^{14}$ initial state) gives the $4f_{7/2\,5/2}$ final state doublet, while the BIS spectrum of Yb gives no structure from 4f states [2.39]

were appropriate. (In Chaps. 3 and 7 the problems of relaxation in d band metals will be treated).

In RE metals the situation is very different. Except for the case of cerium, one can assume that in the RE metals the 4f electrons are quite localized. Thus in a PE experiment they will behave very much like core electrons. This means that one measures a final state which effectively differs by one f electron from the initial state. Thus the PE spectra of the RE metals or RE metallic compounds can be interpreted completely in a final state picture. Figure 2.26 [2.39] gives PES and BIS spectra of the two electronically most simple RE metals, namely the one with no 4f electrons (La) and the one with fourteen 4f electrons (Yb). A combination of the two techniques PES and BIS allows one to map the density of both occupied and empty states in the final state. In the XPS-PE spectrum of La only the $(5d6s^2)$ valence band near the Fermi energy is observed. The empty density of states gives the $4f^1$ peak at 5.5. eV above the Fermi energy where the spin–orbit interaction cannot be resolved. For Yb, on the other hand, the XPS spectrum shows the f^{13} final state (which is equal to a one-hole state f^{-1}) with the spin–orbit splitting into the $^2F_{7/2}/^2F_{5/2}$ doublet clearly resolved. (Note that Yb metal is divalent and therefore has a configuration $4f^{14}6s^2$ in the ground state). The BIS spectrum of Yb metal shows no additional 4f feature.

Figure 2.27 depicts, in the same sense, a combination of an XPS and a BIS spectrum of Er metal ($4f^{11}$). The XPS spectrum has the signature of a $4f^{10}$ final state and the BIS spectrum that of a $4f^{12}$ final state. The bars in the spectra show the positions and relative intensities of the expected final states. These final states can be taken, for example, from optical spectroscopy [2.64, 2.65].

The RE metals can also be used to demonstrate the detailed information that can be obtained from a high-resolution PE spectrum. Figure 2.28 shows UPS spectra of γ-Ce $[4f^1(5d6s^2)]$ and of its low temperature phase α-Ce $[4f^{0.9}(5d^{1.1}6s^2)]$ in the region from the Fermi energy to 0.6 eV below it [2.66]. Because of the variation of the cross section with $\hbar\omega$, the 40.8 eV spectrum shows the weak contributions of the 4f electrons more intensely than the 21.2 eV spectrum. Thus the small difference between the two spectra should give a fair representation of the 4f contribution to the valence-band spectra of γ-Ce and α-Ce.

The spectra reveal a spin–orbit splitting of 280 meV between the $^2F_{5/2}$ and the $^2F_{7/2}$ states. Furthermore, the fact that 4f emission occurs near E_F indicates that one observes a final state not too different from the initial state, and one which can be considered to stem from the following process [2.42]

$$4f^1 5d6s^2 + \hbar\omega \Rightarrow (4f^0 5d6s^2)^* \Rightarrow 4f^0 5d6s^2[4f] \ . \tag{2.25}$$

The photoionization thus leads to an excited state (indicated by the star), which is screened by a 4f electron (indicated by [4f]). Clearly, this state is similar to the ground state. Note that screening can also take place via a 5d conduction electron, leading to a $4f^0 5d6s^2$ [5d] final state located about 2 eV below E_F. It is not visible in the present spectra (see, however, Fig. 3.20).

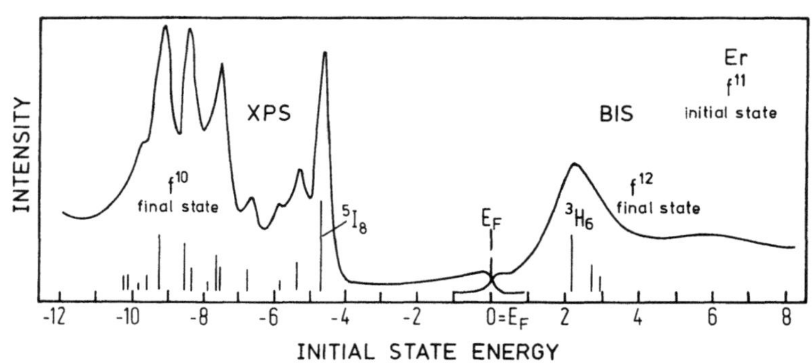

Fig. 2.27. XPS and BIS spectrum of Er metal ($4f^{11}$ initial state) showing the multiplet splitting in the $4f^{10}$ (XPS) and $4f^{12}$ (BIS) final states [2.39]

The width seen in the $^2F_{5/2}$ and $^2F_{7/2}$ final states is mostly due to instrumental and thermal broadening. The transition from γ-Ce to α-Ce is thought to arise from an increase of the f-d hybridization [2.27, 2.40], which leads to a larger 4f itineracy in α-Ce than in γ-Ce. This is also reflected in Fig. 2.28 where one sees that in α-Ce the (excited) $^2F_{7/2}$ state decays faster (gives a smaller signal) than in γ-Ce, a consequence of the increased 4f itineracy. The researchers themselves give a different interpretation of the spectra in terms of the work of Gunnarsson and Schönhammer [2.43, 2.44], see also Chap. 3; the question of whether the simple screening arguments and the elegant many-body calculation are in essence equivalent, is not yet fully settled [2.49, 2.67].

In this context it is worth mentioning that PES has been used extensively to investigate the wave functions of so-called "intermediate valence" compounds. These are RE compounds in which the ground-state wave function

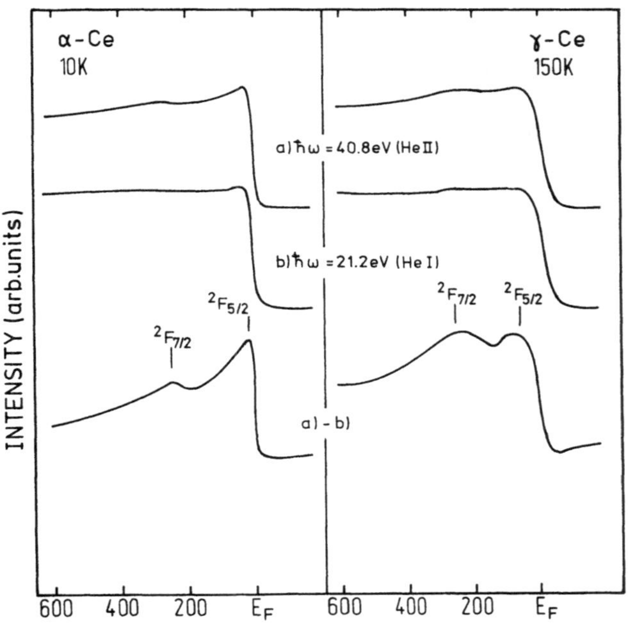

Fig. 2.28. UPS spectra of $\alpha-$Ce and $\gamma-$Ce [2.66]. The $\alpha-\gamma$ transition in Ce metal takes place at $\sim 100\,$K. The two top traces give the raw data at $\hbar\omega = 40.8\,$eV and $\hbar\omega = 21.2\,$eV. At $\hbar\omega = 40.8\,$eV the 4f cross section is larger than at $\hbar\omega = 21.2\,$eV and therefore the subtraction of the two spectra gives essentially the 4f intensity Because of the high resolution, the spin–orbit splitting is visible. The intensity ratio of the $^2F_{7/2}$: $^2F_{5/2}$ peaks is reduced in $\alpha-$Ce with respect to $\gamma-$Ce, probably because of the faster Auger decay rate of the $^2F_{7/2}$ state in $\alpha-$Ce, which has greater itineracy of the 4f electrons

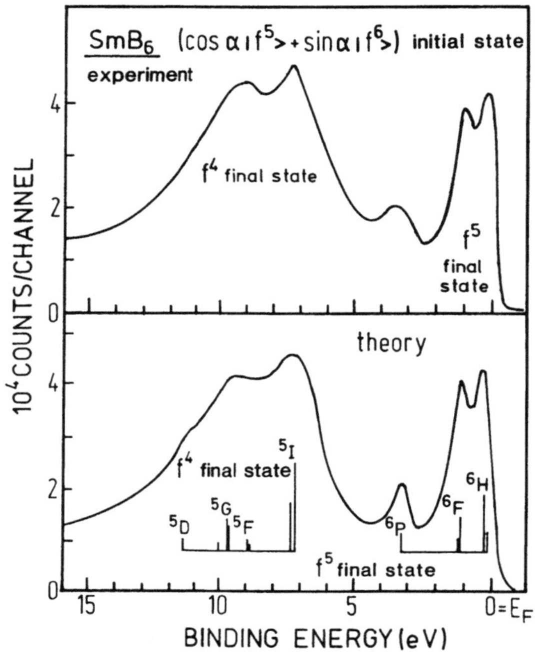

Fig. 2.29. Measured and calculated XPS spectrum of SmB_6. This is an intermediate valence compound with a wave function $\cos\alpha|4f^5 5d6s^2\rangle + \sin\alpha|4f^6 6s^2\rangle$ in the initial state [2.68]. The final-state structure $\cos\alpha|4f^4 5d6s^2\rangle + \sin\alpha|4f^5 6s^2\rangle$ is clearly visible, and a calculated spectrum using the final-state curves of Fig. 2.30 agrees quite well with the measured one [2.69]. *Note*: In the initial state the energies of the two configurations are degenerate and are located near the Fermi energy; therefore a fluctuation between the two electronic configurations can be induced by the thermal energy. In the final state one 4f electron is missing and this shifts the $4f^4$ final state by $\sim 7\,\mathrm{eV}$ away from the Fermi energy

is composed of configurations with different f-counts (this is rather unusual because generally the rare-earth ions have an integer 4f count in solids, too), and therefore can be written in the following way

$$\cos\alpha|4f^n 5d\rangle + \sin\alpha|4f^{n+1}\rangle\ . \tag{2.26}$$

If, in the photoemission process, no relaxation involving f electrons takes place, the signature in the final-state spectra reflects the existence of the two initial-state configurations.

The power of this method can be seen by looking at the PE spectra (Fig. 2.29) of SmB_6 which is known from other measurements to be an intermediate valence compound with an f-count between five and six. As an introduction to the problem Fig. 2.30 first shows PE valence-band spectra of NdSb, SmSb and SmTe. NdSb serves to indicate the location of the valence band, in a trivalent RE antimonide, which is mainly composed of 5d

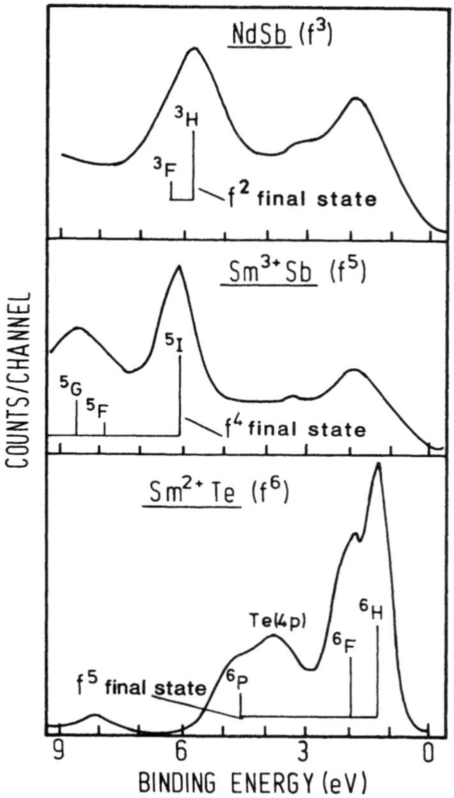

Fig. 2.30. XPS valence-band spectra of (*from top to bottom*): $Nd^{3+}Sb$ (initial state $4f^3$), the $4f^2$ final state is well separated from the valence band; $Sm^{3+}Sb$ (initial state $4f^5$), the $4f^4$ final state is well separated from the valence band near the Fermi energy; $Sm^{2+}Te$ (initial state $4f^6$), here the $4f^5$ final state is near the Fermi energy due to the additional 4f electron [2.68]

electrons. The next two spectra in Fig. 2.30 are given to indicate where one has to expect the final state of f^5 and f^6 initial states, because they are both supposedly present in SmB_6. In SmSb, samarium is in a 3^+ state; thus the initial configuration is f^5 and the final configuration is f^4, as can also be seen from the spectrum. In contrast, in SmTe, samarium is in a 2^+ configuration which is f^6 and thus we expect a f^5 final state and this is again borne out by the experimental findings.

The wave function of SmB_6 may be written

$$\cos\alpha|\mathrm{Sm}^{3+}, 4\mathrm{f}^5 5\mathrm{d}\rangle + \sin\alpha|\mathrm{Sm}^{2+}, 4\mathrm{f}^6\rangle \ . \tag{2.27}$$

Photoemission without f electron relaxation then leads to the following final state:

$$\cos\alpha|\mathrm{Sm}^{4+}, 4\mathrm{f}^4 5\mathrm{d}\rangle + \sin\alpha|\mathrm{Sm}^{3+}, 4\mathrm{f}^5\rangle \ , \tag{2.28}$$

which is a combination of f^4 and f^5 final states (see also Fig. 2.30). The intensities are now determined by the initial-state wave-function weights.

A PE spectrum of SmB_6 is shown in Fig. 2.29. It contains contributions from both, the f^4 and f^5 final states. This is demonstrated in the lower panel of this figure which depicts a "calculated" spectrum where the final-state structure for the f^5 and f^4 states was taken from the data of Fig. 2.30 to synthesize the PE spectrum SmB_6. There is good agreement between theory and experiment [2.68–2.70].

Similar effects have been observed in the valence bands of ionic 3d transition metal compounds. However, in these cases one cannot assume that no relaxation involving d electrons takes place. Therefore, in Chap. 3, we will first have to inspect the relaxation processes in 3d ionic compounds before the PE spectra of these compounds can be analyzed.

State-of-the-art one-electron band structure calculations of solids such as using density-functional theory in the local-density approximation yield ground-state properties. A comparison of such calculations with PES and IPES valence- and conduction-band data is therefore made with the implicit assumption that the final state in the experiment has been screened such that it is not too different from the initial state. In narrow-band systems such as the rare earths this assumption is obviously wrong: We have just seen that (with the exception of Ce, Pr, and, to a lesser degree Nd), photoionization of 4f electrons leads to a final state that differs from the initial state by one electron, and that IPES leads to a final state with one additional f electron. However, it has become apparent that a modification of standard band-structure calculations can be used with considerable success to model excited state properties as observed by PES or IPES. In brief, this method consists of removing one electron from an ion (just as the photoemission process does) and making the ion neutral again by adding the appropriate screening electron. For the rare earths this means that 4f PE from a $4f^n(5d6s)^3$ ion leads to a neutralized $4f^{n-1}(5d6s)^4$ ion. This ion now acts as an impurity in the solid and thereby destroys the translational symmetry. This computational problem, however, can be solved by using the impurity ion and an appropriate number of unperturbed ions surrounding it as a *supercell* of an infinite crystal, thus restoring the convenient periodic boundary conditions [2.71]. It was found that this method is not very sensitive to the size of the supercell. In fact, it already yields quite reasonable results if only the impurity (= photoionized) ions are considered.

This approach has been used to calculate PES and IPES data for rare earths [2.41, 2.64, 2.72–2.75], and also in an exploratory calculation for NiO [2.76]. Results are given in Fig. 2.31 [2.75] (triangles) and compared to the experimental values of Lang et al. [2.39] (circles). For a comparison with thermodynamic data and renormalized atom calculations, see Fig. 2.13. The supercell calculations are based on "first principles" procedures and do not require any experimental information. Considering this fact, the agreement between theory and experiment is remarkable, although there are some deviations in the second half of the series. Thus it looks as if this might provide an excellent method to calculate PES and IPES spectra for narrow-band materials. Calculations within the supercell approach have also shown that the two-

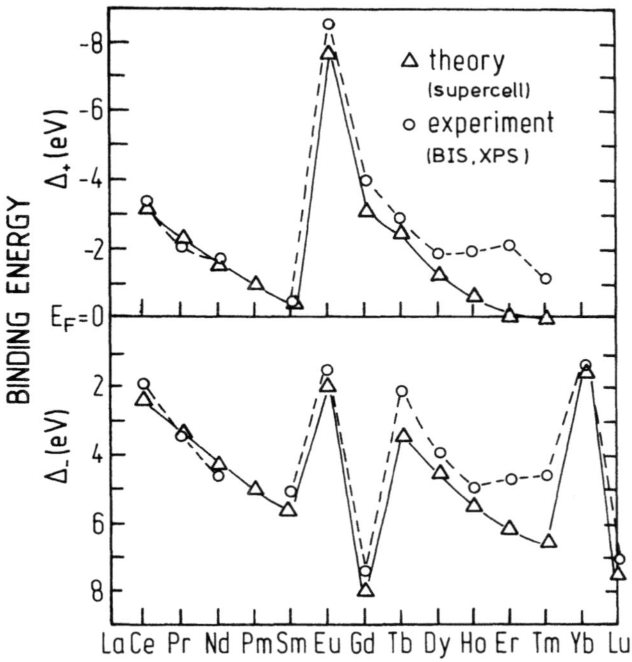

Fig. 2.31. Results for the 4f ionization energy Δ_-, and the 4f addition energy Δ_+ for rare earth metals as calculated by the supercell method [2.72], compared with the experimental data [2.39] (see also Fig. 2.13, for a comparison with thermodynamic data). Note, that Δ_+ and Δ_- are identical to the quantities $E_B(3)$ and $E_B(2)$ as defined in (2.22, 2.23) and Figs. 2.11,2.14, respectively

peaked structure seen in the valence-band PES of Ce (and to a lesser degree in Pr and Nd) can be explained by two different screening channels [2.42, 2.64], namely a sd-screened peak and a f-screened peak (closer to E_F).

2.5 Vibrational Side Bands

So far, in our discussion of the core levels, it has been implied that photoemission takes place in a rigid lattice. However, we shall see later in Chap. 5, where valence-band photoemission is discussed, that especially in free molecules the coupling to the lattice vibrations (phonons) can often not be neglected. This has now also become apparent in a recent high-resolution core-level study of CO adsorbed on Ni(100) (Fig. 1.30) [2.77]. In a situation where the vibrational degrees of freedom have to be taken into account, the photoemission process can most easily be described in a diagram like that given in Fig. 5.2 or Fig. 5.5. In short, one describes the ground state and the excited state (after photoionization) by a potential energy parabola onto which the vibrational energy levels are superimposed. Obviously these two potential energy curves are different, due to the difference in the initial and final state charge distribution. Photoemission can now be thought of as a vertical transition in this potential energy diagram (Frank–Condon principle), which means, that the vibrational state of the system does not change during the photoemission process. This, however, also implies that the photoemission signal does not necessarily have its maximum at the energy connecting the lowest vibrational level of the ground state with that of the excited state, because the potential energy parabolas in the ground- and excited state are displaced with respect to each other. The resulting O 1s and C 1s photoemission spectra of CO on Ni(100) in the c(2 × 2) overlayer structure are given in Fig. 1.30, and they can be fitted by a progression of vibrational (the stretching frequency of CO) side bands, accompanying the so-called adiabatic transition, which is the electronic transition connecting the lowest electronic level in the ground- and excited state, or the transition without any vibrational side bands. Especially with respect to the O 1s line, there is a distinct energy difference between the maximum of the photoemission line and the adiabatic transition energy. The side-band structure is different for the O 1s and the C 1s line, because the vibrations couple differently to these two ions. One has to note that for the CO adsorption it is the C-atom which bonds to the metal.

Similar data have now been observed in Be metal where not a single phonon but the whole phonon spectrum couples to the Be 1s core line. A careful analysis of a spectrum containing the bulk level and four shifted core levels shows that each line is not a simple Lorentzian but a Lorentzian convoluted with the phonon spectrum of Be [2.96].

2.6 Core Levels of Adsorbed Molecules

Because of the small electron escape depth in a solid, one can also investigate the electron energy levels of molecules or atoms adsorbed on a surface [2.78]. The chemical bonding to the surface will, of course, be reflected mainly in the valence orbitals (Sect. 5.10) but the core-level PES can also give some useful information.

The simplest application of core-level spectroscopy from adsorbed species is for chemical analysis. The most common experiment is a determination of the elemental composition. A typical example is given in Fig. 2.32, where an XPS spectrum of a slightly oxidized Al film is reproduced. The two spectra correspond to two detection geometries, where the angle between the direction of the outgoing electrons and the plane of the surface is changed. At 51.5° the chemically shifted oxide peak is hardly evident at all. However, at small detection angles the surface sensitivity is drastically increased and the oxide peak is clearly visible [2.79].

The oxidation of Al has been studied in more detail [2.76], as indicated in Fig. 2.33. Here synchrotron radiation ($\hbar\omega = 130\,\text{eV}$) was employed, giving

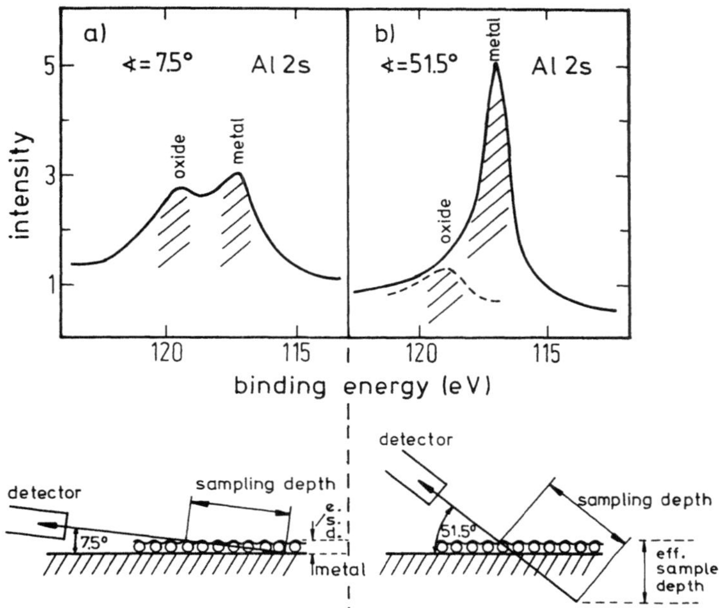

Fig. 2.32a,b. Surface sensitivity of XPS, demonstrated by changing the electron detection angle relative to the surface for a slightly oxidized surface of Al. At 7.5°, the Al 2s signals from Al metal and oxidized Al have the same magnitude, while at 51.5° the oxide signal is hardly visible [2.75]

increased surface sensitivity due to a smaller escape depth than with XPS, and also better resolution. One can now see that two steps occur in the oxidation process. The first one leads to an Al 2p binding-energy shift of 1.4 eV and is due to a chemisorbed state. Chemisorption is obviously a prerequisite for the final oxidation to Al_2O_3, which is characterized by a binding-energy shift of 2.7 eV.

A completely different system is that of a noble-gas mono- and multilayer on metal surfaces [2.81] and as an example the data for Xe on Pd(001) will be briefly discussed. The nice feature of this system is that one can resolve the core levels from successive overlayers as shown in Fig. 2.34. This figure shows the Xe 4d level emission for monolayer, bilayer and four-layer coverage of Xe on Pd(001), where the Xe layers are produced by keeping the substrate at a temperature of $T \approx 40\,\mathrm{K}$. As expected, at monolayer coverage one observes one spectrum (the splitting is due to spin–orbit interaction). At bilayer coverage two superposed spectra are clearly observed, with that of the outermost (second) layer being the more intense one. The spectrum from the four-layer sample can likewise be decomposed into four individual spectra. The chemical shifts for the various layers, with respect, e.g., to the binding

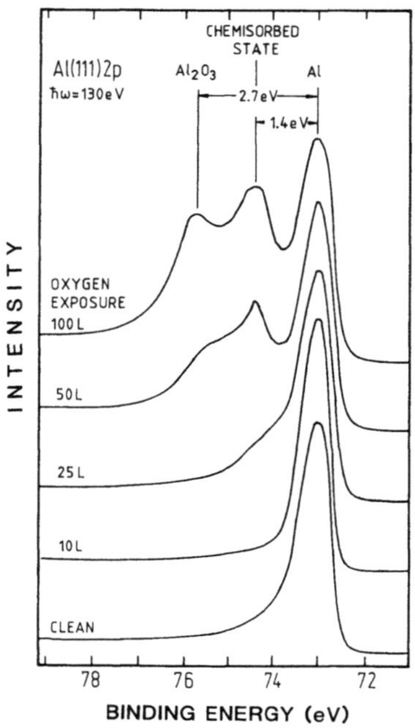

Fig. 2.33. Oxidation of a (111) surface of Al monitored via the Al 2p level with $\hbar\omega = 130\,\mathrm{eV}$ synchrotron radiation [2.80]. The development of a chemisorbed state and subsequent oxide formation can be observed

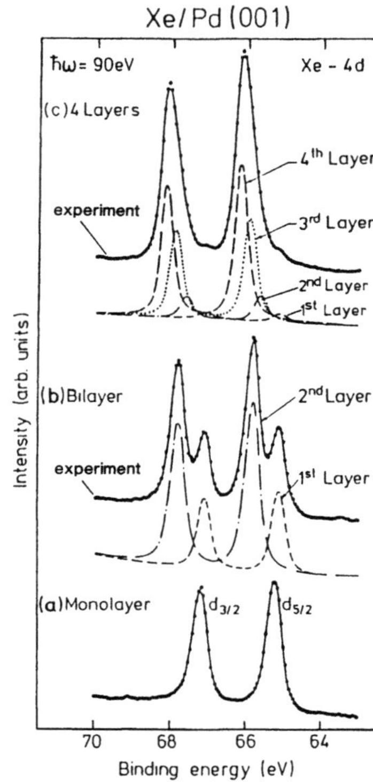

Fig. 2.34. Xe 4d core-level PE spectra for (**a**) a monolayer, (**b**) a bilayer and (**c**) four layers of Xe on Pd(001). The binding energies are with respect to the vacuum level of the adsorbate covered substrate. The solid curves are the result of a least-squares fit of the experimental data (full dots) to (**a**) one spectrum, (**b**) two spectra (1st and 2nd layer) and (**c**) 4 spectra (1st to 4th layer) [2.81]

energy of the free Xe atom, can be interpreted by a model which assumes that the photoionized atom stays charged after the ejection of the photoelectron. Screening is then produced by the induction of an image charge in the metal substrate (and not by any charge transfer), where all other polarization terms are neglected. The screening energy of the charge on the Xe^+ ion outside the metal is given by $e^2/4z$ where z is the distance between the charge e and the image plane. On the basis of this simple model, the observed layer-dependent chemical shifts can be interpreted very well (Table 2.4).

Table 2.4. Xe 4d core-level shifts ΔE_{B} relative to the binding energy of the Xe atom for a monolayer, bilayer and four layers of Xe on Pd(001). All energies are in eV [2.81]

Xe/Pd(001)	layer	ΔE_{B} (exp)	ΔE_{B} (theory)
Monolayer	1	−2.32	−2.61
Bilayer	1	−2.39	−2.70
	2	−1.68	−1.76
Four layers	1	−2.47	−2.70
	2	−1.89	−1.96
	3	−1.62	−1.79
	4	−1.38	−1.51

We note that the chemical shifts observed in metallic clusters on insulating substrates relative to the pure metal [2.82–2.84] can also be explained by the charging of the cluster because of the loss of the photoemitted electron.

2.7 Quantitative Chemical Analysis from Core-Level Intensities

The PE intensity of core levels can be employed as a means to determine the concentration of an element in a sample. This is, however, not straightforward. For details of the complicated procedures we refer the reader to the review by Grasserbauer et al. [2.85].

The experimental intensity of a core line is the sum of the intrinsic (primary) spectrum and the extrinsic (secondary) spectrum. The separation of these two contributions is not always easy and is described in Sect. 4.3 for a homogeneous sample. For simplicity, it is assumed here that the primary spectrum consists of a Lorentzian centered at $E(i,X)$ where i is the index of the core level and X that of the element. If the total integrated intensity of the line is $I(i,X)$ one has [2.86]:

$$I(i,X) = B\sigma(\hbar\omega,\alpha,E(i,X))\lambda(E(i,X))T(i,X)n(X)\,, \tag{2.29}$$

where B is the instrumental factor which contains properties such as the total X-ray flux onto the sample, the angular acceptance of the electron spectrometer (generally small compared to 4π), and the total transmission factor of the spectrometer. The other quantities are defined as follows:

$$\sigma(\hbar\omega,\alpha,E(i,X)) = \sigma(\hbar\omega,E(i,X))f(i,X,\alpha)\,, \tag{2.30}$$

with

$$f(i,X,\alpha) = 1 + \frac{\beta(i,X)}{4}(1-3\cos^2\alpha)\,,$$

where $\beta(i, X)$ is the asymmetry parameter as calculated by Reilman et al. [2.87, 2.88], $\sigma(\hbar\omega, E(i, X))$ is the total photoabsorption cross section of the core shell i of the element X, calculated by Scofield [2.88, 2.89] for free atoms at $\hbar\omega = 1487\,\mathrm{eV}(\mathrm{AlK}_\alpha)$ and $1254\,\mathrm{eV}(\mathrm{MgK}_\alpha)$, α is the angle between the incident photon and the emitted electron, and $\lambda(\mathrm{E}(i, X))$ the inelastic mean free path of the photoexcited electrons according to Penn [2.90] given by $\lambda(E(i, X)) = (E(i, X)/\{a[\ln(E(i, X)) - b]\}$. Here, a and b depend on the electron concentrations and are tabulated in [2.90]. $\lambda(E(i, X))$ can be approximated by $CE(i, X)^{0.71}$, C being a constant weakly dependent on the electronic properties of the sample. $T(i,X)$ is the transmission coefficient of the electrons through the surface, and $n(X)$ the density of element X. If one is interested only in relative concentrations of the constituents in a sample, then one can form the ratios $I(i, X)/I(j, Y)$ (j being a core level in element Y) and thus eliminate the factor B.

While the procedure just outlined is a good starting point, it may be inaccurate to a considerable degree mainly because of elastic scattering of the electrons on their way from the point of photoexcitation to the surface of the sample (see Chap. 1 and [2.91–2.95]). Taking elastic scattering into account accurately is, however, computationally hard and the reader is therefore referred to the original literature [2.91–2.95]. If one works with a small electron acceptance angle and near normal emission and, in addition, chooses the angle between the incoming photons and the outgoing electrons to be $\alpha = 54.7°$ (magic angle; at this angle one has $f(i, X, \alpha) = 1$) one can get an absolute accuracy in the determination of the elemental concentration of $\sim \pm 30\,\%$, even if elastic scattering (and backscattering) are neglected. Under less favorable conditions the accuracy can deteriorate.

It is perhaps useful to point out in addition that the application of core-line intensities for analytical purposes is possible with good accuracy only if one has a symmetric line. This more or less implies a single final state. If the core-line spectrum is made up of multiple final states (satellites, plasmons, electron–hole pair excitations), the analysis becomes very complicated and less precise.

References

2.1 A good description of the breakdown of the one-electron picture in PES has been given in G. Wendin: *Structure and Bonding* **45** (Springer, Berlin, Heidelberg 1981)

2.2 K. Siegbahn, C. Nordling, R. Fahlman, R. Nordberg, K. Hamrin, J. Hedman, G. Johansson, T. Bergmark, S.-E. Karlsson, I. Lindgren, B. Lindberg: In Nova Acta Regiae Sci. Soc., Upsaliensis, Ser IV, **20** (1967)
K. Siegbahn, C. Nordling, G. Johansson, J. Hedman, P.F. Heden, K. Hamrin, U. Gelius, T. Bergmark, L.O. Werme, R. Manne, Y. Baer: *ESCA-Applied to Free Molecules* (North-Holland, Amsterdam 1969)

2.3 W.C. Price: In *Electron Spectroscopy: Theory, Techniques and Applications I*, ed. by C.R. Brundle, A.D. Baker (Academic, New York 1977)

2.4 W.L. Jolly: In *Electron Spectroscopy: Theory, Techniques and Applications I*, ed. by C.R. Brundle, A.D. Baker (Academic, New York 1977)
2.5 H.W.B. Skinner: Proc. Roy. Soc. (London) A**135**, 84 (1932)
2.6 C.E. Moore: Ionization potentials and ionization limits derived from the analysis of optical spectra. Nat. Stand. Ref. Data Sev., NBS, **34** (1970); and Atomic energy levels. Nat. Stand. Ref. Data Sev., NBS **35/I** (1971) and US Nat. Bur. Stand. Circ. No.467 (1958) Vols.**I-III**
2.7 E.M. Baroody: J. Opt. Soc. Am. **62**, 1528 (1972)
2.8 L. Pauling: The *Nature of the Chemical Bond* (Oxford Univ. Press, Oxford 1950)
2.9 P. Finn, R.K. Pearson, J.M. Hollander, W.L. Jolly: Inorg. Chem.**10**, 378 (1971)
2.10 L. Hedin, A. Johansson: J. Phys. B**2**, 1336 (1969)
2.11 W.L. Jolly: Discuss. Faraday Soc. **54**, 13 (1972)
2.12 D.W. Davis, D.A. Shirley: Chem. Phys. Lett. **15**, 185 (1972)
2.13 W.L. Jolly, W.B. Perry: J. Amer. Chem. Soc. **95**, 5442 (1973)
W.L. Jolly, W.B. Perry: Inorg. Chem. **13**, 2686 (1974)
2.14 J.C. Slater: *Quantum Theory of Molecules and Solids*, Vol.IV (McGraw Hill, New York 1974)
2.15 M. Born: Verh. dt. phys. Ges. **21**, 13 and 679 (1919)
F. Haber: Verh. dt. phys. Ges. **21**, 750 (1919)
2.16 C.S. Fadley, S.B.M. Hagstroem, J.M. Hollander, M.P. Klein, D.A. Shirley: J. Chem. Phys. **48**, 3779 (1968)
2.17 N.F. Mott, R.W. Gurney: Electronic Processes in Ionic Crystals (Dover, New York 1964) p.56
2.18 P.H. Citrin, T.D. Thomas: J. Chem. Phys. **57**, 4446 (1972)
2.19 G.K. Wertheim, D.N.E. Buchanan, J.E. Rowe, P.H. Citrin: Surf. Sci. Lett. **319**, L41 (1994). This work shows that in the core levels of alkali halide crystals noticeable surface core-level shifts can be observed
2.20 M. Cardona, L. Ley: In *Photoemission in Solids I*, ed. by M. Cardona, L. Ley, Topics Appl. Phys., Vol. 26 (Springer, Berlin, Heidelberg 1978) Chap.1
2.21 R.C. Martin, D.A. Shirley: In *Electron Spectroscopy: Theory, Techniques and Applications I*, ed. by C.R. Brundle, A.D. Baker (Academic, New York 1977)
2.22 C.S. Fadley: In *Electron Spectroscopy: Theory, Techniques and Applications II*, ed. by C.R. Brundle, A.D. Baker (Academic, New York 1978)
2.23 T. Koopmans: Physica **1**, 104 (1933)
2.24 J.C. Slater, J.B. Mann, T.M. Wilson, J.H. Wood: Phys. Rev. **184**, 672 (1969)
2.25 U. Gelius: J. Electron Spectrosc. **5**, 985 (1974)
2.26 D.A. Shirley: Chem. Phys. Lett. **16**, 220 (1972)
2.27 B. Johansson: J. Phys. F **4**, L169 (1974)
2.28 B. Johansson, N. Mårtensson: Phys. Rev. B **21**, 4427 (1980)
2.29 B. Johansson: Phys. Rev. B **20**, 1315 (1979)
2.30 P. Steiner, S. Hüfner: Solid State Commun. **37**, 79 1981); Acta Met. **29**, 1885 (1981)
2.31 P. Steiner, S. Hüfner, N. Mårtensson, B. Johansson: Solid State Commun. **37**, 73 (1981)
2.32 P. Steiner, S. Hüfner: Solid State Commun. **37**, 279 (1981)
2.33 C.P. Flynn: J. Phys. F **10**, L 315 (1980)
2.34 W.C. Martin, R. Zalubas, L. Hagan: Atomic energy levels – The rare earth elements, Nat. Bur. Stand. (U.S.) Ref. Data Ser. 60 U.S.GPO, Washington, DC (1978)
2.35 U.L. Van der Sluis, L.J. Nugent: J. Opt. Soc. Am. **64**, 687 (1974)
2.36 J.F. Herbst, D.N. Lowry, R.E. Watson: Phys. Rev. B **6**, 1913 (1972)
2.37 J.F. Herbst, R.E. Watson, J.W. Wilkins: Phys. Rev. B **13**, 1439 (1976)

2.38 J.F. Herbst, R.E. Watson, J.W. Wilkins: Phys. Rev. B **17**, 3089 (1978)
2.39 J.K. Lang, Y. Baer, P.A. Cox: J. Phys. Metal. Phys. **11**, 121 (1981)
2.40 S. Hüfner, P. Steiner: *Valence Instabilities*, ed. by P. Wachter, H. Boppart (North-Holland, Amsterdam 1982) p.263
2.41 M.R. Norman, D.D. Koelling, A.J. Freeman, H.J.F. Janssen, B.I. Min, T. Oguchi, Y. Ling: Phys. Rev. Lett. **53**, 1673 (1984)
2.42 S. Hüfner, P. Steiner: Z. Physik B **46**, 37 (1982)
2.43 O. Gunnarsson, K. Schönhammer: Phys. Rev. Lett. **50**, 604 (1983)
2.44 O. Gunnarsson, K. Schönhammer: Phys. Rev. B **28**, 4315 (1983)
2.45 D.M. Wieliczka, C.G. Olson, D.W. Lynch: Phys. Rev. Lett. **52**, 2180 (1984)
2.46 G. Kalkowski, E.V. Sampathkumaran, C. Laubschat, M. Domke, G. Kaindl: Solid State Commun. **55**, 977 (1985)
2.47 R.D. Parks, S. Raan, M.L. Den Boer, M.L. Chang, G.P. Williams: Phys. Rev. Lett. **52**, 2176 (1984)
2.48 S. Hüfner: J. Phys. F **16**, L31 (1986)
2.49 O. Gunnarsson, K. Schönhammer: In *Handbook on the Physics and Chemistry of Rare Earths*, Vol.10, ed. by K.A. Gschneidner, L. Eyring, S. Hüfner (Elsevier Science, New York 1987)
2.50 J.H. van Vleck: Phys. Rev. **45**, 405 (1934)
2.51 G.K. Wertheim, S. Hüfner, H.J. Guggenheim: Phys. Rev. B **7**, 556 (1973)
2.52 S. Hüfner, G.K. Wertheim: Phys. Rev. B **7**, 2333 (1973)
2.53 R.L. Cohen, G.K. Wertheim, A. Rosencwaig, H.J. Guggenheim: Phys. Rev. B **5**, 1037 (1972)
2.54 D.A. Shirley: In *Photoemission in Solids*, ed. by M. Cardona, L.Ley, Topics Appl. Phys., Vol.26 (Springer, Berlin, Heidelberg 1978) Chap.4
2.55 F.R. McFeely, S.P. Kowalczyk, L. Ley. D.A. Shirley: Phys. Lett. A **49**, 301 (1974)
2.56 B.W. Veal, A.P. Paulikas: Phys. Rev. Lett. **51**, 1995 (1983)
2.57 B.W. Veal, A.P. Paulikas: Phys. Rev. B **31**, 5399 (1985)
2.58 D.K.G. de Boer, C. Haas, G.A. Sawatzky: Phys. Rev. B **29**, 4401 (1984)
2.59 V. Kinsinger, I. Sander, P. Steiner, R. Zimmermann, S. Hüfner: Solid State Commun. **73**, 527 (1990)
K. Okada, A. Kotani, B.T. Thole: J. Electron Spectrosc. Relat. Phenom. **58**, 325 (1992)
S.J. Oh, G.H. Gweon, J.G. Park: Phys. Rev. Lett. **68**, 2850 (1992)
V. Kinsinger, R. Zimmermann, S. Hüfner, P. Steiner: Z. Physik B **89**, 21 (1992)
P.S. Bagus, G. Pacchioni, F. Parmigiani: Chem. Phys. Lett. **207**, 569 (1993)
K. Okada, A. Kotani, V. Kinsinger, R. Zimmermann, S. Hüfner: J. Phys. Soc. Jpn. **63**, 2410 (1994)
2.60 K. Karlsson, O. Gunnarsson, O. Jepsen: J. Phys. Condens. Mat. **4**, 895 (1992); **4**, 2801 (1992)
2.61 J.C. Parlebas, M.A. Khan, T. Uozumi, K. Okada, A. Kotani: ISSP Report, Ser. A, No. 2902 (1994)
2.62 K. Okada, A. Kotani: J. Electron Spectrosc. Relat. Phenom. **54**, 313 (1990)
2.63 B. Hermsmeier, C.S. Fadley, M.O. Krause, J. Jimenez-Mier, P. Gerard, S.T. Manson: Phys. Rev. Lett. **61**, 2592 (1988)
2.64 A.J. Freeman, B.I. Min, M.R. Norman: In *Handbook on the Physics and Chemistry of Rare Earths*, ed. by K.A. Gschneidner, L. Eyring, S. Hüfner (Elsevier Science, New York 1987) Vol.10.
2.65 S. Hüfner: *Optical Properties of Transparent Rare Earth Compounds* (Academic, New York 1978)

2.66 F. Patthey, B. Delley, W.D. Schneider, Y. Baer: Phys. Rev. Lett. **55**, 1918 (1985)
2.67 S. Hüfner, L. Schlapbach: Z. Physik B **64**, 417 (1986)
2.68 M. Campagna, G.K. Wertheim, Y. Baer: In *Photoemission in Solids II*, ed. by L. Ley, M. Cardona, Topics Appl. Phys., Vol.27 (Springer, Berlin, Heidelberg 1979) Chap.4
M. Campagna, G.K. Wertheim, E. Bucher: *Structure and Bonding* **30**, 99 (Springer, Berlin, Heidelberg 1976)
2.69 J.N. Chazaviel, M. Campagna, G.K. Wertheim, P.Y. Schmidt: Solid State Commun. **19**, 725 (1976)
2.70 P.A. Cox: *Structure and Bonding* **24**, 59 (Springer, Berlin, Heidelberg 1975)
2.71 A. Zunger, A.J. Freeman: Phys. Rev. B **15**, 2901 (1977)
2.72 M.R. Norman, D.D. Koelling, A.J. Freeman: Phys. Rev. B **31**, 6251 (1985)
2.73 M.R. Norman: Phys. Rev. B **31**, 6261 (1985)
2.74 M.R. Norman, D.D. Koelling, A.J. Freeman: Phys. Rev. B 32, 7748 (1985)
2.75 B.I. Min, H.J.F. Jansen, T. Oguchi, A.J. Freeman: J. Mag. Mag. Mat. **61**, 139 (1986)
2.76 M.R. Norman, A.J. Freeman: Phys. Rev. B **33**, 8896 (1986)
2.77 A. Fröhlisch, N. Warsdahl, J. Hasselströhm, O. Karis, D. Menzel, N. Mårtensson, A. Nielsson: Phys. Ref. Lett. **81**, 1730(1998)
2.78 M. Scheffler, A.M. Bradshaw: In *The Chemical Physics of Solid Surfaces and Heterogeneous Catalysis*, ed. by D.A. King, D.P. Woodruff (Elsevier, Amsterdam 1983)
2.79 S. Kowalczyk: Thesis, Univ. of California, Berkeley, Calif. (1976)
2.80 S.A. Flodstroem, C.W.B. Martinson, R.Z. Bachrach, S.B.M. Hagstroem, R.S. Bauer: Phys. Rev. Lett. **40**, 907 (1978)
2.81 T.C. Chiang, G. Kaindl, T. Mandel: Phys. Rev. B **33**, 695 (1986)
2.82 G.K. Wertheim, S.B. DiCenzo, D.N.E. Buchanan: Phys. Rev. B **33**, 5384 (1986)
2.83 S.B. DiCenzo, G.K. Wertheim: Comments Solid State Phys. **11**, 203 (1985)
2.84 G.K. Wertheim: Z. Phys. B **66**, 53 (1987)
2.85 *Angewandte Oberflächenanalyse*, ed. by M. Grasserbauer, H.J. Dudek, M.F. Ebel (Springer, Berlin, Heidelberg 1986)
2.86 P. Steiner, H. Höchst, S. Hüfner: In *Photoemission in Solids II*, ed. by L.Ley, M. Cardona, Topics Appl. Phys., Vol.27 (Springer, Berlin, Heidelberg 1979)
2.87 R.F. Reilman, A. Msezane, S.T. Manson: J. Electr. Spectr. **8**, 389 (1976)
2.88 J.J. Yeh, I. Lindau: Atomic Data and Nuclear Data Tables **32**, 1 (1985)
2.89 J.H. Scofield: J. Electron Spec. **8**, 129 (1976)
2.90 D.R. Penn: J. Electron Spec. **9**, 29 (1976)
2.91 J. Electron Spectrosc. Relat. Phenom. **100** (1999)
2.92 V.I. Nefedov in Ref. [2.91]
2.93 A. Jablonski, C.J. Powell in Ref. [2.91]
2.94 M.P. Seah in Ref. [2.91]
2.95 D. Briggs, M.P. Seah (Eds.), Practical Surface Analysis (Wiley, Chichester 1990), Vols. I and II
2.96 J.N. Andersen, T. Balasubramanian, C.-O. Almbladh, L.I. Johansson, R. Nyholm: Phys. Rev. Lett. **86**, 4398 (2001)

3. Charge-Excitation Final States: Satellites

Atoms, molecules and solids are many-electron systems. Since the electrons interact with each other via the Coulomb and exchange interaction, the emission of one electron after the photoexcitation process can and must lead to excitations in the remaining system. These excitations require energy and therefore lead to signals in the PE spectrum with a smaller kinetic energy (larger binding energy) than the signal corresponding to the ground state of the system after the PE process. This means that the PE spectrum must consist of the *"main" line* (representing the ground state after photoexcitation) and a number of *"extra"* (*satellite*) lines representing the excited states.

In solids there is yet another source which gives rise to additional lines in the PE spectrum. Here the photoexcited electron travels a relatively long way through a volume which is occupied by other electrons, and therefore has the possibility of exciting them. This excitation energy delivered to the sample must again be supplied from the kinetic energy of the outgoing electron and therefore this photoelectron may be observed with a kinetic energy smaller than that expected otherwise. Of course, there are also electrons which travel through the solid without any loss. These are responsible for the so-called *no-loss line* (or main line), while those experiencing an inelastic collision on their way from the site of the photoexcitation to the surface constitute the "inelastic tail", which generally consists of a featureless part and (broad) discrete peaks.

Thus we see that there are, in principle, two sources of "extra" structure in the PE spectrum (besides the "main" line), an *intrinsic* part (created in the photoemission process) and an *extrinsic* part (created by the photoexcited electron in a solid via interaction with other electrons during its travel to the surface). In this chapter, only the intrinsic part will be dealt with and in Chap. 4, for the particularly simple case of the nearly-free-electron metals, a discussion of intrinsic and extrinsic structure in the PE spectra will be given.[1]

The principal reason for the creation of intrinsic satellites has been given in Sect. 1.6. (We shall henceforth ignore the word "intrinsic" because it is

[1] The separation of the PE spectrum into an intrinsic and an extrinsic part is, in a sense, an artefact of the three-step model (Chap. 6). In a strictly one-step model this distinction is not very meaningful. Since the three-step model is, however, useful for the interpretation of actual data, the separation of the spectra into an intrinsic and an extrinsic part is also very convenient.

rarely used and we have made clear that only satellites created in the photoemission process will be dealt with, not those created by inelastic scattering events during the travel of the photoelectron to the surface.) If a photoelectron is ejected from an orbital of a system (atom, molecule or solid) and that system is then still close to its ground state (meaning that the eigenfunctions of the system before and after the photoemission process are quite similar), the PE spectrum will consist mainly of one line, see (1.28), (1.31). Very often, however, the photoexcitation process can lead to excited states of the system and, as mentioned already, these excited states will be reflected in the spectrum by (extra) satellite lines, see (1.32). The intensity of these lines is proportional to the overlap integral of their initial- and final-state wave functions.

The analysis of satellites is a complex matter and much space would be needed to deal with it in detail [3.1]. We shall therefore restrict the discussion to a simple case of satellites in solids, namely that of the open shell compounds (compounds containing 3d, 4d, 5d, 4f or 5f elements) and also only the most elementary concepts will be discussed

The simplicity of the analysis of these systems stems from the fact that in many cases the d or f electrons in them are still quite localized even in metallic solids (in insulators and semiconductors this statement is even more general). For such situations, quite successful simple models have been devised. The simplest approximation is a molecular orbital approach which represents the d (or f) part of the system by one orbital, and the ligand or valence band by another one. At the next level of sophistication, clusters are employed to represent a solid, for example NiO_6^{10-} for NiO, and their energy levels are calculated "exactly" using ligand field or configuration interaction theory. Finally a model which has been successful for many years for the description of localized states interacting with extended states, namely the Anderson Hamiltonian, can be employed to analyze the excitation spectra of solids containing localized electrons.

All these approaches have been used extensively and successfully [3.2]. In order to keep this chapter to a reasonable length, only the basic facts about satellites in open shell systems will be presented.

3.1 Copper Dihalides; 3d Transition Metal Compounds

3.1.1 Characterization of a Satellite

The problem to be discussed can be illustrated by the XPS spectra of the 2p lines of copper difluoride, copper dichloride and copper dibromide, shown in Fig. 3.1 [3.6]. Each of the two spin–orbit split components ($2p_{3/2}$ and $2p_{1/2}$) consists of two lines instead of one. The main line, which is that with the smaller binding energy, shows large energy differences for the different

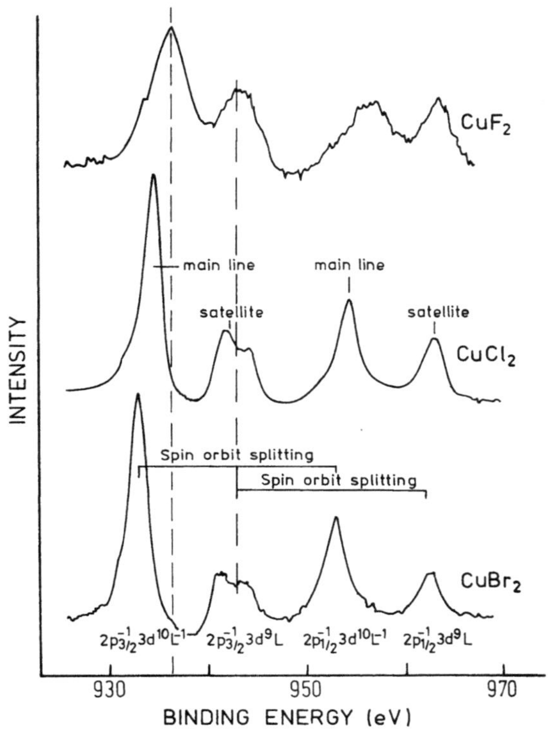

Fig. 3.1. XPS spectra of the 2p core levels of CuF_2, $CuCl_2$ and $CuBr_2$. Cu^{2+} has a $3d^9$ initial-state configuration. The satellite-line energies roughly coincide in the three compounds while the main-line energies show large differences. This makes the assignment of $2p_{3/2,1/2}$ $3d^9$ to the satellite lines likely, leaving the assignment $2p_{3/2,1/2}3d^{10}$ L^{-1} for the main line; their energies differ because of the different ligand binding energies [3.6]

halides, while the so-called satellite line has roughly the same binding energy in all three compounds.

This can be explained using a molecular orbital description, which for an insulating solid is a valid approximation. The main and satellite lines can then be labelled as indicated in the figure. In their ground state the copper dihalides have one hole in the d shell ($3d^9$ configuration) and a filled ligand shell L. This is written as $3d^9L$ where $3d^9$ refers to the Cu^{2+} ion and L represents the highest occupied ligand shell which is 2p for F^-, 3p for Cl^- and 4p for Br^-. The photoexcitation on the Cu site can lead to two final states: one, where after the creation of the 2p core hole the ground-state configuration $3d^9L$ is left roughly intact; another one, where after the core-hole creation one electron is transferred from the ligand shell L into the d shell, leading to a $3d^{10}L^{-1}$ configuration. (Sometimes the hole is indicated in the literature by a bar under the symbol, i.e. $\underline{L}$ instead of L^{-1}, but since this notation has given rise to confusion we avoid it here.) If we assume that the latter configuration applies to the main line we can label the Cu $2p_{3/2}$ photoexcited final state by $2p_{3/2}3d^{10}L^{-1}$. One can then write [3.6]:[2]

$$E_{\mathrm{main}} = E(\mathrm{c}^{-1}\mathrm{d}^{10}\mathrm{L}^{-1}) \simeq E(\mathrm{c}^{-1}\mathrm{d}^{10}) + E(\mathrm{L}^{-1}) . \tag{3.1}$$

The portioning of the $3d^{10}L^{-1}$ configuration into $3d^{10}$ and a L^{-1} assumes that there is only weak hybridization between the metal ion and the ligands.

By the same reasoning the satellite line is described by

$$E_{\mathrm{sat}} = E(\mathrm{c}^{-1}\mathrm{d}^{9}\mathrm{L}) . \tag{3.2}$$

From this argumentation it is evident why, to a first approximation, the satellite line energy does not depend on the ligand, while the main line energy shows a strong correlation with the ligand: the energy needed for the production of a hole in the ligand valence orbital depends on the nature of the ligand ion.

The term "satellite" in PE spectra generally applies to all lines that cannot be explained in a single-ion picture. Or, stated differently, a satellite occurs if more than one final state can be reached in the PE process. In the example of the Cu dihalides one final state leaves the valence configuration largely unaltered, and this leads to the line called the "satellite" line. The second final state, which obviously has a lower energy, is reached if in the final state one electron has been transferred from the valence orbitals onto the metal orbital, and this state gives rise to the "main" line.

This situation can also be viewed slightly differently. Photoionization of a core electron on the metal ion (Cu) produces an extra positive charge on the metal ion [$(Z+1)$-approximation]. This constitutes an additional Coulomb interaction for the electrons with orbital radii larger than those of the photoionized shell. The additional Coulomb interaction then lowers the energy of the ion on which the hole resides, and thus it lowers particularly the energy of the 3d valence shell. In the ground state of transition metal compounds the 3d shell is situated in energy above the ligand valence orbitals. However, the additional Coulomb interaction (U_{cd}, meaning the Coulomb interaction between the core hole c and the 3d shell d) may pull it below the top of the ligand valence band. This is an unstable situation which does not correspond to the ground state but to the satellite. It is then energetically possible, and indeed more favorable, for the system to transfer an electron from the ligand band into the empty 3d orbital, whereby this is filled and the ion attains its ground state (main line).

---

[2] In order to write the final state in a correct way one would have to indicate the hole in the $2p_{3/2}$ shell by $2p_{3/2}^{-1}$; this is hardly ever done and therefore we also use the common nomenclature which implicitly assumes that a hole state is meant. We distinguish c for a core level and c^{-1} for a core hole only in the general discussion.

In the aforegoing description the main line and the satellite line correspond, in principle, to different screening situations in the final state. The language employed, however, was that of molecular orbitals (cluster approach). Although this neglects the "infinite" nature of a solid it seems to be a reasonable approach for an insulating solid, and due to the localized nature of the d electrons it is often as good or even better than the band-structure approach.

The idea that satellites in open f or d shell ions are caused by different screening channels was proposed by Kotani and Toyozawa [3.7, 3.8] and verified in experiments on Ni metal by Hüfner and Wertheim [3.9]. The essence of the arguments is shown schematically in Fig. 3.2, on the top for a conductor and at the bottom for an insulator. In both cases a metal ion was assumed with an unoccupied d level in the initial state.

In the photoemission process this empty d level is pulled below the top of the valence band, and (since it is unoccupied) formally represents a positive charge on the photoionized ion, creating what is generally called a two-hole state containing the photohole and the hole provided by the empty d level.

In the metal it is assumed that the two holes can be screened by charge moving in rapidly from the broad sp conduction band, and this produces the final state labelled (1) in Fig. 3.2. In a different process, charge is transferred from the sp band into the lowered d level, thereby producing the second final state (2). Since the localized d level is more efficient in screening than the extended sp band, this second final state has a smaller binding energy than the first one. The localized level, which was assumed here to be a d level, may of course equally well be an f level.

In insulators the situation is similar, as shown at the bottom of Fig. 3.2 where the metal and the ligand energy levels have been drawn separately. In the initial state the metal ion has an unoccupied d level and an unoccupied broad sp band, which are both pulled down by the core hole to below the top of the ligand valence band (e.g., the F 2p level for a transition metal fluoride). The screening can now occur by charge transfer out of the occupied ligand levels into the metal-ion's sp band [satellite, final state (1) in Fig. 3.2] or into the metal d level [main line, final state (2)].

In order to quantitatively apply the models presented in Fig. 3.2 one needs, in principle, a calculation of the band structure for the one-hole final state, which may be hard to come by. Therefore, for insulators, one resorts to simpler models, and the most successful of these is the representation of the solid (with regard to the metal) by a single metal ion and a cluster of ligands surrounding it. This makes the system tractable in the molecular-orbital configuration interaction approach and offers considerable simplification. The situation can be further simplified by representing the valence orbitals by just one metal and one ligand level which interact to form the two corresponding molecular orbitals (this leaves out the complication of the sp band of the metal ion). In this manner the analysis of the Cu dihalide data (Fig. 3.1) was

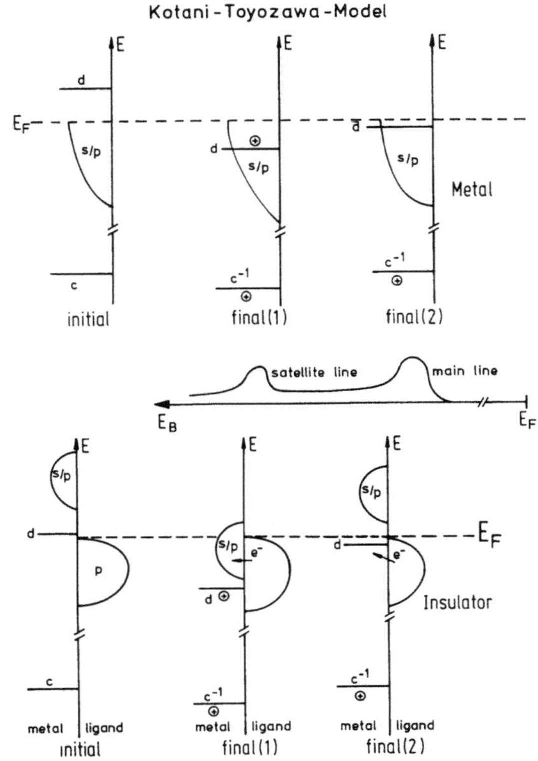

Fig. 3.2. Schematic representation of the Kotani–Toyozawa model for PES from a core level in a metal and an insulator [3.8]. In the initial state the metal d level is above the Fermi energy; in the insulator it is above the ligand p band. In the final state the d level is pulled down below the Fermi energy indicated by a hole ⊕ and can now, in principle, accommodate an (additional) d electron. Screening in this situation can be achieved by sp conduction electrons (final (1), *satellite line*) or if the d-hole is filled by an electron, whereby the ground state is attained (final (2), *main line*)

performed. In that case the initial state configurations are $3d^9L$ and $3d^{10}L^{-1}$ (with some mixing). These are the only one-hole states that can be produced from a configuration with one hole in the metal ($Cu^{2+} : 3d^9 = 3d^{10-1}$) and a filled ligand configuration ($p^6 \equiv L$).

Note that these configurations correspond to two states of which one is filled (mostly $3d^9L$) and the other is empty (mostly $3d^{10}L^{-1}$) in the initial state. These two states can be populated in the final state creating a main line (mostly $3d^{10}L^{-1}$) and a satellite line (mostly $3d^9L$).

This picture is similar to the one in Fig. 3.2 but not identical to it. In both approaches the main line is obtained by screening in the d level. The metal sp conduction band is missing in the simple version of the cluster approach and therefore cannot become populated for screening of the satellite line in this picture. However, since it represents a relatively wide band, its overlap with the ligand p band will not actually accumulate much charge on the photoionized metal ion. This situation is reflected by the term *"non-local screening"* which is frequently used in the literature for this particular screening channel. In this respect the error made in the cluster approximation by neglecting the metal sp band does seem acceptable. It should be added that the metal sp electrons can, of course, be incorporated into the cluster approach, but the simple two-level picture is thereby destroyed.

The simple model above, which uses only a metal d and a ligand L orbital, also sheds light on another terminology used in this context. Referring again to the copper dihalides, the line with the d^9L final state is often called the *"unscreened"* line (because in this simplified picture no charge has been "moved") while the $d^{10}L^{-1}$ final state is called a *"well-screened"* line (because charge has been moved onto the metal ion).

The example just dealt with is the best documented case of satellite creation, but it is not the only one [3.1, 3.10, 3.11]. Generally speaking, any excitation of other transitions during the PE process can give rise to one or more additional lines (satellites). It is, however, not always easy to obtain a full understanding of the transitions involved. In order to give the reader the possibility to look into the original literature, some references are given [3.2]. Satellites in 3d, 4d and 5d compounds have been investigated in [3.1, 3.10–3.57]. Work on 4f elements is reported in [3.58–3.66] and on 5f elements in [3.67–3.74].Theoretical treatments of the satellite problem are given in [3.1, 3.2] and references therein.

3.1.2 Analysis of Charge-Transfer Satellites

Let us now treat the case of the copper dihalides in more detail. The approach used will be the molecular-orbital model, as applied first by Asada and Sugano [3.75]. Similar work was reported by a number of other researchers [3.2, 3.6, 3.10–3.22, 3.46, 3.52–3.57, 3.76–3.83].

We shall describe the case of divalent copper compounds in some detail because it is the simplest possible situation and nevertheless provides considerable insight into the problem of the charge-transfer satellites. The extension of the treatment to more complicated systems like the divalent manganese compounds is straightforward. The problem one has to deal with is the condensation of the mechanism sketched in Fig. 3.2 into a simple tractable model. To that end in Fig. 3.3 we give two possible valence-band excitations in a transition metal compound which will be important in the following discussion. The first one is the excitation of a d electron from a metal ion (configuration d^n) onto the d^n configuration of another (distant) metal ion, a process which needs the energy $U[U = E(d^{n+1}) + E(d^{n-1}) - 2E(d^n)]$. The second one is the excitation of an electron from a ligand ion (configuration L) onto the d^n configuration of a (distant) metal ion, a process which needs the energy

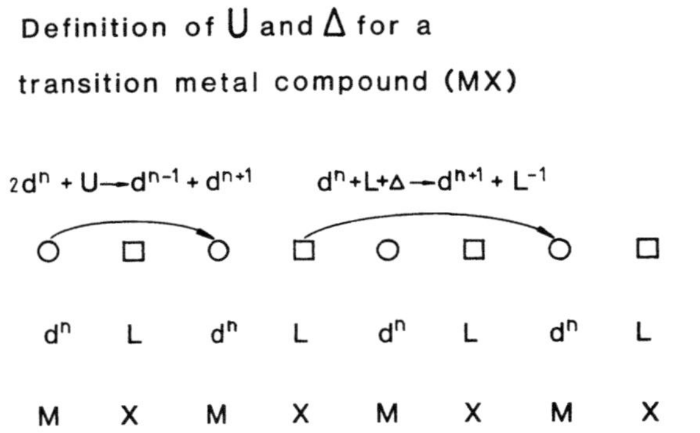

Fig. 3.3. Sketch of a transition metal compound as an ensemble of metal (M, d^n) and ligand (X, L) ions, without any coupling between the ions. This situation defines the parameters $U[2E(\mathrm{d}^n) + U \rightarrow E(\mathrm{d}^{n+1}) + E(\mathrm{d}^{n-1})]$ and $\Delta[E(\mathrm{d}^n) + E(\mathrm{L}) + \Delta \rightarrow E(\mathrm{d}^{n+1}) + E(\mathrm{L}^{-1})]$

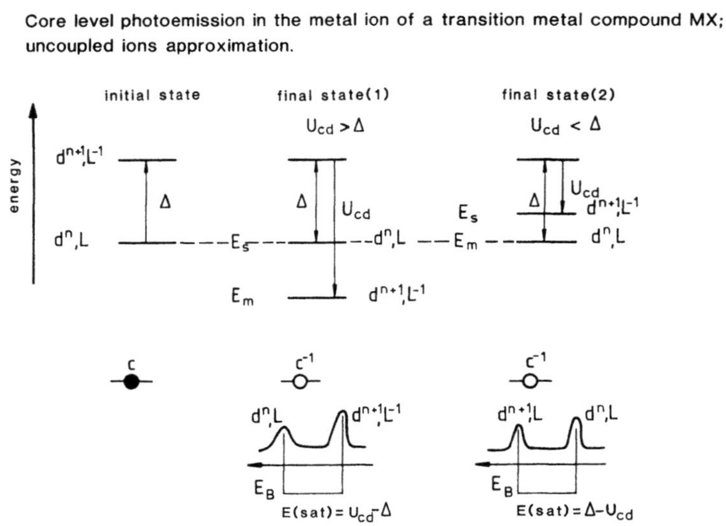

Fig. 3.4. Valence configurations used to interpret the core-level photoemission in a transition metal compound under the assumption that there is no coupling between the metal and ligand ions. It is assumed that the optical gap is of the charge transfer type (which is Δ in the uncoupled ion approximation!). The core hole (c^{-1}) on the metal ion lowers the energy of the d^{n+1} configuration with respect to the d^n configuration. This lowering is described by U_{cd}. Depending on the relative magnitude of U_{cd} and Δ either the $\mathrm{d}^{n+1}\mathrm{L}^{-1}$ configuration is the lowest final state ($U_{\mathrm{cd}} > \Delta$) or the $\mathrm{d}^n\mathrm{L}$ configuration is the lowest ($\Delta > U_{\mathrm{cd}}$). Note that in this approximation the main line-satellite separation is $E(\mathrm{sat}) = U_{\mathrm{cd}} - \Delta$ or $E(\mathrm{sat}) = \Delta - U_{\mathrm{cd}}$, respectively

$\Delta[\Delta = E(\mathrm{d}^{n+1}) + E(\mathrm{L}^{-1}) - E(\mathrm{d}^n) - E(\mathrm{L})]$. A very simple model derived from the ideas of Fig. 3.2 is then shown in Fig. 3.4. It is assumed that the metal and ligand ions are almost uncoupled, such that the energies are not influenced much by the weak coupling (small hybridization). Transitions of electrons between the two kinds of ions are, however, possible. The zero of the total energy in the initial state is placed at the state $\mathrm{d}^n\mathrm{L}$ and the first excited state in this approximation is $\mathrm{d}^{n+1}\mathrm{L}^{-1}$. Creation of a core hole on the metal ion shifts the $\mathrm{d}^{n+1}\mathrm{L}^{-1}$ configuration with respect to the $\mathrm{d}^n\mathrm{L}$ configuration, which is expressed as a lowering of the $\mathrm{d}^{n+1}\mathrm{L}^{-1}$ configuration by an amount U_{cd} under the assumption that the configuration $\mathrm{d}^n\mathrm{L}$ remains at the zero of energy. The two valence configurations $\mathrm{d}^n\mathrm{L}$ and $\mathrm{d}^{n+1}\mathrm{L}^{-1}$ are possible final state configurations after the creation of the core hole. We also see that depending on the relative magnitude of U_{cd} and Δ, the lowest valence configuration can be $\mathrm{d}^{n+1}\mathrm{L}^{-1}[U_{\mathrm{cd}} > \Delta\,, E(\mathrm{sat}) = U_{\mathrm{cd}} - \Delta]$ or $\mathrm{d}^n\mathrm{L}[U_{\mathrm{cd}} < \Delta\,, E(\mathrm{sat}) = \Delta - U_{\mathrm{cd}}]$ where $E(\mathrm{sat})$ is the difference between the satellite line and the main line in the core-level photoemission spectrum. The model depicted in Fig. 3.4 is oversimplified, because it leaves out the important coupling (hybridization) between the metal ions and the ligand ions. It allows, however, to gain insight into the basic physics that is responsible for the mechanism of the creation of the charge-transfer satellites. This model comes closest to reality in the case of the divalent copper compounds ($\mathrm{d}^9\mathrm{L}$ valence configuration) and was applied to them (and the divalent nickel) by Larsson [3.41, 3.78] and by Asada and Sugano [3.75]. We shall now first outline the reasoning of these researchers, as applied to the simple situation of divalent copper compounds, in order to show that the model is even quantitatively correct. We then give a more general description which is applicable to any d^n configuration [3.2, 3.6].

The energy-level diagram for an analysis of the charge transfer satellites in divalent copper compounds is given in Fig. 3.5. It is derived from that in Fig. 3.4 by including the metal ligand interaction (for simplicity only the final state case $U_{\mathrm{cd}} > \Delta$ is given).[3] In the initial state, in molecular-orbital language, the ground state is a bonding level (b) and the first excited state an antibonding level (a). These two states are connected by the optical gap energy (E_{g}) neglecting excitonic effects.

One introduces the core hole c^{-1} and calculates the energy between the two states possible in this situation, assuming that the antibonding level now also can be populated. Thus the energy difference of the main (E_{m}) and the satellite (E_{s}) lines, in terms of the matrix elements in the basis of bonding and antibonding states is [3.75]:

[3] The coupling between the metal and ligand ions is taken into account only in the form of hybridization. The Coulomb interaction U_{dp} (which is the Coulomb interaction between a valence d and a valence p electron) is generally, though not always, neglected. Since the treatment of Asada and Sugano [3.75] uses the bonding and antibonding wave functions as the basis set, it employs slightly different parameters (X_{O}, X_{c}, V instead of Δ, $U_{\mathrm{cd}}, T_{\mathrm{m}}$) to those given in the figures.

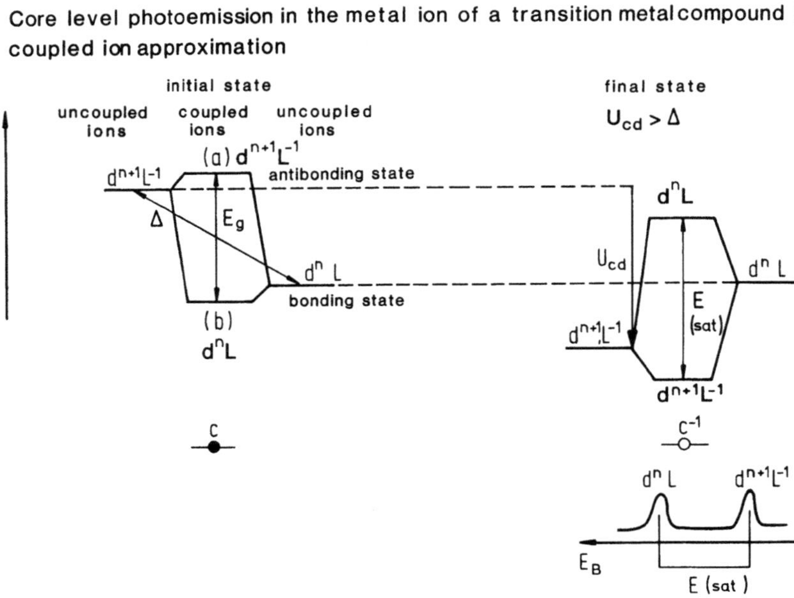

Fig. 3.5. Valence configurations used to interpret the core-level photoemission in a transition-metal compound under the assumption that a coupling exists between the metal (M, d^n) and ligand (X, L) ions. One may describe the two possible valence-band states as bonding and antibonding states in chemical notation. Only the case $U_{\mathrm{cd}} > \Delta$ (applying to divalent copper compounds) is shown

$$E_{\mathrm{s,m}} = \sqrt{X^2 + 4V^2} \, . \tag{3.3}$$

Here X is the diagonal and V the off-diagonal matrix element of the two-level system consisting of the bonding and antibonding states

$$X = \langle a|H|a\rangle - \langle b|H|b\rangle \, , \quad V = \langle a|H|b\rangle \, , \quad H = H_0 + H_{\mathrm{c}} \, . \tag{3.4}$$

In this notation H_0 is the Hamiltonian of the system without the core hole and H_{c} is the Hamiltonian of the core hole. The matrix elements in the presence of the core hole are given in linear approximation by

$$X = X_0 - X_{\mathrm{c}} \, , \quad V = V_0 + V_{\mathrm{c}} \, . \tag{3.5}$$

where X_0 and V_0 are the matrix elements in the absence of the core hole. Since the bonding and antibonding states are eigenfunctions of H_0 one has $V_0 = 0$. X_{c} is the difference in the energy reduction of the metal and the ligand orbital created by the core hole. Since the core hole resides primarily on the metal ion, it will have a greater effect on the metal orbital and therefore the net effect is to lower the energy of the d orbital with respect to the ligand level.[4] We then arrive at

[4] In the uncoupled situation one has $X_{\mathrm{c}} = U_{\mathrm{cd}}$, see Figs. 3.4 and 3.5.

$$E_{\mathrm{s,m}} = \sqrt{(X_0 - X_\mathrm{c})^2 + 4V_\mathrm{c}^2} \quad \text{with} \quad E(\mathrm{sat}) = E_\mathrm{s} - E_\mathrm{m} = E_{\mathrm{s,m}} \ . \tag{3.6}$$

Asada and Sugano [3.75] give arguments to show that the mixing matrix element can be related to the Coulomb energy by the following equation:

$$V = V_\mathrm{c} \simeq K X_\mathrm{c} \tag{3.7}$$

where K is a constant, and

$$X_0 = E_\mathrm{g} \ , \tag{3.8}$$

E_g being the optical gap energy which connects the highest occupied (bonding) and lowest unoccupied (antibonding) levels.

We shall now show how, using independent experimental information, the separation between satellite and main lines measured in the copper dihalides (Fig. 3.1) can be interpreted.

For the analysis of the experimental satellite data in terms of (3.6) one needs V_c, X_c and E_g. Asada and Sugano [3.75] estimated from NMR chemical-shift data a value of $V_\mathrm{c} \simeq 3\,\mathrm{eV}$. In order to estimate X_c one approximates copper with one core hole by zinc. In zinc dihalides the $3\mathrm{d}^9$ final state is about 12 eV below the top of the valence band. In the copper dihalides the top of the valence band ($3\mathrm{d}^9$ final states) is roughly at the Fermi level. This gives of the order of 12 eV for the Coulomb energy X_c based on the equivalent core approximation. E_g is taken from optical experiments [3.84, 3.85].

In the process of analyzing the data it was found that the agreement between theory and experiment can be improved if one assumes that the Coulomb energy decreases from 12 eV in copper difluoride to 10 eV in the hypothetical copper diiodide. This trend is reasonable because the ligand covalency increases in that order. Using these values, and the experimental optical energy gaps [3.84, 3.85], one achieves good agreement between theory and experiment, as shown in Fig. 3.6. The measured optical gaps E_g are joined by an unbroken straight line, and the approximate Coulomb integrals X_c are indicated by the dotted straight line. The calculated separations, $E_{\mathrm{s,m}}$ of the main line and the satellite line are shown by the dashed line and agree quite well with the measured energies.

We have dealt with the analysis of the satellites in the copper dihalides in some detail because it is the simplest and best understood system of charge-transfer satellites [3.6, 3.75].

We shall briefly outline a more general treatment which is, however, also based on the model depicted in Figs. 3.4, 3.5 [3.6]. Sawatzky et al. [3.6] start with the uncoupled wave functions $|3\mathrm{d}^9\mathrm{L}\rangle$ and $|3\mathrm{d}^{10}\mathrm{L}^{-1}\rangle$ (which are the only one hole state wave functions possible) as the basis set of their calculation, then introduce the coupling in order to proceed as Asada and Sugano [3.75].

The energy difference between the two initial states of the valence band is

$$\Delta = E(3\mathrm{d}^{10}\mathrm{L}^{-1}) - E(3\mathrm{d}^9) \ , \tag{3.9}$$

while the mixing matrix element between these states is given by

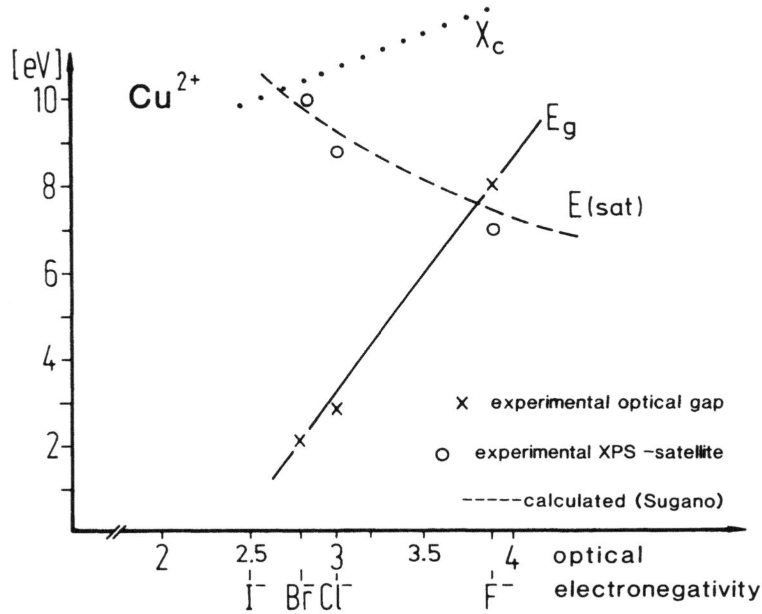

Fig. 3.6. Optical gaps E_g and measured and calculated satellite-main line separations $E_{s,m} = E(\text{sat})$ in Cu dihalides as a function of the optical electronegativity [3.54, 3.84]

$$T_m = \langle 3d^9 | H | 3d^{10} L^{-1} \rangle \, , \tag{3.10}$$

If one neglects overlap, the energies can be calculated to first order as

$$E_{1,2} = E_0 + \frac{1}{2}\Delta \pm \frac{1}{2}\sqrt{\Delta^2 + 4T_m^2} \, . \tag{3.11}$$

(E_0 is the energy for Δ, $T_m = 0$). These two energies belong, in the language of [3.75], to the *bonding and antibonding orbitals.*

Taking overlap into account would introduce an additional matrix element. Without overlap the optical gap E_g is the difference in energy between states E_1 and E_2:

$$E_g = \sqrt{\Delta^2 + 4T_m^2} \, . \tag{3.12}$$

We see that E_g is the energy needed for a charge transfer from the ligand onto the d-ion.

In the final state there is an additional hole in the core level on copper. Thus one has to take into account the Coulomb interaction between this hole and the charges on the copper atom and the ligand. In a first approximation one can neglect the Coulomb interaction between the hole on the copper ion and the ligand. Therefore only the Coulomb interaction between the hole and the d-configuration (designated by U_{cd}) is taken into account. For the energy of the resulting two states one then obtains

$$E_{1,2} = E_0' + \frac{1}{2}\Delta \pm \frac{1}{2}\sqrt{(\Delta - U_{\mathrm{cd}})^2 + 4T_{\mathrm{m}}^2} \, . \tag{3.13}$$

(E_0' is the energy for Δ, U_{cd}, $T_{\mathrm{m}} = 0$). Again we find two final states and the energy separation between main line and satellite is given by

$$E_{\mathrm{s,m}} = \sqrt{(\Delta - U_{\mathrm{cd}})^2 + 4T_{\mathrm{m}}^2} \equiv E(\mathrm{sat}) \, . \tag{3.14}$$

The relation between the Asada and Sugano [3.75] approach and that of Sawatzky et al. [3.6] is obtained by rewriting (3.14) with the help of (3.12) in order to introduce the optical gap energy explicitly

$$E_{\mathrm{s,m}} = \sqrt{E_{\mathrm{g}}^2 - 2\Delta U_{\mathrm{cd}} + U_{\mathrm{cd}}^2} \, , \tag{3.15}$$

$$E_{\mathrm{s,m}} = \sqrt{(E_{\mathrm{g}} - U_{\mathrm{cd}})^2 + 2U_{\mathrm{cd}}(E_{\mathrm{g}} - \Delta)} \, , \tag{3.16}$$

meaning that (assuming $X_{\mathrm{c}} = U_{\mathrm{cd}}$)

$$4V_{\mathrm{c}}^2 = 2U_{\mathrm{cd}}(E_{\mathrm{g}} - \Delta) \, . \tag{3.17}$$

Thus it looks as if the difference between the two formulations is a stronger dependence of the mixing matrix element on the ligand in the work of Sawatzky et al. [3.6] than in that of Asada and Sugano [3.75] because $(E_{\mathrm{g}} - \Delta)$ is strongly varying with the ligand.

The data in Fig. 3.6 provide evidence that the model presented describes correctly the satellites in divalent copper compounds. Since the two formulas (3.6) and (3.14) are essentially equivalent, the latter being the more general one, it is therefore appropriate to use (3.14) to interpret the energetics of the satellites. The parameters entering this description are Δ, U_{cd} and T. It is by now an accepted procedure to use the measured satellite main line separation and intensity ratios, as well as the experimental optical gap energies, to describe the core-level spectra in terms of the charge-transfer model. An example for the degree of agreement which then can be obtained for the $2\mathrm{p}_{1/2}$ lines in CuF_2 and $CuCl_2$ is given in Fig. 3.7 where Fig. 3.8 gives the corresponding energy level diagrams in the spirit of Figs. 3.4, 3.5. Figure 3.8 shows for $CuCl_2$ only an abbreviated diagram for the initial and final state, while for CuF_2 also the connection to the more detailed diagrams of Fig. 3.5 is indicated.

Okada and Kotani [3.86] developed a more complete analysis for satellite spectra by taking into account the 3d-multiplet coupling and the crystal-field interaction.

It was indicated earlier that the relative size of Δ and U_{cd} determines whether in a transition metal compound with a $\mathrm{d}^n\mathrm{L}$ configuration in the initial state, after photoionization, this same $\mathrm{d}^n\mathrm{L}$ configuration is the lowest final state or whether the $\mathrm{d}^{n+1}\mathrm{L}^{-1}$ configuration is the lowest final state. We have just seen that for divalent copper compounds the $\mathrm{d}^{10}\mathrm{L}^{-1}$ state is the lowest final state after core-level photoemission. This is, however, not generally the case for all 3d transition metal compounds. To illustrate the situation

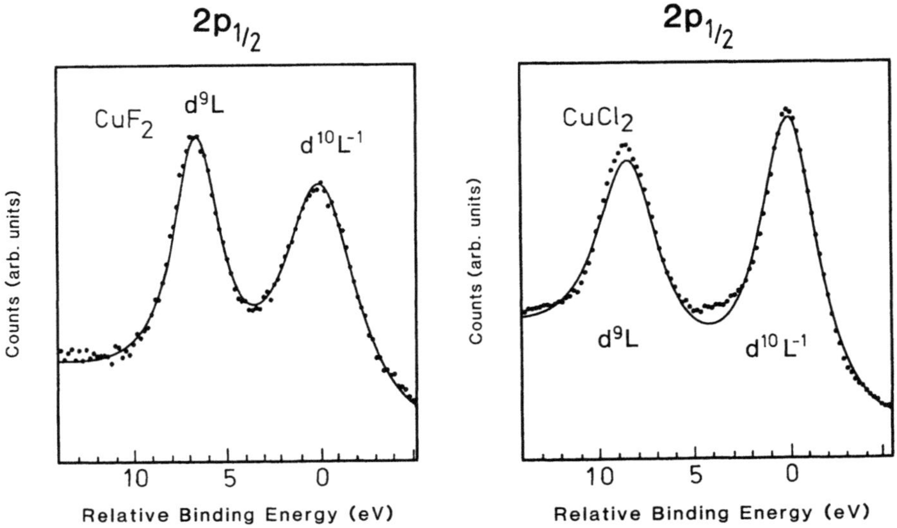

Fig. 3.7. $2p_{1/2}$ core-level lines of CuF_2 and $CuCl_2$ [3.29]. The full lines are fits to the experimental spectra with $\Delta = 7.0\,eV$, $U_{cd} = 9.0\,eV$, $T_{initial} = 2.1\,eV$, $T_{final} = 3.2\,eV$ for CuF_2 and $\Delta = 3.0\,eV$, $U_{cd} = 9.0\,eV$, $T_{initial} = 2.1\,eV$, $T_{final} = 3.0\,eV$ for $CuCl_2$

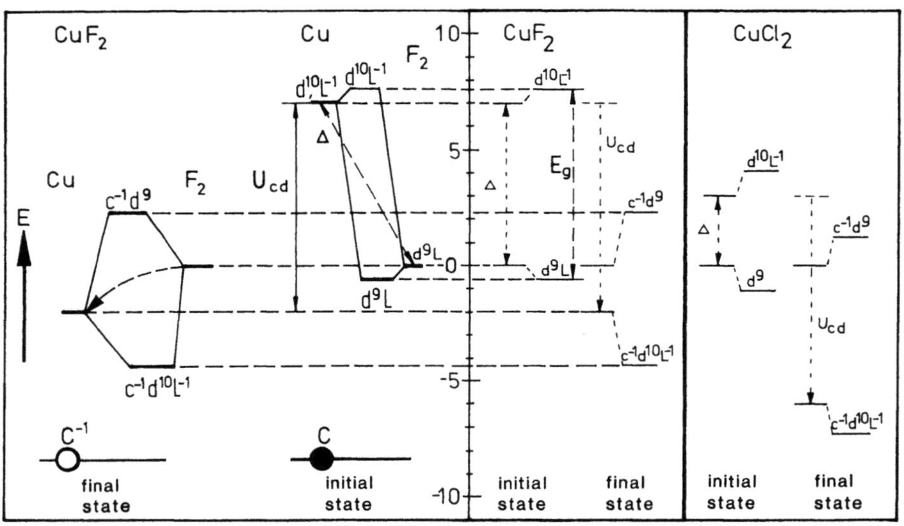

Fig. 3.8. Energy-level diagrams for the valence bands of CuF_2 and $CuCl_2$ with the parameters given in the caption of Fig. 3.7. In the two panels on the right side for the initial and final states the energies of the unhybridized and the hybridized case are given. In the left panel the diagrams for CuF_2 have been redrawn in the spirit of Fig. 3.5 to indicate the way in which they develop

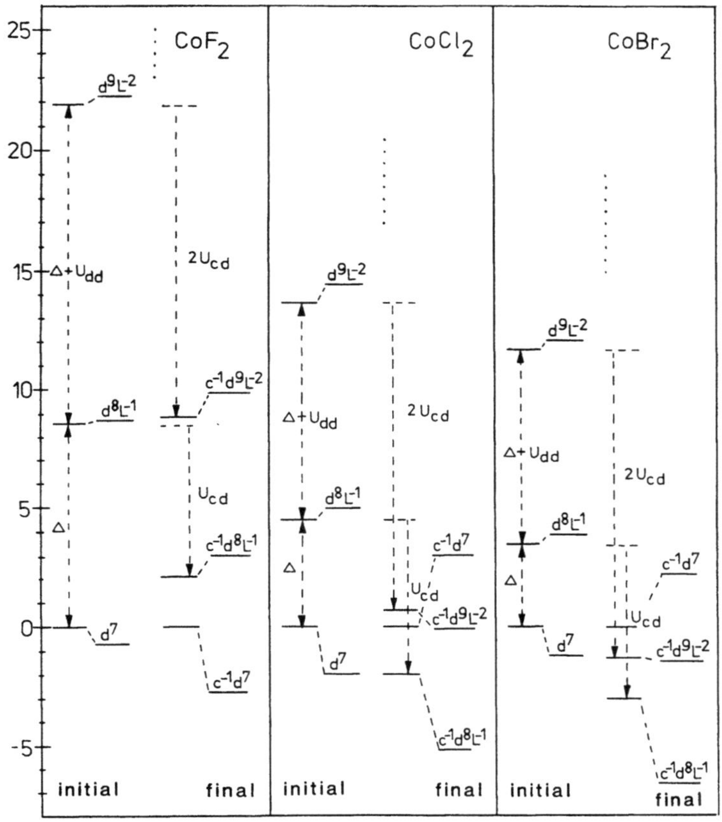

Fig. 3.9. Energy-level diagrams for CoF_2, $CoCl_2$ and $CoBr_2$ obtained from an analysis of the 2p core lines of these compounds. Parameters used are for CoF_2, $CoCl_2$ and $CoBr_2$: U_{dd} = 4.70, 4.70, 4.70; Δ = 8.60, 4.50, 3.50; U_{cd} = 6.50, 6.50, 6.50; $T_{initial}$ = 1.40, 1.95, 1, 30; T_{final} = 1.85, 1.55, 1.65; all numbers are in eV [3.29]

we give in Fig. 3.9 the initial state and final state energy-level diagrams for some cobalt dihalides and in Fig. 3.10 those for the corresponding manganese dihalides. We see that while in $CoCl_2$ and $CoBr_2$ one still has $\Delta < U_{cd}$ and therefore a d^8L^{-1} final ground state (d^7L initial state) for CoF_2 the situation is reversed ($\Delta > U_{cd}$) leading to a d^7L final ground state. In the manganese dihalides one has for all the dihalides $\Delta > U_{cd}$ and therefore a d^5L final ground state. However, in $MnBr_2$ the difference between Δ and U_{cd} is small and therefore the d^5L and d^6L^{-1} states are heavily mixed [3.47].

Note that for the divalent cobalt and manganese compounds not all four ($d^7, d^8L^{-1}, d^9L^{-2}, d^{10}L^{-3}$) and six possible final hole states are shown. In this case also the Coulomb correlation energy U_{dd} between the valence shell d electrons enters the calculations. This is an additional parameter which

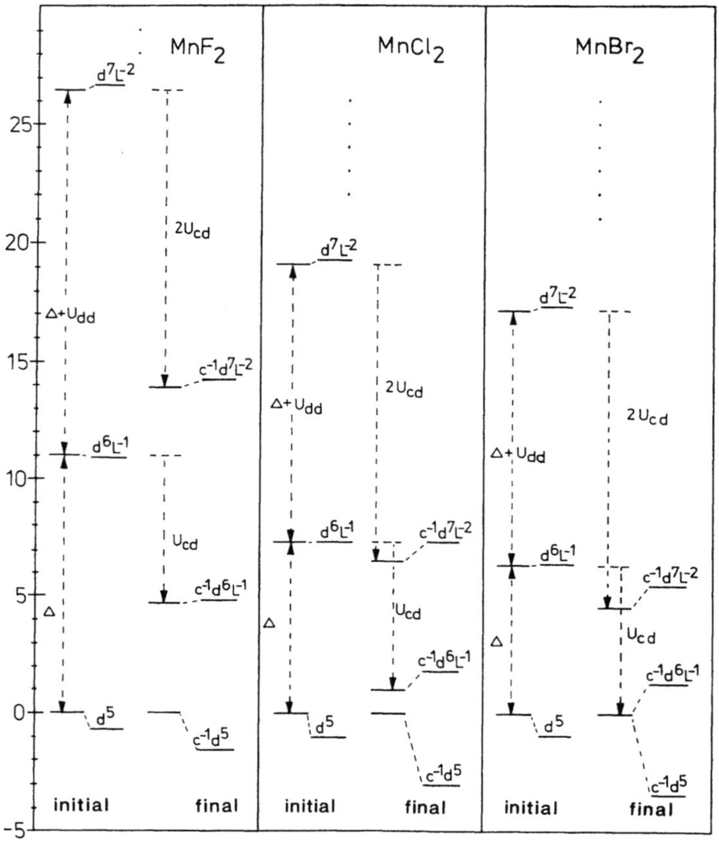

Fig. 3.10. Energy level diagrams for MnF_2, $MnCl_2$ and $MnBr_2$, obtained from an analysis of the 2p core lines of these compounds. Parameters used are for MnF_2, $MnCl_2$ and $MnBr_2$: $U_{dd} = 4.50$, 4.50, 4.50; $\Delta = 11.00$, 7.30, 6.30; $U_{cd} =$ 6.30, 6.30, 6.30; $T_{initial} = 1.30$, 1.25, 1.10; $T_{final} = 1.30$, 1.25, 1.15; all numbers are in eV [3.29]

has to be fixed, where one usually sets $U_{cd} = 1.4\, U_{dd}$, a value consistent with a large body of experimental results. We see from this discussion that only for divalent copper compounds the core-hole final ground state is $d^{n+1}L^{-1}$, while for ions with smaller d occupation number down to manganese, the possibility exists that d^nL is the lowest final state. For $n \leq 5$ probably for all ligands, the final state after core photoionization will be d^nL. These findings are in agreement with the results of the preceding chapter (Fig. 2.23).

These observations form the basis for the so-called *hole doping* in divalent nickel and copper oxides and, in particular, the high temperature superconductors [3.87, 3.88]. The latter compounds are copper oxide based materials like, e.g., $La_{2-x}Sr_xCuO_4$. In the base material La_2CuO_4 the copper ion is divalent. In the doped material $La_{2-x}Sr_xCuO_4 (x > 0)$ some of the copper ions are thought to be trivalent in order to achieve charge neutrality. The PES results show, however, that the situation is different. Since the d electrons in transition metal compounds are quite localized, the valence d electron photoionization can be described in the same framework as just given for the core levels. This implies, e.g., for La_2CuO_4 that the valence-band ground state after photoemission of a d electron is not d^8 ("trivalent copper") but d^9L^{-1} ("divalent copper" with a hole in the oxygen ligand band). If one depletes the valence band of La_2CuO_4 by chemical means of one electron (converting divalent copper into trivalent copper by doping with Sr) the same d^9L^{-1} final state is reached. Therefore the trivalent copper in $La_{2-x}Sr_xCuO_4$ has a hole in the oxygen ligand band and the same d^9 configuration as divalent copper in La_2CuO_4. We also see that such hole doping is obviously possible only for those compounds with a d^nL initial state which after core photoionization have a $d^{n+1}L^{-1}$ final state as the ground state.

Screening by d electrons in the final state can, in principle, also be observed in compounds where no d electrons are present in the initial state. A case in point is CaF_2, which corresponds most closely to the model in Fig. 3.2. In Fig. 3.11 the Ca 2p XPS spectrum is reproduced [3.52] where the loss region (larger binding energies) is also shown on an expanded intensity scale. If the structure accompanying the main line ($E_B > 355$ eV) were exclusively an extrinsic (inelastic electron scattering) loss structure, one should observe identical loss features behind any other line in CaF_2, e.g., near the F 1s line. To check this hypothesis the Ca 2p spectrum was simulated (trace b in Fig. 3.11) by a 2p doublet with the theoretical intensity ratio and a loss structure taken from the F 1s line and folded into the 2p lines. This simulation, while successful for the 2p lines in KF (K is the element adjacent to Ca) is not satisfactory for CaF_2, as seen in Fig. 3.11. However, one can reach almost perfect agreement between the measured and the simulated spectrum (Fig. 3.12) if a satellite at 10.9 eV larger binding energy (see the arrows in Fig. 3.12) is added, with an intensity of 4 % of the original 2p intensity. The resulting new simulation is shown in Fig. 3.12, trace b, and yields almost perfect agreement with the experimental spectrum. Employing the $(Z+1)$-approximation, calcium with a core hole can be approximated by scandium, which has an electronic configuration $3d^1 4s^2$. This means that in photoionized Ca^{2+} the 3d state can, in principle, be close to the valence band and therefore possibly accessible for the screening process [3.10].

Such satellites are not seen in KF in the 2p levels of K, and one can argue that this is because the 3d states in K, even in the photoionized state, are far above the Fermi level [3.52].

The correct assignment of the satellite to the Ca 2p levels seems, however, not yet settled. While Veal and Paulikas [3.52] attribute it to the participation of the d-orbital, de Boer et al. [3.10] favor a p $\rightarrow$ s charge transfer on the ligand (exciton satellite). This means that for the main line (in contrast to

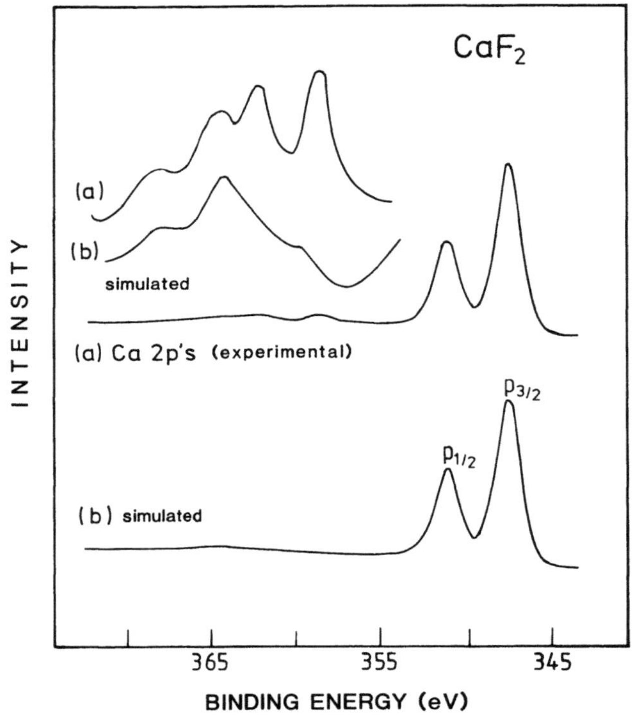

Fig. 3.11. Analysis of the Ca 2p spectrum in CaF_2 using only the loss structure obtained from the F 1s spectrum in this compound [3.52]. (a) measured spectrum, (b) simulated spectrum obtained by superimposing on the Ca $2p_{1/2,3/2}$ doublet the F 1s loss structure. The lower traces are the full spectra. Topmost curves: blown-up versions of bottom curves (a) and (b)

the charge transfer model) the final valence states are similar to those of the initial state; the satellite line then reflects additional charge excitation from an occupied p state into an empty s state on the ligand. On the other hand, there exists other evidence to support the Veal–Paulikas [3.52] interpretation. Ca metal (like Sr, Ba and Rb) shows strong satellites, which probably do not originate from plasmons but from d-screening [3.90,3.91] (see also Sect. 3.2.2).

We add that the systematics presented here for the 3d transition metal compounds (Figs. 2.23 and 3.9, 3.10) favor an interpretation of the satellite in CaF_2 as an exciton satellite [3.10] rather than a charge-transfer satellite.

3.1.3 Non-local Screening

The discussion so far has been based on the electronic structure and the screening processes depicted in Fig. 3.4. This type of analysis has been quite successful in explaining many core-level spectra of open shell ions [3.11–3.20]. However, some details were not explained by this model, for example, the fine

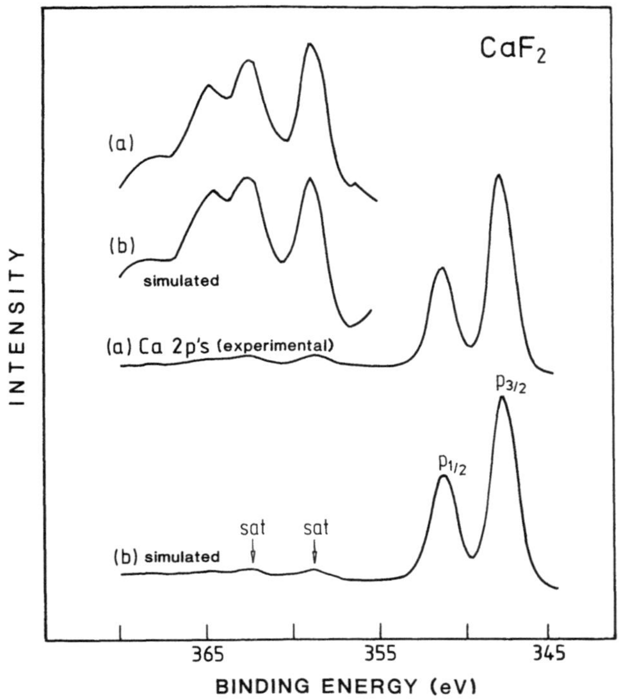

Fig. 3.12. Analysis of the Ca 2p spectrum of CaF_2 employing extrinsic loss structure and a satellite [3.52]. (a) measured spectrum, (b) simulated spectrum with the loss structure of the F 1s line and a satellite doublet (arrows) displaced from the main lines by 10.9 eV and given 4% of the intensity of the main lines. The bottom trace gives the full simulated spectrum

structure in the $2p_{3/2}$ core line of NiO and some of the Cu-based high temperature superconductors [3.92–3.96]. After many failed attempts to explain this additional structure, it now seems to be agreed that it results from what is generally called non-local screening [3.92]. What is meant by this term is indicated in the diagram contained in Fig. 3.13 and also in more detail in Fig. 3.14. With the help of Figs. 3.13 and 3.14 the ground state of NiO can be written as $3d^8L$ and the lowest final state after $2p_{3/2}$ photoionization is $2p_{3/2}^{-1}$ $3d^9L^{-1}$. This means that the charge neutralization or the screening on the photoionized Ni site is produced by transferring an electron from the nearest ligand atoms onto the photoionized Ni atom ($d^8L \rightarrow d^9L^{-1}$). However, there is also another way to neutralize the photoionized Ni atom. This consists of transferring an electron not from one of the six nearest neighbor ligand atoms but from one of the ligand atoms in the next nearest coordination shell. This results in a final state which can formally be written $3d^9L(3d^8L^{-1})$. This process is also shown in Fig. 3.14. In general terms, it

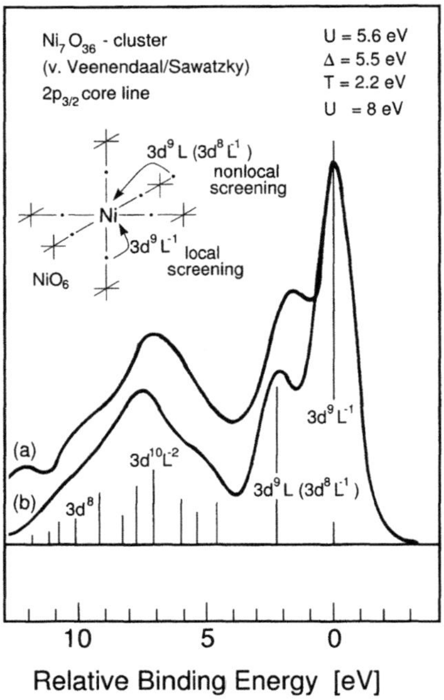

Fig. 3.13. Photoemission spectrum of the $2p_{3/2}$ core line in NiO (a) and calculated spectrum in the non-local screening approximation (b). The inset shows a sketch of the mechanism of the local and the non-local screening processes; the splitting of the main line is due to two screening channels, a local one (d^9L^{-1}) and a non-local one $3d^9L$ ($3d^8L^{-1}$) [3.92]

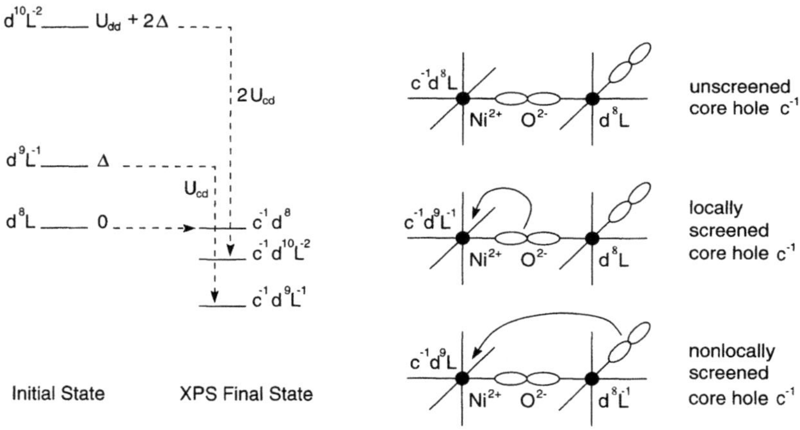

Fig. 3.14. Energy level diagram for the initial and final state photoemission of a nickel core line in NiO. In addition, the processes leading to an unscreened core hole $c^{-1}d^8L$, a locally screened core hole $c^{-1}d^9L^{-1}$, and a non-locally screened core hole $c^{-1}d^9L(d^8L^{-1})$ are shown [3.92]

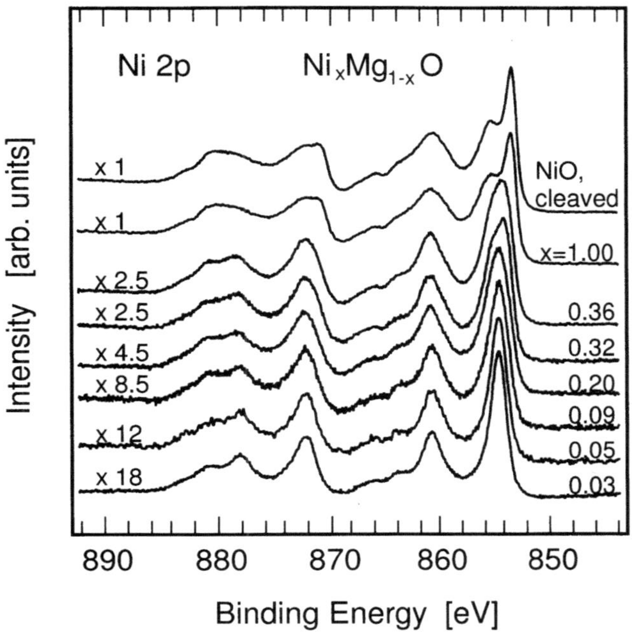

Fig. 3.15. Ni 2p photoemission spectra of $Ni_xMg_{1-x}O$ and for comparison a spectrum of a cleaved NiO surface. The satellite in the main Ni $2p_{3/2}$ line disappears with increasing Mg content indicating that it is caused by a non-local screening channel [3.93]

means that an electron is taken from a far away oxygen site and transferred by hopping to the photoionized Ni site. Calculations based on this concept have been performed by van Venendaal and Sawatzky [3.92], and the resulting spectrum is compared to a measured spectrum in Fig. 3.13. It can be seen that good agreement between theory and experiment is achieved.

One can rationalize the different energetic positions in the locally screened state and in the non-locally screened state in the following way: The core hole produced by the photoionization and the screening electron can be viewed as an exciton, and its binding energy is obviously an inverse function of the separation of the two charges. Therefore the locally screened state has the lowest energy, because its exciton has the highest energy, whereas the non-locally screened state has a small exciton energy and therefore does not represent the ground state. In other words, the two charges in the non-locally screened state are far away from one another, but if they come closer together, as in the locally screened state, they become energetically more stable: the locally screened state is the one with the lowest energy.

These hypotheses can be verified by experiment. If one compares the 2p photoemission spectra of NiO and of NiO dissolved in MgO [3.93] (MgO has the same crystal structure with almost the same lattice constant as NiO), or of one monolayer of NiO on MgO [3.94], one finds that the shoulder at 1.5 eV visible at the $2p_{3/2}$ line in pure NiO is missing (Fig. 3.15). This is expected for a peak that can only be screened locally, because in NiO dissolved in MgO there are no d^8L states available as next nearest neighbors to a particular Ni ion, which could provide the non-local screening. (The same argument also applies, although less strictly, to NiO on MgO).

The 2p photoemission spectra of CuO and of $Sr_2CuO_2Cl_2$ have been interpreted in a similar manner. In all these cases non-local screening effects were definitely identified. However, the details of the spectra could not be explained quantitatively in all cases [3.95, 3.96].

We want to mention a recent paper [3.201] that reports the 2p photoelectron spectrum of atomic Ni. This spectrum, which is similar to the 2p spectrum of NiO, can be explained in a configuration interaction calculation with $2p^53d^84s^2$ and $2p^53d^94s$ final states (the groundstate of atomic Ni is $2p^63d^94s$ and $2p^63d^84s^2$). The authors suggest that a similar explanation should be true for the 2p photoelectron spectrum of NiO thus invalidating an interpretation using the non-local screening effect. Only further work can solve this problem.

3.2 The 6-eV Satellite in Nickel

Another very well investigated and now reasonably well understood example of a satellite is the 6-eV satellite in nickel metal [3.9, 3.10, 3.97]. The problem can be introduced by Fig. 3.16, which shows all core-line spectra of nickel metal with binding energies smaller than 1000 eV. In each case the main line is accompanied by a satellite line, at about 6 eV larger binding energy. The $3p_{1/2,3/2}$ (the $3p_{1/2} - 3p_{3/2}$ spin–orbit splitting is barely visible in the main line) spectra even show a further satellite at about 12 eV larger binding energy. For the 6-eV satellite various interpretations have been presented over the years. They are reviewed briefly in order to show how explanations for extra lines or satellites can be dismissed:

- Impurities: This interpretation can be ruled out on the grounds of the present-day possibilities of high quality sample preparation and chemical analysis of the surface, e.g., by XPS.
- Band-structure effects: These can also be dismissed due to our knowledge of the band structure as determined by angle-resolved photoemission spectroscopy (Chap. 7).
- Plasmon sidebands: In a metal this would seem the most obvious explanation for a satellite at higher binding energy with the same displacement in energy from every core line. However, the loss function (i.e., $\mathrm{Im}\{-1/\varepsilon\}$), as measured by optical means or by electron energy loss experiments, does not show any structure at an energy around 6 eV. Therefore this possibility can also be dismissed [3.98, 3.99].

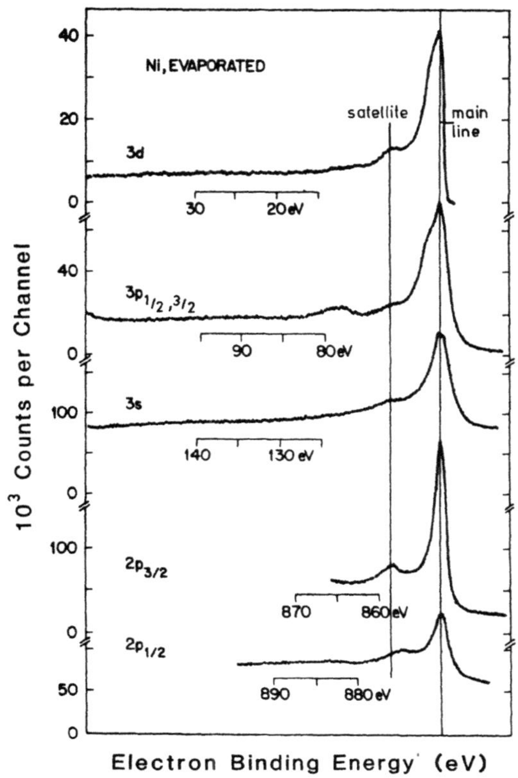

Fig. 3.16. XPS spectra of the 3d, 3p, 3s, $2p_{/2}$ and $2p_{1/2}$ levels of Ni metal [3.9]. The main lines have been lined up to demonstrate the constant distance of the satellite position (even for the 3d valence band)

Thus, only an explanation in terms of final state effects is left. In the following, the picture and the terminology of Kotani and Toyozawa [3.8] will be used. A number of other explanations which differ in detail, but not in substance, have also been given. The physics is summarized in Fig. 3.17, which is an adaptation of Fig. 3.2 to the case of Ni metal.

Figure 3.17 shows a schematic band structure of nickel metal in the initial state, where the $3d^9$ states lie within the 4s band (for simplicity, integral electron occupation numbers are used). In the final state one has two possibilities for screening: the additional Coulomb attraction produced by the core hole (which can be viewed approximately as an additional nuclear charge) can now pull the conduction band so far below E_F that an additional charge can be put into the 3d orbitals leading in a first approximation to a $3d^{10}$ configuration. This is essentially the ground state and is very similar to the ground state in copper metal, to which it corresponds in the $(Z+1)$-approximation. Thus

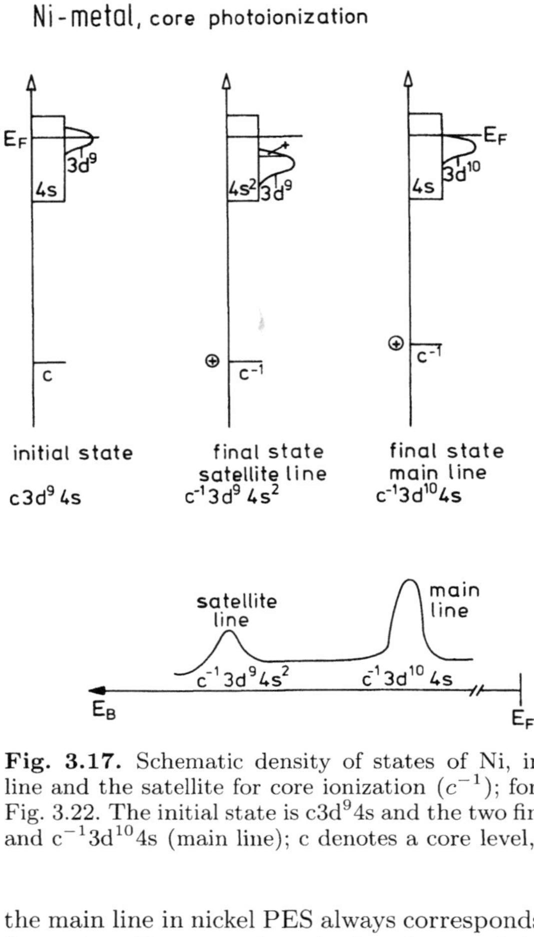

Fig. 3.17. Schematic density of states of Ni, indicating the origin of the main line and the satellite for core ionization (c^{-1}); for valence-band ionization see also Fig. 3.22. The initial state is $\mathrm{c3d^94s}$ and the two final states are $\mathrm{c^{-1}3d^94s^2}$ (satellite) and $\mathrm{c^{-1}3d^{10}4s}$ (main line); c denotes a core level, $\mathrm{c^{-1}}$ a core hole

the main line in nickel PES always corresponds to the $3d^{10}$ valence-band configuration. However, there is also the possibility that the 3d band, although now completely below E_F is not filled by a screening electron, and that the screening is instead produced within the wide 4s band. Effectively, one then has a two-hole state: one hole is produced by the PE process in the core level, and the second, which is also quite localized, is the hole in the d shell. This two-hole configuration is not the ground state. It is an excited state and the excitation energy is roughly 6 eV, in agreement with the experimental results. For more sophisticated treatments of this problem see [3.100–3.116].

Note that the 6 eV satellite is also observed in the valence-band spectrum (Fig. 3.16, top panel). In this case the main "line" corresponds to the $3d^94s$ configuration and the satellite to the $3d^84s^2$ configuration in atomic notation.

3.2.1 Resonance Photoemission

The most convincing evidence to support the two-hole assignment for the 6-eV satellite in the valence band of Ni metal comes from *resonance PE* data [3.100]. In fact, resonance experiments on nickel demonstrated the usefulness of this technique for the investigation of valence-band features in solids. "Resonance" PE means, that one excites photoemission with photons of an energy $\hbar\omega$ very near to the absorption threshold of a core level (index c). Then the direct photoemission of valence-band electrons (v) can interfere with Auger CVV electrons that are emitted in a Super Koster–Kronig process. The intensity for the feature in the valence band for which the direct PE process and the Auger emission overlap is given as a function of the photon energy $\hbar\omega$ by the so-called Fano lineshape [3.117]

$$N(\hbar\omega) \simeq \frac{(\varepsilon + q)^2}{\varepsilon^2 + 1} , \qquad \varepsilon = \frac{\hbar\omega - \hbar\omega_{\mathrm{j}}}{\Delta(\hbar\omega_{\mathrm{j}})/2} \tag{3.18}$$

where $\hbar\omega_{\mathrm{j}}$ is a photon energy equal to the binding energy of a core level (near which the resonance is investigated), q is a parameter for the particular core level and $\Delta(\hbar\omega_{\mathrm{j}})$ is the width (FWHM) of that core level.

In the CVV Super Koster–Kronig Auger process a two-hole state is produced in the valence band. Non-resonant direct PE produces a one-hole state if screening takes place by a d electron (main line), and a two-hole state if screening takes place in the sp band. Thus tuning $\hbar\omega$ through such a resonance enhances the two-hole signal because for this one the direct photoemission current and the CVV Auger current have the same energy and add up to the total observed signal. Therefore, sweeping, e.g., the photon energy through that of the 3p core-level binding energy is a very nice way to demonstrate that the valence-band satellite in Ni metal is indeed a two-hole state.

The coherent superposition of the directly photoemitted electrons and the (CVV) Auger electrons is only possible exactly at resonance; only then is the excitation from the 3p level into the empty d levels possible. At higher photon energies only the non-resonant ("normal") Auger electrons are observed.

It has to be emphasized, that the intensity enhancement is in the satellite (two-hole state!) and not in the main line. This fact has very often been overlooked and has led to various misinterpretations. The main line shows only a weak "dip" in the intensity at resonance. This can be interpreted rather simplistically as being caused by a reduction in the number of photons available for the direct photoemission process; an appreciable number of them are absorbed at resonance by the 3p $\rightarrow$ 3d excitation.

The experimental spectra of Petroff and his group [3.100] are shown in Fig. 3.18. Figure 3.19 gives the normalized satellite intensity as a function of the photon energy. There is a clear resonance at the 3p binding energy $E_{\mathrm{B}} = 67\,\mathrm{eV}$ of Ni metal. At this photon energy direct photoemission from

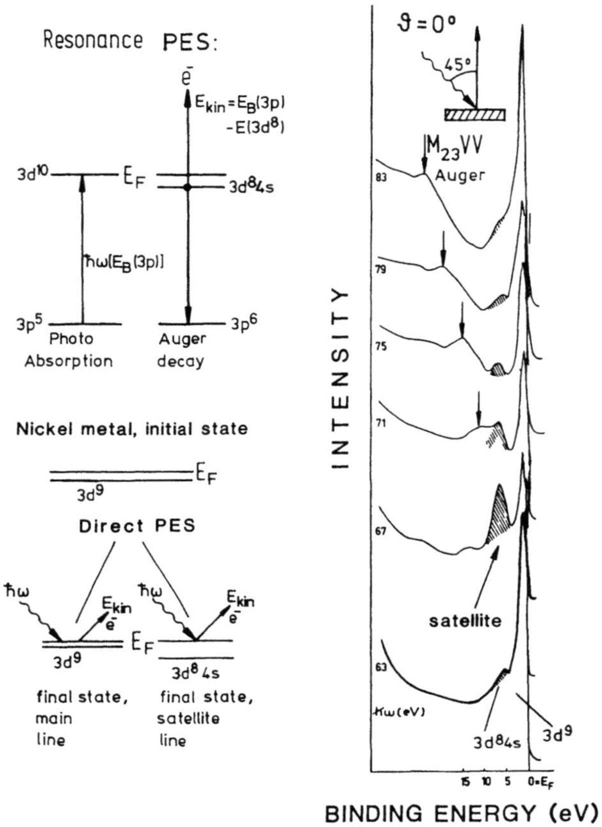

Fig. 3.18. Valence-band PE of Ni metal with photon energies around the 3p core-level binding energy ($E_B = 67\,\text{eV}$) [3.100]. The resonance enhancement of the satellite (*hatched area*) ($E_B \simeq 6\,\text{eV}$) is clearly seen at $\hbar\omega \simeq 67\,\text{eV}$. The diagrams on the left-hand side indicate the process involved in direct PES (*bottom*) and resonance PES (*top*)

the valence band (assuming screening by a 4s electron) may be described by (omitting the work function)

$$3d^9 + \hbar\omega \to 3d^8[4s] + e^-[E_{\text{kin}} = \hbar\omega(E_{3p}) - E_{\text{sat}}(\text{valence})] \qquad (3.19)$$

([4s] is a 4s screening electron). This is the 6-eV satellite.

In addition, the photon energy at the 3p resonance can excite an electron out of the 3p shell into the 3d hole just above the Fermi energy in Ni metal. The excited state resulting from this process can decay by an Auger process. This leads to the same final state ($3d^8[4s]$) and gives electrons with the same kinetic energy as direct photoemission leading to the satellite's final state (omitting the work function)

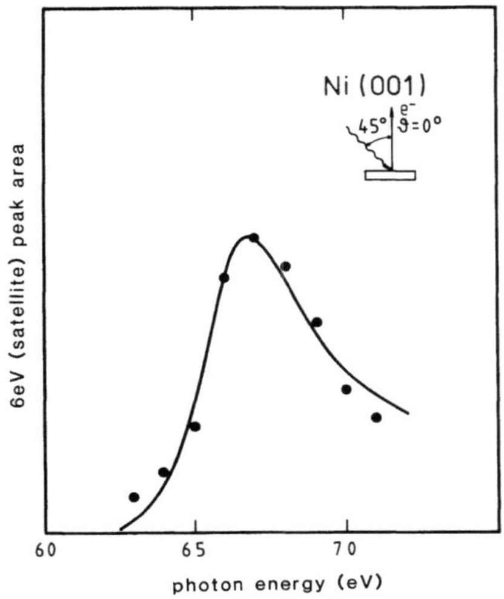

Fig. 3.19. Intensity of the satellite 6 eV below the valence band as a function of photon energy ($\hbar\omega$) [3.100]. Data from Fig. 3.18

$$
\begin{aligned}
&3p^6 3d^9 + \hbar\omega \quad \rightarrow \\
&\qquad 3p^5 3d^{10} \overset{\text{Auger}}{\rightarrow} 3p^6 3d^8 [4s] + e^- [E_{\text{kin}} = E_{3p} - E_{\text{sat}}(\text{valence})] \ .
\end{aligned} \tag{3.20}
$$

The two processes (3.19) and (3.20) overlap coherently at the threshold energy, giving rise to the resonant enhancement of the satellite. At higher photon energy the $M_{23}VV$ Auger electrons move away from the satellite because they appear with fixed kinetic energy [$E_{3p} - E_{\text{sat}}$ (valence)] while the kinetic energy of the photoelectrons increases with $\hbar\omega$. For a detailed treatment of the resonance photoemission in Ni see [3.115].

Resonance photoemission has been investigated in a number of systems [3.100–3.107, 3.118–3.131]. Of particular interest are measurements in atomic 3d metals because they allow one to verify the theory of this technique in considerable detail [3.118–3.121]. Here we describe data for atomic Ni in order to allow a comparison with the data for Ni metal which were just presented.

A spectrum of Ni vapor taken at $\hbar\omega = 68.6$ eV (near the 3p resonance) is shown in Fig. 3.20 [3.130]. The lines have been numbered 1 through 8. In the initial state, at the elevated temperatures ($\simeq 1500°$ C) used to produce the beam of Ni atoms, both the $3d^8 4s^2\ {}^3F$ and the $3d^9 4s\ {}^3D$ state are populated. Therefore, the final state after d electron emission must consist of the configurations $3d^7 4s^2$ and $3d^8 4s$. With the help of atomic data [3.131] the assignment given in Table 3.1 can be made. It shows that the two final states are well separated and only line 5 contains contributions from both the $3d^7 4s^2$ and the $3d^8 4s$ final state configurations. The intensity resonances with $\hbar\omega$ for two typical lines of the different final state configurations (line 3

Table 3.1. Experimental binding energies E_B of the states of free Ni^+ giving rise to the photoemission lines in Fig. 3.20. The tabulated binding energy of the $(3d^8\ ^3F)4s\ ^4F$ state has been used as reference [3.130]

Line	E_B [eV]	State in Ni^+
1	21.8 ± 0.15	$3d^74s^2\ ^2D$
2	19.1 ± 0.15	$3d^74s^2\ ^2F$
3	17.2 ± 0.1	$3d^74s^2\ ^2H,^2D$
4	16.4 ± 0.15	$3d^74s^2\ ^4P,^2G$
5	14.1 ± 0.15	$3d^74s^2\ ^4F$
		$3d^84s\ ^2S$
6	11.7 ± 0.1	$(3d^8\ ^1G,^3P)4s\ ^2G,^2P$
7	10.6 ± 0.1	$(3d^8\ ^1D,^3P)4s\ ^2D,^4P$
8	8.7 ± 0.1	$(3d^8\ ^3F)4s\ ^4F$

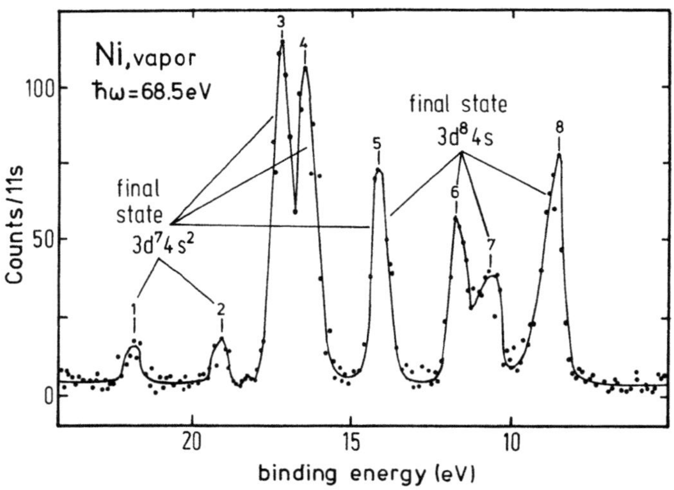

Fig. 3.20. Valence-orbital PE of Ni vapor at the 3p resonance energy ($\hbar\omega = 68.5$ eV) [3.130]. The initial states of Ni in the vapor are $3d^84s^2$ and $3d^94s$ and are almost degenerate. Therefore both final states produced from them, $3d^74s^2$ and $3d^84s$, show up in the spectrum

– $3d^74s^2\ ^2H$, ^2D and line 6 – $3d^84s\ ^2P$, ^2G) are shown in Fig. 3.21. In this figure the full line is a fit to the data with a Fano lineshape (3.18) and the dashed line is the result of a calculation by Davis and Feldkamp [3.124] for $3p^63d^8 \rightarrow 3p^53d^9$ and $3p^63d^9 \rightarrow 3p^53d^{10}$ excitations respectively, and subsequent Auger decay. These channels interfere with the direct valence-electron excitations and thereby produce a Fano [3.117] profile.

The resonance enhancement for lines 1 to 4 (Fig. 3.20) is described by

$$3p^6 3d^8 4s^2\ {}^3F \overset{\hbar\omega}{\rightarrow} \{3p^5 3d^9 4s^2\ {}^3F(64.3\,eV); {}^3D, {}^3P(68.5\,eV)\} \overset{\text{Auger}}{\rightarrow} 3p^6 3d^7 4s^2\ .$$

Since the excitation leaves the 3d shell unfilled in the intermediate state, multiplet splitting leading to a ^3F state (64.3 eV excitation energy) and to a ^3D or ^3P state (68.5 eV excitation energy) causes a splitting of the excitation profile of $\approx 4\,eV$, as is observed.

The resonance enhancement of lines 6–8 is due to the reaction channel:

$$3p^6 3d^9 4s\ {}^3D \overset{\hbar\omega}{\rightarrow} 3p^5 3d^{10} 4s\ {}^3P \overset{\text{Auger}}{\rightarrow} 3p^6 3d^8 4s\ .$$

Since here the 3d shell is filled in the intermediate state, one observes only one resonance maximum. The spin–orbit splitting of the 3p level (1.81 eV) has been ignored in these considerations.

The resonance enhancement of the 6 eV satellite in Ni metal follows quite closely the enhancement measured for line 6 in atomic Ni. This corroborates the assignment of this satellite in the metal to a two-hole (atomic) state, nearly identical to the atomic $(3d^8\ {}^1G)\,4s$ state in atomic Ni. The weak satellite $\simeq 3\,eV$ below E_F in Ni metal may then be attributed to the $(3d^8\ {}^3F)4s$ state in atomic Ni.

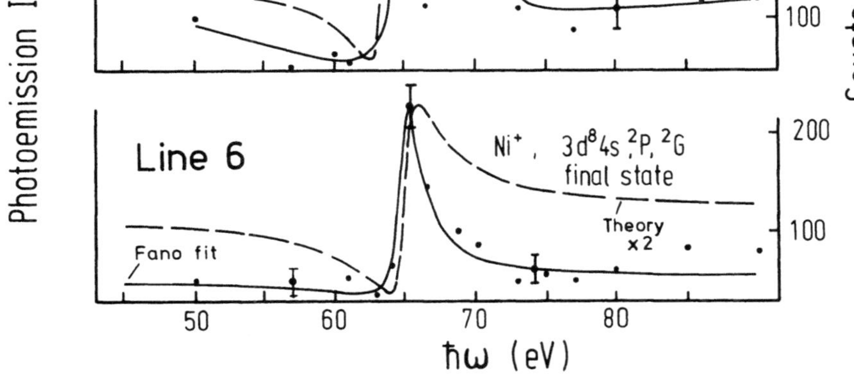

Fig. 3.21. Resonance PE in Ni vapor, for a line with a $3d^7 4s^2$ final state (line 3 of Fig. 3.20) and a line with a $3d^8 4s$ final state (line 6 of Fig. 3.20). The full line is a fit of the experimental data with a Fano lineshape: $N(\hbar\omega) \simeq (\varepsilon + q)^2/(\varepsilon^2 + 1)$. The dashed lines are calculations by Davis and Feldkamp [3.124]

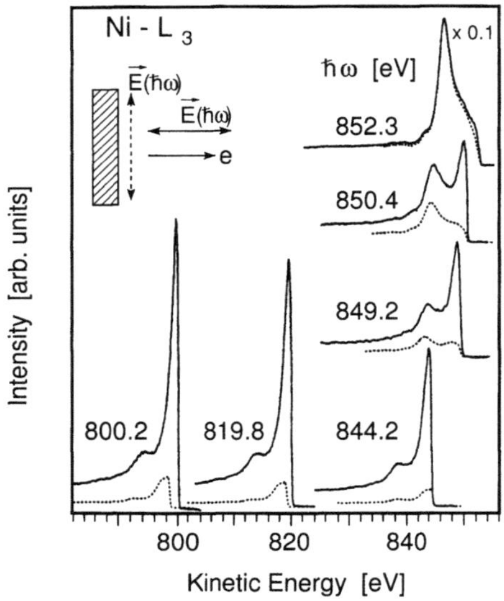

Fig. 3.22. Resonance photoemission spectra of the valence band of Ni metal at the $L_3(2p_{3/2})$ threshold. The dashed curves are for the electric vector of the incident light parallel to the surface (Auger geometry), and the full lines are for the electric vector perpendicular to the surface (photoemission geometry) [3.132]

More detailed investigations of resonance photoemission in Ni metal at the 2p absorption edge have revealed a picture that is not as simple as the one just outlined [3.132, 3.133]. Besides the coherent resonant Auger decay channel, there is also an incoherent Auger decay channel, which turns out to be dominant for excitation energies at and above the threshold (threshold here means the maximum in the photon absorption for the core level, which is used to detect the resonant photoemission signal). For the enhancement of the satellite signal below the threshold a new process has been identified, namely Radiationless Resonant Raman Scattering (RRRS) or Radiationless Resonant Auger Scattering (RRAS). These terms mean that the observed phenomena are the equivalent of resonant Raman scattering in the optical regime.

We now address the question of whether the intensity modulation in the 6 eV satellite in Ni metal, as the photon energy is swept through a core level, 3p or 2p for all practical purposes, is due to simple incoherent Auger processes or whether there is a coherent resonance process present. These two processes can in principle be distinguished by the fact that the ordinary Auger process is an incoherent two-step process, where in the first step photoionization takes place (in the experiment under discussion a photoexcitation of the 2p electron to the Ni valence band) and subsequently an Auger process occurs, involving two valence electrons, by which the core hole is filled and the resulting energy is carried away by an additional valence electron. Conversely,

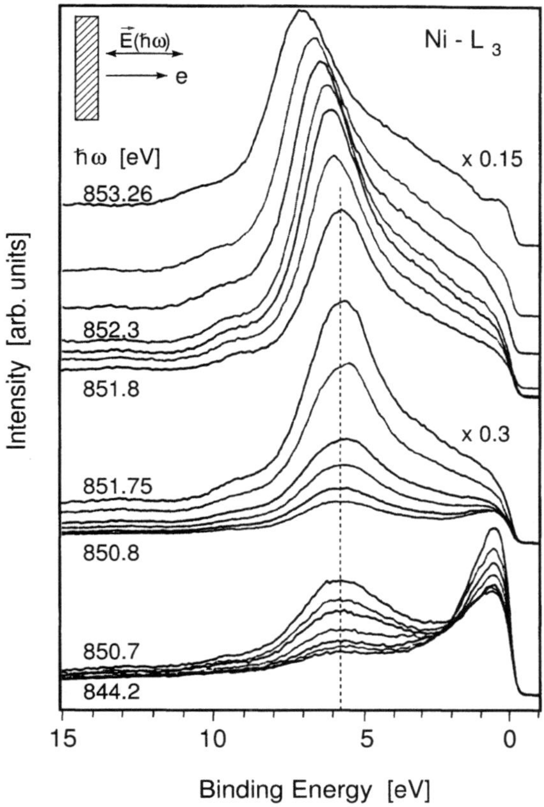

Fig. 3.23. Resonance photoemission spectra at the Ni L_3 threshold with near the threshold photon energies in the Auger geometry. At the threshold of 852.3 eV the Auger signal starts to move to larger binding energies [3.132]

in the resonance process, an electron from the core level is transferred to an intermediate state, from which it decays back to its original state, and an additional electron carries away the excess energy. The important aspect here is that the second (resonance) process has to be viewed as a one-step process, where the intermediate state is virtual. Data from the experiment of Weinelt et al. [3.132, 3.133] are shown in Figs. 3.22 and 3.23. They are taken in small energy increments below the photoabsorption maximum (852.7 eV) and slightly above it. The energetic position of the satellites as a function of photon energy derived from these states is shown in Fig. 3.24. Here one can make an interesting observation. Below the resonance maximum of the $2p_{3/2}$ line the satellite tracks with the photon energy, while at the $2p_{3/2}$ resonance energy it is converted into a constant kinetic energy feature now moving in a binding energy scale signaling its Auger behavior, as also evident in Fig. 3.23. The behavior at the $2p_{1/2}$ resonance is slightly different.

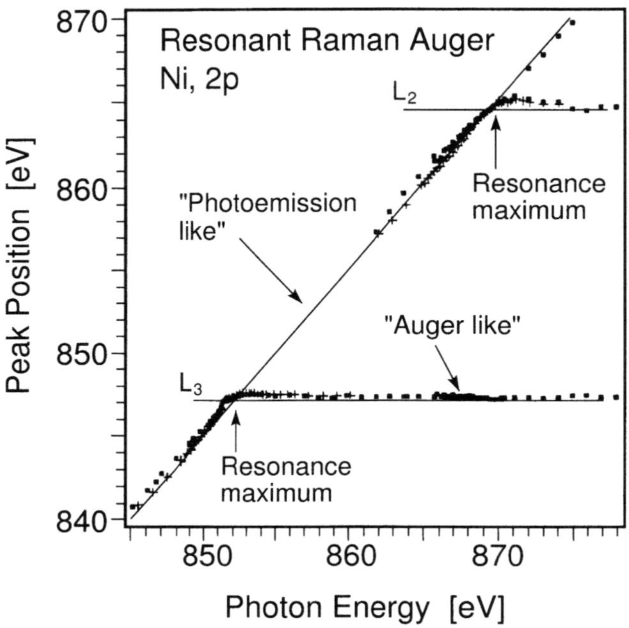

Fig. 3.24. Peak position (kinetic energy) of the valence-band satellite in Ni metal as a function of the exciting photon energy. The transition from the resonant Raman Auger effect below the threshold, to the Auger effect above it are clearly seen at the L_3 threshold, but a slightly different behavior is seen at the L_2 threshold [3.132]

The $2p_{3/2}$ data imply that below the resonance maximum a coherent resonant process exists. This is generally called the resonant Raman Auger regime, while above it one is observing mostly a pure Auger regime. Due to the large intensity, the analysis of the data at and above the resonance maximum is difficult. A detailed analysis of the intensity shows that there is resonance photoemission with interference effects particularly below the resonance maximum.

The spectra in Fig. 3.22 have been taken in two different geometries: one with the vector of radiation polarization normal to the surface, which gives the strongest photoemission intensity, and one with the vector of radiation polarization in the surface (or almost in the surface), which gives small photoemission intensity and therefore enhances the weight of the Auger intensity. The data show that below the resonance the photoemission intensity dominates the intensity. At the resonance (and above it) both geometries result in the same spectral intensity, which implies that much of the intensity now comes from the incoherent Auger process, as shown in Fig. 3.24.

If, as indicated by the data in Fig. 3.24, there is a Raman process below the threshold (resonance maximum), then a behavior similar to that in Ni metal should also be observable in other materials, especially in other transition metals, as is indeed the case. To demonstrate this, we first give a brief description of the RRRS or the RRAS process and subsequently present data for Cr and Fe, which will indeed prove the conjecture, and also indicate that the observation of the satellite is actually quite a general phenomenon in the Raman regime.

Figure 3.25 shows a general energy diagram that can be used to explain RRAS, also for other cases [3.134]. A beam of photons with energy around $\hbar\omega_0$ and with a distribution function G (width, FWHM $= \Gamma_\omega$) impinges on a sample and excites an electron. The system has an excited state with an energy E_r near the Fermi energy, and a distribution function of that state L_r with width Γ_r. The distribution of excited states is now $P_\mathrm{r} = G \cdot L_\mathrm{r}$, from which a decay into a lower state of the system, with energy ε_A^0 relative to the Fermi energy, takes place. The remaining energy is carried away by an Auger electron (therefore the term radiationless). For our purposes the state with energy ε_A^0 is the initial ground state. However, it is also possible that the state from which the electron is initially liberated is filled during the process, creating a vacancy in a higher state.

One can now read from Fig. 3.25 the following relations (note for the RRAS process the condition $\Gamma_\omega < \Gamma_\mathrm{r}$ must be fulfilled):

$$\varepsilon_\mathrm{max} = \hbar\omega_0 - E_\mathrm{f} \tag{3.21}$$

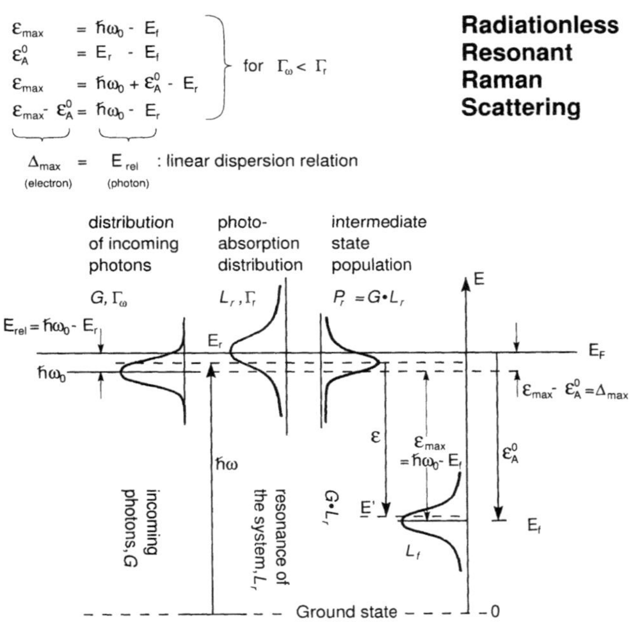

Fig. 3.25. Schematic diagram showing the energies involved in radiationless resonant Raman scattering, also called resonant Raman Auger scattering [3.134]

$$\varepsilon_{\mathrm{A}}^{0} = E_{\mathrm{r}} - E_{\mathrm{f}} \tag{3.22}$$

which lead to:

$$\varepsilon_{\max} = \hbar\omega_0 + \varepsilon_{\mathrm{A}}^{0} - E_{\mathrm{r}} \tag{3.23}$$

$$\varepsilon_{\max} - \varepsilon_{\mathrm{A}}^{0} = \hbar\omega_0 - E_{\mathrm{r}} \tag{3.24}$$

or

$$\Delta_{\max} = E_{\mathrm{rel}} \quad \text{with} \quad \varepsilon_{\max} - \varepsilon_{\mathrm{A}}^{0} = \Delta_{\max},\ \hbar\omega_0 - E_{\mathrm{f}} = E_{\mathrm{rel}}\ , \tag{3.25}$$

where $\Delta_{\max} = E_{\mathrm{rel}}$ is often called the linear dispersion relation characterizing the RRAS process. It means that the Auger kinetic energy, relative to the binding energy of the core level into which the electron decay takes place, linearly follows the photon energy relative to that at the resonance. This is exactly what is observed for the 6 eV satellite in Ni for photon energies below the resonance maximum. Starting at the absorption maximum, the Auger kinetic energy stays constant with increasing photon energy, as it should in an ordinary Auger process. In the literature the regime where the Auger kinetic energy tracks the photon energy is frequently called the photoemission regime (for obvious reasons, because in the photoemission case one has $E_{\mathrm{kin}} = \hbar\omega - \phi - |E_{\mathrm{B}}|$, see (1.2); the photoelectron kinetic energy therefore tracks the photon energy), and the regime where the kinetic energy is constant is called the Auger regime.

Resonance data near the $2p_{3/2}$ edge for Cr, Fe and Ni metal [3.135] are compared in Fig. 3.26. The zero of the photon energy is always the maximum of the photon absorption curve. For Cr and Fe this does not agree with the $2p_{3/2}$ binding energy (EB in the figure) while in Ni metal it does. The details of the crossover from the RRAS regime (photoemission regime) to the Auger regime are different for each metal and not understood at this point. However, the data suggest that it is possible to measure the two-hole satellite not only in Ni but also in Cr and Fe, and there is no reason to believe that it will not also be visible in other systems.

Recently a detailed investigation of resonance photoemission in Cu metal has appeared [3.200]. These authors find deviations in the constant kinetic energy dependence of the resonance signal (satellite at 14.5 eV below E_{F}, belonging to a $2p^6\,3d^8[4s^2]$ final state) which was previously observed, e. g., in Cr, Fe and Ni metal (see Fig. 3.26). They show that these deviations come from a resonance Raman effect even above the absorption threshold.

Resonance photoemission can be used with advantage to enhance certain features in the valence-band spectra. Very often, however, proper care has not been exercised in this procedure. As was seen, it is the satellite that is enhanced (in solids) and not the main line, a fact that has to be kept in mind in analyzing resonance photoemission data.

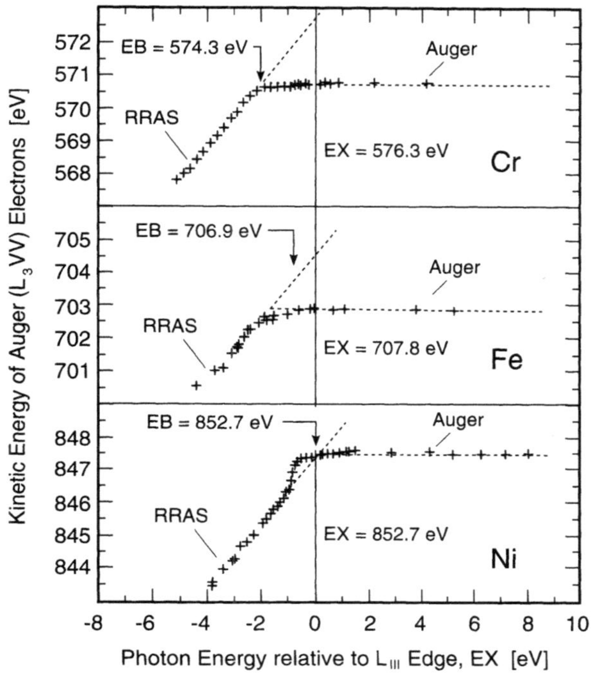

Fig. 3.26. Kinetic energy of the L_3VV electrons as a function of relative photon energy (measured relative to the photo-absorption maximum) at the L_3 edge for Ni, Fe and Cr metal. The behavior is slightly different for each of the three materials [3.135], but the reason is currently not understood

3.2.2 Satellites in Other Metals

Resonance photoemission has also been used to identify satellites or two-hole final states in other transition metals. The best known case is that of cerium metal and cerium compounds. We show in Fig. 3.27 simplified diagrams which compare the valence-band photoemission of Ni metal and Ce metal [3.136]. A valence-band XPS spectrum of Ce metal is given in Fig. 3.28 [3.137].

The X-ray photoemission spectrum of cerium metal [3.137] consists of a broad band which can be attributed (by comparison to the XPS spectrum of lanthanum, Fig. 2.26) to the $5d6s^2$ conduction bands. Sitting on top of this are two sharp signals, one at a binding energy of about 2 eV and the other very close to the Fermi energy (a high-resolution spectrum of the peak near E_F was given in Fig. 2.28, where even the spin–orbit splitting was resolved). The most obvious explanation for these two narrow signals is one which follows closely the case of nickel shown in Fig. 3.17. This means that the signal at about 2 eV binding energy comes from a final state where the screening occurs in the $5d6s^2$ conduction band, whereas the signal at the Fermi energy comes from a final state where the screening is in the f band. This implies that there is delocalization of the f electrons [3.138] which is not so easy to understand, since f electrons are assumed to be similarly localized to core electrons. One has to note, however, that at the beginning of the rare earth series, the f electrons have a radial distribution which is more extended than for the heavy rare earth elements and that, for cerium, a small itinerant character of the f electrons is not unreasonable. The simple picture just given is only meant to demonstrate the essence of the interpretation. Very elaborate calculations have given a quantitative analysis of the measured spectra [3.139, 3.140] (Sect. 3.3), where, however, the exact interpretation of the two "f-related peaks" is also slightly different from the simple picture just presented.

There exist two modifications of metallic cerium compounds. The so-called γ-form is paramagnetic with a Curie–Weiss law. This form is assumed to contain the cerium ions in a $4f^1 5d6s^2$ configuration, and therefore cerium in this modification is often called trivalent cerium. The second modification is α-cerium which is non-magnetic. Therefore it was assumed for a long time that cerium in this form has the valence configuration $4f^0 5d^2 6s^2$, meaning that the magnetic 4f electron has been transformed into a delocalized 5d electron in the $\gamma \rightarrow \alpha$ transition. It was, however, suggested [3.138] that this transformation $4f \rightarrow 5d$ does not take place, and that rather the $\gamma \rightarrow \alpha$ transition is one in which predominantly the 4f electron makes a transition

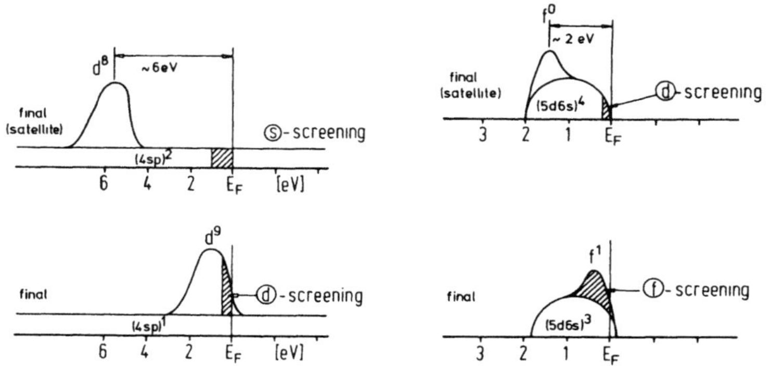

Fig. 3.27. Density of states representation of the mechanism that produces the valence-band satellites in PE from metallic Ni and Ce. In Ni the two screening channels are via sp or d electrons and in Ce they are via $5d6s^2$ or 4f electrons. In Ce the $(5d6s)^4$ screened final state is a single 4f hole [3.136]

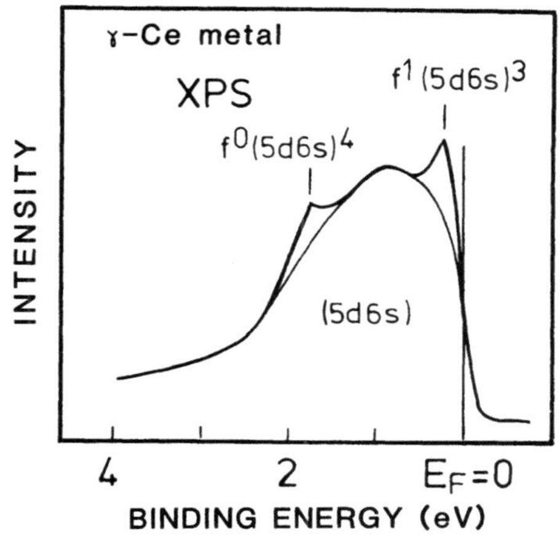

Fig. 3.28. XPS valence-band spectrum of Ce metal [3.137] at 300 K. The $4f^1$ and the $4f^0$ final state are seen to ride on the emission from the $(5d6s)^3$ valence band. The data in Fig. 2.28 comprise only the region of the $4f^1$ final state, within less than 1 eV below the Fermi energy

from a localized into a more delocalized state with little 4f → 5d transfer. This would mean that the $\gamma \to \alpha$ transition is accompanied by a valence transition of the type $4f^15d6s^2 \to 4f^{1-x}5d^{1+x}6s^2$ ($x < 0.5$), where one has in the α-phase a fractional occupation of the 4f level. Photoemission and inverse photoemission spectroscopy have played an important role in showing that this latter picture is correct. Although it is now established that α-type cerium compounds have a sizeable 4f occupation, the valence state of cerium in them is still very often called tetravalent.

To understand the features that prove the present picture of the electronic structure of α- and γ-type cerium compounds we begin the discussion with a typical trivalent Ce system. In γ-cerium, which is the room temperature modification of cerium metal, it can be assumed (e.g., on the basis of magnetic measurements) that the 4f state is occupied by roughly one electron. The $3d_{5/2}$ core-level spectrum of γ-cerium [3.141] shows two lines of unequal intensity, as can be seen in Fig. 3.29. The same interpretation for these two core lines is used here as in the case of nickel metal (Fig. 3.16). The line with the larger binding energy (884 eV) is attributed to a configuration where an additional 4f state is pulled below the Fermi energy but is not occupied and the screening is instead produced in the $5d6s^2$ band ($3d^94f^15d6s^2$ [5d] final state). The second much weaker line, which has smaller binding energy (878 eV), corresponds to the ground state. Here the additional 4f state pulled below the Fermi energy by the Coulomb interaction with the core hole has been filled and leads to a $3d^94f^15d6s^2$ [4f] final state configuration (the screening electrons are indicated in square brackets).

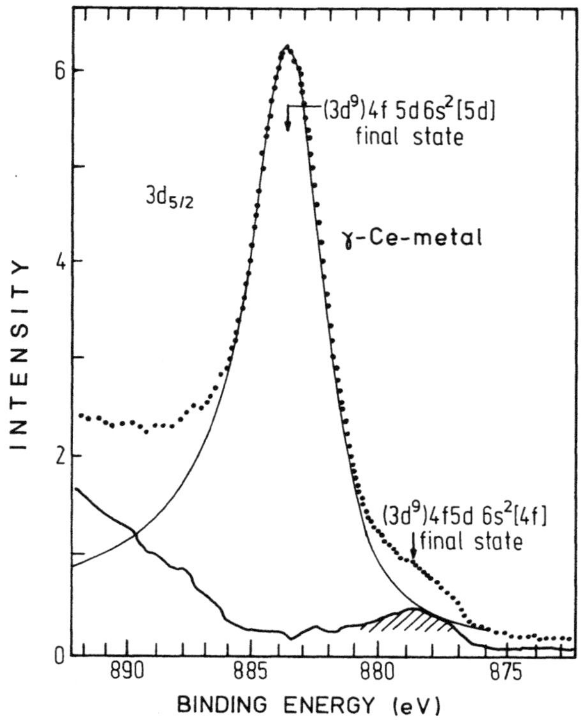

Fig. 3.29. XPS $3d_{5/2}$ line in Ce metal [3.141]. As in the case of Ni metal (Fig. 3.16), two final states are seen, one produced by $(5d6s)^4$-screening and the other produced by $4f(5d6s)^3$-screening. Unfortunately, the signature of satellite and main line given in the literature is reversed as compared to that used for Ni, since generally the $(5d6s)^4$-screened state is called the main line and the $4f(5d6s)^3$-screened state the satellite

The situation is more complicated for α-type compounds where, from thermodynamic and magnetic data, a fractional occupation of the 4f level can already be postulated such as in $CePd_3$. In this compound one can assume a wave function $\sqrt{0.9}|4f^1\rangle + \sqrt{0.1}|4f^0 5d\rangle$ giving a 4f occupation of $n_f = 0.9$. The 3d core-level spectrum of $CePd_3$ is shown in Fig. 3.30 where the final state configurations have also been indicated [3.142]. For comparison, this figure also depicts a 3d spectrum of a trivalent compound (CeSe) and it is evident that the two spectra are different. Their analysis will now be presented.

While in the initial state of $CePd_3$ the configurations $|4f^1\rangle$ and $|4f^0 5d\rangle$ are degenerate, they are separated by $\delta_0 \simeq 10\,\mathrm{eV}$ in the final state (one additional nuclear charge) and the intensities of the two final state signals give approximately the strength of the initial state wave functions. These two states are to be seen in the final state under the assumption that screening takes place in the $5d6s^2$ band. If, however, the screening takes place in the f band, an additional $4f^2$ signal is produced (this implies that the initial state 4f electron and the screening 4f electron are combined) which can also be observed in the figure. Thus the core levels in this case show three final states: $4f^0 5d^2 6s^2[5d]$, $4f^1 5d6s^2[5d]$ and $4f^1 5d6s^2[4f]$. In contrast, the 3d spectrum of trivalent CeSe (with a $4f^1$ initial state) has only two final states ($3d^9 4f^1 5d6s^2[5d]$ and $3d^9 4f^1 5d6s^2[4f]$). Note that to a very good approximation the line assigned to a $3d^9 4f^0 5d^2 6s^2[5d]$ configuration is a measure of the fractional 4f occupancy in the α-type compound $CePd_3$, where if we assume that the initial state is $4f^{1-x} 5d^{1+x} 6s^2$, x determines the intensity of that line.

The conclusion from these observations is that PE, as a complicated many body process, does not lead to a simple final state. In particular, it is by no means obvious how the initial state configuration, which one is generally interested in, can be inferred from a PE signal. Thus care has to be exercised in analyzing PE spectra either in the core regime or in the valence regime.

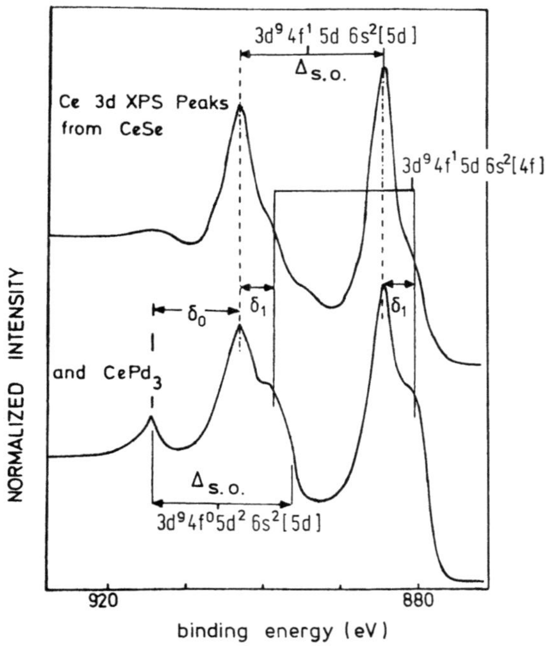

Fig. 3.30. XPS $3d_{5/2,3/2}$ lines in $Ce(4f^1)Se$ and $Ce(4f^{0.8})Pd_3$ [3.142]. For CeSe, where Ce has a $4f^1$ initial state the $4f^1(5d6s)^4$ and the $4f^2(5d6s)^3$ final states appear as in Ce metal (Fig. 3.24), with an energy separation of δ_1. In $CePd_3$ the initial Ce configuration is $4f^{0.8}(5d6s)^{3.2}$ which for instructive purposes is better written as $\cos\alpha|4f^1(5d6s)^3\rangle + \sin\alpha|4f^0(5d6s)^4\rangle$. Thus in addition to the final states observed in the CeSe spectrum, a third final state $4f^0(5d6s)^5$ can now also appear which for the $3d_{3/2}$ line stands out clearly from the other lines. The intensity of this line is a good measure of the initial state 4f occupation, with $I(4f^0) = \sin^2\alpha$ or if one writes the initial state as $4f^{n_f}$ one has $1 - n_f = I(4f^0)/I(4f^1 + 4f^2)$

If for valence levels comparison with calculated valence-band structures is intended, corrections have to be made for the many-body excitations (or put more simply, for the final state effects) [3.143] (Fig. 2.28). On the other hand, the presence of the complicated final-state structure in the PE spectra gives information on the excitation spectrum and the excitation probabilities in the solid, and is thus of considerable interest in itself and allows, using the proper theoretical tools, to infer the initial-state wave functions from the measured final-state spectra.

3.3 The Gunnarsson–Schönhammer Theory

So far we have dealt with satellites such as those observed in narrow band metallic systems (e.g., Ni and Ce) on a phenomenological basis, the basic physics having been given in Fig. 3.2. Within this framework the core (or valence) hole has essentially the following effects. It acts as an additional attractive potential for the other electrons on the photoionized atom and thus, in simple terms, "pulls down" the whole energy level scheme. This can also lower hitherto unoccupied narrow conduction-band states to below the Fermi level. The positive potential also acts as a scatterer for the conduction electrons and can thereby scatter electrons into the narrow empty state produced below the Fermi energy by the hole potential. One can also say that the hybridization coupling between the narrow (correlated) states and the wide conduction-band states leads to a filling of the narrow states which were pulled by the core potential below the Fermi energy. Thus, in a very simple picture, two final states are produced: one where the narrow state below E_F is empty (*poorly screened state, unscreened state, conduction electron screened state*), and another where the narrow state below E_F is filled (*well screened state, screened state, d- or f-screened state*, etc.) which has smaller binding energy than the first one.

This simplified picture indicates the physics of the problem but does not give quantitative details; these can only be obtained from elaborate calculations. The basic problem of the scattering of conduction electrons by a suddenly switched-on local potential has been solved by Mahan [3.144], and Nozières and DeDominicis [3.145]. These researchers, however, worked within a free-electron model and thus naturally did not find the effects caused by the electron–electron correlations. Some of these (core-hole valence-electron interactions) were added to the Mahan–Nozières–DeDominicis treatment by Kotani and Toyozawa [3.7, 3.8, 3.146] (for other work see [3.147–3.155]).

Later it was realized that in narrow-band materials the correlations between the electrons in that band are an important aspect of the problem. These correlations were incorporated as a self-energy starting from a band picture for the interpretation of the 6 eV satellite in Ni [3.108–3.110] and for an analysis of rare-earth PE spectra by starting with a single impurity model.

Important parameters in the calculations by Gunnarsson and Schönhammer [3.139,3.140] are the hybridization interaction H (coupling constant between the "localized" f electrons and the "itinerant" $5d6s^2$ electrons) and the degeneracy N_f of the f level which of course in principle is 14, but was used in this treatment as an expansion parameter [3.155–3.158]. Since N_f is large, one can work in the limit $N_f \to \infty$ which simplifies the calculations (leaving $HN_f = \text{const.}$).

Figure 3.31 shows a calculated valence-band density of states [3.140] for a model system that should correspond closely to Ce metal. One recognizes the two-peak structure also found in the experiment (Fig. 3.28) where one peak is at 2 eV and the other close to the Fermi energy. Thus the calculations can quantitatively reproduce the experimental results.

This same type of analysis has been performed for BIS data of cerium compounds: an example is given in Fig. 3.32 [3.159]. In this case one is dealing with $CeNi_2$ which is non-magnetic and was, like other so-called tetravalent Ce compounds, therefore regarded for some time to have a valence-band configuration $4f^05d^26s^2$. To test this contention a BIS experiment is very suited because the spectrum should consist of a $4f^15d^16s^2$ final state, which should show up as one peak near the Fermi energy. This peak is indeed seen in Fig. 3.32, however, there is in addition a strong signal from a $4f^26s^2$ final state. We conclude therefore, that the initial state must have a wave function of the form $\cos\alpha|4f^15d6s^2\rangle + \sin\alpha|4f^05d^26s^2\rangle$ with $\cos\alpha \simeq 0.9$ (Table 3.2, where $n_f \propto \cos^2\alpha$) as derived from the intensity ratios of the $4f^1$ and $4f^2$

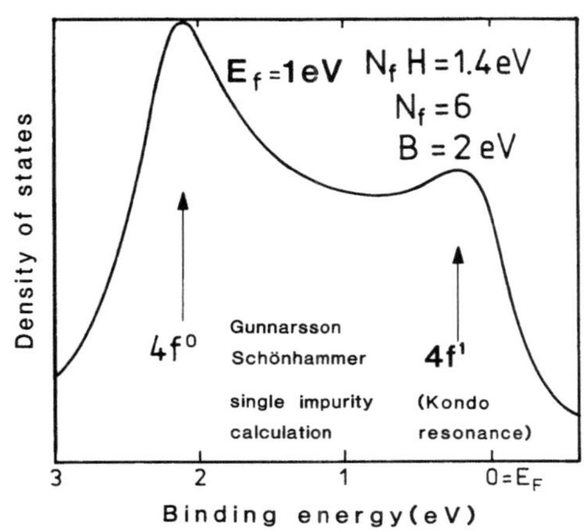

Fig. 3.31. Valence-band density of states of Ce metal calculated by Gunnarsson and Schönhammer [3.140] in the impurity Anderson model. N_f is the degeneracy of the 4f level, H the 4f conduction electron hybridization interaction (π times the square of the coupling matrix element assumed to be constant over the bandwidth B), and B the width of the conduction band. A 0.5 eV Lorentzian broadening was applied to simulate the instrumental resolution. The hybridization width is HN_f

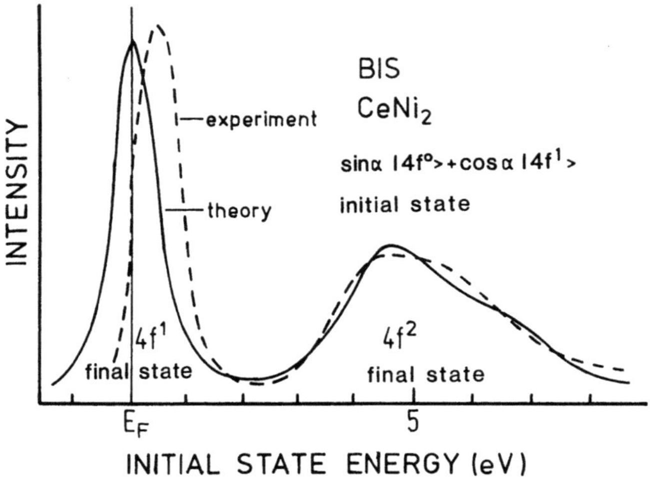

Fig. 3.32. Experimental and theoretical valence-band BIS data for $CeNi_2$ [3.159], which has an initial state 4f occupation of $n_f \simeq 0.8$. Therefore in the BIS spectrum (electron addition spectrum) the $4f^1(5d6s)^3$ and the $4f^2(5d6s)^2$ final states appear

final state peaks. The same result is obtained by the rigorous theoretical treatment which is given in Fig. 3.32 by the solid line [3.159].

Equally suited for the analysis of the valence of Ce compounds by the Gunnarsson–Schönhammer theory is core-level spectroscopy, as was indicated above for $CePd_3$. An example for the nominally tetravalent system $CeRu_2$ is given in Fig. 3.33 [3.160]. $CeRu_2$ in addition to being non-magnetic is also superconducting.

In principle, the $CeRu_2$ valence band should, like that of $CeNi_2$, be described by an electronic configuration $4f^05d^26s^2$. However, core-level spectra such as those in Fig. 3.33 contradict this contention: they show only a rela-

Table 3.2. The f level hybridization H and the f occupancy n_f as deduced from the core-level photoemission (XPS), the 3d → 4f absorption (XAS), and the bremsstrahlung isochromate spectra (BIS) and the static $T = 0$ susceptibility (χ) [3.160] in α-type Ce compounds

	H[eV] XPS	n_f XPS	n_f 3d XAS	n_f BIS	n_f χ
$CeRu_2$	0.10	0.83		$\simeq 0.7$	0.78
$CePd_3$	0.11	0.91	0.86	0.97	0.82
$CeNi_2$	0.10	0.84	0.79	0.78	0.74
$CeNi_5$	0.09	0.79	0.81	$\simeq 0.8$	

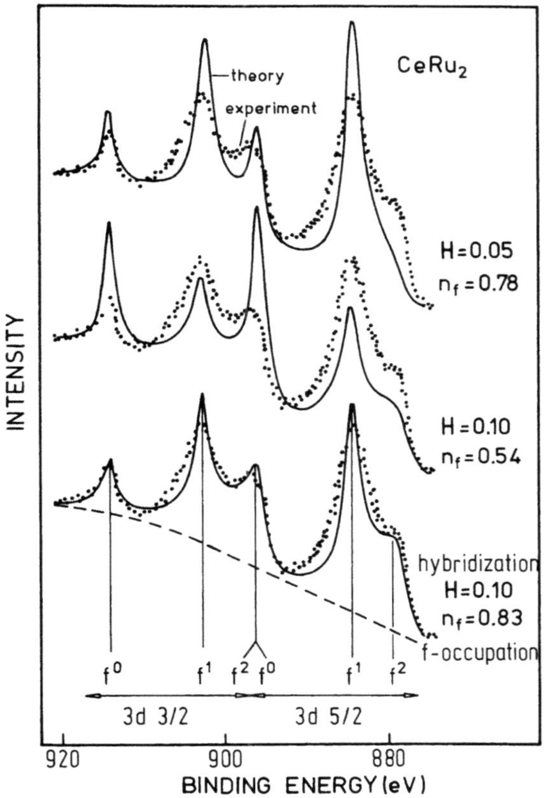

Fig. 3.33. Measured and calculated 3d(3/2, 5/2 spectrum of $Ce(4f^{0.8})\ Ru_2$ [3.160] with n_f being the 4f occupation ($4f^{n_f}$) and H the 4f conduction electron hybridization strength. The fit with $H = 0.10$ eV and $n_f = 0.83$ reproduces the data fairly well. The dashed line is an assumed background

tively weak signal at the anticipated position of the $4f^0 5d^2 6s^2$[5d] final state, while strong signals are observed at the $4f^1 5d 6s^2$[5d] and $4f^1 5d 6s^2$[4f] final state positions. A convenient and often used way of characterizing these different final states is to include the screening electron in the actual configuration and then characterize the final state by the number of 4f electrons contained in it: f^0, f^1 and f^2 final state. Note in particular, that the f^0 final state is the one with the highest energy. This fact makes it unlikely that the PE process alone causes the f^1 and f^2 states to become so strongly populated by shake down processes as to make the observed spectra understandable.

It is much more likely that the f^0 strength in the final state ($\simeq 20\,\%$ of the total intensity) essentially reflects a contribution of the $4f^0$-wave function to the initial state of similar strength. Thus $CeRu_2$ has to be described in the valence band not by a configuration of $4f^0 5d^2 6s^2$, but rather by $4f^{0.8} 5d^{1.2} 6s^2$.

The calculations of Gunnarsson and Schönhammer [3.160] presented by the solid lines in Fig. 3.33 come to just that conclusion. In Fig. 3.33 n_f means the occupation of the f level. The best fit to the experimental spectrum yields a hybridization interaction $H = 0.1\,\mathrm{eV}$ and $n_\mathrm{f} = 0.83$.

Table 3.2 summarizes some data relevant to so-called tetravalent Ce compounds which have been analyzed by the same theoretical approach (XAS stands for X-ray absorption spectroscopy, χ is the magnetic susceptibility). One can see that hybridization widths of $\simeq 0.10\,\mathrm{eV}$ are obtained and that the occupation of the f level is $\simeq 0.8$ of an electron [3.160].

Finally we comment briefly on the satellites observed in Ca, Sr and Ba metal [3.90, 3.91, 3.161, 3.162] which are quite strong especially for Ca. Their obvious interpretation would be in terms of extrinsic (electron energy loss) and intrinsic (photohole-conduction electron coupling) plasmons; see Chap. 4 for details. However, if their total intensity has to be accounted for in this way, very high intrinsic plasmon excitation rates are required [3.161]. These, however, are not very likely in the light of the results to be presented for Na, Mg and Al in Chap. 4.

An interpretation of the strong satellites in Ca in terms of different screening channels is therefore a distinct possibility (similar considerations hold for Sr and Ba). In the $(Z+1)$-approximation a core-ionized Ca ion can be replaced by a Sc ion. While Ca has the valence state atomic configuration $4s^2$ that of Sc is $3d^1 4s^2$ and in the metallic state the 3d electron count of Sc is probably even higher [3.163, 3.164].

Thus one may assume that in the final state of core-ionized Ca an empty d state can be shifted below the Fermi energy and become populated. This raises the possibility that the strong satellite in Ca may be interpreted in the context of the Kotani–Toyozawa theory [3.8] (Figs. 3.2, 3.17) where the satellite is a (poorly) sp-screened state and the main line corresponds to a state screened by a 3d electron. Additional support for this interpretation (Sect. 3.1.2) comes from a multiplet analysis of the Ca 3p line [3.90, 3.91]. If this analysis is correct, it would yield additional support for the interpretation of the spectra from CaF_2 as given by Veal and Paulikas [3.52].

3.4 Photoemission Signals and Narrow Bands in Metals

The satellites observed in Ni and Ce metallic compounds are a direct consequence of the fact that in these materials the d- and f electrons move in relatively narrow energy bands. This means that on the time scale of the experiment, the system cannot totally relax and therefore a complicated excitation spectrum is observed in the PES experiment. While in a material like Ni, with a 3d bandwidth of $\simeq 3\,\mathrm{eV}$, the experimental EDC near E_F will roughly reflect the renormalized one-electron density of states in a polycrystalline sample, in Ce metal with a 4f bandwidth of the order of $0.1\,\mathrm{eV}$ or less, the meaning of the EDC near E_F is less obvious.

Of special interest in this respect are compounds which have a particularly high density of states at the Fermi energy and thus exhibit some very interesting electronic properties. Such systems have coefficients of specific heat of $\gamma \simeq 10^3 \mathrm{mJ} \cdot (\mathrm{K}^2 \cdot \mathrm{mol})^{-1}$, very high as compared to $1\ \mathrm{mJ} \cdot (\mathrm{K}^2 \cdot \mathrm{mol})^{-1}$ in normal metals like Cu. These systems are called heavy fermion systems [3.165–3.173]; if one describes them in the normal free-electron formalism, the large γ means a very large effective mass [3.172, 3.173]. A typical heavy fermion system is $CeAl_3$ [3.165], with $\gamma = 1600\ \mathrm{mJ} \cdot (\mathrm{K}^2 \cdot \mathrm{mol})^{-1}$; its XPS valence-band PES is shown in Fig. 3.34. The structure about 0.8 eV wide near E_F is the "relaxed" 4f peak [coherent part of the spectral function (1.61)] and therefore reflects the 4f excitation spectrum as close to the ground state as it can be measured in a PES experiment with low resolution.

This spectrum is compared in Fig. 3.34 to the results of a point-contact measurement on the same material [3.166, 3.167].[5,6] It is believed that these experiments measure approximately the renormalized single-particle density of states around the Fermi energy,[7] which in this case has a width of only a few meV. A comparison of the two types of measurements shows convincingly that for such materials one cannot measure the proper excitation function (1.61) with standard (low resolution) PES techniques. However, from the core levels one can obtain indirect information on the occupation of the "f level" in the heavy fermion systems employing analysis along the lines given in Sect. 3.3.

On the other hand, advanced (high resolution) photoemission experiments can compete in resolution with the point-contact measurements and contain probably more information. The combination of low temperatures with high resolution UPS equipment has yielded quite detailed electronic structure information in highly correlated systems like cerium compounds. In order to see the progress made in PES over the years we show in Fig. 3.35 (panel b and c) high resolution UPS data on $CeAl_2$ (a compound similar to $CeAl_3$, which also displays heavy fermion behavior; high resolution data on $CeAl_3$ are unfortunately not available) to compare these with the spectra in Fig. 3.34 [3.170]. As in the case of Ce metal (Fig. 2.28) the $\hbar\omega = 40.8\,\mathrm{eV}$ spectra enhance the 4f emission relative to the 21.2 eV spectrum. The $4f^0$ final state (at around 2.5 eV binding energy) and the $4f^1$ final state near the Fermi energy are visible (compare also Fig. 3.28). The spin–orbit splitting of the $4f^1$ final state also

[5] The point-contact spectroscopy is a particular mode of the tunnel spectroscopy. The tunnel junction is produced by pressing a metal tip onto the material which is to be investigated. The interpretation of the tunnel data in Fig. 3.34 in terms of the electronic structure is not obvious.

[6] A tunneling spectroscopy experiment on Ce on an Ag(111) surface which is similar to a Ce compound has yielded a width for the structure near the Fermi energy of $\sim 40\,\mathrm{meV}$, a value similar to the results presented in Fig. 3.34 [3.168]; see Fig. 3.39.

[7] For a proper description of the electronic states near the Fermi energy in materials with large electron–electron correlation, see, e.g., [3.169].

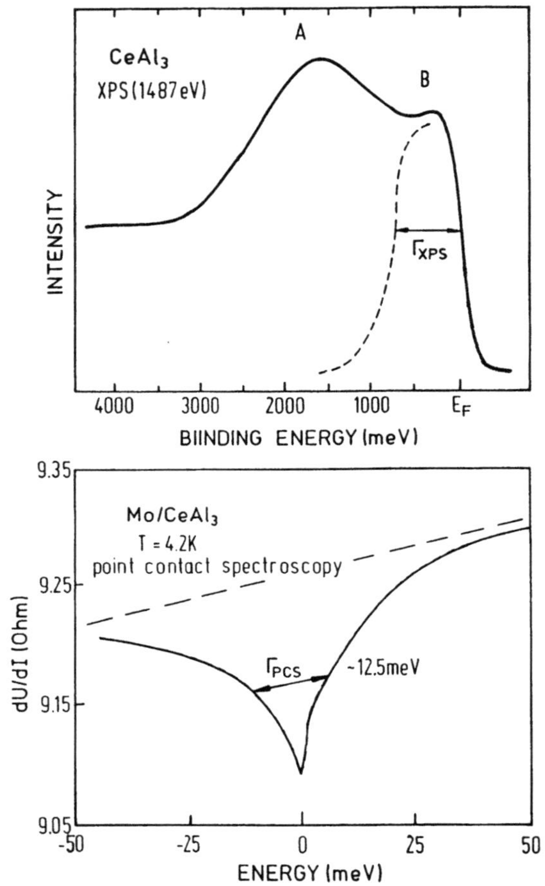

Fig. 3.34. Top: XPS valence-band spectrum of $CeAl_3$ which shows a 4f "bandwidth" $\Gamma_{XPS} = 0.8\,eV$ [3.165]. Bottom: Point-contact spectrum of $CeAl_3$ which indicates a 4f "bandwidth" of $\simeq 12.5\,meV$ [3.166, 3.167]. A comparison with the high resolution PES data of Ce metal (Fig. 2.28) and with the data in Fig. 3.30 indicates that the point-contact technique may measure roughly the "true" 4f bandwidth

appears. The resolution ($\simeq 20\,meV$ according to [3.170]) is not yet sufficient to resolve the crystal-field splitting. The calculated spectrum (panel d) was obtained by deriving the 4f contribution within the formalism of Gunnarsson and Schönhammer (Sect. 3.3). The broad valence-band background was taken from the 21.2 eV spectra. These spectra have only a small 4f contribution because of the low 4f photo-cross-section at this photon energy [3.174–3.176].

In panel (a) of Fig. 3.35 a low resolution spectrum of $CeAl_2$ taken with synchrotron radiation ($\hbar\omega = 122\,eV$) is given. This spectrum obtained by the technique of resonance photoemission contains predominantly the 4f contribution. A comparison with the spectrum in panel (c) reveals the improvement in resolution. In particular, the spin–orbit splitting is not visible in the spectrum of panel (a).

We shall now touch upon a subject which was controversial for some time but is settled now [3.177–3.199] for Ce systems with $T_K < 100\,K$. In Figs. 3.28, 3.31, 3.35 the peak around $E_B = 2\,eV$ in the valence-band spectra of Ce containing systems is ascribed to a $4f^0$ state screened by valence electrons (see also Fig. 3.27). The interpretation of the 4f intensity observed in the energy regime $0 \leq E_B \leq 500\,meV$ is less clear. In the framework of the simple Kotani–Toyozawa model, the peak near E_F is a hole screened by a 4f electron, a situation close to that of the ground state [3.184–3.186]. In the calculations of Gunnarsson and Schönhammer [3.139, 3.140] (Fig. 3.31) a single impurity model is used, where the coupling strength between this impurity and the conduction band (H) is explicitly taken into account. At $T = 0$ this model leads to the so-called Kondo resonance near the Fermi energy. This is a many-body state with a binding energy $k_B T_K$ (T_K: Kondo temperature), which piles up density of states near E_F. The term Kondo

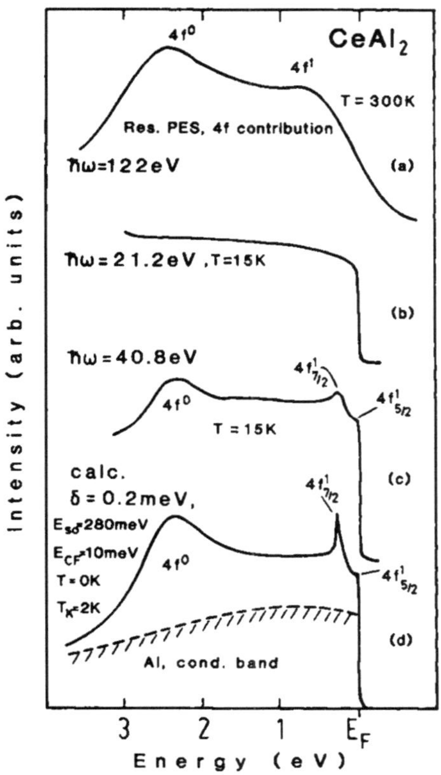

Fig. 3.35. UPS spectra of $CeAl_2$ [3.170]. (a) Synchrotron-radiation spectrum (122 eV), giving an enhanced 4f contribution (resonance enhancement): (b) He I (21.2 eV) spectrum which gives a representation of the valence band, (c) He II (40.8 eV) spectrum which enhances the 4f contribution; the 4f peaks at $\simeq$ 2.5 eV binding energy and near the Fermi energy are visible, (d) simulated spectrum; the 4f contribution was calculated in the Gunnarsson–Schönhammer theory and the broad valence band (*dashed line*) was taken from the spectrum in (b)

resonance was sometimes used for all the intensity observed near the Fermi energy in valence-band PES of cerium compounds.

This terminology has, however, a problem. The Kondo resonance is weakened for $T \gg T_K$. This condition holds, e.g., for $T = 300\,K$ in γ-Ce ($T_K = 70\,K$) or $T = 300\,K$ in $CeAl_2$ ($T_K = 2\,K$). In the simplest Gunnarsson–Schönhammer calculation an extension from $T = 0\,K$ to $T = 300\,K$ would therefore weaken the Kondo resonance and thereby also the intensity near E_F. This, however, is not quite in agreement with the experimental data [Figs. 3.28, 3.35 panel (a)] which show intensity near E_F in the $T = 300\,K$ data. This problem can be solved by realizing that the impurity model in its simplest form neglects the intra 4f (single ion) excitations and it leaves out the aspect that, in a solid, the 4f electron has to be described in principle in a band picture (although this band is narrow because of the high degree of localization of the 4f electron; see [3.169] for details on this point). In a band "single-particle" excitations exist which can contribute to the peak near E_F in Ce compounds. Patthey et al. [3.174–3.176] have introduced the single-ion (intra 4f) excitations (which can be responsible for much of the intensity observed near E_F for $T \gg T_K$) into the impurity model by taking into account the spin–orbit interaction (E_{SO}, this was already considered by Gunnarsson and Schönhammer [3.140]) and the crystal-field interaction (E_{CF}). This procedure yields excitations which are easy to calculate (but cannot account for 4f band excitations). A model calculation [3.174] in the single-impurity model with spin–orbit interaction (E_{SO}) and crystal-field interaction (E_{CF}) for a system with a Kondo temperature of $T_K = 35\,K$ ($CeSi_2$) is depicted in Fig. 3.36. The energy scale of Fig. 3.36 gives only the region around E_F and leaves out the 2 eV feature. At $T = 15\,K$, which is below T_K, the top spectrum has three features: a $4f^1_{7/2}$ peak at 280 meV, a crystal-field peak at 35 meV and the Kondo peak at E_F. This spectrum indicates that the Kondo peak contributes only little intensity to the total signal for $E_B \leq 500\,meV$ for a small-T_K material. If an instrumental broadening (E_{inst}) of 18 meV is introduced, the crystal-field peak and the Kondo resonance are no longer distinguishable. By increasing the temperature to $T \gg T_K$ the Kondo resonance is diminished, yet the total intensity for $400\,meV \geq E_B \geq E_F$ stays roughly constant because the bulk of this intensity is given by the crystal-field and spin–orbit excitations. Note that these can be viewed as "single-particle" excitations within the 4f manifold; they are contained in the simple picture outlined in Fig. 3.27 [3.184–3.186]. This suggests that the simple screening picture given in Fig. 3.27 captures the basic high temperature physics. However, for actual quantitative calculations one has to resort to a manageable model and here the impurity model has had a considerable success [3.140].

We add two points. The screening picture contains the hybridization between the f electrons and the valence band [3.82]. This gives an intensity near E_F proportional to $(T_m/E_f)^2$, E_f being the (unhybridized) ionization energy of the f electron and T_m the hybridization matrix element between

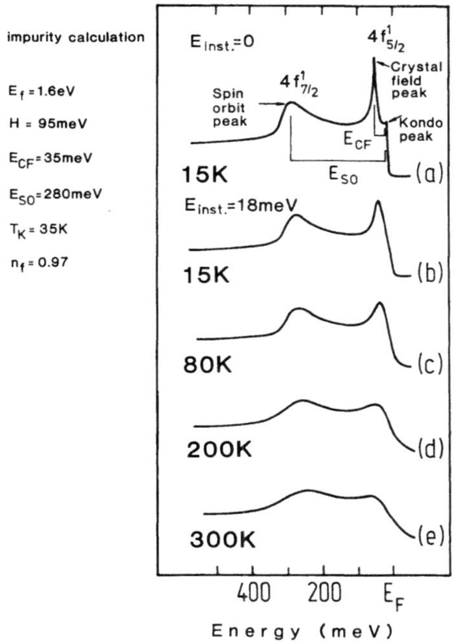

Fig. 3.36. Model calculation for a cerium valence-band spectrum in the energy region 400 meV $\geq E_{\mathrm{B}} \geq E_{\mathrm{F}}$ in the Anderson impurity model [3.174]. Model parameters are $E_{\mathrm{f}} = 1.6\,\mathrm{eV}$, $H = 95\,\mathrm{meV}$, $E_{\mathrm{CF}} = 35\,\mathrm{meV}$, $E_{\mathrm{so}} = 280\,\mathrm{meV}$, $T_{\mathrm{K}} = 35\,\mathrm{K}$, $n_{\mathrm{f}} = 0.97$ (a) $T = 15\,\mathrm{K}$, $E_{\mathrm{inst}} = 0$(no instrumental broadening). (b–e): $E_{\mathrm{inst}} = 18\,\mathrm{meV}$ and $T = 15\,\mathrm{K}$, 80 K, 200 K and 300 K, respectively

the localized f electrons and the valence band according to (3.10). Then, we mention that a temperature-independent intensity near E_{F} is obtained in the Gunnarsson–Schönhammer calculation from the mixing of the f^2 state into the initial state [3.140], in addition to the contributions produced by the crystal-field and spin–orbit excitations.

Another confirmation of the analysis of the electron-spectroscopy data on cerium compounds in terms of the *single-impurity model* has been given by Malterre et al. [3.187]. The BIS (inverse PES using a fixed light frequency, Sect. 1.3) experiment on $\mathrm{CePd_3}$ ($T_{\mathrm{K}} = 240\,\mathrm{K}$) shows a decrease of the $\mathrm{f}^0 \to \mathrm{f}^1$ intensity with increasing temperature. The schematic energy-level diagram in Fig. 3.37 can be used to explain this finding. At $T = 0$ only the ground state of $\mathrm{CePd_3}$ is populated which is a mixture of an f^1 and an f^0 state. The f^0 weight in the ground state gives the $\mathrm{f}^0 \to \mathrm{f}^1$ intensity in the BIS spectrum. At a finite temperature the f^1 excited states are populated (meaning a destruction of the singlet ground state). From the f^1 excited state only transitions to the f^2 final state are possible by BIS. Thus, the destruction of the Kondo state (population of f^1 states) shifts the BIS spectral weight from the $\mathrm{f}^0 \to \mathrm{f}^1$ to the $\mathrm{f}^1 \to \mathrm{f}^2$ transition, as observed.

Liu et al. have demonstrated that within the Gunnarsson–Schönhammer approach for Ce metal a consistent description of the PES data ("high energy scale") and the thermodynamic data ("low energy scale") can be obtained [3.188].

Now we want to approach the question where the differences and similarities are between the simple screening description of the photoemission final state (Fig. 3.27) and that in terms of the spectral function (Fig. 3.31). Cerium seems well suited for such a comparison because with one 4f electron the observed spectra contain only few features. In addition, there exists a simplified version [3.189] of the Gunnarsson–Schönhammer calculation, which makes the physics of this sophisticated theory transparent.The main difference between the description of the photoemission final state in the Green's function formulation and the screening picture is in the number of electrons. The Green's function calculates the spectrum of the hole state while the screening picture assumes that in a metallic environment the hole state is screened by a charge that moves in, to provide a neutral final state. In a one-electron band (no correlations) these descriptions lead, however, to the same result, because the final state in a one-electron system does not depend on the number of additional electrons present. For a one-electron band the screening picture and the hole-state spectrum calculated by the Green's function are thus identical.

In the Gunnarsson–Schönhammer theory [3.148–3.154] each 4f state is described by a local impurity with a magnetic moment (which makes the inclusion of correlations possible), however, the coupling to the conduction band by hybridization is to a reservoir of uncorrelated electrons [3.190]. This has interesting consequences for the photoemission process, as will be shown in an example. The ground state of Ce^{3+} in Ce metal can, in a shorthand notation, be written as a hybridized $4f^1(5d6s)^3$ configuration. Photoemission of

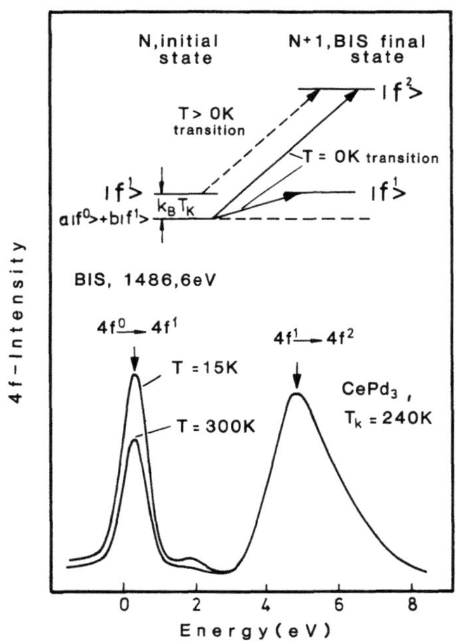

Fig. 3.37. Inverse PES data (BIS mode) of the Kondo system $CePd_3$ ($T_K = 240$ K) at a temperature above and below the Kondo temperature. The insert shows the energy-level diagram explaining the shift of spectral weight from the $f^0 \rightarrow f^1$ transition to the $f^1 \rightarrow f^2$ transition upon increasing the temperature. In the initial state a triplet f^1 state is at $k_B T_K$ above the singlet ground state ($a|f^0\rangle + b|f^1\rangle$). The *solid arrows* give the transitions possible at $T = 0$ K, while the dashed arrow shows the additional transition produced by the occupation of the $|f^1\rangle$ excited state

this state in the Gunnarsson–Schönhammer theory leads to two final states, namely $4f^0$ $(5d6s)^3$ (ionization peak, $4f^0$ peak) and $4f^1(5d6s)^2$ ($4f^1$ peak). Now the $4f^1$ $(5d6s)^2$ state equals the ground state $4f^1(5d6s)^3$ if the $(5d6s)^3$ band is considered a one-electron band. In this sense, also the Gunnarsson–Schönhammer calculation interprets the PES feature observed near the Fermi energy as a state, close to the (initial) ground state. One may even say that in this sense the Gunnarsson–Schönhammer calculation contains a one-electron aspect, where the (correlated) 4f electrons couple to the one-electron band. We note that the screening picture (which leaves the electron numbers conserved in the PE process) leads to $4f^0(5d6s)^4$ and $4f^1(5d6s)^3$ final states, which, because of the one-electron nature of the (5d6s) band, are equal to the final state calculated from the Green's function.

Naturally, the screening approach does not yield the Kondo peak for $T = 0\,\mathrm{K}$. However, for all low T_K materials ($T_K \ll 300\,\mathrm{K}$) this is a small contribution to the observed spectrum near E_F especially if measured at room temperature. The largest contributions come from intra $4f^1$ excitations (spin–orbit, crystal field) that are contained in all models.

We suggest that all the approaches to explain the PE spectra of the Ce system are not too different, namely the screening picture [3.184–3.186], the molecular-orbital-type calculations [3.66, 3.80–3.83, 3.191] (which are most intimately connected with the screening picture) and the calculations based on the Anderson impurity model [3.148–3.154, 3.190, 3.192, 3.193] which yield for low T_K materials the best agreement with experiment.

However, because of the discussion [3.180–3.183, 3.194–3.196] on the interpretation of spectra from Ce- and Yb-based systems this point has to be briefly elaborated upon. Lawrence and co-workers [3.194–3.196] found from their valence-band PE experiments that the single-impurity model predictions by Gunnarsson and Schönhammer [3.148–3.154] fail to account for the experimental spectra because, among other shortcomings, the calculations predict a too wide $4f^0$ peak and are unable to describe the observed dispersion effects in the near E_F structures. On the other hand the Gunnarsson–Schönhammer calculations describe correctly the core-level spectra of Ce-based systems (Fig. 3.33) and the BIS data (Fig. 3.32). They also give a reasonable representation of the valence-band data as found by other authors [3.170, 3.181, 3.199] (Figs. 3.35, 3.38, 3.40, 3.42).

A point of concern is the analysis of the resonance photoemission data that are used frequently to test the Gunnarsson–Schönhammer approach. It is not obvious that this experimental technique is understood with the accuracy with which it is now employed.

The Gunnarsson–Schönhammer assumption of the 4f state as a single impurity state is simple and band aspects have to be introduced (particularly in large-T_K materials) into the model in order to come to a more accurate theoretical description of the PE spectra of Ce and Yb systems. This is in line with the reasoning introduced in this chapter and contained in Fig. 3.27.

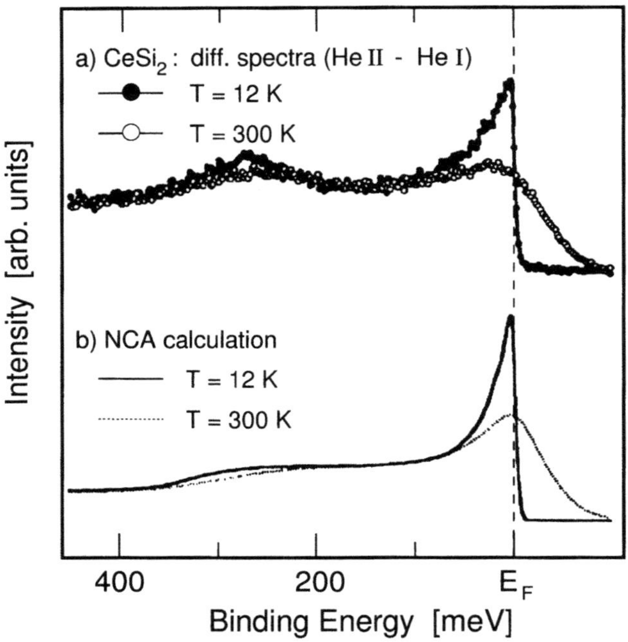

Fig. 3.38. Photoemission difference spectra (He II − He I), in order to bring out more clearly the 4f contribution, for $CeSi_2$ at $T = 12\,K$ and $T = 300\,K$. The Kondo temperature is about 35 K. The lower trace shows the result of an NCA (non-crossing approximation) calculation demonstrating the ability of the Kondo model to explain the data [3.181]

Following this argument one may want to analyze the PE data of Ce and Yb systems in the way employed for the spectra of Ni [3.108–3.116]. In this approach one starts with a single-electron band structure and augments it with a large electron–electron correlation. This leads to renormalized one-electron bands [coherent part of the spectral function (1.61)] and a satellite [incoherent part of the spectral function (1.61)]. Whether such a procedure will ultimately lead to a better agreement between theory and experiment, in particular for more band-like systems like, e.g., $CeRh_3$ [3.197], remains, however, to be seen.

It has to be kept in mind that, for a truly quantitative analysis of photoemission data in terms of models in this field, the spectra have to be separated into their bulk and surface contributions, because they have a different electronic structure [3.198].

Finally, to give the reader the chance to form his own opinion on the quality of comparison of theory (single-impurity Anderson model) and experiment, an example for the spectral function near the Fermi energy in a low T_K (35 K) material ($CeSi_2$) as is presented in Fig. 3.38 [3.181–3.183]. In that figure NCA stands for non-crossing approximation, a formalism that al-

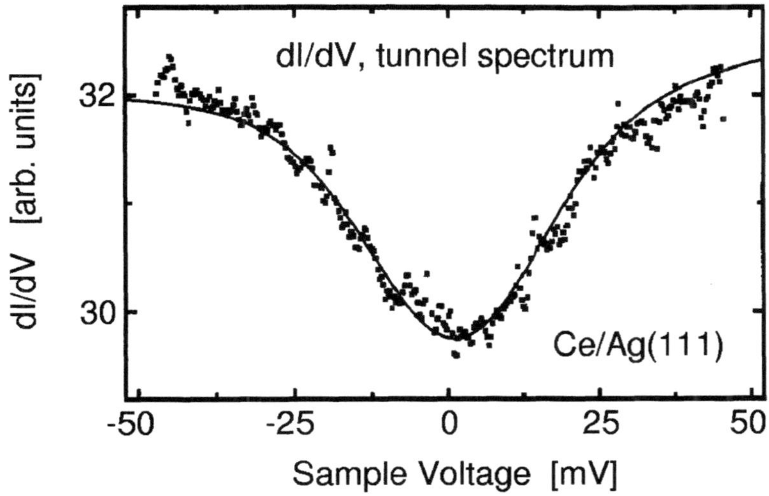

Fig. 3.39. Tunneling current for a Ce impurity on a Ag(111) surface indicating the Kondo resonance [3.168], but also indicating its local character

lows one to extend the $T = 0$ single-impurity Anderson model calculations to finite temperatures. Other examples will be presented later.

The high resolution investigation of $CeSi_2$ shown in Fig. 3.38 (the two peaks are the spin–orbit split Kondo resonance; the incoherent peak or the f^0 resonance is at about 2 eV binding energy and is not visible in these data) shows a distinct temperature dependence and this dependence is very well reproduced by the NCA calculation. The assumption of a density of states model for the description of these data yields a considerably poorer agreement with experiment. This we think is a proof of the validity of the Gunnarsson–Schönhammer theory for the interpretation of these data, within the reservations mentioned.

Recently, an interesting extension of the investigation of the Kondo resonance has been made by applying tunnel spectroscopy. Figure 3.39 shows low temperature tunnel spectra for a Ce impurity on Ag(111). These nicely reproduce the narrow Kondo resonance which is at the heart of the Gunnarsson–Schönhammer theory [3.168].

In the tunneling experiment, however, the Kondo resonance appears to be symmetric with respect to the Fermi energy. This is not quite in agreement with theoretical expectations, which place the maximum of this resonance at $\sim T_K$ above the Fermi energy for cerium systems. This observation can however be explained by a detailed analysis of the tunnel process [3.168]. On the other hand, a high resolution PE spectrum of the heavy Fermion superconductor $CeCu_2Si_2$ ($T_K \sim 10\,K$) measured at 10 K shown in Fig. 3.40 together with the Fermi functions for $T = 10\,K$ can be analyzed such as to yield the Kondo resonance above E_F.

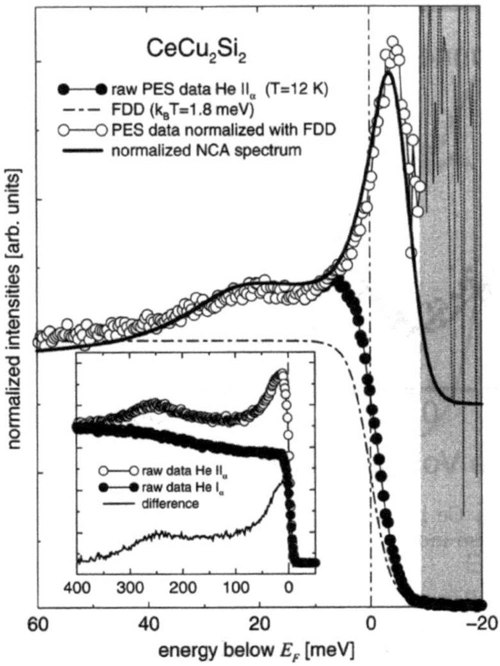

Fig. 3.40. High resolution photoemission results for $CeCu_2Si_2$ at $T = 12\,K$. Shown are the data points (the structures in the data at $E - E_F \approx -30\,meV$ probably reflect crystal-field states; the spin–orbit sideband is seen in the extended energy window shown in the inset), the Fermi function at 12 K (*dash-dotted line*) and the data divided by the respective Fermi–Dirac function, yielding the spectral function, which in the present case is the Kondo resonance; the full line is the result of a calculation in the non-crossing-approximation (NCA)

This kind of analysis will now be presented briefly. It is evident from (1.62) that the photocurrent is given by the spectral function multiplied (apart from momentum- and energy-conservation terms, and transition probabilities) by the Fermi-Dirac function. Therefore, if one divides the measured photocurrent in a narrow energy interval around the Fermi energy by the Fermi-Dirac function, one obtains, in principle, the spectral function, assuming that the other factors are constant over this energy range.

Data to this effect for the heavy fermion system $CeCu_2Si_2$ are shown in Fig. 3.40. Dividing the measured photocurrent by the Fermi function yields a resonance above the Fermi energy in agreement with the expectations.

The question of whether the Gunnarsson–Schönhammer theory is valid for the interpretation of photoemission spectra of Ce compounds, arose probably mostly from the fact that, in the high resolution photoemission spectra, one was only observing the tail of the Kondo resonance. This made a quantitative comparison with theory not easy. On the other hand, inverse photoemission

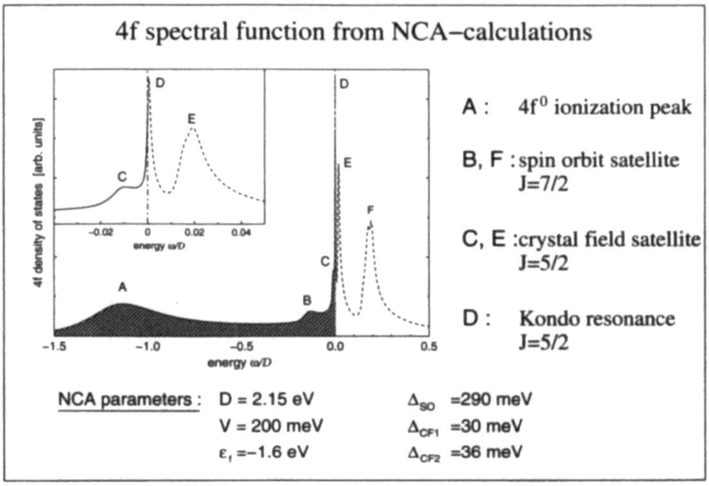

Fig. 3.41. Spectral function for a f^1 system calculated in the NCA approximation. The meaning of the peaks is indicated and also the magnitude of the NCA parameters; D is the bandwidth, V is the 4f conduction electron interaction and ε_f is the bare 4f energy; taken from [3.199]

spectra, which sample the Kondo resonance almost completely (because the resonance is situated with its peak above the Fermi energy) also allow no detailed comparison between experiment and theory due to their low energy resolution (which was about a factor of 100 worse than the width of the Kondo resonance.

The problem has been solved by very high resolution photoemission experiments, in which the Fermi distribution was divided out from the measured spectra to yield the bare spectral function, displaying the Kondo resonance above the Fermi energy [3.199]. In this case one needs very high resolution PES data with good statistical accuracy, because only an energy range of roughly $5\,k_B T$ above the Fermi energy can be covered by this procedure. However, for low T_K materials this is sufficient to analyze even the temperature dependence of the Kondo resonance, which can be compared to calculations using the non-crossing approximation (NCA). From these results it is evident that the Gunnarsson–Schönhammer theory in its temperature dependent form gives a good representation of spectra from Ce compounds also in the energy range around the Fermi energy.

Figure 3.41 gives a calculated spectrum for a Ce system in the NCA, where the peaks A to F have the following meanings:

A: Ionization peak, namely the $4f^1 \rightarrow 4f^0$ transition
D: Kondo resonance, situated above the Fermi energy
B, F: Spin–orbit satellites ($J = 7/2$) of the ground state
C, E: Crystal-field satellites for the ground state 4f level ($J = 5/2$)

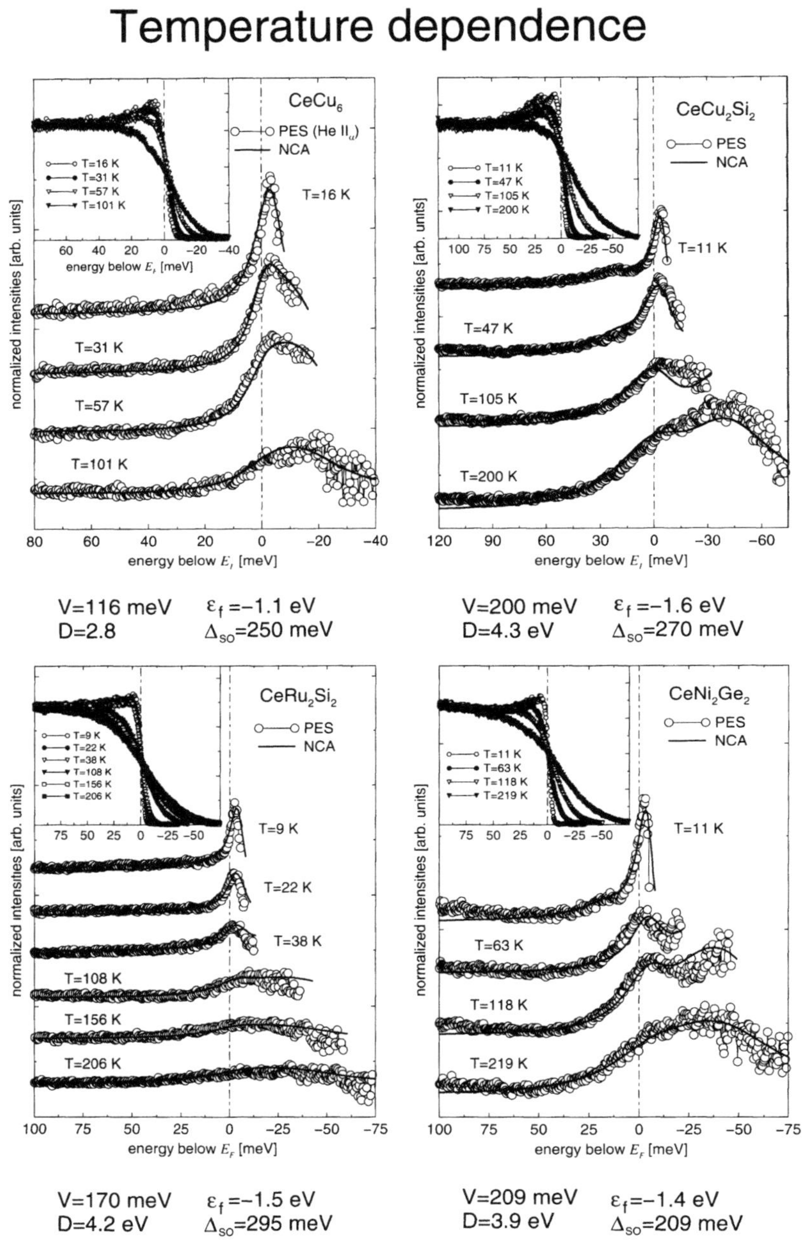

Fig. 3.42. High resolution photoemission spectra of different temperature of the Kondo systems $CeCu_6$, $CeCu_2Si_2$, $CeRu_2Si_2$ and $CeNi_2Ge_2$. The insets show the raw spectra taken at different temperatures. The data presented are the raw spectra divided by the Fermi function of the measuring temperature. The full line is the NCA calculation with the parameters given below each data panel [3.199]

The inset in Fig. 3.41 shows a slightly enlarged version of the spectral range around the Fermi energy, displaying the Kondo resonance (D) and its two crystal-field satellites (C and E).

In what follows, we focus on the area around the Fermi energy displayed in the inset in Fig. 3.41 because the observation of the ionization peak and the spin–orbit satellite in the occupied part of the spectrum is well documented (see Figs. 3.30 and 3.31) in the literature.

Figure 3.42 shows experimental results for the Kondo systems $CeCu_6$ (T_K = 3–7 K), $CeCu_2Si_2$ (T_K = 4.5–10 K), $CeRu_2Si_2$ (T_K = 10–20 K) and $CeNi_2Ge_2$ (T_K = 30 K). These spectra are analyzed in terms of NCA calculations and the agreement between the spectra and the theoretical simulations is very good. Below the data of each compound the parameters used for the NCA calculations are given, where V is the 4f conduction band mixing matrix element, D is the width of the conduction band, ε_f is the bare 4f ionization energy and Δ_{so} is the spin–orbit interaction energy. The raw spectra of each compound, for the different temperatures at which the data were taken, are given in the inset and show the familiar Fermi step. If the Fermi function for the measuring temperature is now divided out from the experimental data one clearly obtains just the spectral function. These are the data that are displayed in the figure, and they show the anticipated behavior, namely a Kondo peak above the Fermi energy, which broadens and shifts to higher energies with increasing temperature due to the population of the crystal-field levels. This is a very nice manifestation of the validity of the Gunnarsson–

Table 3.3. Kondo temperatures and crystal-field parameters for $CeCu_6$, $CeCu_2Si_2$, $CeRu_2Si_2$ and $CeNi_2Ge_2$. The comparison of the data taken from various other sources and those obtained in the present investigation by fitting the photoemission data to the NCA calculation is very gratifying and shows the consistency of photoemission data with results from various other sources, something which has long been questioned [3.199]

	from alternative measurements		from PES and NCA	
	T_K	CEF	T_K	CEF
$CeCu_6$	2.9 . . . 7 K	$\Delta_{1\to2} = 3.6\ldots9.2$ meV $\Delta_{1\to3} = 6.8\ldots20.8$ meV	4.6 K	$\Delta_{1\to2} = 7.2$ meV $\Delta_{1\to3} = 13.9$ meV
$CeCu_2Si_2$	4.5 . . . 10 K	$\Delta_{1\to2} = 12\ldots30$ meV $\Delta_{1\to3} = 34\ldots36$ meV	6 K	$\Delta_{1\to2} = 32$ meV $\Delta_{1\to3} = 37$ meV
$CeRu_2Si_2$	10 . . . 22 K	$\Delta_{1\to2} = 19$ meV $\Delta_{1\to3} = 34$ meV	16.5 K	$\Delta_{1\to2} = 18$ meV $\Delta_{1\to3} = 33$ meV
$CeNi_2Ge_2$	29 K	$\Delta_{1\to2} = 4$ meV $\Delta_{1\to3} = 34$ meV	29.5 K	$\Delta_{1\to2} = 26$ meV $\Delta_{1\to3} = 39$ meV

Schönhammer theory in the non-crossing approximation for the analysis of the photoemission spectra of Ce systems.

Table 3.3 finally shows some relevant parameters obtained from the fits to the photoemission spectra by using the NCA procedures. These results are compared with neutron scattering data and the agreement, even with respect to the hard to measure crystal-field levels, is striking. In summary, the theory of Gunnarsson and Schönhammer in the NCA extension gives a very good description of photoemission spectra from low T_K systems.

References

3.1 G.A. Wendin: *Structure and Bonding* **45** (Springer, Berlin, Heidelberg 1981)
3.2 J. Kanamori, A. Kotani (eds.): *Core-Level Spectroscopy in Condensed Systems*, Springer Ser. Solid-State Sci., Vol.81 (Springer, Berlin, Heidelberg 1988)
3.3 C.S. Fadley: In *Electron Spectroscopy, Theory, Techniques and Applications*, Vol.II, ed. by C.R. Brundle, A.D. Baker (Academic, New York 1978) p.1
3.4 G.K. Wertheim: In *Electron Spectroscopy, Theory, Techniques and Applications*, Vol. II (Academic, New York 1978)
3.5 R.L. Martin, D.A. Shirley: In *Electron Spectroscopy, Theory, Techniques and Applications*, Vol.I, ed. by C.R. Brundle, A.D. Baker (Academic, New York 1977)
D.A. Shirley: In *Photoemission in Solids I*, ed. by M. Cardona, L. Ley, Topics Appl. Phys., Vol.26 (Springer, Berlin, Heidelberg 1978) Chap.4
3.6 G. van der Laan, C. Westra, C. Haas, G.A. Sawatzky: Phys. Rev. B **23**, 4369 (1981); an extension of this work, which takes non-local screening into account, has been given by M.A. van Veenendaal, G.A. Sawatzky: Phys. Rev. B **49**, 3473 (1994)
3.7 A. Kotani, Y. Toyozawa: J. Phys. Soc. Jpn. **35**, 1073 and 1082 (1973)
3.8 A. Kotani, Y. Toyozawa: J. Phys. Soc. Jpn. **37**, 912 (1974)
3.9 S. Hüfner, G.K. Wertheim: Phys. Lett. **51**A, 299 (1975)
3.10 D.K.G. de Boer, C. Haas, G.A. Sawatzky: Phys. Rev. B **29**, 4401 (1984)
G.A. Sawatzky has informed us that a calculation by Okada and Kotani favours the d-screening model, see T. Uozumi, K. Okada, A. Kotani, O. Durmeyer, J.P. Kappler, E. Beaurepaire, J.C. Parlebas: Europhys. Lett. **18**, 85 (1992)
3.11 J.C. Parlebas: J. Phys. (France) **2**, 1369 (1992)
3.12 K. Okada, T. Uozumi, A. Kotani: Jpn. J. Appl. Phys. **32**, Suppl.32-2, 113 (1993)
3.13 K. Okada, A. Kotani: J. Electron Spectrosc. Relat. Phenom. **62**, 131 (1993)
3.14 J.C. Parlebas, M.A. Khan, T. Uozumi, K. Okada, A. Kotani: J. Electron Spectrosc. Relat. Phenom. **71**, 117 (1995)
3.15 A. Kotani: J. Electron Spectrosc. Relat. Phenom. **100**, 75 (1999)
3.16 S. Hüfner: Adv. Phys. **43**, 183 (1994)
3.17 A.E. Bocquet, A. Fujimori: J. Electron. Spectrosc. Relat. Phenom. **62**, 141 (1993)
3.18 A.E. Bocquet, T. Mizokawa, A. Fujimori, S.R. Barman, K. Maiti, D.D. Sarma, Y. Tokura, M. Onoda: Phys. Rev. B **53**, 1161 (1966)
3.19 A.E. Bocquet, T. Mozokawa, A. Fujimori, M. Matoba, S. Anzai: Phys. Rev. B **52**, 13838 (1995)

3.20 M. Imada, A. Fujimori, Y. Tokura: Rev. Mod. Phys. **70**, 1039 (1998)
References [3.11–3.20] review the theory of many-body effects in valence- and core-level spectra of transition metal compounds
3.21 S. Larsson: Chem. Phys. Lett. **40**, 362 (1976)
3.22 S. Larsson, M. Braga: Chem. Phys. Lett. **48**, 695 (1977)
3.23 A. Rosencwaig, G.K. Wertheim, H.J. Guggenheim: Phys. Rev. **27**, 479 (1971)
3.24 G.K. Wertheim, S. Hüfner: Phys. Rev. Lett. **28**, 1028 (1972)
3.25 A. Rosencwaig, G.K. Wertheim: J. Electron Spectrosc. Relat. Phenom. **1**, 493 (1973)
3.26 T.A. Carlson, J.C. Carver, L.J. Saethre, F. Garcia Santibanez, G.A. Vernon: J. Electron Spectrosc. Relat. Phenom. **5**, 247 (1974)
3.27 G.A. Vernon, G. Stucky, T.A. Carlson: Inorg. Chem. **15**, 278 (1976)
3.28 T. Novakov, R. Prins: Solid State Commun. **9**, 1975 (1972)
3.29 V. Kinsinger: Untersuchung der elektronischen Struktur von Dihalogeniden der 3d-Übergangsmetalle, Thesis, Universität des Saarlandes (1990)
3.30 H. Chermette, P. Pertosa, F.M. Michel-Calendini: Chem. Phys. Lett. **69**, 240 (1980)
3.31 J.A. Tossell: J. Electron Spectrosc. Relat. Phenom. **8**, 1 (1976)
3.32 T.A. Carlson, J.C. Carver, G.A. Vernon: J. Chem. Phys. **62**, 932 (1975)
3.33 B. Wallbank, I.G. Main, C.E. Johnson: J. Electron Spectrosc. Relat. Phenom. **5**, 259 (1974)
3.34 A. Gupta, J.A. Tossell: J. Electron Spectrosc. Relat. Phenom. **26**, 223 (1982)
3.35 J.R. Waldrop: J. Vac. Sci. Technol. B **2**, 445 (1984)
3.36 B. Wallbank, C.E. Johnson, I.G. Main: J. Phys. C **6**, L340 (1973)
3.37 D.C. Frost, A. Ishitani, C.A. McDowell: Mol. Phys. **24**, 861 (1972)
3.38 D.C. Frost, C.A. McDowell, I.S. Woolsey: Chem. Phys. Lett. **17**, 320 (1972)
3.39 D.C. Frost, C.A. McDowell, B. Wallbank: Chem. Phys. Lett. **40**, 189 (1976)
3.40 J.S.H.Q. Perera, D.C. Frost, C.A. McDowell: J. Chem. Phys. **72**, 5151 (1980)
3.41 S. Larsson: J. Electron Spectrosc. Relat. Phenom. **8**, 171 (1976)
3.42 B. Wallbank, J.S.H.Q. Perera, D.C. Frost, C.A. McDowell: J. Chem. Phys. **69**, 5405 (1978)
3.43 L.J. Matienzo, L.I. Yin, S.O. Grim, W.E. Swartz: Inorg. Chem. **12**, 2762 (1973)
3.44 I. Ikemoto, K. Ishii, H. Kuroda, J.M. Thomas: Chem. Phys. Lett. **28**, 55 (1974)
3.45 K.S. Kim: Phys. Rev. B **11**, 2177 (1975)
3.46 K. Okada, A. Kotani, B.T. Thole: J. Electron Spectrosc. Relat. Phenom. **58**, 325 (1992)
3.47 J. Park, S. Ryu, M. Han, S.-J. Oh: Phys. Rev. B **37**, 10867 (1988)
3.48 M. Brisk, A.D. Baker: J. Electron Spectrosc. Relat. Phenom. **6**, 81 (1975)
3.49 S.K. Sen, J. Riga, J. Verbist: Chem. Phys. Lett. **39**, 560 (1976)
3.50 M. Scrocco: J. Electron Spectrosc. Relat. Phenom. **19**, 311 (1980)
3.51 G. Lee, S.-J. Oh: Phys. Rev. B **43**, 14674 (1991)
3.52 B.W. Veal, A.P. Paulikas: Phys. Rev. B **31**, 5399 (1985)
3.53 B.W. Veal, D.E. Ellis, D.J. Lam: Phys. Rev. B **32**, 5391 (1985)
3.54 S. Hüfner: Solid State Commun. **47**, 943 (1983)
3.55 S. Hüfner: Solid State Commun. **49**, 1177 (1984)
3.56 J. Zaanen, G.A. Sawatzky: Phys. Rev. B **33**, 8074 (1986)
3.57 J. Zaanen, C. Westra, G.A. Sawatzky: Phys. Rev. B **33**, 8060 (1986)
3.58 G.K. Wertheim, R.L. Cohen, A. Rosencwaig, H.J. Guggenheim: In *Electron Spectroscopy*, ed. by D.A. Shirley (North-Holland, Amsterdam 1972)
3.59 A.J. Signorelli, R.G. Hayes: Phys. Rev. B **8**, 81 (1973)
3.60 L. Schlapbach, S. Hüfner, Th. Riesterer: J. Phys. C **19**, L63 (1986)

3.61 S. Suzuki, T. Ishii, T. Sagawa: J. Phys. Soc. Jpn. **37**, 1334 (1974)
3.62 C.K. Jorgensen, H. Berthou: Chem. Phys. Lett. **13**, 186 (1972)
3.63 H. Berthou, C.K. Jorgensen, C. Bonnelle: Chem. Phys. Lett. **38**, 199 (1976)
3.64 I. Nagakura, T. Ishii, T. Sagawa: J. Phys. Soc. Jpn. **33**, 754 (1972)
3.65 C.K. Jorgensen: Struct. Bonding (Berlin) **24**, 1 (1975)
3.66 A. Kotani, T. Jo, J.C. Parlebas: Adv. Phys. **37**, 37 (1988)
3.67 J.J. Pireaux, J. Riga, E. Thibaut, C. Tenret-Noel, R. Caudano, J.J. Verbist: Chem. Phys. **22**, 113 (1977)
3.68 B.W. Veal, D.J. Lam, H. Diamond, H.R. Hoekstra: Phys. Rev. B **15**, 2929 (1979)
3.69 C. Keller, C.K. Jorgensen: Chem. Phys. Lett. **32**, 397 (1975)
3.70 R.J. Thorn: J. Phys. Chem. Solids **43**, 571 (1982)
3.71 Wei-Yean Howng, R.J. Thorn: Chem. Phys. Lett. **56**, 463 (1978)
3.72 G. M. Bancroft, T.K. Sham, S. Larsson: Chem. Phys. Lett. **46**, 551 (1977)
3.73 G.C. Allen, J.A. Crofts, M.T. Curtis, P.M. Tucker, D. Chadwick, P. Hampson: J. Chem. Soc. Dalton Trans. 1296 (1974)
3.74 T.A. Carlson: *Photoelectron and Auger Spectroscopy* (Plenum, New York 1975)
3.75 S. Asada, S. Sugano: J. Phys. Soc. Jpn. **41**, 1291 (1976)
3.76 B.W. Veal, A.P. Paulikas: Phys. Rev. Lett. **51**, 1995 (1983)
3.77 S. Hüfner: Z. Phys. B **58**, 1 (1984)
3.78 S. Larsson: Chem. Phys. Lett. **32**, 401 (1975)
3.79 M. Braga, S. Larsson: Int'l J. Quantum Chem. Symp. **11**, 61 (1977)
3.80 A. Fujimori: Phys. Rev. B **27**, 3992 (1983)
3.81 A. Fujimori: Phys. Rev. B **28**, 2281 (1983)
3.82 A. Fujimori: Phys. Rev. B **28**, 4489 (1983)
3.83 A. Kotani, H. Mizuta, T. Jo: Solid State Commun. **53**, 805 (1985)
3.84 C.K. Jorgensen: *Progress in Inorg. Chemistry* Vol.12, ed. by S.J. Lippart (Interscience, New York 1970) p.101
3.85 H. Onuki, F. Sugawara, Y. Nishihara, M. Hirano, Y. Yamaguchi, A. Ejiri, H. Takahashi, A. Abe: Solid State Commun. **20**, 35 (1976)
3.86 K. Okada, A. Kotani: J. Phys. Soc. Jpn. **60**, 772 (1991); **58**, 2578 (1989); **58**, 1095 (1989); J. Electron Spectrosc. Relat. Phenom. **53**, 319 (1990)
3.87 J.G. Bednorz,K.A. Müller: Z. Physik B **64**, 189 (1986)
For a review on PES data of high temperature superconductors see
3.88 P.A.P. Lindberg, Z.X. Shen, W.E. Spicer: Surf. Sci. Rpts. **11**, 1 (1990);
3.89 Z.X. Shen, D.S. Dessau: Phys. Reports **253**,1 (1995)
3.90 J.A. Leiro, E.E. Minni: Phys. Rev. B **31**, 8248 (1985)
3.91 A. Fujimori, J.H. Weaver, A. Franciosi: Phys. Rev. B **31**, 3549 (1985)
3.92 M.A. van Veenendaal, G.A. Sawatzky: Phys. Rev. Lett. **70**, 2459 (1993)
3.93 S. Altieri, L.H. Teng, A. Tanaka, G.A. Sawatzky: preprint to be published in Phys. Rev. B **61**, 13403 (2000)
3.94 D. Alders, F.C. Voogt, T. Hibma, G.A. Sawatzky: Phys. Rev. B **54**, 7716 (1996)
3.95 T. Böske, K. Maiti, O. Knauff, K. Ruck, M.S. Golden, G. Krabbes, J. Fink, T. Osafune, N. Motoyama, H. Eisaki, S. Uchida: Phys. Rev. B **57**, 138 (1998)
3.96 K. Okada, A. Kotani: Phys. Rev. B **52**, 4794 (1995)
3.97 S. Hüfner, G.K. Wertheim: Phys. Lett. A **51**, 301 (1975)
3.98 H. Raether: *Springer Tracts Mod. Phys.* **88** (Springer, Berlin, Heidelberg 1980)
3.99 H. Raether: *Springer Tracts Mod. Phys.* **38**, 84 (Springer, Berlin, Heidelberg 1965)

3.100 C. Guillot, Y. Ballu, J. Paigne, J. Lecante, K.P. Jain, P. Thiry, R. Pinchaux, Y. Petroff, L.M. Falicov: Phys. Rev. Lett. **39**, 1632 (1977)
3.101 O. Björneholm, J.N. Andersen, C. Wigren, A. Nilsson, R. Nyholm, N. Mårtensson: Phys. Rev. B **41**, 10408 (1990)
3.102 L.H. Tjeng, C.T. Chen, J. Ghijsen, P. Rudolf, F. Sette: Phys. Rev. Lett. **67**, 501 (1991) resonance photoemission employing the 2p level in CuO
3.103 L.H. Tjeng, B. Sinkovic, N.B. Brookes, J.B. Goedkoop, R. Hesper, E. Pellegrin, F.M.F. Groot, S. Altieri, S.L. Hulbert, E. Shekel, G.A. Sawatzky: Phys. Rev. Lett. **78**,1126 (1997), have extended these investigations by using spin detection; this enabled them to detect directly the singlet state at the top of the valence band of CuO (Zhang-Rice) singlet.
3.104 J.W. Allen: In *Synchrotron Radiation Research, Advances in Surface and Interface Science*, ed. by R.Z. Bachrach (Plenum, New York 1992) Vol.1 (for a review on resonance PES)
3.105 M.F. Lopez, C. Laubschat, A. Gutierrez, G. Kaindl: Z. Phys. B **94**, 1 (1994)
3.106 W. Eberhardt, E.W. Plummer: Phys. Rev. B **21**, 3245 (1980)
3.107 R. Clauberg, W. Gudat, E. Kisker, G.M. Rothberg: Phys. Rev. Lett. **47**, 1314 (1981)
3.108 D.R. Penn: Phys. Rev. Lett. **42**, 921 (1979)
3.109 L.A. Feldkamp, L.C. Davis: Phys. Rev. Lett. **43**, 151 (1979)
3.110 G. Treglia, F. Ducastelle, D. Spanjaard: Phys. Rev. B **21**, 3729 (1980)
3.111 A. Liebsch: Phys. Rev. Lett. **43**, 1431 (1979)
3.112 A. Liebsch: Phys. Rev. B **23**, 5203 (1981)
3.113 S.M. Girvin, D.R. Penn: J. Appl. Phys. **52**, 1650 (1981)
3.114 G. Treglia, F. Ducastelle, D. Spanjaard: J. Phys. (Paris) **43**, 341 (1982)
3.115 L.C. Davis, L.A. Feldkamp: Phys. Rev. B **23**, 6239 (1981); ibid. A **17**, 2012 (1978)
3.116 H. Eckardt, L. Fritsche: J. Phys. F **17**, 1795 (1987)
3.117 U. Fano: Phys. Rev. **124**, 1866 (1961)
3.118 R. Bruhn, E. Schmidt, H. Schroeder, B. Sonntag: J. Phys. B **15**, 2807 (1982)
3.119 R. Bruhn, E. Schmidt, H. Schroeder, B. Sonntag: Phys. Lett. A **90**, 41 (1982)
3.120 R. Bruhn, B. Sonntag, H.W. Wolff: Phys. Lett. A **69**, 9 (1978)
3.121 R. Bruhn, B. Sonntag, H.W. Wolff: J. Phys. B **12**, 203 (1979)
3.122 J. Barth, G. Kalkoffen, C. Kunz: Phys. Lett. A **74**, 360 (1979)
3.123 R. Clauberg, W. Gudat, W. Radlik, W. Braun: Phys. Rev. B **31**, 1754 (1985)
3.124 L.C. Davis, L.A. Feldkamp: Solid State Commun. **19**, 413 (1976)
3.125 L.C. Davis, L.A. Feldkamp: Phys. Rev. B **15**, 2961 (1977)
3.126 L.C. Davis, L.A. Feldkamp: Solid State Commun. **34**, 141 (1980)
3.127 M. Iwan, F.J. Himpsel, D.E. Eastman: Phys. Rev. Lett. **43**, 1829 (1979)
3.128 N. Mårtensson, B. Johansson: Phys. Rev. Lett. **45**, 482 (1980)
3.129 S.H. Oh, J.W. Allen, I. Lindau, J.C. Mikkelsen: Phys. Rev. B **26**, 4845 (1982)
3.130 E. Schmidt, H. Schröder, B. Sonntag, H. Voss, H.E. Wetzel: J. Phys. B **16**, 2961 (1983)
3.131 C. Corliss, J. Sugar: J. Phys. Chem. Ref. Data **10**, 1097 (1981)
3.132 M. Weinelt, A. Nilsson, M. Magnuson, T. Wiell, N. Wassdahl, O. Karis, A. Fröhlisch, N. Mårtensson, J. Stöhr, M. Samant: Phys. Rev. Lett. **78**, 967 (1997)
3.133 B. Sinkovic, L.H. Tjeng, N.B. Brookes, J.B. Goedkoop, R. Hesper, E. Pellegrin, F.M.F. de Groot, S. Altieri, S.L. Hulbert, E. Shekel, G.A. Sawatzky: Phys. Rev. Lett. **79**, 3510 (1997) have used spin detection to separate the d^8 singlets and triplets.
3.134 G.B. Armen, H. Wang: Phys. Rev. A **51**, 1241 (1995)

3.135 S. Hüfner, S.H. Yang, B.S. Mun, C.S. Fadley, J. Schäfer, E. Rotenberg, S.D. Kevan: Phys. Rev. B **61**, 12582 (2000)
3.136 S. Hüfner, P. Steiner: *Valence Instabilities*, ed. by P. Wachter, H. Boppart (North Holland, Amsterdam 1982)
3.137 J.K. Lang, Y. Baer, P.A. Cox: J. Phys. F **11**, 121 (1981)
3.138 B. Johansson: Phil. Mag. **30**, 469 (1974)
3.139 O. Gunnarsson, K. Schönhammer: Phys. Rev. Lett. **50**, 604 (1983)
3.140 O. Gunnarsson, K. Schönhammer: Phys. Rev. B **28**, 4315 (1983); B **31**, 4815 (1985); in *Handbook on the Physics and Chemistry of Rare Earths*, ed. by K.A. Gschneidner, L.R. Eyring, S. Hüfner (North-Holland, Amsterdam 1987) Vol.10; and J. Magn. Mag. Mater. **63/64**, 481 (1987)
J.W. Allen, S.J. Oh, O. Gunnarsson, K. Schönhammer, M.B. Maple, M.S. Torikachvili, I. Lindau: Adv. Phys. **35**, 275 (1986)
3.141 G. Crecelius, G.K. Wertheim, D.N.E. Buchanan: Phys. Rev. B **18**, 6519 (1978)
3.142 R. Lässer, J.C. Fuggle, M. Beyss, M. Campagna, F. Steglich, F. Hulliger: Physica B **102**, 360 (1980)
3.143 F. Patthey, B. Delley, W.D. Schneider, Y. Baer: Phys. Rev. Lett. **55**, 1918 (1985)
3.144 G.D. Mahan: Phys. Rev. **163**, 612 (1967)
3.145 P. Nozières, C.T. DeDominicis: Phys. Rev. **178**, 1097 (1969)
3.146 A. Kotani, Y. Toyozawa: J. Phys. Soc. Jpn. **46**, 488 (1979)
3.147 N.D. Lang, A.R. Williams: Phys. Rev. B **16**, 2408 (1977)
3.148 K. Schönhammer, O. Gunnarsson: Solid State Commun. **23**, 691 (1977)
3.149 K. Schönhammer, O. Gunnarsson: Solid State Commun. **26**, 399 (1978)
3.150 K. Schönhammer, O. Gunnarsson: Z. Physik B **30**, 297 (1978)
3.151 K. Schönhammer, O. Gunnarsson: Phys. Rev. B **18**, 6608 (1978)
3.152 K. Schönhammer, O. Gunnarsson: Surf. Sci. **89**, 575 (1979)
3.153 O. Gunnarsson, K. Schönhammer: Solid State Commun. **26**, 147 (1978)
3.154 O. Gunnarsson, K. Schönhammer: Phys. Rev. Lett. **41**, 1608 (1978)
3.155 T.B. Ramakrishnan: In *Valence Fluctuations in Solids*, ed. by L.M. Falicov, W. Hanke, M.B. Maple (North-Holland, Amsterdam 1981)
3.156 T.B. Ramakrishnan, K. Sur: Phys. Rev. B **26**, 1798 (1982)
3.157 P.W. Anderson: In *Valence Fluctuations in Solids*, ed. by L.M. Falicov, W. Hanke, M.B. Maple (North-Holland, Amsterdam 1981)
3.158 N. Grewe: Solid State Commun. **50**, 19 (1984)
3.159 F.U. Hillebrecht, J.C. Fuggle, G.A. Sawatzky, M. Campagna, O. Gunnarsson, K. Schönhammer: Phys. Rev. B **30**, 1777 (1983)
3.160 O. Gunnarsson, K. Schönhammer, J.C. Fuggle, F.U. Hillebrecht, J.M. Esteva, R.C. Karnatak, B. Hillebrand: Phys. Rev. B **28**, 7330 (1983)
3.161 L. Ley, N. Mårtensson, J. Azoulay: Phys. Rev. Lett. **45**, 1516 (1980)
3.162 H. van Doveren, J.A. Verhoevan: J. Electron Spectrosc. Relat. Phenom. **21**, 265 (1980)
3.163 D. Weighman: Rep. Prog. Phys. **45**, 753 (1982)
3.164 A.R. Williams, N.D. Lang: Phys. Rev. Lett. **40**, 954 (1978)
3.165 H.R. Ott, Y. Baer, K. Andres: *Valence Fluctuations in Solids*, ed. by L.M. Falicov, W. Hanke, M.B. Maple (North-Holland, Amsterdam 1981)
3.166 M. Moser, P. Wachter, F. Hulliger, J.R. Etourneau: Solid State Commun. **54**, 241 (1985)
3.167 M. Moser, F. Hulliger, P. Wachter: Physica **130** , B21 (1985)
3.168 J. Li, W.-D. Schneider, R. Berndt, B. Delley: Phys. Rev. Lett. **80**, 2893 (1998)

3.169 P. Fulde: *Electron Correlations in Molecules and Solids*, Springer Ser. Solid State Sciences, Vol. 100 (Springer, Heidelberg, 1995) 3rd edition
3.170 F. Patthey, W.D. Schneider, Y. Baer, B. Delley: Phys. Rev. B **35**, 5903 (1987)
3.171 F. Steglich, J. Aarts, C.D. Bredl, W. Lieke, D. Meschede, W. Franz, J. Schaefer: Phys. Rev. Lett. **43**, 1892 (1979)
3.172 G.R. Stewart: Rev. Mod. Phys. **56**, 755 (1984)
3.173 G. Zwicknagel: Adv. Phys. **41**, 203 (1992)
3.174 W.D. Schneider, B. Delley, E. Wuilloud, J.-M. Imer, Y. Baer: Phys. Rev. B **32**, 6819 (1985)
3.175 Y. Baer, W.D. Schneider: *In Handbook on the Physics and Chemistry of Rare Earths*, ed. by K.A. Gschneidner, L. Eyring, S. Hüfner (North-Holland, Amsterdam 1987) Vol.10
3.176 F. Patthey, J.-M. Imer, W.D. Schneider, H. Beck, Y. Baer, B. Delley: Phys. Rev. B **42**, 8864 (1990)
3.177 J.J. Joyce, A.J. Arko, J. Lawrence, P.C. Canfield, Z. Fisk, R.J. Bartlett, J.D. Thompson: Phys. Rev. Lett. **68**, 236 (1992)
3.178 J.J. Joyce, A.J. Arko, P.S. Riseborough, P.C. Canfield, J.M. Lawrence, R.I.R. Blyth, R.J. Bartlett, J.D. Thompson, Z. Fisk: Physica B **186**, 31 (1993)
3.179 M. Grioni, D. Malterre, P. Weibel, B. Dardel, Y. Baer: Physica B **186**, 38 (1993)
3.180 A.J. Arko, J.J. Joyce: Phys. Rev. Lett. **81**, 1348 (1998)
3.181 M. Garnier, D. Purdie, K. Breuer, M. Hengsberger, Y. Baer, B. Delley: Phys. Rev. Lett. **78**, 1349 (1998); M. Garnier, D. Purdie, M. Hengsberger, K. Breuer, Y. Baer:Physica B **259-261**, 1095 (1999)
3.182 M. Garnier, K. Breuer, D. Purdie, M. Hengsberger, Y. Baer, B. Delley: Phys. Rev. Lett. **78**, 4127 (1997)
3.183 D. Malterre, M. Grioni, Y. Baer: Adv. Phys. **45**, 299 (1996)
3.184 S. Hüfner, L. Schlapbach: Z. Physik B **64**, 417 (1986)
3.185 S. Hüfner: Z. Physik B **86**, 241 (1992)
3.186 S. Hüfner: Ann. Physik **5**, 453 (1996)
3.187 D. Malterre, M. Grioni, P. Weibel, Y. Baer: Phys. Rev. Lett. **68**, 2656 (1992)
3.188 L.Z. Liu, J.W. Allen, O. Gunnarsson, N.E. Christensen, O.K. Andersen: Phys. Rev. B **45**, 8934 (1992)
3.189 J.-M. Imer, E. Wuilloud: Z. Physik B **66**, 160 (1987)
3.190 P.W. Andersson: Phys. Rev. **124**, 41 (1961)
3.191 K.H. Park, S.J. Oh: Phys. Rev. B **48**, 14833 (1993)
3.192 A. Kotani, H. Ogasawara: J. Electron Spectrosc. Relat. Phenom. **60**, 257 (1992)
3.193 A. Kotani: in *Handbook on Synchrotron Radiation*, ed. by G.V. Marr (North-Holland, Amsterdam 1987) Vol.2, p.611
3.194 J.M. Lawrence, A.J. Arko, J.J. Joyce, R.I.R. Blyth, R.J. Bartlett, P.C. Canfield, Z. Fisk, P.S. Riseborough: Phys. Rev. B **47**, 15460 (1993)
3.195 A.B. Andrews, J.J. Joyce, A.J. Arko, J.D. Thompson, J. Tang, J.M. Lawrence, J.C. Hemminger: Phys. Rev. B **51**, 3277 (1995)
3.196 A.B. Andrews, J.J. Joyce, A.J. Arko, Z. Fisk, P.S. Riseborough: Phys. Rev. B **53**, 3317 (1996)
3.197 E. Weschke, C. Laubschat, R. Ecker, A. Höhr, M. Domke, G. Kaindl, L. Severin, B. Johansson: Phys. Rev. Lett. **69**, 1792 (1992)
3.198 E. Weschke, C. Laubschat, T. Simmons, M. Domke, O. Strebel, G. Kaindl: Phys. Rev. B **44**, 8304 (1991)
3.199 R. Reinert, D. Ehm, S. Schmidt, G. Nicolay, S. Hüfner: Phys. Rev. Lett. **87**, 106401 (2001)

3.200 A. Föhlisch, O. Karis, M. Weinelt, J. Hasselström, A. Nilsson, N. Mårtensson: Phys. Rev. Lett. 88, 027601 (2002)
3.201 K. Godehusen, T. Richter, P. Zimmermann, M. Martins: Phys. Rev. Lett. **88**, 217601 (2002)

4. Continuous Satellites and Plasmon Satellites: XPS Photoemission in Nearly Free Electron Systems

As already emphasized, PES is a very valuable technique for understanding the electronic properties of solids. However, its value is intimately connected with the understanding of the screening of the photohole. In metals screening leads to three types of phenomena:

(i) The positive charge of the photohole is neutralized by an electron which moves onto the site of the photoemission process;[1] the energy of the state from which photoemission has occurred is thereby changed (Sect. 2.2.3). In metals with narrow bands (f and d band metals) the photohole neutralization can take place through various channels (with sp-, d- or f-screening) which leads to satellites (Chap. 3).

(ii) In the screening process the positive photohole produces excitations in the Fermi sea of conduction electrons. In principle, possible excitation energies range between zero (directly at E_F) and the bandwidth of the metal under investigation.

(iii) The deexcitation of the photohole leads to quantized excitations in the conduction-electron system, namely to the creation of plasmons. These are called *"intrinsic"* plasmons, because they are an intrinsic property of the photoemission process. In contrast, *"extrinsic"* plasmons are excited somewhere else in the solid by the outgoing photoelectron during its travel from the place of the photoexcitation process to the surface.[2]

In this chapter we shall be concerned with the phenomena classified under (ii) and (iii) which lead to an asymmetric lineshape of the PE spectrum [4.1–4.9] of metals (more precisely: of systems with free electrons at the Fermi energy), and to the so-called *multi-plasmon sidebands* [4.4, 4.9].

These two effects cannot, however, be easily isolated in the PE spectrum of a solid because they interfere with other excitations. As already mentioned, the photoelectron on its journey to the surface can excite extrinsic plasmons at a well defined energy loss but it can also lose "unquantized" energy by electron–electron and electron–ion collisions. The latter effects lead

[1] The final relaxation of the photohole takes place by an Auger process. This leads to the photoexcited Auger spectra.

[2] The processes (ii) and (iii) cannot be separated rigorously because they occur simultaneously. However, for the sake of an easy presentation we shall separate them here.

to a "featureless", smooth background in the PE spectra. During the penetration through the surface, surface plasmons can be excited. All these different mechanisms contribute to the observed photoemission spectrum, and the intrinsic and extrinsic bulk plasmons are particularly difficult to separate. Finally there are interference effects between the creation of intrinsic and extrinsic plasmons, which complicate the matter further [4.10–4.12].

Because of the complexity of the phenomena it is desirable to study as simple a system as possible and we therefore focus on the simple metals.[3]

For these, the problems indicated have been investigated intensively and our discussion here will be limited to this class of substances. (For a study of core-line asymmetries in other systems see, e.g., [4.2, 4.7, 4.13–4.16]).

The emphasis will be on the two phenomena that are inherent to the photoemission technique, namely the asymmetric lineshape and the intrinsic plasmon creation rate. The other effects, inelastic (extrinsic) electron scattering leading to quantized (plasmons) and unquantized excitation in the electron gas have already been dealt with in great detail elsewhere in the literature [4.17, 4.18].

The interaction between a photohole and a conduction electron system was first documented and dealt with in X-ray spectroscopy [4.19–4.25]. It was realized that for certain transitions in X-ray emission or absorption spectroscopy the threshold behavior (enhancement of the intensity) could not be explained by a single-electron picture but that rather a many-electron description was appropriate.

In simple terms, the creation of a photohole in a core state may be treated as an instantaneously switched-on strong local potential. Since the conduction-electron system has many degrees of freedom, and has a wide spectrum of energetic excitation possibilities, the hole–electron interaction will lead to excitations (Fig. 4.1) which, from intuitive arguments (the matrix element for the electron–hole pair excitation goes as $1/\Delta E$, ΔE being the energy transfer), will be the stronger, the smaller the energy transfer. Thus the excitations will approach a divergence at zero energy transfer (namely at E_F) and this is the origin of the so-called threshold singularity in X-ray spectroscopy, first investigated in detail theoretically by Mahan [4.23]. In a real experiment the $1/E$ singularity is always convoluted with lifetime and instrumental broadening which tend to smear it out.

There is, however, another mechanism that works against the excitations, called the Anderson *"orthogonality catastrophe"*, which again can be rationalized in very simple terms [4.24]: the initial state of an X-ray absorption experiment consists of the core-electron state and the many conduction-electron states. In the final state an electron is moved from the core state into the conduction-electron system. The core-hole potential, however, slightly mod-

[3] The so-called simple metals or nearly free electron metals have a valence band that consists mainly of s and p electrons; examples are the alkali metals or earth alkaline metals.

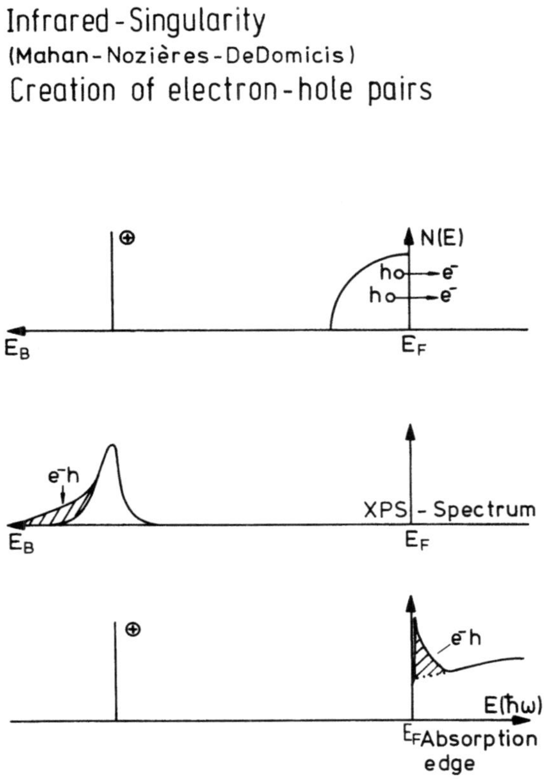

Fig. 4.1. Schematic representation of the Mahan–Nozières–DeDominicis effect [4.23, 4.25] for a core hole in a metal. The core hole acts as a potential which scatters the conduction electron, with a singularity at the Fermi energy. This leads to a tail in the PES spectrum (*shaded*) and an absorption singularity in the X-ray absorption spectrum (*shaded*)

ifies the wave functions of all the conduction electrons such that now their initial and final states are orthogonal to each other [4.24]. This means that the matrix element connecting the initial and final state (which contains the overlap of the initial state of the core electron and the final state into which it is excited, but also the overlap of all the initial-state valence-state functions with the final-state valence-state functions) tends to zero as the number of states approaches infinity. Thus the X-ray-absorption intensity at threshold is the balance of an enhancing and a depressing mechanism.

A rigorous calculation of the X-ray absorption/emission cross section was first carried out by Nozières and DeDominicis [4.25]. They found that near the threshold the absorption cross section has the following form [4]

[4] From here until the end of Sect. 4.1.5 the energy will be represented by ω because this makes a comparison with the literature easier.

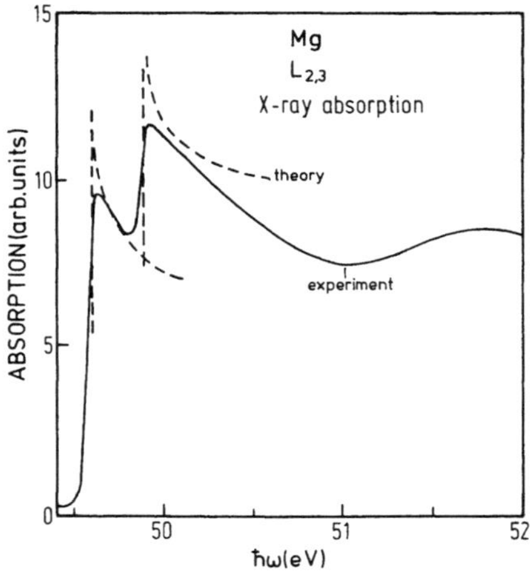

Fig. 4.2. Mg $L_{2,3}$ absorption spectrum taken with synchrotron radiation [4.26,27]. The enhancement of the intensity near the threshold is clearly visible. The splitting of the edge is due to the $L_{2,3}$ spin–orbit splitting

$$A(\omega) = A_0 \left(\frac{\zeta}{\omega - \omega_0} \right)^{\alpha_\ell} , \tag{4.1}$$

where A_0 is a normalization constant, ω_0 the X-ray frequency at the absorption threshold, ζ a parameter of the order of the bandwidth, and ℓ the angular momentum of the conduction state into which the core electron is excited. The threshold exponent α_ℓ is given by

$$\alpha_\ell = (2\delta_\ell/\pi) - \alpha \tag{4.2}$$

with

$$\alpha = 2 \sum_\ell (2\ell + 1)(\delta_\ell/\pi)^2 . \tag{4.3}$$

The phase shift δ_ℓ for angular momentum ℓ can be related to the charge Z screened in the process (unity in the present case) by the Friedel condition

$$Z = 2 \sum_\ell (2\ell + 1)(\delta_\ell/\pi) . \tag{4.4}$$

In the equation for the threshold exponent α_ℓ the first term gives the enhancement of the absorption strength [4.23] due to the creation of electron-hole pairs. The second one reduces it, which is an effect of the orthogonality catastrophe [4.24]. This counterplay has always hampered the interpretation of X-ray absorption and emission data. Figure 4.2 shows the L_{23} soft X-ray absorption spectrum of Mg and one can clearly see the enhancement of the intensity near the absorption edge [4.26, 4.27] which is well accounted for by the theory.

For XPS core-level lines in metals the situation is simplified. The outgoing photoelectron is fast and therefore measures the spectral function of the core hole. This case has been treated by Doniach and Sunjic [4.1]. Their result for the XPS lineshape, obtained by convoluting a $1/\omega^{1-\alpha}$ singularity with a lifetime width of 2γ, is

$$f(\omega) = \frac{\Gamma(1-\alpha)\cos[\pi\alpha/2 + (1-\alpha)\arctan(\omega/\gamma)]}{(\omega^2+\gamma^2)^{(1-\alpha)/2}}\,, \tag{4.5}$$

where Γ denotes the Γ-function. For $\alpha = 0$, (4.5) is reduced to a Lorentzian and for $(\omega/\gamma) \gg 1$ the form $1/\omega^{1-\alpha}$ is obtained.

The lineshape function fits the core lines of simple metals extremely well over an energy range of about the bandwidth. Unfortunately, it has the disadvantage that its integral diverges, which is of course unphysical. A lineshape function that is very similar to the Doniach–Sunjic lineshape but which has a finite integral has been derived by Mahan [4.26](Sect. 4.1.2).

Figure 4.3 gives the Mg $L_{2,3}$ photoemission spectrum (more commonly called the $2p_{1/2,3/2}$ spectrum) in which the spin–orbit splitting is not resolved [4.9]. However, the asymmetry of the spectrum, i.e. its skewing to higher energies, is clearly visible, and the fit of the line with the lineshape of (4.5) is very good.

Besides the asymmetric lineshape, the theory [4.23, 4.25] leads to *plasmon sidebands* of a core-level line [4.28–4.30]. These plasmons, like the electron–hole pairs discussed so far, serve to deexcite the hole state. They can be seen in Fig. 4.4 which shows an extended spectrum of Mg 2p and the Mg 2s lines. The amount of intrinsic contribution to the first bulk plasmon accompanying the 2s line, which was derived from a theory presented later [4.31] (Sect. 4.1.3), is indicated.

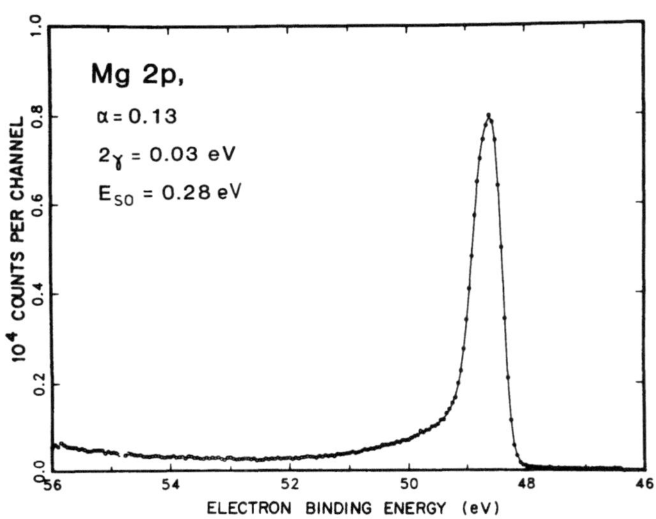

Fig. 4.3. High resolution XPS $2p_{1/2,3/2}$ spectrum of Mg metal [4.9]. The full curve is a Doniach–Sunjic lineshape (4.5) and it fits the data points almost perfectly

These *intrinsic plasmons* are much harder to identify than the asymmetric lineshape, because they coincide with the *extrinsic plasmons*. However, the relative intensities of the successive plasmons are different for the two production mechanisms (intrinsic rate b, and extrinsic rate a), a feature which has allowed an unambiguous identification of intrinsic plasmons first in Be [4.32] and later also in Na, Mg and Al [4.31, 4.33].

Another method of demonstrating the existence of intrinsic plasmons involves Auger electron spectra. The principle of the experiment is given in Fig. 4.5. Here one sees, again as an example for Mg, an initial state which is a Mg atom photoionized in the 1s level. This 1s hole excites (in its neighborhood) intrinsic plasmons, as indicated in the figure. Now, in the deexcitation process Auger electrons can be liberated; these, besides their Auger kinetic energy, also carry the plasmon energy $\hbar\omega_p$. Thus one can observe plasmon-gain satellites in the spectra. Such plasmon-gain spectra have indeed been observed and an example is shown in Fig. 4.6 [4.34, 4.35].

Plasmon sidebands are also observed in the valence-band spectra of the simple metals [4.4,4.32,4.36–4.39]. The question arises of whether the intrinsic rate of plasmon production b in the valence band is the same as in a core

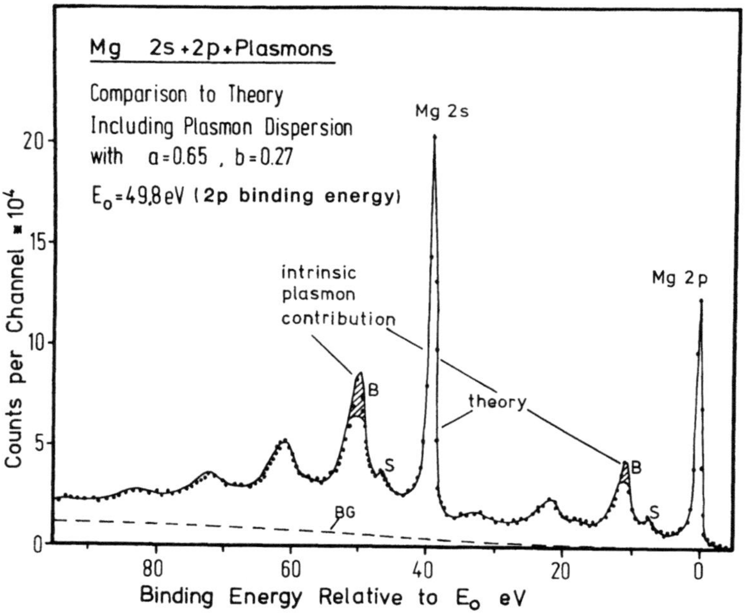

Fig. 4.4. XPS $2p_{1/2,3/2}$ and 2s spectra of Mg metal over an extended energy range. The Mg 2p and the Mg 2s line are both accompanied by a series of bulk (B) and surface (S) plasmons [4.37]. The calculation of the full line, taking the main line asymmetry and the intrinsic (b) and extrinsic (a) plasmon creation into account, is explained in the text. The intrinsic contribution to the first bulk plasmon accompanying the 2s line is indicated and is seen to be by no means negligible

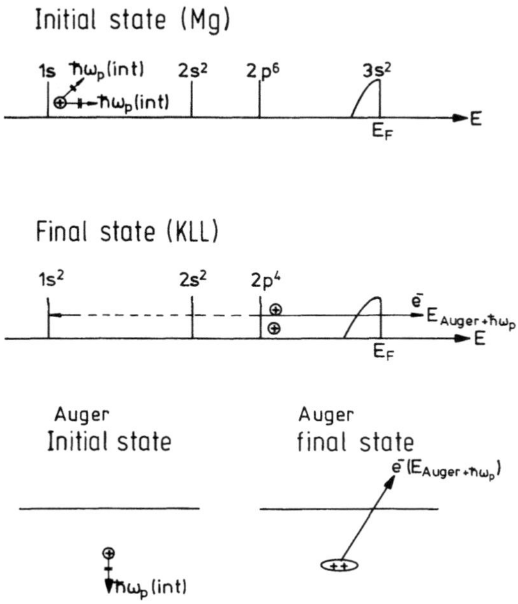

Fig. 4.5. Diagram to explain the production of plasmon-gain satellites observed in Auger spectra of simple metals. The initial state is a 1s hole, which has created intrinsic plasmons. The final state can therefore contain electrons that have kinetic energy $E_{\mathrm{kin}}(\mathrm{KLL}) + \hbar\omega_{\mathrm{p}}$ which is the plasmon gain satellite

level. It is easy to give arguments that the intrinsic plasmon creation rate in the valence band should be smaller than that in a core level: The potential that a valence-band hole creates is more delocalized than that produced by a core hole, with a resulting smaller coupling to the conduction electron system. Calculations indicate a slight decrease of the rate of intrinsic plasmon creation b between the core- and valence-band hole [4.40]. Results, both experimental and theoretical for b_{c} (index: core level) and b_{VB} (index: valence band), are summarized in Table 4.1.

On the experimental side Höchst et al. [4.37, 4.41] find that b_{c} and b_{VB} are similar, where however the errors in the measurements would easily allow a difference of 20 % between the core level and the valence band. In contrast another group [4.38] found a 50 % reduction in b for Al, in going from the core level to the valence band, which seems slightly outside the findings of Höchst et al. [4.37] even allowing for considerable error bars in both cases.

Upon closer inspection of the experimental situation one may reach the conclusion that b should be similar in valence and core levels. In a valence band, electrons are not "smeared out" in energy but rather "sit" in energy levels that are highly dispersive. This is borne out by the fact that UPS measurements on single crystal faces of simple metals [4.42, 4.43] give lines

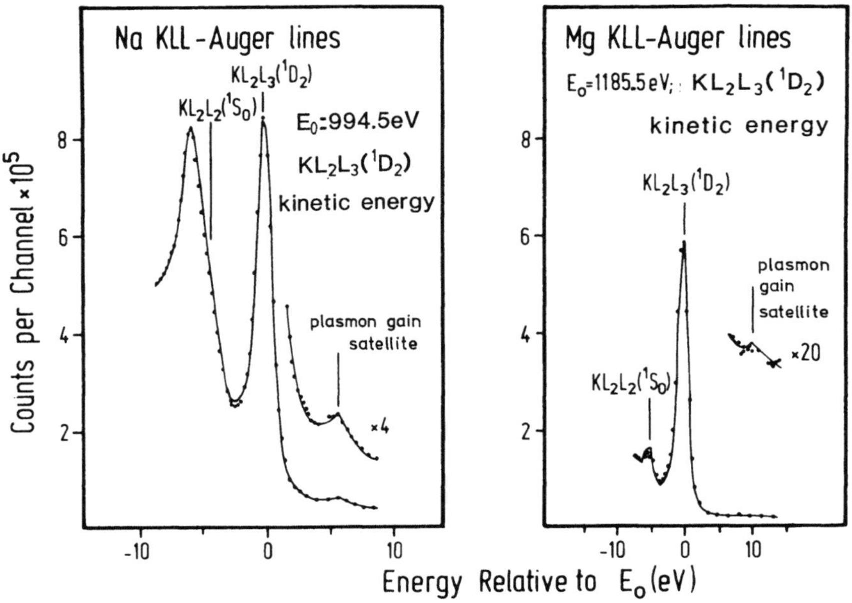

Fig. 4.6. KLL-Auger lines in Na and Mg showing the (weak) plasmon-gain satellite [4.35]. This demonstrates the existence of intrinsic plasmons in the initial (one-hole) state

Table 4.1. Intrinsic production rate of plasmons for core levels b_c and for valence bands b_{VB} for some simple metals

Metal	b_c (exp.)	b_{VB} (exp.)	b_c (theor.)	b_{VB} (theor.)
Na	0.53[a]	0.53[a]	0.66[c]	0.47[c]
Mg	0.27[a]	0.27[a]	0.44[c]	0.35[c]
Al	0.11[a]	0.11[a]	0.34[c]	0.29[c]
	0.21[b]	0.11[b]		

[a] [4.37], [b] [4.38], [c] [4.40]

(for the transitions between a valence and a conduction band) with a width of the order of 1 eV. This width, if attributed completely to the hole, reflects a lower limit of the lifetime of the hole. It is then of the same order of magnitude as that of core levels (the 2s level of Al) to which the Mahan–Nozières–DeDominicis [4.23, 4.25] theory is generally believed to apply. The question to be asked is not whether this theory applies as well to core as to valence levels, but at what magnitude of the lifetime relative to the valence bandwidth, delocalization effects start to become significant.

4.1 Theory

4.1.1 General

We shall first derive the equation for the primary PES spectrum (asymmetric line plus intrinsic plasmons) and then treat the escape of the electrons through the material and the surface, which adds the *secondary spectrum* to the primary one. The *primary spectrum* is created by the photoexcitation of an electron. It is generally a lifetime-broadened Lorentzian, modified by the immediate response of the conduction-electron system to the positive photohole. The theories predict a spectrum consisting of an asymmetric core line at the kinetic energy ω_0 and a series of plasmon lines at energies $\omega_0 - \omega_\mathrm{p}$, $\omega_0 - 2\omega_\mathrm{p} \ldots$ exactly as is observed in the experiment (Fig. 4.4) [4.1, 4.10–4.12, 4.23, 4.25, 4.26, 4.28, 4.30, 4.33, 4.44–4.54].

The primary photoelectron spectrum can be described by (see Sect. 1.5)

$$A(\omega) = \frac{1}{2\pi}\int_{-\infty}^{+\infty} \mathrm{d}t\mathrm{e}^{\mathrm{i}\omega t} f(t) , \tag{4.6}$$

where the Fourier transform $f(t)$ of $F(\omega)$ is given by

$$F(t) = \exp[-m(t)] \tag{4.7}$$

with [4.26, 4.28]

$$m(t) = \int_{-\infty}^{+\infty} \mathrm{d}\omega \frac{\varrho(\omega)}{\omega}(1 - \mathrm{e}^{\mathrm{i}\omega t}) \tag{4.8}$$

for a free-electron metal. The dynamic form factor $\varrho(\omega)$, which includes the interaction of the conduction electrons with the core hole is given in the random-phase approximation by (neglecting lifetime broadening) [4.26]

$$\varrho(\omega) = \frac{\Theta(\omega)}{\pi\omega}\sum_q \frac{|V_q|^2}{v_q}\mathrm{Im}\left\{-\frac{1}{\varepsilon(q,\omega)}\right\} , \tag{4.9}$$

where one has

$$v_q = \frac{4\pi\mathrm{e}^2}{q^2} \quad \text{and} \quad V_q = \int \mathrm{d}^3 r \mathrm{e}^{\mathrm{i}\boldsymbol{q}\cdot\boldsymbol{r}} V(r) . \tag{4.10}$$

$V(r)$ is the sudden change in the Coulomb potential through the photoexcitation process as seen by the conduction electron system. $\Theta(\omega)$ is the step function. $\varrho(\omega)$ peaks sharply at the plasma frequency ω_p and therefore $m(t)$ contains an oscillatory component of frequency ω_p. This oscillatory component when inserted into (4.7) yields components of frequency $\omega_\mathrm{p}, 2\omega_\mathrm{p} \ldots$, which indicates the possibility of intrinsic multiple plasmon production with a production rate b. Under the assumption of a Lorentzian of width $2\gamma_\mathrm{p}$ for $\varrho(\omega)$ one finds

$$f(t) = \sum_{n=0}^{\infty} \mathrm{e}^{-b} b^n [\exp(-\mathrm{i}n\omega_\mathrm{p} t - \mathrm{i}n\gamma_\mathrm{p} t)]/n! . \tag{4.11}$$

We shall now deal separately with the core-line spectrum and with its plasmon tail.

4.1.2 Core-Line Shape

A form for the core-line function which gives a *non-diverging* lineshape, was first suggested by Mahan [4.26]. He used the following equation for the dynamic form factor $\varrho(\omega)$ (the subscript "zero" indicates that one is dealing with the energy regime around the core line):

$$\varrho_0(\omega) = \alpha e^{-\omega/\xi}; \quad \alpha = 2\sum_{\ell}(2\ell+1)(\delta_\ell/\pi)^2 . \tag{4.12}$$

Here ξ is the cut-off parameter with a magnitude of the order of the Fermi energy E_F (which is the width of the occupied band), which is either taken exactly as the Fermi energy or adjusted to experiment. Thus we obtain

$$f_0(t) = (1 + it\xi)^{-\alpha} \tag{4.13}$$

and for the spectrum[5]

$$A_0(\omega) = \frac{1}{\Gamma(\alpha)} \frac{e^{\omega/\xi}}{|\omega/\xi|^{1-\alpha}} \Theta(-\omega) \tag{4.14}$$

$A_0(\omega)$ is the spectral function of the core line, cf (4.6). The expression is similar to the much-used Doniach–Sunjic lineshape [4.1] given by (4.5), which has the numerical disadvantage that it diverges. Therefore it is preferable to use the Mahan lineshape. From (4.14) the Doniach–Sunjic lineshape (4.5) is obtained in the limit $\xi \to \infty$, yielding $A_0(\omega) = \Theta(-\omega)/[\Gamma(\alpha)|\omega|^{1-\alpha}]$. In actual applications the two lineshapes are very similar and can only be distinguished far out in the tail.

Now we discuss the meaning of the Mahan lineshape (4.12), in terms of the dynamic form factor of the electron (4.9). For this purpose one needs an approximation for the potential V_q produced by the photohole.

In order to make the problem tractable we make the approximation $\omega = 0$ which is valid for excitations near to the core line. In addition, the Thomas-Fermi approximation will be used for the calculation of the dielectric constant.

The screened Coulomb potential is approximated by

$$V_q = c_q v_q \tag{4.15}$$

where c_q is the effective charge of the photohole seen by the conduction electrons. For $\omega = 0$ we have

$$\varepsilon(q) = \varepsilon_1(q) + i\varepsilon_2(q) , \tag{4.16a}$$

[5] Here the kinetic energies ω of the photoelectrons are given with respect to the core-level energy.

$$\mathrm{Im}\{-1/\varepsilon(q)\} = \varepsilon_2(q)/|\varepsilon(q)|^2 \, . \tag{4.16b}$$

Using the Thomas-Fermi approximation we get

$$\varepsilon_2(q) = \frac{\pi}{2}\frac{\omega}{v_\mathrm{F}}\frac{k_\mathrm{TF}^2}{q^3} \, , \tag{4.17a}$$

$$|\varepsilon(q)| = 1 + \frac{k_\mathrm{TF}^2}{q^2} \, , \tag{4.17b}$$

where the Thomas–Fermi wave vector is given by

$$k_\mathrm{TF}^2 = \frac{6\pi n e^2}{E_\mathrm{F}} \tag{4.18}$$

with v_F and E_F being, respectively, the Fermi velocity and the Fermi energy of the electron gas with density n.

Inserting (4.10,4.15-4.18) into (4.9) one obtains for $\omega = 0$

$$\varrho(\omega = 0) = 0.083 r_\mathrm{s} c_q^2 \, , \tag{4.19}$$

where $r_\mathrm{s} = [(3/4\pi)n^{-1}]a_0^{-1}$ is the Wigner–Seitz radius (in units of the Bohr radius a_0) of the electron gas with the density n. Comparing (4.19) with (4.12) for $\omega = 0$ one finally gets for the asymmetry parameter

$$\alpha_{\mathrm{Doniach-Sunjic}} = 0.083 r_\mathrm{s} c_q^2 \, . \tag{4.20}$$

This then produces the relation for the dynamic form factor of the electron at zero energy transfer and the exponent in the Doniach–Sunjic lineshape. Table 4.2 gives a compilation of experimental values for α and the values of c_q determined from (4.20). One sees that c_q is always smaller than one, indicating the validity of the procedure.

Table 4.2. Wigner–Seitz radius r_s, experimental Doniach–Sunjic parameter α_exp, and charge c_q of a photohole in the core region for several simple metals

	r_s	α_exp	c_q
Be	1.9	0.05	0.56
Na	4.0	0.19	0.76
Mg	2.6	0.12	0.75
Al	2.1	0.11	0.79

4.1.3 Intrinsic Plasmons

To calculate the plasmon lineshape, one has to know the dynamic form factor $\varrho(\omega)$ in the plasmon region. In order to render the calculations simple, we now neglect, in a first step, the q dependence of the dielectric constant. Of course, this gives a form for the dielectric constant different from (4.17a, 4.17b) where the approximation for $\omega = 0$ and finite q was used. For a free-electron gas the dielectric function is given by ($q = 0$)

$$\varepsilon(\omega) = 1 - \frac{\omega_{\mathrm{p}}^2}{\omega^2 + \mathrm{i}\gamma_{\mathrm{p}}\omega} , \tag{4.21}$$

where $\omega_{\mathrm{p}} = (4\pi n e^2/m)^{1/2}$ is the plasma frequency and $2\gamma_{\mathrm{p}}$ the full width at half maximum of the plasma resonance. For small values of γ_{p} one obtains

$$\varepsilon_1(\omega) = 1 - \omega_{\mathrm{p}}^2/\omega^2 , \tag{4.22}$$

$$\varepsilon_2(\omega) = \gamma_{\mathrm{p}}\omega_{\mathrm{p}}^2/\omega^3 . \tag{4.23}$$

Therefore the loss function $\mathrm{Im}\{-1/\varepsilon(\omega)\}$ is given approximately by a Lorentzian

$$\mathrm{Im}\left\{\frac{-1}{\varepsilon(\omega)}\right\} \simeq \frac{1}{2}\frac{\omega^2}{\omega_{\mathrm{p}}}\frac{\gamma_{\mathrm{p}}}{(\omega-\omega_{\mathrm{p}})^2+\gamma_{\mathrm{p}}^2} \quad \text{for} \quad q < q_{\mathrm{c}} \tag{4.24a}$$

and

$$\mathrm{Im}\left\{\frac{-1}{\varepsilon(\omega)}\right\} = 0 \quad \text{for} \quad q > q_{\mathrm{c}} . \tag{4.24b}$$

Now (4.24) is inserted into (4.9) in order to calculate $m(t)$ via (4.8). We use $V_q = c_q \cdot v_q$ and eliminate the q dependence of v_q by integrating to q_{c}, where $q_{\mathrm{c}} = \omega_{\mathrm{p}}/v_{\mathrm{F}}$ is the so-called *cut-off wave vector* [4.17, 4.18]. With

$$b = (e^2 c_q^2)/v_{\mathrm{F}} = 2\alpha \tag{4.25}$$

this gives

$$m(t) = \ln(1 + \mathrm{i}t\zeta)^{\alpha} + b(1 - \mathrm{e}^{-\omega_{\mathrm{p}}t - \gamma_{\mathrm{p}}t}) . \tag{4.26}$$

Now (4.26) is inserted into (4.7) and the result convoluted with the zero-loss line of width $2\gamma_0$ which yields

$$f(t) = \sum_{n=0}^{\infty} \frac{\mathrm{e}^{-b}b^n}{n!} \frac{\mathrm{e}^{-\mathrm{i}n\omega_{\mathrm{p}}t - \mathrm{n}\gamma_{\mathrm{p}}|t| - \gamma_0|t|}}{(1+\mathrm{i}t\xi)^{\alpha}} = \sum_{n=0}^{\infty} f_n(t) . \tag{4.27}$$

In terms of the photoelectron spectrum one has

$$A(\omega) = \sum_{n=0}^{\infty} A_n(\omega) , \tag{4.28}$$

where $A_n(\omega)$ is the Fourier transform of $f_n(t)$.

Thus, if the dispersion of the plasmons is neglected,[6] the intrinsic spectrum is given as a sum of Mahan lines at $E_n = \hbar\omega_n$ with half widths $\Gamma_n = 2\hbar\gamma_n$, and

[6] The inclusion of the plasmon dispersion into the calculations is difficult and has been discussed in [4.36, 4.56].

$$\omega_n = \omega_0 - n\omega_\mathrm{p} \ , \tag{4.29}$$

$$\gamma_n = \gamma_0 + n\gamma_\mathrm{p} \ . \tag{4.30}$$

The intensities I_n are given by

$$I_n = \mathrm{e}^{-b}\frac{b^n}{n!} \ . \tag{4.31}$$

4.1.4 Extrinsic Electron Scattering: Plasmons and Background

So far we have only considered what is called the *intrinsic spectrum.* In order to arrive at the total (measured) spectrum, we have to add that part of the spectrum (secondary spectrum) which is created during the traveling of the photoelectron from the site of its creation to the vacuum.

The transport of the photoelectrons to the surface can be described by a transport equation, as given by Wolff [4.55]. With the assumption that the electron scattering is mainly in the forward direction, which is realistic for the XPS regime, one finds for the spectrum $P(\omega)$ inside the sample

$$P(\omega) = \lambda_\mathrm{tot}(\omega)A(\omega) + \int_{\omega'>\omega} \mathrm{d}\omega' g(\omega',\omega)P(\omega') \tag{4.32}$$

with

$$g(\omega',\omega) = \Theta(\omega'-\omega)\lambda_\mathrm{tot}(\omega)/\lambda(\omega',\omega) \ , \tag{4.33}$$

$$\lambda^{-1}(\omega',\omega) = S(\omega',\omega)/v(\omega') \ , \tag{4.34}$$

$$1/\lambda_\mathrm{tot}(\omega) = \int_{\omega'<\omega} d\omega'\lambda^{-1}(\omega,\omega') \ . \tag{4.35}$$

Equation (4.32) gives the exact separation of the measured spectrum $P(\omega)$ into *primary spectrum* (first term representing the main line plus intrinsic excitations) and the *secondary spectrum* (second term representing the inelastic events after the photoexcited electron has left the site of the excitation process). In the most common nomenclature one calls the first part of the right-hand side of (4.32) the primary spectrum and the second term the background (which is another name for the secondary spectrum).

The exact evaluation of the background function will be given in Sect. 4.3. In the simple metals the loss function $g(\omega',\omega)$ is dominated by plasmon scattering [4.18] and a separation of the form

$$g(\omega',\omega) = g_\mathrm{p}(\omega',\omega) + g_\mathrm{e}(\omega',\omega) \tag{4.36}$$

where $g_\mathrm{p}(\omega',\omega)$ contains solely the plasmon part and $g_\mathrm{e}(\omega',\omega)$ the other inelastic processes, seems possible.

If we assume that the plasmon lineshape is given by a Lorentzian $L(\omega' - \omega_\mathrm{p}, \gamma_\mathrm{p})$ we have

$$g_{\mathrm{p}}(\omega',\omega) = \frac{\lambda_{\mathrm{tot}}(\omega)}{\lambda_{\mathrm{p}}(\omega)} L(\omega' - \omega_{\mathrm{p}}, \gamma_{\mathrm{p}}) \tag{4.37}$$

yielding for the spectrum due to bulk-plasmon scattering (extrinsic plasmons)

$$P_{\mathrm{b}}(\omega) = \frac{\lambda_{\mathrm{tot}}(\omega)}{\lambda_{\mathrm{ep}}(\omega)} \int \mathrm{d}\omega' P(\omega+\omega') L(\omega' - \omega_{\mathrm{p}}, \gamma_{\mathrm{p}}) \; . \tag{4.38}$$

The functions are defined as follows:

$\lambda_{\mathrm{tot}}(\omega)$	:	total mean free path,
$L(\omega' - \omega_{\mathrm{p}},\ \gamma_{\mathrm{p}})$	:	normalized lineshape of a plasmon resonance described by a Lorentzian at the energy ω_{p} with the linewidth $2\gamma_{\mathrm{p}}$,
$v(\omega')$	:	velocity of electron with energy ω',
$S(\omega',\omega)$	:	probability that an electron is scattered from ω' to ω,
$A(\omega)$	:	primary photoelectron spectrum, spectral function,
$\lambda^{-1}(\omega',\omega)$	:	differential scattering length,
$\Theta(\omega' - \omega)$	:	step function,
$\lambda_{\mathrm{p}}(\omega)$	:	attenuation length due to plasmon creation.

If the weak ω-dependence of λ_{tot} and λ_{p} is neglected one can write

$$\frac{\lambda_{\mathrm{tot}}(\omega)}{\lambda_{\mathrm{p}}(\omega)} = a \; , \tag{4.39}$$

which leads to

$$P_{\mathrm{b}}(\omega) = \sum_{n} P_{\mathrm{b}n}(\omega) \tag{4.40}$$

with

$$P_{\mathrm{b}n}(\omega) = a \int \mathrm{d}\omega' P_{\mathrm{b}\,n-1}(\omega+\omega') L(\omega' - \omega_{\mathrm{p}}, \gamma_{\mathrm{p}}) \; . \tag{4.41}$$

For the part of the background produced by electron–electron and electron–ion scattering (i.e., all inelastic processes except plasmon scattering), one obtains the intensity

$$\bar{B}(\omega) = \frac{1}{1-a} \int_{\omega'>\omega} \mathrm{d}\omega' g_e(\omega',\omega) P(\omega') \; . \tag{4.42}$$

($\bar{B}(\omega)$ is the background intensity without the plasmons, whereas $B(\omega)$ is the total background intensity, i.e., the *secondary spectrum*).

Assuming in the evaluation of $B(\omega)$ that $S_{\mathrm{e}}(\omega,\omega')$ [which is inserted into (4.42) via (4.33, 4.34)] is proportional to the density of states in phase space at energy $\hbar\omega$ one finds:

$$\bar{B}(\omega) = \frac{3}{2\bar{\omega}} \int_{\omega'>\omega} \mathrm{d}\omega' P(\omega') \; , \tag{4.43}$$

where $\bar{\omega}$ is the mean energy of the photoelectrons for the spectrum under consideration.

Finally, the electron passing through the surface can also excite surface plasmons.[7] One takes them into account in a way which is analogous to the extrinsic bulk-plasmon contribution

$$P_{\mathrm{s}}(\omega) = \sum_{n} a_{\mathrm{s}} P_{\mathrm{s}n} \tag{4.44}$$

which gives the surface-plasmon contributions with relative intensities a_{s}.

4.1.5 The Total Photoelectron Spectrum

The complete photoelectron intensity spectrum $P_{\mathrm{tot}}(\omega)$, neglecting interference effects between the probability amplitudes of the various mechanisms, can then be written as

$$P_{\mathrm{tot}}(\omega) = \bar{B}(\omega) + P_{\mathrm{s}}(\omega) + P_{\mathrm{b}}(\omega) + A_{\mathrm{i}}(\omega) + A_{0}(\omega) \; , \tag{4.45}$$

where the partial intensities are

$\bar{B}(\omega)$: background (excluding plasmons),
$P_{\mathrm{s}}(\omega)$: surface plasmons,
$P_{\mathrm{b}}(\omega)$: extrinsic bulk plasmons
$\sum_{n=1}^{\infty} a^n \int \mathrm{d}\omega' A_0(\omega+\omega') L(\omega' - n\omega_{\mathrm{p}}, n\gamma_{\mathrm{p}})$,
$A_{\mathrm{i}}(\omega)$: intrinsic bulk plasmons
$\sum_{n=1}^{\infty} \frac{b^n}{n!} \int \mathrm{d}\omega' A_0(\omega+\omega') L(\omega' - n\omega_{\mathrm{p}}, n\gamma_{\mathrm{p}})$,
$A_0(\omega)$: zero-loss line (Doniach–Sunjic or Mahan lineshape), as obtained in the experiment. It therefore contains implicitly as a factor the attenuation $(1-a)^{-1}$ due to extrinsic and e^{-b} due to intrinsic plasmon creation.

4.2 Experimental Results

4.2.1 The Core Line Without Plasmons

The PE process from the core level of an atom produces a Lorentzian line, where the linewidth is determined by the lifetime of the state. In an insulator, this Lorentzian is hard to observe because of the tendency for differential charging which adds a Gaussian (symmetric) contribution to the linewidth.

In metals, the situation is inherently different because such charging does not occur. As was seen, the PE process in a core level of a metal produces a Lorentzian (with a width determined by the lifetime of the level), which

[7] Surface plasmons are collective excitations in the electron gas, similar to the bulk plasmons, which are, however, restricted in their spatial extension to the last atomic layer of the surface [4.17, 4.18].

is, however, "dressed" with the spectrum that is produced by the creation of electron–hole pairs. This is an intrinsic process, and the intrinsic lineshape is given by the asymmetric Doniach–Sunjic or Mahan lineshape. In principle, one always has to add the phonon broadening but this represents, except for Li metal, a small contribution for the metals investigated so far [4.9].

Looking at XPS core-level spectra of metals, it seems that the effect of the asymmetric lineshape is very obvious, especially in transition metals where the effect is particularly pronounced [4.2, 4.13, 4.14]. Due to the instrumental limitations of early PE experiments, experimenters did not have sufficient confidence in their data; thus only some time after the work of Doniach and Sunjic was the observation of this effect reported [4.2, 4.3]. The exact agreement of the analytical Doniach–Sunjic lineshape with a measured spectrum, taking lifetime and instrumental resolution into account [4.7], was then taken as proof of the Mahan–Nozières–DeDominicis [4.23, 4.25] effect in metals, also demonstrating the existence of the Anderson [4.24] orthogonality catastrophe.

The most thoroughly investigated group of metals with respect to the many-body excitations are the simple metals [4.4, 4.9] because for these comparison can be made with the many-body effects observed by other spectroscopic techniques like X-ray emission and absorption and electron scattering. A typical example of XPS lines (1s and 2s of Mg metal) fitted to (4.14) is shown in Fig. 4.7. It is immediately evident that the agreement between theory and experiment is good, lending strong support to the underlying theory (see also Fig. 4.3). Table 4.3 gives a compilation of a number of parameters obtained from fits to PES spectra for Na, Mg and Al.

Table 4.3. Parameters describing the core-line spectra of Na, Mg, and Al. ($2\gamma_0$: linewidth (FWHM), α: singularity index, a: creation rate for extrinsic plasmons, b: creation rate for intrinsic plasmons)

	$2\gamma_0$	$2\gamma_0$	α	α	α	a	a	a	b	b	b	b
Na 2s	0.14[a]	0.28[b]	0.19[a]	0.21[b]	0.21[c]	0.67[a]		0.58[d]	0.53[a]		0.41[e]	0.66[h]
Na 2p	0.05[a]	0.02[b]	0.19[a]	0.20[b]	0.21[c]	0.67[a]		0.58[d]	0.53[a]		0.41[e]	0.66[h]
Mg 2s	0.48[a]	0.46[b]	0.12[a]	0.13[b]	0.13[c]	0.65[a]	0.67[f]	0.66[d]	0.27[a]	0.19[f]	0.36[e]	0.44[h]
Mg 2p	0.05[a]	0.03[b]	0.12[a]	0.14[b]	0.13[c]	0.65[a]	0.67[f]	0.66[d]	0.27[a]	0.19[f]	0.36[e]	0.44[h]
Al 2s	0.76[a]	0.78[b]	0.10[a]	0.12[b]	0.11[c]	0.66[a]	0.62[g]	0.71[d]	0.11[a]	0.21[g]	0.26[e]	0.34[h]
Al 2p	0.05[a]	0.04[b]	0.10[a]	0.12[b]	0.11[c]	0.66[a]	0.62[g]	0.71[d]	0.11[a]	0.21[g]	0.26[e]	0.34[h]

[a] [4.56], [b] [4.9], [c] [4.48], [d] [4.17, 4.18, 4.57], [e] [4.33], [f] [4.38], [g] [4.39], [h] [4.40]

With respect to the linewidth $2\gamma_0$ and the singularity parameter α, there is good agreement between the results of two groups [4.4, 4.9] that have used quite different methods of data analysis. It can also be seen that different core levels have the same α, as expected since at no point in the discussion of

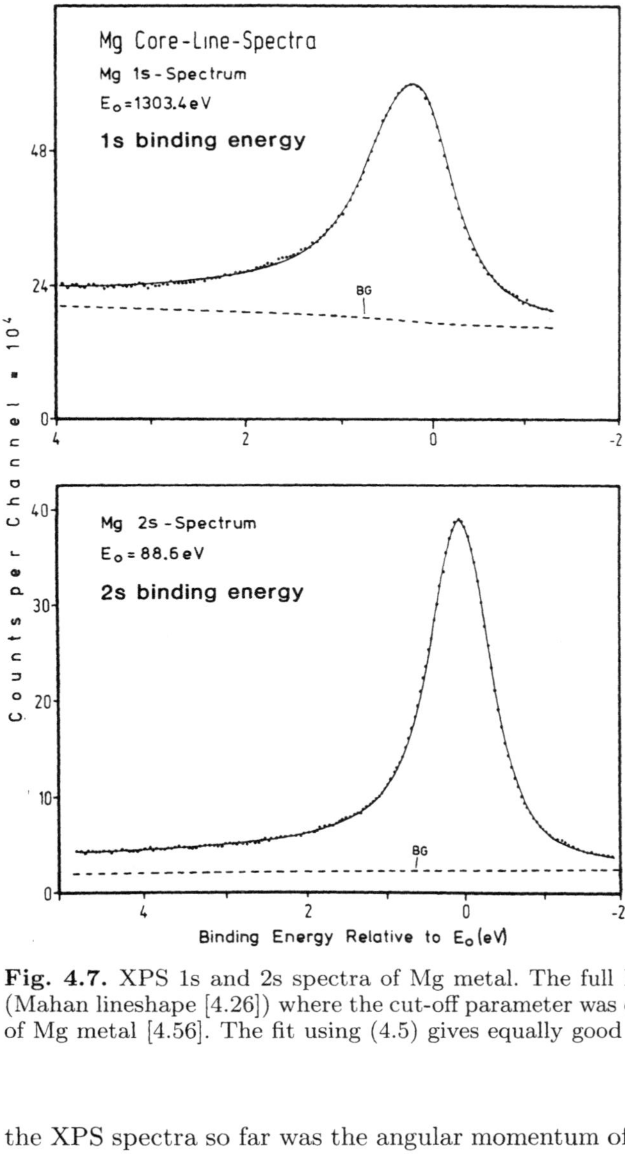

Fig. 4.7. XPS 1s and 2s spectra of Mg metal. The full lines are fits using (4.12) (Mahan lineshape [4.26]) where the cut-off parameter was equal to the Fermi energy of Mg metal [4.56]. The fit using (4.5) gives equally good results. BG: background

the XPS spectra so far was the angular momentum of the core hole used, see (4.5). There have been a number of calculations of α for the simple metals. The results deviate slightly but not significantly from one another. Table 4.3 contains the results of a representative calculation [4.28] and again one notices good agreement with the experimentally determined numbers [4.9].

Fits to the experimental spectrum with a Mahan [4.26] or Doniach–Sunjic [4.1] lineshape can only be made over a limited energy range away from the center of the line, because the background and the plasmons become important further out. Therefore a full fit to a spectrum including the plasmon tail and the background not only helps to obtain data for the core lines but also gives additional information on the intrinsic plasmon creation rate. A careful data analysis thus helps to confirm the validity of our physical picture about the photoemission process in a metal.

4.2.2 Core-Level Spectra Including Plasmons

The question of whether the plasmon satellites seen in the core-level spectra are intrinsic or extrinsic has aroused much controversy [4.58–4.64]. While it was evident from theoretical work that the intrinsic contribution to the plasmon creation rate must exist [4.10–4.12, 4.45, 4.46, 4.65–4.68], its magnitude was debated. Since the extrinsic and the intrinsic plasmons cannot easily be distinguished from one another, the demonstration of intrinsic plasmon creation is a problem of quantitative analysis of the intensities of adjacent plasmons. This involves a considerable amount of data analysis and therefore is subject to some uncertainty.

A most elegant experiment to settle the question had been designed by Fuggle et al. [4.63]. They compared the core-level spectra of Mn and Al in overlayer samples of Al on Mn. Whereas the core spectra from Al always showed their plasmon structure, even for thick (40 Å) layers of Al on Mn, no Al plasmons could be seen on the core levels from Mn, which should have been the case if the plasmons were predominantly extrinsic. On the basis of this experiment the researchers judged that, at least in Al, the plasmon satellites are mostly of intrinsic origin. This rather nicely designed experiment produced, however, the wrong results because, as will be demonstrated below, in Al the plasmons are mostly extrinsic and have only a small intrinsic contribution. One can only speculate that the Al overlayers in this experiment consisted of large, not connected islands, making the missing plasmons in the spectra of the overlayer samples understandable. Pardee et al. [4.64] analyzed the core-level spectra of Mg, Al and Na in order to find out whether the intensities of the plasmon satellites are produced by intrinsic or extrinsic processes. If intrinsic these should follow a Poisson distribution as a function of n [$(b^n/n!)e^{-b}$ in (4.31)], while the nth extrinsic contribution is obtained as a pure power law (a^n). Pardee et al. [4.64] came semiquantitatively to the conclusion of small intrinsic plasmon contributions in Al and Mg and a measurable one in Na.

The first fully quantitative analysis was that of Höchst et al. [4.32] for the 1s spectrum of Be, proving conclusively the existence of a large intrinsic contribution to the plasmon spectrum. Figure 4.8 shows the core-level spectra of the Be 1s line [the points are the raw data and the full lines are the fits to the data with (4.45) including the plasmon tail]. The intrinsic contribution to the plasmon creation is most easily seen in the first bulk plasmon loss because of the rapid decay of the intensity of this term following the $(b^n/n!)e^{-b}$ law. The indicated "intrinsic area" in Fig. 4.8 shows that a purely "extrinsic" analysis would clearly be in error.

From Fig. 4.8 one sees that the plasmons contain a great deal of intensity. One can also observe that the theoretical curve fits the measured spectrum quite well, giving support to the underlying theory. Note also that Fig. 4.4 shows an analysis of the region of the 2p and 2s spectra of Mg metal and that there again good agreement between theory and experiment is evident.

Parameters for the intrinsic and extrinsic plasmon creation rates are also given in Table 4.3. Unfortunately only two cases exist where independent experimental determinations of b have been made, and the values show a considerable scatter [4.37–4.39, 4.56]. More experimental work is needed to clarify the situation. Results of the plasmon analyses of some simple metals are given in Fig. 4.9, where the normalized intensities W_n have been calculated from the relation

$$W_n = b_n + aW_{n-1} \ , \tag{4.46}$$

with

$$b_n = \frac{b^n}{n!}, \quad W_n = \frac{I_n}{I_0}, \quad I_0 = \mathrm{e}^{-b} \ ,$$

and are compared to the measured intensities. Again one sees good agreement between the calculated curve and the measured intensities.

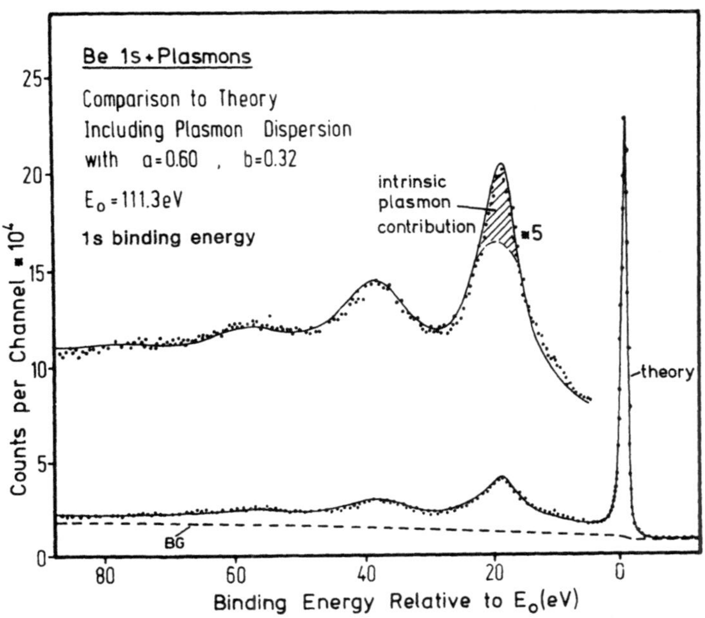

Fig. 4.8. XPS 1s spectrum of Be metal [4.32] showing the main line and the plasmon "sidebands". The latter have an intrinsic contribution as indicated (*shaded*) at the first bulk plasmon. The full line is a fit using (4.45)

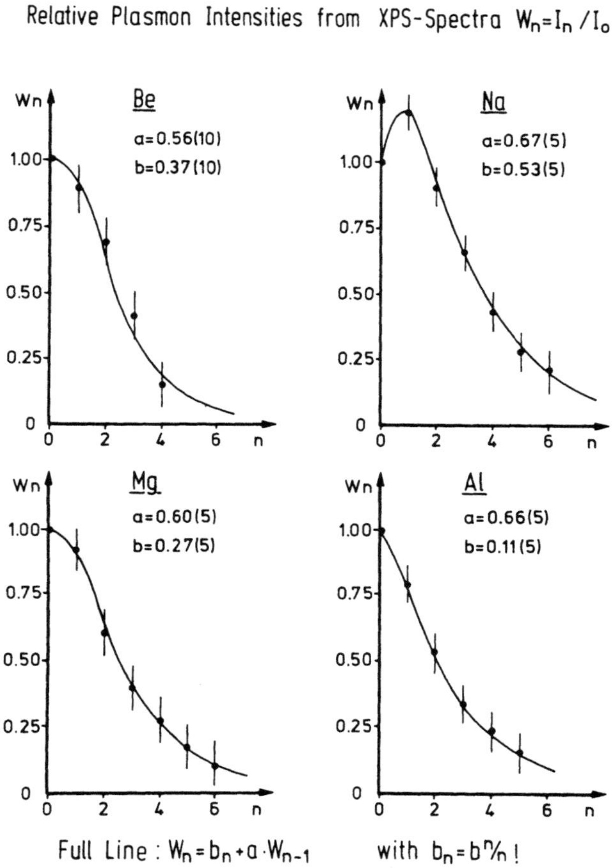

Fig. 4.9. Consistency check on the plasmon creation rates in Be, Na, Mg, Al. Plotted for each metal is the function $W_n = (b^n/n!) + aW_{(n-1)}$ with $W_n = I_n/I_0$ (I: intensity) and the a and b values as given in the figure. The points are extracted from the measured spectra. We observe reasonable agreement between measured and calculated intensities [4.56]

Steiner et al. [4.4,4.56] employed the line shape given by Mahan [4.26] for analyzing the core lines. The singularity exponent α obtained in this way does not differ much from the one obtained in the core-line analysis by Wertheim and Citrin [4.6,4.8,4.9,4.69] using the simpler Doniach–Sunjic lineshape [4.1]. The cut-off parameter ξ is also found from the fits [4.4,4.56] but contains large uncertainties; it is, however, gratifying to realize that it is of the order of the Fermi energy (and the bulk plasmon energy) as it should be [4.70,4.71].

The plasmon linewidths and energies follow the anticipated behavior as shown in Fig. 4.10 for the 2s plasmons in Mg metal: they are a linear function of the excitation number n (4.26). This indicates that no plasmon–plasmon coupling is present. The linewidth plotted in Fig. 4.10 is not the true plasmon linewidth, but one which is augmented by the fact that the large angle of the measurement integrates over the plasmon dispersion. If this contribution is subtracted from the measured width, the results are in agreement with those from direct energy-loss measurements [4.17, 4.18]. The same statement holds for the measured plasmon energies. They are slightly larger than those from the direct energy-loss measurements [4.17] which is again the result of averaging over the plasmon dispersion in the XPS experiment because of its large acceptance angle.

The values for the parameter a in Table 4.3 can be used to determine the electron mean free paths. The constant a is given by $a = \lambda_{\mathrm{tot}}/\lambda_{\mathrm{ep}}$ where λ_{tot} is the total mean free path of an electron of kinetic energy $\hbar\omega$ and λ_{ep} is the mean free path due to extrinsic plasmon creation. Now $\lambda_{\mathrm{ep}} = 4a_0 \times (\omega/\omega_{\mathrm{p}})/\ln(\omega/E_{\mathrm{F}})$, where a_0 is the first Bohr radius, ω_{p} is the bulk plasmon energy and E_{F} is the Fermi energy. From these relations the total mean free paths given in Table 4.4 are calculated. They show good agreement with theoretical predictions of Penn [4.57].

One question that remains to be solved is that of the interference between the extrinsic and intrinsic plasmons [4.48, 4.49]. Various estimates indicate that there should be considerable interference, especially at low kinetic energies of the photoelectrons. These interference effects would reduce the measured values of a and b relative to those calculated. They are, however, strongly dependent upon the kinetic energy of the photoelectron and could amount to as much as about 50 % for the Na 1s level and to much less for the Na 1s and 2p levels. However, experimentally the a and b parameters obtained for these three levels seem to be the same to within experimental

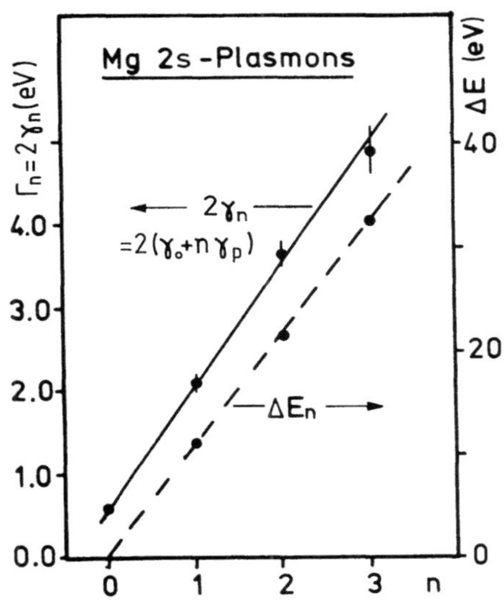

Fig. 4.10. Half width ($2\gamma_n$) and energy separation ΔE_n of the nth plasmon accompanying the 2s line in Mg metal [4.56]

Table 4.4. Electron attenuation length λ_{tot}^{exp}[Å] calculated from the bulk plasmon creation rate a and calculated values λ_p [Å] of the plasmon scattering length. λ_{ei}[Å] is the contribution of the electron–electron and electron–ion interaction to λ_{tot}. λ_{tot}^{theory} are the values obtained in the theoretical treatment of [4.57]

	E_{kin} [eV]	a	λ_p [Å]	λ_{tot}^{exp} [Å]	λ_{tot}^{theory} [Å]	λ_{ei} [Å]
Be 1s	1376	0.55	32	18	24	41
Na 1s	416	0.67	31	21	18	65
Na 2s	1423	0.67	86	58	49	178
Na 2p	1456		86	58	50	158
Mg 2s	1398	0.65	52	34	32	98
Mg 2p	1437		53	35	33	103
Al 2s	1369	0.66	39	25	26	70
Al 2p	1414		40	26	27	74

accuracy which can indicate that interference phenomena between extrinsic and intrinsic plasmon creation processes are not very important.

An interesting aspect of the available data concerns the computer fitting of XPS core lines. The 2p lines of Na, Mg and Al are all spin–orbit split doublets, and the spin–orbit splitting had been inserted in the fitting routine,

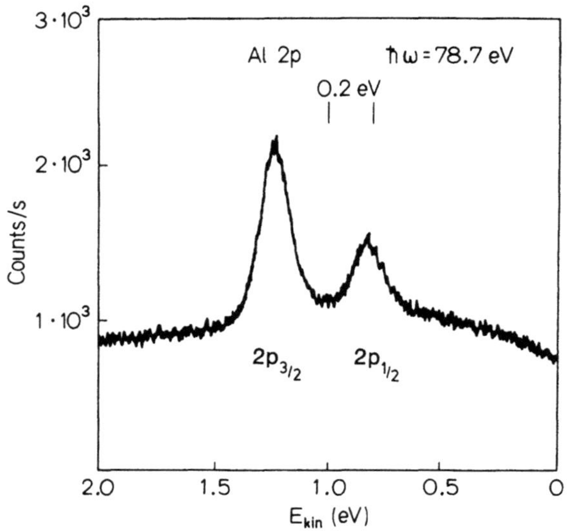

Fig. 4.11. High resolution Al 2p spectrum from Al metal taken with synchrotron radiation. The spin–orbit splitting is well resolved [4.72]

although in the data it was hardly resolved, except for Al. For illustrative purposes Fig. 4.11 shows the Al 2p spectrum measured at threshold with synchrotron radiation by Eberhardt et al. [4.72]. It clearly displays the spin–orbit splitting of 0.4 eV into the $2p_{1/2}$ and $2p_{3/2}$ lines, a value also found by fitting an unresolved line doublet. However, also in XPS experiments with very high resolution [4.69] the $2p_{1/2,3/2}$ doublet of Al can be resolved.

The consistent analysis of the core lines demonstrated by the data in Table 4.3 shows that successful computer fitting of unresolved lines can be performed [4.73].

4.2.3 Valence-Band Spectra of the Simple Metals

Theoretical and experimental results for DOS curves of simple metals have been available now for more than 40 years. However, if for a particular element one compares the various experimental results with one another and with the theoretical predictions, the agreement is not as good as one might hope [4.37].

Therefore it was tempting to use the method of PES to obtain more reliable experimental data. Since the simple metals are in principle free-electron-like, which means that features induced by the crystal potential and by electron correlations are small (leading, e.g., to an almost spherical Fermi surface) one might think that it should be easy to determine, e.g., the bandwidth (or in other words the Fermi energy) of the simple metals by XPS or UPS from a polycrystalline sample. This has the added advantage that one can work with vapor-deposited samples, which is the easiest method to produce clean surfaces of these very reactive materials. The outcome of this attempt has been somewhat disappointing, however, as can be seen from the data of Table 4.5. Here we compare the results for the Fermi energy of a few simple metals as determined by XPS [4.37] and ARUPS [4.42, 4.43] experiments. It is apparent that ARUPS gives smaller Fermi energies than XPS in every case. The ARUPS values are also consistently smaller than predicted theoretically (by a one-electron calculation). The origin of the difference in the XPS and ARUPS data is not evident. One should point out, however, that the data analysis is more difficult in the XPS case than in the ARUPS case. The analysis of the XPS data will now be outlined briefly. This is done with the intention of giving some idea of how XPS data can be used to derive Density-Of-States (DOS) results. The latter are useful in understanding the electronic structure of a compound and are sometimes the only data available (see also Sect. 7.4.3).

We would like to reemphasize here that PE measures a many-electron final state. This can be denoted as an electron removal spectrum, an Energy Distribution Curve (EDC) or, more formally, the imaginary part of the one particle Green's function. If these data are analyzed properly they yield a many-electron band structure. The state-of-the-art band-structure calculations (local-density approximation) are, however, one-electron calculations. Only if correlations are negligible can one compare electron removal spectra

Table 4.5. Fermi energies of simple metals in eV [4.4, 4.37, 4.42, 4.43, 4.74]

	E_F free electron	E_F band theory	E_F ARUPS	E_F XPS
Be	14.3	11.2	11.1	11.9
Na	3.2	3.15	2.5	3.2
Mg	7.1	6.8	6.1	6.9
Al	11.7	11.1	10.6	11.2

with the results of the one-electron calculations in good faith. Alternatively one can use the discrepancy between an EDC and a one-electron band structure to estimate the strength of the electron–electron interactions which are usually expressed by the effective mass.

It seems fair to state at this point that the problem of the bandwidth of the simple metals is not yet solved completely neither with respect to the differences found in the results of different measuring techniques nor with respect to the comparison of experimental and theoretical (one-electron) results. In this situation we can only define the open questions as clearly as possible.

In principle, a photoemission spectrum obtained in an XPS experiment yields information about the total PE density of states. However, a number of corrections have to be applied to the PE spectrum in order to compare it directly with a theoretical (correlated) density of states curve. These corrections take into account the different transition probabilities of electrons with different angular momentum, the correction introduced by the photohole–conduction–electron coupling, and the correction for inelastic scattering of the photoexcited electron. In performing an actual experiment and analyzing the data, it is difficult to recover the "bare" PE density of states function from the measured spectrum. Therefore one usually applies the aforementioned corrections to a theoretical single-particle density of states and compares this synthesized spectrum with the experimental data. This procedure biases the final result towards the theoretical one-electron density of states used and may be one source of the discrepancies between the XPS and ARUPS data seen in Table 4.5.[8]

To illustrate the procedure, the results for Be metal will be discussed. Figure 4.12 shows an XPS spectrum of polycrystalline Be and, as an insert, the total single-particle DOS curve for this material [4.75]. Several authors have calculated these DOS curves, but the deviations between the various calculations are not large. The agreement between the XPS spectrum and

[8] In principle, one would have to employ a correlated band structure to construct the synthesized spectrum.

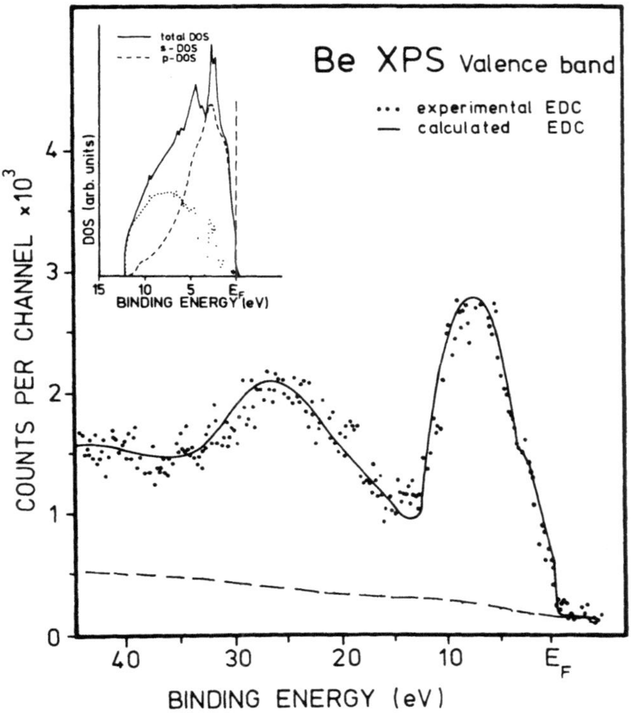

Fig. 4.12. XPS valence-band spectrum of Be metal and its analysis [4.4,4.37]. The insert gives the calculated total density of states [4.75] and the s- and p-partial density of states [4.76,4.77]. The background is indicated by the dashed line

the calculated total single-particle density of states in the insert is not good. One notes especially that the XPS curve shows a peak at about 8 eV (below the Fermi energy), whereas the calculated total DOS curve shows a peak around 2 eV. A clue to this puzzling difference can be obtained from the SXE (Soft X-ray Emission) spectrum of Be metal [4.76]. Since this spectrum is obtained from a transition to an s level (valence band to 1s transition), it measures, due to the $\Delta\ell = \pm 1$ selection rule, only the p partial DOS curve of this material. If one decomposes the total calculated DOS curves of Be metal by using this p contribution [4.77], one obtains the s partial DOS, as also shown in the insert of Fig. 4.12. It is seen that the peak in the s-partial DOS curve roughly coincides with the peak in the experimental XPS spectrum. This is an indication that the XPS spectrum is dominated by the s density of states.

The separation of the maxima in the s- and p-partial densities of states by 6 eV is in agreement with conclusions reached from Auger electron spectra [4.78, 4.79]. In order to generate a theoretical XPS spectrum, the theoretical s- and p-partial density of states were folded with the Doniach–Sunjic function [4.1] with an asymmetry taken from the core-hole spectrum and also its surface and bulk plasmon tail. The s- and p-partial densities of states modified in this way were then added to produce the "theoretical" XPS spectrum. The only free parameter in this procedure was the relative contributions of the p- and s-partial density of states to the observed spectrum, which was adjusted by a fit to the experimental XPS spectrum. The best fit between theoretical and experimental XPS spectrum is given as the full curve in Fig. 4.12. It was obtained with a ratio of the s to p photoexcitation cross sections of $\sigma_s : \sigma_p = 5 : 1$. The agreement between the measured and the generated curves is quite good, indicating the validity of the procedure and the accuracy of the theoretical density of states curve. Interestingly, the theoretical DOS curve of Herring and Hill [4.80], which was obtained in 1940, gives results not very different from those shown in Fig. 4.12.

A point of concern in the calculation of the theoretical photoemission spectrum is the use of the electron–hole coupling strength, as deduced from the core level. This is only justified if the photohole produced in a valence-band photoemission process stays highly localized. There are experimental data that support this assumption. Photoelectron diffraction measurements [4.81, 4.82] have given evidence for a strong hole-state localization in valence-band photoemission measurements. In addition, the use of electron–hole coupling constants for the valence-band interpretation different from those derived for core levels leads to worse agreement between the calculated and the measured photoemission spectra.

The XPS spectra of Na, Mg, and Al have been analyzed in a similar way [4.37,4.56]. The results for the Fermi energies are summarized in Table 4.5 and compared with numbers from other sources.

Some comments about the discrepancies in Table 4.5 seem to be in place. We shall concentrate on Na, Mg and Al because these metals can be described most easily as free-electron-like.

The difference between the ARUPS data and the theoretical results may have (at least in part) a simple origin, connected with the electron–hole pair creation by the photohole. The following reasoning is not meant as a rigorous theory but rather an indication of how the many-body interactions may influence even UPS experiments [4.83]. For the theoretical data, the free-electron results will be used because the various band-structure calculations are not in complete agreement. The sum rule [4.84–4.87] for the total intensity of a PE spectrum $A(E)$ fixes its center of gravity at the Koopmans' binding energy. If $A(E_B)$ is the PE spectrum, one has [4.54, 4.84–4.88]

$$\int_{-\infty}^{+\infty} A(E)E\mathrm{d}E = -\varepsilon_K \; , \tag{4.47}$$

where the index B in the binding energy has been dropped and ε_K is the Koopman's binding energy (Chap. 1). Note that in the original literature, the photon energy is explicitly retained. Neglecting the work function for the moment, setting $-E_{\mathrm{B}} = E_{\mathrm{kin}} - \hbar\omega$, and dropping the index "kin", this yields the usual formula

$$\int_{-\infty}^{+\infty} A(E - \hbar\omega) E \mathrm{d}E = \varepsilon_{\mathrm{K}} \ . \tag{4.48}$$

The Koopmans' binding energy of a state at the bottom of the conduction band is, by definition, the Fermi energy. We now calculate the center of gravity of a single line spectrum at a binding energy equaling the Fermi energy which has the asymmetric form given by the Mahan line shape function (4.14). Setting $\xi = E_{\mathrm{F}}$ one obtains:[9]

$$\int_{-\infty}^{+\infty} A(E) E \mathrm{d}E = \alpha E_{\mathrm{F}} + E_{\mathrm{F}} \tag{4.49}$$

with $A(E)$ given by (4.14).

The sum rule (4.47,4.48), however, fixes the right-hand side of (4.49) to E_{F}! This means that one has to start out with a line at energy $E_{\mathrm{F}} - \alpha E_{\mathrm{F}}$ which then is the position of the experimentally observed bottom of the conduction band or the experimentally observed Fermi energy $E_{\mathrm{F}}^{\mathrm{exp}}$ satisfying

$$E_{\mathrm{F}}^{\mathrm{exp}} = E_{\mathrm{F}} - \alpha E_{\mathrm{F}} \ . \tag{4.50}$$

Using α values from Table 4.3, the "experimental" Fermi energies that should be observed in ARUPS experiments have been calculated and are given in Table 4.6 together with the measured ones. The data at least suggest that the reasoning presented contains some elements of truth. With respect to the use of α values derived from core levels in the calculation of valence-band properties, we remark, as before, that judging from the ARUPS experiments the valence-band holes have a lifetime similar to that found in core levels for which the Mahan/–Nozières–DeDominicis theory has been successfully applied, justifying their use also for the valence bands. In addition, there is other experimental evidence for strong valence-band hole localization [4.81, 4.82] lending support for our procedure.

If this analysis is correct, one may say that the discrepancy between the photoemission Fermi energy (second column of Table 4.6) and the free-electron Fermi energy (first column of Table 4.6) is caused by a PE final-state effect (electron–hole pair creation). This would imply that the one-electron calculations give the correct band structure for the free-electron like metals.

A much more elaborate explanation for the apparent reduction in bandwidth in the ARUPS data, as compared to the free-electron value, has been given by Shung and Mahan [4.89, 4.90]. Their conclusion is that "the band structure of the simple metals is indeed simple, although it does not always reveal itself in a simple way". For a further discussion of the ARUPS data in relation to band structures, see [4.74].

9 ξ in (4.14) is a cut-off parameter that is generally taken to be of the order of the Fermi energy (or bandwidth).

Table 4.6. Comparison of the free-electron Fermi energies for Na, Mg and Al, with the UPS-measured Fermi energy and a theoretical Fermi energy corrected for the electron–hole asymmetry; α_c is the asymmetry parameter determined for the core levels [4.8]

Metal	E_F free electron [eV]	E_F^{exp} (measured) [eV]	α_c	αE_F [eV]	$E_F^{exp} = E_F - \alpha E_F$ (calculated) [eV]
Na	3.2	2.5[a]	0.19	0.6	2.6
Mg	7.1	6.1[b]	0.12	0.9	6.2
Al	11.7	10.6[c]	0.10	1.2	10.5

[a] [4.43], [b] [4.42], [c] [4.4]

The differences in Fermi energies deduced from ARUPS and angle integrated XPS (for which we have no explanation) indicate the problems encountered in deducing accurate (experimental) bandwidths from XPS spectra where the transitions from the bottom of the conduction band are obscured by Doniach–Sunjic "tails" and plasmons (Fig. 4.12).

Data such as in Fig. 4.12 show that it is close to impossible to use the XPS spectrum alone to define a bottom of the conduction band. On the other hand, the evaluation of the data, as indicated for Be metal, contains so many parameters that a considerable uncertainty in the determination of the bandwidth from photoemission curves as those in Fig. 4.12 is unavoidable.

4.2.4 Simple Metals: A General Comment

The reader may wonder why a relatively complicated formalism [4.26] has been described at some length. We feel that this is necessary in order to demonstrate that for the simple metals a long-standing problem in photoemission experiments has now been solved, at least for the XPS regime: namely, the full and quantitative theoretical description of an experimental core-line EDC (Figs. 4.4, 4.8), including its quite structured loss tail [4.56]. This theory relies on the three-step model and takes into account the plasmon dispersion as well as the creation of plasmons by intrinsic and extrinsic processes. The analysis of the core lines gives quantitatively the relative magnitudes of the intrinsic and extrinsic plasmon creation rates. For Be and Na they are of similar magnitude, for Mg and Al the intrinsic contributions are small. Neglecting plasmon contributions, a form for the background correction in the XPS spectra has been derived: at energy $\hbar\omega$ it is proportional to the number of electrons between $\hbar\omega$ and $\hbar\omega_0$ (where $\hbar\omega_0$ is the zero-loss-energy). This form should also be applicable to the spectra of other metallic samples.

With respect to the valence bands from polycrystalline samples, the results are twofold. First, as demonstrated most clearly by the data for Be metal, PE data cannot be directly compared with DOS curves but have to be analyzed with respect to the contributions of different angular momentum wave functions. Second, and more important in the case of strong plasmon tails, the analysis of an XPS valence band spectrum is obscured by the tailing out due to electron–hole pair creation and by the plasmons. This introduces a barely solvable problem in the data analysis and leaves the XPS data with much uncertainty. Fortunately, for transition metals the plasmon creation rate is small, and therefore in these cases the analysis of XPS spectra can be performed more reliably.

4.3 The Background Correction

PE spectra of solids (and to a lesser degree also those of free molecules and atoms) always contain a sizable background, and, in order to recover the "primary spectrum", this background has to be removed by suitable procedures. In this connection, one can distinguish two general cases. Usually one is interested in line positions and for these a correction of the form given in (4.43), which has been used successfully in many cases, is sufficient. This equation assumes a background at kinetic energy E which is proportional to the weight of the primary spectrum for all kinetic energies $E' > E$.[10] For a Lorentzian-type primary spectrum one obtains a steplike background correction (Fig. 4.8). This way of treating the background correction is generally used for core levels. For valence levels a second less sophisticated procedure is often more appropriate. One simply fits the flat regions of the spectrum "before" and "behind" the "region of interest" to a polynomial (second or third order is almost always sufficient) and by subtracting this structureless background over the entire range one hopes to get a fair representation of the primary spectrum.

The choice of an appropriate background correction is of course of considerable importance if PES is to be used for quantitative purposes [4.91–4.94]. Indeed, the accuracy of the separation of the primary and the secondary (background) spectrum determines the usefulness of PES as an analytical tool. For such application of PES a more elaborate background correction is necessary, and will now be outlined for the simple case of a homogeneous system neglecting surface properties.

The relevant equation for the separation of the primary spectrum A(E) from the secondary spectrum is (4.32) which reads, with respect to $A(E)$:[11]

[10] Except for the simple metals and a few other sp metals, the plasmon contributions are small and can be neglected. Thus, to a good approximation, (4.43) represents the total background.

[11] In contrast to (4.38) kinetic energy is now designated by E instead of ω.

$$A(E) = \frac{1}{\lambda_{\text{tot}}(E)} \times \left[P(E) - \int_{E'>E} \mathrm{d}E' \lambda_{\text{tot}}(E) \Theta(E' - E) S(E, E') P(E') / v(E') \right] . \tag{4.51}$$

In order to rewrite this equation in the form generally found in the literature, we notice that $A(E)$ as defined in (4.6) is the primary spectrum per unit volume and thus $A(E) \cdot \lambda_{\text{tot}}(E) = \bar{A}(E)$ is the primary spectrum per unit area of the sample which is the quantity usually quoted; furthermore, the effect of the step function is trivial and it is therefore deleted; and $S(E, E')/v(E') = \bar{S}(E, E')$ will now be called the scattering functions. We thus obtain

$$\bar{A}(E) = P(E) - \int_{E'>E} \mathrm{d}E' \lambda_{\text{tot}}(E) \bar{S}(E, E') P(E') . \tag{4.52}$$

This is the form of (4.51) given, e.g., by Tougaard et al. [4.91–4.94]. If in (4.51) it is assumed that the scattering is independent of the energy transfer $(E' - E)$, then one recovers essentially the background given in (4.43) namely

$$\bar{A}(E) = P(E) - \lambda_{\text{tot}} \bar{S}(E) \int_{E'>E} P(E') \mathrm{d}E' , \tag{4.53a}$$

where the second term represents the background.

We reiterate that in (4.53a) the second term contains all inelastic events (plasmons and "unstructured" terms) while in (4.43) the plasmon contribution has been subtracted out because it was of particular interest there. However, for most purposes the plasmons also constitute "background" and therefore the separation of primary and secondary spectrum, as given in (4.51–4.53a), is the generally accepted one.

In order to estimate the magnitude of the background one can rewrite (4.52) as

$$\bar{A}(E) = P(E) - \int_{E'>E} \mathrm{d}E' \frac{\lambda_{\text{tot}}(E)}{\lambda(E, E')} P(E') . \tag{4.53b}$$

The differential path length $\lambda(E, E')$ for the scattering of an electron of energy E to energy E' is obtained by integrating (1.4) over all momentum transfers q yielding

$$\lambda^{-1}(E, E') = \frac{1}{\pi a_0 E} \ln \left(\frac{1 + (E'/E)^{1/2}}{1 - (E'/E)^{1/2}} \right) \operatorname{Im} \left(\frac{-1}{\varepsilon(E - E')} \right) , \tag{4.54}$$

where a_0 is the Bohr radius and $\varepsilon(E - E')$ is the dielectric function. Thus one can use the measured dielectric constant of a medium $\varepsilon(E - E')$ to calculate the inelastic part of the measured spectrum $P(E)$ and thus recover the primary spectrum $\bar{A}(E)$ [4.92]. Note that this background subtraction does involve the phonon scattering processes only if they are included in the dielectric constant ε.

An analysis of an XPS spectrum for the Cu $2p_{1/2,3/2}$ spectrum in Cu metal is shown in Fig. 4.13 [4.91], where the procedure described by (4.53b and 4.54) was applied together with the known loss function $\mathrm{Im}\{-1/\varepsilon(q, E - E')\}$ as given in Fig. 4.14 [4.95]. This analysis suggests that the background rises more slowly "behind" the main line than is generally accepted and that the spectrum of a material like Cu contains "behind" the main line an additional intrinsic (primary) contribution of considerable strength. This is important for quantitative analyses. But the question of the extra intrinsic structure still needs further study.

Figure 4.14 also shows the loss function of Al for a small q value [4.18]. The loss function for Al differs significantly from that of Cu and shows hardly any strength beyond the well-known plasmon resonance at $E_p = 15.5\,\mathrm{eV}$. This is a good justification for the separation of the inelastic part of the PE spectrum of the simple metals into a plasmon part and a featureless "background" part [Sect. 4.1.4 and (4.35)]. In addition, this loss function shows why it is possible to analyze the core lines of simple metals in the range $E \ll E_p$ without any background correction at all.

For practical applications, the question arises of how one can perform a reliable background correction without indulging in lengthy calculations. To illustrate the possibilities, we shall briefly discuss three different types of background correction for the $2p_{1/2,3/2}$ lines of Ni metal. These lines are known to be accompanied by a considerable background (much stronger than,

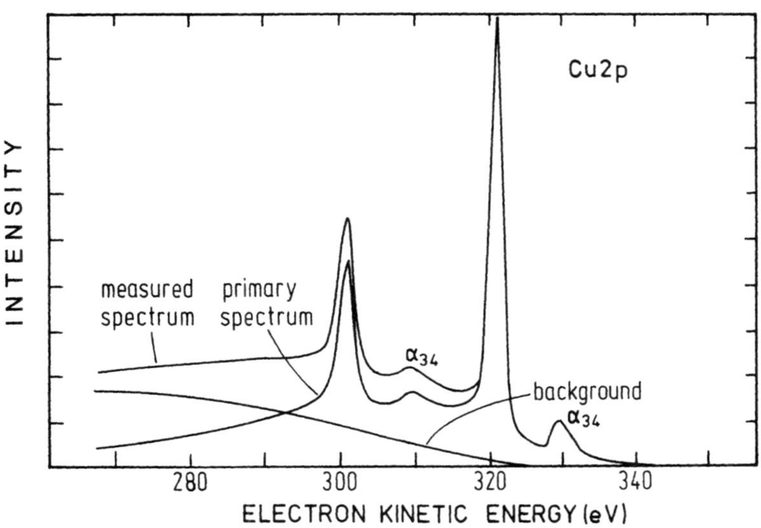

Fig. 4.13. Analysis of the Cu $2p_{1/2,3/2}$ spectrum in Cu metal [4.91]. The measured spectrum is deconvoluted using (4.53b and 4.54) employing the loss function $\mathrm{Im}\{-1/\varepsilon(0, E - E')\}$ from Fig. 4.14

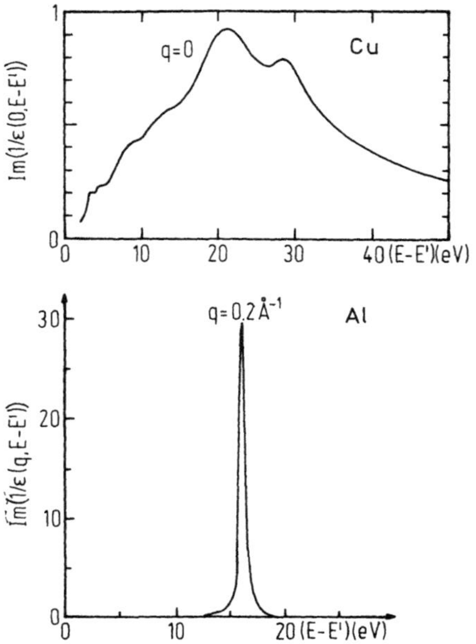

Fig. 4.14. Loss function $\mathrm{Im}\{-1/\varepsilon\}$ for Cu [4.93] and Al [4.18]

e.g., that of the $2\mathrm{p}_{1/2,3/2}$ spectrum in Cu) and therefore seem suited for such a study.

As a first approximation for the background correction we shall use that given in (4.53a) which assumes that the scattering function $\bar{S}(E, E')$ is a constant leading to a background of the form[12]

$$B_1(E) = A \int_{E'>E} \mathrm{d}E' [P(E') - P_0] \tag{4.55}$$

(P_0 is the intensity "before" the line for which the background correction is to be performed). Figure 4.15 shows an original XPS spectrum ($\hbar\omega =$ 1487 eV) of the 2p region of Ni metal, with the background calculated from (4.55) indicated as B_1 and the "primary" spectrum, $P(E) - B_1$ obtained by subtracting the background from the measured spectrum.

In order to calculate a more realistic background along the lines of (4.53b and 4.54) Tougaard [4.92] proposed, guided by an analysis of a large number of energy-loss spectra of d metals and noble metals, the following "universal" form for the scattering function

$$\lambda_{\mathrm{tot}}(E)\lambda^{-1}(E - E') = B \frac{E - E'}{[C + (E - E')^2]^2} \tag{4.56}$$

with $C = 1643\,(\mathrm{eV})^2$ and $B = 2866\,(\mathrm{eV})^2$. But similar to the factor A in (4.55) the factor B is often fitted to the experiments so that the background-corrected spectrum vanishes at about 30 to 50 eV above the

[12] This background is frequently called the "Shirley" background.

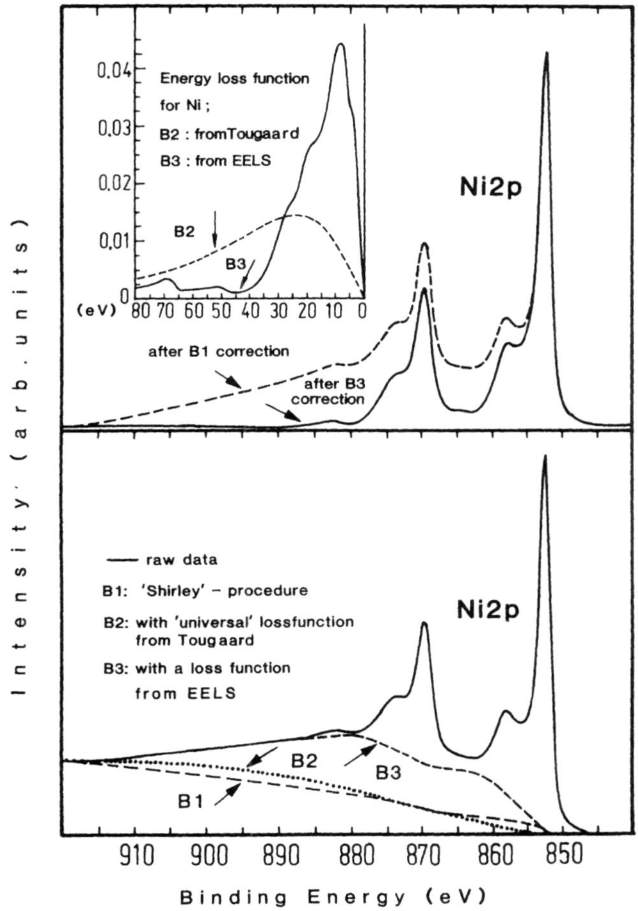

Fig. 4.15. Background subtraction for the $2p_{1/2,3/2}$ lines of Ni metal. The lower part depicts the raw data, together with the background obtained by (i) the "Shirley" procedure (B1, [4.96]), (ii) using the "universal" loss function (B2) of Tougaard [4.92], and (iii) with the loss function deduced from an Electron Energy-Loss Spectroscopy (EELS) experiment with a primary electron energy of 700 eV. In the upper part the intrinsic Ni $2p_{1/2,3/2}$ spectra obtained from procedure (i) and (iii) are compared; [procedure (ii) gives results very similar to (i); it is not shown here]. The energy loss function for procedures (ii) and (iii) is depicted in the insert in the upper part. It is a constant for the "Shirley" procedure (i)

main line. The best procedure is, of course, to use directly the loss function $\lambda_{tot}(E)\lambda^{-1}(E,E')$ as, e.g., obtained in an Electron Energy Loss Experiment (EELS) with a comparable primary energy of 500 to 1000 eV. This has been done for Ni metal. The three procedures outlined above are displayed in Fig. 4.15. In the lower part the original data after subtraction of the constant P_0 in (4.55) are shown, together with the three background curves.

The insert depicts the "universal" loss function of Tougaard in comparison to the one experimentally determined from an electron energy-loss experiment with 700 eV primary electrons on the same Ni sample. (The "Shirley" loss function is, of course, a constant). While in this case the "Shirley" correction and Tougaard's universal procedure give very similar results, the use of the correct loss function reveals remarkable differences. This is displayed in Fig. 4.15 in the corrected spectra. This example demonstrates that it is very necessary to apply a good background-correction procedure if reliable analytic information is to be obtained. This is especially necessary when one wants to analyze the intrinsic loss structure over a larger energy range in more detail.

In recent years Tougaard has considerably refined the analysis of the inelastic background in PE spectra and tackled the related problem of the separation of the intrinsic (primary) and the secondary spectrum [4.97]. Therefore the treatment in this chapter can only be regarded as an introduction to what, mathematically too, is quite a complex field. Readers wishing to perform such analysis in any detail are referred to [4.97] and the many interesting references therein.

In addition it should be mentioned that Hedin [4.98, 4.99] has given a comprehensive treatment of the theory of photoemission, which includes in one step the intrinsic and extrinsic contributions to the spectra.

References

4.1 S. Doniach, M. Sunjic: J. Phys. C **3**, 285 (1970)
4.2 S. Hüfner, G.K. Wertheim, D.N.E. Buchanan, K.W. West: Phys. Lett. A **46**, 420 (1974)
S. Hüfner, G.K. Wertheim, D.N.E. Buchanan: Chem. Phys. Lett. **24**, 527 (1974)
S. Hüfner, G.K. Wertheim: Phys. Lett. A **47**, 349 (1974)
4.3 P.H. Citrin: Phys. Rev. B 8, 5545 (1973)
4.4 P. Steiner, H. Höchst, S. Hüfner: In *Photoemission in Solids II*, ed. by L. Ley, M. Cardona, Topics Appl. Phys., Vol.27 (Springer, Berlin, Heidelberg 1979) Chap.7
4.5 Y. Baer, P.H. Citrin, G.K. Wertheim: Phys. Rev. Lett. **37**, 49 (1976)
4.6 P.H. Citrin, G.K. Wertheim, Y. Baer: Phys. Rev. Lett. **35**, 885 (1975)
4.7 G.K. Wertheim, S. Hüfner: Phys. Rev. Lett. **34**, 53 (1975)
4.8 P.H. Citrin, G.K. Wertheim, M. Schlueter: Phys. Rev. B 20, 4343 (1979)
4.9 G.K. Wertheim, P.H. Citrin: *In Photoemission in Solids I*, Vol.26, ed. by M. Cardona, L. Ley, Topics Appl. Phys., Vol.26 (Springer, Berlin, Heidelberg 1978) Chap.5
4.10 M. Sunjic, D. Sokcevic: Solid State Commun. **15**, 165 (1974)
4.11 M. Sunjic, D. Sokcevic, A.A. Lucas: J. Electron Spectrosc. Relat. Phenom. **5**, 963 (1974)
4.12 M. Sunjic, D. Sokcevic: Solid State Commun. **18**, 373 (1976)
4.13 S. Hüfner, G.K. Wertheim: Phys. Rev. B **11**, 678 (1975)
4.14 S. Hüfner, G.K. Wertheim, J.H. Wertnick: Solid State Commun. **17**, 417 (1975)

4.15 S. Hüfner, G.K. Wertheim: Phys. Rev. B **11**, 5197 (1975)
4.16 S. Hüfner, G.K. Wertheim: Phys. Lett. A **51**, 301 (1975)
4.17 H. Raether: *Solid State Excitations by Electrons*. Springer Tracts Mod. Phys. **38**, 84 (Springer, Berlin, Heidelberg 1965)
4.18 H. Raether: Springer Tracts in Modern Physics, Vol.88 (Springer, Berlin, Heidelberg, 1980)
4.19 H.W.B. Skinner: Phil. Trans. Roy. Soc. London A **239**, 95 (1940)
4.20 W.M. Cady, D.H. Tomboulian: Phys. Rev. **59**, 381 (1941)
4.21 S. Crisp, S.E. Williams: Phil. Mag. **6**, 625 (1961)
4.22 D.H. Tomboulian: In *Handbuch der Physik*, Vol.30, ed. by S. Flügge (Springer, Berlin, Heidelberg 1957)
4.23 G.D. Mahan: Phys. Rev. **163**, 612 (1967); and Solid State Physics **29**, 75 (Academic, New York 1974)
4.24 P.W. Anderson: Phys. Rev. Lett. **18**, 1049 (1967)
4.25 P. Nozières, C.T. DeDominicis: Phys. Rev. **178**, 1097 (1969)
4.26 G.D. Mahan: Phys. Rev. B **11**, 4814 (1975)
4.27 C. Kunz, R. Haensel, G. Keitel, P. Schreiber, B. Sonntag: *Electronic Density of States*, ed. by L.H. Bennet, Nat. Bur. Stand. Spec. Publ. 323 (USGPO, Washington DC, 1971)
4.28 P. Minnhagen: Phys. Lett. A **56**, 327 (1976)
4.29 D.C. Langreth: Phys. Rev. B **1**, 471 (1970)
4.30 L. Hedin, B.I. Lundquist, S. Lundquist: J. Res. Nat. Bur. Stand. A **74** (1970)
A first-principles calculation of the plasmon satellites in the valence bands of Na and Al has been published by
F. Aryasetiawan, L. Hedin, K. Karlson: Phys. Rev. Lett. **77**, 2268 (1996)
4.31 P. Steiner, H. Höchst, S. Hüfner: J. Phys. F **7**, L105 (1977)
4.32 H. Höchst, P. Steiner, S. Hüfner: Phys. Lett. A **60**, 69 (1977)
4.33 D.R. Penn: Phys. Rev. Lett. **38**, 1429 (1977)
4.34 J.C. Fuggle, L.M. Watson, D.J. Fabian, S. Affrossman: J. Phys. F **5**, 375 (1975)
4.35 P. Steiner, F.J. Reiter, H. Höchst, S. Hüfner: Phys. Stat. Sol. (b) **90**, 45 (1978)
P. Steiner, F.J. Reiter, H. Höchst, S. Hüfner, J.C. Fuggle: Phys. Lett. A **66**, 229 (1978)
4.36 D.R. Penn: Phys. Rev. Lett. **40**, 568 (1978)
4.37 H. Höchst, P. Steiner, S. Hüfner: Z. Physik B **30**, 145 (1978)
4.38 P.M.Th.M. van Attekum, J.M. Trooster: Phys. Rev. B **18**, 3872 (1978)
4.39 P.M.Th.M. van Attekum, J.M. Trooster: Phys. Rev. B **20**, 2335 (1979)
4.40 P. Longe, S.M. Bose: Solid State Commun. **38**, 527 (1981)
4.41 H. Höchst, P. Steiner, S. Hüfner: J. Phys. F **7**, L309 (1977)
4.42 E.W. Plummer: Surf. Science **152/153**, 162 (1985)
B.S. Itchkawitz, In-Whan Lyo, E.W. Plummer: Phys. Rev. B **41**, 8075 (1990)
4.43 E. Jensen, E.W. Plummer: Phys. Rev. Lett. **55**, 1912 (1985)
4.44 D.C. Langreth: Theory of Plasmon Effects in High Energy Spectroscopy, presented before the Nobel Symposium, Sweden, 1973, in *Nobel Foundation Series Nobel Symposia, Medicine and Natural Sciences*, Vol.24 (Academic, New York)
4.45 J.J. Chang, D.C. Langreth: Phys. Rev. B **8**, 4638 (1973)
4.46 G.D. Mahan: Phys. Stat. Sol. (b) **55**, 703 (1973)
4.47 D.C. Langreth: Phys. Rev. Lett. **26**, 1229 (1971)
4.48 S.M. Bose, P. Kiehen, P. Longe: Phys. Rev. B **23**, 712 (1981)
4.49 S.M. Bose, St. Prutzer, P. Longe: Phys. Rev. B **27**, 5992 (1983)
4.50 G.D. Mahan: Phys. Rev. B **2**, 4334 (1970)

4.51 W.I. Schaich, N.W. Ashcroft: Solid State Commun. **8**, 1959 (1970)
4.52 W.I. Schaich, N.W. Ashcroft: Phys. Rev. B **3**, 2452 (1971)
4.53 G.K. Wertheim, L.R. Walker: J. Phys. F **6**, 2297 (1976)
4.54 J.W. Gadzuk: *In Photoemission and the Electronic Properties of Surfaces*, Vol.I, ed. by B. Feuerbacher, B. Fitton, R.F. Willis (Wiley, Chichester 1978)
4.55 P.A. Wolff: Phys. Rev. **95**, 56 (1954)
4.56 P. Steiner, H. Höchst, S. Hüfner: Z. Physik B **30**, 129 (1978)
4.57 D.R. Penn: J. Electron Spec. **9**, 29 (1976)
4.58 Y. Baer, G. Busch: Phys. Rev. Lett. **30**, 280 (1973)
4.59 S.P. Kowalczyk, L. Ley, F.R. McFeely, R.A. Pollak, D.A. Shirley: Phys. Rev. B **8**, 3583 (1973)
4.60 J. Tejeda, M. Cardona, N.J. Shevchik, D.W. Langer, E. Schoenherr: Phys. Stat. Sol. (b) **58**, 189 (1973)
4.61 A. Barrie: Chem. Phys. Lett. **19**, 109 (1973)
4.62 R.A. Pollak, L. Ley, F.R. McFeely, S.P. Kowalczyk, D.A. Shirley: J. Electron Spectr. **3**, 381 (1974)
4.63 J.C. Fuggle, D.J. Fabian, M.L. Watson: J. of El. Spectr. **9**, 99 (1976)
4.64 W.J. Pardee, G.D. Mahan, D.E. Eastman, R.A. Pollak, L. Ley, F.R. McFeely, S.P. Kowalczyk, D.A. Shirley: Phys. Rev. B **11**, 3614 (1975)
4.65 J. Chang, D.C. Langreth: Phys. Rev. B **5**, 3512 (1972)
4.66 P. Minnhagen: J. Phys. C **8**, 1535 (1975)
4.67 B. Gumhalter, D.M. Newns: Phys. Lett. A **53**, 137 (1975)
4.68 J. Harris: Solid State Commun. **16**, 671 (1975)
4.69 P.H. Citrin, G.K. Wertheim, Y. Baer: Phys. Rev. Lett. **41**, 1425 (1978); ibid. **35**, 885 (1975)
4.70 The present situation with respect to the experimental verification of the Mahan-Nozières-DeDominicis theory was discussed in:
P.A. Bruhwiler, S.E. Schnatterly: Phys. Rev. B **41**, 8013 (1990)
E. Zaremba, K. Sturm: Phys. Rev. Lett. **66**, 2144 (1991)
G.K. Wertheim, D.N.E. Buchanan: Phys. Rev. B **43**, 13815 (1991)
4.71 G.K. Wertheim: J. Phys. Soc. Jpn. **64**, 4023 (1995).
This paper demonstrates the complete compatibility of the parameters for the Mahan/Nozières/DeDominicis theory obtained from analyzing PES data with those derived from X-ray emission and absorption spectra
4.72 W. Eberhardt, G. Kalkoffen, C. Kunz: DESY Rpt. SR 79/15 (1978)
4.73 D.M. Riffe, G.K. Wertheim, P.H. Citrin: Phys. Rev. Lett. **67**, 116 (1991)
D.M. Riffe, G.K. Wertheim, D.N.E. Buchanan, P.H. Citrin: Phys. Rev. B **41**, 6216 (1992)
W. Theis, K. Horn: Phys. Rev. B **47**, 16060 (1993)
These publications demonstrate the effect of phonons on the core lines of simple metals.
4.74 In-Whan Lyo, W.E. Plummer: Phys. Rev. Lett. **60**, 1558 (1988)
4.75 S.T. Inoues, J. Yamashita: J. Phys. Soc. Jpn. **35**, 677 (1973)
4.76 G. Wiech: In *Soft X-Ray Band Spectra and the Electronic Structure of Metals and Materials*, ed. by D.J. Fabian (Academic, New York 1968) p.59-70 4.75
4.77 P.O. Nilsson, G. Arbmann, T. Gustafson: J. Phys. F **4**, 1937 (1974)
4.78 R.G. Musket, R.J. Forther: Phys. Rev. Lett. **26**, 80 (1971)
4.79 H.G. Maguire, P.D. Augustus: Phil. Mag. (GB) **30**, 95 (1974)
4.80 C. Herring, A.G. Hill: Phys. Rev. **58**, 132 (1940)
4.81 J. Osterwalder, T. Greber, S. Hüfner, L. Schlapbach: Phys. Rev. Lett. 64, 2683 (1990)
4.82 J. Osterwalder, T. Greber, S. Hüfner, L. Schlapbach: Phys. Rev. B **41**, 495 (1990)

4.83 S. Hüfner: Solid State Commun. 59, 639 (1986). It was most regrettably overlooked in this paper that an earlier, much more complete work by L. Hedin exists [Physica Scripta **21**, 477 (1980)] which provides a rigorous derivation of the estimates given by Hüfner
4.84 B.I. Lundquist: Phys. kondens. Mat. **6**, 193 (1967)
4.85 B.I. Lundquist: Phys. kondens. Mat. **6**, 206 (1967)
4.86 B.I. Lundquist: Phys. kondens. Mat. **7**, 117 (1968)
4.87 B.I. Lundquist: Phys. kondens. Mat. **9**, 236 (1969)
4.88 D.A. Shirley: In *Photoemission in Solids I*, ed. by M. Cardona, L. Ley, Topics Appl. Phys., Vol.26 (Springer, Berlin, Heidelberg 1978) Chap.4
4.89 K.W.K. Shung, G.D. Mahan: Phys. Rev. Lett. **57**, 1076 (1986)
4.90 K.W.K. Shung, B.E. Sernelius, G.D. Mahan: Phys. Rev. B **36**, 4499 (1987)
4.91 S. Tougaard, B. Jorgensen: Surf. Sci. **143**, 482 (1984)
4.92 S. Tougaard: Surface and Interface Analysis **11**, 453 (1988); this paper contains references on the background correction in PES
4.93 S. Tougaard, P. Sigmund: Phys. Rev. B **25**, 4452 (1982)
4.94 S. Tougaard, A. Ignatiev: Surf. Sci. **124**, 451 (1983)
4.95 C. Wehenkel: J. Physique **36**, 199 (1975)
4.96 D.A. Shirley: Phys. Rev. B **5**, 4709 (1972)
4.97 A.C. Simonsen, F. Yubero, S. Tougaard: Surf. Sci. **436**, 149 (1999)
4.98 L. Hedin, J. Michiels, J. Inglesfield: Phys. Rev. B **58**, 15565 (1998)
4.99 L. Hedin: J. Phys. Condens. Matter **11**, R 489 (1999

5. Valence Orbitals in Simple Molecules and Insulating Solids

It has been shown in the preceding chapters that PES measures the binding energies of electrons in an atom, a molecule or a solid. These measured binding energies, however, are not single-particle energies, but energies which contain many-body relaxation contributions. The energy levels of atoms, molecules and solids, at about 10–20 eV below the vacuum level are those which are responsible for chemistry in the widest sense and thus determine to a considerable degree the appearance of our environment. The study of valence orbitals has therefore always played a prominent role in PES. In this chapter we will discuss the kind of information that can be obtained from PES on atoms, molecules and insulating solids. Note that besides molecules in their ordinary state, it is also possible to study them as adsorbates on a surface. In addition, relatively short-lived molecules, so-called transient species, have also been investigated by this technique and their orbital energies have been determined, but will not be treated in this chapter.

The description of the photoemission spectra of atomic gases and fee molecules is very brief in this volume and can only serve as an introduction to that field. It has to be noted in particular that advances in technology have now given resolutions in that field of the order of 0.01 meV or slightly better. This opens a new perspective for photoemission spectroscopy in free molecules. For an introduction into that field and interesting new literature we refer the reader to the article of Hollenstein et al. [5.1]. Examples of this kind of spectroscopy have been given in Figs. 1.32, 1.33 and 1.34 and an additional one will be presented in Figs. 5.6 and 5.7. The spectrum in Fig. 5.6 is a conventional photoemission spectrum of N_2. With the best now available techniques much more details can be observed, and as an example, a ZEKE spectrum of the line at 15.57 eV is shown in Fig. 5.7. Again one realizes the considerably improved resolution, where now this resolution allows to detect the rotational structure present in the spectra.

For a long time XPS and UPS measurements on molecules were equally popular. However, the information obtained from the core-level studies by XPS is rather limited, and nowadays core-level spectroscopy is performed mostly for analytical purposes. Thus UPS spectroscopy, particularly with synchrotron radiation, is the technique which has found the widest application in the study of molecules. A special advantage of UPS is that it has a much higher resolution than XPS. Also, neglecting for the moment the cases of the rare-earth and transuranium elements, UPS generally enjoys higher photoionization cross-sections than XPS. Therefore, in this chapter, for the most part, UPS data will be discussed. (For reviews see [5.2–5.14]).

In two cases, the present chapter, despite its title, will also deal with metallic compounds; namely, it will treat the high-temperature superconductors and also discuss the difference between a Fermi liquid and a Luttinger liquid. In both cases the materials concerned are transition metal compounds, and since their electronic structure is described in detail in this chapter, we will also include here a discussion of these compounds in their metallic state. The more conventional metals, such as Na or Cu, are treated in Chap. 7.

5.1 UPS Spectra of Monatomic Gases

Gases consisting of single atoms (e.g., the noble gases) produce the simplest photoelectron spectra, because they contain no vibrational excitations. As an example, Fig. 5.1a shows the photoionization of argon with He I radiation. The reaction equation is the following:

$$3p^6 + \hbar\omega \rightarrow 3p^5 \equiv (3p)^{-1}\,. \tag{5.1}$$

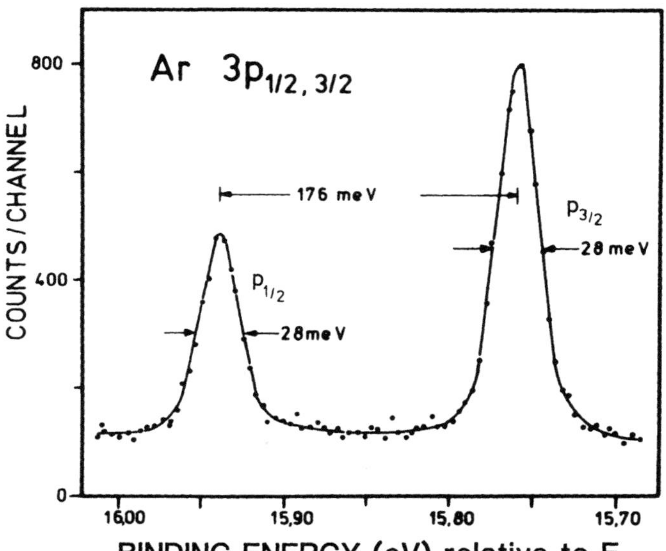

Fig. 5.1a. UPS spectrum (He I, 21.2 eV) of gaseous argon showing the $3p_{1/2,3/2}$ doublet. The linewidth is an indication of the resolution (almost exclusively determined by the electron spectrometer). Spectra like this one are often used to measure the resolution of a PE spectrometer

Thus the final state is that of one hole in the 3p shell, which gives a spectrum equivalent, apart from the sign, to that of one 3p electron. This spectrum contains two lines, the $p_{3/2}$ and $p_{1/2}$ lines, at $E_B = 15.755\,eV$ and $E_B = 15.933\,eV$, where the separation of the two lines is given by the atomic spin–orbit interaction. The binding energy in free atoms and molecules is measured relative to the vacuum level. In these systems, the binding energy is often referred to as the ionization energy. Fig. 5.1b shows a He I spectrum of the Xe $5p_{3/2}$ line at high resolution. This type of spectrum is often used these days to determine the resolution of a PE spectrometer.

The intensity ratio of the two lines, as estimated by the occupation numbers of the emitting orbitals, should be $(2J_1 + 1)/(2J_2 + 1) = 4/2 = 2$. This prediction is approximately borne out by the experimental findings. The spectrum of this experiment represents the "valence levels" of argon and is typical for an atom. One sees that atomic valence orbital spectra are

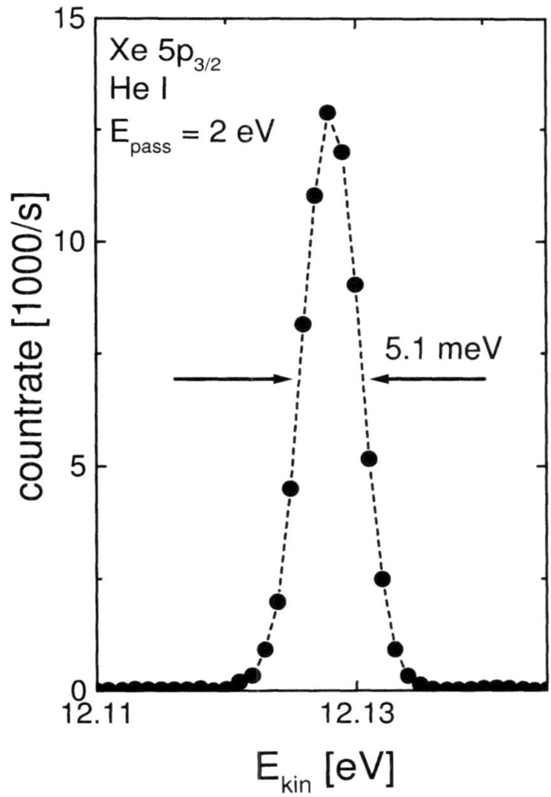

Fig. 5.1b. Xenon $5p_{3/2}$ spectrum at high resolution measured with a path energy of 2 eV. The measured width is 5.1 meV, which can be taken as a good indication of the resolution of the system

very similar to the core-level spectra. The problem of satellites is also encountered in the spectra of atoms and has been discussed in considerable detail [5.11, 5.15, 5.16]. Because of the limited applications of photoelectron spectra from atoms, these will not be considered further, and we move on to a discussion of somewhat more complex systems.

5.2 Photoelectron Spectra of Diatomic Molecules

The simplest example of a valence orbital photoemission spectrum of a diatomic molecule is the He I spectrum of molecular hydrogen shown in Fig. 5.2 [5.17].

Such a molecule, along with its electronic degrees of freedom, has the additional possibility of absorbing energy by rotation or vibration of the two nuclei with respect to each other. The relation between the photon energy and the kinetic energy of the escaping electron is therefore

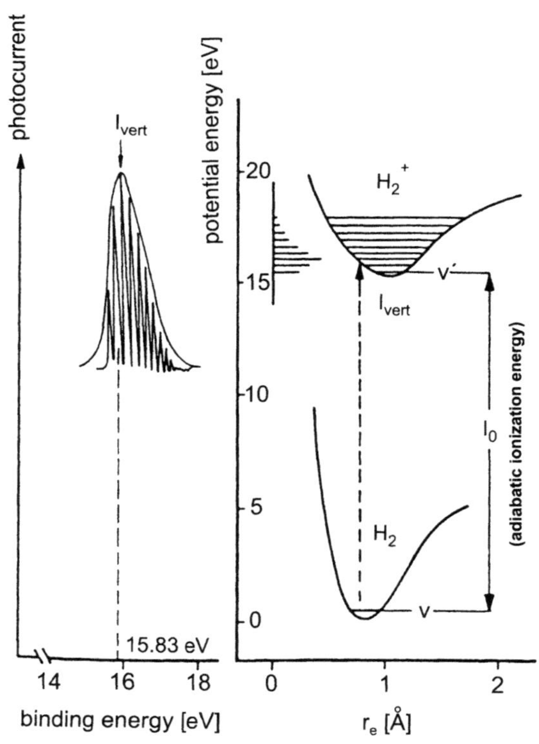

Fig. 5.2. UPS spectrum of H_2 [5.17] and the potential energy curves for H_2 and H_2^+ indicating the possible transitions. The vertical ionization energy I_{vert} is indicated as occurring when the transition takes place without any change in the internuclear coordinate (r_e)

$$\hbar\omega - E_{\mathrm{kin}} = I_0 + \Delta E_{\mathrm{vib}} + \Delta E_{\mathrm{rot}} \, . \tag{5.2}$$

Here I_0 is the so-called adiabatic ionization energy, while ΔE_{vib} and ΔE_{rot} are the changes in vibrational and rotational energy of the molecule caused by the photoionization. In Fig. 5.2 the potential-energy curves of molecular hydrogen (in the ground state) and ionized molecular hydrogen are shown in order to demonstrate the photoemission process [5.17].

Photoionization of the neutral H_2 molecule leads to the H_2^+ molecule. In order to describe this transition in the diagram of Fig. 5.2, it is assumed that the electronic transition is fast compared to a vibrational period. This means that the transition can be described by a vertical line in Fig. 5.2 (Frank–Condon principle). The potential energy curve for H_2^+ is not identical to that of H_2. Since the two electrons of the H_2 molecule are in a bonding orbital, the removal of one of these electrons (in the creation of H_2^+) leads to a less tightly bound molecule, which is reflected in Fig. 5.2 by the fact that the minimum of the potential energy for H_2 occurs at a larger internuclear distance r_e than that in H_2. Only the lowest vibrational energy level ($v = 0$, v is the vibrational quantum number) is generally occupied in the H_2 molecule, whereas after the transition the H_2^+ molecule can end up in an excited vibrational state ($v' > 0$).

The strength of the photocurrent, or the number of photoemitted electrons can be calculated under the assumption that the electronic wave function varies slowly with internuclear distance. Then the electronic part ψ_e and the vibrational part ψ_v of the molecular wave function can be separated (meaning that the total wave function can be written as a product of the two), which leads to the following equation for the photocurrent (Born–Oppenheimer approximation):

$$I = |M_e|^2 \left| \int \psi_v \psi_{v'}^* \, \mathrm{d}\tau \right|^2 \tag{5.3}$$

for a transition from $H_2(v)$ into $H_2^+(v')$.

The transition strength is therefore governed by the vibrational overlap integral between the ground state H_2 and the excited state H_2^+ (provided, $|M_e|^2$ is constant). As can be seen from Fig. 5.2 transitions are possible out of the ground state of $H_2(v = 0)$ to a number of excited vibrational states ($v' > 0$) of the H_2^+ molecule. The maximum of the envelope of the various vibrational transitions as shown in Fig. 5.2 gives the so-called vertical ionization energy (I_{vert}). In contrast, the ionization energy connecting the $v = 0$ and $v' = 0$ vibrational states is called the adiabatic ionization energy (I_0).

Figure 5.2 illustrates an additional benefit of PE experiments on free molecules: besides measuring electronic energies one can also determine vibrational energies in excited states, especially for vibrations involving light atoms.

The vibration structure of the PES of H_2 is shown in greater detail in Fig. 5.3 [5.2]. To extract the vibrational frequency and the dissociation limit

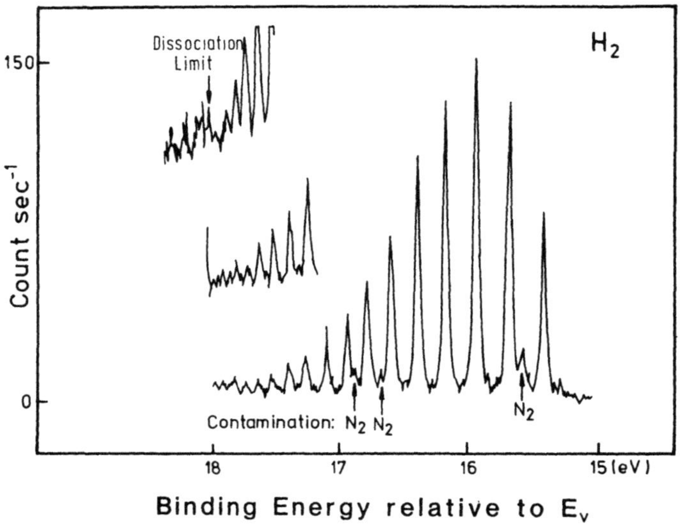

Fig. 5.3. UPS spectrum ($\hbar\omega = 21.2\,\text{eV}$) of molecular hydrogen (with a small amount of N_2 impurity) showing the vibrational structure up to the dissociation limit [5.2], E_V is the vacuum level

from such a spectrum, one proceeds as follows: the vibrational energies in a diatomic molecule like H_2 are given by

$$G(v') = (v' + 1/2)\hbar\omega_3 - (v' + 1/2)^2\hbar\omega_3 x \ , \tag{5.4}$$

where ω_3 is the excited state vibrational frequency. v' is the vibrational quantum number and x is the anharmonicity constant. For the spacing between adjacent vibrational levels $\Delta G = [G(v' + 1) - G(v')]$ one finds

$$\Delta G = \hbar(\omega_3 - \omega_3 x) - 2(v' + 1/2)\hbar\omega_3 x \, . \tag{5.5}$$

Thus ΔG is a linear function of v' and indeed a plot of ΔG as a function of v' gives a straight line as shown in Fig. 5.4. This plot can be used to determine the dissociation limit.

Before proceeding to more complicated molecules, we briefly generalize the photoemission process for molecules consisting of two different atoms (AB) in Fig. 5.5. This figure shows the potential energy curves for a molecule (AB) in its ground state (M_0, v) and its photoionized state $(AB)^+$ after removal of a non-bonding electron (M_1^+, v'), an antibonding electron (M_2^+, v'') and a bonding electron (M_3^+, v'''); in addition, the transition into an unstable state (M_4^+) is indicated. Transitions are assumed to take place vertically (Frank–Condon process) and assuming that the initial molecule is in the vibrational ground state ($v = 0$), the shaded area indicates the part in the diagram for which non-zero transition strength can be expected.

If a non-bonding electron is excited, the potential energy curves of the ground state (M_0) and the excited state (M_1^+) will be quite similar. This

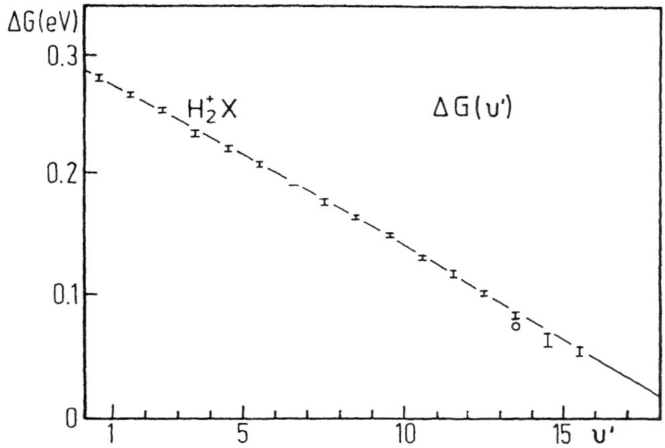

Fig. 5.4. Plot of ΔG (energy separation between two vibrational energy levels) as a function of $v' + 1/2$ (v': vibrational quantum number)(5.5) in H_2^+ (data from Fig. 5.3). The straight line allows the determination of the dissociation limit [5.2]

leads to an almost perfect overlap (5.3) of the $v = 0$ and $v' = 0$ vibrational wave functions and therefore this transition carries the most intensity. The higher $v' > 0$ vibrational bands occur with diminishing strength.

If by contrast an antibonding electron is excited, the internuclear distance of M_2^+ is smaller than in M_0, and if a bonding electron is excited, the internuclear distance of M_3^+ is larger than that of M_0. For these two cases the excitation of the $v = 0 \rightarrow v'' = 0$, and $v = 0 \rightarrow v''' = 0$ transitions carry smaller weight, as is evident from Fig. 5.5.

We note that this discussion gives only a qualitative picture of the PE process in a free molecule. Quantitatively one also has to realize that the excitation of non-bonding electrons leaves the molecule relatively "intact", which leads to little vibrational excitation. On the other hand, the excitation of a bonding electron (and to a lesser degree of an antibonding electron) presents a large disturbance, which will generally result in a large amplitude in the vibrational excitations.

In addition we note with reference to Fig. 5.5 that the excitation into M_3^+ is beyond the dissociation limit and leads to a dissociation of the molecule. The top diagram shows the potential energy curve for an unstable state M_4^+ of the molecule.

To apply these general principles and illustrate some features of the spectra of slightly more complicated diatomic molecules, Fig. 5.6 shows the He I photoemission spectrum of molecular nitrogen [5.2]. One sees three electronic transitions at 18.7 eV, 16.6 eV and 15.5 eV, which can be assigned to transitions out of the σ_u-, π_u- and σ_g orbitals. All these transitions show distinct vibrational structure which can be used for the interpretation of the observed electronic spectra [5.2] in the spirit of Fig. 5.5.

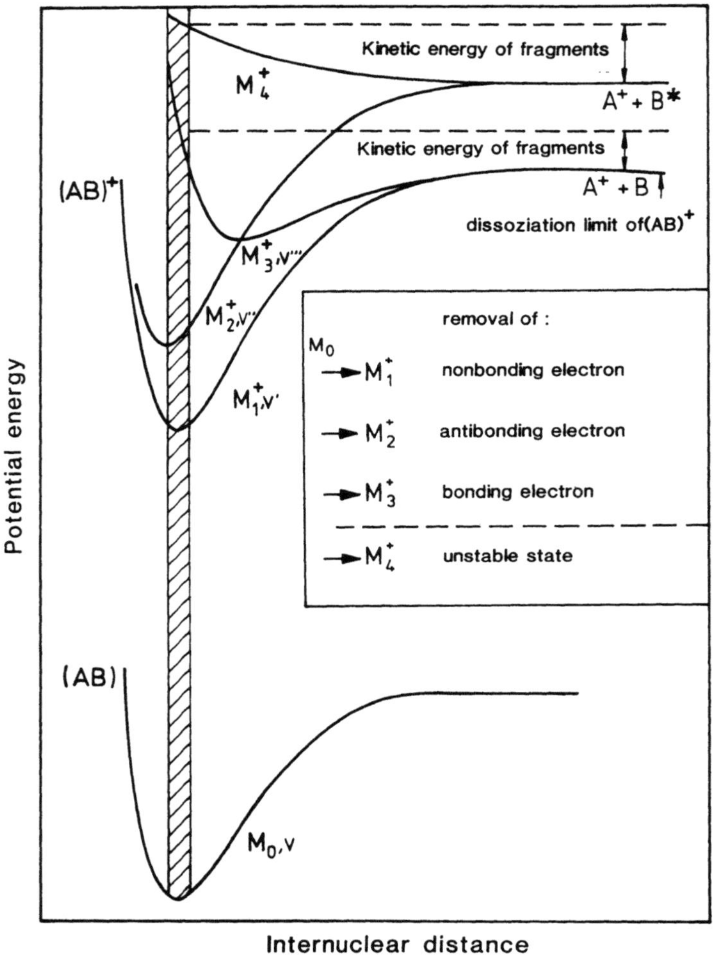

Fig. 5.5. Potential-energy curves for a diatomic molecule AB in the ground state M_0 and for the photoionized molecule $(AB)^+$ in the ground state M_1 and three excited states M_2, M_3 and M_4 [5.2]

The spectra around 18.7 eV ($\hbar\omega_1 = 2340\,\mathrm{cm}^{-1}$; note: $8065\,\mathrm{cm}^{-1} \hat{=} 1\,\mathrm{eV}$) and 15.5 eV ($\hbar\omega_1' = 2100\,\mathrm{cm}^{-1}$) show only weak vibrational structure and are therefore assigned to the excitation from non-bonding σ orbitals. The spectrum around 17 eV ($\hbar\omega_3 = 1850\,\mathrm{cm}^{-1}$) shows distinct vibrational structure and is assigned to the excitation from a bonding π orbital. (In the ground state the vibrational frequency is $\hbar\omega_0 = 2345\,\mathrm{cm}^{-1}$; the comparison with the above numbers emphasizes the bonding nature of the 17 eV state as detailed below).

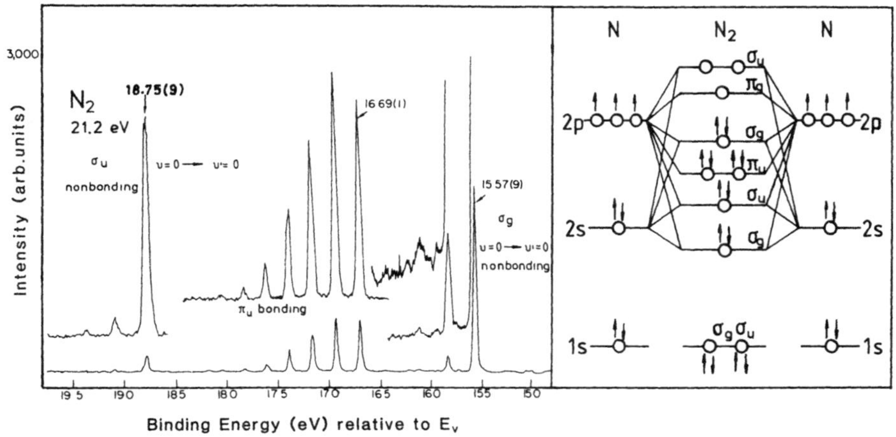

Fig. 5.6. UPS (21.2 eV) spectrum of N_2 [5.2]. The two non-bonding orbitals (σ_u and σ_g) exhibit little vibrational structure, while the bonding orbital (π_u) shows intense vibrational structure

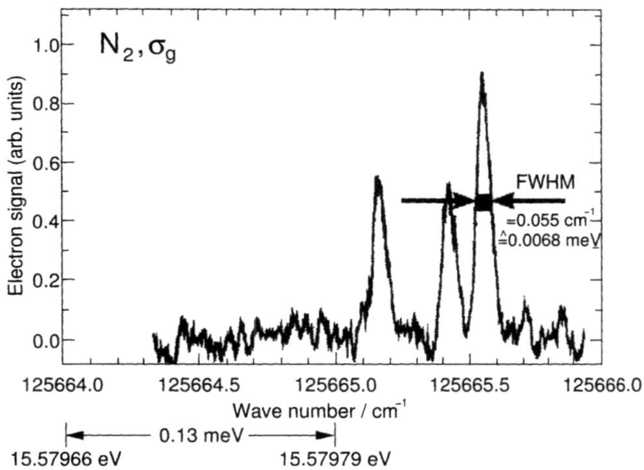

Fig. 5.7. ZEKE photoelectron spectrum corresponding to the photoionization out of the σ_g orbital of N_2. The additional structure relative to the lower resolution spectrum in Fig. 5.6 comes from the rotational energy levels [5.1]

It was just shown how one can differentiate in the PE spectra between bonding and non-bonding electrons. A similar distinction between bonding and antibonding orbitals is more difficult. However, it can be made by comparing the vibrational parameters deduced from the photoemission spectra with those obtained from Raman or infrared spectroscopy, which measure ground-state vibrational properties. If in the PE process a bonding electron is excited, then the bonding of the final state is weaker than that of the initial state (Fig. 5.5) and therefore the vibrational frequency obtained from PES for the excited final state is lower than for a ground-state measurement. If, on the other hand, an antibonding electron is emitted in the PE process, the bonding of the final state is stronger and the vibrational frequency for the excited final state will be correspondingly higher than that measured for the ground state by Raman or infrared spectroscopy. Finally we mention that in the case of non-bonding orbitals the vibrational frequencies measured by PES for the excited final state are slightly lower than those of the ground state because these orbitals actually have a slight bonding character (see data for N_2 above).

These general considerations will now be verified. An example of the change in vibrational structure between bonding and non-bonding orbitals can be found in the valence-band PE spectra of the hydrogen halides [5.2]. Figure 5.8 shows a PE spectrum of hydrogen chloride and as an insert a sketch of the electronic structure of the molecules. The PE spectrum consists of two bands at ~ 16 eV and ~ 13 eV binding energy. The first has pronounced vibrational structure, whereas for the second band it is much weaker. From this fact alone, one can assign the 16 eV band to the σ-bonding orbital and the 13 eV band to the π-non-bonding orbital. Table 5.1 [5.18] gives a compilation of the vibrational frequencies ($v' = 0 \rightarrow v' = 1$ energy separation) of hydrogen chloride, hydrogen bromide and hydrogen iodide, as deduced from the PE spectra of the non-bonding and bonding orbitals [5.2] and also the cor-

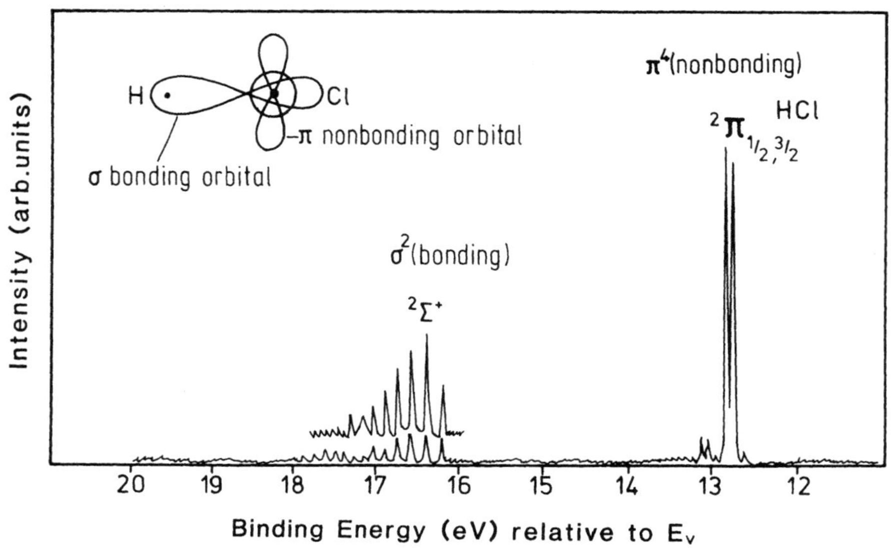

Fig. 5.8. UPS (21.2 eV) spectrum of HCl together with a schematic diagram of its molecular structure [5.2]. The bonding orbital (σ) shows strong vibrational structure, while the non-bonding orbital (π) shows no vibrational structure

Table 5.1. Vibrational frequencies [cm^{-1}] ($1cm^{-1} \hat{=} 1.24 \times 10^{-4}$ eV) of the hydrogen halides; ν_π (non-bonding) and ν_σ (bonding) are the frequencies associated with the π- and σ orbital PE, while ν_0 is the frequency as measured with the Raman effect [5.18]

	HCl	HBr	HI
ν_π (non-bonding)	2660	2420	2100
ν_σ (bonding)	1610	1290	1300
ν_0 (Raman)	2886	2560	2230

Table 5.2. Ratio of the vibrational frequencies of the dihalides, as measured by PES from antibonding π orbitals (ν_π^*) and obtained by Raman spectroscopy in the ground state (ν_0) [5.18]

	F_2	Cl_2	Br_2	I_2
ν_π^*/ν_0	1.18	1.14	1.11	1.02

responding vibrational frequencies obtained for the groundstate by Raman or infrared measurements. One sees that the vibrational frequencies determined from the Raman data and from the non-bonding PE orbital roughly agree although the latter is slightly smaller indicating that the nominally non-bonding orbital also takes part slightly in the bonding. However, the vibrational frequency determined from the PE signal of the bonding orbital is markedly lower in all cases in agreement with the above contention that the molecule becomes less tightly bound if an electron is ejected from a bonding orbital.

An example for the vibrational structure of antibonding orbitals can be found in the antibonding π orbitals of the dihalides. Table 5.2 lists the ratio of the frequency ν_π^* of the vibrations as determined from the PE spectra to that determined from ground-state Raman measurements (ν_0). One sees that in each case the frequency obtained from the PE spectrum is larger than that measured for the ground state. This confirms that the removal of an antibonding electron indeed increases the interatomic force constant and thereby the vibrational frequency.

5.3 Binding Energy of the H_2 Molecule

The photoemission spectrum of the H_2 molecule has its vertical ionization energy (Fig. 5.2) at 15.8 eV. One can rationalize this number in the following way [5.17]: the vertical ionization potential consists of two parts: (1) the ionization potential IP(H) of the hydrogen atom, which is 13.595 eV, and (2) one half of the binding energy of the H_2 molecule $E_b(H_2)$, which can be determined by calorimetric methods: thus $I_{\text{vert}} = IP(\text{H}) + 1/2E_{\text{b}}(\text{H}_2)$. The factor of one half takes into account that two electrons take part in the bonding. This second term amounts to 2.23(9) eV. Thus a total vertical ionization potential of 15.83 eV is predicted, in very good agreement with the measured value ($I_{\text{vert}} = 15.83\,\text{eV}$).

In Sects. 5.4–5.7 a number of representative spectra of small molecules will be discussed. The selection of examples reflects the preferences of the author.

5.4 Hydrides Isoelectronic with Noble Gases

In this section we consider diatomic molecules which, if their nuclei were united, could be considered as noble gases. Conversely, one can imagine the hydrides HF, H_2O, NH_3 and CH_4 as produced from the noble gas neon by the successive separation of a proton (this very nice compilation has been given by Price [5.17]). The geometrical structures of these molecules are shown in Fig. 5.9. The noble gas neon has a threefold degenerate orbital configuration in the ground state. By the successive partitioning of the protons, the

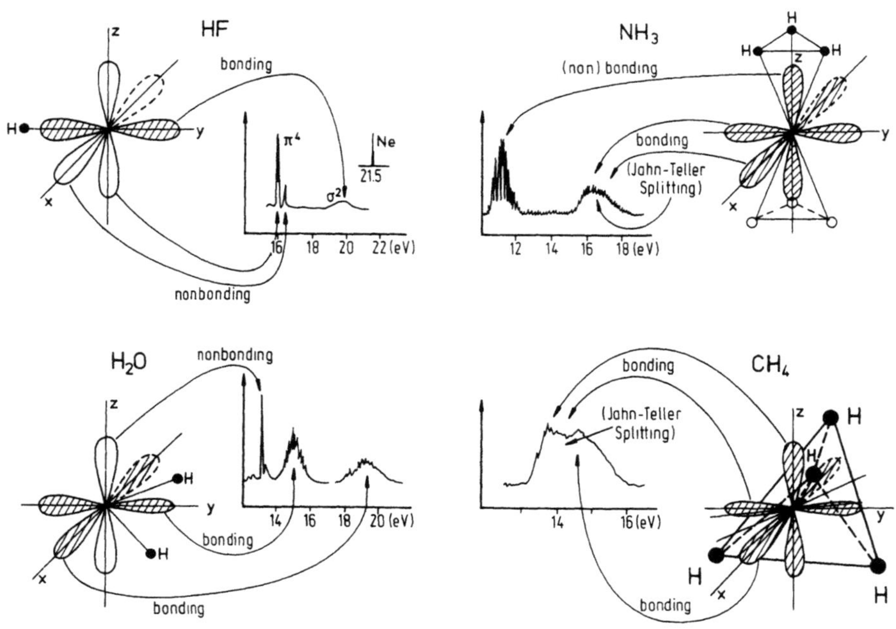

Fig. 5.9. UPS (21.2 eV) spectra of hydrides that are isoelectronic with Ne together with a sketch of their electronic structure (HF, H_2O, NH_3, CH_4) [5.17]. The splitting in the spectrum of CH_4 is due to a Jahn–Teller effect

degeneracy of this p^6-configuration is progressively lifted. This mechanism can explain the appearance of the photoemission spectra which will now be described in more detail (Fig. 5.9).

The electronic configurations of the molecules are: Ne: $2p^6 2s^2 1s^2$; HF: 1s;$2p^5 2s^2 1s^2$; H_2O: 1s; 1s; $2p^4 2s^2 1s^2$; NH_3: 1s; 1s; 1s; $2p^3 2s^2 1s^2$; CH_4: 1s;1s; 1s;1s; $2p^2 2s^2 1s^2$. This somewhat unconventional notation has been chosen to emphasize the partitioning of the protons. The following discussion parallels that of Price [5.17].

Neon (Ne)

The PE spectrum of neon consists of only one level (which is split by the spin–orbit interaction yielding a splitting of 0.1 eV), and is similar to the spectrum shown for argon (Fig. 5.1), except for the smaller spin–orbit splitting.

Hydrogen Fluoride (HF)

For symmetry reasons the axis of this molecule must lie along one p orbital. Thus one has a singly degenerate σ^2 orbital along the bonding axis and a double degenerate π^4 orbital perpendicular to that axis. The π^4 orbital is non-bonding, thus showing weak vibrational structure, and is therefore identified with the photoemission signal at 16 eV. The σ^2 orbital is bonding, and thus has a rich vibrational structure; it is identified as the transition observed at 20 eV [5.17].

Water (H_2O)

If one separates two protons from neon then in principle a linear molecule could result. However, this arrangement is not the most stable, because only the two electrons along the bond axes of the two protons can screen the two protons from the oxygen core. One can obtain additional screening from one of the two orbitals perpendicular to the initial orbital if the two protons deviate from the linear configuration and are located just inside one of the perpendicular electron clouds (the "real" extension of the orbitals is somewhat larger than shown in the schematic illustration of Fig. 5.9). This second electron cloud lies in the same plane as the oxygen nucleus, the two protons and the first electron cloud. It bisects the bond angle of the H_2O molecule, and determines to a large degree the angle of the molecular configuration (Fig. 5.9). The third electron cloud is perpendicular to the plane of the molecule. It does not take part in the bonding, nor does it influence the bond angle; therefore it is a non-bonding orbital. From this it follows that the orbital degeneracy of the p^6-configuration is completely lifted and consequently one expects three photoemission signals, as indeed observed in Fig. 5.9. The three signals can be interpreted as follows:

13 eV: This signal shows no vibrational structure and is therefore attributed to the non-bonding orbital lying perpendicular to the plane of the molecule (p_z).
15 eV: This signal has vibrational structure whose peak separation corresponds to the frequency of the bending mode of the molecule. It is therefore attributed to photoionization out of the orbital which determines the bonding angle of the molecule; this is the orbital which bisects the bond angle of the H_2O molecule (p_y).
19 eV: This photoemission signal also shows intense vibrational structure which corresponds to the stretching frequency of the free H_2O molecule. Therefore this signal is attributed to photoionization out of the orbital which determines the H-O-H-bonding (p_x).

For H_2O the observed vibrational structure has thus helped in the identification of the photoemission signals.

Finally it has to be noted that in the final state the H_2O^+ molecule is linear and therefore has a shape that is distinctly different from that of the H_2O molecule in the ground state.

Ammonia (NH_3)

This molecule has the structure of a pyramid. The p_z orbital along the axis of the pyramid determines the geometrical form of the molecule. The two other orbitals (p_x, p_y) perpendicular to the symmetry axis screen the charge of the protons from the charge of the nitrogen core and thereby determine the bond length of the molecule. In the vibrational structure of the PE signal at 11 eV one finds the frequency of the out of plane bending vibration mode. The intensity of this vibrational structure is unusual for a (largely) non-bonding orbital. However, upon photoionization, the molecule NH_3^+ becomes planar. It is this major change of geometry that is responsible for the intensity of the vibrational structure. The photoemission band at 16 eV shows Jahn–Teller splitting, which indicates that the signal comes from a doubly degenerate state, and it is thus attributed to the two other orbitals which determine the bond length.

Methane (CH_4)

This molecule has the structure of a tetrahedron, with the C atom in the center and a H atom at each corner. In this structure all the orbitals are degenerate. Therefore the valence orbitals should give rise to a single line in the PE spectrum. However, it is energetically favorable to lift the degeneracy of the orbitals by a Jahn–Teller distortion of the molecule which is responsible to the doublet structure in the spectrum.

Figure 5.10 summarizes the photoemission spectra of the neon atom and of the corresponding "hydrides" of F, O, N and C discussed in the preceding

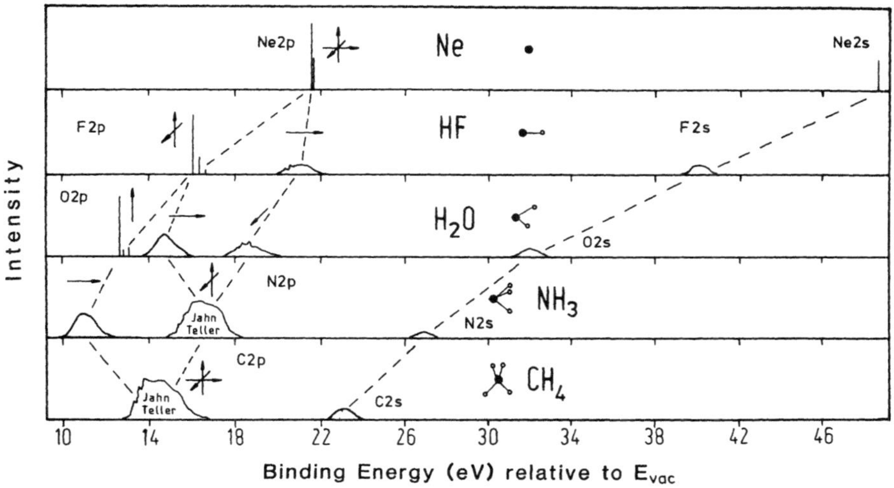

Fig. 5.10. Diagrammatic representation of the data in Fig. 5.9 for Ne and its isoelectronic hydrides, namely HF, H_2O, NH_3 and CH_4 [5.17]

paragraphs [5.17]. It shows more clearly the development of the three valence-band features seen in H_2O from the single (spin–orbit split) line in Ne and the subsequent merging into a single (Jahn–Teller split) feature in CH_4. This figure also shows the 2s lines of Ne, F, O, N and C, which move to smaller binding energy as the atomic number decreases from Ne to C.

5.5 Spectra of the Alkali Halides

The molecules treated in the previous section are examples of more or less covalently bonded molecules. We now consider an example of photoemission spectra from strictly ionically bonded free molecules.

A prototype of such molecules are the alkali halides [5.17, 5.19]. In this case the electrons are quite strongly localized around the respective atoms. In lithium fluoride, for example, one has for lithium a configuration Li^+ ($1s^2$) and for fluorine a configuration $F^-(1s^2 2s^2 2p^6)$. Thus both ions have more or less a noble gas configuration. If the valence orbital of the system M^+X^- (M: metal, X: halide) is photoionized then the final state can be described formally as M^+X^0. This means that the ionic bond is almost completely destroyed and only a weak bonding remains between the M^+ ion and the dipole moment induced by the M^+ ion on the X atom.

On the halide ion X^- the PE process produces a p^{-1} configuration (X^0) which is, in principle, a p state with a spin–orbit splitting where the splitting between the $p_{1/2}$ and $p_{3/2}$ state is $3/2\,\zeta$ (ζ being the spin–orbit coupling

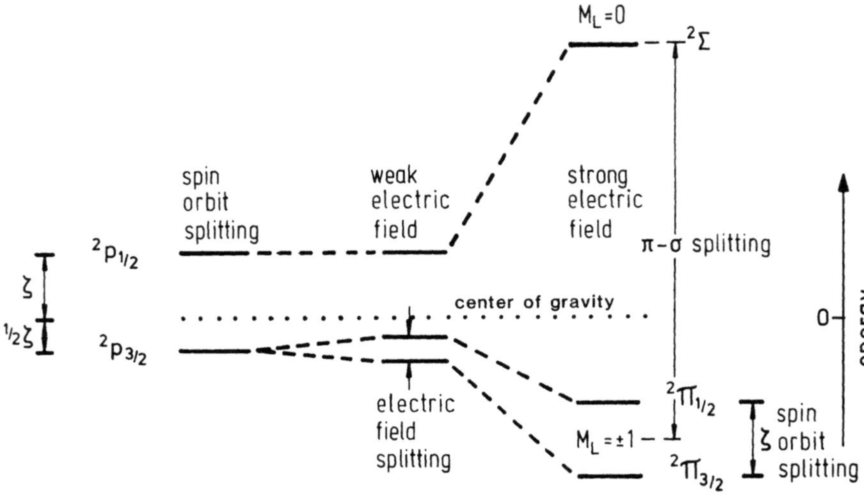

Fig. 5.11. Diagram showing the interplay of spin–orbit interaction and weak and strong electric-field interaction for a $2p_{1/2,3/2}$ state [5.17, 5.20]. On the left there is only the spin–orbit interaction. A weak electric field splits the $2p_{3/2}$ level into two levels while not affecting the $2p_{1/2}$ state. Finally in a strong electric field the 2p level is split into the $M_L = 0$ and the $M_L = \pm 1$ component, where the latter is further split by the spin orbit interaction

constant) provided the electric-field interaction[1] is small compared to the spin–orbit interaction. In addition to the spin–orbit interactions, the X^0 atom feels an electric field produced by the M^+ ion, thus destroying the spherical symmetry experienced by the p electrons in the free X^0 atom; hence the $2p_{3/2}$ state can be split into two states. There are, however, cases in which one has to apply the strong field coupling scheme (electric-field interaction large compared to the spin–orbit interaction) in order to understand the spectrum. This means, that the p-hole ($L = 1$) splits in the electric field into a state $M_L = 0$ ($^2\Sigma$) and another state with $M_L = \pm 1(^2\Pi)$ where the latter of these can be split into states with $S = +1/2$ and $S = -1/2$ by the spin–orbit interaction. The transition from the case of pure spin–orbit splitting to that with an additional weak electric field and finally to that with a strong electric field is shown in Fig. 5.11 [5.17, 5.20].

In order to analyze actual spectra, some consideration of the expected relative trends of spin–orbit splitting versus electric field interaction is necessary [5.19, 5.21]. Table 5.3 gives the spin–orbit splitting parameters for the halogens, both in the atomic state, and in the case where they form molecular orbitals in the $M^+ X^-$ molecule [5.19]. Only for Br and I is this parame-

[1] In the literature this splitting is sometimes called a crystal-field splitting in analogy to the interaction observed in a crystal. This is slightly misleading, however, because one is really dealing with free molecules here.

Table 5.3. Diagonal spin–orbit coupling constants of the uppermost occupied π orbitals of the alkali halide (LiF, ...) molecules and the halide atoms (F, Cl, Br, I) [eV] [5.19]. The numbers in parentheses are estimated from the trends observed in the calculated values

	F	Cl	Br	I
Halide atom	0.033	0.073	0.302	0.628
Li	0.028	0.063	0.263	0.530
Na	0.028	0.063	0.263	0.530
K	0.029	0.063	0.264	0.531
Rb	0.037	(0.068)	(0.273)	(0.546)
Cs	(0.043)	(0.073)	(0.298)	(0.628)

ter large enough that a splitting in the spectra can be observed. Table 5.4 gives the energetic separation of the π and σ orbitals (electric-field splitting parameter) as calculated (and extrapolated) [5.19]. For zero spin–orbit splitting this parameter gives the total splitting observable between the $^2\Pi$ and $^2\Sigma$ levels. In the case where both interactions (spin–orbit and electric field) are present, the corresponding (2×2) secular equation has to be solved. However, the numbers indicate that only for Na and Li (and possibly for K) halides will the electric-field splitting be observable. For a particular metal ion the electric-field splitting increases in going from F to I because the atomic polarizability increases in that order. For a particular halogen X, however, the electric-field splitting increases in the order CsX to LiX because of the decreasing metal radius, which increases the metal–halogen interaction. Table 5.5 lists the splitting of the halogen p state under the combined action of the electric-field interaction (Table 5.4) and the spin–orbit interaction (Table 5.3). This is useful in data analysis [5.21].

Table 5.4. Separation of the π- and σ orbitals (electric-field splitting) derived from the uppermost occupied (halogen p-like) orbitals [eV](the π is least bound in all cases)[a] [5.19] of all alkali halide molecules from LiF to CsI

	F	Cl	Br	I
Li	0.591	0.754	0.770	(0.8)
Na	0.362	0.453	0.472	(0.5)
K	0.215	0.248	(0.27)	(0.3)
Rb	0.026	(0.1)	(0.15)	(0.19)
Cs	(0)	(0.03)	(0.05)	(0.1–0.14)

[a]Values have been obtained assuming Koopman's theorem. The quantities in parentheses have been estimated from the trends observed in the calculated values.

Table 5.5. Splitting of the halogen p level of alkali halide ions at internuclear distance r_e of the neutral molecule [eV] [5.21]

	F			Cl			Br			I		
	$^2\Pi_{3/2}$	$^2\Pi_{1/2}$	$^2\Sigma_{1/2}$	$^2\Pi_{3/2}$	$^2\Pi_{1/2}$	$^2\Sigma_{1/2}$	$^2\Pi_{3/2}$	$^2\Pi_{1/2}$	$^2\Sigma_{1/2}$	$^2\Pi_{3/2}$	$^2\Pi_{1/2}$	$^2\Sigma_{1/2}$
Li	0.0	0.027	0.606	0.0	0.060	0.788	0.0	0.213	0.951	0.0	0.337	1.238
Na	0.0	0.027	0.377	0.0	0.058	0.489	0.0	0.183	0.684	0.0	0.254	1.041
K	0.0	0.026	0.232	0.0	0.054	0.289	0.0	0.136	0.530	0.0	0.173	0.924
Rb	0.0	0.012	0.070	0.0	0.043	0.160	0.0	0.087	0.472	0.0	0.892	0.116
Cs	0.0	0.065	0.0	0.0	0.037	0.121	0.0	0.476	0.026	0.0	0.989	0.076

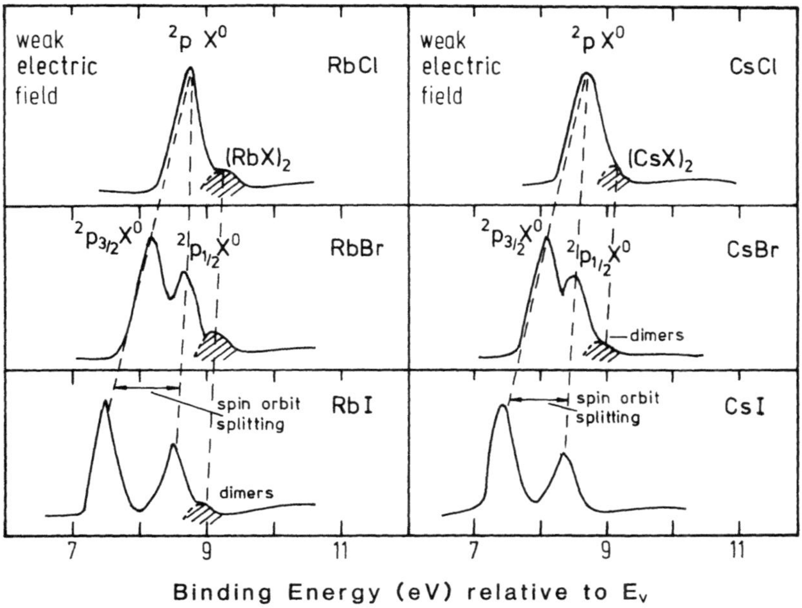

Fig. 5.12. UPS ($\hbar\omega = 21.2\,\mathrm{eV}$) of Rb halides and Cs halides. The spectra are characteristic for the $X^0(X^0 = \mathrm{Cl, Br, I})$ 2p spin–orbit interaction indicated by the increase in splitting from the chloride to the iodide. An additional small line is caused by molecular dimers, $(MX)_2$ [5.17, 5.20]

Figure 5.12 shows the spectra of the Rb and Cs halides. As expected from the numbers in Tables 5.3–5.5 they correspond to the weak electric-field case [5.20]. This means that the splitting observed is roughly the spin–orbit interaction, the splitting being due to the halogen ion, and it is the same for both series. The intensity ratio of the two lines observed is approximately 2:1 as expected for a $\mathrm{p}_{3/2}$, $\mathrm{p}_{1/2}$ doublet.

The spectra for the Na halides in Fig. 5.13 are different [5.20]. The splitting is slightly larger than in the corresponding Rb and Cs halides, and it does not change too much in going from Na Cl to Na I. In Na I, the $^2\Pi$ peak shows a splitting which indicates that one is approaching the strong coupling regime. The main splitting now is largely, though not entirely, due to electric-field interaction, as seen from the numbers in Tables 5.3, 5.4. Note that the diagram in Fig. 5.11 contains as the strong field case the final limit, which is by no means reached for Na I. Therefore the $^2\Pi_{1/2}$, $^2\Pi_{3/2}$ spin–orbit splitting of Na I is still reduced compared to the spin–orbit splitting as observed directly, e.g., in Rb I and Cs I (Fig. 5.12).

Finally, the binding energies (often called ionization potentials in the chemical literature) of the outer p electrons of the X^- ions have to be commented upon. The relevant diagram is shown in Fig. 5.14 [5.19], in which

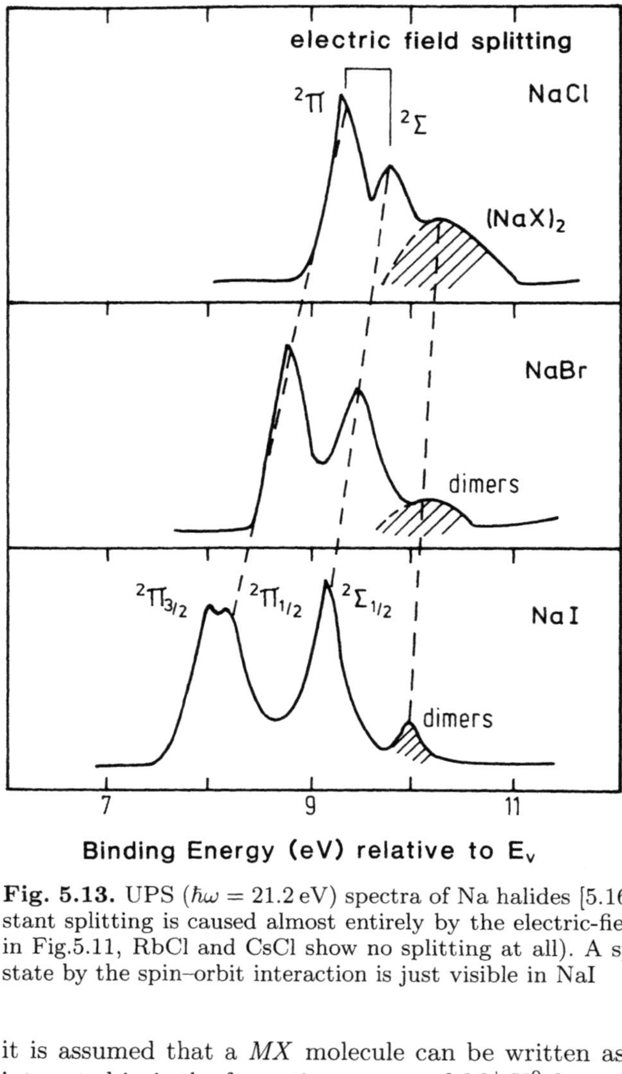

Fig. 5.13. UPS ($\hbar\omega = 21.2\,\mathrm{eV}$) spectra of Na halides [5.16, 5.20]. The roughly constant splitting is caused almost entirely by the electric-field interaction (note that in Fig.5.11, RbCl and CsCl show no splitting at all). A splitting of the $2p_{3/2}(^2\Pi)$ state by the spin–orbit interaction is just visible in NaI

it is assumed that a MX molecule can be written as M^+X^-. What one is interested in is the formation energy of M^+X^0 from M^+X^-. The ionization potentials of the free X^- ions (or rather their Electron Affinities EA) are well known [5.23]. To these one has to add the dissociation energy of the M^+X^- molecule $D_0(M^+X^-)$ to arrive at the total measured ionization potential. If r_e is the equilibrium nuclear separation in the MX molecule, $D_0(M^+X^-)$ can be written as:

$$D_0(M^+X^-) = \frac{e^2}{4\pi\varepsilon_0 r_e}\,, \tag{5.6}$$

where r_e can be taken, e.g., from an analysis of vibrational spectra [5.20]. To be more correct, one has to take into account the bonding between the M^+

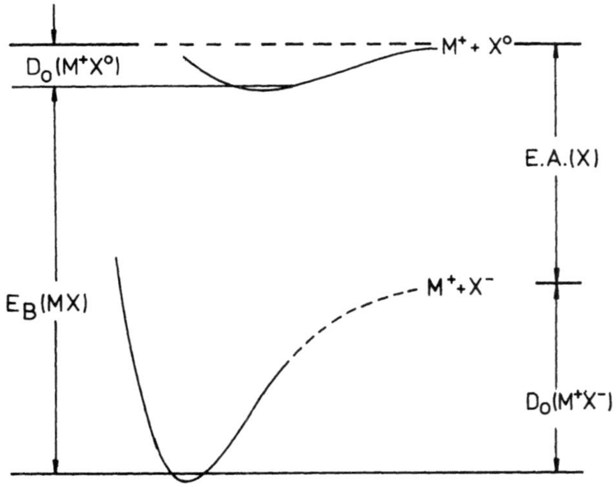

Fig. 5.14. Potential energy diagram for the molecule MX and its photoionized species MX^0 [5.19]. $D_0(M^+X^-)$ is the dissociation energy of M^+X^-. $EA(X)$ is the electron affinity of X. The upper curve is very shallow because the binding in M^+X^0 is only via the polarization of the X^0 ion induced by the surrounding M^+ ions. $D_0(M^+X^0)$ is the (small) dissociation energy of the (M^+X^0) complex

and X^0 species, which, although weak, nevertheless exists. Thus one arrives at the following equation:

$$E_B(MX) = D_0(M^+X^-) + EA(X) - D_0(M^+X^0) . \quad (5.7)$$

Neglecting $D_0(M^+X^0)$ in a first approximation, one obtains quite good agreement with the experimental binding energies, which are taken from the spectra as the peaks with the smallest binding energy. As an example, data for CsX are compiled in Table 5.6.

In the spectra of the alkali halides, sometimes an additional line at larger binding energies is observed. This is due to dimer formation $(MX)_2$ of the respective molecules (Figs. 5.12, 5.13).

Data on solid alkali halides can be found in [5.22].

Table 5.6. Binding energies of halogen p electrons in CsX molecules [eV] [5.19]

	CsF	CsCl	CsBr	CsI
$EA(X)$ [5.23]	3.4	3.6	3.4	3.0
$D_0(M^+X^-)$ [5.24]	5.6	4.8	4.6	4.3
$EA + D_0$	9.0	8.4	8.0	7.3
E_B	8.8	7.8	7.4	7.1

5.6 Transition Metal Dihalides

Compounds such as nickel oxide (NiO) or nickel dichloride ($NiCl_2$) have played a considerable role in the understanding of the ground-state and excited-state electronic structures of solids. In order to reach an understanding of the photoemission spectra of such compounds, it is instructive to look at the PE spectrum of a typical free molecule of this kind, such as manganese dichloride [5.25, 5.26]. The He I and He II spectra of this molecule are shown in Fig. 5.15. The two different radiations are used to distinguish between orbitals with mainly d character and those with mainly p character, since the He II radiation has a larger cross-section for d orbitals, whereas He I radiation is more sensitive to p orbitals (Fig. 1.14). Inspection of Fig. 5.15 thus leads to the conclusion that the signal C at 13.6 eV originates from an orbital with almost pure d character, while the structure designated as B around 12 eV

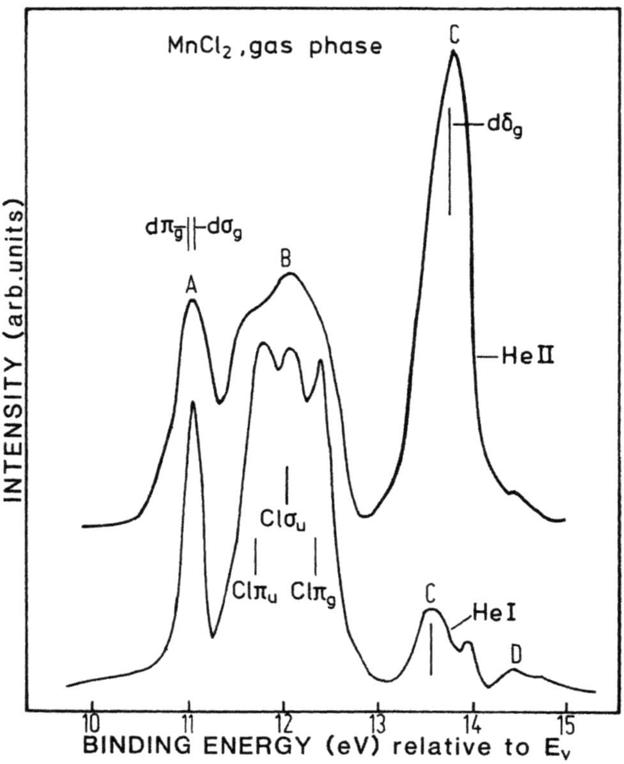

Fig. 5.15. UPS ($\hbar\omega = 21.2\,\text{eV}$ and $\hbar\omega = 40.8\,\text{eV}$) of $MnCl_2$ in the gas phase [5.25, 5.26]. Structures A and C reflect photoionization from orbitals with mostly d character while structure B reflects photoionization from orbitals with mostly p character. The ratio of the d-cross-section to the p-cross-section increases in going from He I (*lower curve*) to He II radiation (*upper curve*)

should be mostly of p character. The structure labelled A can be assigned to a mixed character. The simple Koopmans' theorem calculation leads to the result [5.25, 5.26] that, neglecting relaxation effects, the ionization energy (binding energy) of the d orbitals is always larger than that of the p orbitals, meaning that the d orbitals should always have a lower energy (relative to E_v) than the p orbitals. This is in contrast to the observation in Fig. 5.15 since the $d\pi_g$ and $d\sigma_g$ orbitals occur at smaller binding energies than the p orbitals and only the $d\delta_g$ orbital occurs at its anticipated place below the p orbitals.

For the interpretation of the spectra in Fig. 5.15 one is thus led to the conclusion that relaxation effects are important. The $d\pi_g$ and the $d\sigma_g$ orbitals have counterparts in the chlorine p orbitals with the same symmetry. Therefore an easy charge transfer from the p orbitals into these d orbitals can take place, making relaxation via charge transfer possible. However, there are no p orbitals of $d\delta_g$ character making relaxation difficult for the orbitals of this type. Thus the signal C in the spectrum which is assigned to the $d\delta_g$ orbital is an almost unrelaxed state, whereas the structure A which is assigned to the $d\pi_g$ and $d\sigma_g$ orbitals is a more or less relaxed state due to charge transfer from the p orbitals into the photoionized d orbitals. In this sense the structure C can be interpreted as a "satellite" of the structure A, both coming from the d orbitals. This is in accordance with the solid state spectra of transition metal dihalides (Chap. 3).

5.7 Hydrocarbons

The investigation of hydrocarbons has played an important role in the field of PES of free molecules [5.2, 5.27]. Here it is impossible to treat this field in any depth and we confine ourselves to a brief sketch of two examples in order to demonstrate the kind of information that can be obtained. It is emphasized that spectra from organic molecules are often very complicated and that an understanding of an Energy Distribution Curve (EDC) in terms of the molecular structure is, in many cases, only obtained with the help of molecular-orbital calculations and/or the comparison with analyzed EDCs for related compounds.

As examples we choose acetylene (C_2H_2) [and diacetylene (C_4H_2)] because it is perhaps the simplest molecule involving a C-C bond, and then benzene (C_6H_6) will briefly be dealt with since it serves as a prototype of the *aromatic molecules*.

He I spectra of C_2H_2 and C_4H_2 are given in Fig. 5.16 [5.5, 5.28], where their molecular structure is also indicated. The EDC of C_2H_2 in the energy region given shows three bands with binding energies of 11.4 eV, 16.4 eV and 18.4 eV. In experiments with higher photon energies another band at 23.55 eV can be found. The interpretation of the spectrum can be made in terms of the known electronic structure of C_2H_2 (Table 5.7).

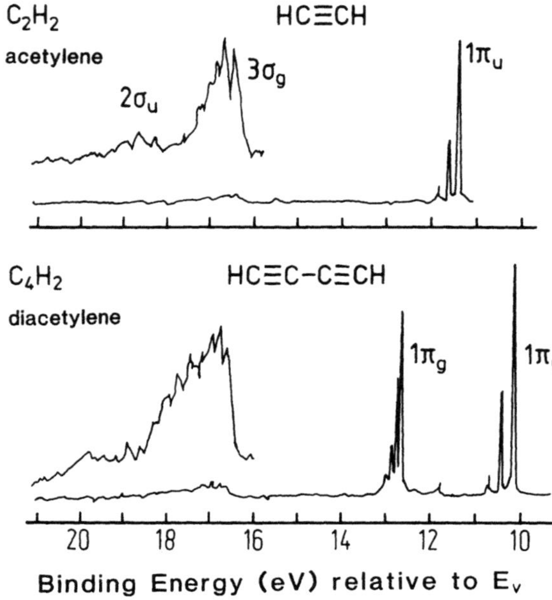

Fig. 5.16. UPS spectra of acetylene (C_2H_2) and diacetylene (C_4H_2) [5.2, 5.5]. In going from $HC \equiv CH$ to $HC \equiv C - C \equiv C - H$ an additional $C \equiv C$ bond appears which clearly shows up in the PE spectrum

Because C_2H_2 is a linear molecule, the π and σ bands are well separated from each other, which simplifies the interpretation of the spectra. The band with the smallest binding energy stems from the π-bonding of the two carbon atoms. This assignment can also be deduced from the vibrational structure of the band at 11.4 eV, because, as shown in Fig. 5.17 [5.2], the vibrational structure contains the frequency $\nu_2 = 1830\,\mathrm{cm}^{-1}$ which is very close to the $C \equiv C$ stretching frequency of the C_2H_2 free molecule ($\nu_2 = 1983\mathrm{cm}^{-1}$ [5.2]).

Table 5.7. Schematic electronic structure of C_2H_2 [5.2]

E_B[eV]	290	23.6	18.4	16.4	11.4
orbital	$(1\sigma_g 1s_c)^2(1\sigma_u 1s_c)^2$	$(2\sigma_g)^2$	$(2\sigma_u)^2$	$(3\sigma_g)^2$	$(1\pi_u)^4$
bonds		$\sigma(C-C)$	$\sigma(C-C)$ $\sigma(C-H)$	$\sigma(C-H)$ $\sigma(C-C)$	$\pi(C-C)$
atomic levels	1s core level		C 2s		C 2p, H 1s

The lowering of the frequency in the ion as compared to the molecule indicates that the band is weakly bonding.

The spectrum of diacetylene (C_4H_2) is very similar to that of acetylene except that it shows a symmetric splitting of the π band signal. This is to be expected because in C_4H_2 one has two $C \equiv C$ bonds that interact with each other leading to two π bands with different binding energies. The π_u signal occurs at 10.2 eV and the π_g signal at 12.8 eV. The rest of the spectrum, as expected, is very much the same as that of C_2H_2.

With respect to further applications of PES to the field of hydrocarbons, the reader is referred to the literature [5.2]. However, to illustrate the state of the art a PE spectrum of benzene (C_6H_6), which has been interpreted by a combination of comparisons with analogous compounds and calculations, is discussed. The PE spectrum [5.29] and the assignments of the bands are given in Fig. 5.18.

The interpretation of the spectrum of Fig. 5.18 was first proposed by Lindholm [5.30]. We now illustrate, for one case, how the classification can be verified.

The $1e_{1g}$ level at about 9 eV in benzene is, according to the classification of Lindholm [5.30], a twofold degenerate state. If the symmetry of the benzene ring is destroyed by replacing one hydrogen atom (one electron) by an atom with one hole, e.g., a halide atom, then the symmetry degeneracy of that orbital must be lifted. Figure 5.19 gives the PE spectrum of bromo-

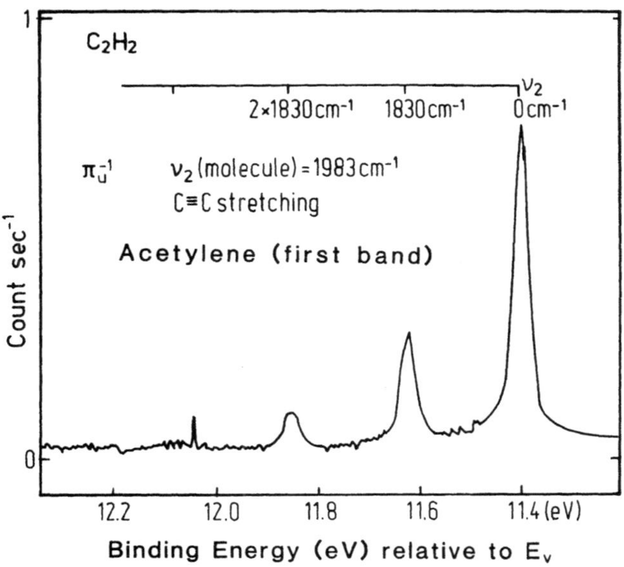

Fig. 5.17. High-resolution UPS spectrum of the π_u photoionization in C_2H_2 (Fig. 5.16) which reveals the vibrational structure ($C \equiv C$ stretching frequency) [5.2]

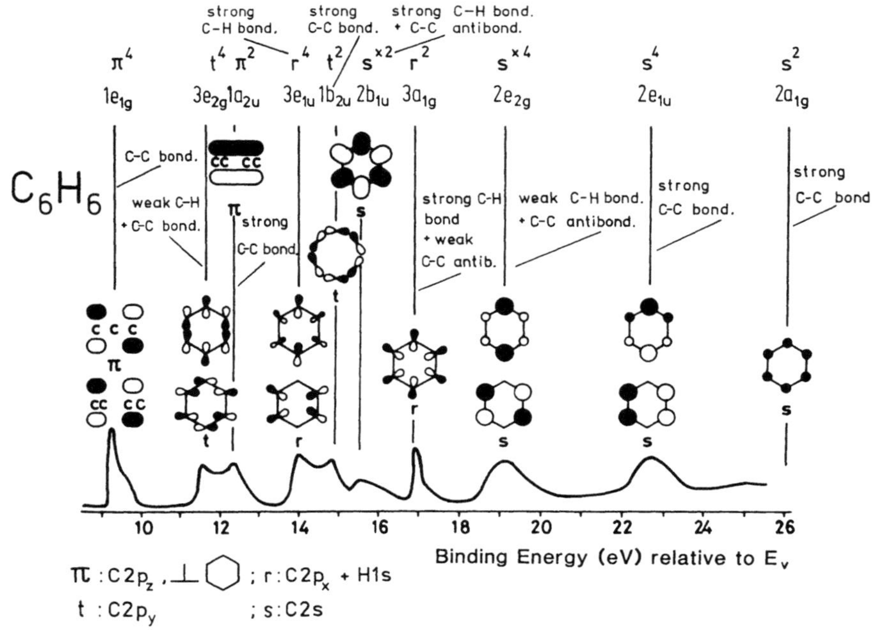

Fig. 5.18. UPS (He II radiation) spectrum and molecular assignment of the bands of benzene C_6H_6 [5.27, 5.29, 5.30].
s: orbitals which are composed of C 2s orbitals;
t: orbitals which are composed of C 2p orbitals and are directed tangentially to the benzene ring;
r: orbitals which are composed of C(2p)+H(1s) orbitals and are directed radially to the ring of the benzene;
π: orbitals which are composed of C 2p orbitals and are perpendicular to the plane of the benzene molecule

benzene [5.2]. The details of the spectrum are not important but one should concentrate on the area around 9 eV. Here, instead of one level as in C_6H_6, one observes in C_6H_5Br two distinct energy levels at 9.05 and 9.67 eV. This is a direct consequence of the lifting of the twofold degeneracy of the $1e_{1g}$ orbital due to the replacement of one hydrogen atom by a bromine atom. This simple example illustrates how substitution can be employed to understand and interpret a photoemission spectrum of a particular molecule.

Finally Fig. 5.20 shows a comparison of the UPS spectra of C_6H_6 in the gas phase and in the condensed form. As expected, the fine structure is washed out in the condensed phase but otherwise the spectrum of the latter is very similar to the gas phase spectrum, except for a small "solid-state" relaxation shift (if both spectra are referenced to the vacuum level) of ≈ 1.15 eV [5.31, 5.32].

Spectra of solid C_{60} are given in [5.33].

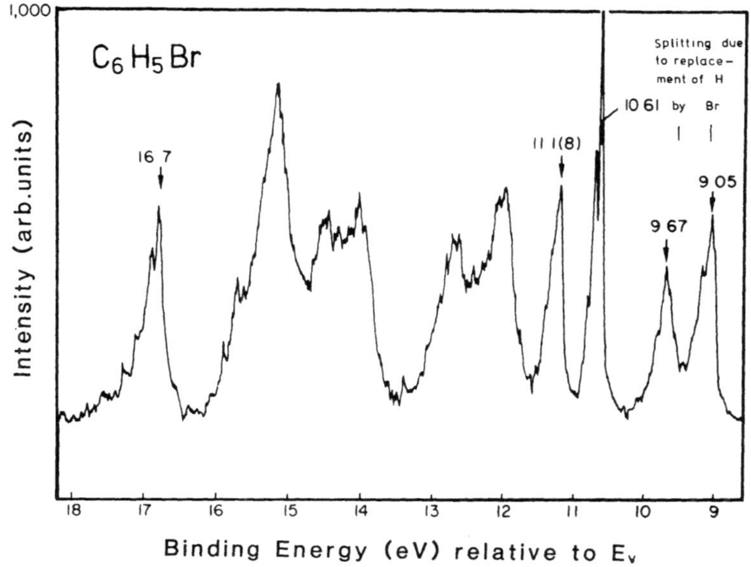

Fig. 5.19. UPS spectrum (He I radiation) of C_6H_5Br [5.2]. Note the splitting of the band with the smallest binding energy (which is degenerate in C_6H_6) due to the distortion of the symmetry of the molecule upon replacing an H ion by a Br ion

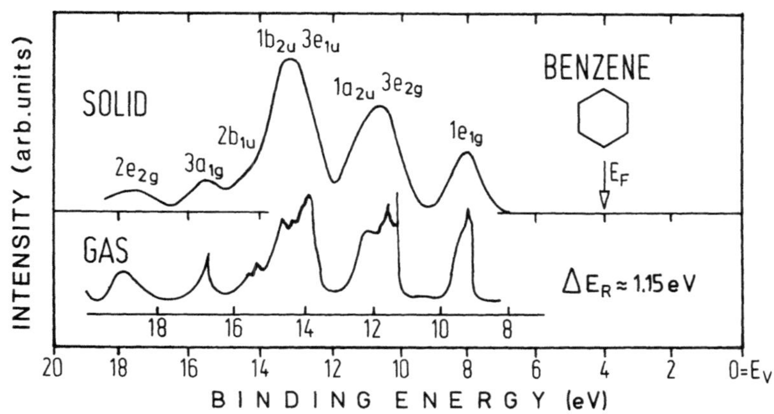

Fig. 5.20. UPS spectra of C_6H_6 (benzene) in the gas phase and the solid phase showing a relaxation shift between the two species [5.31, 5.32]. Except for some broadening in the solid state spectrum the two spectra are quite similar

5.7.1 Guidelines for the Interpretation of Spectra from Free Molecules

It has perhaps become apparent from the material presented on the analysis of EDCs from free molecules that, especially for hydrocarbons with many atoms, the correlation between a measured EDC and the electronic structure of the molecule is not always trivial. Some guidelines that may be of use in analyzing a measured EDC are summarized below.

- Electronic structure calculations for molecules: with the advanced computer codes available today, molecular-structure calculations have achieved a remarkable degree of accuracy and it is certainly always a very good first approach to use a recent calculation in order to reach an understanding of an EDC.
- Vibrational structure analysis (see the N_2 example, Fig. 5.6).
- Comparison within an isoelectronic series (see, e.g., the data on Ne, HF, H_2O, NH_3, CH_4, Figs. 5.5, 5.9, 5.10).
- Chemical comparison and/or chemical trends. This approach can be used, e.g., in analyzing the spectra of CCl_4, CBr_4 and $SiCl_4$ [5.18].
- Lifting of degeneracies by chemical substitutions (see the comparison of C_6H_6 with C_6H_5Br, Figs. 5.18, 5.19).
- Chemical systematics. This can be applied in series such as $(C_2H_2)_n$ or $(C_6H_6)_n$.
- Intensity measurements as a function of photon energies (e.g., $MnCl_2$, Fig. 5.15 or Fig. 1.15).

5.7.2 Linear Polymers

The way in which a solid forms from isolated atoms, through small clusters containing only a few atoms, to an "infinite" solid is still not very well documented because the investigation of small clusters is not an easy matter [5.34]. PES has been used to study supported metal clusters, and one can indeed observe the narrowing of the valence band as the clusters get smaller. Effects on the core levels have also been observed. However, if the present understanding is correct, these are dominated by a final-state effect, namely the production of the positive charge on the cluster by the PE process [5.34]. A rather instructive way to observe the evolution of the valence band of a solid from that of the isolated atomic constituents is by investigating polymers and a case in point are the linear alkanes with the formula C_nH_{2n+2}, starting out with methane (CH_4) and ending up in polyethylene $(CH_4)_x$ [5.32, 5.35, 5.36]. Figure 5.21 shows the XPS spectra of the first four members of this series together with the result of a first principles calculation of their "valence-band" structure [5.32, 5.37]. The valence band is derived from the C 2p, C 2s and H 1s atomic orbitals.

Methane (CH_4) has already been discussed briefly in Sect. 5.4. The electronic structure can be written as [5.2]

$$\begin{matrix} (1s\,a_1)^2 & (2s\,a_1)^2 & (2p\,t_2)^6 \\ C\,1s & C\,2s & C\,2p \end{matrix} ,$$

where the first orbital is C 1s-derived, etc. Because of the symmetry of the molecule, the $2pt_2$ orbitals are all degenerate and the small splitting actually observed is due to a Jahn–Teller effect (Fig. 5.9).

In ethane, C_2H_6, the molecular orbitals are [5.2]:

$$\begin{matrix} (1a_{2u})^2(1a_{1g})^2 & (2a_{1g})^2(2a_{2u})^2 & (1e_u)^4(3a_{1g})^2(1e_g)^4 \\ C\,1s & C\,2s & C\,2p \end{matrix} .$$

Since there are two carbon atoms, the C 1s and C 2s atomic levels are both split into two levels. This splitting can be nicely seen in the C 2s orbitals of Fig. 5.21 ($E_B \approx 20 - 25\,\mathrm{eV}$). There are, as is apparent from the molecular geometry (Fig. 5.22), three C 2p-derived orbitals (one C-C and two different C-H) yielding in principle three peaks in the 2p-derived part of the spectrum

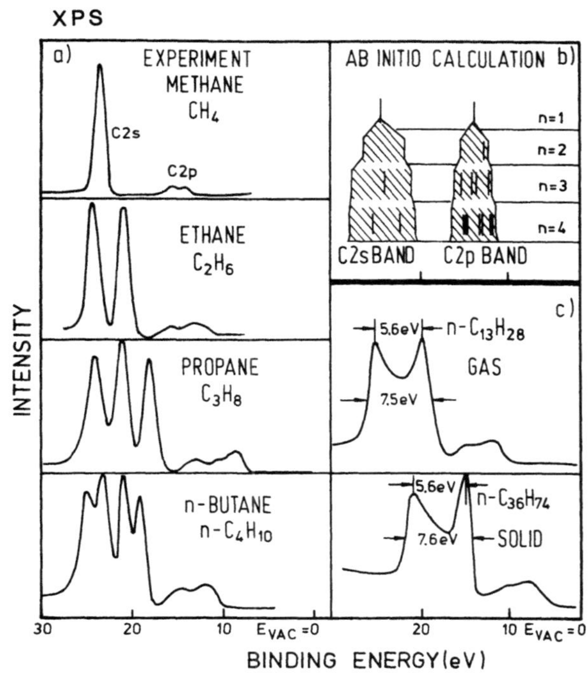

Fig. 5.21. (**a**) EDCs of a number of linear polymers [5.32]: CH_4, C_2H_6, C_3H_8 and C_4H_{10}; (**b**) comparison with the ab initio calculation of their valence bands. The C 2s orbital nicely shows the splitting expected into 1, 2, 3 and 4 levels. (**c**) For higher members of this family the single orbitals originating from the C 2s level can no longer be resolved and lead to a structure seen, e.g., in the gas-phase spectrum of $C_{13}H_{28}$. This spectrum is very similar to the one observed in solid samples such as $C_{36}H_{74}$

(Fig. 5.21) which are not visible in the XPS spectrum. Even a UPS spectrum (Fig. 5.22) shows only indications of these three levels [5.2].

The 2s-derived spectra of the linear C_nH_{2n+2} molecules should always reflect the number of C atoms. This is illustrated in the spectra of propane (C_3H_8) and n-butane (n-C_4H_{10}), which show three and four C 2s levels, respectively. The spectra reproduced in Figs. 5.21, 5.22 were obtained from free molecules.

In this way only data with not very good statistical accuracy are often obtained. Therefore the small molecules or metal clusters are deposited on a substrate in order to have a sample with a higher density of the molecules or clusters. In these samples, however, in addition to the intra-molecular interactions (or the intracluster interactions), also interactions with the substrate are present. These additional interactions are an unwanted side effect and they may obscure the interactions which one is interested in. Therefore one has to compare spectra from free molecules (or clusters) with those obtained from the deposited molecule (or cluster).

Figure 5.23 shows, for instructive purposes, the comparison of spectra of a linear alkane molecule (n-nonane, n-C_9H_{20}) in its (free) molecular form and condensed on a glass substrate [5.38]. Except for a trivial general shift of the two spectra with respect to each other, the only noticeable difference is a slight broadening in the condensed phase spectrum with respect to that of the gas phase, a phenomenon generally observed, for which the most likely

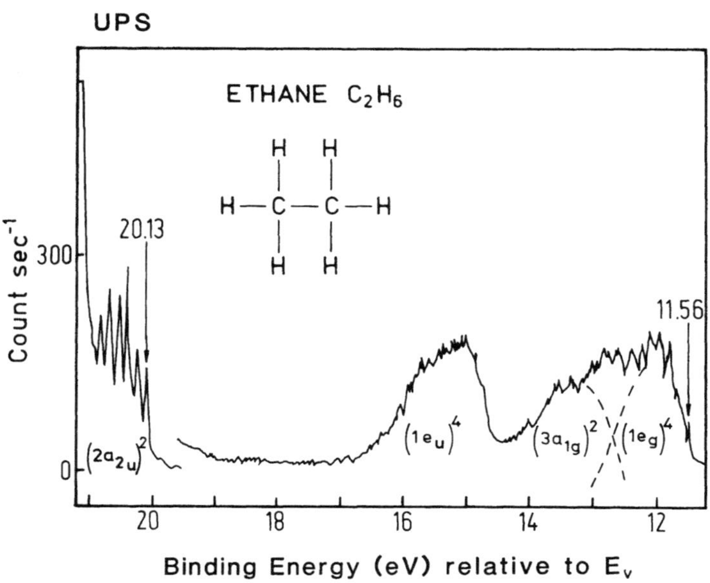

Fig. 5.22. UPS (21.2 eV) spectrum of C_2H_6 to study the structure in the 2p-derived orbitals [5.2]. The three expected orbitals can barely be discerned

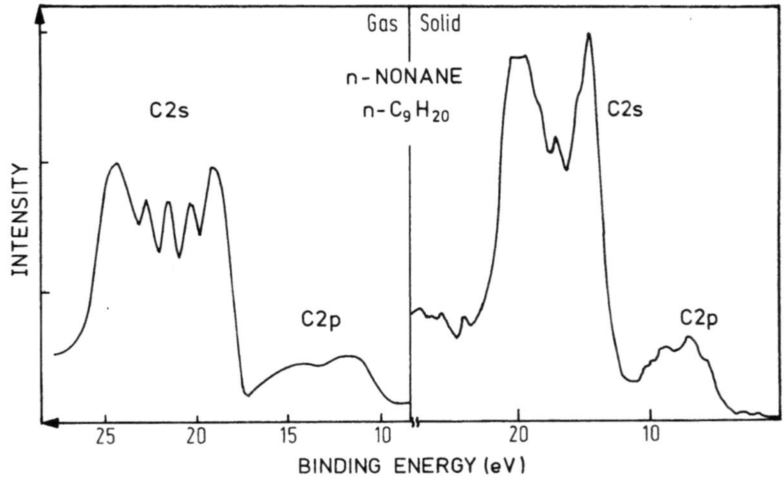

Fig. 5.23. PES spectra of C_9H_{20} in the gas and in the solid phase [5.38]

explanation is differential charging. Furthermore, one can see in Fig. 5.23 how the C 2s part of the spectrum develops into a two-peak structure with increasing chain length, while the 2p part develops into one broad hump.

Figure 5.21, as a further example in this respect, shows a gas phase spectrum of n-$C_{13}H_{28}$ and a spectrum of solid n-$C_{36}H_{74}$. These spectra are almost identical indicating that the gas phase data are also representative for those of the solid state.

These data suggest that, for organic molecules, the inter-molecular interaction and the interaction with the substrate are not very important compared to the intra-molecular interactions. The data at this point seem insufficient to make a similar statement for metallic clusters.

Finally, to close the circle, it is worthwhile studying the "real" solid that is made up of infinite $(CH_2)_n$ chains, namely polyethylene. Its XPS spectrum is given in the upper panel of Fig. 5.24 [5.38, 5.39]. This spectrum is, as expected, quite similar to the solid state spectrum of n-C_9H_{20} (Fig. 5.23) or that of n-$C_{36}H_{74}$ (Fig. 5.21).

The series of data in Figs. 5.21, 5.23, 5.24 shows that even at a quite early stage – namely at $n = 3$ – the bandwidth of the solid form has been largely established and that further lengthening the chain of the polymer increases the bandwidth only slightly, whilst significantly increasing the level density within the band.

In Fig. 5.24 in addition to the XPS spectrum of polyethylene, which can be thought of as an approximately one-dimensionally bonded carbon system, the spectra of graphite, diamond and a fullerene (C_{60}) are shown for comparison. These are (approximately) two-dimensionally, three-dimensionally and spherically bonded extensions of polyethylene (of course, without hydrogen).

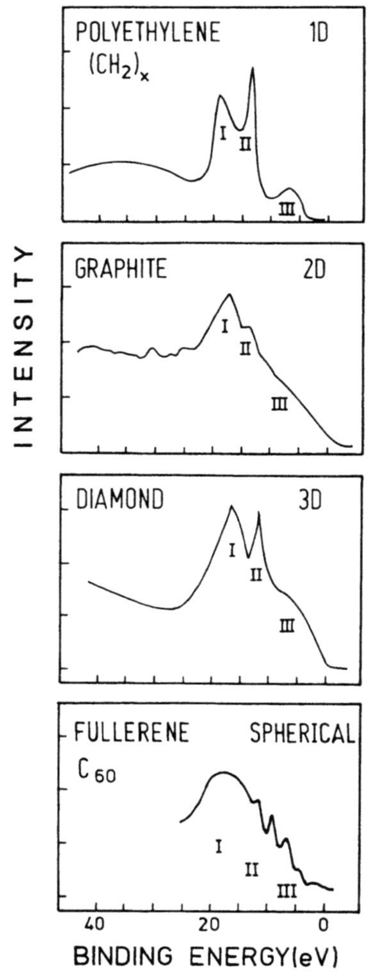

Fig. 5.24. Comparison of the XPS spectra of "one", "two", "three" dimensionally and "spherically" bonded carbon systems [5.38, 5.39]. The spectrum of $(CH_2)_x$, polyethylene, is only marginally different from those of C_9H_{20} and $C_{13}H_{28}$ (Figs. 5.21, 5.23). The similarity to the spectra of graphite, diamond, and the fullerene C_{60} is interesting and indicates the derivation of the regions I, II and III

These spectra which, with reservations concerning the cross sections, represent the corresponding densities of states, are remarkably similar. (An analysis shows that region III is C 2p-derived as expected, region I C 2s-derived and region II C 2p/C 2s mixture).

Figure 5.25 finally gives an XPS spectrum of trans-polyacetylene – $(CH)_x$ – [5.35, 5.40], which is not too different from that of the systems shown in Fig. 5.21. The interpretation of the PES spectra, however, is not yet fully established [5.35, 5.40, 5.41]. In staying with the interpretation of the data in Fig. 5.21 and the literature [5.35, 5.40, 5.41] we can assume that the intensity for $E_B \leq 10\,eV$ reflects emission out of C 2p-derived orbitals while the intensity for $E_B > 10\,eV$ stems from emission out of C 2s-derived orbitals.

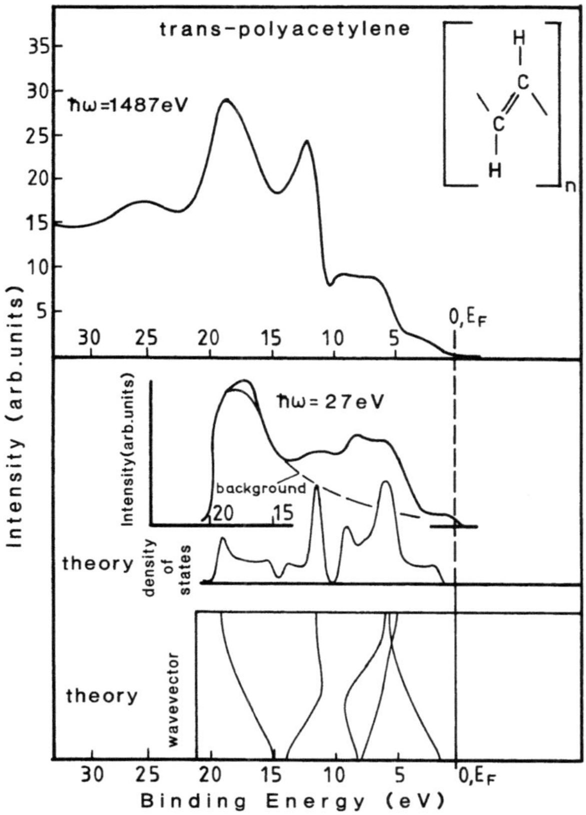

Fig. 5.25. XPS and UPS spectra of trans-polyacetylene [5.40] together with a band structure and a density of states derived from it. The two experimental spectra have been aligned at the Fermi energies (as they show up in the spectra) by the present author

Polyacetylene has found some interest because it can be doped to make it highly conducting [5.42–5.44]. Attempts to obtain information on the electronic structure of doped $(CH)_x$ have been only partly successful [5.45, 5.46]. The dopants only contribute a fraction of an electron to the π-bond of the $(CH)_x$ chain and that gives rise to only a very small signal in the PES spectra. It has not yet been possible to observe the development of a "real" Fermi edge in these conducting materials. This probably is not an effect of bad samples but reflects the one-dimensional nature of these compounds. In one-dimensional systems the excitation spectra give no Fermi step. The interpretation of the core-level spectra, which likewise change upon doping, also remains to be established [5.47].

5.8 Insulating Solids with Valence d Electrons

Valence bands of large gap insulators like NaCl or MgO can, at least as far as their gross features are concerned, be very well understood within the framework that was applied to their core levels. This means that the position of the valence bands can be obtained approximately by a Born–Haber-type calculation (Sect. 2.2.1). The only difference is a larger width of the valence band as compared to the core levels which reflects the dispersion of the bands.

The interpretation of PE spectra becomes more complicated in compounds that contain open shell ions like transition metal or 4f/5f ions. These systems will be the topic of the following discussion, where the emphasis will be placed on 3d transition metal compounds.

The topic of this section is currently a very active research area. This also has to do with the fact that two classes of interesting compounds are transition metal compounds with d electrons in the valence band, namely the high-temperature superconductors (parent material La_2CuO_4) and compounds showing the colossal magneto-resistance (parent compound $LaMnO_3$). Therefore the literature in this field is enormous and we shall touch here only a small fraction of the available material and refer the reader to the original literature.

Figure 5.26 [5.48,5.49] shows XPS valence bands of TiO_2, NiO and Cu_2O. The first is an insulator with no d electrons (TiO_2, with Ti^{4+}, $3d^0$), the second an insulator with eight d electrons (NiO, Ni^{2+}, $3d^8$) and the third a semiconductor ($E_g = 2\,eV$) with a closed d^{10} shell (Cu_2O, Cu^+, $3d^{10}$). TiO_2, which represents insulating oxides with no d electrons only shows the O 2p band and the O 2s band with a separation of $\sim$16 eV. This is the separation also found in atomic oxygen [5.53–5.56] and indicates that solid-state effects hardly influence the relative position of the valence bands in these ionic compounds. Indeed in almost all oxides investigated so far by PES the O 2s–O 2p separation is found to be $\sim$16 eV, as is also seen in Fig. 5.26 for the semiconducting Cu_2O. This constancy can in turn be used to determine the O 2p position in compounds like NiO, where it is not a priori identified in the spectrum.[2]

Nickel oxide is a prototype material for the transition metal compounds, which derive many of their interesting properties from the d electrons [5.57]. Therefore NiO will be discussed at some length in this section. Simple molecular-orbital considerations lead to an energy level ordering that places the d states above the O 2p states, as observed in the data of Fig. 5.26. It will be demonstrated that matters are more complicated.

In the simplest picture of its electronic structure, the d-like valence band in NiO is not completely filled (Ni is Ni^{2+} in NiO, with a $3d^8$ configuration)

[2] The position of the O 2p band in NiO indicated in Fig. 5.26 was deduced from its position in TiO_2 and Cu_2O. The correctness of this positioning has been verified by UPS investigations in which the intensity of the O 2p emission relative to the 3d-emission is larger than in the XPS data of Fig. 5.26.

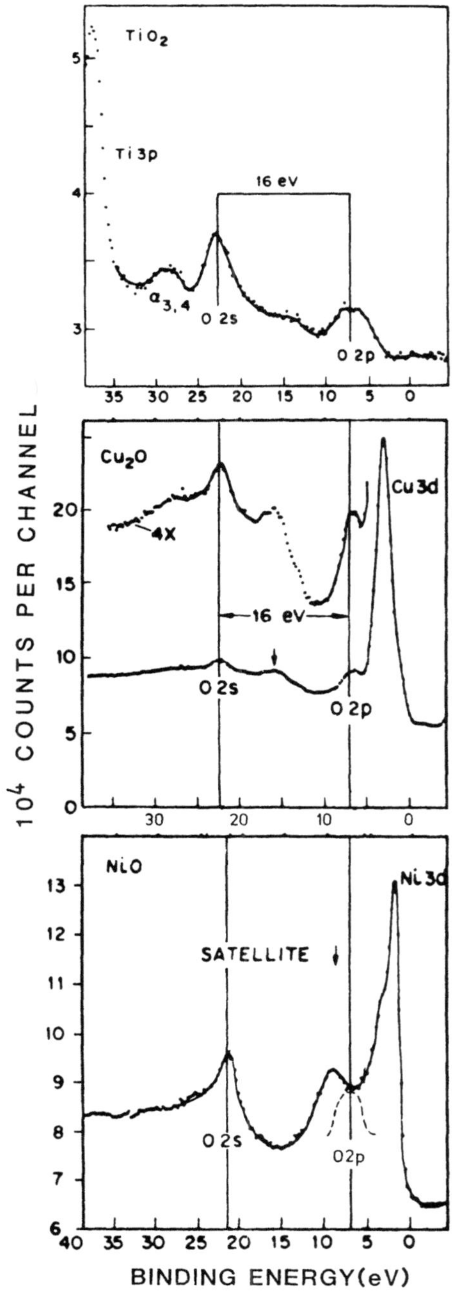

Fig. 5.26. XPS valence-band spectra of TiO_2 ($3d^0$ initial state), Cu_2O ($3d^{10}$ initial state) and NiO ($3d^8$ initial state) [5.48]. The O 2s and O 2p signals line up for TiO_2 and Cu_2O. This suggests that similar structure seen in NiO at around $E_B = 9\,eV$ (see *arrow*) is not due to O 2p photoionization and is therefore termed "satellite" here. The O 2p band is therefore inferred from its positions in TiO_2 and Cu_2O

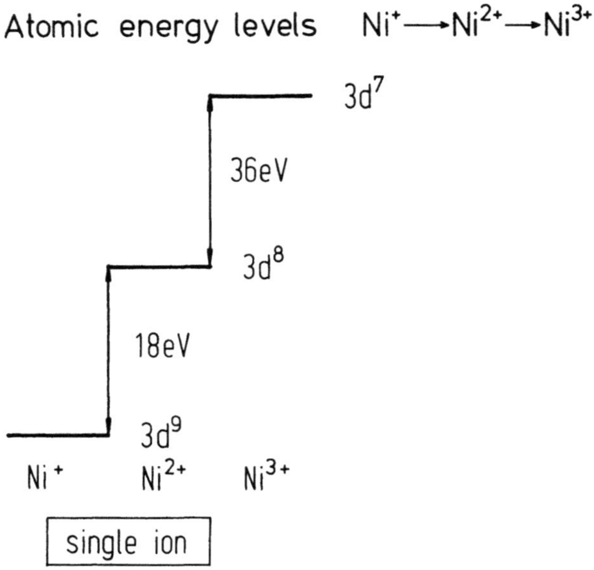

Fig. 5.27. Ionization potentials in Ni^+ and Ni^{2+} [5.53–5.56]

and therefore NiO should be a metal. However, NiO is an insulator with a gap of 4 eV [5.58,5.59]. The insulating nature of a material like NiO is thought to result from the correlation [5.60] between the d electrons, a picture advocated for a long time by Mott [5.61]; compounds like NiO are thus often called Mott insulators.[3] Mott argues that to achieve conduction in the Ni 3d band of NiO one requires the reaction [5.60–5.63]:

$$2\mathrm{d}^8 \rightarrow \mathrm{d}^7 + \mathrm{d}^9 - U\ . \tag{5.8}$$

This reaction [5.64] is defined only in the atomic limit. The relevant energy levels for Ni^+, Ni^{2+} and Ni^{3+} are given in Fig. 5.27, where it is seen that the ionization potential (I) of Ni^{2+} is 36 eV and its electron affinity (EA) is 18 eV. Thus the above reaction corresponds to the creation of a Ni^+ and Ni^{3+} pair out of two (atomic) Ni^{2+} ions which requires an energy of $U = I - EA = 18\,\mathrm{eV}$ (Fig. 5.28). In the solid, the atomic energy levels are broadened by the ion–ion interaction and this can be shown in a diagram of the form of Fig. 5.29 [5.61] where the energy levels of Fig. 5.28 are given as a function of $1/a$ for an ensemble of ions with separation a. One sees that for $a = \infty$ (atomic limit!) one has $U = I - EA$, but for finite a and with the bandwidths of the lower and upper bands denoted by B_1 and B_2 respectively one has

$$E_\mathrm{g} = U - \frac{1}{2}(B_1 + B_2)\ , \tag{5.9}$$

[3] We shall see later that, contrary to general belief, NiO is not a Mott insulator, but a so-called charge-transfer insulator.

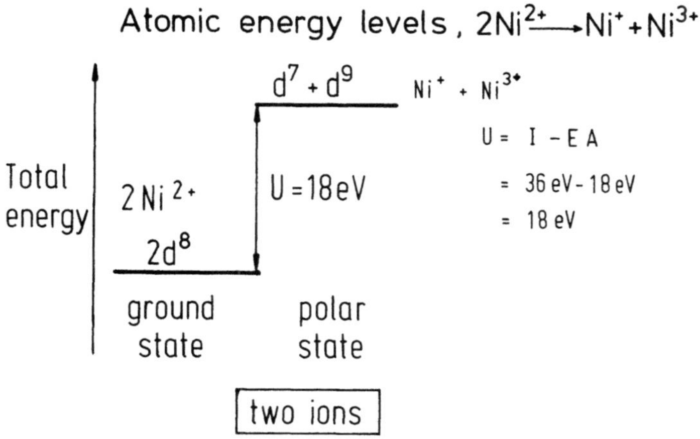

Fig. 5.28. Energy level diagram corresponding to the reaction $2Ni^{2+} \rightarrow Ni^+ + Ni^{3+} - U$ (see Fig. 5.27 for numbers); I is the ionization potential of Ni^{2+}, and EA its electron affinity

where E_g is the energy gap between the valence and conduction band. In this picture U is always defined as the atomic value; the effect of the solid is taken into account by B_1 and B_2.

However, one can also perform the reaction (5.8) in the solid and then define U in the solid as the energy needed to create a distant Ni^+, Ni^{3+} pair from a Ni^{2+}, Ni^{2+} pair. Then U will be smaller than in the atomic limit because of the various screening mechanisms acting in the solid [5.61].

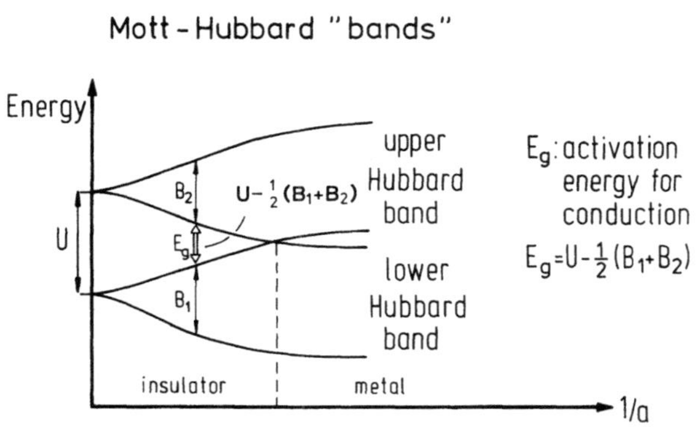

Fig. 5.29. Mott–Hubbard [5.61] energy level diagram for the d band (not filled) in a solid. (Energy as a function of $1/a$, a = interatomic distance). The separation at $1/a = 0$ is U from Fig.5.27 ($U = I - EA$). With increasing $1/a$ the two energy levels broaden and their separation is the gap energy $E_g = U - \frac{1}{2}(B_1 + B_2)$ where B_1 and B_2 are the widths of the two bands

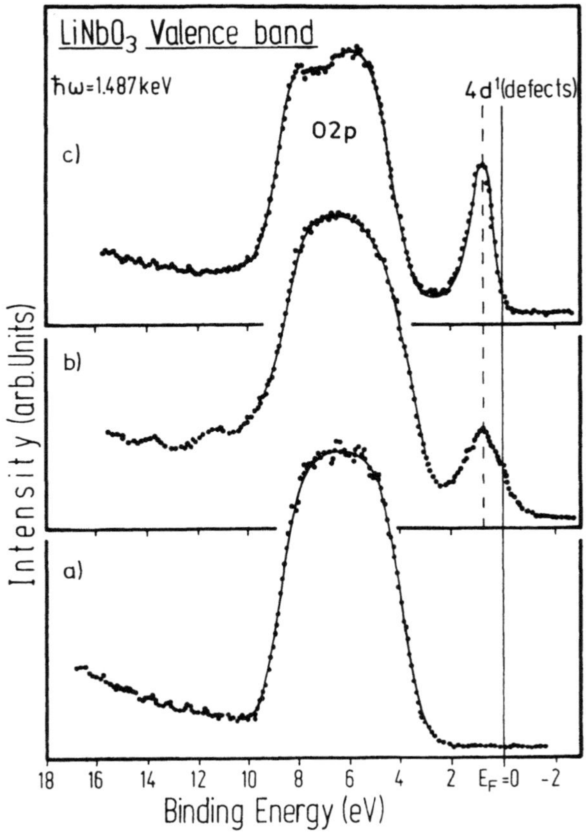

Fig. 5.30. XPS valence band of $LiNbO_3$ (Nb^{5+} $4d^0$ initial state) [5.65], obtained after different surface treatments. (**a**) Clean sample; (**b**) after argon ion sputtering; (**c**) after electron bombardment. The irradiation produces predominantly O defects which in turn lead to Nb^{4+} ions with a $4d^1$ initial state

If U is larger than $W = \frac{1}{2}(B_1 + B_2)$, metallic conduction is not possible. Since for Ni^{2+} the atomic limit is $U = 18\,\text{eV}$ [5.53–5.56] and $W \simeq 1 - 3\,\text{eV}$ (see Fig. 5.25 for NiO), even with a considerable reduction of U by solid-state screening the insulating nature of NiO seems understandable in this simple picture. This is the reason why NiO for a long time has been considered to be the prototype of a Mott insulator.

The above-mentioned basic level ordering in transition metal oxides with the d band above the O 2p band (Fig. 5.26) can most easily be seen in systems that have two oxidation states, one with no d electrons and another with one d electron, for example Ti^{4+} ($3d^0$) and Ti^{4+} ($3d^1$) or Nb^{5+} ($4d^0$) and Nb^{4+} ($4d^1$) because in these compounds, the d states and the O 2p band are well separated. Figure 5.30a reproduces the valence-band spectrum of

$LiNbO_3$ ($Li^+Nb^{5+}O_3^{2-}$), which shows the O 2p band separated by the gap energy (3 eV) from the Fermi level [5.65]. In the experiment the sample was irradiated with a flux of low energy electrons during the measurement which pinned the Fermi energy at the bottom of the conduction band (top of the energy gap). Subsequently argon ion sputtering was used to produce oxygen defects in the sample. This leads to a reduction of some of the Nb^{5+} ions to Nb^{4+} with a $4d^1$ configuration. One can see the emergence of this $4d^1$ state in the gap in Figs. 5.30b, c verifying the assumed level ordering.

Based on these observations, the valence bands of transition metal compounds were interpreted for some time in the following way [5.66]. While the ligand p band was assumed identical to that of non-transition metal compounds (the O 2p band for NiO was approximated by that of MgO or TiO_2), the "d structure" on top of the p band was interpreted in a local way by assuming that it reflected the d^{n-1} final state structure produced by the PE process out of the d^n initial state. For the case in point, namely NiO, this means that the structure observed above the O 2p band was assumed to reflect the $3d^7$ final-state structure and the successful interpretation of many data seemed to support this assumption. (We shall see shortly that the situation is more complicated and by no means clear in all details).

A basis for this interpretation of the d part of the PE spectra was derived from a comparison of valence-band spectra of trivalent Cr compounds and divalent iron compounds, as shown in Figs. 5.31a, 5.8b. A schematic energy diagram is given in Fig. 5.32 [5.66,5.67]. In Cr^{3+} compounds ($3d^3$ initial state) all the d electrons are in a spin-up state and therefore, irrespective of which electron is photoemitted, the final state will always be a triplet, and indeed in trivalent chromium compounds the d structure always consists of a single peak (Fig. 5.31a). This argument assumes that the crystal field interaction is small. In divalent iron compounds, however, staying within this simplified picture, one can emit the one spin-down electron resulting in a sextet final state or one can eject one of the five spin-up electrons which will result in a quartet final state. These two final states will be separated by the quartet–sextet exchange interaction. The spectrum of FeF_2, shown in Fig. 5.8b, seems to support this assumption because it shows in the d band a weak structure (sextet final states) above a stronger structure (quartet final states) where the intensity ratio is roughly 1:5 as expected from the simple picture just outlined. Data for a number of Fe^{2+} compounds that always showed this very characteristic two-peak structure seemed to support the interpretation just given. This interpretation has been applied to many cases including the series of oxides measured by Eastman and Freeouf [5.68] (Fig. 5.33) where the bars reflect final-state structure calculated under the d^{n-1} final state assumption [5.69, 5.70].

It will now be shown that matters are more complicated and, for convenience, we summarize the contents of the next sections:

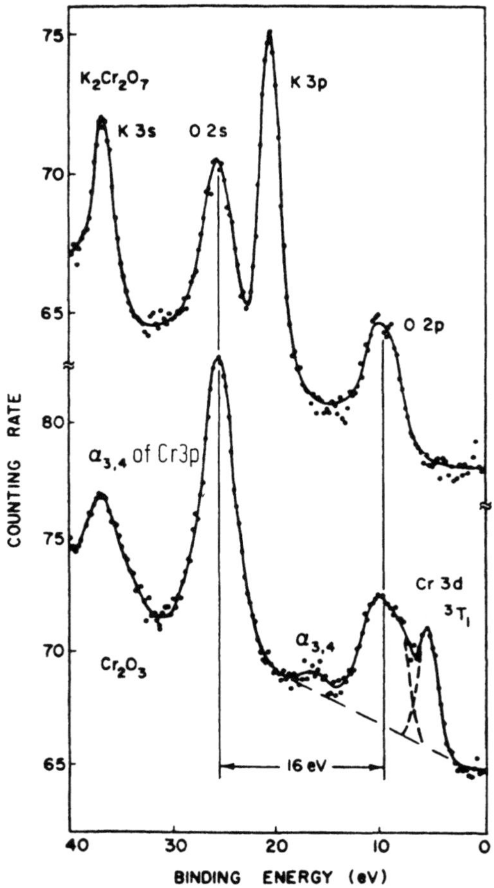

Fig. 5.31a. XPS valence bands of $K_2Cr_2O_7$ (initial state $3d^0$) and Cr_2O_3 (initial state $3d^3$) [5.66]. The O 2s- and O 2p- bands can be lined up accurately with respect to each other. For Cr_2O_3, the Cr 3d spectrum is visible above the O 2p spectrum

The basic problem with the interpretation of the d electron photoemission just given stems from the fact that it neglects the very large 3d ionization energies, as is apparent, e.g., in Fig. 5.27, of the (late) transition metal ions, and the hybridization of the metal d electrons with the ligand p electrons.

The latter fact means that the d-like states of the valence band are in fact a mixture of ligand (L) p electrons and metal d electrons and can be written as d^nL. This valence band leads to two final states after photoemission, namely $d^{n-1}L$ and d^nL^{-1}, where the first corresponds to the one we have just been dealing with, however, the second, not considered so far, has been produced by a ligand to metal charge transfer out of the photoionized state $d^{n-1}L$. The d^nL^{-1} and the $d^{n-1}L$ state contain the same number of electrons (or

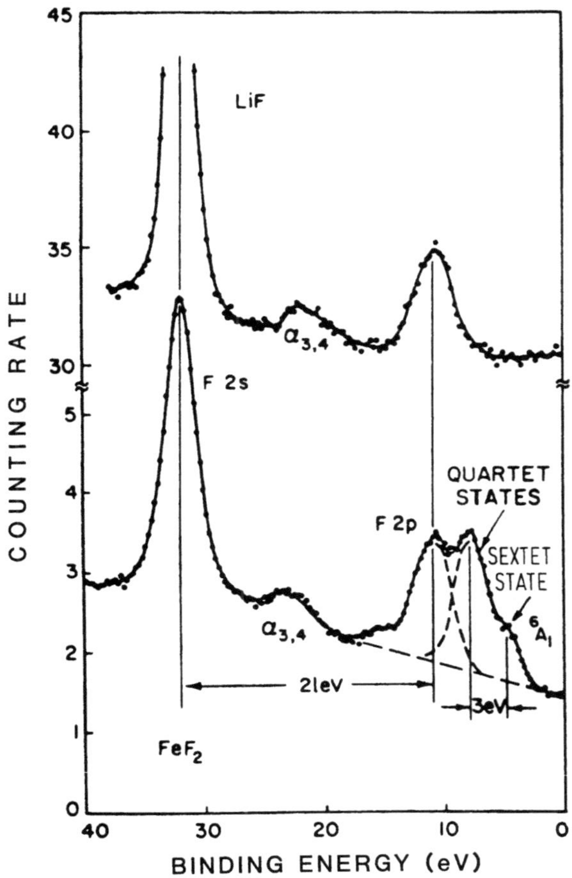

Fig. 5.31b. XPS valence-band spectra of LiF and FeF_2 [5.66]. The LiF spectrum serves to identify the F 2s and F 2p states. In the FeF_2 spectrum, a doublet structure produced by the $3d^5$ final state is visible above the F 2p state. The assignment follows from the diagram in Fig. 5.33

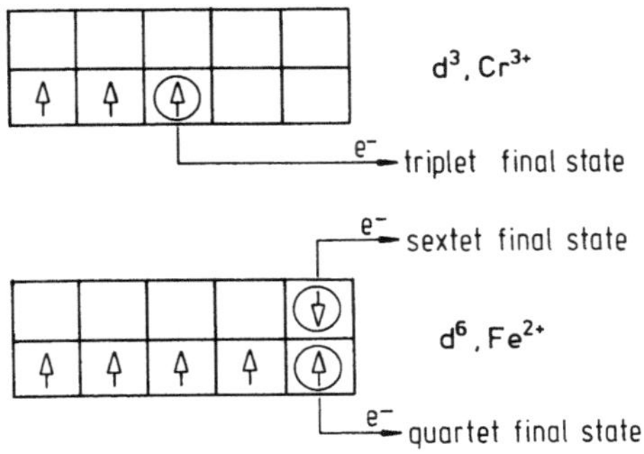

Fig. 5.32. Hund's rule energy level diagrams for photoionization from Cr_2O_3 ($3d^3$ initial state) and FeF_2 ($3d^6$ initial state). For Cr_2O_3 only one 3d final state is possible, while for FeF_2 two are possible. In a full treatment the crystal-field interaction has to be included [5.70]

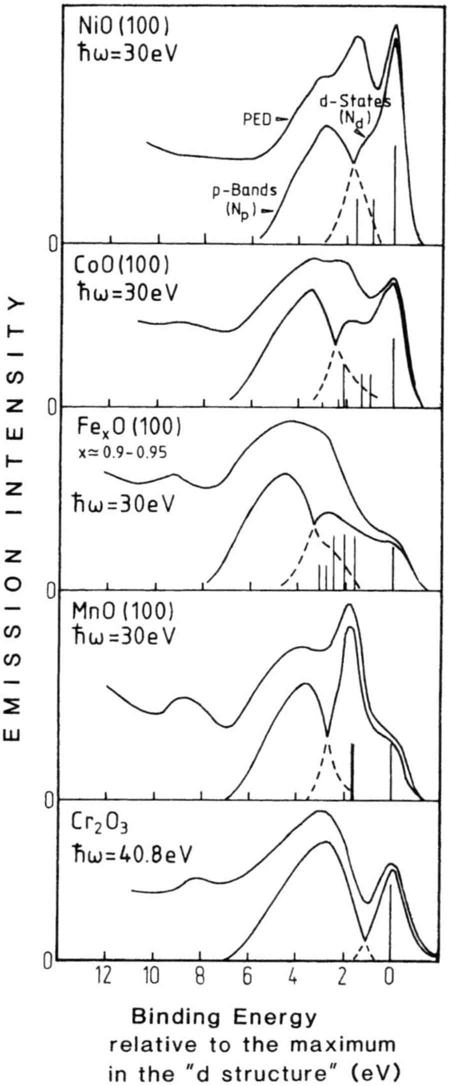

Fig. 5.33. UPS spectra (synchrotron radiation) of NiO ($3d^8$ initial state), CoO ($3d^7$), Fe_xO ($3d^6$), MnO ($3d^5$) and Cr_2O_3 ($3d^3$) [5.68]. The bars in the d bands represent final states calculated from the multiplets of Sugano et al. [5.70] with the intensities given by Cox [5.69] for d^{n-1} final states. The spectra have been lined up with respect to the maximum of the d structure with the smallest binding energy

holes) and therefore mix. As will be shown, the bar diagrams in Fig. 5.33 only contain one half of the "truth" by leaving out the d^nL^{-1} component. For the particular case of NiO the d^8L^{-1} state produces most of the weight near the Fermi energy while the d^7L state produces most of the weight in the satellite at $E_B \simeq 9\,eV$, which is not seen in the data presented in Fig. 5.33 but is clearly visible in the XPS data of Fig. 5.26. We thus realize that there are two types of electronic excitations possible in the valence band of a transition metal compound (Fig. 3.3). The excitation of a d electron from a

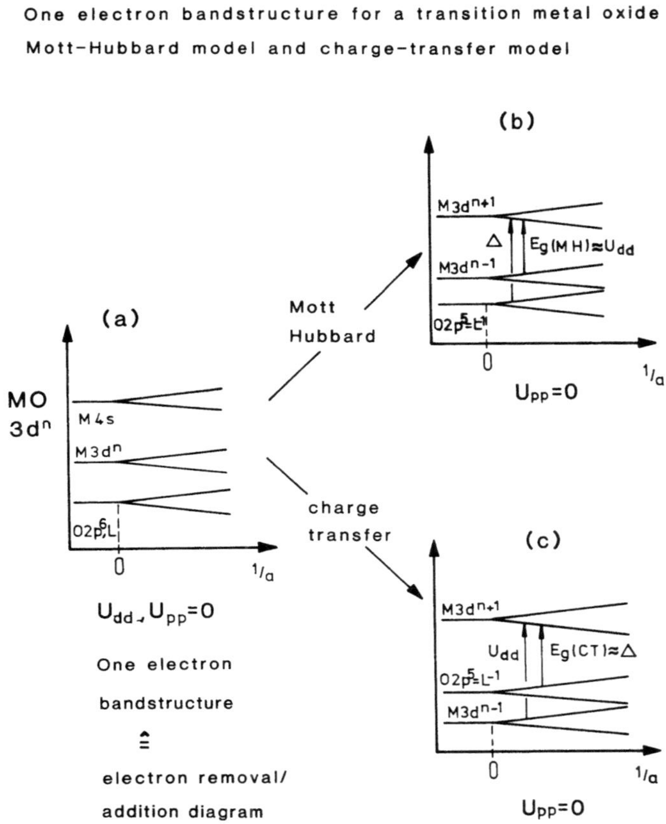

Fig. 5.34. (**a**) One-electron diagram of a transition metal compound (oxide). The energies reflect also the electron removal spectrum (PES) and the energy addition spectrum (IPES); (**b**) Mott–Hubbard diagram for a transition metal compound ($U_{\mathrm{dd}} < \Delta$) for $U_{\mathrm{pp}} = 0$; (**c**) charge-transfer diagram for a transition metal compound ($U_{\mathrm{dd}} > \Delta$) for $U_{\mathrm{pp}} = 0$

transition metal ion and its transfer to another distant transition metal ion which is governed by the energy U in the spirit of Fig. 5.29 and, in addition, there is the possibility of the excitation of a ligand electron onto a transition metal ion (often described by the energy Δ). The relative magnitude of these two energies determines whether a compound is a Mott insulator (or, more generally, a Mott compound) or a *charge-transfer compound.*

For the sake of the further discussion we compare in Fig. 5.34 a one-electron diagram for a transition metal oxide with the excitation spectra in the Mott–Hubbard and the charge-transfer case. The energies are plotted as a function of $1/a$ (a is the interatomic distance in the solid) in order to indicate the evolution of the energy levels of the solid out of those of the atoms. In a one-electron description the electron–electron interactions are zero ($U_{\mathrm{dd}} = 0$; $U_{\mathrm{pp}} = 0$). In this case the one-electron band structure (density of states) reflects also the electron removal spectrum (PE spectrum) and the electron addition spectrum (IPE spectrum).

In the right-hand side of Fig. 5.34 the correlation energy between the 3d electrons has been switched on ($U_{\mathrm{dd}} > 0$). For simplicity we make the crude assumption that the correlation energy between the oxygen p electrons is zero ($U_{\mathrm{pp}} = 0$). In addition the metal 4s states are neglected. Now the excitation spectra are different from the one-electron states. The metal 3d states are split into what is generally called the upper ($\mathrm{M3d}^{n+1}$) and lower ($\mathrm{M3d}^{n-1}$) Hubbard band. Depending on the relative magnitude of U_{dd}, the correlation energy between the 3d electrons and Δ, the charge-transfer energy for the transfer of an electron from the oxygen p states to the metal d states one has the Mott–Hubbard case (top) or the charge-transfer case (bottom).

With respect to the examples already mentioned, we shall see that TiO can approximately be described as a Mott compound (as can Cr_2O_3) and that NiO is a charge-transfer compound. In FeF_2 the two possible final states, namely d^5 and $d^6\underline{L}^{-1}$ are strongly mixed and therefore neither of the two classifications applies to it.

5.8.1 The NiO Problem

(a) Outline

In this section the question whether NiO is indeed a Mott insulator will be discussed. Phrased differently, it will be analyzed whether the energy level diagram of this compound can be described in terms of the diagram given in Fig. 5.29. This means that we have to elucidate the nature of the optical gap transition. The reader is reminded of the analyses of the metal core-level spectra of copper and nickel compounds in Chap. 3. There it could be shown that for a compound with an initial state $3d^n\mathrm{L}$ after core-metal ionization, two final states for the valence band are possible, namely $3d^n\mathrm{L}$ and $3d^{n+1}\mathrm{L}^{-1}$. It will be seen that, in a very similar way, valence photoionization of those compounds will also lead to two final states.

(b) Analysis of the Photoelectron Spectra of NiO

A successful interpretation of the PE spectra of NiO (Fig. 5.26) must provide an explanation for the so-called *satellite* [5.71]. Figure 5.35 shows the valence-band photoemission spectrum of nickel oxide taken with X-ray excitation and also with UV excitation [5.68, 5.71].

These spectra exhibit three contributions. The structure with the smallest binding energy is frequently called the *main 3d emission*. It is followed in binding energy by emission out of the oxygen 2p band. This is discernible in

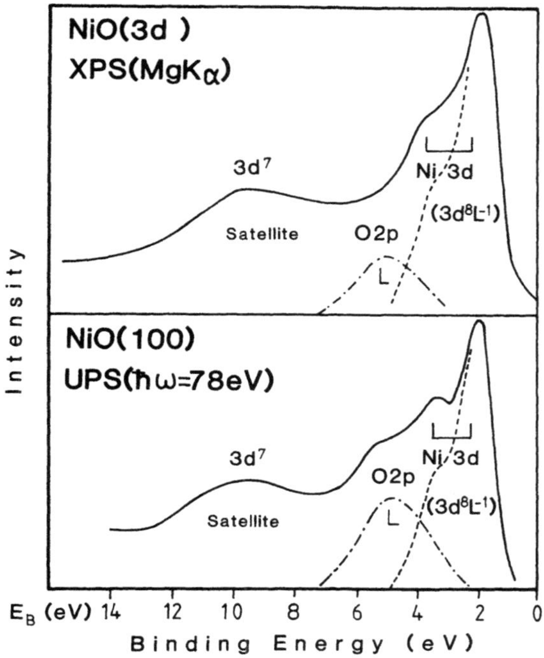

Fig. 5.35. Comparison of XPS and UPS spectra of NiO [5.71]. In the XPS spectrum the O 2p contribution was estimated from a comparison of XPS spectra with those of other transition metal oxides (see also Fig.5.25); in the UPS spectrum the O 2p band is visible. In both cases a distinct satellite is evident. Its assignment to a $3d^7$ final state (NiO has a $3d^8$ initial state) is discussed in the text [5.71]. The zero is the experimental Fermi energy, which is close to the top of the valence band

the UV photoemission spectrum but only barely visible in the XPS spectrum. Finally at about 7 eV below the d emission one sees the so-called satellite, the interpretation of which was for a long time a matter of controversy. As mentioned already, it was assumed for some time that the d-emission near the Fermi energy is essentially a structure produced by a d^{n-1} final state if d^n is the initial state configuration. For nickel oxide this means that the emission near the zero of energy would be representative of a d^7 final state. The satellite was attributed to a multi-electron excitation, where the exact nature of the excitation was left open.

In the light of the previously mentioned interpretation of the core-level spectra (Sect. 3.1.2) this designation of the different PE signals cannot be correct. Formally and in keeping with the notation of Sect. 3.1.2, the valence-band configuration of nickel oxide in the ground state can be written d^8L. Here the L stands for the full 2p band configuration of oxygen. Valence-band photoemission can now lead to only two states, namely d^{n-1}L and d^nL^{-1}. If one used the explanation successfully applied to the interpretation of the

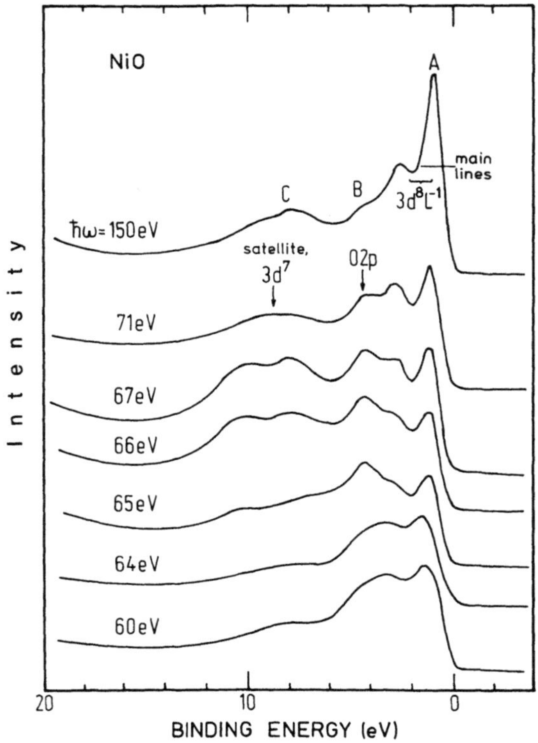

Fig. 5.36. Resonance PE spectra of NiO [5.72,5.73]. A is the "main 3d structure", B the O 2p valence band and C the satellite. Resonance enhancement of the satellite occurs via the reaction: $3p^6 3d^8 + \hbar\omega \rightarrow 3p^5 3d^9 \rightarrow 3p^6 3d^7$, indicating that the satellite is a $3d^7$ final state

satellites in the core-level spectra, one could assume that the main d emission corresponds to a final-state configuration $d^8\underline{L}^{-1}$, whereas the satellite would be a configuration d^7L.

This can be checked by an inspection of the resonance photoemission (Sect. 3.2.1) data of nickel oxide [5.72,5.73]. Resonance photoemission means that if one observes EDCs from the valence band of a particular material with variable photon energy, one particular feature in the valence-band emission is intensified when the photon energy sweeps through the binding energy of a core level of the system under investigation (Sect. 3.2.1). For NiO (Fig. 5.36), this means that if the photon energy approaches the 3p binding energy of nickel ($E_B \simeq 65$ eV), the satellite (C) which is about 7 eV below the main d-emission, gets enhanced with respect to the other features. The intensity of the spectral regions A, B and C of Fig. 5.36, as a function of photon energy, is shown in Fig. 5.37. The resonance enhancement is observed in structure C and a dip in the intensity is visible for structure A. In a simple picture the

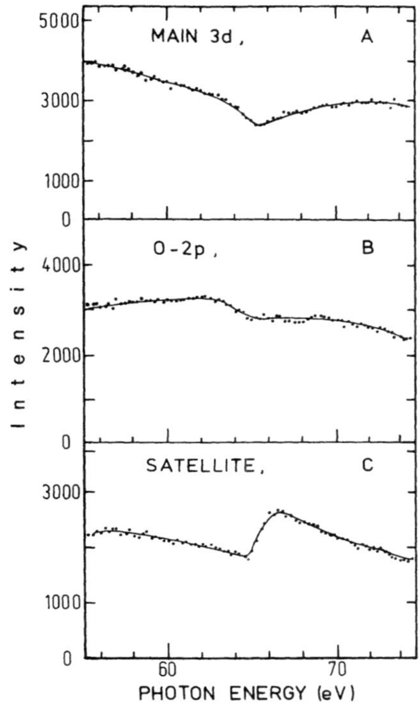

Fig. 5.37. Intensities of structures A, B and C from Fig. 5.36 as a function of the photon energy $\hbar\omega$ used to measure the PE spectra [5.72]

resonance behavior is explained as a superposition of the following processes. In the first step, the photon energy is used to excite an electron out of the $3p^6$ shell of nickel ($E_B \simeq 65\,eV$) into its $3d^8$ configuration to form a $3d^9$ state. This metastable configuration decays via an Auger process back into a $3p^6 3d^7$ configuration in the following reaction equation (neglecting the work function):

$$3p^6 3d^8 + \hbar\omega \rightarrow 3p^5 3d^9 \rightarrow 3p^6 3d^7 + e^- [E_{kin} = E(3p^5 3d^9) - E(3p^6 3d^7)] . \tag{5.10}$$

This reaction can be compared with the direct photoemission using a photon energy that matches the binding energy of the 3p level of nickel in NiO ($\sim 65\,eV$). This leads to (neglecting the work function)

$$3p^6 3d^8 + \hbar\omega \rightarrow 3p^6 3d^7 + e^- [E_{kin} \hat{=} \hbar\omega(3p^5 3d^9) - E(3p^6 3d^7)] . \tag{5.11}$$

Equations (5.10, 5.11) show that both processes lead to the same final state and thus the same kinetic energy of the liberated electron. Or, expressed differently, at the resonance energy which is the binding energy of the 3p level, two channels lead to electrons with the same kinetic energy and thus the final state signal, which the electrons leave behind, is enhanced. This is the $3d^7$ final state. This interpretation of the valence-band PES spectrum is in line with the conclusions drawn from the analysis of the core lines.

Resonant photoemission data on FeO [5.74] and Fe_2O_3 [5.75] show that their valence-band PE spectra can be interpreted in a way analogous to that just given for NiO. However, in these compounds the mixing between the $d^{n-1}L$ and the d^nL^{-1} is very substantial. It is therefore no longer possible to attribute one of the wave functions predominantly to the main d emission and the other to the satellite.

The information obtained from the analysis of the core level and the resonant photoemission spectra is supported by the Auger spectrum of NiO. Figure 5.38 shows the L_3 VV Auger spectrum together with an X-ray photoemission spectrum of NiO. The Auger spectrum shows a two-peak structure. The L_3 VV Auger spectra of transition metals display only one dominant line with some weaker structure produced by multiplet coupling. In view of the classification given in Figs. 5.35 and 5.36 these two final states are both of mixed (by hybridization) $d^7 + d^8L^{-1}$ character. The Auger transition projects out the d^7 contribution in each of these states. This is an indication that this Auger decay leads in NiO to two different final states. The Ni L_3 VV Auger energy is given by

$$E_{\mathrm{kin}}(\mathrm{L_3\,VV}) = E_{\mathrm{B}}(2\mathrm{p}_{3/2}^{-1}3\mathrm{d}^9\mathrm{L}^{-1}) - E(3\mathrm{d}^7\mathrm{L}^{-1})\,. \tag{5.12}$$

The energy of interest for the interpretation of the PE valence-band spectrum is that of the $3d^7$ final state. If one assumes small hybridization between $3d^7$ and $3d^7L^{-1}$ one has

$$E(3\mathrm{d}^7\mathrm{L}^{-1}) \simeq E(3\mathrm{d}^7) + E(\mathrm{L}^{-1})\,, \tag{5.13}$$

where $E(L^{-1})$ is the unhybridized ligand binding energy. This is roughly 5 eV if the separation between the O 2s and the O 2p level is assumed to be 16 eV which is the value for the free ion. $E(L^{-1}) \simeq 5$ eV is certainly an overestimate because there is strong p-d hybridization present in NiO. This can be seen from the following comparison. The lowest-energy final state in 3p photoemission (main line) is $3p^{-1}3d^9L^{-1}$ with a binding energy of 67.5 eV. On the other hand, the final state of 3p photoabsorption is $3p^{-1}3d^9$ which has an energy of 65.5 eV [5.73]. For this case the energy separation of the states $3d^9L$ and $3d^9L^{-1}$ is only 2 eV instead of the 5 eV expected for zero hybridization. Since there is no other information we use $E(L^{-1}) \simeq 2$ eV in order to obtain the energy $E(3d^7)$ from $E(3d^7L^{-1})$ as measured in the Auger spectrum, and get

$$E_{\mathrm{B}}(3\mathrm{d}^7) \simeq E_{\mathrm{B}}(2\mathrm{p}^{-1}3\mathrm{d}^9\mathrm{L}^{-1}) - E_{\mathrm{kin}}(\mathrm{L_3\,VV}) - 2[\mathrm{eV}] \tag{5.14}$$

with [5.67, 5.71]

$$E_{\mathrm{B}}(2\mathrm{p}^{-1}3\mathrm{d}^9\mathrm{L}^{-1}) = 854.5\,\mathrm{eV} \tag{5.15}$$

$$E_{\mathrm{kin}}(\mathrm{L_3\,VV}) = 849.5,\ 843\,\mathrm{eV} \tag{5.16}$$

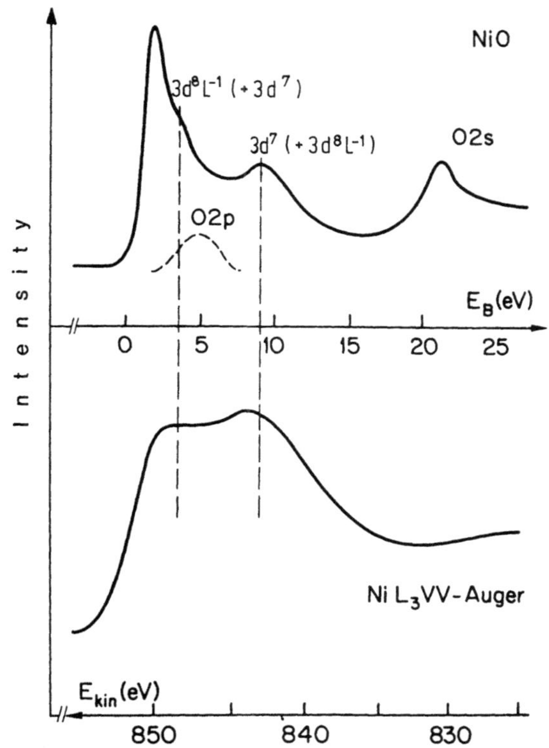

Fig. 5.38. XPS valence-band spectrum of NiO superimposed on a L_3VV Auger spectrum [5.71]. The L_3VV Auger spectrum leads to a $3d^7$ final state assuming the L_3 hole has a $2p^5 3d^9 L^{-1}$ configuration. This final state occurring at $E_{kin} = 843\,eV$ lines up with the satellite in the XPS spectrum. The second feature in the Auger spectrum must then be due to a "screened" $3d^7$ state, i.e. a state that can be approximately described by $3d^7 + (L \rightarrow d) \simeq 3d^8L^{-1}$. The "conventional" way of interpreting the PES d structure as only a $3d^7$ final state was given in Fig. 5.34 [5.68]; see also [5.74, 5.75]

one then obtains

$$E_B(3d^7) \simeq 3,\ 9.5\,eV\ . \tag{5.17}$$

These last numbers have to be compared with the structures at 2 eV and 9 eV respectively found in the XPS spectrum of NiO (Figs. 5.26, 5.35), with the zero placed at the zero of the photoemission signal, which is presumably the top of the valence band. The energy separation of 2 eV between the state $E(3d^7)$ and $E(3d^7L^{-1})$ has also been used in the alignment of the photoemission and the Auger emission spectrum in Fig. 5.38.

The Auger spectra lead to the conclusion that d^7 weight occurs at two places in the valence-band EDCs of NiO. In other words there is hybridization in the valence band of nickel oxide and the various states do not have pure wave functions. Yet the resonance photoemission data (Figs. 5.36, 5.37) indicate that the main wave function component in the satellite is $3d^7$, not however by a wide margin as the Auger data and cluster calculations show.

The so-called *main d emission* is then a d^7 final state, but screened by a ligand electron transferred into an excited many-body state of the d configuration. This results in a configuration which can be written as d^8L^{-1}. Otherwise the large energy separation between the satellite and the main d emission would be hard to understand.

The separation between satellite and main line is similar in the valence band and the core levels. Since in the core levels the two configurations attributed to the satellite and main line differ by one d electron, one can assume a similar situation for the valence band, which is a further argument for the above interpretation of the valence-band PE spectrum.

A theoretical analysis of the PES and BIS data yields the same results. Fujimori et al. [5.76, 5.77] have performed a calculation by representing the NiO crystal as a Ni d-ion with its 6 nearest oxygen neighbors which yields a NiO_6^{10-} cluster. Although this is a molecular calculation it should represent the Ni d levels well because these are localized even in the solid. Figure 5.39 gives some of the results of that work. The left part displays the energy levels of the d electrons in the unhybridized and hybridized situations for the initial cluster (NiO_6^{10-}) and for the cluster with one electron removed or added, which represents the PES or BIS spectra. For the initial state, the wave function was chosen as $\sin\alpha|3d^8\rangle + \cos\alpha|3d^9L^{-1}\rangle$ omitting the third two-hole state namely $|3d^{10}L^{-2}\rangle$. This leads in PES to three final states: $|3d^7\rangle$, $|3d^8L^{-1}\rangle$, and $|3d^9L^{-2}\rangle$, and in BIS to two final states: $|3d^9\rangle$, and $|3d^{10}L^{-1}\rangle$, and their energetic positions relative to the ground state are given. The right part of Fig. 5.39 compares a measured XPS valence-band spectrum of NiO with one constructed from the cluster calculation, where the crystal-field interaction is now also taken into account. The lower part shows that there is a strong mixing of the wave functions, in line with the results from the Auger spectrum (Fig. 5.38). (Note that the d^7 contribution is of about equal strength at $E_B \approx 2\,eV$ and $E_B \approx 8\,eV$, nicely correlating with the Auger spectrum of Fig. 5.38.)

The interpretation of the valence-band spectra given for NiO also holds for other Ni compounds (e.g., $NiCl_2$, [5.77]) and for Cu compounds [5.78]. At present, however, it is not completely clear to what extent it can be applied to 3d elements to the left of Ni. The resonance PE data for the 3d transition metal chlorides indicate [5.79] that the mixing between the two final states $d^{n-1}L$ and d^nL^{-1} (if d^n is the initial ground state) gets stronger with decreasing number of d electrons. This conclusion stems from the fact that while in NiO the resonance enhancement is in the satellite and the main line shows a dip, for $CoCl_2$ both the main line and the satellite show enhancement, and in $MnCl_2$($3d^5$ initial state) the enhancement is largely in the main line, which indicates that it has to be interpreted as the d^4 final-state structure.

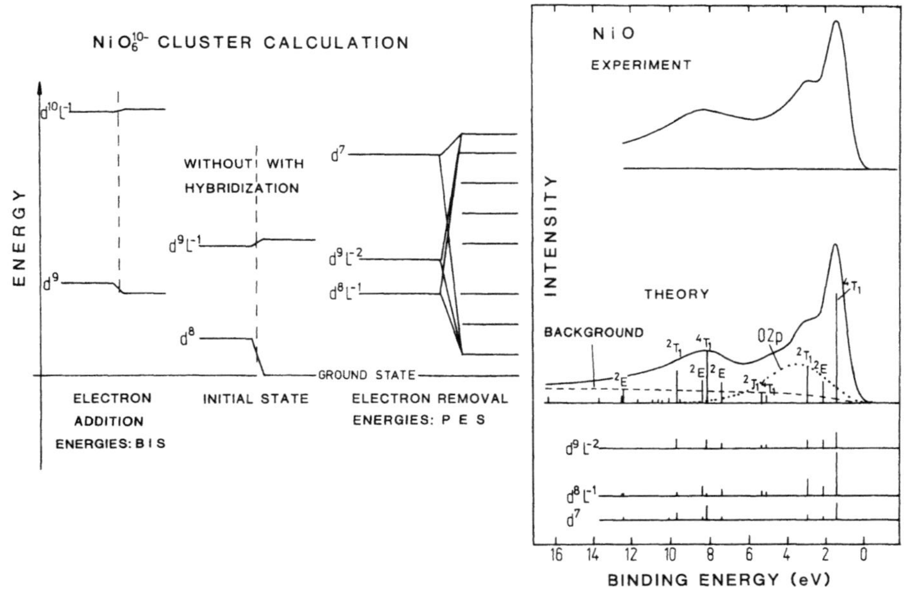

Fig. 5.39. *Left*: Schematic energy level diagram derived for a NiO_6^{10-} cluster in the initial state (d^8, d^9L^{-1}), in the PE situation where one electron has been removed (d^7, d^8L^{-1}, d^9L^{-2}) and in the BIS situation where one electron has been added (d^9, $d^{10}L^{-1}$). The energies on the left always refer to the unhybridized case, whereas for those on the right hybridization has been taken into account. In the hybridized PE situation the crystal-field interaction is also indicated. *Right*: Analysis of the valence-band XPS spectrum of NiO in terms of the calculation for the NiO_6^{10-} cluster (left). The multiplet lines were lifetime broadened by a width increasing with increasing binding energy and finally convoluted with a Gaussian resolution function. The dotted line represents the O 2p emission, and the dashed line approximates the background. In the bottom panels the decomposition into the d^7, d^8L^{-1} and d^9L^{-2} final states is given; it shows a strong mixing of these three contributions [5.76, 5.77]

We note, however, that Fujimori et al. have analyzed data on FeO ($3d^6$ initial state) [5.74] and Fe_2O_3 ($3d^5$ initial state) [5.75] in very much the same way that proved successful for NiO, finding strong mixtures of $d^{n-1}L$ and d^nL^{-1} configurations in the valence-band PE spectra.

More work is necessary to reveal the details of the valence-band PES of transition metal compounds. However, it is established that the Coulomb correlation energy U decreases (though not monotonically) in going from $Cu^{2+}(3d^9)$ to $Ti^{3+}(3d^1)$. This means that at the beginning of the 3d series (for Ti, V and Cr compounds) the d structure nearest to the Fermi energy corresponds to transitions of the type $3d^n \rightarrow 3d^{n-1}$ while, as was shown for NiO, for the late 3d transition metal compounds, the structure near to the Fermi energy corresponds to transitions of the type $3d^n \rightarrow 3d^nL^{-1}$.

For the early 3d metal compounds the three relevant interaction energies in the valence band (the d-d Coulomb correlation energy U, the p-d charge transfer energy Δ and the)p-d hybridization energy T are of similar magnitude (of the order of 4 eV), which makes it difficult to interpret the PES data in terms of the local models employed here.

There is probably an area (ranging from Mn to Co compounds) where the $d^{n-1}L$ and $3d^nL^{-1}$ final states are so strongly mixed that a distinction between these two cases is not very meaningful (Fig. 2.23). Also for one particular 3d ion, the energy level scheme can be strongly dependent on the nature of the ligand, as was seen, e.g., in the core level case for Co^{2+} compounds (Fig. 3.9). The interpretation presented here is roughly in line with the photoemission spectra of molecular dihalides of transition metals (Sect. 5.6).

We finally note that the PE data on the *high-temperature superconductors*, e.g., $La_{2-x}Sr_xCuO_4$ [5.80] and $Y_1Ba_2Cu_3O_{7-x}$ [5.81], show that the Cu ions in these compounds are mostly in a 2^+ state [5.82] and that their valence-band spectra and core-level spectra are similar to those of CuO. This indicates a large correlation energy U for the Cu d electrons (of the order of 6–8 eV). Therefore conductivity in these systems cannot be produced by a breakdown of Mott insulation but is the result of introducing charge carriers (usually holes) into hitherto unoccupied copper-oxygen orbitals [5.82] (Sect. 5.9.)

(c) The Optical Gap in NiO

The optical gap in insulating NiO has a magnitude of 4 eV [5.59]. While chemists agree that the charge transfer energy gaps in transition metal compounds are of the type ligand-p→metal-d [5.83], for NiO a number of different possible transitions had been advocated [5.84]

(I) metal $3d^8 \rightarrow$ metal $3d^7 4s$
(II) metal $3d^8 \rightarrow$ metal$(3d^7 + 3d^9)$
(III) ligand 2p $\rightarrow$ metal $3d^8 4s$
(IV) ligand 2p $\rightarrow$ metal $3d^9$.

Of these possibilities, in view of the foregoing discussion and for intensity reasons [5.85], cases (I) to (III) can be dismissed (see [5.86] for a discussion of the optical gap in Ni:MgO).

We shall briefly outline some problems connected with the determination of the optical gap in NiO from PES and IPES data. In principle, if one knows the lowest ionization state and the lowest affinity state of NiO and their atomic origin it is possible to settle the question of the nature of the optical energy gap. The combination of PES and IPES [5.87,5.88] on the same sample with the same reference point therefore measures the optical gap (Sect. 1.3, Chap. 9).

Figure 5.40 presents a combined UPS/BIS (BIS is a special mode of doing IPES) spectrum of NiO, where the NiO had been grown as a thin layer on

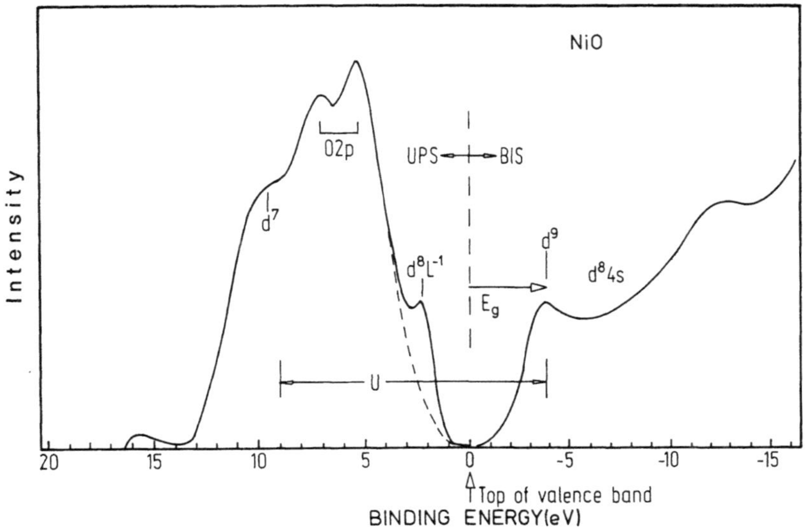

Fig. 5.40. Combined UPS ($\hbar\omega = 21.2\,\mathrm{eV}$) and BIS ($\hbar\omega = 9.7\,\mathrm{eV}$) spectrum of NiO. The approximate labeling of the structures is indicated [5.76, 5.77, 5.87, 5.88]. It is assumed that the experimental Fermi energy is near to the top of the valence band, and that the d^8L^{-1} structure has an excitation energy (relative to E_F) of $\sim 1.8\,\mathrm{eV}$. The optical gap E_g is then due to a p $\rightarrow$ d (ligand-to-metal) transition [5.87, 5.88]

Ni by oxidation in air at 800° C [5.87]. This sort of sample gives spectra similar to those obtained from bulk NiO samples but has the advantage that one does not encounter charging problems. Since NiO generally exhibits Ni^{2+} vacancies the most likely charge compensating defects are O^- ions which can be regarded as holes and thus make NiO a p-type material [5.89, 5.90]. It is therefore safe to assume that in the NiO sample under investigation the Fermi edge is pinned at (or close to) the top of the valence band. In Fig. 5.40 the PE spectrum is similar to the XPS spectrum shown earlier (Figs. 5.26, 5.35, 5.38) in displaying the d^7 state, the O 2p valence band and the d^8L^{-1} state. The d^8L^{-1} state has an energy of $\sim 2\,\mathrm{eV}$ below the Fermi edge. This indicates that the d structure seen at about 2 eV binding energy is not close to the ground state but an excited state with a sizable binding energy.

If the electronic states of a system can be described in the one-electron approximation the PE final states of the valence band are similar to those of the initial state, which generally is the ground state. The fact that the lowest energy state in photoemission is at 2 eV binding energy is a manifestation of electron–electron correlation.

Note that the placement of the gap transition from the (experimental) Fermi energy to the maximum of the $d^8 \rightarrow d^9$ final state contains a certain amount of personal judgement. Other researchers [5.88] have placed this

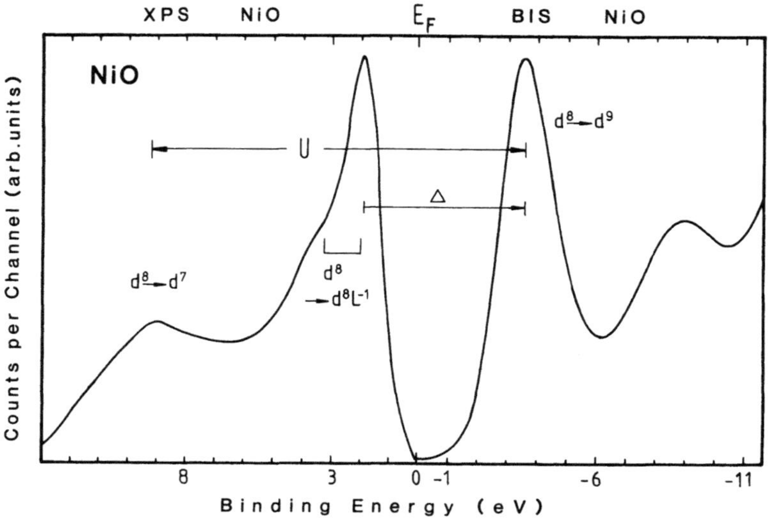

Fig. 5.41. Combined XPS (1254 eV) and BIS ($\hbar\omega = 1254$ eV) spectrum of NiO. The three main features have been labelled according to the transitions from which they are derived. The intensities are very different from those in Fig. 5.40 because of the different photon energies

transition between the half height of the $d^8 \rightarrow d^8L^{-1}$ and $d^8 \rightarrow d^9$ final states.

In order to show the differences of PES and IPES at UV energies and X-ray energies Fig. 5.41 gives a combined XPS/BIS spectrum for a sample of NiO on Ni metal. Now the d excitations are enhanced with respect to the p excitations. However, the resolution of the spectra is worse. We would also like to point out that the PE spectra in Figs. 5.40, 5.41 have been measured with different analyzers which explains the higher background in the former compared to the latter one.

The nature of the optical gap in NiO can be inferred from Fig. 5.40. The transitions responsible for this gap are from the Fermi energy to the lowest empty states. The Fermi energy is the energy zero in Fig. 5.40, and corresponds to a position roughly at the top of the valence band because NiO is a p type conductor. The first empty state is d^9 and therefore the optical gap in NiO is a p $\rightarrow$ d transition ($d^8L^{-1} \rightarrow d^9$). These considerations seem to be in agreement with activation measurements [5.91, 5.92] and photoconductivity data. As yet, calculations give no consistent picture of the situation [5.93–5.97]. For instruction purposes we show in Fig. 5.42 the result of a local-density band-structure calculation of some transition metal oxides which indicates a small gap for NiO [5.93], which occurs between the d states. We have seen, however, that the gap is actually large (4 eV) and corresponds to a ligand-to-metal charge transfer. The deficiency that local-density theory

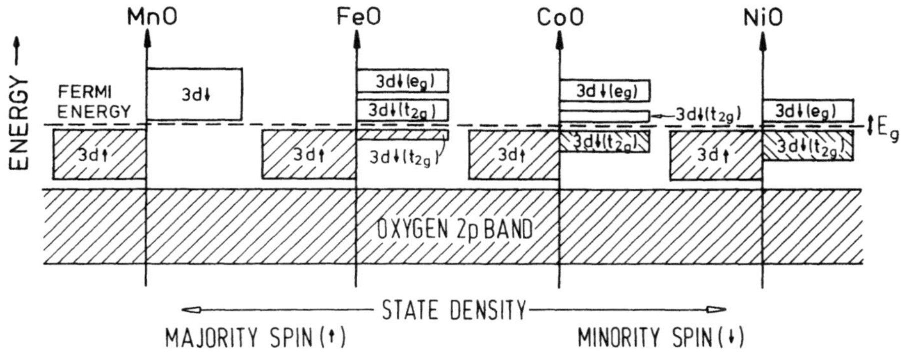

Fig. 5.42. Results of a local-density band-structure calculation for MnO, FeO, CoO and NiO [5.93, 5.94]. This calculation yields a small gap in the d band of NiO (0.4 eV) produced by the $e_g\uparrow - e_g\downarrow$ exchange splitting (see, however, [5.95])

gives too small a gap (or none at all) may have now been remedied [5.98] in systems like NiO; these scientists calculated gaps for transition metal oxides in good agreement with experiment. The procedure of these researchers is, however, not truly a priori and therefore can only be regarded as a first step to a solution of the gap problem in local-density theory.

(d) Comparison of the Valence- and Conduction-Band Data of NiO and NiS

The interpretation of the optical gap in NiO as an O 2p → Ni 3d transition makes use of the assumption that NiO is a p-type conductor, and that the Fermi energy is near the top of the valence band. A problem in this context is the fact that in insulators like NiO the Fermi edge is not clearly visible in the experimental spectra.

Therefore a comparison with a compound that is metallic, but otherwise similar to NiO, can be useful. Such a compound is NiS, which has a nonmetal–metal transition at 260 K (this may actually be a metal-to-metal transition (Fig. 5.54), a fact which does not invalidate the following argument) and is thus metallic at room temperature. Figure 5.43 gives a comparison of the UPS (21.2 eV) valence band of NiS and NiO [5.99–5.102]. In the NiS spectrum the Fermi energy is visible. One can also see that in NiS the $3d^8L^{-1}$ final-state structure is displaced from the Fermi energy (the peak position is at 1.3 eV). In metallic NiS the screening in the final state is more effective than in insulating NiO. Therefore it is not surprising that the peak in the $3d^8L^{-1}$ structure in NiO is found at 2 eV, as compared to 1.3 eV in NiS.

Further support for this interpretation is obtained from a comparison of the BIS spectra of NiO and NiS (Fig. 5.44). In NiS one can see the d^9 state at the Fermi energy, a finding which is in accord with the metallic nature of this compound. This also shows that conduction in this compound takes

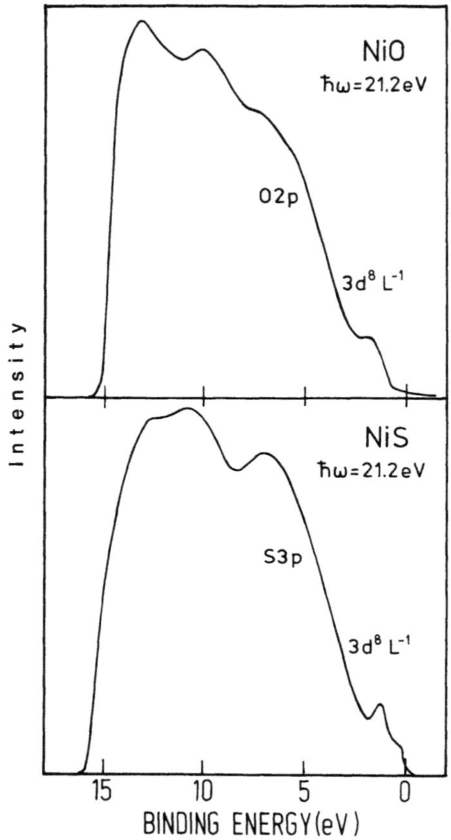

Fig. 5.43. Comparison of UPS ($\hbar\omega = 21.2\,\mathrm{eV}$) spectra of NiO and NiS [5.99, 5.101,5.102]. In the NiS spectrum (NiS is a metal) the Fermi edge is visible and the $3d^8L^{-1}$ structure is observed away from the Fermi energy

place with an active participation of the Ni 3d electrons. The BIS spectrum of NiO displays the first electron-affinity peak, i.e., the d^9 structure, at 4 eV, which corresponds to the optically measured gap energy. Thus, if the first ionization state of NiO is d^8L^{-1} and the first conduction-band state is $3d^9$ the optical gap transition must be of the O 2p $\rightarrow$ Ni $3d^9$ type in NiO, where this statement is not meant to have a single-particle meaning.[4] Finally we present in Fig. 5.45 a combined PES/BIS diagram ($\hbar\omega = 21.2\,\mathrm{eV}/9.7\,\mathrm{eV}$) for NiS, where also the most likely interpretation of the various features in terms of the transitions from which they originate is given. There is still some uncertainty with respect to the interpretation of the spectrum right at the

[4] The optical gap in a highly correlated material like NiO cannot be described in a single-particle picture. The optical gap is the energy difference between the smallest electron affinity energy and the smallest electron ionization energy. The first is obtained by the transition $d^8L \rightarrow d^9L$. The latter is deduced from the $d^8L \rightarrow d^8L^{-1}$ transition. Here the final state d^8L^{-1} is not just a state with a free hole in the valence band but a highly correlated state.

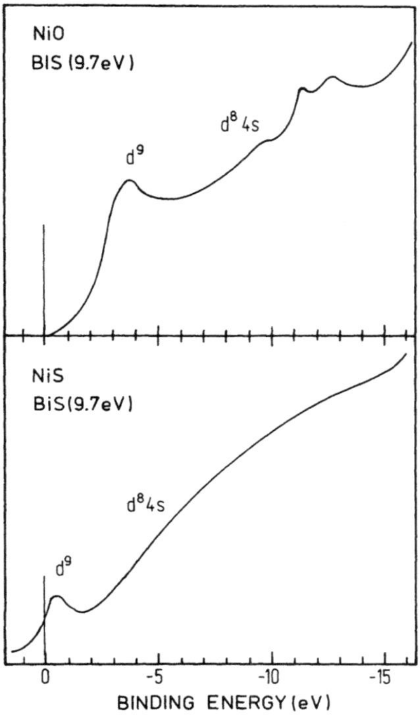

Fig. 5.44. Comparison of BIS spectra ($\hbar\omega = 9.7\,eV$) of NiO and NiS [5.99,5.101]. In NiS the d^9 structure intersects the Fermi energy, while in NiO it is removed by $\sim$ 4 eV (the optical gap energy) from it

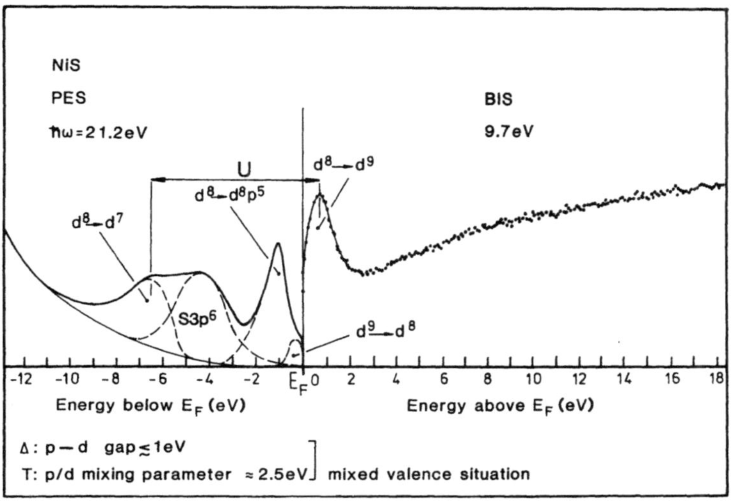

Fig. 5.45. Combined UPS ($\hbar\omega = 21.2\,eV$) and BIS ($\hbar\omega = 9.7\,eV$) of NiS. The interpretation of the UPS spectrum as indicated must be regarded as tentative. The much reduced background with respect to the data in Fig.5.42 is the result of the use of different spectrometers

Fermi energy. It is here suggested that at the Fermi energy, or slightly below it, one is observing S 3p density of states and a $d^9 \rightarrow d^8$ transition strength by PES. The latter is possible because the d-occupation number in NiS exceeds eight. In addition, one realizes from the BIS part of the data that the d^9 band intersects the Fermi energy, which suggests the presence of d^9 states in the valence band.

Note again, by a comparison with Figs. 5.43, 5.44, that different instruments can yield very different background signals. This is the reason for the improved quality of the data in Fig. 5.45.

5.8.2 Mott Insulation

At this point we return to the question posed in Sect. 5.8.1a of whether NiO is a Mott insulator [5.61] meaning that its energy levels can be described by the diagram of Fig. 5.29. This raises the further question of the magnitude of the Coulomb correlation energy U, given in (5.8).

In principle, the procedure to obtain U is simple. One has in the combined PES/BIS valence and conduction-band spectrum the position (Fig. 5.40, 5.41) of the d^7 state and the d^9 state. According to (5.8) their energy separation is U, giving $\sim 13\,\text{eV}$. This is an oversimplification, however, because the energy levels in NiO are hybridized. This hybridization tends to increase the energy splitting and if one makes a proper correction for hybridization a limit of $6\,\text{eV} \leq U_{\text{eff}} \leq 9\,\text{eV}$ is obtained [5.87]. From the PES spectra one also finds that the bandwidth is of the order of 1 eV or less [5.103] and thus clearly $U > W$ such that, in principle, NiO could be called a Mott insulator in the sense of Fig. 5.29.

In order to gain more insight, a look at the case of Ce metal is instructive. Fig. 5.46 shows a combined XPS/BIS spectrum of Ce metal [5.104] and Fig. 5.47 gives the relevant energies in a bar diagram [5.105]. In this case one has $U = 6\,\text{eV}$ and an application of Fig. 5.29 then implies that conduction including the f electrons is impossible. One knows, however, [5.106, 5.107] that in the so-called α-type Ce compounds the f electrons are indeed in an itinerant state which expresses itself as a f^1-peak in the BIS spectrum of such α-Ce systems at the Fermi energy (Fig. 3.27) [5.108].

This shows that the U inferred from a combined PES/BIS experiment does not necessarily describe, e.g., the conductivity. In the case of the α-Ce systems, one can describe this formally by saying that for electrons at the Fermi energy $U = 0$ (by definition!) [5.109] or in slightly more physical terms, that a charge fluctuation in the f band is compensated (screened) by the readjustment of many other charges such that the energy provided by any one of them is only small.

With these considerations in mind, we reinspect the XPS/BIS results for NiO for which Fig. 5.48 gives a bar diagram.

In comparing Fig. 5.48 with Fig. 5.29 one can see that NiO cannot be termed a Mott insulator [5.58, 5.105, 5.110–5.113] because it is not only the

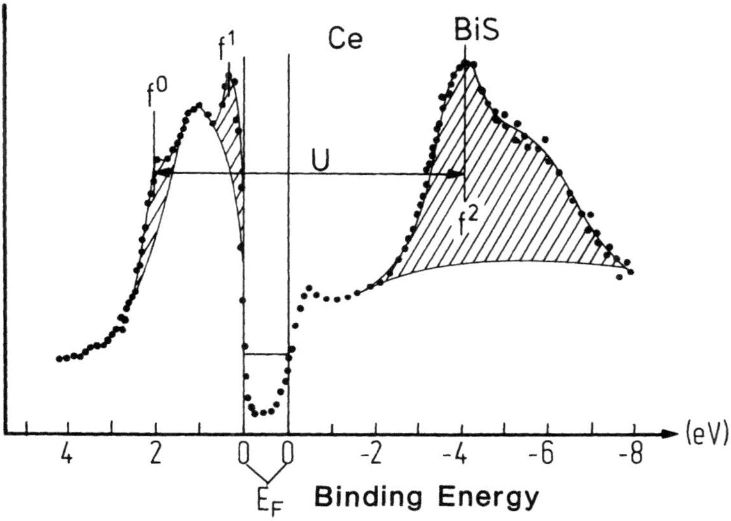

Fig. 5.46. Combined XPS and BIS spectrum of Ce metal, showing the f^0, f^1 and f^2 final states [5.104]

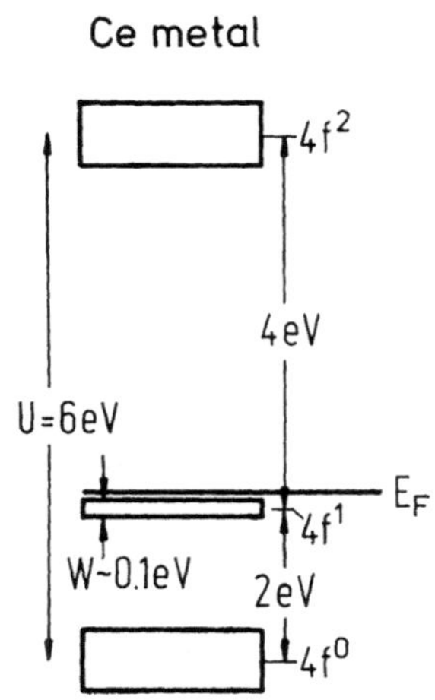

Fig. 5.47. Schematic representation of the data of Fig.5.45 [5.105]. The value of U (derived from the equation $2f^1 \rightarrow f^0 + f^2 - U$) is read off from this diagram

splitting of the d bands which makes NiO insulating. The gap that separates the valence and the conduction band (4 eV) is between the $3d^8L^{-1}$ state and the $Ni3d^9$ band. This gap determines the electrical nature of the compound and, since the gap is of the *charge-transfer type*, one may call NiO a *charge-transfer insulator* rather than a *Mott insulator* [5.58, 5.105, 5.110–5.113].

These findings are emphasized for NiS which has a phase transition at 260 K (the transition at 260 K in NiS may be a metal-to-metal transition and not a nonmetal-to-metal transition as widely believed). Its energy level

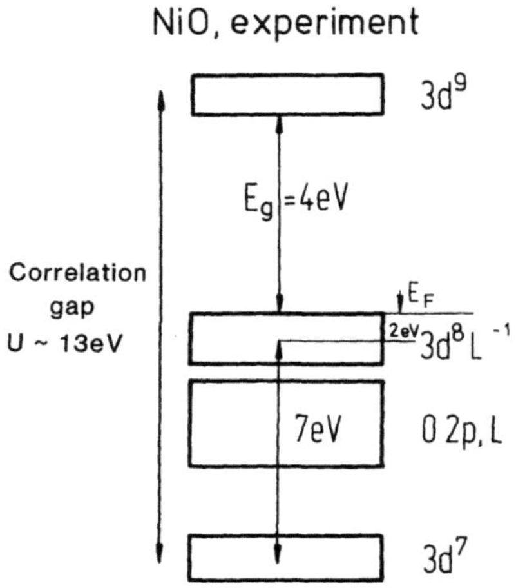

Fig. 5.48. Bar diagram in the spirit of Fig. 5.47 for NiO (data taken from Fig. 5.40). The correlation gap ($3d^7 - 3d^9$ separation) contains states, namely the O 2p band and the $3d^8L^{-1}$ final state [5.105]; U is not corrected for hybridization

diagram obtained from a combined PES/BIS experiment [5.101,5.102] is given in Fig. 5.49 [5.105]. These data show a $3d^9$-$3d^7$ energy separation of 7 eV, leading, after a hybridization correction to $U \simeq 5$ eV, which would make this material an insulator. However the S 3p band and the Ni $3d^9$ band are separated by at best a very small gap (see Fig. 5.54), which can be closed by a crystallographic transformation making NiS a metal for $T \geq 260$ K. (Note that the diagram in Fig. 5.49 corresponds to the non-metallic state at $T < 260$ K, where a small gap exists). The situation in CoS is treated in [5.114]. See next section for detailed temperature dependent PE spectra of NiS.

One now has to ask whether the situation found in NiO and NiS holds generally for transition metal compounds or whether there are also "genuine" Mott compounds for which the electrical properties are governed by the Mott–Hubbard bands in the simple sense of Fig. 5.29. For this purpose Fig. 5.50 compares the valence-band EDCs of a number of transition metal oxides, namely Ti_2O_3 (semimetal, $E_g = 0.1$ eV), V_2O_3 (metal), Fe_2O_3 (semiconductor, $E_g = 2$ eV) and NiO (insulator, $E_g = 4$ eV) [5.103]. (These spectra differ slightly with respect to intensities from those presented previously be-

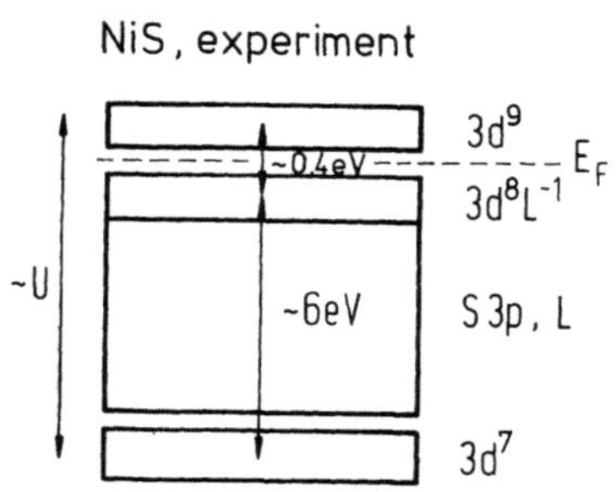

Fig. 5.49. Bar diagram in the spirit of Fig. 5.47 for NiS. Data are taken from Figs. 5.43– 5.45; U is not corrected for hybridization; recent PES data (Fig. 5.54) suggest that NiS has no gap

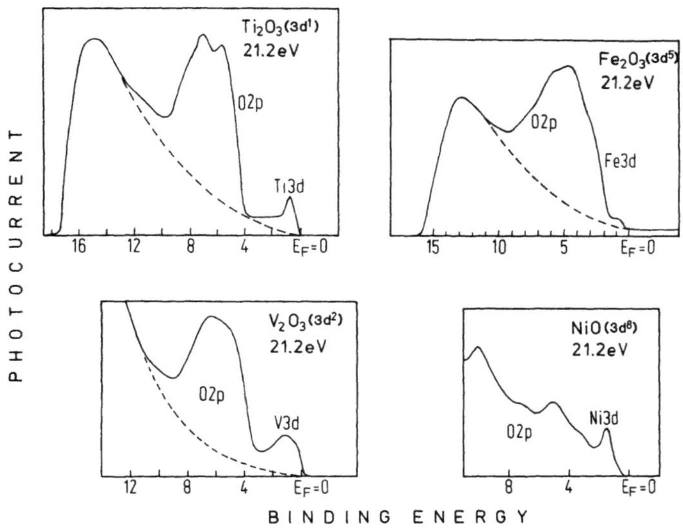

Fig. 5.50. UPS spectra ($\hbar\omega = 21.2\,\mathrm{eV}$) of Ti_2O_3 ($3d^1$ initial state), V_2O_3 ($3d^2$ initial state), Fe_2O_3 ($3d^5$ initial state) and NiO ($3d^8$ initial state) [5.103]. With increasing 3d occupation the 3d band and oxygen 2p band move closer to each other, indicating increasing hybridization. This suggests that Ti_2O_3 and V_2O_3 are approximately Mott compounds with a correlation gap between occupied and unoccupied d states (Fig. 5.28) while Fe_2O_3 and NiO are charge-transfer compounds [5.110–5.112] where the correlation gap contains other states

cause they have been obtained from single crystals. The energy positions are however the same as obtained from polycrystalline samples).

One can see that in Ti_2O_3 (and to a lesser degree in V_2O_3) the metal 3d band and the O 2p band are very well separated, although there is large metal–oxygen hybridization (Fig. 5.51). In this situation conduction by holes in the O 2p band is very unlikely and must therefore take place in the d band. This would make Ti_2O_3 (or V_2O_3) a system to which the diagram of Fig. 5.29 is approximately applicable. Resonance photoemission data (Fig. 5.51) on a very similar system, $SrTiO_3$ with defects associated with a $3d^1$ initial state, show that the Ti $3d^0$ final state is near to the Fermi energy [5.110] but due to strong hybridization also in the O 2p band. The defect structure of $SrTiO_3$ contains O^{2-} vacancies which are compensated by Ti^{3+} ions where in pure stoichiometry one has only Ti^{4+} ions. The position of the $3d^1$ defect band in $SrTiO_3$ is therefore very similar to that of the $3d^1$ band in Ti_2O_3. Since in Ti_2O_3 the $3d^2$ band must be just above the Fermi energy because this compound is a metal, Ti_2O_3 can be called a Mott compound [5.111]. The same situation can be assumed for V_2O_3 where the V 3d band and the O 2p band are still well separated. Thus V_2O_3 can also be called a Mott compound in the sense of Fig. 5.29. This is emphasized by the PES data on the similar compound VO_2 in going through the Mott transition [5.115, 5.116]. The

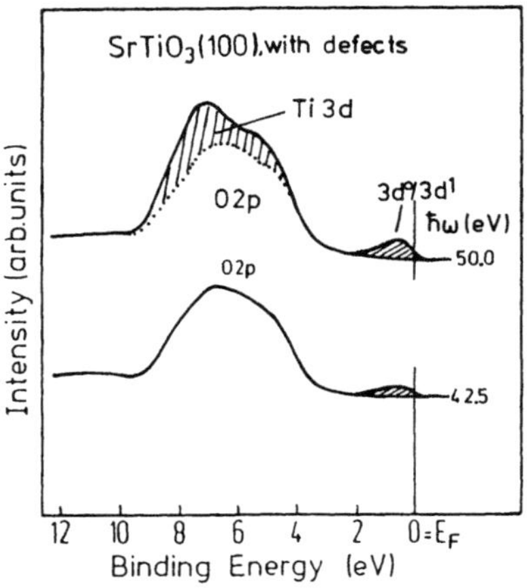

a)

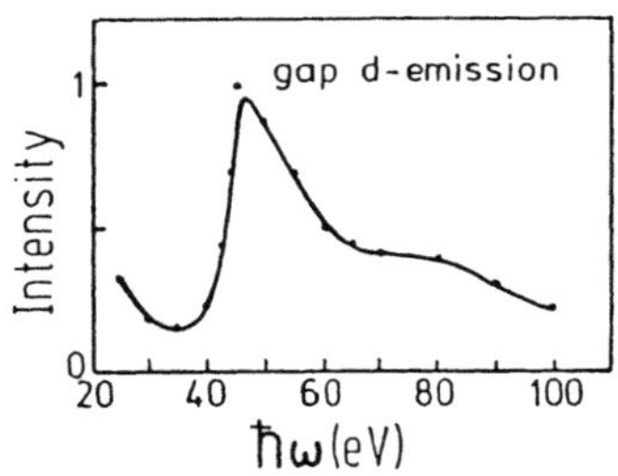

b)

Fig. 5.51. (**a**) UPS (synchrotron radiation) spectra of a defect-rich $SrTiO_3$ (100) surface, showing the $3d^1$ state at the Fermi energy; the resonance PE data indicate that the observed final state also contains the $3d^0$ component (resonance PES means: $3p^63d^1 + \hbar\omega \rightarrow 3p^53d^2 \rightarrow 3p^63d^0$). The resonance enhancement in the O2p band signals strong p-d mixing due to hybridization. (**b**) Resonance enhancement of the d emission in the gap near the Fermi energy

separation of the V 3d band and the O 2p band is seen and the closing of the Mott–Hubbard gap on passing through the phase transition is also visible in the data (Fig. 5.52) [5.112].

These different behaviors can be put into a simple diagram derived from Fig. 5.34 (Fig. 5.53) which plots the gap energy (in other words, the energy in a compound needed to make the lowest energy transitions between the occupied and the empty bands) with respect to the difference of the ligand p-ionization energies and metal d ionization energies [5.110–5.112].

If the ligand ionization energy is larger than the metal 3d ionization energy, one has the classical Mott case, and this is shown in the left part of the diagram. It is easy to realize that then the optical gap is given by the energy it takes to create charge fluctuations in the d-manifold, meaning that one transfers an electron from one metal ion to another. In the reverse case, namely that the ligand ionization energy is smaller than the metal ionization energy, the optical band gap is influenced by all three properties, namely the correlation energy and the metal and ligand ionization energy, as shown in the right part of the diagram. We have indicated in that diagram typical

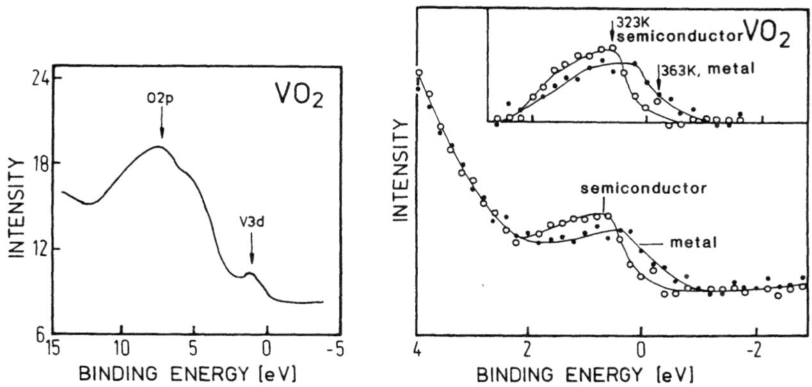

Fig. 5.52. XPS valence-band spectra of the metal–nonmetal transition in VO_2 [5.115,5.116]. The left panel gives the complete valence band. The right panel gives the 3d structure above (363 K), and below (323 K) the metal–nonmetal transition ($T_N = 340$ K). In going through the transition the 3d band moves towards the Fermi energy to intersect it

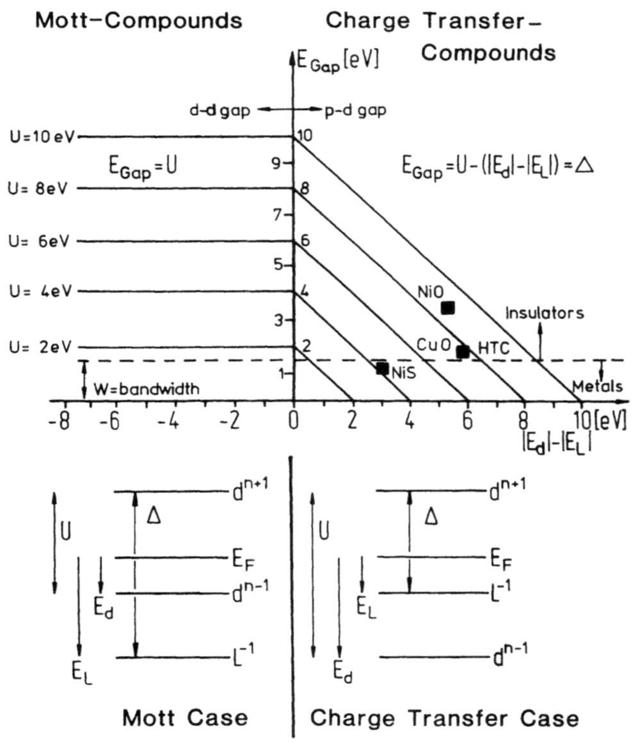

Fig. 5.53. Diagram showing the magnitude of the optical gap energy (E_g), which separates the highest occupied from the lowest empty states, as a function of the relative position of the d electron (E_d) and ligand electron (E_L) ionization states

materials and have also given a not untypical bandwidth of 1.5 eV, meaning that if the gap energy is smaller than this bandwidth, metallic behavior should be observed. Note that CuO and the high-temperature superconductors (Sect. 5.9) are charge-transfer compounds.

5.8.3 The Metal–Insulator Transition; the Ratio of the Correlation Energy and the Bandwidth; Doping

Interest in the investigation and understanding of the metal–insulator transition has a long tradition [5.49–5.52] and has, in addition, been stimulated recently by the discovery of two classes of very interesting compounds: The high-temperature superconductors, which can be derived by electron and hole doping from an antiferromagnetic parent compound (e.g., La_2CuO_4 in the simplest case) and the substances showing so-called colossal magnetoresistance, which exhibit a very large magnetoresistance around the magnetic transition temperature, and for which the parent compound $LaMnO_3$ is also an antiferromagnetic insulator. The interesting compounds exhibiting the large magnetoresistance are produced by various kinds of doping of the insulating parent compounds. So far it seems fair to state that, although the accuracy with which the metal–insulator transition can be investigated has reached a considerable degree of sophistication, the exact nature of this phase transition in these systems is not well understood. Therefore we show a few examples in this field to give the reader the possibility to judge the nature of the data presently available and to give a starting point for deeper study of this field.

The study of the metal–insulator transition was an early subject of photoemission spectroscopy [5.115,5.116] because in principle this method should yield detailed data on the change of the electronic structure of a system in going from the metallic to the insulating (semiconducting) state. Figure 5.52 shows an investigation on a prototype material, namely VO_2 from the early days of photoemission spectroscopy. This example was used to demonstrate the development of the Mott–Hubbard gap.

Data like those in Fig. 5.52 suggest that in this case the metal–insulator transition (or rather metal–nonmetal transition, a nomenclature more precise because it also contains the metal–semiconductor transition) can be explained in a diagram as given in Fig. 5.29. In the insulating state the gap is of the Mott–Hubbard type. The correlation between the d electrons is responsible for the gap. With increasing temperature the upper Hubbard band becomes populated, the correlations are increasingly screened and, finally, at the Mott temperature, the gap collapses and a metallic state is produced. This is probably an adequate interpretation of the data for VO_2 (Fig. 5.52).

It is unlikely, however, that this reasoning also explains the nonmetal–metal transitions in the compounds that display high-temperature superconductivity in their metallic state. In the insulating state these materials have a charge-transfer gap and not a Mott–Hubbard gap and the gap energies are of the order of 1.5–2 eV, which makes a collapse of the gap by a "conventional" doping process unlikely. Instead, much evidence points to the creation of additional states in the gap by the doping procedure [5.49], although the exact mechanism that leads to the nonmetal–metal transition in these systems is not yet established.

To introduce this topic we shall show some data on the change of the PES spectra by doping in the prototype material NiO, and also demonstrate the doping dependence of the PE spectra of the simplest parent compound for the high-temperature superconductors namely La_2CuO_4. We hope that these few data will demonstrate that PES, with high resolution and on carefully prepared samples, may eventually produce new insights into the nonmetal–metal transition, e.g., in the high-temperature superconductors.

A material, that has played a considerable role in the investigation of the metal–insulator transition is NiS, which has been dealt with in the previous sections. In this material the transition temperature is near 260 K; however, it has never really been established what kind of metal–insulator transition occurs in this compound. The temperature-dependent resistivity data suggest a small insulating gap (0.4 meV) as depicted in Fig. 5.49. High-resolution photoemission spectra of this system are displayed in Fig. 5.54 and the inset shows the density of states near the Fermi energy. The latter can in principle be "extracted" from these data by fitting them to a product of the density of states shown with a Fermi function and by convoluting this product with a Gaussian resolution function appropriate for the experiment [5.117]. It can

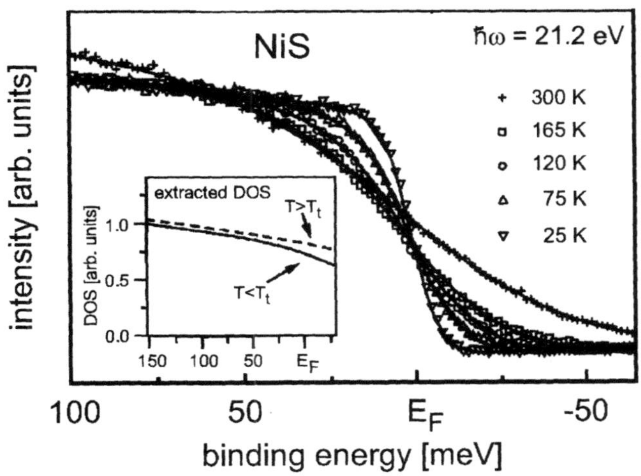

Fig. 5.54. Temperature dependence of the photoemission spectra of NiS near the Fermi energy. The inset shows the density of states above and below the transition temperature, which is used together with a Fermi function to describe the data (the transition temperature is around 260 K). The data can be well described by the theory indicating that the transition in NiS is a metal-to-metal transition [5.117]

be seen that a good fit of the data is achieved by assuming that the density of states changes only little in going through the transition temperature, and that, in principle, one can describe the transition at 260 K in NiS by a metal-to-metal transition and probably not by a metal–insulator transition. Therefore the established picture of the density of states of NiS below the transition temperature of 260 K (Fig. 5.49) will probably have to be revised. These data demonstrate how high-resolution temperature-dependent photoemission experiments can help to reveal changes in the electronic states in compounds.

In Mott–Hubbard compounds such as VO_2 the physics of the metal–insulator transition should be understandable along the lines of the diagram in Fig. 5.29. This diagram indicates that the relevant parameter that drives the metal–insulator transition can be considered either as a^{-1} (where a is the interatomic distance) or U/W, where W is the bandwidth and U the Mott–Hubbard correlation energy. For zero bandwidth and finite U one always has an insulating state. For small U and large bandwidth W the correlation cannot open a gap and one has a metal. Thus, with decreasing U/W one goes from an insulating to a metallic state. This has been demonstrated by an investigation of a series of compounds, that have different U/W ratios, as shown in Fig. 5.55. Here $YTiO_3$ is a ferromagnetic insulator, which has a large value of U/W, $LaTiO_3$ is an antiferromagnet, which is weakly metallic, $SrVO_3$ and the metallic phase of VO_2 are ferromagnetic metals and ReO_3 is an uncorrelated metal, with a small value of U/W. The figure shows the photoemission spectra and the result for the density of states obtained in the local density approximation, which neglects U. If the bandwidth W is large compared to the correlation energy U, as in ReO_3, band theory and photoemission spectrum agree quite well, as anticipated. On the other hand, for large U/W one has an insulator, and in this case the photoemission spectrum represents the $d^1 \to d^0$ excitation spectrum and the density of states is not visible in the spectrum. The other three cases (VO_2, $SrVO_3$, $LaTiO_3$), which can be considered correlated metals, exhibit a mixture of band behavior and local behavior in that they show intensity near the Fermi energy, which can be considered a density of states, and a final state $d^n \to d^{n-1}$, which can be related to the satellite, e.g., in nickel metal, as discussed extensively in Chap. 3 [5.119].

Next we show how the spectra change in a particular system when the correlation energy U, namely the energy needed to put an extra electron in a state d^n, changes [5.120]. Typical examples where this phenomenon has been investigated are $CaVO_3$ and $SrVO_3$, where the former has a larger U/W ratio, W being similar in both compounds. The spectra in Fig. 5.56 show that by increasing U/W, the peak at the Fermi energy, which is usually related to the single particle spectrum (or sometimes called the coherent part of the spectrum) is depressed in intensity with respect to the incoherent spectrum or the so-called satellite. This is in agreement with expectations. For small

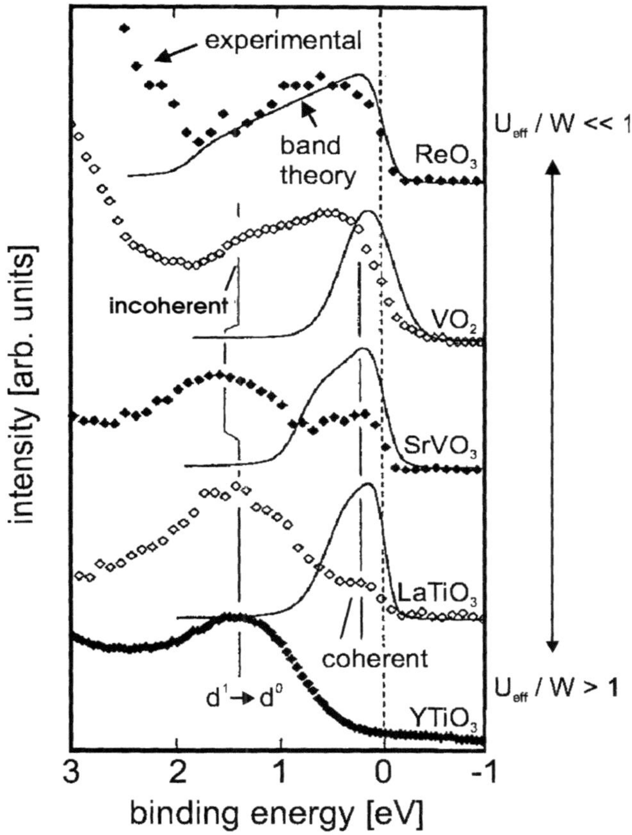

Fig. 5.55. Valence-band photoemission spectra of ReO_3 (uncorrelated metal), VO_2 (correlated metal), $SrVO_3$ (correlated metal), $LaTiO_3$ (weak metal) and $YTiO_3$ (insulator). The ratio of the correlation energy to the bandwidth (U/W) increases in that order. The full lines are LDA density of states calculations. They show agreement with the part of the spectrum near the Fermi energy (often called the coherent part of the spectrum), while the satellite or incoherent part of the spectrum is not described by the LDA calculation [5.119]

U/W, namely for small correlations, the single particle density of states is observed in a PES experiment and there is a small incoherent part of the spectrum or a small satellite in the language of Chap. 3. For a large U/W ratio the final PES state is missing one electron, and one observes mostly the satellite or the incoherent part, or the lower Hubbard band and the single-particle part of the spectrum is reduced. The figure also contains the LDA density of states of the two materials, which are quite similar, as they should be due to the similarity in crystal structure and atomic constituents.

It was just shown how, in systems with a low 3d-count, a change in the U/W ratio can lead from an insulating to a metallic state and vice versa.

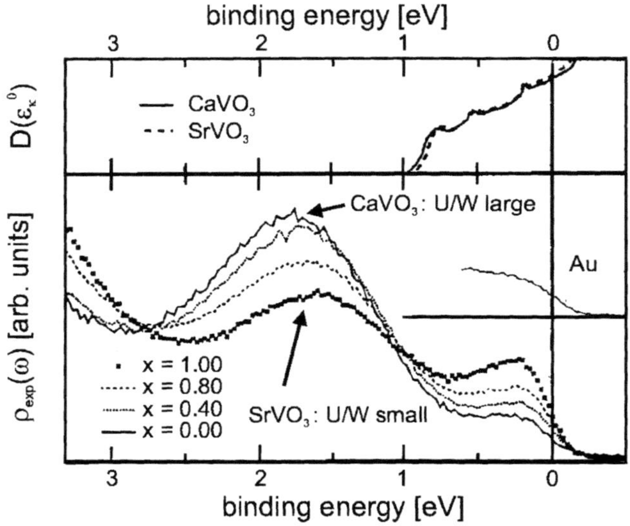

Fig. 5.56. Valence-band photoemission spectra of $Sr_xCa_{1-x}VO_3$. With increasing Ca concentration the ratio of U/W gets larger. This is reflected by an increase of the satellite intensity and a decrease in the intensity of the valence band. The LDA density of states is almost identical for the two materials [5.120]

This mechanism is not responsible for the nonmetal–metal transition in the high-temperature superconductors as achieved by doping. While the details of this process are not understood, we shall nevertheless give some information obtained by PES and IPES on the change of the electronic structure produced by doping in a prototype material, namely NiO. Unfortunately in this system the metallic state is not reached. However, since NiO has been treated in some detail in this volume, for the ease of presentation it will be used again.

So-called hole doping can be produced in two ways in NiO. The first method is by "alloying" it with Li which, in a one-dimensional picture, leads to the following configuration: $-O^{2-} - Ni^{2+} - O^{2-} - Li^{+} - O^{-} - Ni^{2+} -$. This means that if Li^+ ions are incorporated into the NiO structure, a neighboring O^{2-} ion has to give up an electron to maintain charge neutrality. A missing electron is usually called a hole in semiconductor language, because this site can accept an electron. Another way to produce holes in NiO is by a Ni^{2+} vacancy ($\Box$). Such a vacancy gets neutralized by two O^- sites: $O^{2-} - Ni^{2+} - O^{-} - \Box - O^{-} - Ni^{2+}$. The Ni^{2+} vacancies occur "naturally" in NiO and mean that this material is always slightly hole doped.

Sometimes it has been stated in the literature that Li^+ doping creates Ni^{3+} sites in NiO. In looking at Fig. 5.35 one sees that this is an unlikely possibility: the d^7 configuration of Ni^{3+} is at 9 eV below E_F while the d^8L^{-1}

configuration, corresponding to the missing electron on the O^{2-} site, is at 2 eV below E_F showing, that "oxygen holes" are strongly energetically favored over Ni^{3+} sites as charge compensation centers for Li^+.

The question now arises as to the energetic position of the (Li^+O^-) hole centers. Naively, one could argue that a Li^+ ion produces a d^8L^- configuration on any of the adjacent $[NiO_6]^{10-}$ clusters and its energy should be equivalent to the d^8L^{-1} final state, which is at 1.8 eV below the Fermi energy.

However, a closer look reveals a different picture. In a one-dimensional model the final state of O^{2-} valence photoionization can be written as: $-O^{2-}-Ni^{2+}-O^{-}-Ni^{2+}$. Conversely a measurement of the hole state energy produced by Li doping can be performed by a technique that places an electron into this hole, e.g., photoabsorption spectroscopy or IPES. Considering for the moment the IPES experiment, it results, in putting an electron into the O^- site, in the following final state: $\ldots - O^{2-} - Li^{+} - O^{2-} - Ni^{2+} - \ldots$ This shows that the two final states, the PE final state of NiO and the IPE final state in Li:NiO, represent similar charge disturbances, but with opposite signs with respect to the charge deficiencies. Therefore, if the PES final state is observed at 1.8 eV below E_F (Fig. 5.35), the signal from the hole state, as measured by BIS (or IPES) should show up at about the same energy but above E_F.

These anticipations are in agreement with the experimental findings displayed in Fig. 5.57. Here a combination of XPS/BIS valence-band spectra are shown for pure NiO and for NiO doped with 10 % Li. The Li doping has two effects on the spectra. It shifts the Fermi energy by 0.4 eV to lower energies. Since in the acceptor (Li) doped sample, E_F must be at the top of the valence band, E_F is at 0.4 eV into the gap in the undoped sample. The pinning centers in the undoped material are not known. The second feature produced by the Li doping is a state at 1.4 eV above E_F, which should be identified with the acceptor position. It occurs, as predicted, as the mirror image of the PES final state. Thus, Li doping produces states in the gap of NiO (Fig. 5.58d).

A combination of different data related to the gap in NiO is given in Fig. 5.58. Part (a) shows the combination of an IPES and an X-ray absorption spectrum [5.121] on similarly Li-doped NiO samples, which both display the acceptor signal above E_F. Part (b) gives an XPS/BIS spectrum of lightly Li-doped NiO such that the Fermi energy is pinned at the top of the valence band. This diagram again demonstrates the creation of states in the gap of NiO by Li doping. Part (c) gives an optical absorption spectrum of NiO, where the energy normalization has been performed in such a way that the first maximum of the optical absorption coincides with the maximum of the d^9 final state in the first two parts.

A similar investigation on the development of the states produced by hole doping in the high-temperature superconductor ($La_{2-x}Sr_xCuO_4$) is shown in Fig. 5.59. At this point these data are not very detailed, however, they

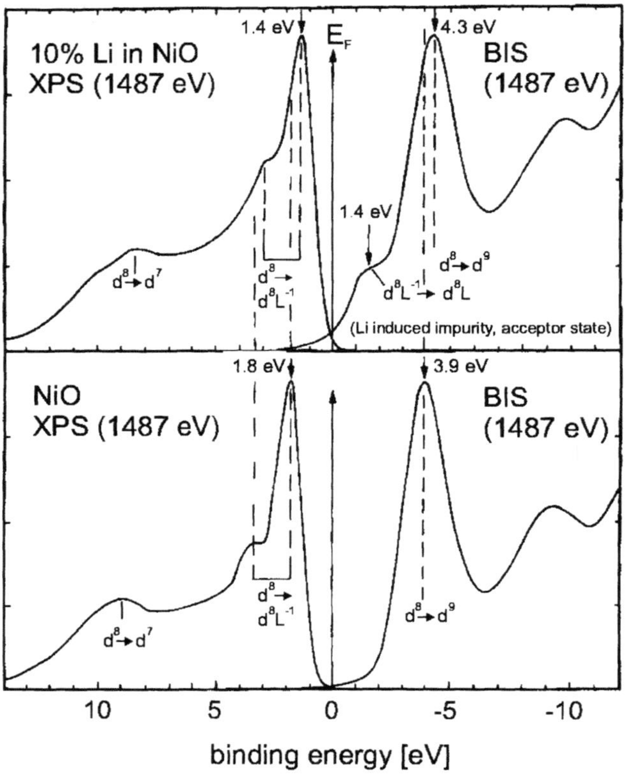

Fig. 5.57. XPS Photoemission and inverse photoemission spectra of NiO and NiO doped with 10 % Li. The lithium induced state in the gap is clearly visible in the inverse photoemission spectra. The doping leads to a shifting of the Fermi energy into the gap by 0.4 eV [5.50]

indicate that, as in the case of Li doping of NiO, hole doping produces states in the charge transfer gap [5.49].

Next we want to move on to some manganites. Figure 5.60 shows the phase diagrams for the $LaMnO_3$ and $NdMnO_3$ based manganites, which exhibit various phases, namely a paramagnetic insulator (PI), a paramagnetic metal (PM), a ferromagnetic metal (FM), a ferromagnetic insulator (FI), an antiferromagnetic insulator (AFI), a spin-canted insulator (CI) and a charge-ordered insulator (COI). We do not show PES data for $LaMnO_3$, because they are not very convincing, but we instead give in Fig. 5.61 data for $Nd_{1-x}Sr_xMnO_3$. For $x = 0.47$ in the temperature range measured, namely 170 K to 20 K, the sample is in a ferromagnetic metallic state and the spectra for all compounds are very similar. They display a small finite density of states at the Fermi energy indicating that this system is indeed a metal. For $x = 0.50$ the phase diagram in Fig. 5.60 indicates that between 140 K and

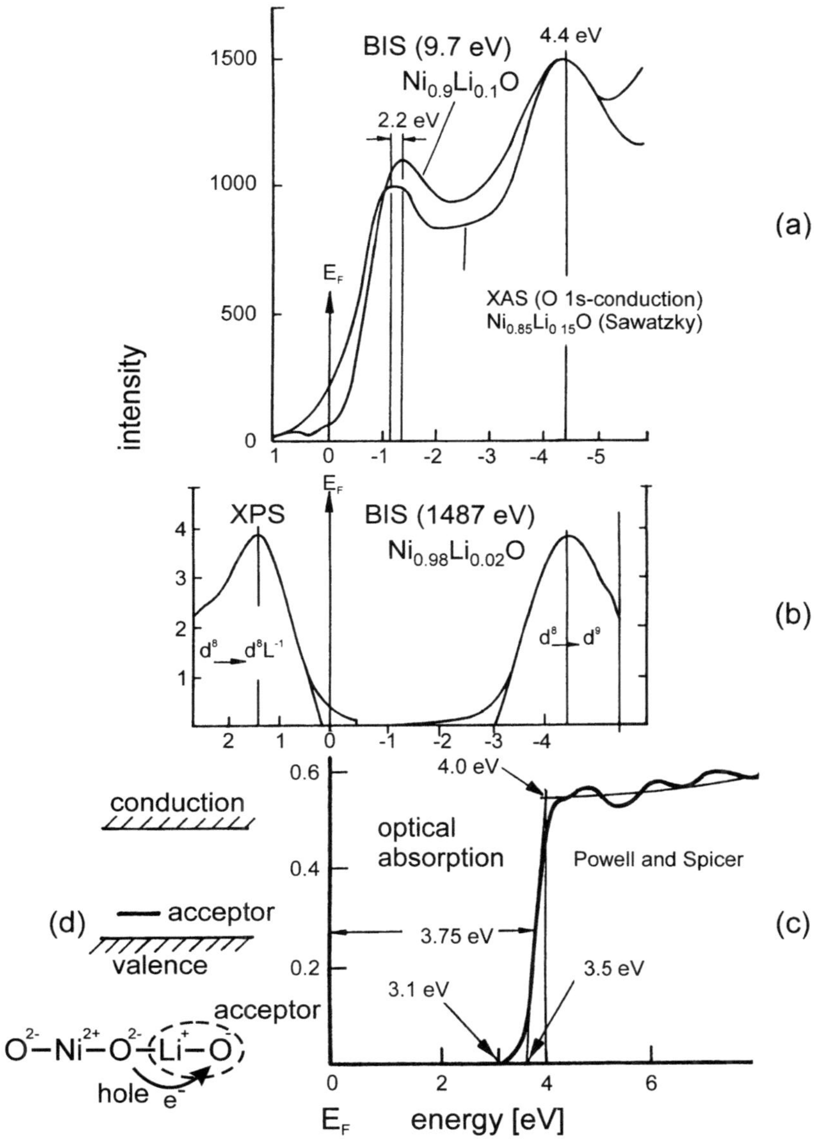

Fig. 5.58. (**a**) X-ray absorption and inverse photoemission data of $Ni_{0.9}Li_{0.1}O$ showing the doping induced state in the gap. (**b**) Photoemission and inverse photoemission spectra of $Ni_{0.98}Li_{0.02}O$ showing in principle the occupied and the unoccupied valence data as measured by photoemission; the small doping only pins the Fermi energy. (**c**) Optical absorption of undoped NiO indicating that the band gap in NiO is a transition from the top of the valence band into the conduction band. (**d**) Semiconductor picture for the doping of NiO by Li [5.121]

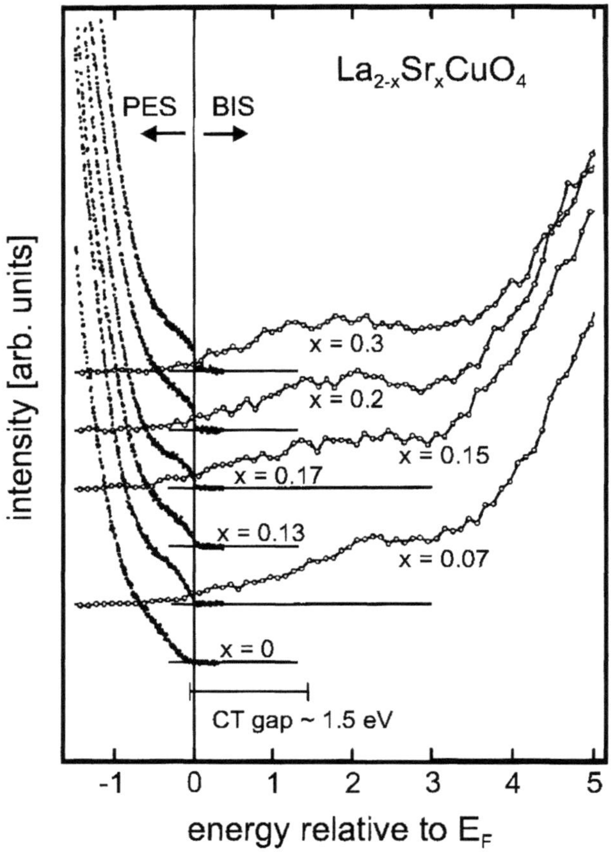

Fig. 5.59. Photoemission and inverse photoemission data of $La_{2-x}Sr_xCuO_4$, which can be considered a parent compound of the high-temperature superconductors. The doping leads to the creation of states in the charge-transfer gap [5.49]; the charge-transfer gap as determined from optical data is 1.5 eV

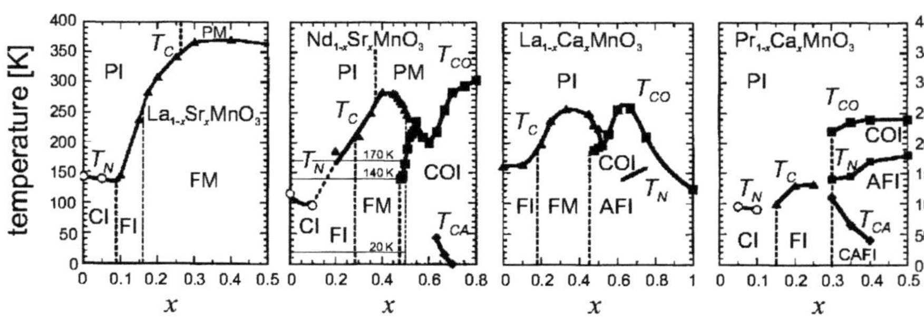

Fig. 5.60. Possible phase diagrams for the manganites, which show the colossal magneto resistance [5.49, 5.122]

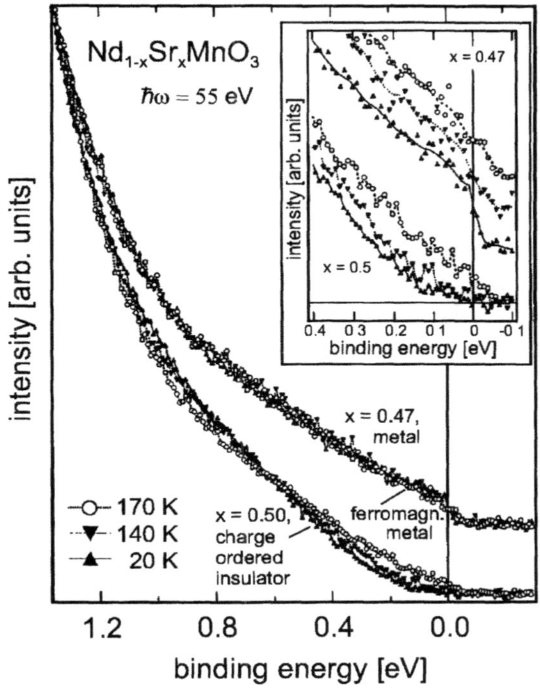

Fig. 5.61. Photoemission spectra as a function of temperature for $Nd_{1-x}Sr_xMnO_3$ at two concentrations of Sr. A slight increase in the intensity near the Fermi energy in the $x = 0.5$ sample seems to indicate a transition into the metallic state; this region is amplified in the inset [5.49, 5.122]. According to the phase diagram (Fig. 5.59) the sample with $x = 0.47$ is always in the metallic state, which is substantiated by the small Fermi step at all temperatures

170 K the system undergoes a transition from the charge ordered insulator to the ferromagnetic metallic state and indeed a slight increase in intensity in the photoemission spectrum is observed in Fig. 5.61 for the spectra from this sample [5.49, 5.122].

5.8.4 Band Structures of Transition Metal Compounds

The electronic states in a solid are described by the many-body band structure. For the valence and conduction band one generally expects sizable dispersion which can be measured by PES and IPES (Chap. 7). In order to make this part on the transition metal compounds reasonably complete, we discuss one example of a band structure, namely that for NiO (Fig. 5.62). The method of determining the experimental band structure from angular-resolved PE data will be discussed in Chap. 7.

Figure 5.62 shows that the calculated one-electron O 2p bands in NiO are very similar to those calculated for MgO. The Ni 3d bands are situated,

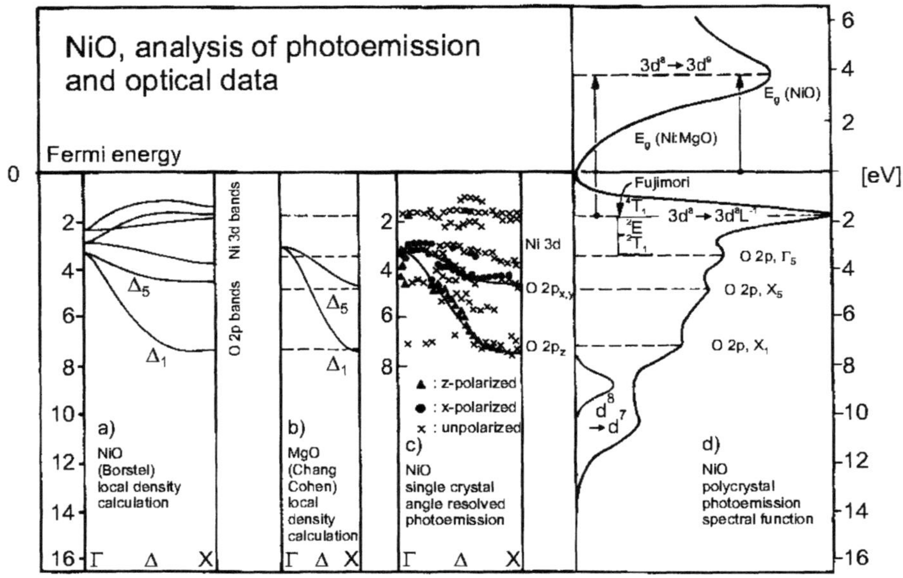

Fig. 5.62. (**a**) Band structure of NiO in the ΓX-direction. (**b**) Band structure of MgO in the ΓX-direction. (**c**) Experimental excitation spectra measured by angular-resolved photoemission spectroscopy of NiO in the ΓX-direction. (**d**) Combined photoemission and inverse photoemission experiment on a polycrystalline sample of NiO

according to calculations, on top of the O 2p bands near the Fermi energy. The experimental data obtained from a NiO single crystal show that the O 2p bands agree quite well in theory and experiment while the experimental "d structures" show only little dispersion less than the one-electron bands. We are inclined to interpret this finding by assuming that the broad O 2p bands provide enough degrees of freedom for relaxation so that the PES experiment measures approximately the one-electron band structure. On the other hand the localized 3d bands which have high correlations cannot "relax during the measurement" and therefore the final states are excited states, which we have written so far as $3d^8L^{-1}$. These excitations have smaller dispersions than the one-electron bands from which they are derived.

In a more formal way the data in Fig. 5.62 can be interpreted by stating that the correlations between the O 2p electrons are so small relative to the O 2p bandwidth that their bands can still be described by one-electron theory. For the Ni 3d electrons the correlations are so large that a one-electron description of their band structure is no longer valid.

Finally (Fig. 5.62d) a comparison with a PES/IPES measurement for a polycrystalline sample allows a tentative interpretation of the PES peaks in terms of critical points in the O 2p part of the band structure (which

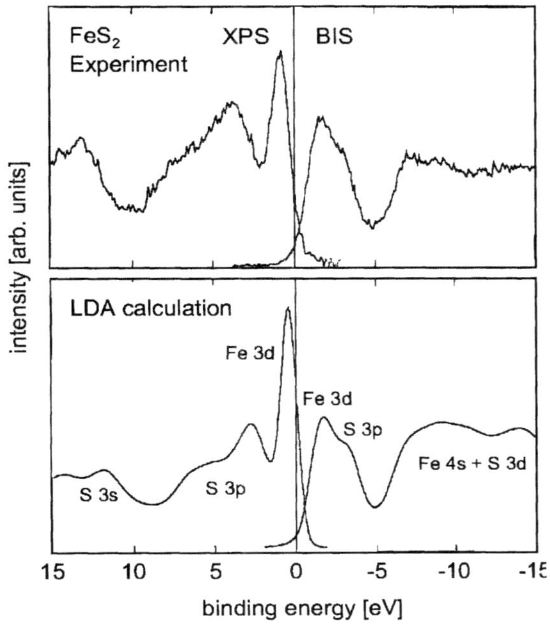

Fig. 5.63. Photoemission and inverse photoemission spectra of FeS_2 compared with the results of a local-density calculation. The agreement is good indicating that the ratio of the correlation energy to the bandwidth (U/W) is small for this compound [5.51]

is one-electron-like). In addition, we have indicated the possibilities for the transitions that are responsible for the optical gap in NiO (4.0 eV) and in Ni:MgO (6.2 eV). The system Ni:MgO represents the model that is used in the single-impurity calculation. Therefore, the optical gap in Ni:MgO should approximately represent the $d^8L^{-1} - d^9$ energy difference which is the parameter Δ.

The diagrams in Fig. 5.62 should be treated with caution and could even mislead the unwary since they combine one-electron and many-electron states. Therefore we emphasize that the one-electron band structure [part (a)] does not represent the (many-electron) d states correctly, especially since it gives practically no gap (see also Fig. 5.42), while the optical gap in NiO is 4 eV. Whether this deficiency has been cured by the LDA+U method remains to be seen. Part (d) also contains one-electron language (critical points in the O 2p density of states) and many-electron processes (optical gap in NiO) in one diagram.

For compounds with 3d ions containing fewer d electrons, for which the U/W ratio is smaller and hence band effects larger, the one-electron LDA method can do reasonably well in reproducing the PES spectra.

A case in point are data from FeS_2 in Fig. 5.63 that show a combined XPS/BIS spectrum of FeS_2 and its successful comparison with the result of a LDA calculation of Imada et al. reported in [5.51].

5.9 High-Temperature Superconductors

We want to state at the outset that this section is written with some reservations. One is dealing here with a field that is still in rapid development and, therefore, more than for any other subject treated in this volume, there is the danger that some of the material that will be presented is out of date by the time that this book is published. In order to reduce this possibility to the absolute minimum, we have decided to restrict this chapter to a very few investigations and especially to some from which we hope that their results still prove to be correct and meaningful in the future. Also, more than in any other section of this book, the choice of the literature here is a very personal one and since, so far, thousands of papers have already appeared, no attempt is made to cover the existing literature in any detail [5.49,5.122–5.126,5.187].

The first question to be asked is why the detection of superconductivity in ceramic CuO-based materials by Bednorz and Müller [5.80] has been greeted with so much enthusiasm. This can perhaps be illustrated by the diagram in Fig. 5.64 which shows a plot of the highest known superconducting transition temperature as a function of year. One can see that such a plot produces roughly a linear relationship, up to 1985 when the maximum transition temperature reached was still below 25 K. We mention that already before 1986 superconductivity in oxides had been detected, however, with transition temperatures that did not exceed those of the known intermetallic compounds. Since 1986 the slope of the curve of the transition temperatures versus time has changed dramatically and now the highest confirmed transition temperature is 133 K which is well above the boiling temperature of liquid nitrogen and therefore represents a dramatic scientific achievement with considerable technological potential. The surprise with which the detection of the ceramic high-T_c superconductors was greeted came also from the fact that these are materials which, although it is known that they can be metals, generally are considered insulators or, at best semiconductors and poor metals.

Very soon after the detection of the high-T_c superconductors it was realized that, although they may obey the rules given by the BCS theory, probably the electron phonon coupling mechanism is not the only one responsible for the coupling interaction resulting in the Cooper pairs. At this point in time it is not clear what the detailed coupling mechanism looks like. However, most likely a mechanism of an electronic (or magnetic) nature will contribute to the coupling. This makes the understanding of the electronic structure of these compounds of importance for a final understanding of the nature of the superconductivity in the high-T_c materials. Therefore PES and inverse

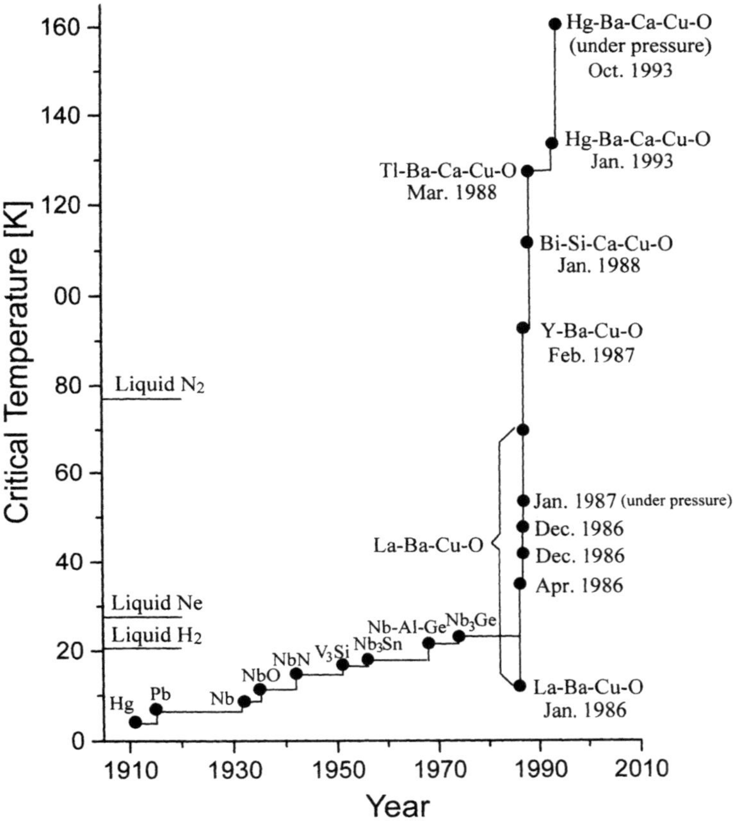

Fig. 5.64. Superconducting transition temperatures of selected compounds as a function of the year of their first realization; we thank Y. Xu (Saarbrücken) for this figure

PES may have considerable value in revealing at least some basic features governing the nature of the electronic structures of these compounds.

5.9.1 Valence-Band Electronic Structure; Polycrystalline Samples

In principle, the basic facts about the electronic structure of the class of compounds, to which so far all of the high-T_c superconductors belong, have been outlined in Sect. 5.8 of this book (see also [5.49, 5.105, 5.110–5.112]). There it was shown that in the late 3d transition metal compounds, large *Coulomb correlation energies* (5 eV to 10 eV) between the d electrons are active, which, in principle, localize the d electrons to a considerable degree. However, charge fluctuations in these compounds can be produced by a charge transfer from the ligands onto the empty d states and the energies for these charge transfers are in many cases smaller than the Coulomb correlation energies. Therefore these compounds can be named charge-transfer compounds. It was also demonstrated for the test material of NiO, that these compounds, if they contain "excess" oxygen, are p type conductors, which means that the mobile carriers are holes in the oxygen p-bands [5.89, 5.90]. This behavior is different from that found for the early 3d transition metal compounds like, e.g., Ti_2O_3. In these compounds, the Coulomb correlation energy can be of the same order of magnitude as the charge-transfer energy for an electron transition out of the ligand band onto the metal ion. Therefore, for these systems, the relative magnitude of the Coulomb correlation energy and the 3d bandwidth can govern the electronic nature, which is near to the classical Mott–Hubbard case [5.61].

These facts have been summarized in the diagram of Fig. 5.53. The important consequence of this diagram is that the cuprate-based high-T_c superconductors (HTC), as far as their overall electronic structure goes, can be regarded as similar to CuO. Therefore in what follows we will give a discussion of the electronic structure of CuO and also Cu_2O. This discussion will be supplemented by some additional information on the high-T_c superconductors, showing where they differ from the parent materials. So far all superconductors with $T_c > 77\,K$ which have been found, contain copper as the transition metal ion.

(a) Cu_2O and CuO

We start the discussion of the electronic structure of the high-T_c materials with an inspection of the situation for Cu_2O (see P. Steiner et al. in ref. [5.125]). Although this oxide has little to do with the high-temperature superconductors, much can be learned from the investigation of the electronic structure of this compound by PES. Fig. 5.65 gives PE spectra taken with 21.2 eV, 40.8 eV and 1254 eV radiation. The ratio of the photoelectron cross sections multiplied by the number of electrons per atom are O 2p/Cu 3d = 2.0, 1.0, 0.03 for He I, He II and XPS energies, respectively. This means that the XPS spectrum gives a rather good representation of the d excitation spectrum and the He II spectrum gives an equally good representation of the combined O 2p and Cu 3d excitation spectrum. The spectra in Fig. 5.65 obtained with He I, He II and Mg-K_α radiation indicate that the 3d and 2p part of the excitation spectrum of Cu_2O are well separated. Of particular interest is a comparison with the results of a band structure calculation employing the local-density approximation, as shown in Fig. 5.66. This figure shows in the top three panels the oxygen part of the theoretical density of states (d), the copper part of the density of states (c) and the total density of states (b), which has been smoothed by a 0.3 eV resolution function. A comparison of the density of states curve (b) with the measured He II spectrum (a) is shown in the lowest panel (e) where the calculated and measured spectra have been

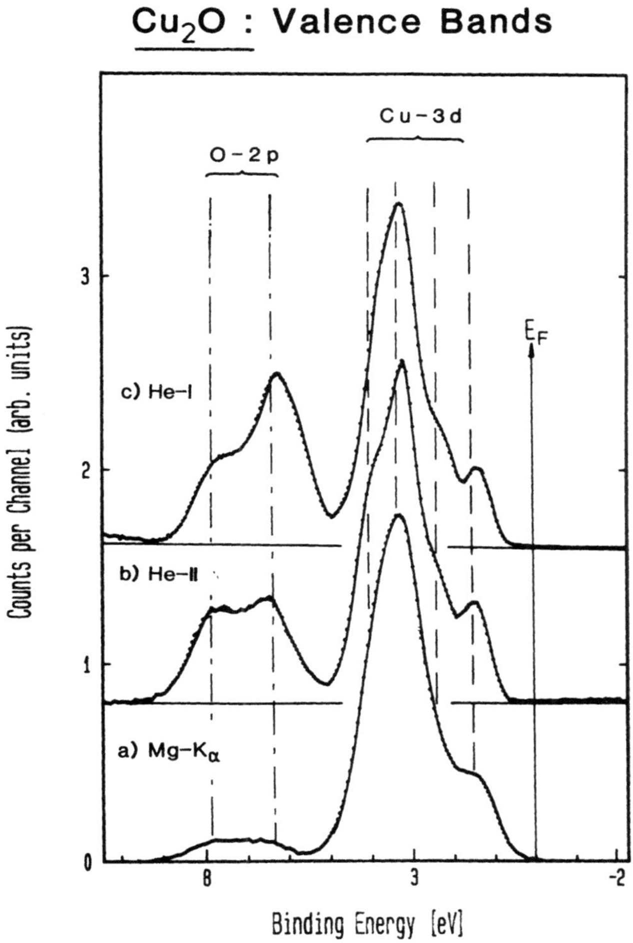

Fig. 5.65. Valence bands of polycrystalline samples of Cu_2O taken with He I, He II and Mg-K_α radiation. The relative O 2p to Cu 3d cross sections are 2.0, 1.0 and 0.03 for He I, He II and Mg-K_α radiation, respectively. This indicates that the XPS spectrum reflects almost exclusively a d density of states, whereas the He II spectrum gives a fair representation of the combined O 2p and Cu 3d density of states

shifted with respect to each other. This shift is needed because Cu_2O is a semiconductor and therefore it is not unexpected that the calculated (top of the valence band) and the experimental (in the gap) Fermi energy do not coincide.

The remarkable fact, which can be seen from the lowest panel of Fig. 5.66, is the almost perfect agreement between the calculated (ground-state) occupied one-electron density of states and the measured excitation spectrum.

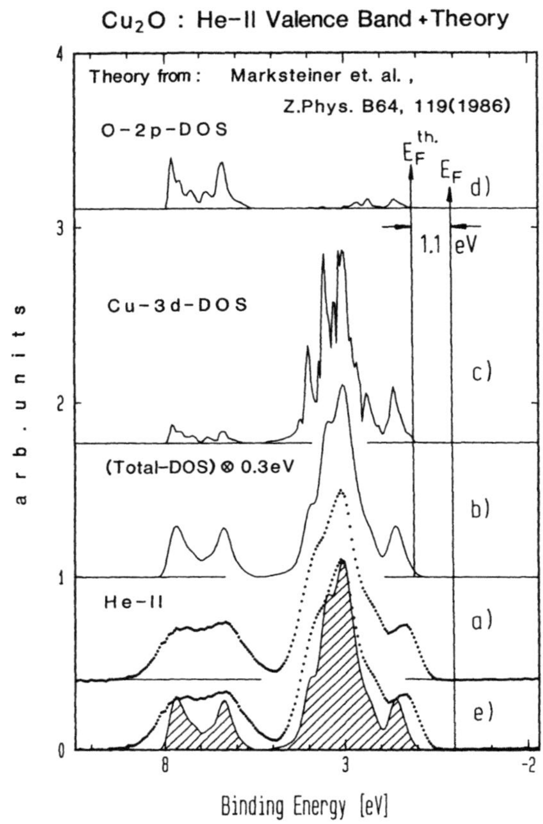

Fig. 5.66. Comparison of the experimental photoelectron density of states and a density of states calculated in the local-density approximation for Cu_2O. (a) He II photoelectron spectrum of Cu_2O with a resolution of 0.3 eV and a background subtracted. (b) Total density of states of Cu_2O calculated by the local-density approximation and smoothed with a 0.3 eV resolution function. (c) Cu 3d part of the calculated density of states (d) O 2p part of the calculated density of states. (e) Comparison of the measured He II photoelectron density of states and the broadened calculated density of states, where the Fermi energies have been shifted in order to give an optimal agreement between the two curves [5.125]

This means that the excitation spectrum has relaxed to a degree which makes it very similar to the ground state. On the other hand, it gives indications that the local-density approximation is successful in providing a ground-state band structure and density of states of materials like Cu_2O. While this good agreement is not unexpected for a broad band of weakly correlated electrons like that of the oxygen 2p electrons, it cannot be taken for granted for the 3d electron part. One realizes from the atomic energy levels [5.53–5.56] that the $3d^{10}$ and $3d^94s$ energies for Cu^+ have a separation of only 3 eV, a value which will probably be reduced in the solid. This means that an excitation in the d part of the spectrum of Cu_2O can be screened out by d electron transfer leading to a situation which is close to the ground state. This, in turn, is described quite well by the one-electron local-density calculation.

In more formal language one can say that the d^9 final state configuration produced by PE out of the d^{10} initial state configuration is a one-hole state. This one-hole state does not correlate with itself and therefore one-electron band theory will give a good description of the final state of Cu_2O produced by PES.

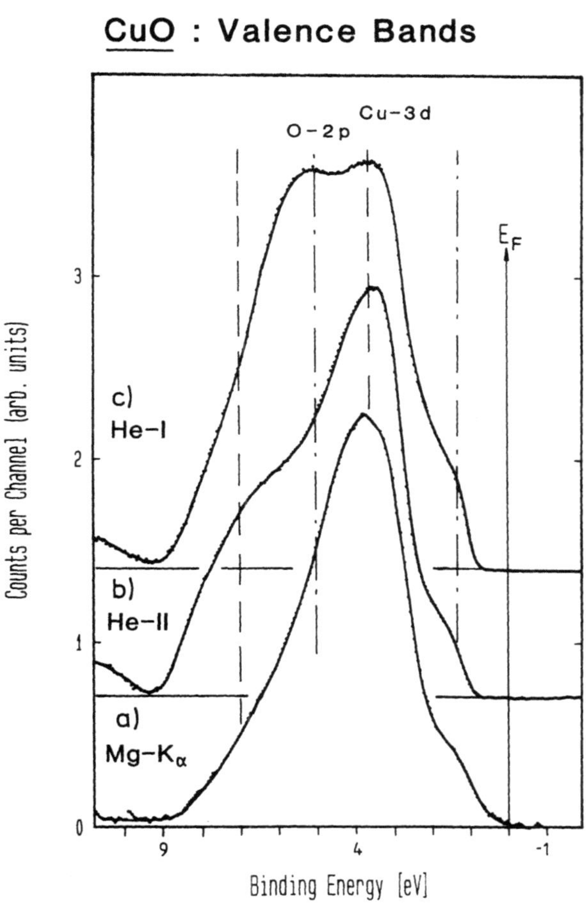

Fig. 5.67. Photoclectron valence-band studies of CuO with He I, He II and Mg – K_α radiation. In this case, in contrast to Cu_2O, the separation of the O 2p and Cu 3d orbitals is no longer so pronounced, indicating a much stronger hybridization between the ligands the metal d electrons [5.125]

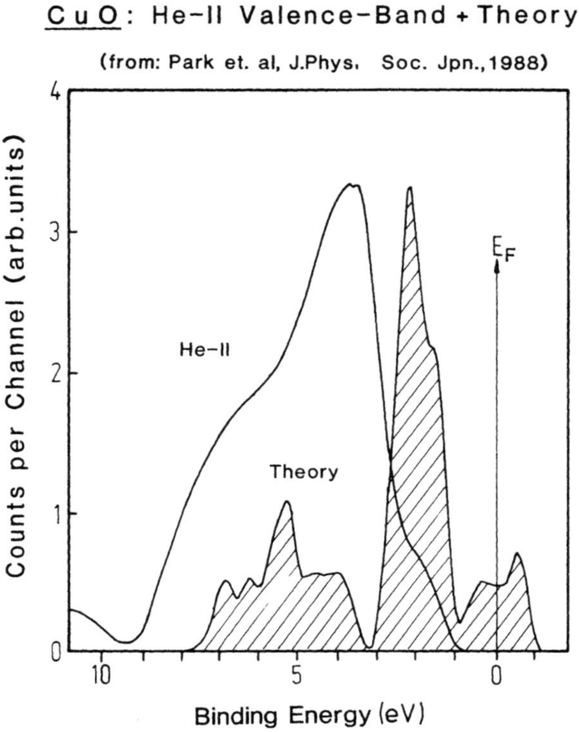

Fig. 5.68. He II valence-band photoelectron density of states compared to the density of states calculated by the local-density approximation in CuO. No adjustment of the calculated Fermi energy relative to the experimental one has been performed. Note that the density of states is one for the groundstate, while the photoelectron spectrum represents an excitation spectrum

With these results for Cu_2O in mind we move on to CuO (which can be considered the parent material for the high-T_c compounds containing Cu^{2+}) in order to see what kind of information can be obtained from a PE investigation here. Fig. 5.66 gives PE spectra taken with He I, He II and Mg-K_α radiation for CuO [5.125]. The material used for these spectra showed only one O 1s line, indicating that it had no additional contaminant phases. For CuO there is no longer such a clear separation between O 2p and the Cu 3d part of the excitation spectrum as seen for Cu_2O. This indicates a stronger hybridization between the ligand 2p and metal 3d wave functions. Now we compare in Fig. 5.67 the He II spectrum from the preceding figure and a density of states from a local-density calculation. The disagreement between the PE spectrum and the local-density curve is apparent.

This disagreement is gratifying because one is comparing in Fig. 5.68 two different things, namely a one-electron density of states obtained for a ground-state calculation, and an excitation spectrum which, in this case, does not reflect the ground state. The reason why a similar comparison for Cu_2O was so successful in Fig. 5.66 lies in the fact that there relaxation produced a state close to the ground state in the PE experiment. This is apparently not the case for CuO. The reason for this different behavior is as follows. Cu_2O has a $3d^{10}$ configuration in the initial state leading to a $3d^9$ PE final state configuration. This one-hole state cannot show correlations. This is another way of saying that (in the atom) the $3d^{10}$ and $3d^9 4s$ configurations are close to each other. However, CuO has a $3d^9$ initial state configuration and a $3d^8$ final state configuration. The configuration $3d^8$ contains two holes which correlate and therefore produce an excitation spectrum [5.126] that deviates from the ground-state one-electron density of states as given, e.g., by the local-density calculation.

This can be emphasized by an inspection of Fig. 5.69 which gives a PE spectrum of CuO at the so-called 3p resonance photon energy (resonance PE, Sect. 3.2.1) [5.73], a technique which enhances the signal that stems from the

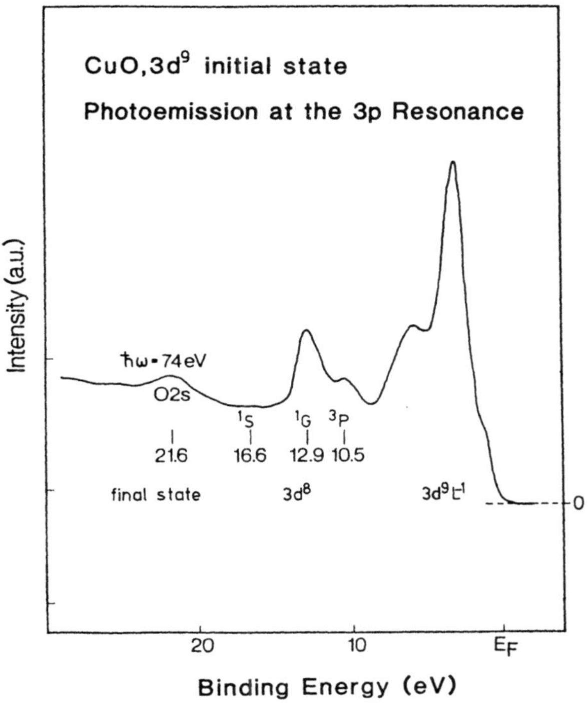

Fig. 5.69. Photoelectron spectrum of CuO with a photon energy at the resonance for 3p excitation. From the theory of resonant photoemission one knows that this type of excitation emphasizes the $3d^8$ states which are created out of the $3d^9$ states of the groundstate in CuO. The d structure near the Fermi energy is then one, created out of the $3d^8$ state by a charge transfer of an O 2p electron into the excited $3d^8$ configuration, leading to a $3d^9L^{-1}$ state [5.73]

final state created by the d electron photoemission. For CuO with a $3d^9$ initial state this is the $3d^8$ final state and it is seen that it gives considerable intensity over about a region of 6 eV. The main 3d intensity near the Fermi energy is then a final state of the type $3d^9L^{-1}$, and the total photoemission spectrum can be described by a superposition of these two final states and, in addition, the O 2s and O 2p final states. Since one is dealing here with localized d electrons, a calculation using a cluster approach, or one using the Anderson Hamiltonian, can describe the d excitation spectrum well, as shown by Fujimori and Sawatzky [5.126].

This situation need not mean a complete breakdown of local-density theory to describe certain aspects of the electronic structure of a system like CuO in the ground state (like, e.g., the total energy). Rather it signals that local-density theory does not do well in representing the excitation spectrum of a compound in which this spectrum is determined to a large degree by the d-d Coulomb correlation energy. This fact is indicated by the occurrence of the $3d^8$ configuration, an aspect not contained in the local-density approach. Thus local-density theory must be used with care in order to give meaningful results for the interpretation of PE (excitation) spectra. On the other hand this indicates that one has to be careful if from PE results conclusions about the possible low-energy excitations (which of course govern the supercon-

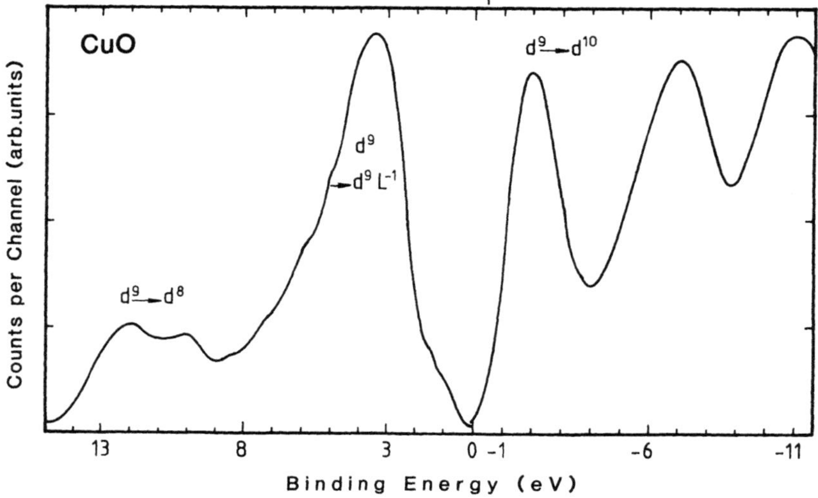

Fig. 5.70. Combined XPS ($\hbar\omega = 1254\,\text{eV}$) and BIS ($\hbar\omega = 1254\,\text{eV}$) spectra for CuO. The origin of the most important features in terms of the transitions from which they originate is indicated. Note that the highest occupied state, after ejection of an electron, is d^9L^{-1}, and the lowest unoccupied state after the addition of an electron is d^{10}, indicating the charge transfer nature of the optical gap

ducting behavior) in a material like CuO (or the high-T_c materials) are to be deduced.

In order to complete the information on the electronic structure of CuO, we show in Fig. 5.70 a combined PES/BIS spectrum taken with soft X-rays for a sample of oxidized Cu. It is similar to the same sort of diagram for NiO (Fig. 5.41) giving the lowest electron-affinity state as d^{10} and the smallest ionization state as d^9L^{-1}. This signals in the context of Fig. 5.53 that CuO and also all high-temperature superconductors are compounds in which the conductivity gap is of the charge-transfer nature.

(b) High-Temperature Superconductors

The data presented for CuO and Cu_2O allow some general statements with respect to the high-temperature superconductors for which CuO can be considered the base material. All these superconductors are in the undoped form charge-transfer semiconductors with a gap of the order of 2 eV. As in a semiconductor, also for the superconductors two forms of doping are possible, namely hole-doping (which is the common form of doping in cuprate superconductors) and electron doping (which has so far only been observed in two types of high-temperature superconductors, namely $Nd_{2-x}Ce_xCuO_4$ and $Sr_{1-y}Nd_yCuO_2$).

It is not yet established how in the high-temperature superconductors doping creates the metallic state out of the semiconducting one [5.127].

Optical and photoemission data show that by hole or electron doping, intensity is created in the (intensity free) gap region of the undoped material. This cannot be taken as evidence for the creation of "genuine" states in the gap by the doping process. It has been demonstrated [5.128] that in correlated materials doping leads to a transfer of spectral weight from above the optical gap into the optical gap.

For a qualitative discussion it looks as if nevertheless the hole-doping process in the high-temperature superconductors can, with respect to the energetics, be described by a picture derived from semiconductor physics (see also data given for Li:NiO earlier in this chapter). There one is of course dealing with one-electron arguments, which are not valid for correlated materials. However, we shall see that in the hole-doped materials the Fermi energy shifts only slightly upon doping. This can be interpreted by assuming a pinning near the top of the valence band by impurities in the undoped sample and by the dopants in the doped sample. The creation of the holes at the top of the valence band by the dopants can be measured by X-ray absorption or inelastic electron scattering experiments. These data give, in addition, the position of the (empty) conduction band, and the energy difference between the hole states (which are assumed to be at the top of the valence band) and the conduction band is in close agreement to the measured optical gap energy.

The semiconductor-derived view makes it understandable why in the metallic state of the high-temperature superconductors the measured bands near the Fermi energy agree with those calculated by one-electron (local-density) theory (except for some renormalization). If in the doped, but not yet metallic state, the Fermi energy is near the top of the valence band it is at least intuitively appealing that when the charge carriers become metallic, the local-density band structure describes the topology of the Fermi surface adequately.

These remarks are not meant to play down the importance of correlations for the electronic structure of the high-temperature superconductors. However, it seems that some aspects of that electronic structure can still be discussed by using the semiconductor language.

On the other hand, we shall see that a comparison of the spectroscopic data of $La_{2-x}Sr_xCuO_4$ and $Nd_{2-x}CuO_4$ in terms of the semiconductor picture leads to inconsistencies. This obviously signals the limitations of using one-electron language for the interpretation of spectroscopic data from correlated materials.

We shall see in what follows, that in the hole conducting high-temperature superconductors the superconducting state is determined by wave functions of the type $3d^9L^{-1}$ (many people have chosen to term this trivalent copper which, however, is slightly misleading as we will show, because the extra electron missing with respect to divalent copper is taken from the oxygen 2p reservoir rather than from the Cu 3d reservoir). This $3d^9L^{-1}$ wave function is exactly the one which is produced by PE out of the $3d^9$ initial state and subsequent ligand-to-metal charge transfer. Therefore the PE signal is a representation of those parts of the wave functions which are intimately connected with the superconducting properties.

Finally we want to point out a fact which emerges from Fig. 5.66. There it was demonstrated that for Cu_2O one has good agreement between the experimental density of states and the theoretical one calculated in the local density approximation. This also means that the theoretical one-electron dispersion curves should be a good approximation of the experimental ones (which so far have not been measured). These theoretical dispersion curves have a magnitude of the dispersion which is only about a factor of two smaller than that found in Cu metal (Chap. 7). This is also reflected by the bandwidth of the Cu part of the density of states in Fig. 5.65, which is approximately 3 eV wide. Therefore any theory that deals with the electronic structure of Cu_2O (or CuO, or the high-temperature superconductors) must take into account these large dispersions. It also means that strictly local models like the Anderson impurity model or models working with small clusters probably will have difficulties in capturing all the physics present in these compounds.

In order to go one step further in sophistication, we take a look at the PE results from La_2CuO_4. This material contains all the essential properties

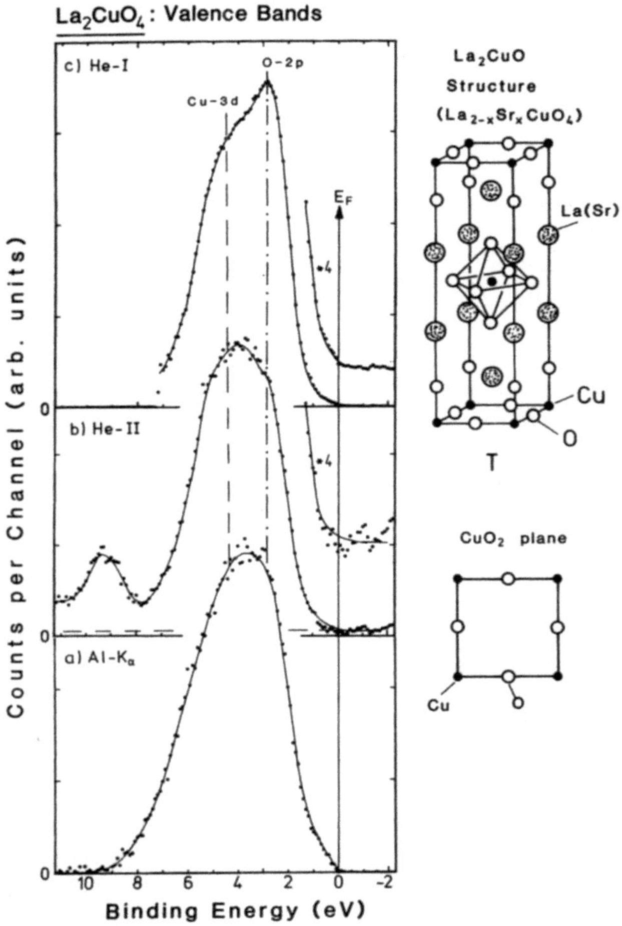

Fig. 5.71. Valence-band photoemission spectra of La_2CuO_4 taken with He I, He II and Al-K_α radiation. The peak at 9 eV in the He II spectrum is produced by an impurity [5.125]

of the high-temperature superconductors in having CuO_2 planes which are separated by LaO planes.

The crystallographic structure of La_2CuO_4 is shown in Fig. 5.71. There we have also indicated the essential feature of that structure, namely the CuO_2 planes which are found in all Cu containing high-temperature superconductors. Stoichiometric La_2CuO_4 is an insulator of the charge-transfer type. If some of the trivalent La-ions are replaced by divalent Sr or Ba ions, the material becomes a high-temperature superconductor (in fact this was the first cuprate superconductor discovered by Bednorz and Müller [5.80]). This replacement can be compensated either by a conversion of Cu^{2+} to

Cu^{3+} (going from $3d^9$ to $3d^8$) or by converting O^{2-} to O^{1-} (going from $2p^6$ to $2p^5 = L^{-1}$). Since it is well known [5.129] that in these materials energy considerations favor oxygen holes, the second possibility is the one that takes place. This can also be inferred from the data in Fig. 5.69. There it can be seen that the $3d^8(Cu^{3+})$ configuration is 10 eV above the $3d^9L^{-1}$ configuration. Thus in a CuO type material after the ejection of one electron, the occurrence of the $3d^9O^-$ configuration is favored by a wide margin over the $3d^8O^{2-}$ configuration [5.90].

PE spectra of La_2CuO_4 from a ceramic sample with three representative photon energies are shown in Fig. 5.71. We see that a trend already noticed in going from Cu_2O to CuO is further emphasized, namely the complete merging of the O 2p and Cu 3d signals, which are hardly distinguishable in the data. This is also brought out by the band-structure calculations which are shown in Fig. 5.72 [5.130] where one can see that it is hardly any longer possible to separate the O 2p from the Cu 3d density of states.

We should be aware of the fact that the comparison of a PE spectrum of a correlated material La_2CuO_4 and a density of states of a one-electron calculation can be very misleading in view of what has been demonstrated in this chapter. However, Fig. 5.72 shows a strong hybridization of Cu 3d and O 2p states in the local-density calculation and we have seen that O 2p bands come out experimentally as calculated in a one-electron local-density calculation. At least with respect to the O 2p part, a comparison between the PE spectrum and the local density result is not completely meaningless. Since the data do not indicate a separation of the O 2p part and the Cu 3d part also in the final state, the strong p-d hybridization seems to be maintained. We add that the PE spectra of the high-temperature superconductors do not change much in going from the insulating to the metallic state. The experiments have shown that in the metallic state the measured bands agree (except for some renormalization) with those calculated in the local-density approximation. This indicates the degree to which the local-density results can be used for an interpretation of the electronic structure of the high-temperature superconductors.

Thus we realize one important feature valid for all the high-temperature superconductors, namely a very strong hybridization between the oxygen 2p and the copper 3d electrons.

The spectra in Fig. 5.71 have been taken on poor samples. As evidence for this claim we point to the 9 eV peak in the He II spectrum of Fig. 5.71. As far as one knows, this peak is produced by an impurity of not completely known origin. In spectra taken from well-prepared single-crystal surfaces this peak is absent. This brings out one problem of the PE data obtained in the field of high-temperature superconductivity. Much of the work reported so far has been performed on ceramic (polycrystalline) samples and they are not as well characterized as one might wish.

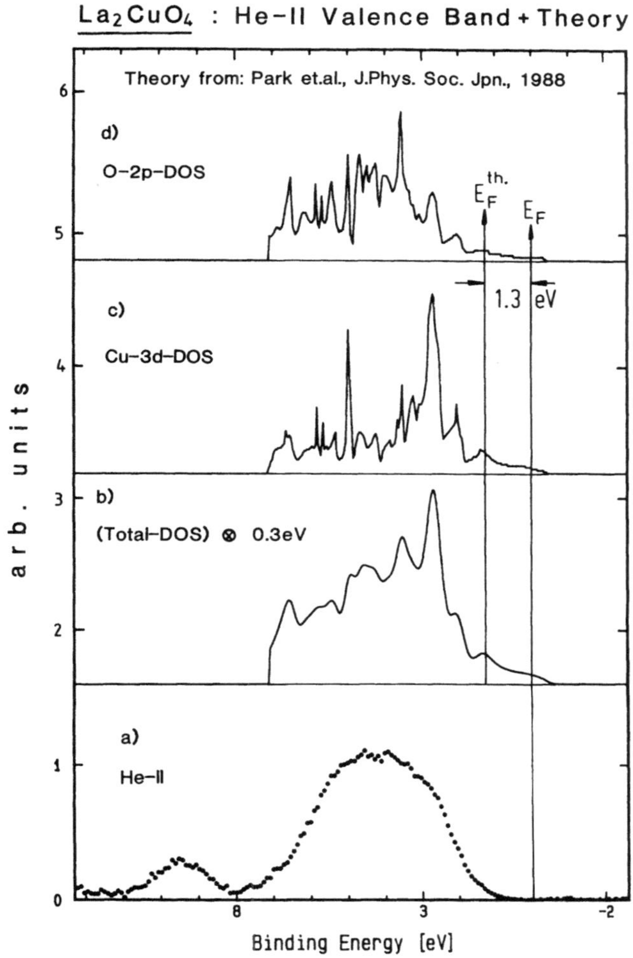

Fig. 5.72. Comparison of the He II spectrum of La_2CuO_4 with the results of a local-density calculation. The calculated spectra have been shifted by 1.3 eV relative to the measured spectrum. The partial densities of states for the Cu 3d electrons and the O 2p electrons indicate that they overlap completely and are thus completely mixed, a result which is in agreement with the PE density of states of Fig. 5.62 [5.125]

Therefore in many cases the definite data will have to wait until work with good single crystals has been reported. However, for the conclusion that was drawn from the spectra in Fig. 5.71, namely a very strong mixing between the Cu 3d and O 2p orbitals, the quality of the data presented is sufficient.

Next we briefly deal with the question of the nature of the charge carriers in these compounds. As was pointed out already, the base materials, such as La_2CuO_4, are semiconductors with a gap energy of $E_g \approx 2$ eV. Unfortunately so far one has not been able to dope the same base material with holes or with electrons, which would be highly interesting. A series of cuprate superconductors are known which can be doped by holes (like in $La_{2-x}Sr_xCuO_4$, $Y_1Ba_2Cu_3O_{7-x}$) and two materials have been found which could be doped with electrons ($Nd_{2-x}Ce_xCuO_4$ and $Sr_{1-y}Nd_yCuO_2$).

As already mentioned, from thermodynamic arguments [5.90,5.129] it can be inferred that the most likely defects in these materials are Cu defects which are charge compensated by holes in the oxygen 2p band. A doping of similar nature is achieved if, e.g., in La_2CuO_4 some of the La ions are replaced by Sr ions. Then the charge neutrality in the system is again not achieved by creating an equal number of Cu^{3+} ions out of the Cu^{2+} ions, but rather by forming some O^{1-} ions out of the O^{2-} ions which, in other words, means the production of holes in the oxygen 2p band.

These holes can in principle be seen by investigating excitations with the O 1s level as an initial state. Such experiments have been successfully performed by X-ray absorption and inelastic electron scattering experiments, for which we show representative data in Fig. 5.73 [5.131].

In order to explain the measurement, an energy level diagram containing the O 1s level (from which the electron is excited) and the valence and conduction band in the doped and undoped material is also given in Fig. 5.73. In the undoped material ($x = 0$) transitions out of the O 1s level can only take place into the d^{10} conduction band (and higher empty bands). In the hole doped material ($x > 0$) transitions out of the O 1s level into the empty states in the valence band (hole states) are possible. They have indeed been observed (putting an electron into the d^9L^{-1} hole state leads to a $[d^9L + d^{10}L^{-1}]$ state as indicated in the spectrum). The energy separation between the $[d^9L + d^{10}L^{-1}]$ state and the d^{10} state is 2 eV which corresponds to the known gap energy of La_2CuO_4.

So far only small effects on the PE spectra produced by doping have been found. This means, e.g., that the spectra shown in Fig. 5.71 are changed only slightly if instead of La_2CuO_4 a sample of $La_{1.9}Sr_{0.1}CuO_4$ is used.

Even more puzzling is the comparison of spectra from a hole-doped material ($La_{1.8}Sr_{0.2}CuO_4$) and an electron-doped material ($Nd_{1.85}Ce_{0.15}CuO_4$), as given in Fig. 5.74 [5.125]. These spectra are similar. Note that both base materials, namely Nd_2CuO_4 and La_2CuO_4 have a gap energy of $\sim 2\,eV$. In $La_{1.8}Sr_{0.2}CuO_4$ (hole doped) the Fermi energy can be assumed (in using the experience with NiO) to be positioned near the top of the valence band. In $Nd_{1.85}Ce_{0.15}CuO_4$ (electron doped) the Fermi energy would be expected at the bottom of the conduction band if this system would behave like an "ordinary" semiconductor. This, however, would mean that the peak in the PE density of states of $Nd_{1.85}Ce_{0.15}CuO_4$ would be shifted by $\sim 2\,eV$ to larger binding energy with respect to its position in $La_{1.8}Sr_{0.2}CuO_4$. This is obviously not the case, indicating that also in the electron-doped material the Fermi energy may be near the top of the valence band, if the semiconductor-

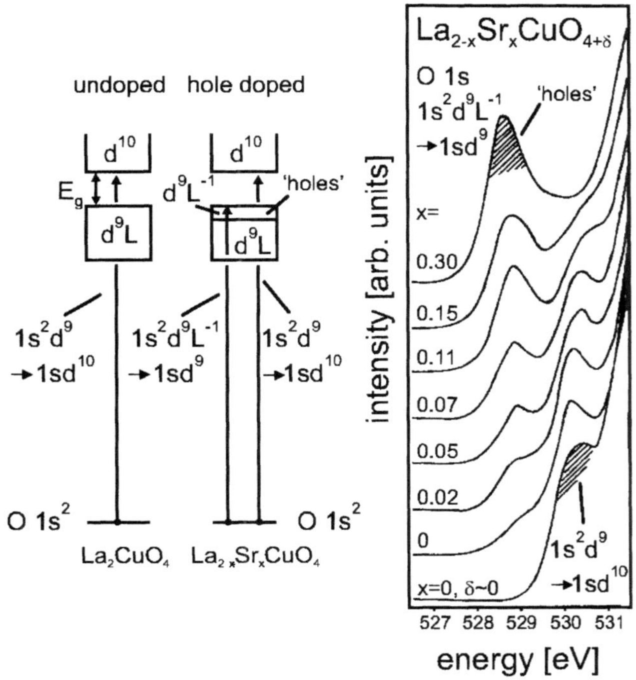

Fig. 5.73. Inelastic electron scattering cross section near the O 1s threshold for $La_{2-x}Sr_xCuO_4$ ($x \geq 0$) and energy level diagram relevant to the experimental data. For $x = 0$ the material is non-conducting (and non-superconducting) and absorption is only possible into the empty d^{10} conduction band. For $x > 0$ charge neutrality in the compound $La_{2-x}Sr_xCuO_4$ can only be obtained if an appropriate number of Cu^{3+} ions are created or if, alternatively, some O^- ions are produced. The intensity of the O 1s excitation preceding the conducting band absorption representing transitions into the valence band are strong evidence for the creation of O^- ions, which represent holes in the O 2p band. [5.131]

derived picture would be correct. One could rationalize this behavior by realizing that in electron-doped Nd_2CuO_4, a Cu $3d^{10}$ configuration is produced, in which there is no correlation energy, leading to a breakdown of the optical gap.

There exists, however, also the possibility that in the hole-doped and the electron-doped superconductor the Fermi energy is pinned in the middle of the gap. We add that La_2CuO_4 and Nd_2CuO_4 have different crystallographic structures and consequently different electronic structures. This may invalidate any comparison of the two spectra in Fig. 5.74. However, both structures contain the CuO_2 planes and therefore a certain similarity of the electronic structures can be anticipated making the comparison not useless.

One can of course take the position that in a system with highly correlated electrons any one-electron-based discussion is meaningless. Then the

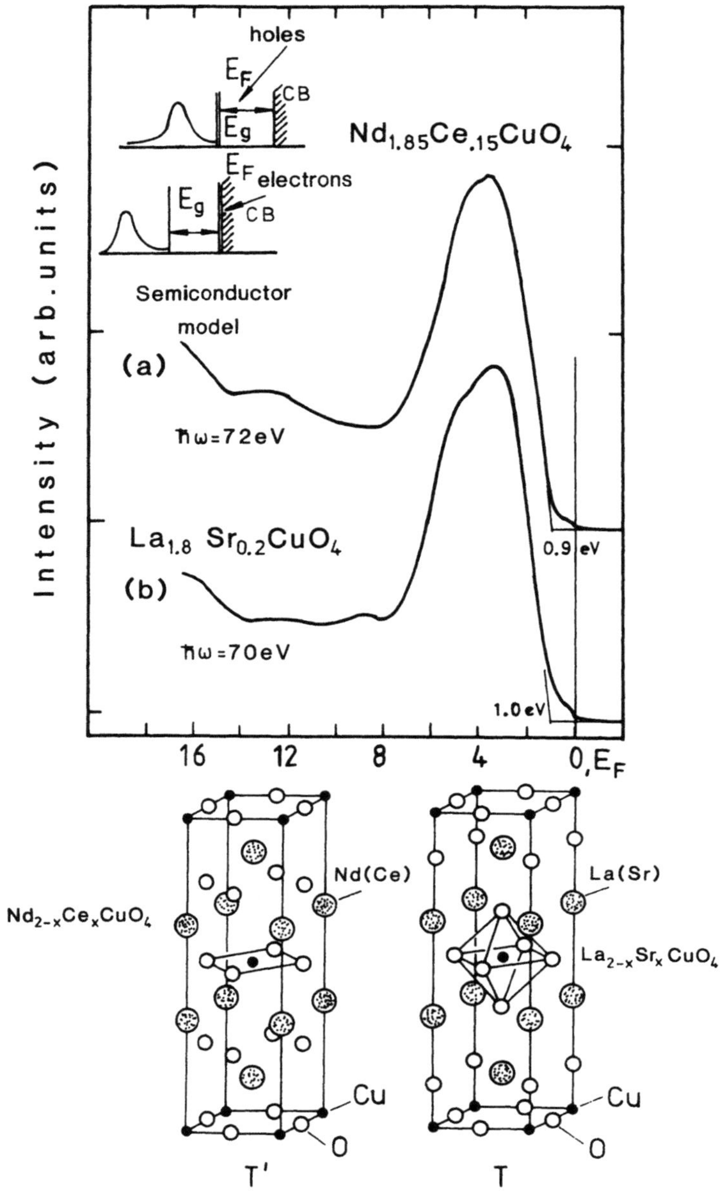

Fig. 5.74. UPS spectra of the hole superconductor $La_{1.8}Sr_{0.2}CuO_4$ and the electron superconductor $Nd_{1.85}Ce_{0.15}CuO_4$. Data were obtained from ceramic samples, and the 9 eV peak indicates that impurities were present in the samples. The O 2p/Cu 3d valence-bands overlap, which is not in agreement with the semiconductor model shown for hole and electron doping. Note that the optical gap in the base materials La_2CuO_4 and Nd_2CuO_4 is ~ 2 eV [5.125]. The lower part gives the crystal structures of $Nd_{2-x}Ce_xCuO_4$ and $La_{2-x}Sr_xCuO_4$

problems just discussed with respect to the two curves in Fig. 5.74 are simply a manifestation of correlation.

5.9.2 Dispersion Relations in High Temperature Superconductors; Single Crystals

Since the high-temperature superconductors are metals, and in general terms a metal is a system with itinerant electrons, such a system has a band structure and a Fermi surface. This makes this section (Sect. 5.9) slightly inconsistent because so far this chapter has dealt mostly with localized levels, which by definition have no dispersion and no Fermi surface. However, in order to treat all the material on the high-temperature superconductors in one section, we shall deal here with the dispersion relations instead of discussing them in Chap. 7.

A metal is often called a Fermi liquid, which means that its properties have a one-to-one-correspondence to those of a free-electron metal. It was speculated early on in the investigation of the high-temperature superconductors that they, because of their unusual properties, may not be Fermi liquids but rather so-called marginal Fermi liquids. However, we will not deal with this question because it is not yet settled.

Early PE studies of high-temperature superconductors were plagued by the absence of a clearly defined Fermi edge, which one would expect to be present in a metallic material. Such Fermi edges, which are typical for metallic samples, were first seen in the Bi-containing high-temperature superconductors, as is demonstrated in Fig. 5.75. Both the He I and the He II spectrum of $Bi_2Sr_2Ca_1Cu_2O_8$ present a clear Fermi edge, which is particularly apparent in the expanded scale spectra. The sample used in this experiment was polycrystalline [5.132].

With increasing sophistication in sample preparation, Fermi edges were observed for all high-temperature superconductors. Fig. 5.76a gives an example for $Nd_{2-x}Ce_xCuO_4$ displaying XPS spectra measured from a well prepared polycrystalline sample [5.133]. It is evident from the data that while the semiconducting sample shows no Fermi edge, this is clearly visible in the conducting and superconducting material. The conducting sample $Nd_{1.85}Ce_{0.15}CuO_4$ is made superconducting by reducing it slightly to $Nd_{1.85}Ce_{0.15}CuO_{4-x}$. Similar data taken with UPS energies and correspondingly higher resolution are displayed in Fig. 5.76b [5.134].

In understanding the nature of the superconducting state in the cuprate high-temperature superconductors it is important to have a knowledge of their correlated electronic bands near the Fermi energy. An important question in this respect is whether the band structure as obtained by PES has any relation to that derived from one-electron calculations.

The exact determination of an experimental band structure from a PE experiment will be dealt with later (Chap. 7). Here it suffices to say that in order to measure the energy of a band as a function of the crystal momentum

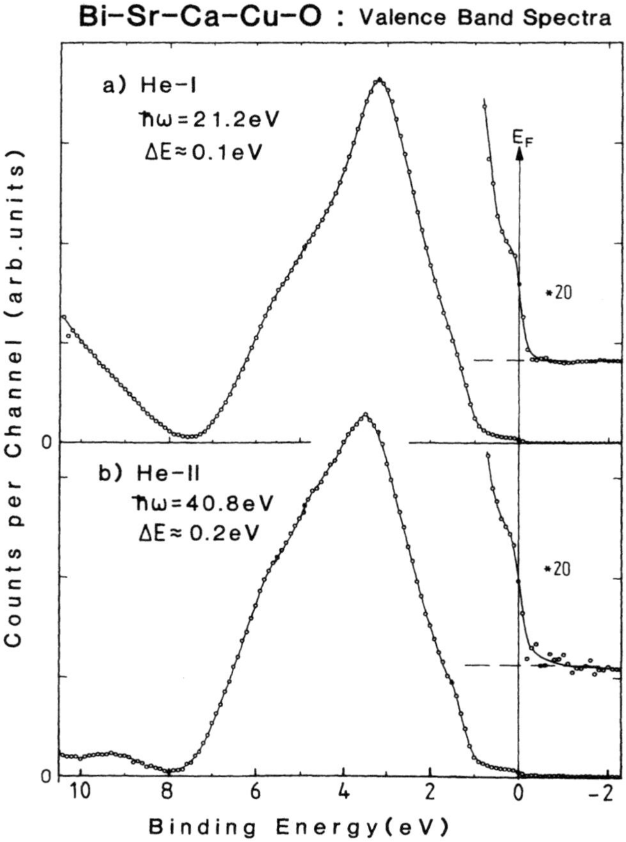

Fig. 5.75. UPS spectra of $Bi_2Sr_2Ca_1Cu_2O_{8+x}$ taken with He I and He II radiation and a resolution of $\Delta E = 0.1\,\mathrm{eV}$ and $0.2\,\mathrm{eV}$ respectively. A Fermi edge is clearly visible in both spectra [5.132]

$\hbar \boldsymbol{k}$, the initial state energy is measured together with the momentum of the electron photoemitted from it, which, in turn, allows one to determine (by some not always simple procedures) the momentum of the initial state. The momentum of the photoemitted electron is a function of its direction relative, e.g., to the crystal normal. Thus in measuring spectra as a function of the electron detection angle, one can determine the energy of the initial state as a function of its momentum, which is nothing else but an experimental band structure determination.

The results of such a measurement [5.134] for a sample of $Nd_{2-x}Ce_xCuO_4$ are displayed in Fig. 5.77. The wave vector has been changed by varying at fixed frequency of the exciting radiation the electron detection angle relative to the surface normal. A dispersion of a state A is observed. The insert gives the small part of the one-electron (local-density) band structure relevant to

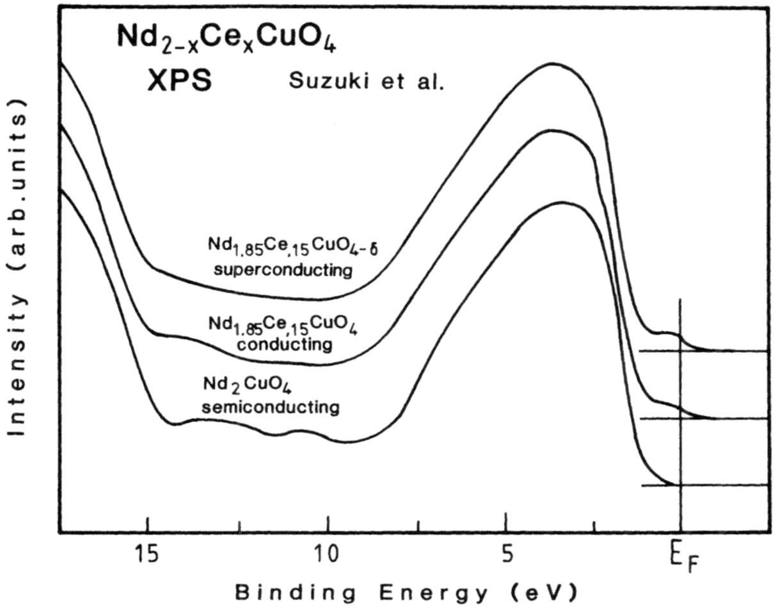

(a)

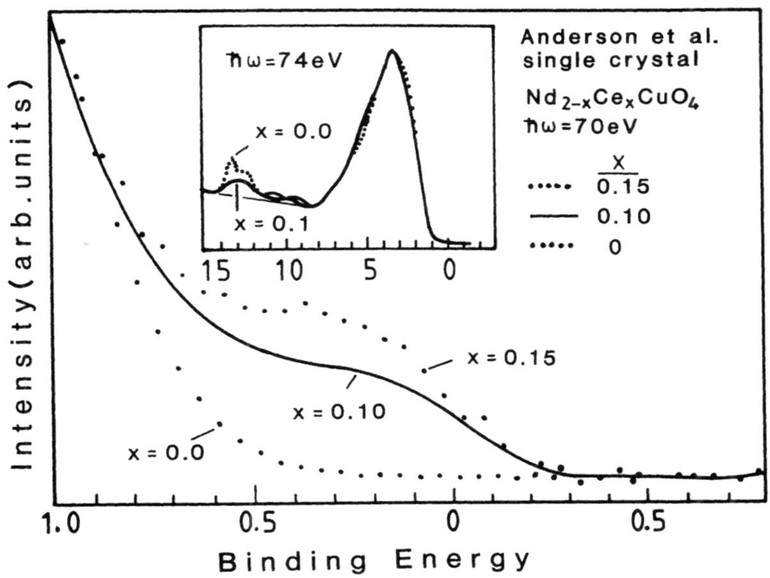

(b)

Fig. 5.76. (**a**) XPS spectra of the electron conducting material $Nd_{2-x}Ce_xCuO_4$. For $x = 0$ the material is a semiconductor displaying no Fermi edge. For $x = 0.15$ (unreduced) the sample is metallic (non-superconducting) and shows a Fermi edge, as does the same sample after reduction which makes it superconducting [5.133]. (**b**) UPS spectra of $Nd_{2-x}Ce_xCuO_4$ showing the development of the Fermi energy as the doping is increased from $x = 0.0$ to $x = 0.15$ [5.134]

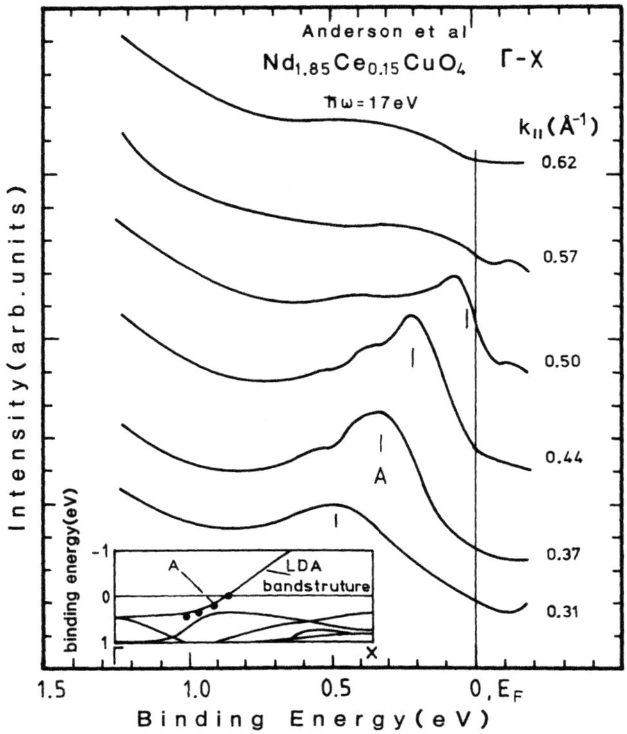

Fig. 5.77. UPS spectra of $Nd_{2-x}Ce_xCuO_4$. The spectra have been taken at fixed photon energy by changing the electron detection angle relative to the crystal normal. The insert shows a portion of the local-density band structure of $Nd_{2-x}Ce_xCuO_4$ with the data points inserted [5.134]

the experiment where the data points have been inserted by full dots. The results from a series of similar experiments are compiled in Fig. 5.78. This figure gives a cut through the Fermi surface of $Nd_{1.85}Ce_{0.15}CuO_4$ in the basal plane (see Fig. 5.74 for the crystal structure).

The open circles denote points in $\boldsymbol{k}$-space where angle resolved PE spectra have been measured. The full circles indicate the Fermi level crossing of a structure in the EDC. These full points agree with the contour of the Fermi surface in the basal plane calculated by a one-electron (tight binding) calculation (full curve). The picture that emerges from Figs. 5.77 and 5.78 is in a sense representative for the results of similar measurements on a number of high-temperature superconductors. Weakly dispersing bands have been found experimentally above and below the Fermi energy. The dispersions are, however, often smaller than predicted by (one-electron local-density) theory (they are renormalized due to correlations). In all high-temperature superconductors investigated so far by PES the experimental Fermi surface topology has been found to agree in the main features with that calculated by a one-

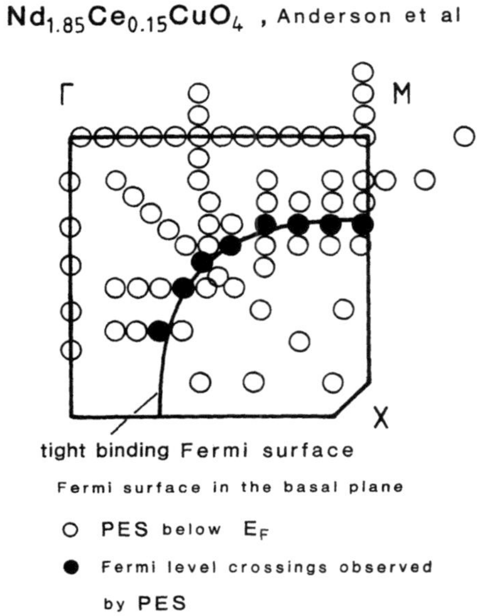

Fig. 5.78. Cut through the Fermi surface of $Nd_{2-x}Ce_xCuO_4$ in the basal plane. Open circles: points in $\boldsymbol{k}$ space where PE spectra have been taken; full circles: structure in the PE spectrum at E_F. The full line gives the Fermi surface contour in the basal plane obtained in tight binding calculation [5.134]

electron local-density calculation (Luttinger theorem). One may rationalize this in simple terms. At the Fermi energy the correlation energy vanishes and therefore for this energy the one-electron states are recovered even in a system with large correlations.

We now summarize the main features of the electron dispersion relations near the Fermi energy of all the high-temperature superconductors, these being the states that are largely responsible for the unusual electronic properties of these materials. Most important in this respect is the momentum dependence of the electronic states at the Fermi energy, which is what defines the Fermi surface [5.135, 5.136].

It should be mentioned that the details of the Fermi surface, although they have been investigated quite intensively, are still not documented beyond doubt.

Figure 5.79 gives some basic features which are useful to understand the Fermi surface of the high-temperature superconductors. In part (A) the position of the ions in the CuO_2 plane is indicated. These are the planes responsible for the properties of the high-temperature superconductors. The Cu ions sit at the corners and the oxygen ions on the connecting lines between the corners. Part (C) gives the band structure of such a system in the tight binding approximation. Note that there is a saddle point at the X-point of the Brillouin zone and another Fermi level crossing of the band between Γ and M. Part (B) gives the crystal's first Brillouin zone and the Fermi surface obtained from the band structure in part (C). The points (a) and (b) are indicated in the band structure and also in the Fermi surface. It is evident

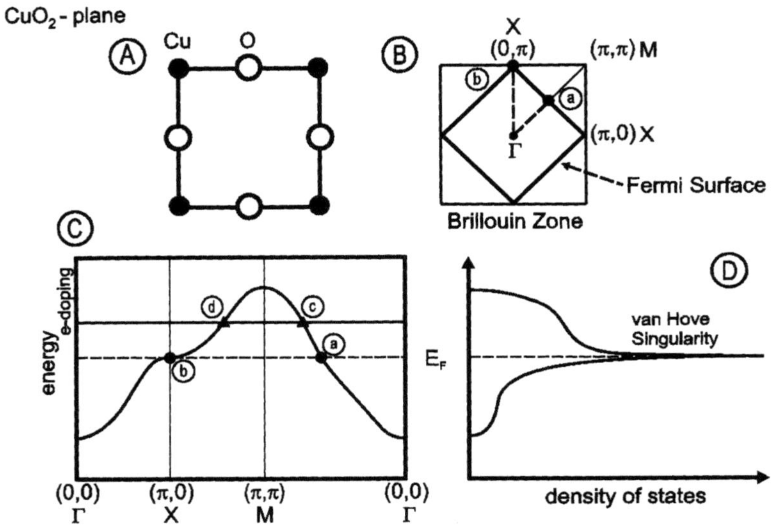

Fig. 5.79. The electronic properties of the CuO_2 plane. (**A**) Geometric structure. (**B**) Fermi surface. (**C**) Energy band structure obtained by a tight binding calculation. The dashed curve is the Fermi energy in the normal state, whereas the full curve shows the Fermi energy in the electron-doped state. (**D**) Density of states in the normal state, showing a van Hove singularity created by the saddle point at (b) [5.135]

that in the density of states [part (D)] there is a van Hove singularity at the Fermi energy.

Unfortunately the various high-temperature superconductors have different crystal structures and although the basic feature, namely the CuO_2 plane, is the same in all the materials, the labeling of the Brillouin zone is different in the different materials, and in particular there is an exchange of X and Y with $\bar{M}$ and M, which has to be noted in comparing results from different papers (see Fig. 5.80). If one starts from the band structure in Fig. 5.79 and adds a slight electron doping, the Fermi surface in the extended zone scheme becomes that shown in Fig. 5.81, and, in many papers at least, it looks as if this Fermi surface reflects the general features of all the Fermi surfaces measured so far.[5] In the simple model of Fig. 5.79 there are closed orbits around the M-points (note that these are the X- and Y-points in the Bi-containing high-temperature superconductors, which are the most frequently investigated ones) and this Fermi surface is generally called a hole-type Fermi

[5] Most of the high-temperature superconductors are hole-doped systems. The electron doping is used in this discussion because only by this type of doping the model band structures of Fig. 5.79 yields the Fermi surface of Fig. 5.81 which is close to the observed Fermi surface. The band structure of a real high-temperature superconductor (see Fig. 5.82) is close to that of the model superconductor shown in Fig. 5.81.

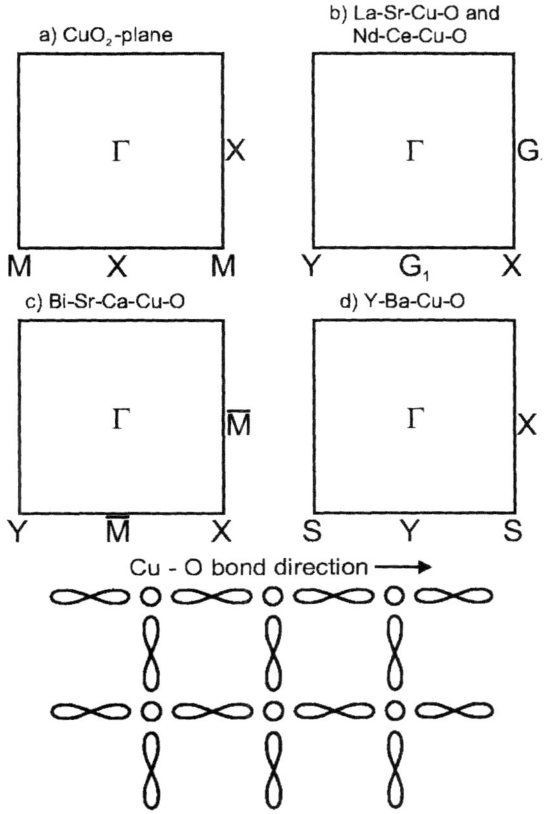

Fig. 5.80a–d. Nomenclature designating the points of the Brillouin zone for the various types of high-temperature superconductors [5.135]

surface for obvious reasons. So far this Fermi surface has not only been found in the electron-doped materials but also for almost all hole-type materials. A nice general view of the Fermi surface has been obtained by using a technique described in Chap. 11 of this book, which samples all the electrons in a 2π geometry over the surface of these quasi two-dimensional systems. In principle this maps the two-dimensional Fermi surface, although obviously not with the high resolution desirable. The results of this experiment are given in Fig. 5.82, where (a) gives the raw data and (b) the pieces of the Fermi surface deduced from these data. They are compared with the results of a local-density calculation (Fig. 5.82) and, except for the electron pockets around $\bar{M}$, there is generally good agreement between experiment and theory.

Since all the high-temperature superconductors have as their main building blocks the CuO_2 planes, one can try to map all the results from the very different compounds into a Brillouin zone like that shown in Fig. 5.79. This has been done by Shen et al. [5.135] and the result is given in Fig. 5.83. One

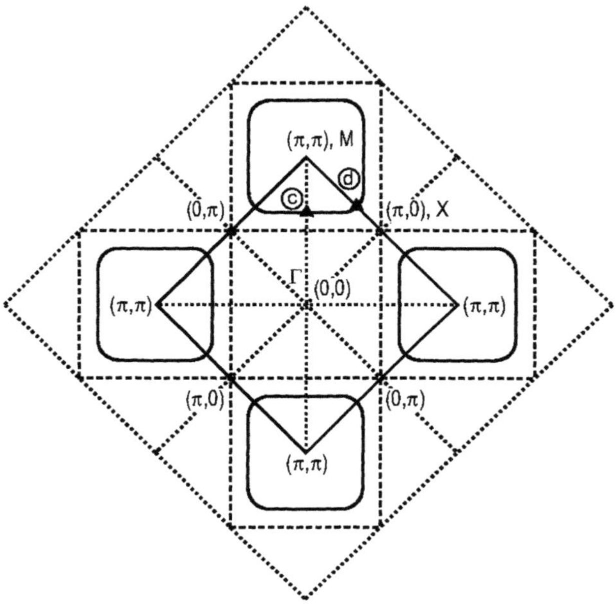

Fig. 5.81. Brillouin zone and Fermi surface (full lines around π, π for the electronic structure given in Fig. 5.79. The points (c) and (d) from Fig. 5.79 are also indicated [5.135]

now sees that basically, at least as far as the general features are concerned, and not unexpectedly, the data fall on top of each other and that the general form of the Fermi surface seems to be one with holes where the orbits are centered around M in the basic CuO_2 structure, around X and Y in the BiCaSrCuO structure, or around (π, π) in the most general notation.

The band structure near the Fermi energy and the Fermi surface of the high-temperature superconductors is still under active investigation. There are still many open questions in that field and therefore here only some important references are given [5.179–5.185]

5.9.3 The Superconducting Gap

Finally, we address the question of whether the modifications in the electronic states, as predicted by the *BCS theory* can be observed by PE spectroscopy.[6]

[6] Here the term BCS density of states is frequently used in the literature. In the context of the present chapter we point out that this density of states is actually an excitation spectrum of a correlated electron system (Bose condensate).

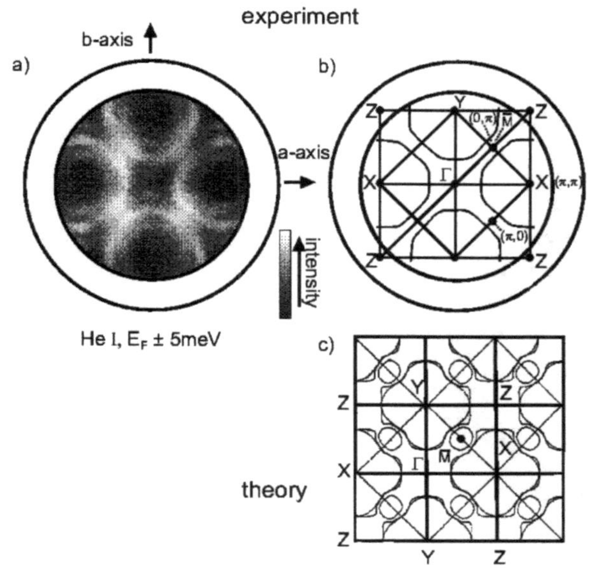

Fig. 5.82. Measurement of the Fermi surface of a $Bi_2Sr_2CaCu_2O_8$ superconductor and its comparison with the results of a local-density calculation. (**a**) Experimental data. (**b**) Schematic drawing of the experimental data. (**c**) Theory [5.136]

Note that these modifications are of the order of the transition temperature ($T_c \simeq 100\,K \simeq 10\,meV$), and therefore can only be observed in high-resolution experiments. In addition, since they take place most noticeably at the Fermi energy, they can only be detected in a narrow energy interval around that energy. Thus, they require a good Fermi step in the experimental data together with high experimental resolution.

The first data on this phenomenon [5.137] are depicted in Fig. 5.84 (other groups have meanwhile obtained similar results). The middle panel gives the change of the excitation spectra produced by the superconducting pairing energy of the electrons. In short, this gives rise to the opening of an energy gap of magnitude $2\Delta_{BCS}$ with singularities in the excitation spectra above and below E_F, respectively. This has been nicely observed in high-resolution PE experiments, as the data in the top panel reveal [5.137]. They clearly indicate the enhancement of the intensity below the Fermi energy in the superconducting state relative to the non-superconducting state. The lower panel gives a best fit which, by the way, results in a ratio of $2\Delta_{BCS}/k_BT_c = 8$. The corresponding BCS value is 3.5, indicating that one is here in the strong coupling limit.

We may conclude by stating that PES has demonstrated that the CuO based superconducting compounds discovered so far, are so-called *charge-*

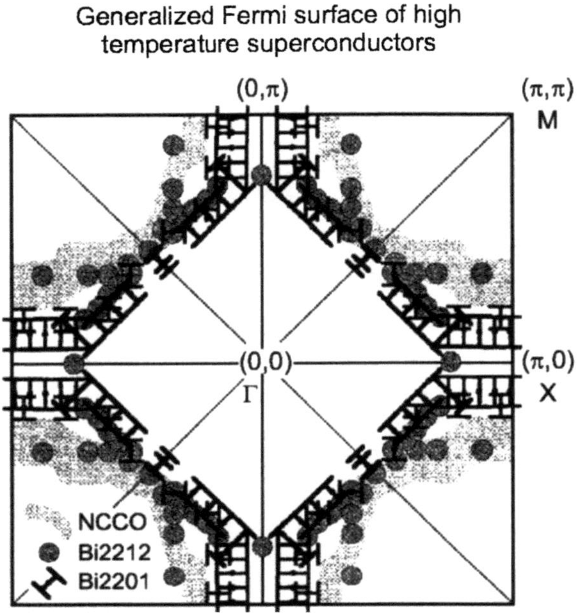

Fig. 5.83. Summary of the Fermi surface data of three different types of high-temperature superconductors in a normalized Brillouin zone, showing the close correspondence of the data from the various sources as expected from the fact, that the main electronic properties of the system come from the CuO_2 planes [5.135]. NCCO = $Nd_{2-x}Ce_xCuO_4$, Bi2212 =$Bi_2Sr_2CaCu_2O_8$, Bi2201 =$Bi_2Sr_2CuO_6$

transfer compounds, which have strong electron–electron correlations. The charge carriers in these systems are either holes or electrons in the two-dimensional copper-oxygen antibonding bands. While in the hole-doped systems holes seem to be necessary for conductivity and superconductivity to occur, there are also compounds (like, e.g., $Y_{1-x}Zn_xBa_2Cu_3O_7$ for $x \approx 0.1$) which do have the holes and which are neither superconducting nor conducting. Employing high resolution and low temperatures one has been able to detect the modification of the excitation spectra, as predicted by the BCS theory, although a higher value for the ratio of the gap and the superconducting temperature than predicted by BCS has been observed.

Since the first convincing measurement of the gap in a high temperature superconductor, many additional experiments have been performed, showing, e.g., the anisotropy of the gap. This will be discussed in the next section.

5.9.4 Symmetry of the Order Parameter in the High-Temperature Superconductors

There has been a long discussion about whether the order parameter in the high-temperature superconductors is isotropic, which would mean that it has

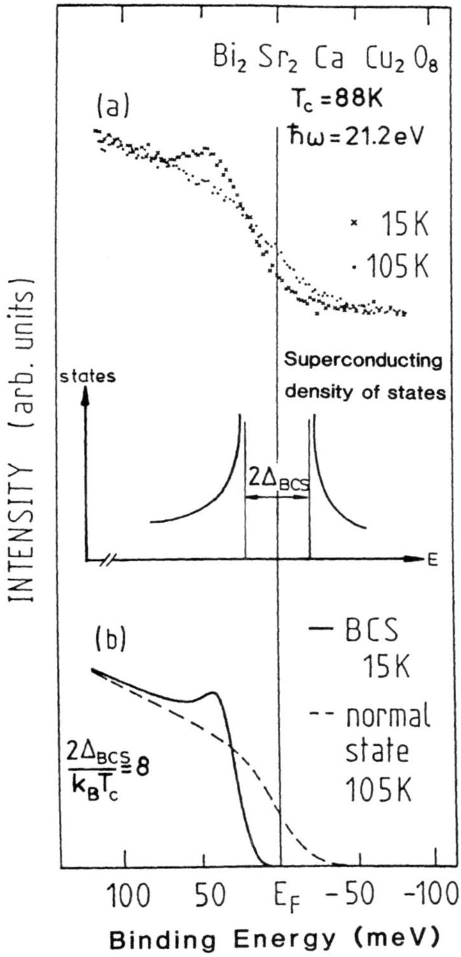

Fig. 5.84. Measurement of the superconducting density of states by PES. (**a**) PES spectra of $Bi_2Sr_2CaCu_2O_8$ ($T_c = 88\,K$) below T_c ($T = 15\,K$) and above T_c ($T = 105\,K$). (**b**) Simulation of the experimental spectra from the superconducting density of states, which has a divergence in the occupied (at $-\Delta_{BCS}$) and the empty (at $+\Delta_{BCS}$) density of states. The normal state spectrum is sloping and smeared out at E_F because of the instrumental broadening. If this normal state spectrum is convoluted with the superconducting density of states the full line in part (**b**) is obtained which resembles closely the experimental spectrum taken at 15 K [5.137]

s-type symmetry, or, as suggested by the bonding property in the CuO_2 plane, whether it is of the $d_{x^2-y^2}$ type. A photoemission experiment has at least indicated that it is of the latter type. Data pertaining to this effect are shown in Fig. 5.85, which compares photoemission spectra above and below the superconducting transition temperature at two points of the Brillouin zone, as depicted in the inset. The two sets of data correspond to point A in the Brillouin zone, which is situated along the Cu-O bond direction and to point B, which is on a direction at 45° to the bond direction. It is apparent from the data that the spectra taken at points A and B in the Brillouin zone, change in a different way with temperature. There is no gap opening at point B, but a large gap opening at point A. If one plots the size of the gap, as given by the difference in spectral height at the Fermi energy between the high and the

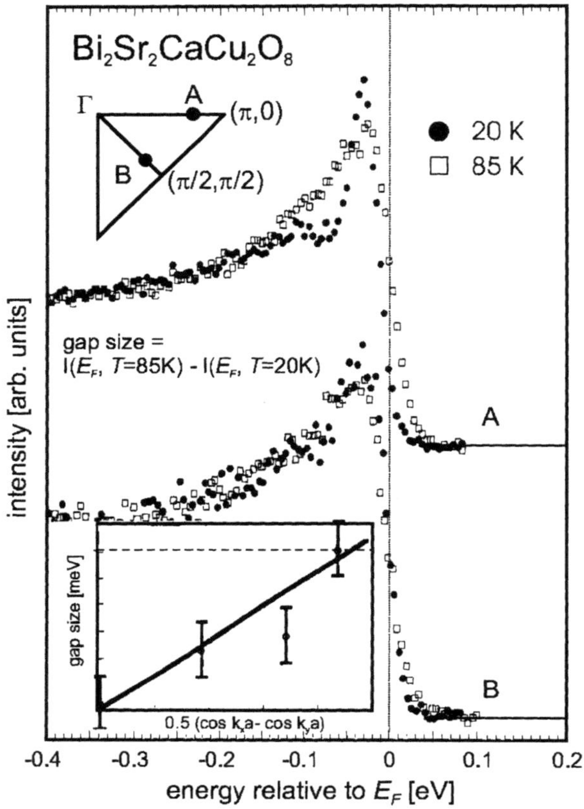

Fig. 5.85. Measurement of the anisotropy of the band gap of a high-temperature superconductor. The gap has been measured at two points in the Brillouin zone as indicated in the insert and plotted as a function of 0.5 $(\cos k_x a - \cos k_y a)$, where the straight line indicates that the gap of the $d_{x^2-y^2}$ type [5.135]

low temperature, as a function of 0.5 $(\cos k_x a - \cos k_y a)$, one gets the straight line shown in the inset, where the two extreme points are given by the data at points A and B. These data are consistent with the assumption that the order parameter is of the $d_{x^2-y^2}$ type in the high-temperature superconductors. One should note, however, that a very anisotropic s-type order parameter could in principle also show a similar behavior, so the final proof that one is indeed observing a $d_{x^2-y^2}$ order parameter in the high-temperature superconductors has to come from other experiments. So far, these all support the finding of the elegant PES experiment by Shen's group [5.135].

5.9.5 Core-Level Shifts

Intimately connected with the question of the charge carriers in the superconducting compound is that of the so-called *valency of the copper ions*. As a matter of fact one may argue that the question of the holes in these compounds and the question of the valency of the copper ions are identical and represent only different views of the same problem [5.82].

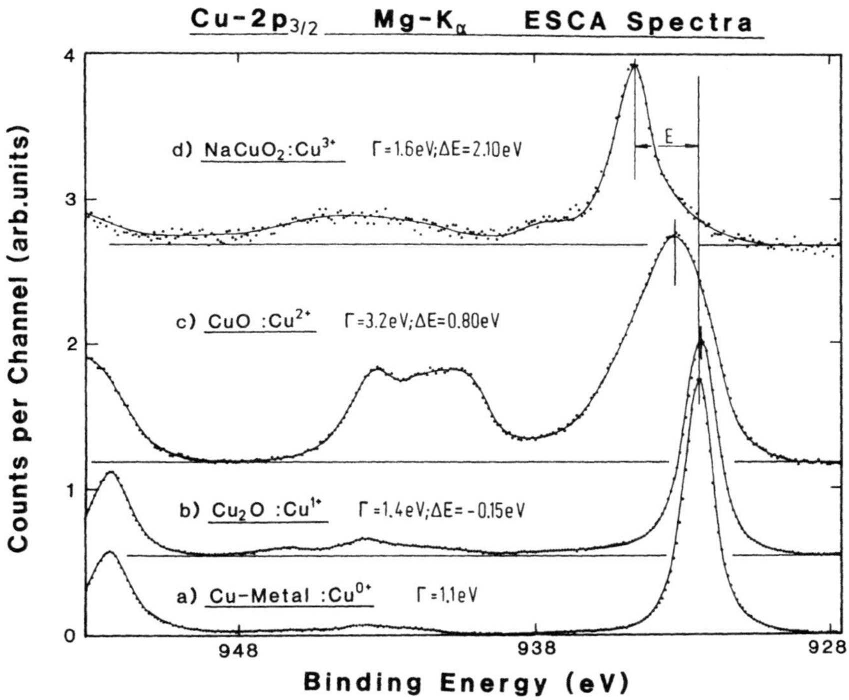

Fig. 5.86. Cu $2p_{3/2}$ spectra taken with Mg-K_α radiation for Cu metal (Cu^{0+}), Cu_2O (Cu^{1+}), CuO (Cu^{2+}) and $NaCuO_2$ (nominally Cu^{3+}). The chemical shifts are given with respect to the energy in Cu metal

In PES some information about the valency can be inferred from the core levels (Sect. 2.1). To that end Fig. 5.86 shows some relevant Cu 2p-core levels, namely those of Cu metal (Cu^{0+}), $Cu_2O(Cu^{1+})$, $CuO(Cu^{2+})$, and $NaCuO_2(Cu^{3+}$; it has and will be shown that in $NaCuO_2$ the Cu ion has the electronic configuration d^9L^{-1}. The frequently used term Cu^{3+} is therefore quite unfortunate). We see that the core-level shifts (ΔE measured with respect to the metal) are small but noticeable; however, detailed analysis has to be performed if $Cu^{2+}(3d^9)$ is to be distinguished from $Cu^{3+}(3d^9L^{-1})$. Such an analysis is given in Fig. 5.87 for the $2p_{3/2}$ level in $Y_1Ba_2Cu_3O_{7-x}$.

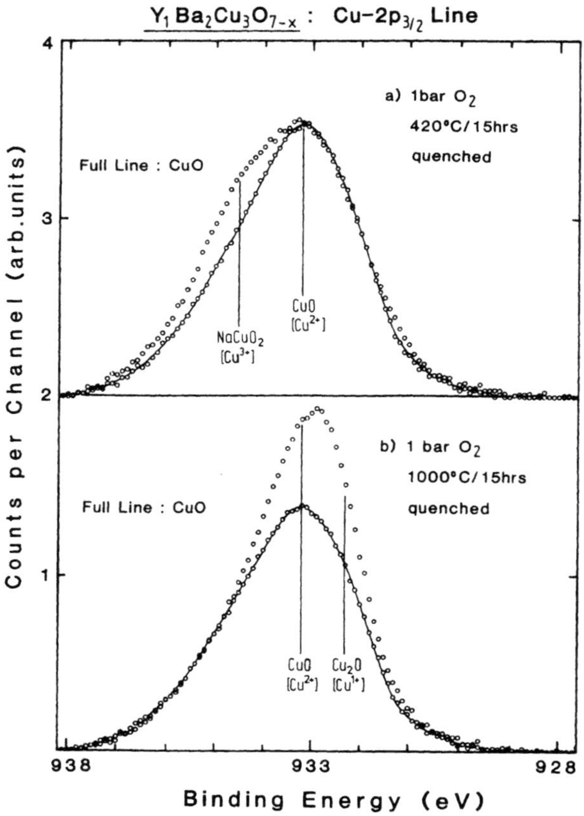

Fig. 5.87. Cu $2p_{3/2}$ spectra of two differently annealed samples of $YBa_2Cu_3O_{7-x}$. The sample for spectrum (**a**) was treated in 1 bar O_2 at 420° C for 15 h and subsequently quenched to room temperature. This procedure should produce a superconducting sample $x \simeq 0$. The Cu $2p_{3/2}$ line of this sample, if compared to that obtained for CuO, gives an extra signal at a position where it is expected from the position in $NaCuO_2$ (nominally Cu^{3+}, which, however, should rather be regarded as $Cu^{2+}O^{-}$). The sample (**b**) was treated in 1 bar O_2 at 1000° C for 15 h. This procedure should produce an oxygen content of 6.2. The comparison of the line with that obtained for CuO shows that additional weight is present near to the position expected for Cu^{+} [5.82]

It is well known that the oxygen content in $YBa_2Cu_3O_{7-x}$ can be controlled by the annealing conditions of the ceramic samples. In short, a high-temperature anneal at 900° C and 1 bar O_2 tends to reduce the oxygen content in the sample, leading to samples with $x > 0.5$ and these samples are known to be nonmetallic and non-superconducting. If a formal valence analysis of spectra of such samples is made, as in Fig. 5.87b, they should consist of divalent and monovalent copper ions. A superposition of lines from the calibration compounds as given in the preceding figure allows a very good fit of the actual observed line, indicating that the annealing procedure has indeed produced Cu^{1+}. Conversely, annealing at lower temperatures, namely 420° C leads to a high oxygen content $x \simeq 0$ in the sample, which then becomes conducting and superconducting. A formal valence count for this sample leads, as we have already indicated, to a certain amount of Cu^{3+}. Indeed intensity in the Cu 2p line is observed at the energy where the line occurs in the nominally pure trivalent compound $NaCuO_2$. One is also able to fit this line as a superposition of the divalent and trivalent compound, and thus, in principle, able to extract the amount of trivalent copper in the superconducting ceramic sample (Fig. 5.87a).

We emphasize again that here the term trivalent, although frequently used, is slightly misleading. In a simplistic view, one may regard Cu^{3+} as almost identical to Ni^{2+} as far as the valence electron shell is concerned. This would mean that Cu^{3+} in the superconducting compounds has a ground-state electronic configuration of $3d^8$. This, however, cannot be the case as, e.g., the data in Fig. 5.69 clearly demonstrated. They show that the $3d^8$ configuration is 12 eV above the ground state and therefore cannot be populated at room temperature. The ground state configuration of Cu^{3+} must therefore be $3d^9L^{-1}$, which, as is also shown in Fig. 5.69, is the lowest energy state in the PE of CuO. This corresponds to the creation of a hole in a $Cu^{2+}L$ initial-state configuration. This can be viewed as the production of Cu^{3+}, but instead of leading to a $3d^8$ configuration it leads to a $3d^9L^{-1}$ configuration, which represents the holes at the Fermi energy.

5.10 The Fermi Liquid and the Luttinger Liquid

Metals contain itinerant electrons, whose number is of the order of 10^{23} in a typical sample. It is obvious that these electrons, averaged over time, can be quite near to each other and therefore interact with each other. The description of 10^{23} interacting electrons is a demanding problem and therefore has not been solved. What is fairly well understood is the so-called free-electron problem, in which one neglects the interaction between the electrons and treats the system of electrons in a metal as a gas of non-interacting particles. One might think that this is a very crude approximation, which indeed it is in some cases, but it works surprisingly well in a great many other cases. The reason is revealed by a "construction" by Landau, which these days is called a "Fermi liquid". A Fermi liquid is a system of interacting electrons, whose properties can be mapped onto those of a system of non-interacting electrons by one-to-one-transformation. This "renormalizes" many properties of the system of non-interacting particles, for example the mass, and therefore the particles in the system of interacting electrons have been given a special name and are called quasi-particles.

The signature of a system of non-interacting particles in photoemission spectroscopy, if the spectrum is plotted versus the momentum, is a succession of δ-functions from the bottom of the conduction band up to the Fermi energy. Because the electrons have no interactions with other electrons or with the lattice, they have an infinite lifetime (see Fig. 5.88, $E^0_{\boldsymbol{k}}$).

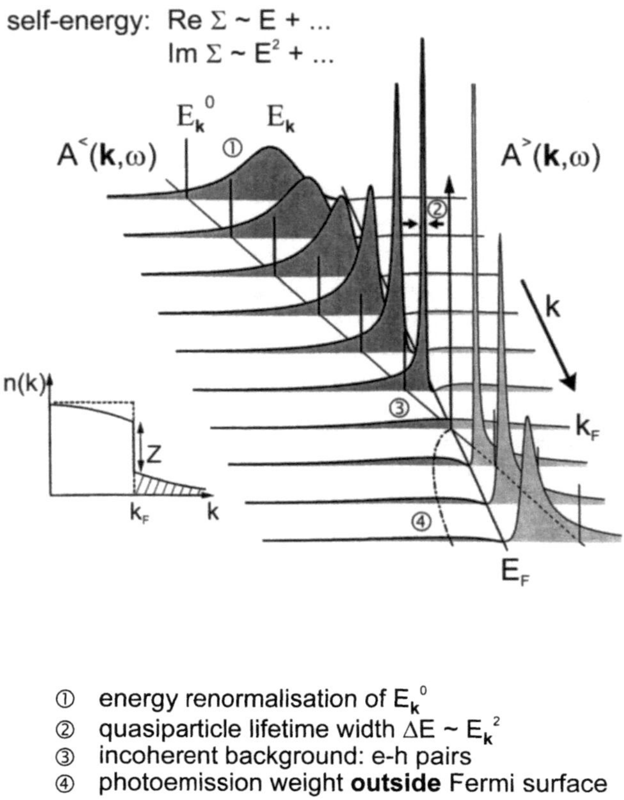

Fig. 5.88. Principle of the photoemission from a Fermi liquid near the Fermi energy. With respect to the non-interacting case there are four distinct differences: (1) an energy renormalization, (2) a finite quasi-particle lifetime, (3) an incoherent background created by electron–hole pairs, (4) photoemission weight outside the Fermi surface [5.140]

The spectral properties, as measured by photoemission spectroscopy, of a system of interacting electrons (Fermi liquid) deviates in four respects from that of an system of non-interacting electrons as shown in Fig. 5.88, and one can spell out the following distinct signatures of a Fermi liquid (see Fig. 5.88:

(1) The energy of the interacting system ($E(\boldsymbol{k})$) is renormalized relative to that of the non-interacting system ($E^0(\boldsymbol{k})$), where the renormalization is given by $\mathrm{Re}\Sigma = aE$, the so-called self-energy.

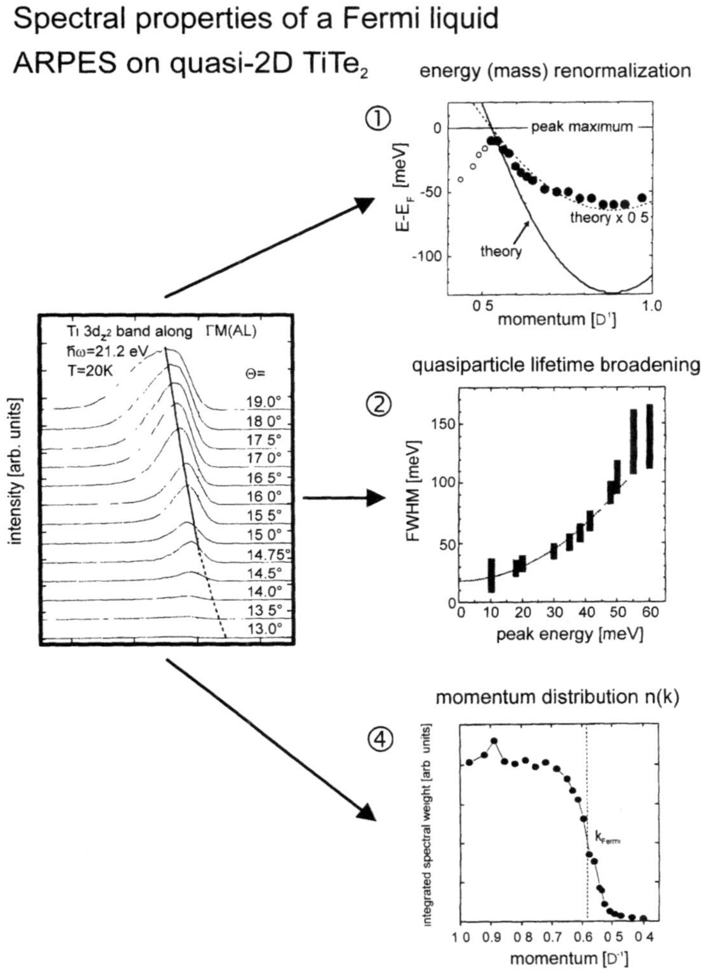

Fig. 5.89. Experimental results on the valence-band structure of $TiTe_2$ indicating the validity of the prediction for Fermi liquid photoemission as given in Fig. 5.88. The data show (1) an energy renormalization, (2) a finite quasi particle lifetime, (4) momentum outside the Fermi wave vector; the incoherent background (3) is barely visible in the spectra (R. Claessen, private communication)

(2) There is a finite quasi-particle lifetime, because of the electron–electron interaction and this lifetime is given by $\mathrm{Im}\Sigma = bE^2$, which is the imaginary part of the self energy. The total self energy is: $\Sigma = aE + \mathrm{i}bE^2$, where E is measured with respect to the Fermi energy.

(3) There is an incoherent background produced by electron–hole pairs.

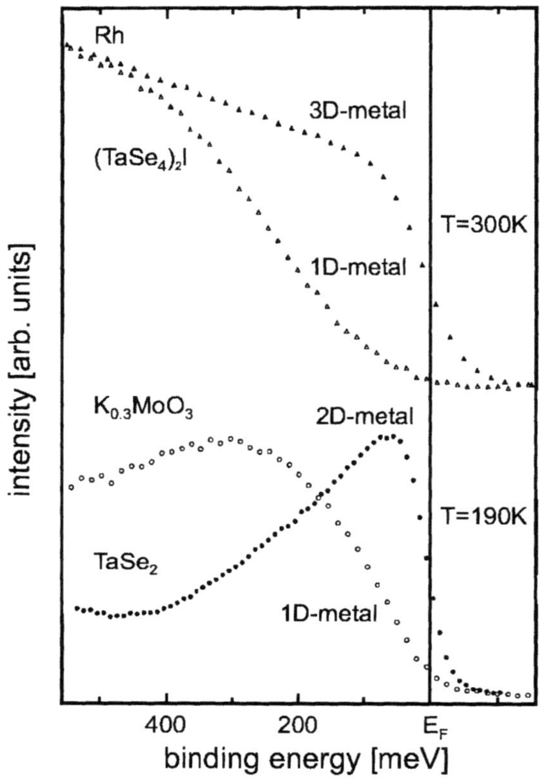

Fig. 5.90. Comparison of the Fermi edge region for a three-dimensional metal (Rh), a two-dimensional metal ($TaSe_2$) and two one-dimensional metals ($(TaSe_4)_2I$ and $K_{0.3}MoO_3$), where the latter two in contrast to the former show no Fermi step indicating Luttinger type behavior [5.141]

(4) There is photoemission weight outside the Fermi surface, which means that for $T = 0$ the height of the Fermi step Z is reduced to $Z < 1$ instead of $Z = 1$.

The derivative of the momentum distribution function $\mathrm{d}n/\mathrm{d}k$ always has a singularity at the Fermi wave vector. As a matter of fact the occurrence of this step at the Fermi wave vector and/or the form of the self-energy $\mathrm{Im}\Sigma = bE^2$ are often used to define a Fermi liquid. The predictions for the Fermi liquid are reasonably well observed for the typical two-dimensional metal $TiTe_2$, as can be seen by the data in Fig. 5.89.[7] Here, one has to use a two-dimensional rather than a three-dimensional metal in order to find the signature of a

[7] High-resolution experiments on $TiTe_2$ have revealed structures in the data of Fig. 5.89 which are not yet understood (Saarbrücken, Stanford, Berkeley collaboration, 2000). A recent paper on $TiTe_2$ [5.186] confirms the findings presented in Fig. 5.89 with very accurate data and a detailed analysis.

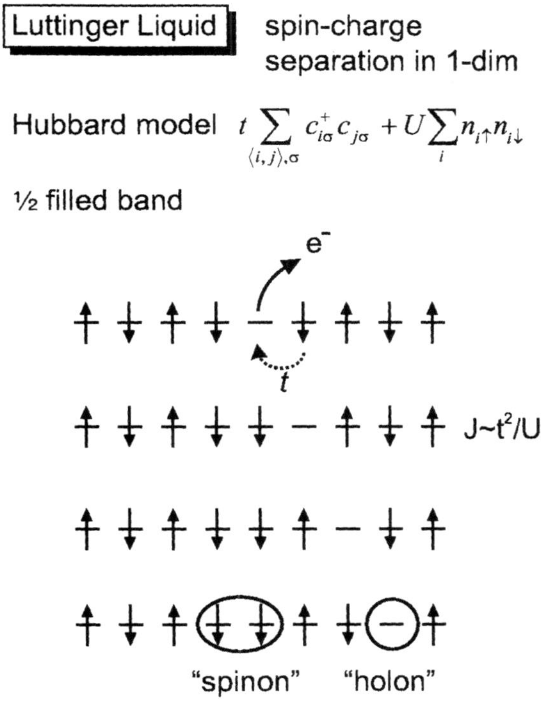

Fig. 5.91. Schematic diagram indicating the creation of a spinon and a holon for a one-dimensional antiferromagnetically coupled system (Luttinger liquid); (R. Claessen, private communication)

Fermi liquid, because kinematics tell us that the width of a line is given by (see Chap. 1):

$$\Gamma_{\text{exp}}(\Theta = 0) = \Gamma_h + \Gamma_e \left| \frac{v_{h\perp}}{v_{e\perp}} \right|$$

and this allows one to measure only the hole lifetime function, which is given by

$$\Gamma_{\text{exp}} = \Gamma_h = 2\text{Im}\Sigma$$

if the velocity of the electrons normal to the surface ($v_{h\perp}$) is zero, which is only the case in two-dimensional and one-dimensional systems.

One may now ask what happens if Z approaches zero. This is the case of the so-called Luttinger liquid [5.138, 5.139]. A Luttinger liquid is a one-dimensional metal with electron–electron interactions. If the spectral properties of such a system are calculated, one finds that $Z = 0$, and this means that there is no singularity in the derivative of the spectral function near the Fermi energy, or, put in another way, instead of a Fermi step at the Fermi

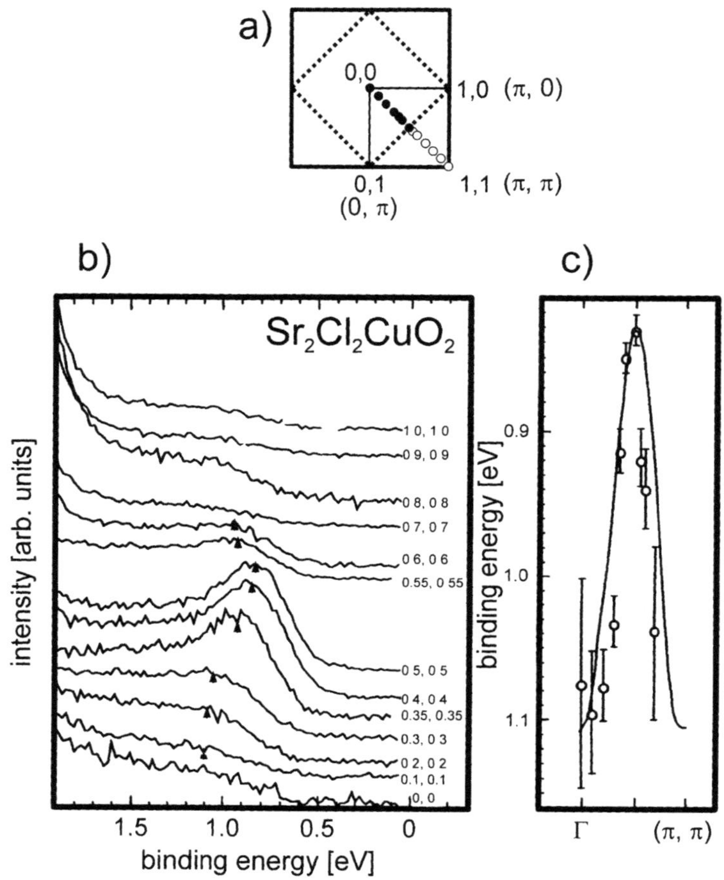

Fig. 5.92. Spin–charge separation as obtained for the insulating compound $Sr_2CuO_2Cl_2$. (**a**) Brillouin zone and path of the measurement. (**b**) Data obtained along the path indicated in (**a**), k values in units of (π, π). (**c**) Holon dispersion taken from the data in (**b**) together with the results of a calculation [5.142]

energy one observes a power-law dependence in the photoemission spectrum. This prediction is verified by photoemission experiments such as that shown in Fig. 5.90. Here the spectral function near the Fermi energy for a three-dimensional metal (Rh) and a two-dimensional metal ($TaSe_2$) are shown and they both exhibit the typical step function expected near the Fermi energy, broadened here by instrumental and temperature effects. If one performs the same kind of experiment for a one-dimensional system ($(TaSe_4)_2I$ and $K_{0.3}MoO_3$) one finds no hint of a step function but instead a power-law dependence is observed, as predicted for a Luttinger liquid [5.140, 5.141].

This statement, however, must be regarded with caution. One-dimensional metals show a Peierls transition, making them insulators below the transition temperature. Above the transition temperature they do not enter a "simple" one-dimensional metallic state, but one with large fluctuations (which, as the temperature is lowered, precede the phase transition into the insulating state). This fluctuating state has a spectral function similar to that of a Luttinger liquid and therefore one must have some reservations when attributing data such as those in Fig. 5.90 to Luttinger type behavior.

We end this section with some additional remarks on the quasi-particle picture in the Fermi liquid. So far in the definition of the Fermi liquid only the electron–electron interaction has been taken into account. In a solid, however, there are also phonons which interact with the electrons. If the electron–phonon interaction is weak, phonons do not influence the electron system. This changes in the case of strong electron–phonon interaction. An example, namely the surface state of Be, was shown in Fig. 1.41. Far away from the Fermi energy, the surface state behaves like a quasi-particle. If, however, the surface state energy is of the order of the phonon energy ($\hbar\omega_{\text{max}} = 70\,\text{meV}$, maximum phonon energy), there is a strong coupling between the electron and phonon system and this leads to two peaks in the excitation spectrum, reflecting mixed electron–phonon states for which the Fermi liquid picture no longer applies. Therefore these excitations can no longer be described as quasi-particles. Only directly at the Fermi energy is the quasi-particle picture recovered for this system.

Finally it is perhaps useful to point out that the expression for the quasi-particle self-energy $\Sigma = aE + \text{i}bE^2$ reflects the first two terms of an expansion. In this sense the quasi-particle picture is valid only for $E = E_{\text{F}}$, and strictly speaking this also holds for the concept of the Fermi liquid.

Another prediction for the spectral response of a Luttinger liquid is a so-called spin–charge separation, which can be visualized by a simple model as given in Fig. 5.91. Here it is assumed that we have a chain of antiferromagnetically coupled ions in one dimension. If at time t_0 an electron is ejected by photoemission, subsequent hopping induces a disturbance separated with respect to the charge and the spin. These separate charge and spin excitations are called holons and spinons, respectively. So far the experimental evidence for these excitations is not overwhelming and the best documented example is for an insulating one-dimensional system, namely $Sr_2Cl_2CuO_2$. Figure 5.92a shows the Brillouin zone and the path along which the data displayed in Fig. 5.92b have been taken. A clear dispersion in the spectrum is observed and these data are displayed in Fig. 5.92c where the full curve is the calculation using the t-J model with a value for the antiferromagnetic coupling constant taken from other data. This at least indicates that the model describes the experiment reasonably well [5.142].

5.11 Adsorbed Molecules

5.11.1 Outline

Since PES is a surface sensitive technique it is especially suited for the investigation of molecules adsorbed on solid surfaces [5.143–5.145]. A one-monolayer coverage of, say CO on Ni, gives an easily detectable signal from the CO molecular orbitals and thus allows one to study the molecule–substrate interaction. This is a very interesting subject for two reasons. Firstly, ordered overlayers on single crystal surfaces can be very good approximations of two-dimensional systems and it is thus highly desirable to study their electronic structure. Secondly, and more importantly, the study of the surface chemical bond is a fascinating area of research in itself and provides in addition a starting point for many highly controversial subjects such as catalysis or corrosion.

Problems in this area stem from three sources. Firstly the analysis of a PES spectrum of an adsorbed species is not always trivial, even if the corresponding spectrum of the free molecule is well documented [5.2] and understood. Secondly, the molecular orbital energy shifts that occur between the two chemical states, the free molecule and the surface adsorbed molecule, are not easy to interpret in terms of chemical bonding information. And finally there is a problem of magnitudes [5.143]. The heat of adsorption of CO on Ni is 1.3 eV (30 kcal/mole) and that of CO on Cu is 0.6 eV (14 kcal/mole), which is also the difference in heats of adsorption of a normally chemisorbed molecule and a weakly chemisorbed one. This difference of 0.7 eV is, however, only slightly more than the error limits of PE data taken by different groups and analyzed by different procedures. Thus two chemically quite different states do not show very distinctive differences in PES, a feature which will become more apparent when looking at actual data.

Rather than accumulating many examples, it seems more appropriate to discuss a typical case in order to demonstrate the potential of the method.

5.11.2 CO on Metal Surfaces

The adsorption of H, O, CO, NO and C_6H_6 on metal substrates [5.143] has been studied by PES in some detail. Since the information on CO seems the most abundant, it will be used as an example here. The XPS spectrum of gaseous CO together with a schematic energy level diagram, is shown in Fig. 5.93 [5.146, 5.147]. As far as the valence orbitals are concerned, CO is isoelectronic with N_2 (see Fig. 5.6 for a spectrum of N_2) and the valence orbital spectra of the two species must be similar (the number of electrons in CO and N_2 is identical, however the symmetry of the two molecules is different) as are their UPS spectra shown in Fig. 5.94 [5.2]. The spectrum of CO consists at the highest binding energies of the O 1s level (543 eV) and the C 1s level (296 eV), which are both essentially atomic-like because they

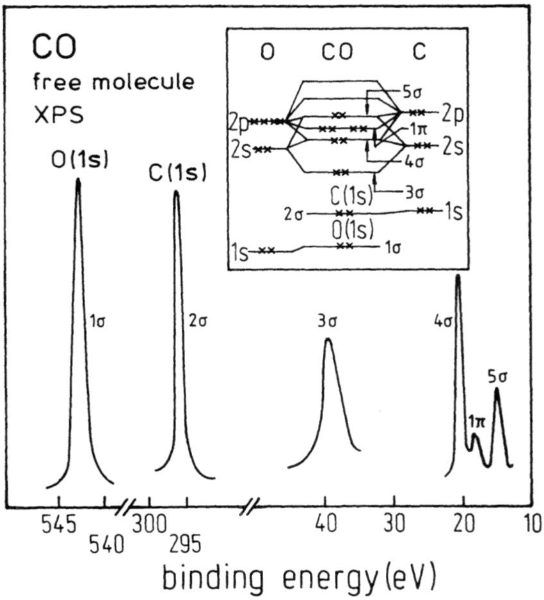

Fig. 5.93. XPS spectrum of gaseous CO [5.146, 5.147] and a molecular orbital energy level diagram

are very far below the valence orbitals. The orbitals at 40 eV and smaller binding energies are the $C(2s^22p^2)$ and the $O(2s^22p^4)$ orbitals which form the 3σ, 4σ, 1π, 5σ molecular orbitals, all of which are detected in the XPS spectrum. Note that the high-resolution UPS spectrum in Fig. 5.94 also gives the vibrational structure, which cannot be resolved in the XPS data. The strong vibrational structure of the second band shows that it corresponds to the bonding π-orbital. The weak vibrational structure of the two other bands indicates that they correspond to the non-bonding 4σ and 5σ orbitals.

Figure 5.95 gives a schematic picture of the outermost molecular orbitals of the CO molecule [5.145] and indicates how it binds to a metal surface such as Ni. The energetically highest occupied orbital and thus also the spatially most extended one is the 5σ orbital which is located mainly on the C atom. Thus it is intuitively understandable that the CO molecule tries to attach its "softest" part to the metal surface in the surface chemical bond and the initial step in the binding is therefore a 5σ to metal charge transfer. This charge transfer is compensated (partlv) bv back donation of charge from the metal surface onto the CO molecule [5.148]. This back donated charge has to go into the lowest unoccupied orbitals of the CO molecules. It is interesting to note that these orbitals are $\sim 2\,\text{eV}$ [5.149, 5.150] above the vacuum level E_v in the free CO molecule. Back donation will certainly lower them and thus one expects to find them below E_v in the chemisorbed case.

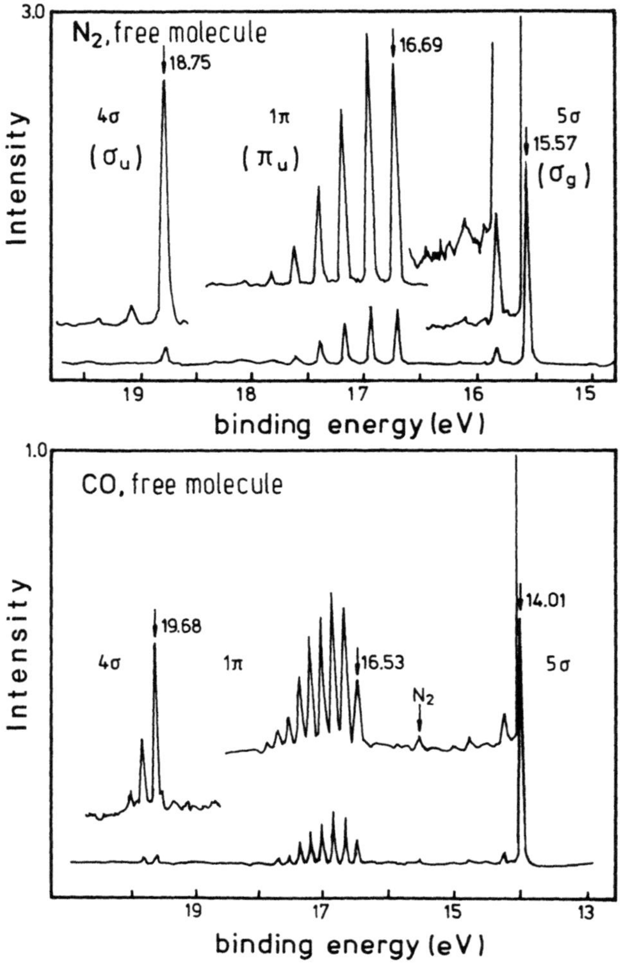

Fig. 5.94. Comparison of the UPS spectra ($\hbar\omega = 21.2\,\text{eV}$) of the isoelectronic free molecules N_2 and CO [5.2]. These gas spectra also display the vibrational structures, which cannot be resolved in the XPS spectrum (Fig. 5.93). The arrows indicate the adiabatic ionization potentials (Sect. 5.2)

At this point, before discussing actual data, we have to address an essential point of relevance for the analysis of the spectra. What we are interested in here is the electronic structure of the chemisorptive (or physisorptive) bond. In other words, one wants to measure the electronic states of "one" "isolated" molecule bound to a surface. Such a system does, of course, not exist and in order to get a reasonable signal, one has to deal with molecule coverages on a surface of the order of one monolayer. This, however, raises the question of the relative magnitude of the adsorbate-substrate interac-

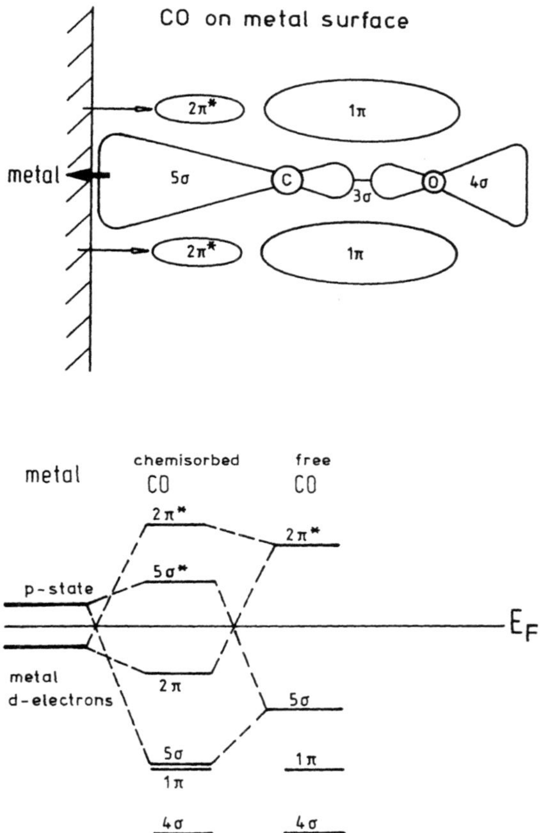

Fig. 5.95. Bonding of a CO molecule to a metal surface. Top part: geometry; bonding (thick arrow) is provided by the most extended orbital of the molecule which is the 5σ orbital. Back donation from the metal into the formerly empty $2\pi^*$ orbitals is indicated [5.142] by thin arrows. Bottom part: schematic energy level diagram. The two outer diagrams give the metal and CO molecule energies respectively. The middle diagram indicates the energies of the adsorbed CO molecule

tion with respect to the adsorbate-adsorbate interaction. We shall see that in chemisorbed systems at a coverage of one monolayer, the adsorbate-adsorbate interaction can be large (1 eV and more) leading to a two-dimensional band structure of the electronic states of the adsorbate system (Chap. 7). To interpret the adsorbate–substrate interaction quite detailed and complicated data analyses are necessary. In particular it has to be noted that the structure of an ordered overlayer on a substrate depends on the crystal face. This means that the adsorbate–adsorbate interactions also depend on the crystal face.

The situation is shown schematically in Fig. 5.96, which is modelled for one monolayer of CO on Ni(110) [5.151]. Column (a) depicts, on a common energy scale, the higher orbitals of a Ni atom (d, p, s orbitals) and those of a (free) CO molecule (4σ, 1π, 5σ, $2\pi^*$). If the Ni atom and the CO molecule are brought into contact (column (b)) an interaction takes place as indicated schematically in Fig. 5.95. The 5σ orbital of the CO molecule interacts with the p-orbital of the Ni atom, whereby its energy is changed and the $2\pi^*$ orbital of the CO molecule interacts with the d/s orbitals of the Ni atom, also leading to energy changes. This situation is also representative for the interaction of one CO molecule with a Ni surface because, in essence, in the bonding process the CO molecule is attached to one Ni atom of the surface. In the metal, of course, the Ni energy levels are bands as indicated in column (a) and (b) by the widths of the states that represent the metal. This is the situation which is relevant to our problem.

Since, however, one usually has a considerable coverage of molecules on the substrate, we indicate the modifications which are introduced by the adsorbate–adsorbate interaction. In a first step one can ignore the adsorbate–substrate interaction altogether and inspect only the energy levels of a two-

One Electron Scheme of an Adsorbate

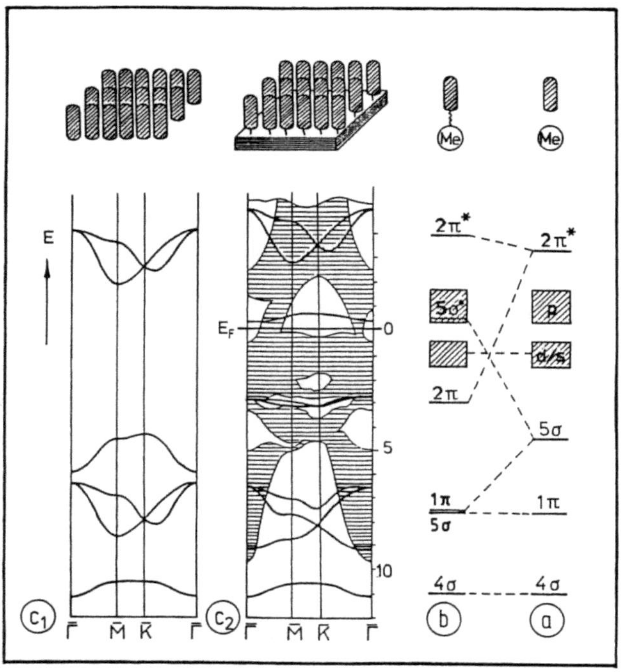

Fig. 5.96. Schematic energy level diagram for the various steps of the adsorbate (CO) substrate (Ni) interaction [5.151]. (a) Energy levels of the free molecule and Ni metal (non-interacting); (b) Interaction of a CO molecule with an Ni metal surface; (c_1) Band structures of a free unsupported layer of CO molecules; (c_2) Energy levels of an ordered overlayer of CO molecules on an Ni metal surface. Shaded area: two-dimensional projected Ni band structure

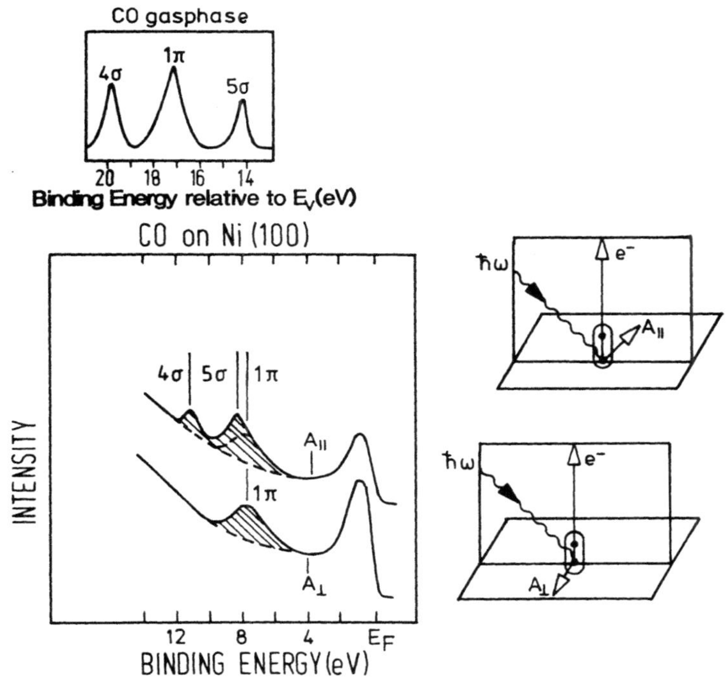

Fig. 5.97. UPS ($\hbar\omega = 21.2\,\mathrm{eV}$) valence-band spectra of CO on Ni [5.143, 5.155] with two different polarizations of the light. The geometry of the experiment is indicated, and also a low resolution spectrum of CO in the gas phase (relative to E_v). The 5σ level does not shift in parallel with the 4σ and 1π level in going from the gas phase to the adsorbed phase. Since the 5σ orbital, which is responsible for the surface bond, undergoes a strong relaxation shift, the initial state shift is hard to estimate from the data

dimensional adsorbate layer: this is (c_1). Here one sees that the energy levels of the free molecule layer have attained considerable dispersion.

Finally we move to the most realistic situation, i.e., that of an (ordered) overlayer of molecules bound to a substrate surface, as indicated in column (c_2). Here the hatched areas are the projection of the bulk band structure of Ni metal on the (110) surface Brillouin zone. Also shown are the adsorbate-induced bands. In the case of the $2\pi^*$- and 4σ-derived bands, only a rigid shift with respect to their position in the free unsupported layer of molecules is observed. However, for the 5σ and 1π energy levels, which are brought close together by the substrate–adsorbate interaction, the bands in the adsorbed overlayer differ from those in the free layer due to strong hybridization.

The actual situation in an adsorbate–substrate system can thus be quite complicated because of competition between adsorbate-substrate and adsorbate–adsorbate interactions.

For simplicity, we shall treat the system of CO on metal surfaces only within the approximation outlined in column (b) of Fig. 5.96, keeping in mind that this will omit a number of details. Later, in Chap. 7, the band-structure aspects of the typical system CO on Ni(110), will be presented.

Spectra for CO on Ni [5.152,5.153] are shown in Fig. 5.97. They were taken with electron detection normal to the surface, and with two different light polarizations, parallel ($A_{||}$) to the plane containing the surface normal and the incident light beam and perpendicular ($A_{\perp}$) to that plane. In the former geometry ($A_{||}$) the vector potential $\boldsymbol{A}$ has a component parallel to the axis of the molecule and one perpendicular to it if we assume that the molecule is oriented normal to the surface, while in the latter geometry ($A_{\perp}$) the light has only a component perpendicular to the axis of the molecule. The spectrum consists essentially of two parts, one near the Fermi energy, which represents the Ni d band, and another consisting of one or two peaks, depending on the polarization, which represent the signal from the adsorbed CO. It is at first sight worrying that this latter part has only two signals because, from a simple-minded comparison with the gas phase spectrum (also shown in Fig. 5.97) one would expect three.

Before addressing this question, we deal with the observed polarization dependence [5.143,5.152–5.156]. The electron detection is for normal emission (along the axis of the CO molecule). Thus the photoemitted electron is in a totally symmetric state with respect to the symmetry operations of the surface normal. Therefore, as will be discussed in Chap. 6, the experiment can only detect electrons in a symmetric ($|+\rangle$) final state (σ symmetry). In the bottom curve of Fig. 5.97 the light is polarized perpendicular to the molecular axis (and to the plane of detection) and has therefore odd mirror symmetry ($|-\rangle$). Therefore the initial state $|i\rangle$ must also have odd symmetry $|-\rangle$ in order to give a non-zero transition matrix element $\langle +| - |-\rangle$. The peak that is still observable under these conditions ($A_{\perp}$) must correspond to photoemission out of a π state. In the other geometry one has components of $\boldsymbol{A}$ both parallel and perpendicular to the molecular axis and thus both σ and π initial states can be seen in the top curve.

The next question concerns the exact assignment of the two observed peaks. An analysis of the shape of the 8 eV peak in the top spectrum of Fig. 5.97 indicates that it contains a further peak and it seems reasonable to assume that the peak at $\approx +8$ eV binding energy contains the 1π and the 5σ state. One must then ask why the 5σ state is so much shifted from its gas phase position relative to the 4σ and 1π states. The naive answer would be that the 5σ state mediates the molecular surface bond and thus is most prone to a change of the wave function. Since the 5σ level will obviously be more tightly bound in the surface configuration than in the free molecule, a lowering in energy is not entirely unreasonable. For a discussion of this point see [5.157–5.163].

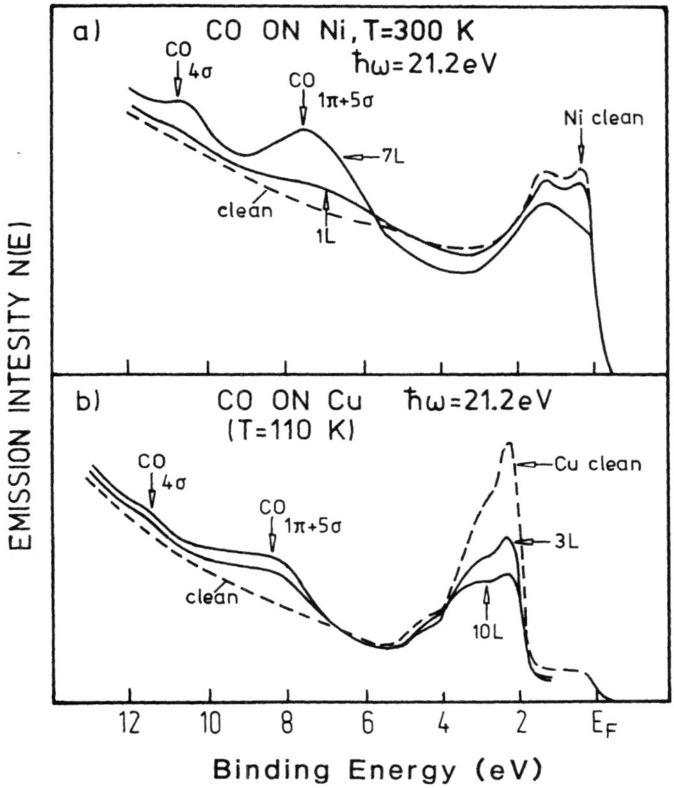

Fig. 5.98a,b. UPS ($\hbar\omega = 21.2\,\mathrm{eV}$) valence-band spectra of CO on Ni and CO on Cu [5.153]. The spectra are remarkably similar considering the fact that in Ni the electrons at the Fermi energy are mostly of d-type while those in Cu are mostly of sp-type. The coverage is given in Langmuirs (L)

To give further insight into the specific case of CO on Ni, Fig. 5.98 shows a comparison of data for CO chemisorbed on Ni and on Cu ($T = 110\,\mathrm{K}$) [5.153]. At first sight it is surprising to see that the adsorbed CO parts of the spectra are quite similar. However, one should note that the difference in heat of adsorption of CO on Cu and on Ni is only $\approx 0.7\,\mathrm{eV}$ [5.132]. It is evident from the data that such a small difference will not be easy to recover from the actual PE spectra.

A summary of PES data for CO as a free molecule, as a solid and deposited on various metals, is presented in Fig. 5.99 [5.164]. These spectra have all been referenced to the vacuum level because this is the only meaningful zero for the gas phase spectrum. To transform data from a reference level at the Fermi energy to one at the vacuum level, the work function of the adsorbate covered surface has to be used. These work functions are strongly coverage dependent and this may explain differences between the data from different authors.

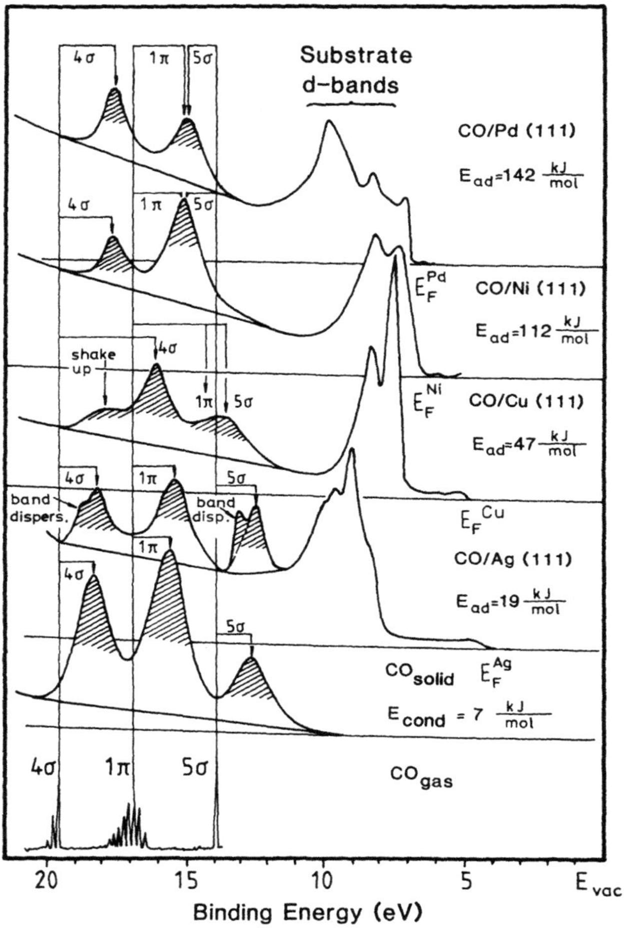

Fig. 5.99. PE spectra of gas phase CO, solid CO and CO on various metal surfaces [5.164]. The spectra for free CO, solid CO and CO (physisorbed) on Ag are quite similar. CO is chemisorbed on Cu, Ni and Pd and therefore binding occurs via the 5σ orbital, which lowers its energy to approximately that of the 1π orbital. Note that the common reference of the energy scales is the vacuum level. The different Fermi energies of the metallic substrates have been indicated

Furthermore, the adsorbate structure depends on that of the substrate and therefore, for the same substrate, differences in the spectra are expected for different crystal faces.

In going from the gas phase spectrum of CO to that of solid CO one finds only an (expected) broadening and a relaxation shift, but no change of the relative energies. The spectrum of the physisorbed species [CO on Ag(111)] again agrees quite well in its gross structure with that of solid CO. The extra structure in the 4σ and 5σ peaks is due to band dispersion effects.

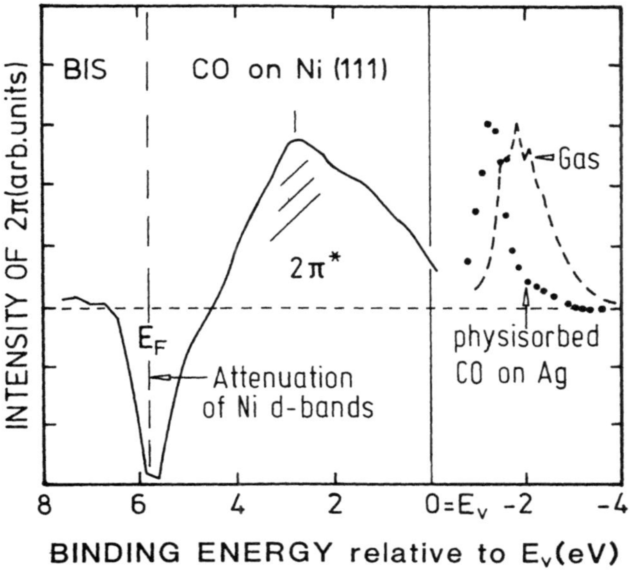

Fig. 5.100. Inverse PE spectrum of CO on Ni(111) [5.165] showing the $2\pi^*$-antibonding state of the CO molecule. For comparison the position of the $2\pi^*$ resonance in gaseous CO and in CO adsorbed on Ag are also given [5.149, 5.150]; in both cases the difference spectra of the adsorbate-covered sample and the clean sample are exhibited

The next three spectra in Fig. 5.99 which are for systems with real chemical bonding, are markedly different from the previous ones (gas, condensed phase, physisorbed phase). The 5σ orbital is now lowered with respect to the 4σ and 1π orbitals, such that it becomes almost degenerate with the 1π orbital. The spectrum of CO on Cu(111) shows an additional shake-up feature (satellite). The spectrum for the weakly chemisorbed system CO/Cu(111) is, however, quite similar to those of strongly chemisorbed systems such as CO/Ni(111) and CO/Pd(111). In particular, with respect to the Fermi energy the 4σ and the $1\pi/5\sigma$ peaks come out at approximately the same binding energy with respect to the vacuum level(!) for all three systems. This indicates that the simple view of CO chemisorption indicated in Fig. 5.95 is a valid approach since the specific nature of the metal surface does not enter into it. We stress the fact that the data show bonding via the 5σ orbital, whereas the 4σ and 1π orbitals seem to be little affected by chemisorption.

Thc adsorption of CO on metals such as Ni takes place via bonding of the 5σ level and back bonding into the $2\pi^*$ states. Therefore it is necessary to study the $2\pi^*$ states for CO adsorbed on Cu and Ni to find possible differences in their position. The $2\pi^*$ state is unoccupied in the free molecule (above E_{v}) and thus with some back donation one might expect it to be below E_{v} in a chemisorbed state, but hardly below E_{F}. The most suitable method to study

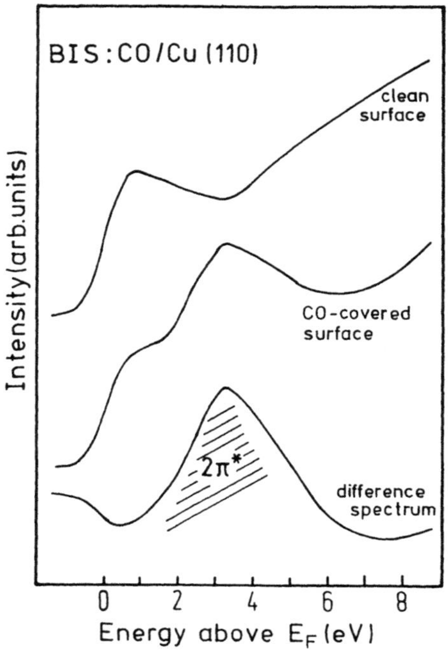

Fig. 5.101. BIS spectra of the CO/Cu(110) system [5.166]. The spectrum of a clean Cu (110) surface is given together with a spectrum of CO on Cu(110) and the difference spectrum clearly showing the $2\pi^*$ state of CO on Cu(110)

this level is thus IPES or BIS (Chap. 9) and this is indeed the technique by which the state has been observed [5.165–5.168]. Figure 5.100 presents a BIS spectrum of CO on Ni(111) [5.165], where the difference of the spectra of the CO-covered sample and the clean sample have been plotted. It shows the $2\pi^*$ peak at $\approx 3.0\,\mathrm{eV}$ above the Fermi energy (2.6 eV below E_v), although the exact position is hard to determine because of the width of the signal.

Also shown are the $2\pi^*$ resonances of the CO gas [5.149] and of CO physisorbed on Ag [5.150]. These are both above the vacuum level as they must be since the $2\pi^*$ level is not occupied in free CO and is thus (probably) also unoccupied in CO in the physisorbed state.

Figure 5.101 shows a BIS measurement for CO on Cu [5.166] and again the signal is seen in the difference spectrum at $\approx 3\,\mathrm{eV}$ above the Fermi energy, very similar to the situation found in CO on Ni.

Looking at Figs. 5.100, 5.101 one notices the considerable linewidth of the $2\pi^*$ level, which suggests that it contains a number of lines reflecting the dispersion produced by the adsorbate–adsorbate interaction.

A more detailed analysis of the BIS spectrum of CO on Ni(111) is given in Fig. 5.102. In this experiment the adsorption of CO was studied at low temperatures ($T = 140\,\mathrm{K}$). The lower temperature must be responsible for the differences observed with respect to the data shown in Fig. 5.101. In Fig. 5.102, the spectrum of the clean Ni(111) surface shows a peak at the

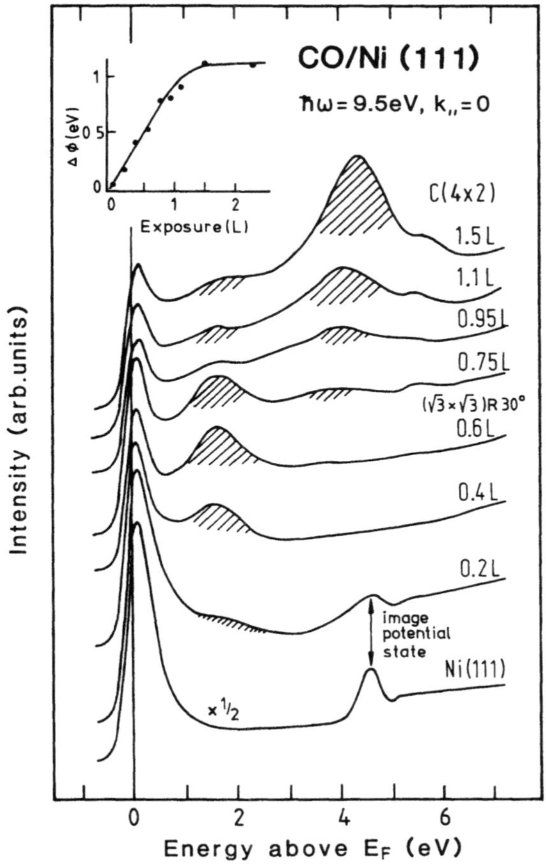

Fig. 5.102. BIS spectra [5.169] of the system CO on Ni(111) for various CO exposures. The clean Ni(111) surface shows the empty d states at E_F and an image potential state at 4.3 eV above E_F. With increasing coverage a first adsorbate state at 1.5 eV above E_F and later a second one at 4.5 eV above E_F develop. The separation of these two peaks is a measure of the dispersion of the two-dimensional band structure of the ordered CO overlayer. The insert shows the work function change induced by CO adsorption

Fermi energy due to the empty d states and a peak at 4.5 eV which stems from an image potential state (Chap. 9).

At low CO coverages a CO-derived state is observed at 1.5 eV above E_F, and at higher coverages a second CO-derived state develops at 4.5 eV above E_F. Note that at 0.6 L and 1.5 L exposure, ordered structures develop, which can give rise to a two-dimensional band structure. The splitting of the $2\pi^*$ state into two levels is thus attributed to a band-structure effect. The center of gravity of the two peaks agrees roughly with the 3 eV measured at room temperature in the experiment shown in Fig. 5.101. The data in Fig. 5.102

emphasize the problems in determining binding energies in terms of simple molecular orbitals when dealing with ordered overlayers, because in this case the adsorbate–adsorbate interactions can dominate the spectra.

The energies measured in a PES or IPES experiment cannot be directly related to chemical properties since they consist of two parts: an initial state shift due to the bonding and a relaxation shift due to screening in the final state. Thus (as before) one has for the shift ΔE_{B} of the binding energy between the free molecule and the molecule adsorbed on a surface two contributions, one which comes from the change in the chemistry and another one which reflects the difference in relaxation energies, i.e.,

$$\Delta E_{\mathrm{B}} = \Delta E_{\mathrm{chemical}} + \Delta E_{\mathrm{relaxation}} \ . \tag{5.18}$$

In a rigorous sense one has

$$\Delta E_{\mathrm{B}} = (E_{\mathrm{B}})_{\mathrm{CO\ molecule}} - (E_{\mathrm{B}})_{\mathrm{CO\ on\ metal}} \ , \tag{5.19}$$

which may be broken up to give approximately

$$\begin{aligned} \Delta E_{\mathrm{B}} &= (E^{n}_{\mathrm{total}} - E^{n-1}_{\mathrm{total}})_{\mathrm{CO\ molecule}} \\ &- (E^{n}_{\mathrm{total}} - E^{n-1}_{\mathrm{total}})_{\mathrm{CO\ on\ metal}} \ . \end{aligned} \tag{5.20}$$

Thus it is not easy (and may be almost impossible), to obtain clear chemical information from the PE data as such, because in some way or another they have to be compared with theory. Such a theory exists but in view of the small energies ($\sim 0.5\,\mathrm{eV}$) it seems that the present accuracy of the theory does not allow a clear analysis of the data [5.162]. (See also [5.171] for a discussion of this point).

The conventional chemical view of CO adsorption on metal surfaces has been questioned by Bagus and collaborators [5.160–5.163, 5.172–5.178]. Their calculations show that the dominant mechanism of the CO metal bonding is a metal-to-ligand charge transfer, while the ligand-to-metal transfer is only weak. The large 5σ shift between the free CO molecule and CO on metal is then seen as an electrostatic initial-state effect reflecting the metal adsorbate bond length. The relaxation shift is found to be approximately the same for all three valence orbitals (1π, 4σ, 5σ). Since discussion on this topic remains active, it seems sensible to defer judgement.

One can perhaps make one general comment. In a small free molecule like CO, relaxation cannot be accomplished easily (compared to a metallic solid) because there is simply not enough charge around to effectively screen the photohole. On the other hand if CO is bound to a metallic surface, the screening charge will be available in the form of conduction electrons. Since one knows that the relaxation shifts are of the order of a few volts, this means that the binding energies of free CO are larger than those of CO on a metal, as indeed is observed. The general magnitude of this shift ($\approx 2\,\mathrm{eV}$) is also not unreasonable. In addition, the fact that it is the 5σ level that shifts relative to the 4σ and 1π levels upon adsorption is also in agreement with expectation. Thus, on a qualitative basis, one can claim that the photoemission results support the picture in which adsorption of CO on metals is achieved by a combination of 5σ bonding and $2\pi^*$ back bonding. Perhaps it is worth pointing out that the $2\pi^*$ energies suggest that the pinning of the unoccupied levels is also relative to the Fermi energy rather than to the vacuum level, indicating a reasonably strong surface chemical bond.

References

5.1 U. Hollenstein, R. Seiler, H. Schmutz, M. Andrist, M. Merkt: J. Chem. Phys. **115**, 5461 (2001)

5.2 D.W. Turner, A.D. Baker, C. Baker, C.R. Brundle: *Molecular Photoelectron Spectroscopy* (Wiley, New York 1970)
Further molecular photoelectron spectra can be found in:
K. Kimura, S. Katsumata, Y. Achiba, T. Yamazaki, S. Iwata: *Handbook of He I Photoelectron Spectra of Fundamental Organic Molecules* (Jpn. Sci. Soc. Press, Tokyo 1981)
Data Bank of He II Photoelectron Spectra:
Pt.I Hydrocarbons (C,H), J. Electron Spectrosc. Relat. Phenom. **10**, 149 (1980)
Pt.II Aza-compounds (C,H,N), ibid. **21**, 93 (1980)
Pt.III Oxo-compounds (C,H,O), ibid. **21**, 175 (1980)
Pt.IV Fluoro-compounds (C,H,F), ibid. **23**, 281 (1981)
Pt.V Hetero-compounds containing first-row elements (C, H, B, N, O, F), ibid. **24**, 293 (1981)
Pt.VI Halogen compounds (C,H,X;X=Cl,Br,I); ibid. **26**, 173 (1982)
Pt.VII Miscellaneous compounds, ibid. **27**, 129 (1982)

5.3 A.D. Baker, D. Betteridge: *Photoelectron Spectroscopy* (Pergamon, Oxford 1972)

5.4 J.H.D. Eland: *Photoelectron Spectroscopy* (Butterworths, London 1974)

5.5 J.H.D. Eland: *Photoelectron Spectroscopy*, 2nd Ed. (Butterworths, London 1984)

5.6 C.R. Brundle, A.D. Baker (eds.): *Electron Spectroscopy, Theory, Techniques and Applications* Vol.1 (Academic, New York 1977)

5.7 C.R. Brundle, A.D. Baker (eds.): *Electron Spectroscopy, Theory, Techniques and Applications*, Vol.II (Academic, New York 1978)

5.8 C.R. Brundle, A.D. Baker (eds.): *Electron Spectroscopy, Theory, Techniques and Applications*, Vol.III (Academic, New York 1979)

5.9 C.R. Brundle, A.D. Baker (eds.): *Electron Spectroscopy, Theory, Techniques and Applications*, Vol.IV (Academic, New York 1981)

5.10 C.R. Brundle, A.D. Baker (eds.): *Electron Spectroscopy, Theory, Techniques and Applications*, Vol.V (Academic, New York 1984)

5.11 T.A. Carlson: *Photoelectron and Auger Spectroscopy* (Plenum, New York 1975)

5.12 H. Bock, B.G. Ramsey: Angew. Chemie, Int'l Ed. **12**, 734 (1973)

5.13 H. Bock: Angew. Chemie, int. Ed. **16**, 613 (1977)
H. Bock, B. Soloniki: Angew. Chemie, int. Ed. **20**, 427 (1981)

5.14 Landolt-Börnstein: New Series, Group III, Vol.23, Subvols. A(1989), B(1994),C1(to appear), C2(1999) ed. by A. Goldmann, E.E. Koch (Springer, Berlin, Heidelberg 1989)

5.15 D.P. Spears, H.J. Fischbeck, T.A. Carlson: Phys. Rev. A **9**, 1603 (1974)

5.16 C.S. Fadley: In [5.7]
5.17 W.C. Price: In [5.6]
5.18 A.F. Orchard: In *Handbook of X-Ray and Ultraviolet Photoelectron Spectroscopy*, ed. by D. Briggs (Heyden, London 1977)
5.19 J. Berkowitz: in [5.6]
5.20 A.W. Potts, T.A. Williams, W.C. Price: Proc. Roy. Soc. London A **341**, 147 (1974)
5.21 J. Berkowitz, J.L. Dehmer, T.E.H. Walker: J. Chem. Phys. **59**, 3645 (1973)
5.22 G.K. Wertheim, J.E. Rowe, D.N.E. Buchanan, P.H. Citrin: Phys. Rev. B **51**, 13675 (1995). This work shows that an appreciable fraction of the PES valence-bandwidth of alkali halides is due to phonon broadening. After this has been accounted for, the agreement between the PES data and the band-structure calculations is good
5.23 R.S. Berry: Chem. Rev. **69**, 533 (1969)
5.24 L. Brewer, E. Brackett: Chem. Rev. **61**, 425 (1961)
5.25 E.P.F. Lee, A.W. Potts, M. Dovan, I.H. Hillier, J.J. Delamy, R.W. Hawksworth, M.F. Guest: J. Chem. Soc., Faraday Trans.II, **76**, 506 (1980)
5.26 A.W. Potts, D. Law, E.P.F. Lee: J. Chem. Soc., Faraday Trans. II, **77**, 796 (1981)
5.27 E. Heilbronner, J.P. Maier: in [5.6]
5.28 C. Baker, D.W. Turner: Proc. Roy. Soc. A **308**, 19 (1968)
5.29 P.A. Clark, F. Brogli, E. Heilbronner: Helv. Chem. Acta **55**, 1415 (1972)
5.30 E. Lindholm: Faraday Disc. Chem. Soc. **54**, 200 (1972)
5.31 K.J. Yu, J.C. McMenahim, W.E. Spicer: Surf. Science **50**, 149 (1975)
5.32 W.D. Grobman, E.E. Koch: In *Photoemission in Solids II*, ed. by L. Ley, M. Cardona, Topics Appl. Phys., Vol.27 (Springer, Berlin, Heidelberg 1979) Chap.5
5.33 G.K. Wertheim: Phys. Rev. B **51**, 10248 (1995)
O. Gunnarsson, H. Handschuh, P.S. Bechthold, B. Kessler, G. Ganteför, W. Eberhardt: Phys. Rev. Lett. **74**, 1875 (1995)
(These communications) References [5.32, 5.33] investigate the valence-band structure of C_{60} and show that phonon broadening is a significant factor in the measured width
5.34 R.M. Nieminen, P. Huttunen: J. Phys. F **16**, L93 (1986)
G.K. Wertheim: Z. Physik B **66**, 53 (1987)
For a review see K.H. Meiwes-Broer: Appl. Phys. A **55**, 430 (1992)
5.35 W.R. Salaneck: CRC Crit. Rev. in Solid State and Mater. Sci. **12**, 267 (1985)
5.36 A. Dilks: In [5.9]
5.37 J.J. Pireaux, S. Svensson, E. Basilier, P.A. Malmqvist, U. Gelius, R. Caudano, K. Siegbahn: Phys. Rev. A **14**, 2133 (1976)
5.38 J.J. Pireaux, R. Caudano: Phys. Rev. B **15**, 2242 (1977)
J.J. Pireaux, J. Riga, R. Caudano, J. Verbist: In *Photon, Electron and Ion Probes of Polymer Structure and Properties*, ed. by D.W. Dwight, T.J. Fabish, H.R. Thomas (Am. Chem. Soc., Washington D.C., 169 1981)
5.39 F.R. McFeeley, S.P. Kowalczyk, L. Ley, R.G. Cavell, R.A. Pollak, D.A. Shirley: Phys. Rev. B **16**, 5268 (1974)
J.H. Weaver, J.L. Martins, T. Komeda, Y. Chen, T.R. Ohno, G.H. Kroll, N. Troullier, R.E. Haufler, R.E. Smalley: Phys. Rev. Lett. **66**, 1741 (1991)
5.40 J. Rasmusson, J. Stafström, M. Lögdlund, W.R. Salaneck, U. Karlsson, D.B. Swanson, A.G. McDiarmid, G.A. Arbuckle: Synth. Met. **41–43**, 1365 (1991)
M.P. Keane, A. Naves de Brito, N. Correia, S. Svensson, L. Karlsson,

B. Wannberg, U. Gelius, St. Lunell, W.R. Salaneck, M. Lögdlund, D.B. Swanson, A.G. McDiarmid: Phys. Rev. B **45**, 6390 (1992)
5.41 C.B. Duke, A. Paton, W.E. Salaneck, H.R. Thomas, E.W. Plummer, A.J. Heeger, A.G. MacDiarmid: Chem. Phys. Lett. **59**, 146 (1978)
W.R. Salaneck: CRC Crit. Rev. Solid State and Mater. Sci. **12**, 267 (1985)
5.42 C.K. Chiang, C.R. Fincher Jr., Y.W. Park, A.J. Heeger, H. Shirakawa, E.J. Louis, S.C. Cau, A.G. MacDiarmid: Phys. Rev. Lett. **39**, 1098 (1977)
5.43 A.G. MacDiarmid, A.J. Heeger: in *Molecular Metals*, ed. by W.E. Hatfield (Plenum, New York 1979)
5.44 T. Skotheim (ed.): *Handbook of Conducting Polymers* (Dekker, New York, 1986)
5.45 W.R. Salaneck, H.R. Thomas, C.B. Duke, A. Paton, E.W. Plummer, A.J. Heeger, A.G. MacDiarmid: J. Chem. Phys. **71**, 2044 (1979)
W.R. Salaneck, H.R. Thomas, C.B. Duke, A. Paton, E.W. Plummer, A.J. Heeger, A.G. MacDiarmid: J. Chem. Phys. **72**, 3674 (1980)
5.46 B. Dardel, D. Malterre, M. Grioni, P. Weibel, Y. Baer, F. Lèvy: Phys. Rev. Lett. **67**, 3144 (1991)
5.47 M. Sasai, H. Fukutome: Solid State Commun. **58**, 735 (1986)
G. Beamson, D. Briggs: *High Resolution XPS of Organic Polymers* (Wiley, Chichester 1992). This volume gives a useful compilation of XPS core and valence bands for a large number of organic polymers
5.48 G.K. Wertheim, S. Hüfner: Phys. Rev. Lett. **28**, 1028 (1972)
S. Hüfner, G.K. Wertheim: Phys. Rev. B **8**, 4857 (1973)
5.49 For an early review of PES in 3d transition metal compounds see L.C. Davis: J. Appl. Phys. **59**, R25 (1986)
More recent reviews:
E. Dagotto: Rev. Mod. Phys. **66**, 763 (1994). Dealing with theoretical ideas and experimental results about electronic structure of the high-temperature superconductors
Z.X. Shen, D.S. Dessau: Phys. Repts. **253**, 1 (1995). Treating the electronic structure and PE studies of late transition metal oxides, Mott insulators and high-temperature superconductors
S. Hüfner: Adv. Phys. **43**, 183 (1994). Dealing with the electronic structure of NiO and related 3d transition metal compounds
A. Fujimori, Y. Tokura (eds.): Spectroscopy of Mott Insulators and Correlated Metals, Springer Ser. Solid-State Sci., Vol.119 (Springer, Berlin, Heidelberg 1994). Proceedings of the Taniguchi Symposium that investigated the nature of highly correlated insulating and metallic states in 3d transition metal oxides
5.50 S. Hüfner: Adv. Phys **43**, 183 (1994)
5.51 M. Imada, A. Fujimori, Y. Tokura: Rev. Mod. Phys. **70**, 1039 (1998); a review on the metal insulator transition
5.52 N.F. Mott: *The Metal Insulator Transition*, taylor and Francis, London (1990)
5.53 C.E. Moore: *Atomic Energy Levels*, Vols.I-III, NBS No.467 (US Governm. Printing Office, Washington, DC 1958)
5.54 C.E. Moore: *Atomic Energy Levels*, Vol.I, NBS No.467 (US Governm. Printing Office, Washington DC 1949)
5.55 C.E. Moore: *Atomic Energy Levels*, Vol.II, NBS No.467 (US Governm. Printing Office, Washington DC 1952)
5.56 C.E. Moore: *Atomic Energy Levels*, Vol.I, NSRDS-NBS 35, Nat'l Bur. Stand., Washington DC (December 1971); and *Ionization Potentials and Ionization Limits Derived from the Analyses of Optical Spectra*, NSRDS-NBS 34, Nat'l Bur. Stand., Washington DC (September 1970)

5.57 J.A. Wilson: Adv. Phys. **26**, 143 (1972)
5.58 S. Hüfner: Solid State Commun. **53**, 707 (1985)
5.59 R. Newman, R.M. Chrenko: Phys. Rev. **115**, 882 (1959)
5.60 A good introduction into the field of correlations can be found in P. Fulde: *Electron Correlations in Molecules and Solids*, 2nd edn., Springer Ser. Solid-State Sci., Vol.100 (Springer, Berlin, Heidelberg 1993)
5.61 N.F. Mott: Proc. Phys. Soc. London Sect. A **62**, 416 (1949)
N.F. Mott: *Metal Insulator Transitions*, (Taylor and Francis, London 1974)
5.62 J. Hubbard: Proc. Roy. Soc. A **176**, 328 (1963)
5.63 B. Brandow: Adv. Phys. **26**, 651 (1977)
5.64 C. Herring: In *Magnetism*, Vol.IV, ed. by G.T. Rado, H. Suhl (Academic, New York 1966)
5.65 R. Courths, P. Steiner, H. Höchst, S. Hüfner: Appl. Phys. **21**, 345 (1980)
5.66 G.K. Wertheim, H.J. Guggenheim, S. Hüfner: Phys. Rev. Lett. **30**, 1050 (1973)
5.67 S. Hüfner: In *Photoemission in Solids II*, ed. by L. Ley, M. Cardona, Topics Appl. Phys., Vol.27 (Springer, Berlin, Heidelberg 1979) Chap. 3
5.68 D.E. Eastman, J.L. Freeouf: Phys. Rev. Lett. **34**, 395 (1975)
5.69 P.A. Cox: In *Structure and Bonding* **24**, 59 (Springer, Berlin, Heidelberg 1975)
5.70 S. Sugano, Y. Tanaka, H. Kamimura: *Multiplets of Transition Metal Ions in Crystals* (Academic, New York 1970)
5.71 S. Hüfner, F. Hulliger, J. Osterwalder, T. Riesterer: Solid State Commun. **50**, 83 (1984)
5.72 S.J. Oh, J.W. Allen, I. Lindau, J.C. Mikkelsen: Phys. Rev. B **26**, 4845 (1982)
5.73 M.R. Thuler, R.L. Benbow, Z. Hurych: Phys. Rev. B **27**, 2082 (1983): NiO; ibid. B **26**, 669 (1982): CuO
5.74 A. Fujimori, N. Kimiyuka, M. Taniguchi, S. Suga: Phys. Rev. B **36**, 6691 (1987)
5.75 A. Fujimori, M. Sacki, N. Kimizuka, M. Taniguchi, S. Suga: Phys. Rev. B **34**, 7318 (1986)
5.76 A. Fujimori, F. Minami: Phys. Rev. B **30**, 957 (1984)
5.77 5A. Fujimori, F. Minami, S. Sugano: Phys. Rev. B **29**, 5225 (1984)
See also J. Zaanen, G.A. Sawatzky: Can. J. Phys. **65**, 1262 (1987)
J. Zaanen, G.A. Sawatzky: Prog. Theo. Phys. (Jpn.), Suppl. 101, 231 (1990)
5.78 G. van der Laan, C. Westra, C. Haas, G.A. Sawatzky: Phys. Rev. B **23**, 4369 (1981)
B.O. Wells, Z.X. Shen, A. Matsuura, D.M. King, M.A. Kastner, M. Greven, R.J. Birgenau: Phys. Rev. Lett. **74**, 964 (1995). This very interesting experiment determined for the first time the dispersion of d-derived valence-band states in an insulating material with large correlations, namely $Sr_2CuO_2Cl_2$. An analysis of these data has been given by Dagotto [5.49]
5.79 A. Kakizaki, K. Sugano, T. Ishii, H. Sugawara, I. Nagakura, S. Shin: Phys. Rev. B **28**, 1026 (1983)
5.80 J.G. Bednorz, K.A. Müller: Z. Phys. B **64**, 189 (1986)
5.81 C.W. Chu, P.H. Hor, R.L. Meng, L. Gao, Y.J. Huang, Y.Q. Wang: Phys. Rev. Lett. **58**, 405 (1987)
5.82 P. Steiner, S. Hüfner, V. Kinsinger, I. Sander, B. Siegwart, H. Schmitt, R. Schulz, S. Junk, G. Schwitzgebel, A. Gold, C. Politis, H.P. Müller, R. Hoppe, S. Kemmler-Sack, H.-G. Kunz: Z. Phys. B **69**, 449 (1988)
5.83 C.K. Jorgensen: Progress in Inorg. Chemistry Vol.12, ed. by S.J. Lippart (Interscience, New York 1970) p.101
5.84 R.J. Powell, W.E. Spicer: Phys. Rev. B **2**, 2182 (1970)

5.85 R. Merlin, T.P. Martin, A. Polian, M. Cardona, B. Andlauer, D. Tannhauser: J. Magn. Magn. Mat. **9**, 83 (1978)
5.86 K.W. Blazey: Physica B **89**, 47 (1977)
S. Hüfner, P. Steiner, I. Sander, F. Reinert, H. Schmitt: Z. Physik B **86**, 207 (1992)
5.87 S. Hüfner, J. Osterwalder, T. Riesterer, F. Hulliger: Solid State Commun. **52**, 793 (1984)
5.88 G.A. Sawatzky, J.W. Allen: Phys. Rev. Lett. **53**, 2339 (1984)
5.89 D. Adler: *Solid State Physics* Vol.21, ed. by F. Seitz, D. Turnbull, H. Ehrenreich (Academic, New York 1968)
5.90 D. Adler, J. Feinleib: Phys. Rev. B **2**, 3112 (1970)
5.91 J.E. Keem, J.M. Honig, L.L. Van Zandt: Phil. Mag. B **37**, 537 (1978)
5.92 J.E. Keem, M.A. Wittenauer: Solid State Commun. **26**, 213 (1978)
5.93 K. Terakura, A.R. Williams, T. Oguchi, J. Kübler: Phys. Rev. Lett. **52**, 1830 (1984)
5.94 T. Terakura, A.R. Williams, T. Oguchi, J. Kübler: Phys. Rev. B **30**, 4734 (1984)
5.95 J. Kübler, A.R. Williams: J. Magn. Magn. Mater. **54**, 603 (1986)
5.96 B. Brandow: Proc. NATO Adv. Workshop "Narrow Band Phenomena" (1987)
5.97 M.R. Norman, A.J. Freeman: Phys. Rev. B **33**, 8896 (1986)
5.98 A. Svane, O. Gunnarsson: Phys. Rev. Lett. **65**, 1148 (1990)
V.I. Anisimov, J. Zaanen, O.K. Andersen: Phys. Rev. B **44**, 943 (1991)
J. Bala, A.M. Oles, J. Zaanen: Phys. Rev. Lett. **72**, 2600 (1994)
V.I. Anisimov, P. Kuiper, J. Nordgren: Phys. Rev. **50**, 8257 (1994)
S. Massida, A. Continenza, M. Posternak, A. Baldereschi: Phys. Rev. Lett. **74**, 2323 (1995)
F. Manghi, C. Calandra, S. Ossicini: Phys. Rev. Lett. **73**, 3129 (1994)
5.99 S. Hüfner, T. Riesterer: Phys. Rev. B **33**, 7267 (1986)
5.100 S. Hüfner, G.K. Wertheim: Phys. Lett. **44** A, 133 (1973)
5.101 S. Hüfner, J. Osterwalder, Th. Riesterer, F. Hulliger: Solid State Commun. **54**, 689 (1985)
5.102 A. Fujimori, K. Terakura, M. Tamiguchi, S. Ogawa, S. Suga, M. Matoba, S. Anzai: Phys. Rev. B **37**, 3109 (1988)
5.103 V.E. Henrich: Reports Prog. Phys. **48**, 1481 (1985); for similar data see S. Shin, S. Suga, M. Taniguchi, M. Fujisawa, H. Kanzaki, A. Fujimori, H. Daimon, J. Ueda, K. Kosuge, S. Kachi: Phys. Rev. B **41**, 4993 (1990)
5.104 J.K. Lang, Y. Baer, P.A. Cox: J. Phys. F **11**, 121 (1981)
5.105 S. Hüfner: Z. Phys. B **58**, 1 (1984)
5.106 B. Johansson: J. Phys. F **40**, L169 (1974)
5.107 B. Johansson: Phil. Mag. **30**, 469 (1974)
5.108 G. Rosina, E. Bertel, F.P. Netzer, J. Redinger: Phys. Rev. B **33**, 2364 (1986)
5.109 J.F. Janak: Phys. Rev. B **18**, 7165 (1977)
5.110 S. Hüfner: Z. Physik **61**, 135 (1985)
5.111 The question whether the early transition metal compounds can be termed "Mott compounds", has not yet been settled, see
S. Hüfner: In [5.49]
R. Heise, R. Courths, S. Witzel: Solid State Commun. **84**, 599 (1992)
K. Okada, A. Kotani: J. Electron Spectrosc. Relat. Phenom. **62**, 131 (1993)
U. Uozumi, K. Okada, A. Kotani: J. Phys. Soc. Jpn. **62**, 2595 (1993)
For $3d^1(Ti^{3+})$ systems one has according to these publications $U_{dd} \approx \Delta \approx T_m \approx 4\,eV$. Therefore a classification as a Mott compound or a charge-transfer compound (Fig. 5.53) is not very meaningful. Probably band-

structure effects are important for these systems
5.112 I.H. Inoue, I. Hase, Y. Aiura, A. Fujimori, Y. Haruyama, T. Maruyama, Y. Nishihara: Phys. Rev. Lett. **74**, 2539 (1995). Describing the systematic development of the spectral function in $3d^1$ Mott–Hubbard systems
5.113 J. Zaanen, G.A. Sawatzky, J.W. Allen: Phys. Rev. Lett. **55**, 418 (1985)
5.114 T. Riesterer, L. Schlapbach, S. Hüfner: Solid State Commun. **57**, 109 (1986)
5.115 C. Blaauw, F. Leenhouts, F. van der Woude, G.A. Sawatzky: J. Phys. C **8**, 459 (1975)
5.116 G.K. Wertheim: J. Franklin Inst. 298, 289 (1974)
5.117 D.D. Sarma, S.R. Krishnakumar, N. Chandrasekharan, E. Weschke, C. Schüssler-Langeheine, L. Kilian, G. Kaindl: Phys. Rev. Lett. **80**, 1284 (1998)
5.118 In another controversial system, namely Fe_3O_4, PES has shown convincingly that the Verwey transition at 122 K is of the metal–nonmetal type: A. Chainani, T. Jokoya, T. Morimoto, T. Takahashi, S. Todo: Phys. Rev. B **51**, 17976 (1995)
5.119 A. Fujimori, I. Hase, H. Namatame, Y. Fujishima, Y. Tokura, H. Eisaki, S. Uchida, K. Takegahara, F.M.F. deGroot: Phys. Rev. Lett. 69, 1796 (1992)
5.120 I.H. Inoue, I. Hase, Y. Ainra, A. Fujimori, Y. Haruyama, T. Maruyama, Y. Nishihara: Phys. Rev. Lett. **74**, 2539 (1995)
5.121 P. Kuiper, P. Kruizinga, G. Ghigsen, G.A. Sawatzky, H. Verwey: Phys. Rev. Lett. **62**, 221 (1989)
5.122 A. Sekiyama, S. Suga, M. Fujikawa, S. Imada, T. Iwasaki, K. Matsuda, T. Matsushita, K.V. Kaznacheyev, H. Fujimori, H. Kuwahara, Y. Tokura: Phys. Rev. B **59**, 15528 (1999)
5.123 P.A.P. Lindberg, Z.-X. Shen, W.E. Spicer, I. Lindau: Surf. Phys. Rpts. **11**, 1 (1990); see also the special issue: Spectroscopy and High-Temperature Superconductors – The Current Understanding of High-Temperature Superconductors Revealed by Spectroscopy. J. Electron Spectrosc. Relat. Phenom. **66** (1994)
F.M. Shamma, J.C. Fuggle: Physica C **169**, 325 (1990)
H. Kuzmany, M. Mehring, J. Fink (eds.): *Electronic Properties of High-T_c Superconductors and Related Compounds*, Springer Ser. Solid-State Sci., Vol.99 (Springer, Berlin, Heidelberg 1990)
5.124 J. Kanamori, A. Kotani (eds.): *Core-Level Spectroscopy in Condensed Systems*, Springer Ser. Solid-State Sci., Vol.81 (Springer, Berlin, Heidelberg 1988)
5.125 J.W. Allen, C.G. Olson, M.B. Maple, J.S. Kang, L.Z. Lui, J.H. Park, R.O. Anderson, W.P. Ellis, J.T. Markert, J. Dolichaouck, L.Z. Liu: Phys. Rev. Lett. **64**, 595 (1990). These researchers interpret the data by assuming that for both materials the Fermi energy is in the middle of the gap; earlier data can be found in P. Steiner, S. Hüfner, A. Jungmann, V. Kinsinger, I. Sander: Z. Physik **74**, 173 (1989); and A. Grassmann, J. Ströbel, M. Klauda, J. Schlotterer, J. Saemann-Ischenko: Europhys. Lett. **9**, 827 (1989)
5.126 A. Fujimori: In *Core-Level Spectroscopy in Condensed Systems*, ed. by J. Kanamori, A. Kotani, Springer Ser. Solid-State Sci., Vol.81 (Springer, Berlin, Heidelberg 1988) p.136
G.A. Sawatzky: In *Core-Level Spectroscopy in Condensed Systems*, ed. by J. Kanamori, A. Kotani, Springer Ser. Solid-State Sci., Vol.81 (Springer, Berlin, Heidelberg 1988) p.99; the photoemission spectrum of CuO has been calculated accurately by J. Ghijsen, L.H. Tjeng, J. van Elp, H. Eskes, J. Westering, G.A. Sawatzky, M.T. Czyzyk: Phys. Rev. B **38**, 11322 (1988); a dif-

ferent calculation is given by B. Brandow: J. Solid State Chem. **88**, 28 (1990)
5.127 J. Zaanen, M. Alouani, O. Jepsen: Phys. Rev. B **40**, 837 (1989)
S. Hüfner: Solid State Commun. **74**, 969 (1990)
5.128 W. Stephan, P. Horsch: Phys. Rev. B **42**, 8736 (1990)
H. Eskes, M.B.J. Meinders, G.A. Sawatzky: Phys. Rev. Lett. **67**, 1035 (1991)
H. Eskes: Some unusual aspects of correlated systems. Dissertation, University of Groningen (1992)
5.129 S.J. Van Houten: J. Phys. Chem. Solids **17**, 7 (1960)
5.130 K.T. Park, K. Terakura, T. Oguchi, A. Yanase, M. Ikeda: Technical Report of ISSP (Japan), Serie A, Nr. 1360 (1988)
5.131 N. Nücker, J. Fink, B. Renker, I. Ewert, C. Politis, P.J.W. Weits, J.C. Fuggle: Z. Physik B **67**, 9 (1987)
H. Romberg, M. Alexander, N. Nücker, P. Adelmann, J. Fink: Phys. Rev. B **42**, 8768 (1990)
5.132 P. Steiner, S. Hüfner, A. Jungmann, S. Junk, V. Kinsinger, I. Sander, W.R. Thiele, N. Backes, C. Politis: Physica C **156**, 213 (1988)
5.133 T. Suzuki, M. Nagoshi, Y. Fukuda, K. Oh-Ishi, Y. Syono, M. Tachiki: Phys. Rev. B **42**, 4263 (1990)
Y. Sakisaka, T. Maruyama, Y. Morikawa, H. Kato, K. Edamoto, M. Okusawa, Y. Aiura, H. Yanashima, T. Terashima, Y. Bando, K. Iijima, K. Yamamoto, K. Hirata: Phys. Rev. B **42**, 4189 (1990)
A summary of Fermi surface properties of the high-temperature superconductors can be found in: J. Phys. Chem. Solids **52**, 11/12 (1991); band-structure calculations using the Quantum Monte Carlo method are given by
W. Hanke, A. Muramatsu, G. Dopf: Phys. Blätter **47**, 1061 (1992); see also Dagotto in [5.49]
5.134 R.O. Anderson, R. Claessen, J.W. Allen, C.G. Olson, C. Janowitz, L.Z. Liu, J.-H. Park, M.B. Maple, Y. Dalichaouch, M.C. de Andrade, R.F. Jardim, E.A. Early, S.-J. Oh, W.P. Ellis: Phys. Rev. Lett. **70**, 3163 (1993)
5.135 Z.X Shen, W.E. Spicer, M.D. King, D.S. Dessau, B.O. Wells: Science **267**, 343 (1995). Reviewing, in simple terms, the present understanding of the band structure of high-temperature superconductors as derived from PES
5.136 P. Aebi, J. Osterwalder, P. Schwaller, L. Schlappbach, M. Shimoda, T. Mochiku, K. Kadowaki: Phys. Rev. Lett. **72**, 2757 (1994)
5.137 J.-M. Imer, F. Patthey, B. Dardel, W.D. Schneider, Y. Baer, Y. Petroff, A. Zettl: Phys. Rev. Lett. **62**, 336 (1988)
Jian Ma, C. Quitmann, R.J. Kelley, H. Berger, G. Margaritondo, M. Onellion: Science **267**, 862 (1995). Discussing the temperature dependence of the superconducting-gap anisotropy in $Bi_2Sr_2CaCu_2O_8$
5.138 J.M. Luttinger: J. Math. Phys. **4**, 1154 (1963)
5.139 S. Tomonaga: Prog. Theo. Phys. **5**, 349 (1950)
5.140 R. Claessen, G.H. Gwear, F. Reinert, J.W. Allen, W.P. Ellis, Z.X. Shen, C.G. Olson, L.F. Schneemeyer, F. Levy: Journ. Electr. Spectrosc. **76**, 121 (1995)
5.141 B. Dardel, D. Malterre, M. Grioni, P. Weibel, Y. Baer: Phys. Rev. Lett. **67**, 3144 (1992)
5.142 B.O. Wells, Z.X. Shen, A. Matsunra, D.M. King, M.A. Kastner, M. Greven, R.J. Birgeneau: Phys. Rev. Lett. **74**, 964 (1995)
A similar investigation on another insulator material, namely NaV_2O_5, was performed by K. Kobayashi, T. Mizokawa, A. Fujimori, M. Isoke, Y. Ueda, T. Tokyama, S. Maekawa: Phys. Rev. Lett. **82**, 803 (1999)
5.143 E.W. Plummer, W. Eberhardt: Advances in Chemical Physics **49**, 533 (1982)
5.144 F.J. Himpsel: Adv. Phys. **32**, 1 (1983)

5.145 M. Scheffler, A.M. Bradshaw: In *The Chemical Physics of Solid Surfaces and Heterogeneous Catalysis*, ed. by D.A. King, D.P. Woodruff (Elsevier, Amsterdam 1983)
5.146 K. Siegbahn, C. Nordling, R. Fahlman, R. Nordberg, K. Hamrin, J. Hedman, G. Johansson, T. Bergmark, S.-E. Karlsson, I. Lindgren, B. Lindberg: *ESCA, Atomic, Molecular and Solid State Structure Studied by Means of Electron Spectroscopy*, Nova Acta Regiae Soc. Sci. Upsaliensis Ser.IV, Vol.20 (1967)
5.147 K. Siegbahn, C. Nordling, G. Johansson, J. Hedman, P.F. Heden, K. Hamrin, U. Gelius, T. Bergmark, L.D. Werme, R. Manne, Y. Baer: *ESCA – Applied to Free Molecules* (North-Holland, Amsterdam 1969)
5.148 G. Blyholder: J. Vac. Sci. Technol. **11**, 865 (1975)
5.149 H. Ehrhardt, L. Langhans, F. Linder, H.S. Taylor: Phys. Rev. **173**, 222 (1968)
5.150 J.E. Demuth, D. Schmeisser, Ph. Avouris: Phys. Rev. Lett. **47**, 1166 (1981)
5.151 H. Kuhlenbeck, H.B. Saalfeld, U. Buskotte, M. Neumann, H.J. Freund, E.W. Plummer: Phys. Rev. B **39**, 3475 (1989)
5.152 R.J. Smith, J. Anderson, G.J. Lapeyre: Phys. Rev. Lett. **37**, 1081 (1976)
E.W. Plummer, T. Gustafsson, W. Gudat, D.E. Eastman: Phys. Rev. A **15**, 2339 (1977)
5.153 D.T. Ling, J.N. Miller, D.L. Weissmann, P. Pianetta, P.L. Stefan, I. Lindau, W.E. Spicer: Surf. Sci. **57**, 157 (1976)
5.154 J. Hermanson: Solid State Commun. **22**, 9 (1977)
M. Scheffler, K. Kambe, F. Forstmann: Solid State Commun. **23**, 7896 (1977)
5.155 J.W. Davenport: J. Vac. Sci. Technol. **15**, 433 (1978); Phys. Rev. Lett. **36**, 945 (1976)
5.156 C.L. Allyn, T. Gustafsson, E.W. Plummer: Chem. Phys. Lett. **47**, 127 (1977)
5.157 I.P. Batra, P.S. Bagus: Solid State Commun. **16**, 1097 (1975)
5.158 G. Borstel, W. Braun, M. Neumann, G. Seitz: Phys. Stat. Sol. (b) **95**, 453 (1979)
5.159 L.S. Cederbaum, W. Domcke, W. von Niessen, W. Brenig: Z. Phys. B **21**, 381 (1975)
5.160 K. Hermann, P.S. Bagus: Phys. Rev. B **16**, 4195 (1977)
5.161 P.S. Bagus, K. Hermann, C.W. Bauschlicher: J. Chem. Phys. **81**, 1977 (1984)
5.162 P.S. Bagus, C.J. Nelin, C.W. Bauschlicher: J. Vac. Sci. Technol. A **2**, 905 (1984)
5.163 P.S. Bagus, C.J. Nelin, C.W. Bauschlicher: Phys. Rev. B **28**, 5423 (1983)
5.164 A.J. Freund, M. Neumann: Appl. Phys. A **47**, 3 (1988)
5.165 Th. Fauster, F.J. Himpsel: Phys. Rev. B **27**, 1390 (1983)
5.166 J. Rogozik, H. Scheidt, V. Dose, K.C. Prince, A.M. Bradshaw: Surf. Sci. **145**, L481 (1984)
5.167 F.J. Himpsel, Th. Fauster: Phys. Rev. Lett. **49**, 1583 (1982)
5.168 J. Rogozik, J. Küppers, V. Dose: Surf. Sci. **148**, L653 (1984)
5.169 K.H. Frank, H.J. Sagner, E.E. Koch, W. Eberhardt: Phys. Rev. B **38**, 8501 (1988)
5.170 D.E. Ellis, E.J. Baerends, H. Adachi, F.W. Averill: Surf. Sci. **64**, 649 (1977)
5.171 S. Krause, C. Mariani, K.C. Prince, K. Horn: Surf. Sci. **138**, 305 (1984)
5.172 P.S. Bagus, K. Hermann, C.W. Bauschlicher: J. Chem. Phys. **80**, 4378 (1984)
5.173 P.S. Bagus, K. Hermann, C.W. Bauschlicher: J. Chem. Phys. **81**, 1966 (1984)
5.174 P.S. Bagus, K. Hermann, M. Seel: J. Vacuum Sci. Technol. **18**, 435 (1981)
5.175 P.S. Bagus, K. Hermann: Surf. Sci. **89**, 588 (1979)
5.176 C.W. Bauschlicher, P.S. Bagus: J. Chem. Phys. **81**, 5889 (1984)
5.177 P.S. Bagus, K. Hermann: Phys. Rev. B **33**, 2987 (1986)
5.178 P.S. Bagus, M. Seel: Phys. Rev. B **23**, 2065 (1981)

5.179 A. Kaminski, M. Randeria, J.C. Campuzano, M.R. Norman, H. Fretwell, J. Mesot, T. Sato, T. Takahashi, K. Kadowaki: Phys. Rev. Lett. **86**, 1070 (2001)
5.180 S.V. Borisenko, M.S. Golden, S. Legner, T. Pichler, C. Dürr, M. Knupfer, J. Fink: Phys. Rev. Lett. **84**, 4453 (2000)
5.181 X.J. Zhou, T. Yoshida, S.A. Kellar, P.V. Bogdanov, E.D. Lu, A. Lanzara, M. Nakamura, T. Noda, T. Kakeshita, H. Eisaki, S. Uchida, A. Fujimori, Z. Hussain, Z.-X. Shen: Phys. Rev. Lett. **86**, 5578 (2001)
5.182 D.L. Feng, A. Damascelli, K.M. Shen, N. Motoyama, D.H. Lu, H. Eisaki, K. Shimizu, J.-I. Shimoyama, K. Kishio, N. Kaneko, M. Greven, G.D. Gu, X.J. Zhou, C. Kim, F. Ronning, N.P. Armitage, Z.-X. Shen: Phys. Rev. Lett. **88**, 107001 (2002)
5.183 H. Ding, J.R. Engelbrecht, Z. Wang, J.C. Campuzano, S.-C. Wang, H.-B. Yang, R. Rogan, T. Takahashi, K. Kadowaki, D.G. Hinks: Phys. Rev. Lett. **87**, 227002 (2001)
5.184 A. Kaminski, J. Mesot, H. Fretwell, J.C. Campuzano, M.R. Norman, M. Randeria, H. Ding, T. Sato, T. Takahashi, T. Mochiku, K. Kadowaki, H. Hoechst: Phys. Rev. Lett. **84**, 1788 (2000)
5.185 D.L. Feng, N.P. Armitage, D.H. Lu, A. Damascelli, J.P. Hu, P. Bogdanov, A. Lanzara, F. Ronning, K.M. Shen, H. Eisaki, C. Kim, Z.-X. Shen: Phys. Rev. Lett. **86**, 5550 (2001)
5.186 L. Perfetti, C. Rojas, A. Reginelli, L. Gavioli, H. Berger, G. Margaritondo, M. Grioni, R. Gaal, L. Forro, F. Rullier Abenque: Phys. Rev. B **64**, 115102 (2001)
5.187 A. Damascelli, Z.-X. Shen, Z. Hussain: Rev. Mod. Phys. (to be published) This is a very good review on photoemission data on the high-temperature superconductors

6. Photoemission of Valence Electrons from Metallic Solids in the One-Electron Approximation

Solids are characterized with respect to the size of the energy gap between valence and conduction band as insulators, semiconductors or metals. In PES and IPES experiments, only metals possess an experimentally accessible (and generally accepted) zero of energy, namely the Fermi energy. The Fermi energy generally shows up as a step in the EDC[1] and therefore it is a very convenient experimental reference point. In insulators and semiconductors (although they of course also have a Fermi energy, it does not usually show up clearly in the PE data) the top of the valence band or the bottom of the conduction band are often taken as the experimental zero of energy. These energy points, however, are not very well defined in the experimental spectra. In order to treat only the simplest possible case, the discussion in this chapter will be restricted to metals. The generalization to insulators or semiconductors is often straightforward.

In the most naive view, a metal can be considered as a sea of conduction electrons. The simplest excitations in this metal are the "plasmons". These are collective excitation modes, which one can view as oscillations of the conduction electrons against the cores of the positively charged ions [6.2]. The best way to investigate these plasma oscillations is via inelastic electron scattering whereby both the energy and the momentum vector can be measured. The photoelectrons on their way to the surface may also produce such excitations and these show up as side bands to the primary photoelectron spectrum (Chap. 4).

In the next step of sophistication the energy states of a metal are characterized by the single-electron energies E and their wave vectors $\boldsymbol{k}$; this $E(\boldsymbol{k})$ relation is called the band structure.[2] In the band-model interpretation PES (IPES) measures transitions between states in occupied (empty) and empty (empty) bands. These transitions are vertical in a reduced zone scheme (energy and wave-vector conservation) and therefore occur without

---

1 This statement is not correct for one dimensional metals [6.1] which do not show a step in the measured spectra but rather a gradual increase in intensity starting at E_{F} (therefore, the valence electrons in these systems are sometimes called Luttinger liquid).

2 $\boldsymbol{k}$ designates the wave vector of the Bloch states in a crystal; $\boldsymbol{K}$ designates the wave vector of the photoexcited electrons within the crystal and $\boldsymbol{p}/\hbar$ designates the wave vector of the photoelectrons outside the crystal in the vacuum.

the participation of other excitations. They are called vertical or direct transitions. Via direct transitions it is possible, in principle, to determine the band structure of the occupied and the unoccupied bands. This is the main topic of this chapter.

A question that has already been posed often in this book concerns the final state in PES. It was shown in Chap. 3 that even in some metals the screening of the photohole is neither instantaneous nor complete. Thus for every set of experimental data and their theoretical interpretation it has to be kept in mind that only if the screening process is accurately known, a comparison between the PES- (or IPES-) deduced band structure and the theoretical band structure is strictly allowed. Since in many cases the screening process is not understood, comparison between theory and experiment has to be made with caution.

In order to give at the outset a working knowledge for the use of PES in the investigations of band structures, the essential rules will be summarized immediately.

Henceforth, we deal with PES experiments. The application to IPES experiments is mostly straightforward if one reverses the time direction. Specific examples applying to IPES experiments will be given in Chap. 9.

In commonly used experimental setups for XPS [XPS in present day jargon usually means $\hbar\omega = 1254\,\mathrm{eV}$ (Mg K_α) or $1487\,\mathrm{eV}$ (Al K_α)], the angular acceptance of the photoelectrons is so large that in the final state the wave vector $\boldsymbol{K}$ of the photoelectrons is smeared out over the whole Brillouin zone.[3] More explicitly, this corresponds to a parallel (to the surface) wave vector of the photoelectron of

$$K_{||} \simeq 0.5\text{Å}^{-1} \sin\vartheta \sqrt{E_{\text{kin}}[\text{eV}]}$$

$$\Delta K_{||} \simeq 0.5\text{Å}^{-1} \cos\vartheta \sqrt{E_{\text{kin}}[\text{eV}]}\Delta\vartheta \, .$$

If one assumes that the angular acceptance of the electron analyzer is $\Delta\vartheta = 2°$ and that the detected electrons are photoexcited at the Fermi energy, one has:

$$\Delta K_{||} \simeq 0.6\text{Å}^{-1}$$

which can be compared with the Brillouin zone dimensions of Cu which are $2\pi/\text{a} \simeq 1.7\text{Å}^{-1}$. There is also a second effect: for electromagnetic radiation of 1.5 keV, the photon wave vector is relatively large, $\kappa \simeq 0.7\text{Å}^{-1}$, and therefore one no longer has a "vertical" transition. This adds to the averaging over the Brillouin zone. In experiments using a polycrystalline sample, the XPS spectrum, therefore, reproduces a density of states of the material, in which the contributions from electrons with different angular momenta are weighted differently because of their different photoelectric cross-sections. In experiments with single crystals this XPS density of states is further modulated

[3] Wave vector and momentum are sometimes measured in the same units, which corresponds to setting $\hbar = 1$.

with an angular projection factor [6.3]. However, experience has shown that single crystal experiments on pure materials do not give much more detailed information than experiments with polycrystalline samples and so the former are not very often performed. In the study of adsorbate systems, however, single crystals are often very useful, even in the XPS regime.

In the UPS regime (photon energies smaller than 100 eV) the spectra are dominated by direct ("vertical") transitions in the reduced zone scheme. In an experiment with a polycrystalline sample one has to average all these transitions over the whole Brillouin zone and the measured property is thus called the Joint Density Of States (JDOS). In an experiment with a single crystal sample (angle-resolved UPS=ARUPS) one investigates direct transitions with well-defined $\boldsymbol{k}$-vectors. This then provides a method to map the band structure $E(\boldsymbol{k})$. Details are discussed in the next section.

6.1 Theory of Photoemission: A Summary of the Three-Step Model

Some of the material presented here has already been given in Chap. 1. However, for the sake of making the text easy to use, a few redundancies seem permissible.

As a starting point we use the one-electron approximation, which makes the spectral function a δ-function (with a finite integral) that will be neglected until later in this chapter. To make the equations as simple as possible, the Fermi function will also be ignored until the end.

Since in this chapter we are interested in questions of momentum conservation and the generation of the photocurrent in the sample the omissions are permissible.

The most commonly used model for the interpretation of photoemission spectra in solids is the so-called three-step model, developed by Berglund and Spicer [6.4]. It is a purely phenomenological approach, which has nonetheless proved to be quite successful. It breaks up the complicated PE process into three steps: the excitation of the photoelectron, its passage through the solid to the surface and its penetration through the surface into the vacuum, where it is detected [6.5].

The three-step model and its relation to the correct one-step model are sketched in Fig. 6.1. In the one-step model one considers the excitation from an initial state (Bloch wave in the crystal) into a damped final state near the surface, the damping taking care of the short mean free path of the electrons in the solid. The one-step model will be discussed later. First the various steps in the three-step model will be treated in more detail.

We note at the outset a problem that arises with respect to the definition of $\boldsymbol{k}$: band structures are usually plotted in the reduced zone scheme, which means that the bands outside the first Brillouin zone are folded back into the first Brillouin zone by adding the appropriate reciprocal lattice vector $\boldsymbol{G}$

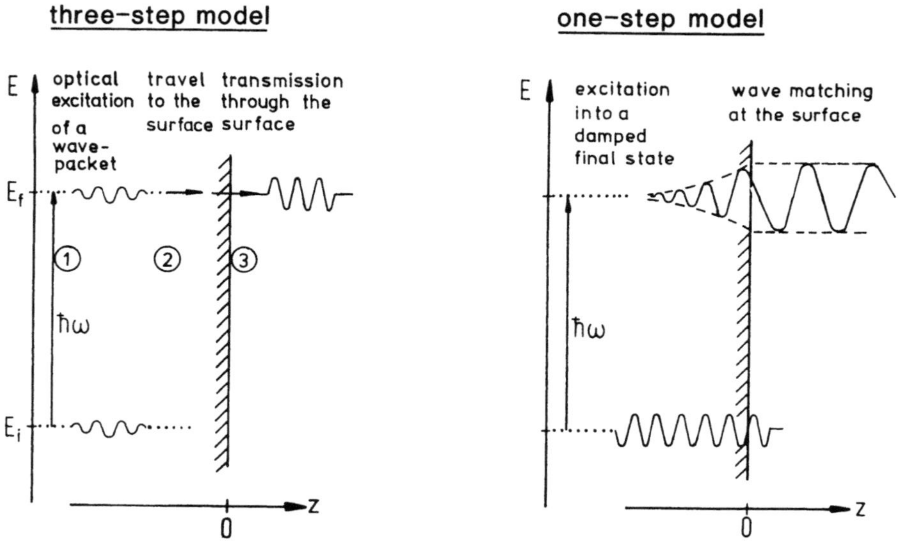

Fig. 6.1. Illustration of the three-step and the one-step model in PES. The three-step model consists of (1) photoexcitation of an electron, (2) its travel to the surface and (3) its transmission through the surface into the vacuum. In the one-step model a Bloch wave electron is excited into a wave that propagates freely in the vacuum but decays away from the surface into the solid

(Fig. 6.8). Thus in the reduced zone scheme an optical transition is "vertical" if $\boldsymbol{k}_i = \boldsymbol{k}_f$, whereas in the extended zone scheme one has $\boldsymbol{k}_f = \boldsymbol{k}_i + \boldsymbol{G}$. Therefore it is useful to distinguish between the wave vector of the crystal states $\boldsymbol{k}$ and that of the photoexcited electron within the crystal $\boldsymbol{K}$: $\boldsymbol{K} = \boldsymbol{k}_i + \boldsymbol{G}$. In cases where the final state energy E_f is given as a function of the wave vector of the photoelectron in the crystal, this momentum will be denoted $\boldsymbol{K}_f$. The wave vector of the photoelectron outside the crystal is $\boldsymbol{p}/\hbar$. A good description of the theory of valence-band photoemission is given by Braun [6.6].

Step 1: Optical excitation of the electron in the solid

The electronic states in the solid are described by bands. Neglecting the momentum of the photon, the optical excitation is a direct (momentum conserving or vertical) transition in the reduced zone scheme (Fig. 6.2). Fig. 6.2 also indicates the energy dispersion relation of the electron in the final state *after* escape into the vacuum.

The internal energy distribution of photoexcited electrons $N_{\text{int}}(E, \hbar\omega)$, where E is the final kinetic energy and $\hbar\omega$ the photon energy, is given in the reduced zone scheme by (this is (1.39), but keeping the momentum conservation in the matrix element)

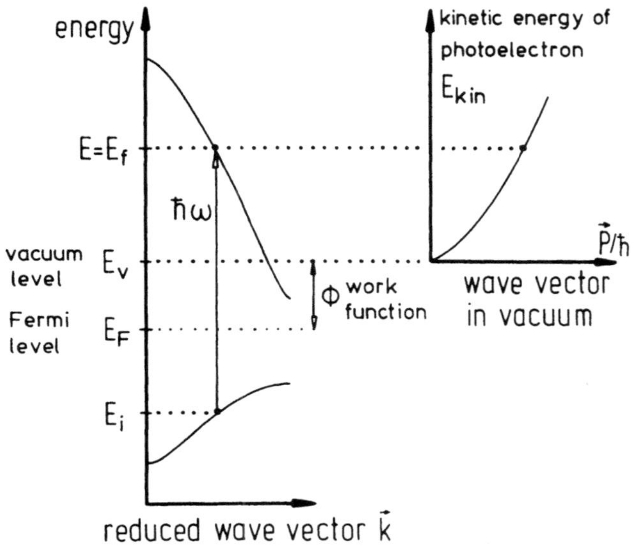

Fig. 6.2. Optical transition between two bands (final state above the vacuum level!) and kinetic energy of the photoelectron in the vacuum as a function of its wave vector $\boldsymbol{p}/\hbar$; note that we have $E_\mathrm{i} = -E_\mathrm{B}$

$$N_\mathrm{int}(E,\hbar\omega) \propto \sum_{f,i} |M^1_{fi}(\boldsymbol{k}_i,\boldsymbol{k}_f)|^2 \delta(E_f(\boldsymbol{k}_f) - E_i(\boldsymbol{k}_i) - \hbar\omega)$$
$$\times \delta(E - [E_f(\boldsymbol{k}_f) - \phi]) \,, \qquad (6.1)$$

where $E_f(\boldsymbol{k}_f)$ and $E_i(\boldsymbol{k}_i)$ denote the energies of the final bands $|f,\boldsymbol{k}_f\rangle$ and the initial bands $|i,\boldsymbol{k}_i\rangle$, respectively.[4,5] $|M^1_{fi}(\boldsymbol{k}_i,\boldsymbol{k}_f)|^2$ is the square of the

[4] In the three-step model the matrix element $M^1_{fi}(\boldsymbol{k}_i,\boldsymbol{k}_f)$ is taken between the Bloch states of the initial and final state inside the crystal. In the one-step theory the initial state is the same as in the three-step model, however, the final state is a so-called reversed LEED state (6.65), which is considered the "correct" final state in a one-step theory. This matrix element between the initial state and the "true" final state will be termed $M_{fi}(\boldsymbol{k}_i,\boldsymbol{k}_f)$.
We will in some cases write out the momentum conservation of the matrix elements explicitly which means the following relations:

$$M^1_{fi}(\boldsymbol{k}_i,\boldsymbol{k}_f) = \tilde{M}^1_{fi}(\boldsymbol{k}_i,\boldsymbol{k}_f)\delta(\boldsymbol{k}_i + \boldsymbol{G} - \boldsymbol{k}_f)$$

and

$$M_{fi}(\boldsymbol{k}_i,\boldsymbol{k}_f) = \tilde{M}_{fi}(\boldsymbol{k}_i,\boldsymbol{k}_f)\delta(\boldsymbol{k}_i + \boldsymbol{G} - \boldsymbol{k}_f)$$

or more explicitly

$$M_{fi}(\boldsymbol{k}_i,\boldsymbol{k}_f) = \tilde{M}_{fi}(\boldsymbol{k}_i,\boldsymbol{k}_f)\delta(\boldsymbol{k}_{i||} + \boldsymbol{G}_{||} - \boldsymbol{k}_{f||})\delta(k_{i\perp} + G_\perp - k_{f\perp}) \,.$$

[5] In the summation over the initial and final states it would be sufficient to sum over i, because the final states f are connected rigorously to the initial states and,

transition matrix element of the interaction operator (neglecting non-linear processes) as given in (1.21)

$$H^{\text{int}} = \frac{e}{mc}\boldsymbol{A} \cdot \boldsymbol{p} , \tag{6.2}$$

where $\boldsymbol{A}$ is the vector potential (not to be confused with the spectral function) of the exciting electromagnetic field and $\boldsymbol{p}_{\text{op}}$ the momentum operator of the electron. The first delta function imposes energy conservation during the first (excitation) step, and the second delta function ensures that the kinetic energy measured outside the sample equals the final state energy inside (Fig. 6.2) minus the work function.

In summary, for the localized optical excitation it can be assumed that the complete volume of the solid within the electromagnetic field penetration depth α^{-1} participates in the PE process. The initial state can be taken to be a wave packet of extension D along z (normal to the surface) such that $a_\perp \ll D \ll \alpha^{-1}$ ($a_\perp$: lattice constant along z). The optical excitation probability for such a wave packet is about equal to the bulk transition probability for Bloch states. On average, the excited wave packet is well localized in space with respect to its distance from the surface (Fig. 6.1). A semiclassical description thus seems to be justified. The validity of such a description will diminish when interactions that limit the inelastic mean free path are included. We note that although the treatment of the PE process in the reduced zone scheme is very convenient and often used, it can be misleading, because the importance of reciprocal lattice vectors is not apparent in this scheme. It will be shown later that one should actually use the extended zone scheme and thus in general $E_f = E_f(\boldsymbol{K}_f) = E_f(\boldsymbol{k}_i + \boldsymbol{G})$.

Step 2: Transport of the electron to the surface

The dominant scattering mechanism that reduces the number of photoexcited electrons reaching the surface with E_f is the electron–electron interaction (except very close to the threshold where phonons dominate).

Assuming that the scattering frequency $1/\tau$ (τ = lifetime) is isotropic and depends only on E, the electron inelastic mean free path $\lambda(E,k)$ is given by

$$\lambda(E,k) = \tau v_{\text{g}} = (\tau/\hbar)\text{d}E/\text{d}k , \tag{6.3}$$

where v_{g} is the group velocity in the final state. Berglund and Spicer [6.4] found in a classical treatment that the transport can be described by a coefficient $d(E,k)$ which describes the fraction of the total number of photoelectrons created within one mean free path λ from the surface and is given by

therefore, the sums over i and f are not independent of each other. However, in order to bring out the fact that the summation over i always also includes one over f we shall give both summation indices.

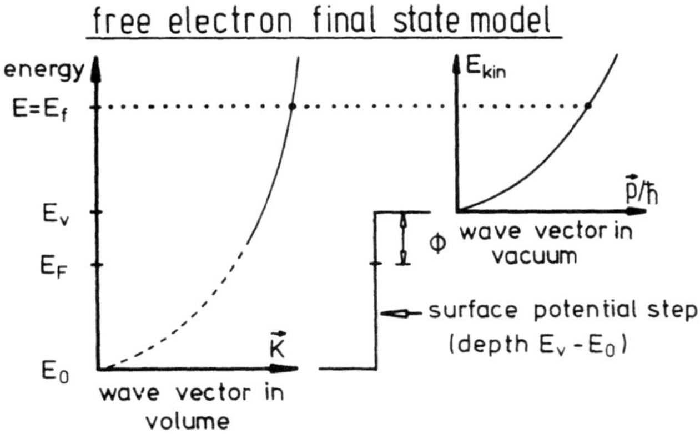

Fig. 6.3. Free-electron final-state model for PES. In the solid it is assumed that the final-state wave vector can be obtained by the intersection of a free-electron parabola with its zero at the bottom of the valence band ($E_0 < 0$) and the final state energy E_f. The wave vector of that electron in the vacuum is then given by the intersection of a free-electron parabola with its zero at the vacuum level and the final-state energy E_f

$$d(E,k) \simeq \frac{\alpha\lambda}{1+\alpha\lambda}\,, \tag{6.4}$$

where α is the optical absorption coefficient of the light ($\alpha^{-1} \sim 100-$ $-1000\,\text{Å}$ in the energy range discussed here). In the limit $\alpha\lambda \ll 1 (\lambda \approx 10$–$20\,\text{Å})$ in which the mean-free path of the electron, λ, is much smaller than the penetration depth of the light α^{-1}, one obtains $d(E,k) \to \alpha\lambda$.

Step 3: Escape of the electron into vacuum

The transmission of the photoexcited electrons into vacuum can be described using the *escape-cone argument* (Figs. 6.3–6.5). The escaping electrons are those for which the component of the kinetic energy normal to the surface is sufficient to overcome the surface potential barrier; the other electrons are totally reflected back into the bulk. Inside the crystal, as shown in Fig. 6.3, the electron travels in a potential of depth $E_\mathrm{v} - E_0$. For escape into the vacuum the electrons must satisfy the condition:

$$(\hbar^2/2m)K_\perp^2 \geq E_\mathrm{v} - E_0 \tag{6.5}$$

where $E_0(<0)$ is the energy of the bottom of the valence band and $K_\perp$ is the component of the wave vector of the excited electron $\boldsymbol{K}$ normal to the surface. The transmission of the electron through the surface leaves the parallel component of the wave vector conserved (Fig. 6.4) such that (assuming emission only into the first Mahan cone takes place, see Sect. 6.2.2)

$$\boldsymbol{p}_{||}/\hbar = \boldsymbol{K}_{||} = \boldsymbol{k}_{||} + \boldsymbol{G}_{||}\,. \tag{6.6}$$

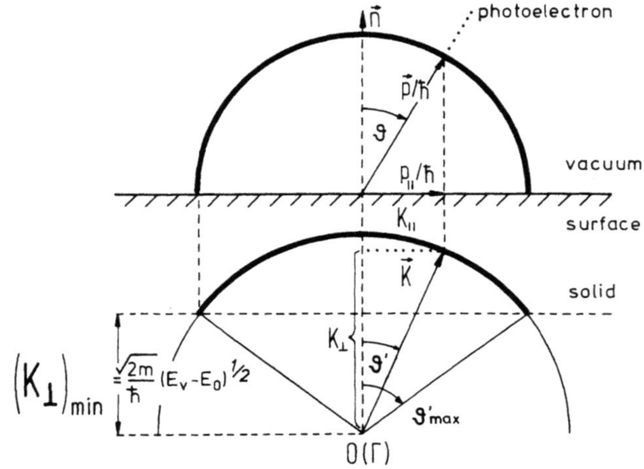

Fig. 6.4. Escape condition for the simple model shown in Fig. 6.3. The thick solid lines show the internal and external escape cones, respectively, for photoexcited electrons having momentum on a circle with radius $|\boldsymbol{K}| = \text{const.}$ and $|\boldsymbol{p}|/\hbar = \text{const.}$, respectively

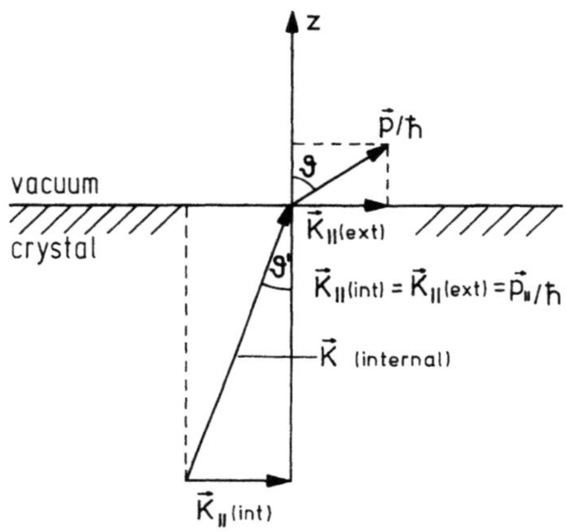

Fig. 6.5. Momentum relations at the solid-vacuum interface. It is assumed that the photoexcited electron has momentum $\boldsymbol{K} = \boldsymbol{k}_f$ inside the solid, and that $\boldsymbol{k}_{f||} = \boldsymbol{K}_{||}$ is conserved in transversing the solid-vacuum interface $\boldsymbol{K}_{||}(\text{int}) = \boldsymbol{K}_{||}(\text{ext}) = \boldsymbol{p}_{||}/\hbar$

Here $\boldsymbol{p}$ is the momentum of the photoelectron in vacuum, $\boldsymbol{K}_{||}$ is the parallel component of the wave vector $\boldsymbol{K}$ of the photoexcited electron. In the free-electron model for the excited electrons, (6.6) simply reflects Snell's law, see Fig. 6.5

$$\boldsymbol{k}_{f||} = \sin\vartheta \left(\frac{2m}{\hbar^2} E_{\text{kin}}\right)^{1/2} = \sin\vartheta' \left[\frac{2m}{\hbar^2}(E_f - E_0)\right]^{1/2} \tag{6.7}$$

(ϑ: angle outside the sample, ϑ': angle inside the sample, with respect to the surface normal).

Equation (6.7) states that the escape direction of the electron makes a *larger* angle with the surface normal *outside* the crystal than inside it. This, however, means that for every final state energy $E_f(\vartheta \leq \pi/2)$ there is a maximum angle $\vartheta'_{\text{max}} < \pi/2$ inside the sample at which the electrons excited in the sample can cross the surface:

$$\sin\vartheta'_{\text{max}} = \left(\frac{E_{\text{kin}}}{E_f - E_0}\right)^{1/2} . \tag{6.8}$$

The angular region $\vartheta' \leq \vartheta'_{\text{max}}$ is called the escape cone.

The situation is sketched in Fig. 6.4. The maximum escape angle outside the crystal is 90°, for which $\sin\vartheta = 1$. An escape angle of 90° obviously corresponds to the maximum internal parallel wave vector that can be detected. This gives a simple construction for the determination of the internal escape cone as shown in Fig. 6.4. One sees that for $K_\perp = (1/\hbar)\times(2m)^{1/2}(E_{\text{v}}-E_0)^{1/2}$ electrons just escape parallel to the crystal surface, meaning that at this point one has reached the threshold, $K_{\perp\text{min}}$.

The kinetic energy outside the crystal is determined by

$$E_{\text{kin}} = \hbar^2/2m[\boldsymbol{K}_{||}^2 + (p_\perp/\hbar)^2] = E_f(\boldsymbol{k}) - E_{\text{v}} , \tag{6.9}$$

where $p_\perp/\hbar$ is the perpendicular component of the electron wave vector in the vacuum.

None of these equations contains $K_\perp$(or $k_\perp$) the perpendicular component of the wave vector inside the crystal [6.5]. Thus an experiment taken at any $\boldsymbol{k}_{||}$ does not allow the determination of the full wave vector $\boldsymbol{k}$ of the crystal state, because $k_\perp$ remains undetermined. The assumption of a step function for modeling the potential change and of a free-electron final state are not accurate enough to allow an exact determination of $K_\perp$. Therefore one has either to restrict work to two-dimensional solids, where only $\boldsymbol{k}_{||}$ matters, or to apply more elaborate methods to determine $k_\perp$.

In a "free electron" solid one would be able to determine the depth of the crystal potential $E_{\text{v}} - E_0$ via the escape cone opening and thereby also measure $\boldsymbol{k}$ [6.7].

In a real solid the wave function $\psi_f(\boldsymbol{k})$ of a state at $E_f(\boldsymbol{k})$ is not a single plane wave but a Bloch wave containing plane-wave contributions with a number of reciprocal lattice vectors $\boldsymbol{G}$:

$$\psi_f(\boldsymbol{k}) = \sum_{\boldsymbol{G}} u_f(\boldsymbol{k},\boldsymbol{G})\mathrm{e}^{\mathrm{i}(\boldsymbol{k}+\boldsymbol{G})\cdot\boldsymbol{r}} . \tag{6.10}$$

Each component has the possibility of matching to an escaping wave outside the crystal, which means that the photoelectron can emerge from the crystal in a number of possible directions determined by (6.7) and (6.9).

Plane wave components of a single Bloch wave of energy E with the same value of $\boldsymbol{k}_{||} + \boldsymbol{G}_{||}$ leave the crystal in the same direction and must be treated coherently. The total transmission factor $|T(E_f, \boldsymbol{K}_{||})|^2$ for such a $(\boldsymbol{k}_{||} + \boldsymbol{G}_{||})$

beam at a particular final state energy E_f is then expressed as the sum of the transmission factors for each plane wave $|t(E_f, \boldsymbol{K}_{||})|^2 |u_f(\boldsymbol{G}, \boldsymbol{k})|^2$:

$$|T(E_f, \boldsymbol{k}_{||})|^2 = |t(E_f, \boldsymbol{K}_{||})|^2 \left| \sum_{(k+G)_\perp > 0} u_f(\boldsymbol{G}, \boldsymbol{k}) \right|^2 \tag{6.11}$$

where the summation is only over components propagating towards the surface, $(k + G)_\perp > 0$. A "classical" expression for the reduced transmission factor $t_f(E_f, \boldsymbol{K}_{||})$ is obtained using $p_\perp/\hbar$ from (6.9)

$$(p_\perp/\hbar)^2 \hbar^2/2m = E_f(\boldsymbol{k}) - E_\mathrm{v} - (\hbar^2/2m)\boldsymbol{K}_{||}^2 \tag{6.12}$$

and inserting $\boldsymbol{K}_{||} = \boldsymbol{k}_{||} + \boldsymbol{G}_{||}$, this leads to the estimate

$$|t(E_f, \boldsymbol{K}_{||})|^2 = \begin{cases} 1 & \text{if} \quad E_f(\boldsymbol{k}) - E_\mathrm{v} > \hbar^2(\boldsymbol{k}_{||} + \boldsymbol{G}_{||})^2/2m \\ 0 & \text{if} \quad E_f(\boldsymbol{k}) - E_\mathrm{v} \le \hbar^2(\boldsymbol{k}_{||} + \boldsymbol{G}_{||})^2/2m \end{cases} . \tag{6.13}$$

This is certainly an overestimate because one knows, from the occurrence of surface plasmons, that electrons undergo inelastic scattering processes at the surface, meaning that $t(E_f, \boldsymbol{K}_{||}) < 1$ under all circumstances.

The final expression for the angle-resolved photoelectron energy spectrum $N(E, \boldsymbol{K}_{||}, \hbar\omega)$ at photon energy $\hbar\omega$ is[6]

$$\begin{aligned} N(E, \boldsymbol{K}_{||}, \hbar\omega) \propto \sum_{f,i} & \left| \tilde{M}^1_{fi}(\boldsymbol{k}_i, \boldsymbol{k}_f) \right|^2 d(E_f, \boldsymbol{k}_f) |T(E_f, \boldsymbol{K}_{||})|^2 \\ & \times \delta(E_f(\boldsymbol{k}_f) - E_i(\boldsymbol{k}_i) - \hbar\omega) \delta(E - [E_f(\boldsymbol{k}_f) - \phi]) \\ & \times \delta(\boldsymbol{k}_i + \boldsymbol{G} - \boldsymbol{K}) \delta(\boldsymbol{K}_{||} - \boldsymbol{p}_{||}(\vartheta, \varphi)/\hbar) \end{aligned} \tag{6.14a}$$

where the last δ function $\delta(\boldsymbol{K}_{||} - \boldsymbol{p}_{||}(\vartheta, \varphi)/\hbar)$ expresses the fact that the component of the momentum parallel to the crystal surface is conserved in the experiment (inside the crystal $\boldsymbol{k}_{i||} + \boldsymbol{G}_{||} = \boldsymbol{K}_{||}$, outside the crystal $\boldsymbol{p}_{||}(\vartheta, \varphi)/\hbar = \boldsymbol{K}_{||}$).

We have used here the function $\delta[\boldsymbol{K}_{||} - \boldsymbol{p}_{||}(\vartheta, \varphi)/\hbar]$ to the more commonly employed $\delta(\boldsymbol{k}_{i||} + \boldsymbol{G}_{||} - \boldsymbol{K}_{||})$ to emphasize the nature of the three-step model. In addition we have indicated the direction (ϑ, φ) of the electron emission to show that we are dealing with an angle-resolved experiment. In the matrix element the $\boldsymbol{k}$-conservation was written out changing M^1 to $\tilde{M}^1$.

The three-step model, taken at face value, violates the uncertainty principle, as has been discussed by Caroli et al. [6.9], because it assumes that the optical excitation takes place at a given point in the solid before propagation and transmission into vacuum, and that the inelastically scattered electrons lose their energy after they have first been optically excited. However, the analysis of a wide variety of experimental data has shown that the three-step model is a very useful and also accurate approximation.

[6] Note that in the three-step model the matrix element contains only the general $\boldsymbol{k}$-conservation rule, while the $\boldsymbol{k}_{||}$-conservation is added in step 3 of the model and therefore shows up explicitly in (6.14a).

This has been a brief outline of the ingredients of the three-step model. It suffices for the reader who only wants to know the basic principles of PE from solids, particularly under $\boldsymbol{k}$-conserving conditions. In the remainder of this chapter we discuss refinements that enable one to determine electronic band structures in solids solely from experimental data. Such band-structure determinations will be the topic of Chap. 7.

6.2 Discussion of the Photocurrent

6.2.1 Kinematics of Internal Photoemission in a Polycrystalline Sample

The Hamiltonian for the interaction between an electron and electromagnetic radiation with the vector potential $\boldsymbol{A}$ can be written (6.2)

$$H^{\text{int}} = \frac{e}{mc}\boldsymbol{A}\cdot\boldsymbol{p}$$

The vector potential $\boldsymbol{A}$ is perpendicular to the propagation direction of the photons. It is also usually assumed that the vector potential, except for reflection and refraction, is not modified by the interaction with the medium into which it penetrates in order to produce the photoemission current. In addition, one can assume that the vector potential is constant over space (because of the large wavelength $\approx \lambda = 10^3$ Å for 10 eV photons; see [6.10–6.12] for deviations from this assumption). The transition probability between initial and final state is given by Fermi's Golden Rule as [see (1.19)]

$$w_{fi} = \frac{2\pi}{\hbar}|\langle f|H^{\text{int}}|i\rangle|^2\delta(E_f - E_i - \hbar\omega)\,. \tag{6.15}$$

This formula leads to the photocurrent in an (exact) one-step theory if for i and f the "true" initial and final states are inserted.

Since we are dealing with a discrete number of band states, but a (semi-) continuous distribution of $\boldsymbol{k}$-vectors, it is convenient to write out the $\boldsymbol{k}$-dependence explicitly in (6.15) leading to

$$w_{fi} = \frac{2\pi}{\hbar}|\langle f,\boldsymbol{k}_f|H^{\text{int}}|i,\boldsymbol{k}_i\rangle|^2\delta(E_f - E_i - \hbar\omega)\,. \tag{6.16}$$

From (6.16) one can derive an expression for the photocurrent for the case that the momentum information is blurred completely. As will be shown later, this is not a bad approximation if one works with a polycrystalline sample and uses X-ray photons for photoexcitation (XPS). Assuming that the matrix element $|\tilde{M}^1_{fi}|^2$ and the density of final states are constant at fixed $\hbar\omega$ one obtains

$$\begin{aligned} N(E,\hbar\omega) &\propto |\tilde{M}^1_{fi}|^2\sum_i \delta(E_f - E_i - \hbar\omega)\delta(E - E_f + \phi) \\ &\propto |\tilde{M}^1_{fi}|^2 DOS(E_i)\delta(E - E_f + \phi) \end{aligned} \tag{6.17}$$

where $DOS(E_i)$ is the density of occupied states.

This equation means that the distribution of photoemitted electrons in the XPS experiment is approximately proportional to the density of initial states $[DOS(E_i)]$. Thus this type of spectroscopy measures a density of states, which however, if electrons with different angular momenta are involved, shows the contributions from these various angular momentum contributions with different strengths because $|\tilde{M}^1_{fi}|^2$ depends on the angular momentum of the initial state. In the UPS regime one has to take into account the energy and momentum (wave vector) dependence of the matrix element.

The simplest model of a metallic solid is that of the free-electron gas. In that model the energy dispersion relation is

$$E = \frac{\hbar^2}{2m}\boldsymbol{k}^2 \,. \tag{6.18}$$

In the following treatment, which runs through to (6.46), the zero of the energy scale is placed at the bottom of the conduction band, E_0. This is done because the literature uses this convention and because certain essential geometrical considerations are thereby made very easy.

It is apparent (Fig. 6.6) that for an infinite crystal with no *periodic potential* this model allows no direct transitions, because the momentum of the photon is too small compared to the momentum of the electron to make a transition from any initial state E_i possible to a final state under the restriction of $\boldsymbol{k}$-conservation (direct transitions). In other words, direct (vertical) transitions are prevented by the lack of appropriate final states. Thus within this free-electron-gas model PE transitions can be only explained by invoking a semi-infinite crystal (which is the actual experimental situation) where the surface barrier provides a coupling to the whole crystal and thus makes a direct transition possible. This case is, however, rarely considered. In any model that takes the crystal potential into account it is obvious that via the periodic potential the lattice can absorb the required change of crystal momentum ($\Delta\boldsymbol{k} = \boldsymbol{G}$), i.e., allows direct transitions in the reduced zone scheme.

A comparison of the PE process in a free-electron system and in a periodic system in which the crystal potential provides a source of wave vector, is sketched in Fig. 6.7. In the free electron system one has

$$E_f - E_i = \frac{\hbar^2}{2m}(\boldsymbol{K}_f^2 - \boldsymbol{k}_i^2)) \tag{6.19}$$

with $\boldsymbol{k}_f = \boldsymbol{k}_i + \boldsymbol{\kappa}$ ($\boldsymbol{\kappa}$ being the wave vector of the photon $\hbar\omega$)

$$(\boldsymbol{K}_f - \boldsymbol{\kappa})^2 = \boldsymbol{K}_f^2 + \boldsymbol{\kappa}^2 - 2|\boldsymbol{K}_f||\boldsymbol{\kappa}|\cos(\boldsymbol{\kappa}, \boldsymbol{K}_f) \tag{6.20}$$

and thus

$$E_f - E_i = \frac{\hbar^2}{2m}\left[2|\boldsymbol{K}_f||\boldsymbol{\kappa}|\cos(\kappa, \boldsymbol{K}_f) - \boldsymbol{\kappa}^2\right] \tag{6.21}$$

$$= E_f\frac{\kappa}{|\boldsymbol{K}_f|}\left[2\cos(\boldsymbol{\kappa}, \boldsymbol{K}_f) - \frac{\kappa}{|\boldsymbol{K}_f|}\right] \approx E_f\frac{\kappa}{|\boldsymbol{K}_f|} \ll \hbar\omega \tag{6.22}$$

(since $\kappa/|\boldsymbol{K}_f| \ll 1$) and therefore an optical transition is not possible.

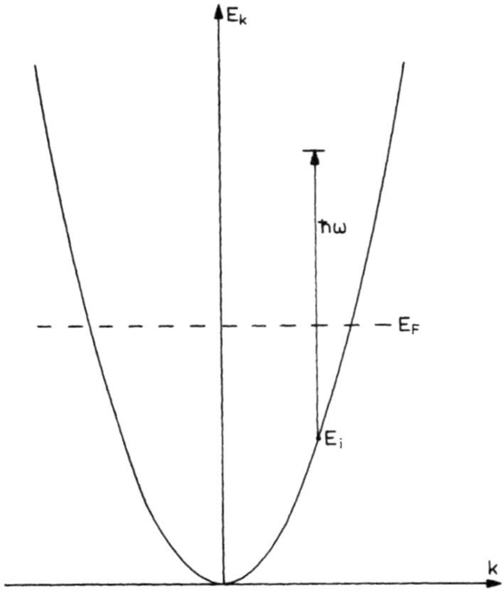

Fig. 6.6. PE process in a "perfectly free-electron" situation. Since photons can only excite an almost vertical transition, a transition between two states on the dispersion parabola cannot take place

In the case where a (weak) crystal potential is present (called the *nearly free-electron approximation*) one neglects the momentum of the photon in the UPS regime. The final-state wave vector $\boldsymbol{K}_f$ is made up of the initial-state wave vector $\boldsymbol{k}_i$ plus a reciprocal lattice vector $\boldsymbol{G}$. The difference between the final and the initial energy is then

$$E_f - E_i = \frac{\hbar^2}{2m}[\boldsymbol{K}_f^2 - \boldsymbol{k}_i^2] \tag{6.23}$$

$$= \frac{\hbar^2}{2m}[\boldsymbol{K}_f^2 - (\boldsymbol{K}_f - \boldsymbol{G})^2] \tag{6.24}$$

$$= \frac{\hbar^2}{2m}[\boldsymbol{K}_f^2 - (\boldsymbol{K}_f^2 + \boldsymbol{G}^2 - 2|\boldsymbol{K}_f||\boldsymbol{G}|\cos(\boldsymbol{K}_f, \boldsymbol{G})] \, . \tag{6.25}$$

with

$$\cos\alpha = \cos(\boldsymbol{K}_f, \boldsymbol{G}) \tag{6.26}$$

we finally obtain

$$E_f - E_i = 2(E_f E_G)^{1/2}\cos\alpha - E_G; \quad E_G = \frac{\hbar^2 \boldsymbol{G}^2}{2m} \tag{6.27}$$

which equals the photon energy $\hbar\omega$ in the case of an optical transition.

From this one can derive the following equations for the final and the initial state energies

$$E_f = \frac{(\hbar\omega + E_G)^2}{4E_G\cos^2\alpha} = \frac{E_0}{cos^2\alpha} \quad \text{with} \quad E_0 = \frac{(\hbar\omega + E_G)^2}{4E_G}$$

$$E_i = \frac{(\hbar\omega - E_G)^2}{4E_G\cos^2\alpha'} \quad \text{with} \quad \cos\alpha' = \cos(\boldsymbol{k}_i, \boldsymbol{G}) \, , \tag{6.28}$$

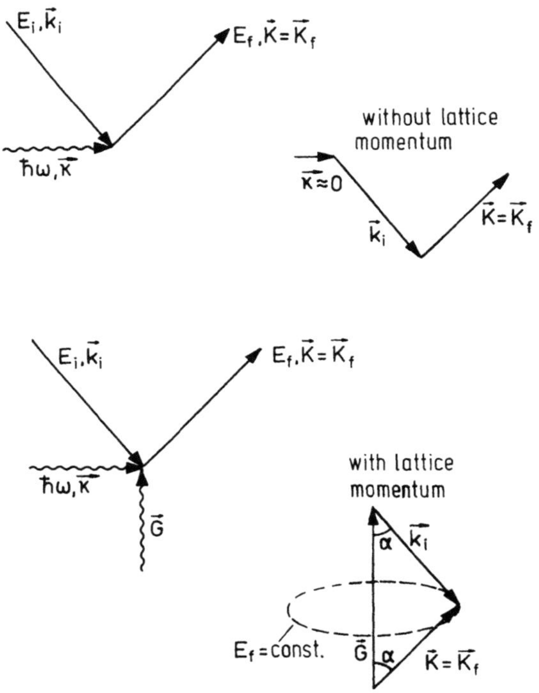

Fig. 6.7. Top: momentum addition for the PE process in a free-electron situation. Because of the smallness of $\boldsymbol{\kappa}$, the momentum of the photon, the momentum triangle cannot be closed – thus making the transition impossible. Bottom: momentum addition for the PE process in a nearly free-electron situation, where the crystal can supply the momentum $\hbar \boldsymbol{G}$. The triangle consisting of $\hbar \boldsymbol{k}_i$, the initial-state momentum, $\hbar \boldsymbol{k}_f = \hbar \boldsymbol{K}$, the final-state momentum, and $\hbar \boldsymbol{G}$, the crystal momentum, can now be closed (the photon momentum $\boldsymbol{\kappa}$ has been neglected in this situation)

where α and α' are the angles between the reciprocal lattice vector and the final and initial state wave vectors respectively (Fig. 6.7). For a specific reciprocal lattice vector and a fixed photon energy, the allowed values of the final and initial state wave vectors lie on a circle about the reciprocal lattice vector (Fig. 6.7). From this, the origin of the "Mahan cones", as discussed above, is immediately apparent.

This situation is demonstrated by an example in Fig. 6.8. It applies to a simple cubic lattice, where the electrons are detected perpendicular to the surface. For this system the reciprocal lattice vector is given by [6.13]

$$\boldsymbol{G} = (2\pi/a, 0, 0)$$

for a surface normal in the x-direction, and it is assumed that the photon energy has a magnitude

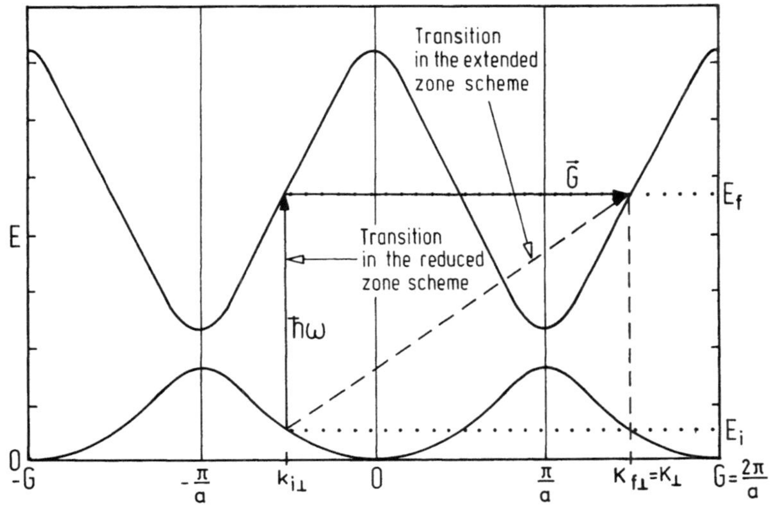

Fig. 6.8. PE transition in a nearly free-electron band structure. The vertical transition in the reduced zone scheme is really a non-vertical transition in the extended-zone scheme, between two points in the band structure connected by a reciprocal lattice vector $\boldsymbol{G}$

$$\hbar\omega = \frac{1}{2}E_G = \frac{1}{2}\left(\frac{\hbar \boldsymbol{G}^2}{2m}\right) . \tag{6.29}$$

For normal emission $\vartheta = 0$, such that $\boldsymbol{k}_{i||} = \boldsymbol{k}_{f||} = 0$, one has the initial state energy.

$$E_i = +\frac{1}{4}\left(\frac{1}{2}E_G\right)^2 \frac{1}{E_G} = \frac{1}{16}E_G; \quad E_f = \frac{9}{16}E_G . \tag{6.30}$$

One now sees that $\boldsymbol{k}_f = k_{f\perp} = 3/4\boldsymbol{G}$ and thus $\boldsymbol{k}_i = k_{i\perp} = k_{f\perp} - \boldsymbol{G} = -1/4\boldsymbol{G}$.

Thus one is observing a transition in which the initial state and the final state have wave vectors differing by one reciprocal lattice vector $\boldsymbol{G}$. In the reduced zone scheme one is thus observing a vertical transition with $\hbar\omega$ just fitting into the band scheme. In the extended zone scheme, however, which is the more physical one, this corresponds to a transition between two points in different Brillouin zones which are just one reciprocal lattice vector apart. Thus, for every photon energy, mostly only one transition in the reduced zone scheme can be observed and therefore if the photon energy is varied, one can map the band structure by ARUPS.[7] This indicates the usefulness of working with synchrotron radiation with its continuously variable energy for the determination of band structures.

[7] For high-lying final states it is possible that there are several $\boldsymbol{G}$'s which make a transition possible.

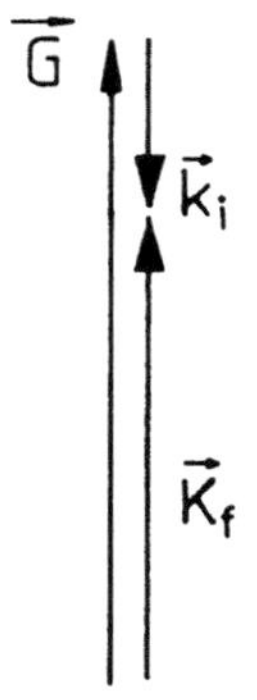

Fig. 6.9. Momentum relation for a normal emission experiment ($\boldsymbol{k}_{i||} = \boldsymbol{K}_{f||} = 0$)

The wave vectors involved in the above example are illustrated schematically in Fig. 6.9. It is assumed that $\alpha = 0$, so that the initial and final momenta and the reciprocal lattice vector are all parallel to each other. Thus the final and initial state energies (6.28) for $\alpha = 0$ are

$$E_f = \frac{(\hbar\omega + E_G)^2}{4E_G} \quad \text{and} \quad E_i = \frac{(\hbar\omega - E_G)^2}{4E_G}\,, \tag{6.28a}$$

both of which have fixed values, and with $K_f = -|\boldsymbol{k}_i| + G$ (which can be read off from Fig. 6.9) one has

$$0 = \frac{\hbar^2}{2m}(-2|\boldsymbol{k}_i|G + G^2) - \hbar\omega \tag{6.28b}$$

which implies that there is just one $|\boldsymbol{k}_i|$ for which the transition is possible.

Therefore, if the photon energy and the emission angle are kept constant, only one transition is detected. This is the signature of the direct transition model in its simplest form.

We now proceed to discuss the more general equation for the final state energy

$$E_f = \frac{(\hbar\omega + E_G)^2}{4E_G\cos^2\alpha} \tag{6.28c}$$

and to compare its predictions with experimental results.

The simplest metallic systems are the alkali metals and we shall attempt to understand the simplest of their spectra, namely those of polycrystalline materials. The band structure of the alkali metals will be approximated by that of free electrons so that we can write the following equations for the initial and final state energies:

$$E_i(\boldsymbol{k}) = \frac{\hbar^2}{2m}\boldsymbol{k}^2$$
$$E_f(\boldsymbol{k}) = \frac{\hbar^2}{2m}(\boldsymbol{k} + \boldsymbol{G})^2\,. \tag{6.31}$$

We are thus assuming that a small lattice potential is present such that $\boldsymbol{G}$ is defined (otherwise photoemission would be impossible) but nevertheless that the potential is so small that one can approximate the energies by free-electron parabolas.

For a fixed $\hbar\omega$ and a polycrystalline sample, the planes in $\boldsymbol{k}$-space that contain the optical transitions are given by the following equation (these are the so-called planes of constant interband transitions):

$$\begin{aligned} \Omega_{if}(\boldsymbol{k}) &= E_f(\boldsymbol{k}) - E_i(\boldsymbol{k}) - \hbar\omega = 0 \\ &= \frac{\hbar^2}{2m}(\boldsymbol{k}^2 + 2\boldsymbol{k}\cdot\boldsymbol{G} + \boldsymbol{G}^2) - \frac{\hbar^2}{2m}\boldsymbol{k}^2 - \hbar\omega = 0 \\ &= \frac{\hbar^2}{2m}(2\boldsymbol{k}\cdot\boldsymbol{G} + \boldsymbol{G}^2) - \hbar\omega = 0\,. \end{aligned} \tag{6.32}$$

We see that $\Omega_{if}(\boldsymbol{k})$ is perpendicular to the $\boldsymbol{G}$-vector that mediates the optical transition. Since the momentum of the initial state is $\boldsymbol{k}_i$, one finds

$$\Omega_{if}(\boldsymbol{k}_i) = \frac{\hbar\omega}{2m}(2\boldsymbol{k}_i\cdot\boldsymbol{G} + \boldsymbol{G}^2) - \hbar\omega = 0. \tag{6.33}$$

Since $\boldsymbol{K}_f = \boldsymbol{k}_i + \boldsymbol{G}$, one finally obtains:

$$\begin{aligned} \Omega_{if}(\boldsymbol{K}_f) &= \frac{\hbar^2}{2m}[2(\boldsymbol{K}_f - \boldsymbol{G})\cdot\boldsymbol{G} + \boldsymbol{G}^2] - \hbar\omega \\ &= \frac{\hbar^2}{2m}(2\boldsymbol{K}_f\cdot\boldsymbol{G} - \boldsymbol{G}^2) - \hbar\omega\,. \end{aligned} \tag{6.34}$$

The maximum initial state energy is the Fermi energy $(E_i)_{\max} = E_{\mathrm{F}}$. It was also shown (6.28) that for the initial state energy one has $E_i = (\hbar\omega - E_G)^2/4E_G\cos^2\alpha'$. Thus the minimum initial state energy ($\alpha' = 0$) is given by

$$(E_i)_{\min} = (\hbar\omega - E_G)^2/4E_G. \tag{6.28d}$$

These relations can be represented in a graphical construction for the possible transition with photon energy $\hbar\omega$ as shown in Fig. 6.10. The various values of $\boldsymbol{k}_i$ are located on a plane perpendicular to $\boldsymbol{G}$. This plane intersects $\boldsymbol{G}$ at the point where $(E_i)_{\min} = (\hbar\omega - E_G)^2/4E_G$ is satisfied. The maximum value of E_i is obtained for the Fermi energy. The possible values of the final-state wave vector $\boldsymbol{K}_f$ are then given by the points where the circle with radius $E_f = E_i + \hbar\omega$, centered at the origin of $\boldsymbol{G}$, intersects the plane perpendicular to $\boldsymbol{G}$ at $(E_i)_{\min}$.

One is interested in the photon energy range over which PE will be observed in a particular experiment. This can be derived from the angular dependence of the internal photoemission, which in turn can be obtained using the following equation for the final state energy:

$$E_f = \frac{\hbar^2\boldsymbol{K}_f^2}{2m} = \frac{(\hbar\omega + E_G)^2}{4E_G\cos^2\alpha}\,. \tag{6.28e}$$

The smallest angle for which photoemission is observed is of course $\alpha_{\min} = 0$, which leads to the smallest value of the initial state energy. The maximum

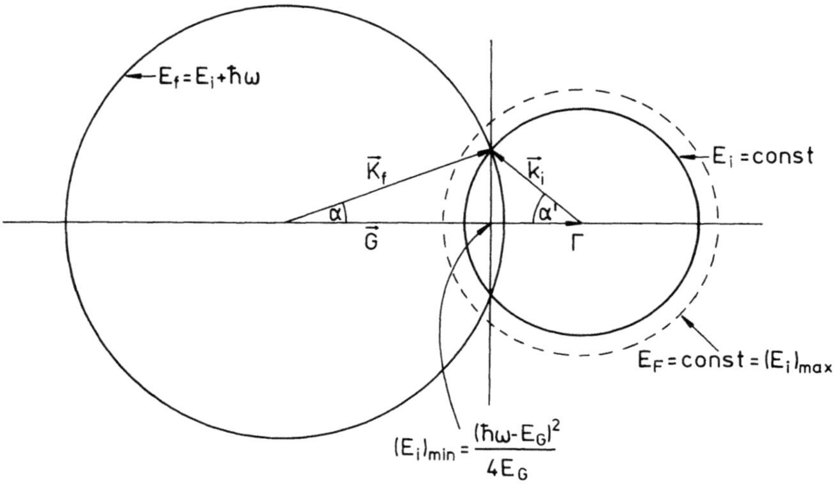

Fig. 6.10. Momentum relation for a PE experiment with off-normal emission. The origins of $\boldsymbol{k}_i$ and $\boldsymbol{K}_f$ are separated by $\boldsymbol{G}$. The minimum initial state energy is $(E_i)_{\min} = (\hbar\omega - E_G)^2/(4E_G)$, see (6.28d). The possible transitions are on the plane normal to $\boldsymbol{G}$ at $(E_i)_{\min}$; $\boldsymbol{k}_i$ and $\boldsymbol{K}_f$ have to meet on that plane. The maximum initial-state energy is the Fermi energy $(E_i)_{\max} = E_{\mathrm{F}}$ [6.6, 6.15]

angle is obtained for the maximum energy of the initial state, which, by definition, is the Fermi energy. This yields

$$(E_i)_{\max} = E_f \quad \text{and} \quad (E_f)_{\max} = E_{\mathrm{F}} + \hbar\omega \tag{6.35}$$

and thus

$$1 \geq \cos^2\alpha = \frac{(\hbar\omega + E_G)^2}{4E_G E_f} = \frac{(\hbar\omega + E_G)^2}{4E_G(E_{\mathrm{F}} + \hbar\omega)} .$$

Therefore, one has

$$\alpha_{\max} = \cos^{-1}\left(\frac{(\hbar\omega + E_G)}{2[E_G(E_f + \hbar\omega)]^{1/2}}\right) . \tag{6.36}$$

The energy range of $\hbar\omega$ for which *internal* photoabsorption is possible is therefore limited and one has

$$1 \geq \frac{(\hbar\omega + E_G)^2}{4E_G(E_{\mathrm{F}} + \hbar\omega)}$$

$$4E_G(E_{\mathrm{F}} + \hbar\omega) \geq (\hbar\omega + E_G)^2 = (\hbar\omega)^2 + 2(\hbar\omega E_G) + E_G^2 , \tag{6.37}$$

$$4E_G E_{\mathrm{F}} + 4E_G\hbar\omega \geq (\hbar\omega)^2 + 2\hbar\omega E_G + E_G^2 ,$$

$$0 \geq \hbar\omega^2 - 2\hbar\omega E_G + E_G^2 - 4E_G E_{\mathrm{F}} ,$$

$$0 \geq (\hbar\omega - E_G)^2 - 4E_G E_{\mathrm{F}} , \tag{6.38}$$

which finally yields

$$\hbar\omega_{\rm min} = E_G - 2(E_G E_{\rm f})^{1/2} \le \hbar\omega \le E_G + 2(E_G E_{\rm F})^{1/2} = \hbar\omega_{\rm max} \,. \quad (6.39)$$

This is the photon energy range for which photoemission is possible for a nearly free electron metal assuming only direct transitions in the reduced zone scheme for a given $\boldsymbol{G}$.

6.2.2 Primary and Secondary Cones in the Photoemission from a Real Solid

It has been demonstrated in the preceding section that photoemission is only possible with the assistance of a reciprocal lattice vector. This led to the wave-vector diagram in Fig. 6.7. In this diagram the values for the wave vectors of the final state are situated on a cone around the reciprocal lattice vector yielding the transition in a nearly free electron approximation. This cone is known [6.7] as the *primary cone.*

So far we have assumed, for simplicity, that the crystal just provides one reciprocal lattice vector, in order to make photoemission possible.

In a real solid the crystal potential, however, mixes wave functions with different $\boldsymbol{G}$ (symmetry permitting). The Bloch function is written

$$\psi_f(\boldsymbol{k}) = \exp(i\boldsymbol{k}\cdot\boldsymbol{r}) \sum_{\boldsymbol{G}\neq\boldsymbol{G}_1} u_f(\boldsymbol{k},\boldsymbol{G}) \exp(i\boldsymbol{G}\cdot\boldsymbol{r}) \,, \quad (6.40)$$

where $\boldsymbol{k} = \boldsymbol{k}_i + \boldsymbol{G}_1$ and $\boldsymbol{G}_1$ is the smallest reciprocal lattice vector that makes photoemission possible. It was taken out of the summation to indicate that in most experiments only the first Mahan cone is observed. Therefore addi-

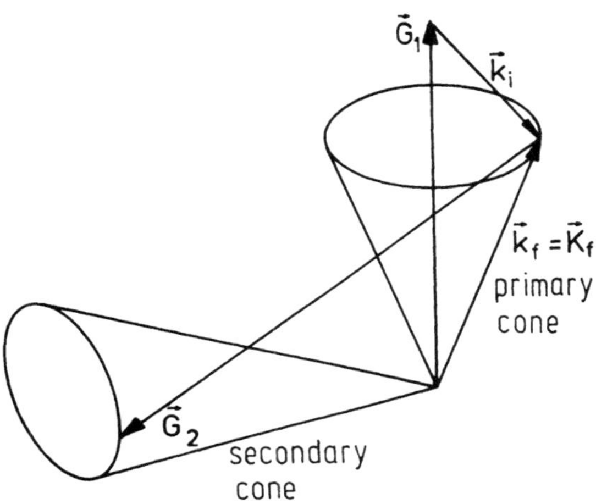

Fig. 6.11. Schematic representation of a primary and a secondary cone [6.7]. The secondary cone is reached from the primary cone by adding an additional reciprocal lattice vector $\boldsymbol{G}_2$ to the momentum triangle; generally the secondary cone is not "conical"

tional eigenfunctions with different wave vectors are mixed into the primary eigenfunction leading to new directions of electron emission. One has

$$\boldsymbol{k} = \boldsymbol{k}_i + \boldsymbol{G}_1 : \text{ primary cone}$$
$$\boldsymbol{k} = \boldsymbol{k}_i + \boldsymbol{G}_1\boldsymbol{G}_2 : \text{ secondary cone, etc.}$$

(Note that the convention "cone" for the higher-order emission directions is misleading because the electron distributions are not conical in the sense as it is the case for the primary cone distributions [6.7]). The situation is sketched in Fig. 6.11. This shows in essence that the photoexcited electrons can propagate in the crystal in any direction for which wave-vector conservation is achieved. This has to be taken into account if single crystal spectra are to be analyzed rigorously.

6.2.3 Angle-Integrated and Angle-Resolved Data Collection

The parameters for a photoemission experiment from a solid are sketched in Fig. 6.12. A photon is incident at an angle ψ with respect to the surface normal and via the photoemission process electrons are liberated from the sample and are detected in a solid angle $\mathrm{d}\Omega(\vartheta, \varphi)$. The measured quantity is $N(E, \hbar\omega, \boldsymbol{k}, \boldsymbol{K})$, being the Energy Distribution Curve (EDC) as a function of both the angular parameters and the energy E of the emitted electrons. One generally distinguishes between two different geometries, namely angle-resolved and angle-integrated measurements.

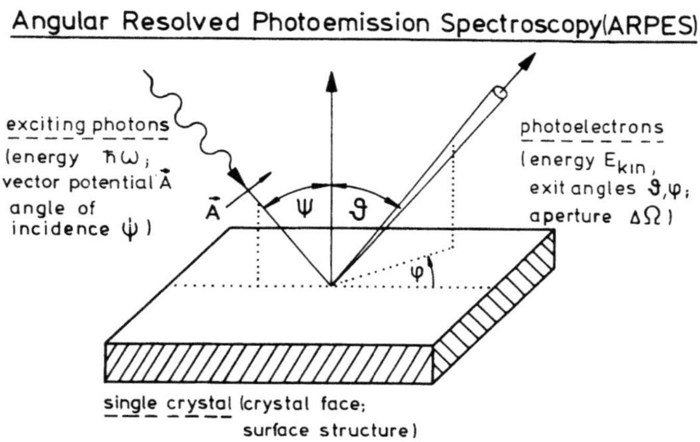

Fig. 6.12. Definition of the parameters in an angle-resolved PE (ARPE) experiment; $\hbar\omega$: photon energy, $\boldsymbol{A}$: vector potential of the radiation, ψ: angle of incidence of photon, ϑ: polar angle of the detected electrons, φ: azimuthal angle of the detected electrons with respect to the crystal axes, $\Delta\Omega$: detector acceptance angle

I. Angle-resolved measurements. In this case single crystal samples are used and the solid angle of the detection is small (preferably $\leq 2°$) which allows exploitation of $\boldsymbol{k}$-conservation by detection in a narrow $\boldsymbol{k}$-interval. For the excitation in the solid both energy and momentum are conserved. The electrons excited in the sample must have a kinetic energy larger than the work function in order to escape from the solid. Four major parameters control the experiment, namely the two angles ϑ, φ, specifying the direction of the electrons, their kinetic energy E_{kin} and the energy of the impinging radiation $\hbar\omega$.

In Fig. 6.13 the results of an angle-resolved experiment are shown schematically. One observes a number of lines (two in the case of Fig. 6.13) as a function of kinetic energy. This spectrum is obtained from (6.16) by separating out the $\boldsymbol{k}$-conservation from $M^1_{fi}(\boldsymbol{k}_f, \boldsymbol{k}_i)$ which yields $\tilde{M}^1_{fi}(\boldsymbol{k}_f, \boldsymbol{k}_i)$ by adding the $\boldsymbol{k}_{||}$-conservation in the spirit of the three-step model and by specifying the electron-detection condition with a particular set of angles (ϑ, φ). We have

$$N(E, \hbar\omega, \boldsymbol{k}, \boldsymbol{k}_{||}) \propto \underbrace{\sum_{i,f} |\tilde{M}^1_{fi}(\boldsymbol{k}_f, \boldsymbol{k}_i)|^2}_{\text{selection rules}} \; \underbrace{\delta(E_f - E_i - \hbar\omega)}_{\text{energy conservation}} \; \underbrace{\delta(\boldsymbol{k}_f - \boldsymbol{k}_i - \boldsymbol{G})}_{\text{momentum conservation}}$$

$$\times \underbrace{\delta(\boldsymbol{K}_{||} - \boldsymbol{p}_{||}(\vartheta, \varphi)/\hbar)}_{\text{momentum conservation parallel to the surface}} \cdot \underbrace{\delta(E - E_f + \phi)}_{\text{detection by an analyzer}} \tag{6.14b}$$

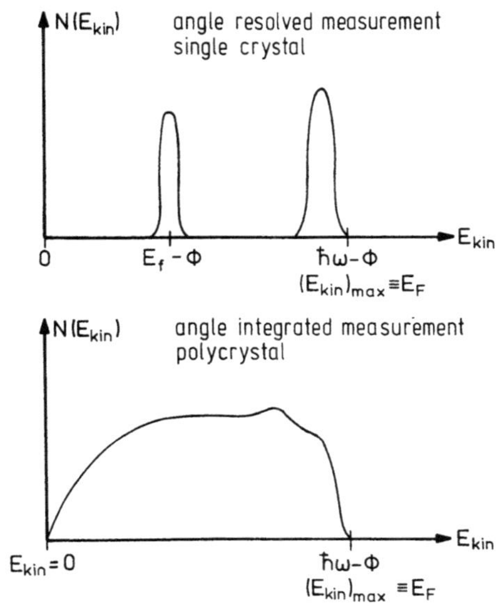

Fig. 6.13. Schematic diagram of the energy distribution curves obtained in an angle-resolved and an angle-integrated experiment

The discrete nature of the spectrum is a result of the various conservation laws and selection rules.

Another way of interpreting the experiment is illustrated in Fig. 6.14, where the top part gives the experiment in real space and the lower part shows the energetics involved. In contrast to the internal PE process, the zero of energy is now important and will be taken as the Fermi energy. This implies that in order for electrons to be observable in normal emission, one requires $E_{\text{kin}} \geq 0$ or $E_f \geq \phi$. Secondly, for electrons escaping at off-normal emission, refraction has to be taken into account in the way outlined in Figs. 6.4 and 6.5. Thus in the equations describing off-normal detection, a correction according to the procedures given in Figs. 6.4 and 6.5 has to be performed. We refer the reader to [6.7] for details.

Refraction is important for experiments in off-normal emission in the ARUPS mode and in the next chapter its effect will be documented. However, if one works with polycrystalline samples even in the UPS mode, while refraction is of course present, it does not show up in the experimental data and in XPS it is hardly noticeable at all. In the following equations we implicitly assume normal electron detection outside the crystal.

II. Angle-integrated measurements. The solid angle of detection is large in this type of experiment (under favorable conditions close to 2π), which means that $\boldsymbol{k}$-information is blurred by the integration over all outgoing wave vectors. Such measurements can be performed on single crystal faces but one more usually employs polycrystalline samples and they yield a joint density of states.

Before proceeding to the actual calculation of a joint density of states, a few useful definitions will be given [6.15], to facilitate a better understanding of the experiment where we write $\boldsymbol{k}_i = \boldsymbol{k}_f = \boldsymbol{k}$ (assuming that $\boldsymbol{k}_f$ is in the first Brillouin zone) and $E_{\text{kin}} = E$.

1. A plane of constant interband energy

If $E_f(\boldsymbol{k})$ and $E_i(\boldsymbol{k})$ represent the energy dispersion relations in the two bands f and i, respectively, the optical transitions at photon energy $\hbar\omega$ are confined to a surface $\Omega_{fi}(\boldsymbol{k})$ in $\boldsymbol{k}$-space; for all points of this surface one has

$$E_f(\boldsymbol{k}) - E_i(\boldsymbol{k}) - \hbar\omega = 0 \,. \tag{6.41}$$

2. The combined (joint) density of states

This quantity is obtained if the $\boldsymbol{k}$-integration is performed for a surface of constant interband energy leading to

$$J(\hbar\omega) = \frac{1}{(2\pi)^3} \sum_{i,f} \int \Omega_{fi}(\boldsymbol{k}) \mathrm{d}^3 k \,. \tag{6.42}$$

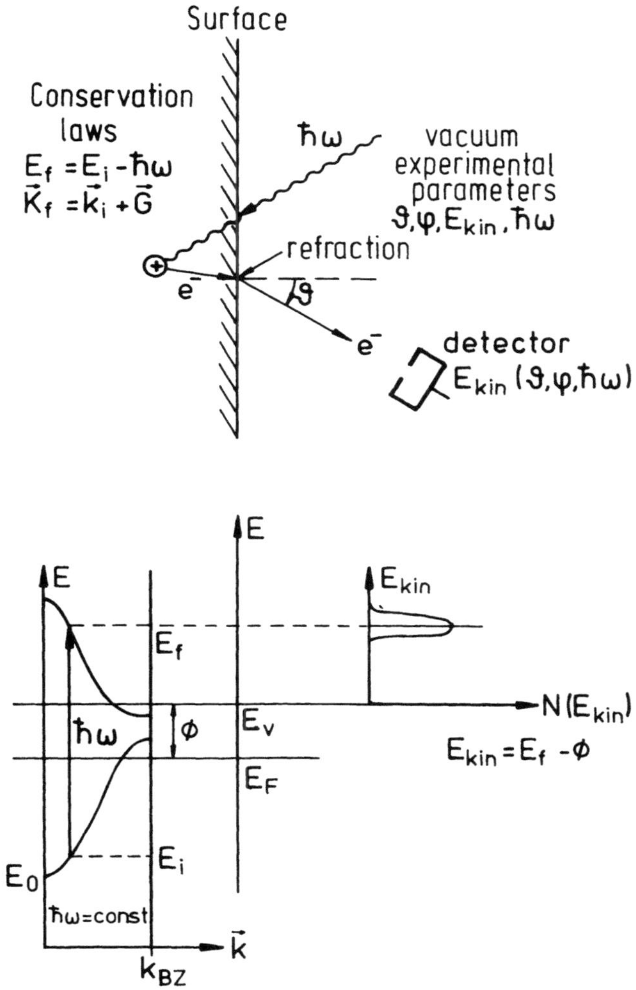

Fig. 6.14. PE experiment on a solid as viewed in real space (*top*) and in momentum space (*bottom*)

The joint density of states gives the total number of direct transitions that are possible for a particular photon energy. It can also be written as [6.16]

$$J(\hbar\omega) = \frac{1}{2\pi^3} \sum_{i,f} \int \frac{\mathrm{d}S_{fi}}{|\nabla_{\boldsymbol{k}} E_f(\boldsymbol{k}) - \nabla_{\boldsymbol{k}} E_i(\boldsymbol{k})|} \tag{6.43}$$

where $\mathrm{d}S_{fi}$ is an element on the surface $\Omega_{fi}(\boldsymbol{k})$.

3. The energy distribution of the combined (joint) density of states

This quantity is defined by

$$D(E,\hbar\omega) = \frac{1}{(2\pi)^3}\sum_{i,f}\int \Omega_{fi}(\boldsymbol{k})\delta(E - E_f + \phi)\mathrm{d}^3k \tag{6.44}$$

where $\delta(E - E_f + \phi)$ selects the transitions that involve the energy E_f.

It can also be written as [6.16]

$$D(E,\hbar\omega) = \frac{1}{(2\pi)^3}\sum_{i,f}\int \frac{\mathrm{d}l_{fi}}{|\nabla_{\boldsymbol{k}}E_f(\boldsymbol{k}) \times \nabla_{\boldsymbol{k}}E_i(\boldsymbol{k})|} \tag{6.45}$$

where the sign $\times$ designates a vector product and $\mathrm{d}l_{fi}$ is an element of the line that is produced by the intersection of the surfaces $E_i = E$ and $E_f = E + \hbar\omega$ (Fig. 6.10).

Note that none of these definitions contains weighting factors meaning that we have assumed that the matrix elements are constant. These would have to be added if we wanted to treat a real example.

4. The Density Of States (DOS)

For completeness we add the definition of the density of states (DOS):

$$D(E) = \frac{1}{(2\pi)^3}\int \frac{\mathrm{d}S_E}{|\nabla_{\boldsymbol{k}}E|} \tag{6.46}$$

where $\mathrm{d}S_E$ is an element on the surface of constant energy and $\nabla_{\boldsymbol{k}}E$ is the gradient of the energy.

An experiment performed with a fixed photon energy $\hbar\omega$ and integration over the total solid angle (which is of course an oversimplification) yields the energy distribution curve $N(E,\hbar\omega)$. We now ask: what is the meaning of $N(E,\hbar\omega)$ in terms of the electronic properties of the sample? Here two limiting cases will be treated, that involving only direct (vertical) transitions, and that involving only indirect transitions.

Direct Transition Case

Here one is considering transitions which are vertical in the reduced zone scheme meaning that if:

> $\boldsymbol{K}_f = \boldsymbol{k}_i + \boldsymbol{k}$ is the final-state wave vector of a photoexcited electron within the solid,

one has

> $\boldsymbol{k}_f = \boldsymbol{k}_i = \boldsymbol{k}$ ($\boldsymbol{k}$ being in the first Brillouin zone) for the wave vector of the final and initial state in the reduced zone scheme.

In order to relate the measured EDC, namely $N(E,\hbar\omega)$, to the theoretical photocurrent, one has to integrate (6.16) over the total solid angle yielding[8]

$$N_{\text{direct}}(E,\hbar\omega) \propto \sum_{i,f} \int \mathrm{d}^3\boldsymbol{k}_f \mathrm{d}^3\boldsymbol{k}_i |M^1_{fi}(\boldsymbol{k}_f,\boldsymbol{k}_i)|^2 \delta(E_f(\boldsymbol{k}_f) - E_i(\boldsymbol{k}_i) - \hbar\omega) \times\delta(E - E_f + \phi) . \tag{6.47}$$

This can now be related to the quantity $\varepsilon_2\omega^2$ ($\varepsilon = \varepsilon_1 + \mathrm{i}\varepsilon_2$ is the dielectric constant) measured in an optical experiment. In the latter one has:

$$\varepsilon_2\omega^2 \propto \sum_{i,f} \int \mathrm{d}^3k |M^1_{fi}(\boldsymbol{k}_f,\boldsymbol{k}_i)|^2 \delta(E_f(\boldsymbol{k}_f) - E_i(\boldsymbol{k}_i) - \hbar\omega) \tag{6.48}$$

and thus, in this approximation, the photoemission measures the imaginary part of the dielectric constant multiplied by ω^2.

If we assume that the matrix element $|M^1_{fi}(\boldsymbol{k}_f,\boldsymbol{k}_i)|^2$ is constant and setting $\boldsymbol{k}_i = \boldsymbol{k}_f = \boldsymbol{k}$, we have

$$N_{\text{direct}}(E,\hbar\omega) \propto \sum_{i,f} \int \mathrm{d}^3k\ \delta(E_f(\boldsymbol{k}) - E_i(\boldsymbol{k}) - \hbar\omega)\delta(E - E_f + \phi) . \tag{6.49}$$

From this equation one sees using (6.15 and 6.44) that, in the case of direct transitions only and for a polycrystalline sample, the PE experiment measures the Energy Distribution of the Joint Density Of States EDJDOS namely $D(E,\hbar\omega)$

$$N_{\text{direct}}(E,\hbar\omega) \propto D(E,\hbar\omega) . \tag{6.50}$$

Indirect-Transition Case

The other limiting case is reached when the experimental conditions are chosen in a way such that transitions from all initial states are possible which means that one can perform the integration over $\boldsymbol{k}_i$ and $\boldsymbol{k}_f$ independently. This is the case most relevant to measurements with XPS photon energies ($\hbar\omega = 1254\,\text{eV}$, $1487\,\text{eV}$). In these experiments, generally, the acceptance angle is so large that transitions in the whole Brillouin zone are possible. Here one has

$$N_{\text{indirect}}(E,\hbar\omega) \propto \sum_{i,f} \int \mathrm{d}^3\boldsymbol{k}_i \mathrm{d}^3\boldsymbol{k}_f |M^1_{fi}|^2 \times\delta(E_f(\boldsymbol{k}_f) - E_i(\boldsymbol{k}_i) - \hbar\omega)\delta(E - E_f + \phi) \tag{6.51}$$

which results in

[8] The additional $\delta(E - E_f + \phi)$ which shows up in (6.47) but not in (6.48) takes care of the fact that in photoemission the energy of the photoexcited electron, E, must be sufficient for it to escape into the vacuum. The optical experiments can also reach final states below the vacuum.

$$N_{\text{indirect}}(E,\hbar\omega) \propto DOS(E_i)DOS(E_f)|\bar{M}^1_{fi}|^2 \,, \tag{6.52}$$

where $|\bar{M}^1_{fi}|$ is an averaged matrix element. Thus, assuming a constant matrix element, this experiment measures a product of the initial and final densities of states. Assuming, as is usually done in XPS experiments, that the final density of states is constant over the interval of the measurement, this type of experiment actually measures the density of initial states provided the matrix element is constant, see (6.17).

The above considerations will now be applied to calculate the EDJDOS of a simple metal. We assume that all energy bands may be represented by the nearly free-electron model. For that purpose one rewrites (6.49) with the help of (6.45) (assuming direct transitions only) as

$$N_{\text{direct}}(E,\hbar\omega) \propto D(E,\hbar\omega) = \sum_{i,f} \frac{1}{(2\pi)^3} \int \frac{dl_{fi}}{|\nabla_{\boldsymbol{k}} E_f(\boldsymbol{k}) \times \nabla_{\boldsymbol{k}} E_i(\boldsymbol{k})|} \,. \tag{6.53}$$

Using the relations

$$E_f = (\hbar^2/2m)(\boldsymbol{k}_i + \boldsymbol{G})^2, \quad E_i = (\hbar^2/2m)\boldsymbol{k}_i^2 \tag{6.54}$$

we have

$$\begin{aligned} |\nabla_{\boldsymbol{k}} E_f \times \nabla_{\boldsymbol{k}} E_i| &= |(2\hbar^2/2m)(\boldsymbol{k}_i + \boldsymbol{G}) \times (2\hbar^2/2m)\boldsymbol{k}_i| \\ &= 4(\hbar^2/2m)^2|\boldsymbol{G} \times \boldsymbol{k}_i| = 4(\hbar^2/2m)^2 k_i G \sin\alpha' \,, \end{aligned} \tag{6.55}$$

where α' is the angle between $\boldsymbol{G}$ and $\boldsymbol{k}_i$ and therefore

$$D(E,\hbar\omega) = \frac{1}{(2\pi)^3} \sum_{i,f} \int \frac{dl_{fi}}{4(\hbar^2/2m)^2 k_i G \sin\alpha'} \,. \tag{6.56}$$

Because of the cylindrical symmetry with respect to $\boldsymbol{G}$, $|\boldsymbol{k}_i|$ and α' are constant along the line of integration, and one can therefore write

$$D(E,\hbar\omega) = \sum_{i,f} \frac{1}{4(\hbar^2/2m)^2(2\pi)^3 k_i G \sin\alpha'} \int \mathrm{d}l_{fi} \,. \tag{6.57}$$

Now

$$\int \mathrm{d}l_{fi} = 2\pi k_i \sin\alpha' \tag{6.58}$$

and therefore

$$D(E,\hbar\omega) = \sum_{i,f} \frac{2\pi}{4(\hbar^2/2m)^2(2\pi)^3 G} \,. \tag{6.59}$$

Since $N_{\text{direct}}(E,\hbar\omega) \propto D(E,\hbar\omega)$ this means that the electron energy distribution is a constant for (nearly) free electron metals where the maximum initial state energy is E_{F} and the minimum one is $(\hbar\omega - E_G)^2/4E_G$ as given in (6.28a). Therefore we have

$$N_{\text{direct}}(E,\hbar\omega) = \begin{cases} \text{const.} \quad \text{for } E_{\min} = \frac{(\hbar\omega - E_G)^2}{4E_G} < E < E_{\max} = E_{\text{F}} \\ 0 \quad \text{otherwise} \end{cases} \,. \tag{6.60}$$

One thus expects a rectangular EDJDOS with

$$E_{\max} - E_{\min} = E_{\mathrm{F}} - \frac{(\hbar\omega - E_G)^2}{4E_G} . \tag{6.61}$$

(This is the negative of the binding energy referenced to E_{F}; for a zero at the bottom of the conduction band this is a positive number and therefore in this section the zero has been placed at the bottom of the conduction band).

The width of the rectangular EDC varies with the square of the frequency of the exciting radiation. It has been assumed in this derivation that $|M^1_{fi}(\boldsymbol{k})|^2 = \mathrm{const}$, and it can be shown that this is a valid approximation for a nearly free-electron system.

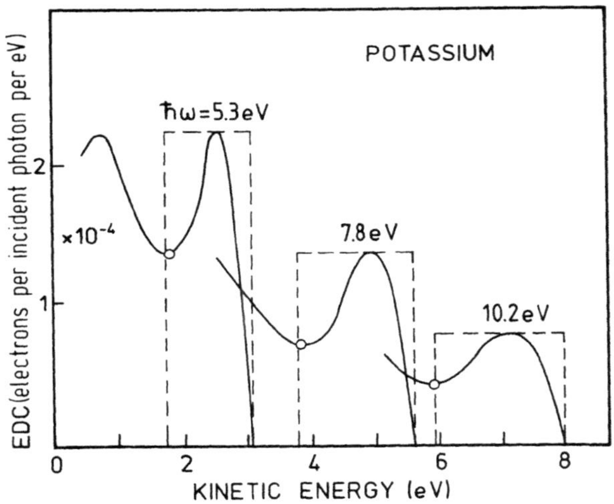

Fig. 6.15. EDCs from polycrystalline K as a function of the kinetic energy for $\hbar\omega = 5.3\,\mathrm{eV}$, $\hbar\omega = 7.8\,\mathrm{eV}$ and $\hbar\omega = 10.2\,\mathrm{eV}$ [6.15]. The EDCs expected from (6.61) (*square shaped*) are indicated

These ideas are roughly supported by experiment [6.15]. Figure 6.15 shows experimental EDCs obtained with three different photon energies, 5.3 eV, 7.8 eV, and 10.2 eV, for a typical free electron metal, namely a polycrystalline potassium sample in the angle-integrating mode. Also indicated are the expected EDCs (6.61) where the heights of the columns have been adjusted to the experimental curves. The agreement between theory and experiment can be considered fair with respect to the width of the EDC. In Fig. 6.16 one finds the experimentally determined minimum energies (open dots in Fig. 6.15) compared to those calculated according to $E_{\min} = (\hbar\omega - E_G)^2/4E_G$ for $\boldsymbol{G} = (110)$ and $\boldsymbol{G} = (200)$, see (6.28a). The results indicate that the transitions observed in the experiment are for $\boldsymbol{G} = (110)$.

In order to avoid confusion, a further remark about the energy zero is necessary. In Fig. 6.15 the EDCs are plotted with respect to the kinetic energies. The high energy cutoffs in these curves give the Fermi energy in the curves for the different photon energies. The kinetic energies are converted to binding energies E_B by the relation: $E_{kin} + \phi + E_B = \hbar\omega$. The work function ϕ can be obtained from the low kinetic energy cutoffs which are not shown in Fig. 6.15. In Fig. 6.16 the binding energies (referenced to E_F) are used in accordance with the usual convention. In (6.28a) the energy is referenced to the bottom of the conduction band E_0 ($E_0 = 2.12\,\text{eV}$ for K). The energies calculated from (6.28a) have been shifted by E_0 in order to plot them as a function of E_B in Fig. 6.16. Alternatively one can, of course, use (6.61) directly; this gives the energies with respect to E_F.

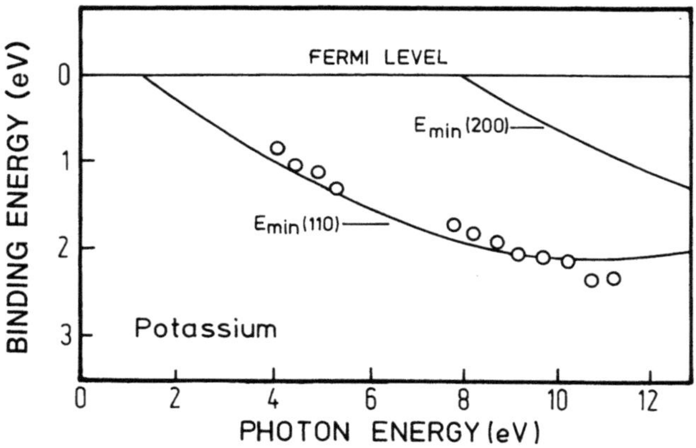

Fig. 6.16. Energy position of the open dots in Fig. 6.15 measured with respect to the Fermi energy (cut-off at the right-hand side of the curves in Fig. 6.15) as a function of the photon energy $\hbar\omega$. Also given is the curve $E_{min} = (\hbar\omega - E_G)^2/4E_G$, for $G(110)$ and $G(200)$ [6.15]

6.3 Photoemission from the Semi-infinite Crystal: The Inverse LEED Formalism

To understand the PE process in a solid, we have so far proceeded stepwise in order to reveal the underlying physics. First the phenomenological three-step model was introduced. The next step was to write down an equation for the photocurrent I or the electron energy distribution $N(E, \hbar\omega)$, see (6.17), using the Fermi Golden Rule (6.15). This equation (6.15) is in principle the correct formulation of the photoemission process and one can also call it a one-step theory provided that the final state is taken to be that of the escaping electron in the vacuum. Note that (6.15) was only applied to the photoabsorption process itself in Sect. 6.2; propagation to the surface and penetration through the surface were added on separately, thus making the theory not truly one step. In evaluating (6.15), wave-number conservation was extracted somewhat artificially. It was then postulated that wave-number and energy conservation determine the information obtained from a PES experiment.

We now consider the extent to which a one-step theory devised along the lines of (6.15) and the three-step model (6.14a, 6.14b) are actually related to one another.

What physical effects have so far been neglected in our treatment of the PE process? The most significant is the relatively short mean free path of the photoexcited electrons and care will now be exercised to take this properly into account. To do this, it is useful to turn to a theory that has treated the problem of the short mean free path of eV electrons in great detail; this is the theory of LEED (Low-Energy Electron Diffraction).

The LEED process is shown schematically in Fig. 6.17. A beam of low energy electrons with velocity $-\boldsymbol{v}$ (the usefulness of the minus sign will become apparent in a moment) impinges on a sample. The direction vector of the beam is $\hat{\boldsymbol{R}} = \boldsymbol{v}/|\boldsymbol{v}|$. At the surface it is split into a beam that penetrates into the sample and a reflected beam. If one now sets the reflected beam to zero, reverses the velocities of the two remaining beams and adds the impinging photon, one has the situation characteristic for a PES experiment. Hence the approach to the PE process employing the LEED theory is called the *inverse LEED theory* of PES.

There have been a number of treatments of PES along these lines [6.7, 6.17–6.27]. Here we follow that of Mahan [6.7] and of Feibelman and Eastman [6.23], since these readily allow the recovery of the three-step model [6.4, 6.30, 6.31]. Later on, a "complete one-step theory" as used by Pendry will be mentioned briefly [6.25, 6.26].

Fairness requires the author to state that he is deeply impressed by the formidable paper of Mahan [6.7] on the theory of PES, the reading of which conveys great insight into this subject.

In the asymptotic form, i.e., at large distances from the solid-vacuum interface, the LEED functions corresponding to Fig. 6.17, can easily be written down. The wave function has to be split up into two parts parallel and perpendicular to the surface. The wave function in the vacuum is

$$\psi^{\mathrm{L}}_{z\to\infty} \Rightarrow \exp(\mathrm{i}\boldsymbol{K}_{\|}\cdot\boldsymbol{\varrho})\{\exp[-\mathrm{i}(p_{\perp}/\hbar)z] + r\exp[\mathrm{i}(p_{\perp}/\hbar)z]\} \tag{6.62}$$

and in the solid

$$\psi^{\mathrm{L}}_{z\to-\infty} \Rightarrow t\exp(\mathrm{i}\boldsymbol{K}_{\|}\cdot\boldsymbol{\varrho} - \mathrm{i}K_{\perp}z)\psi(\boldsymbol{k},z) \tag{6.63}$$

where z is along the surface normal and points out of the surface, and $\psi(\boldsymbol{k},z)$

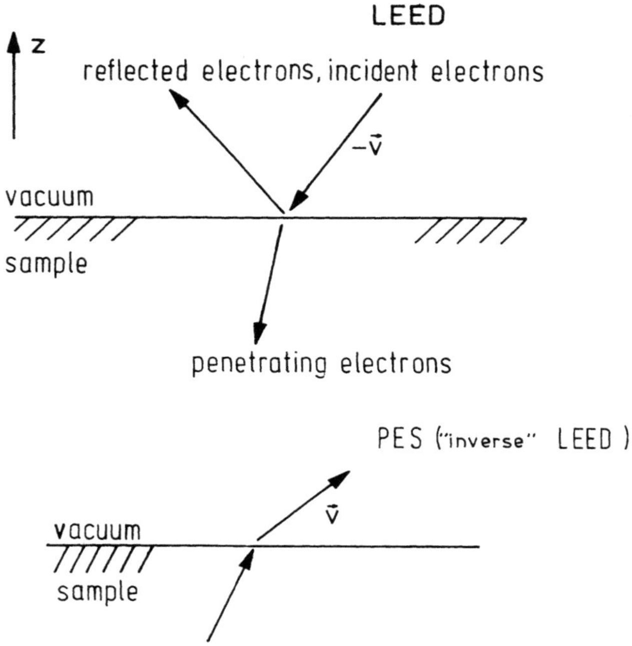

Fig. 6.17. Asymptotic wave functions in LEED (*top*) and PES (*bottom*). In LEED there is an incoming electron beam ($-\boldsymbol{v}$), a reflected beam and a penetrating beam. In PES only the time-reversed penetrating and incoming ($+\boldsymbol{v}$) beams are used

is a bulk Bloch function. The surface is the $x-y$ plane and $\boldsymbol{\varrho} = \boldsymbol{e}_x x + \boldsymbol{e}_y y$ is a vector in the surface; in addition r and t are the reflection and transmission coefficients, respectively. Since the transition matrix element will be evaluated in the crystal, the final state wave function to be used in the inverse LEED formalism will have the form of (6.63).

In treating PES via the *inverse LEED formalism* one takes the LEED wave function ψ^{L} for the final-state electrons in PE with the incident electron beam velocity $-\boldsymbol{v}$ taken equal to the negative of the emitted photoelectron velocity (which is then $\boldsymbol{v}$). For the case in which hole damping can be neglected, and using the independent electron approximation, one obtains the following Golden Rule formula for the EDC along the observation direction $\hat{\boldsymbol{R}} = \boldsymbol{v}/|\boldsymbol{v}|$ [see also (6.16)]:

$$N(E, \hbar\omega, \hat{\boldsymbol{R}}) \propto \boldsymbol{v} \cdot \sum_{\text{occupied},i} (|\langle\psi^{\mathrm{L}}(\boldsymbol{r}, E, \boldsymbol{K})|H^{\mathrm{int}}|\psi_i(\boldsymbol{r}, \boldsymbol{k})\rangle|^2)|M_{fi}|^2 \times\delta(E_f - E_i - \hbar\omega) \tag{6.64}$$

where $E = E_f - \phi$ is the kinetic energy of the photoelectron, $\hat{\boldsymbol{R}} = \boldsymbol{v}/|\boldsymbol{v}|$ is the unit vector along the photoelectron beam, and $\boldsymbol{k} = \boldsymbol{\varrho} + \boldsymbol{e}_z z$. The outgoing electron velocity $\boldsymbol{v}$ is equal to $[2(E)/m]^{1/2}$ and E_i and ψ_i are the eigenvalues and (occupied) eigenstates, respectively, of the Schrödinger equation. This is also the correct expression for the photocurrent in a one-step theory.

The two-dimensionality of the surface manifests itself in the fact that only the wave vector parallel to the surface is conserved. The wave functions ψ^{L} and ψ_i are therefore conveniently expanded in a way that takes this into account and thus one writes them as two-dimensional Bloch functions of the position vector $\boldsymbol{\varrho}$ along the surface. However, an inspection of the electron wave function with respect to the wave-vector components perpendicular to the surface is necessary to recover the direct transition excitation. To describe the wave functions in the solid one may add an imaginary contribution to the potential in the Schrödinger equation (optical potential) which "mimics" the damping of the electrons (lifetime broadening).

The various wave function combinations are given in Fig. 6.18. For the final state one has in situation (a) the matching of a free electron wave in the vacuum to a propagating Bloch wave in the crystal, with small damping. In case (b) the wave in the vacuum is matched such that it ends up in a band gap of the crystal, resulting in what is called an *evanescent wave.* In this terminology propagating means behaving as a Bloch wave for $z \to -\infty$ and evanescent means that the wave is damped exponentially (imaginary wave vector). Case (c) for the final state corresponds to band-to-band photoemission in a system with a small electron escape depth. In summary, the final states in the crystal can be written as sums of propagating and evanescent waves expanded in terms of two-dimensional Bloch functions

$$\psi^{\mathrm{L}} \propto \exp(\mathrm{i}\boldsymbol{k}_{f\|} \cdot \boldsymbol{\varrho}) \sum_m t_m \exp(\mathrm{i}k_{\perp m} z) u_m(\boldsymbol{r}, \boldsymbol{k}_{f\|}, E) \,. \tag{6.65}$$

(For convenience the indices k, E are dropped from the transmission coefficient).

One now has

$$k_{\perp m} = k_{\perp m}^{(1)} + \mathrm{i}k_{\perp m}^{(2)} \tag{6.66}$$

and the magnitude of $k_{\perp m}^{(2)}$ determines whether one is dealing with a propagating ($k_{\perp m}^{(2)} = 0$) or an evanescent (damped) wave.

The $\boldsymbol{k}$-value of a propagating wave is nearly real with a small imaginary part resulting from inelastic scattering, whereas an evanescent wave has an imaginary part both from inelastic scattering and because its energy lies in the band gap. Thus the latter represents a state that is confined to the surface.

The initial state wave function can likewise describe two situations. The first is a propagating wave with one component traveling towards the surface and another which, after reflection, travels away from it and an evanescent wave which reaches out into the vacuum [case (d) in Fig. 6.18]. In the case of an initial surface state the propagating Bloch wave is of course missing, and one has only the state confined to the surface, decaying both into the

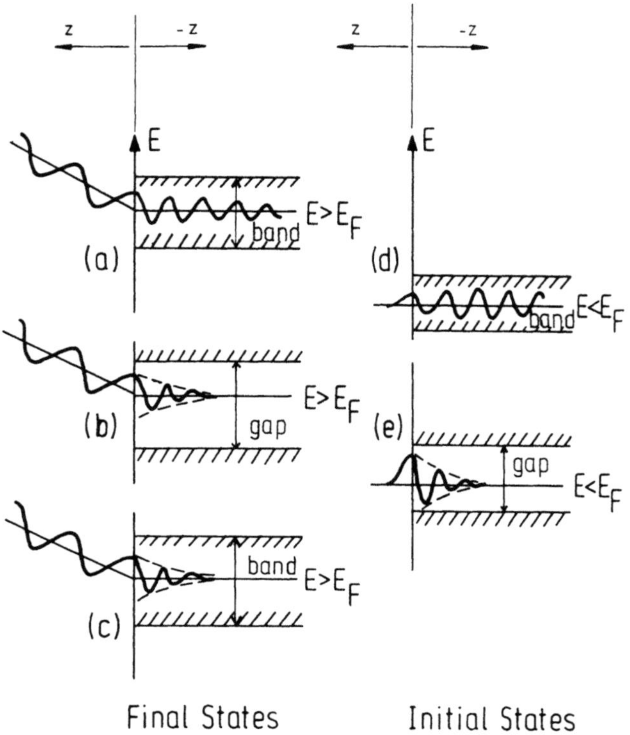

Fig. 6.18. Wave functions involved in PES in the LEED formalism [6.17]. Final states: (**a**) bulk Bloch wave, weakly damped; (**b**) strongly damped gap state (surface state); (**c**) strongly damped Bloch wave (short mean free path). Initial states, (**d**) bulk Bloch wave; (**e**) surface state in the gap

vacuum and into the bulk of the sample; case (e). One can thus write the wave function of the initial state in terms of two-dimensional Bloch functions as

$$\psi_i = \exp(\mathrm{i}\boldsymbol{k}_{i\|} \cdot \boldsymbol{\varrho}) \sum_n C_n \exp(\mathrm{i}k_{\perp n} z) v_n(\boldsymbol{r}, \boldsymbol{k}_{i\|}, E_i) \tag{6.67}$$

and again $k_{\perp n} = k_{\perp n}^{(1)} + \mathrm{i}k_{\perp n}^{(2)}$ in order to describe a traveling wave and a damped state.

At this point we are only interested in bulk band states, in order to recover the three-step model of photoemission from the Golden Rule expression. One therefore sets $k_{\perp n}^{(2)} = 0$; $k_{\perp m}^{(2)}$ is small but finite (weak damping) such that $a_\perp k_{\perp m}^{(2)} \ll 1$, $a_\perp$ being the lattice constant perpendicular to the surface normal.

The matrix element M_{fi} in (6.64) can be split up into one for the surface region M_{fi}^{S} (integration over a thickness which contains the region that deviates from the bulk) and one for the true bulk (M^{B}_{fi}) meaning a region where the properties of the crystal correspond to the situation of the perfect crystal. We neglect M^{S}_{fi} in the following treatment.

However, if one wants to calculate an accurate and complete photoemission spectrum, M^{S}_{fi} has to be taken into account [6.23–6.25]. Only with the inclusion of this term the surface states are reproduced by the calculations (Chap. 8).

The matrix element for the bulk contribution M^{B}_{fi} can be further decomposed by realizing that the crystal is periodic with respect to the unit cell:

$$M^{\mathrm{B}}_{fi} = \sum_{\text{one layer}} \exp[\mathrm{i}(\boldsymbol{k}_{i||} - \boldsymbol{k}_{f||}) \cdot \boldsymbol{\varrho}_\ell] \sum_n \sum_m t_m C_n \times \sum_{\perp \text{to surface}} \exp[\mathrm{i}(k_{n\perp} - k_{m\perp}) z_\perp] \underbrace{\int u^*_m H^{\text{int}} v_n \, d\tau_0}_{\tilde{M}^{\mathrm{B}}_{nm}} \tag{6.68}$$

($\mathrm{d}\tau_0 = \mathrm{d}x_0 \mathrm{d}y_0 \mathrm{d}z_0$: integration in one unit cell). This matrix element M^{B}_{fi} can be simplified by performing the sums in the layer and the one perpendicular to it. In the layer one has

$$\sum_{\text{one layer}} \exp[\mathrm{i}(\boldsymbol{k}_{i||} - \boldsymbol{k}_{f||}) \boldsymbol{\varrho}_\ell] = \delta(\boldsymbol{k}_{i||} - \boldsymbol{k}_{f||} + \boldsymbol{G}_{||}) \tag{6.69}$$

which is the momentum conservation rule parallel to the surface.

Then one obtains (setting $z_\perp = r \cdot a_\perp$, where $a_\perp$ is the lattice constant perpendicular to the surface)

$$\varDelta_{nm} = \sum_{r=0}^{\infty} \exp[\mathrm{i}(k_{\perp n} - k_{\perp m}) r \cdot a_\perp] \, . \tag{6.70}$$

This yields with $k^{(2)}_{\perp m} \approx 0$ the momentum conservation perpendicular to the surface. With

$$\varDelta k^{(1)}_\perp = k^{(1)}_{\perp n} - k^{(1)}_{\perp m}$$

we find

$$\varDelta k^{(1)}_\perp \cdot \left(\frac{1}{2} a_\perp \right) = n\pi, \; k^{(1)}_\perp = n \left(\frac{2\pi}{a_\perp} \right) \quad \text{and} \quad \varDelta k^{(1)}_\perp = n \cdot G_\perp \, , \tag{6.71}$$

where $\boldsymbol{G}_\perp$ is the reciprocal lattice vector perpendicular to the surface.

For small but finite damping one obtains

$$|\varDelta_{nm}|^2 = \left((a_\perp k^{(2)}_{\perp m})^2 + \left[a_\perp (k^{(1)}_{\perp n} - k^{(1)}_{\perp m}) \right]^2 \right)^{-1} \tag{6.72}$$

which yields approximately with (6.71)

$$|\varDelta_{nm}|^2 \approx \left[\pi / \left(a^2_\perp k^{(2)}_{\perp m} \right) \right] \delta \left(k^{(1)}_{\perp n} - k^{(1)}_{\perp m} + G_\perp \right) . \tag{6.73}$$

Therefore, one obtains the following equation for the photocurrent only those factors that are important for the subsequent discussion, unnecessary indices, and adding a $\delta(E - E_f + \phi)$ in order to fix t the energy at the Fermi energy):[9]

$$N(E,\hbar\omega) \propto \sum_{n,m} |t_m C_n|^2 \lambda_m \left|\tilde{M}^{\mathrm{B}}_{nm}\right|^2$$
$$\times \delta(k^{(1)}_{\perp n} - k^{(1)}_{\perp m} + G_{\perp})$$
$$\times \delta(\boldsymbol{k}_{i||} - \boldsymbol{k}_{f||} + \boldsymbol{G}_{||})\delta(E_f - \hbar\omega - E_i)\delta(E - E_f +$$

This corresponds to the three-step model as derived on a phenom basis in (6.14a,6.14b):

1) The matrix element $|\tilde{M}^{\mathrm{B}}_{nm}|^2 = \int u^*_m H^{\mathrm{int}} v_n \mathrm{d}\tau_0$ describes the p tation process; B stands for "bulk".
2) $\lambda_m = (k^{(2)}_{\perp m})^{-1}$ describes the average distance from which an ele travel to the surface without scattering (escape depth).
3) $|t_m C_n|^2$ gives the transmission probability of the electron thr surface. For a relation between the transmission coefficient for (state to the total transmission coefficient of all plane waves in direction, see (6.11).

This formula for the photocurrent (6.74a) should be compared derived in Chap. 1, namely (1.62):

$$I(E,\hbar\omega) \propto \sum_{i,f} \frac{\mathrm{Im}\Sigma(\boldsymbol{k}_i)}{\left(E - E^0(\boldsymbol{k}_i) - \mathrm{Re}\Sigma(\boldsymbol{k}_i)\right)^2 + \left(\mathrm{Im}\Sigma(\boldsymbol{k}_i)\right)^2}$$
$$\times \frac{|\tilde{M}_{i,f}|^2}{\left(k^{(1)}_{i,\perp} - k^{(1)}_{f\perp}\right)^2 + \left(k^{(2)}_{f\perp}\right)^2}$$
$$\times \delta\left(\boldsymbol{k}_{i||} - \boldsymbol{k}_{f||} + \boldsymbol{G}_{||}\right)\delta\left(\boldsymbol{k}_i - \boldsymbol{k}_f + \boldsymbol{G}\right)$$
$$\times \delta\left(E^{\mathrm{l}}(\boldsymbol{k}_f) - E^{\mathrm{l}}(\boldsymbol{k}_i) - \hbar\omega\right)\delta\left(E - E^{\mathrm{l}}(\boldsymbol{k}_f) + \phi\right) \cdot f(E,T)$$

The first term in (6.74b) does not occur in (6.74a), because the la was obtained in the one-electron approximation, where the spectral can be set to unity. The transmission process was not considered ex Chap. 1 and therefore does not show up in (6.74b). The damping fac same in (6.74a) and (6.74b) if one keeps in mind that the form in (6 been derived from (6.72), which contains a form identical to that in The $k_{\perp}$-conservation in (6.74a) is implicitly contained in the $\boldsymbol{k}$-con in (6.74b).

[9] For details of the calculation, see [6.8, 6.23].

The δ-functions contain the momentum and energy conservation laws; note that in (6.74a) the $k_\perp$-conservation is implicitly contained in $\delta(E_f(\boldsymbol{k}) - E_i(\boldsymbol{k}) - \hbar\omega)$.

In (6.64) the momentum conservation is in the matrix element M_{fi}, which is between the initial and final states. The matrix element $\tilde{M}^{\mathrm{B}}_{nm}$ in (6.74a) is, however, between the Bloch functions (6.65 and 6.67). Therefore, the momentum conservation is already taken out and appears in the δ-function.

Finally, some special features of the two frequently used photon energy regimes, namely UPS and XPS, will be mentioned.

6.3.1 Band Structure Regime

Thus far we have assumed for the most part that direct transitions are dominant. This is usually the case provided there is at least one final state available via energy and momentum conservation and provided the photon momentum can be neglected. The latter condition is true for the UPS regime ($\hbar\omega \leq 100\,\mathrm{eV}$) and gives rise to structures in the energy distribution curves (EDC) whose energy and/or intensity varies with photon energy and photoelectron emission angle. In this regime special procedures (Chap. 7) can be used to map out band structures.

Direct transitions with full conservation of $\boldsymbol{k}$ occur if the electron damping is sufficiently weak. This is a necessary but not sufficient condition to observe structures in the EDJDOS. Further necessary criteria are that the number of final states that conserve both energy and wave vector is small (i.e., that the different transitions do not overlap to produce a continuum) and that the hole wave-vector broadening is small [6.23].

Band mapping can also be performed with higher ($\hbar\omega \geq 100\,\mathrm{eV}$) photon energies (see Sect. 7.3.8 for an experiment with $\hbar\omega = 580\,\mathrm{eV}$). In order to perform these experiments successfully one needs high energy and momentum resolution and has to take the momentum of the photon into account for an accurate analysis of the data.

6.3.2 XPS Regime

Another photon energy range with weak electron damping is the XPS regime with photon energies of $\simeq 1.5\,\mathrm{keV}$ ($\mathrm{Mg\,K_\alpha}$ and $\mathrm{Al\,K_\alpha}$ radiation). Here under the usual experimental conditions the $\boldsymbol{k}$-broadening is so large that it enables transitions to occur over virtually the whole Brillouin zone. In addition, at high energies so many reciprocal lattice vectors are available to fold the band structure into the first zone that practically a continuum of final states is available for transitions.

This conclusion can also be reached by a simple argument. Let us assume that an electron is excited by absorbing a 1.5 keV photon, such that the electron has a momentum of $k = (2mE)^{1/2}/\hbar \simeq 20\,\mathrm{\AA}^{-1}$. For Cu, with a lattice constant of $a \simeq 4$ Å, the Brillouin zone has a radius $2\pi/a \simeq 1.7$ Å^{-1}. To have final $\boldsymbol{k}$ values distributed over the whole Brillouin zone such that transitions from the entire first zone are possible, requires that the final state of the electron has a momentum spread of $2k_{\mathrm{BZ}} \simeq 3.4$ Å^{-1}. For a total momentum of 20 Å^{-1} this is achieved with an acceptance angle of $\Delta\vartheta \simeq 3.4/20 \simeq 10°$, a value that is smaller than those often employed in experiments. Thus XPS experiments using Mg K_α and Al K_α radiation will, as far as the $\boldsymbol{k}$-selection rule is concerned, sample states over the whole Brillouin zone.

In experiments with polycrystalline samples (in order to average over anisotropies of the optical matrix element) one thus measures a density of states (except for contributions from surface effects which will be discussed later), modulated in intensity by the matrix elements if initial states with different orbital character contribute to the photocurrent.

In angle-resolved XPS experiments with single crystals, the optical matrix elements have to be taken into account explicitly in order to analyze the measured spectra [6.30]. Since the final state is essentially a plane wave, the matrix element has the form:

$$M_{fi}^1 \propto \langle i|\boldsymbol{A} \cdot \boldsymbol{p}_{\mathrm{op}}|\exp(\mathrm{i}\boldsymbol{K} \cdot \boldsymbol{r})\rangle \; . \tag{6.75}$$

In order to reveal the symmetry properties of the matrix element one can approximate the initial state by a combination of atomic orbitals. One thus has

$$M_{fi}^1 \propto \langle \mathrm{d}_n|\boldsymbol{A} \cdot \boldsymbol{p}_{\mathrm{op}}|\exp(\mathrm{i}\boldsymbol{K} \cdot \boldsymbol{r})\rangle \tag{6.76}$$

where the d_n can be, e.g., the five d-orbitals, d_{xy}, d_{yz}, d_{xz}, $\mathrm{d}_{x^2-y^2}$ and $\mathrm{d}_{3z^2-r^2}$.

The angular intensity distribution of the emitted electrons is thus determined by the symmetry of the d_n. Let us assume, e.g., that $\boldsymbol{K}_{||}$ is along [100] of a fcc crystal and consider normal emission. The final state has a wave function $\sim \exp(\mathrm{i}K_\perp x)$ and this connects only to the two $\mathrm{e_g}$ wave functions $\mathrm{d}_{x^2-y^2}$ and $\mathrm{d}_{3z^2-r^2}$(d_{xy}, d_{xz} and d_{yz} being zero for this direction). Because of the finite acceptance angle of the detector and the high density of final states, we assume complete loss of the $k_\perp$-information. In an XPS spectrum taken from a (100) single crystal surface, one thus measures only the $\mathrm{e_g}$ projected density of states. For normal emission from a (111) single crystal surface one measures the $\mathrm{t_{2g}}$ projected DOS because now the matrix element is only non-zero for the d_{xy}, d_{xz} and d_{yz} states; the two other states $\mathrm{d}_{x^2-y^2}$ and $\mathrm{d}_{3z^2-r^2}$ being zero in this direction. In these experiments the electromagnetic radiation must also fulfil certain conditions. Returning to the case of a (100) single crystal surface with electron detection in the [100] direction, the transition matrix element is

$$\langle \mathrm{d}_n|\boldsymbol{A} \cdot \boldsymbol{p}_{\mathrm{op}}|\mathrm{e}^{\mathrm{i}\boldsymbol{K}\cdot\boldsymbol{r}}\rangle \simeq \boldsymbol{A} \cdot \langle \mathrm{d}_n \boldsymbol{p}_{\mathrm{op}} \mathrm{e}^{\mathrm{i}\boldsymbol{K}\cdot\boldsymbol{r}}\rangle = \boldsymbol{A} \cdot \boldsymbol{e}_x \left\langle \mathrm{d}_n \frac{\partial}{\partial x} \mathrm{e}^{\mathrm{i}k_\perp x} \right\rangle \; . \tag{6.77}$$

This shows that only the A_x component gives a non-zero contribution to the transition.

6.3.3 Surface Emission[10]

When working at UPS photon energies, along with the $\boldsymbol{k}$-conserving direct transitions, additional transitions are observed. By making comparisons with band structures these can be shown to be related to the occupied DOS (one-dimensional DOS, if an angle-resolved experiment is performed [6.30]). This is already evident in the observation that in UPS EDCs in the angle-resolved mode of metal single crystals, a Fermi edge is always visible. This is proof of the fact that "DOS transitions" are always present even in the UPS spectra, although these spectra are usually dominated by direct transitions and analyzed in these terms. (Note that XPS usually measures a DOS, i.e., the spectrum reflects the occupied density of states of the material.)

One of the reasons for the presence of "DOS transitions" is the occurrence of phonon-assisted indirect transitions, whose strength depends on the temperature (Sect. 6.4).

The main cause of indirect transitions is the damping of the electron final state (finite $k_\perp^{(2)}$), which is responsible for a relaxation of the strict $\boldsymbol{k}$-conservation rule. This leads to emission via excitation into evanescent states in gaps and is called gap emission or surface emission because the gap states are localized near the surface (already the step in the potential at the surface causes a violation of $k_\perp$-conservation).

If in an actual experiment, the final state energy is chosen to be in a bulk band gap, no transmitted beams are possible, in principle, since the final states are evanescent and thus no PE signal should be observed. However, it is a matter of common experience that EDCs in the UPS regime are also observed when the final state is in the gap. This means that if EDCs are measured in a particular area of the Brillouin zone, emission peaks do not suddenly cease when a gap is reached. The intensity may become weaker, but spectra are observed nevertheless. A case in point is the PE from Cu(110) [6.31], the results of which are shown in Fig. 6.19. The spectra (left panel) are EDCs taken with synchrotron radiation and two different polarization directions of the light. The polarization of $\boldsymbol{A}$ (the vector potential of the incoming light) was always in the plane of incidence, defined by the propagation vector of the incoming light and the vector of the outgoing electrons.

[10] The remainder of this chapter contains experimental material taken mostly from Cu metal. It will involve some nomenclature which will only be explained in detail in Chap. 7. The energy bands of a metal like Cu or Ag are given as $E(\boldsymbol{k})$ and are three-dimensional entities. Published band structures generally give the $E(\boldsymbol{k})$ curves along the principle directions such as (100), (110) and (111). These directions, the points where they intersect the Brillouin zone, and other principle lines in the Brillouin zone are designated in a specific way and are shown in Fig. 7.5. The band structures of Cu and Ag can be found in Figs. 7.7, and 7.24.

For the full curve the crystal was adjusted such that the [001] direction was also in the plane of incidence whereas for the broken curve the crystal was rotated such that the plane of incidence contained a [1$\bar{1}$0] direction.

The four peaks observed in a typical spectrum are numbered consecutively. The right part of Fig. 6.19 gives (in the bottom panel) the final state dispersion of Cu(110) as taken from Janak's band structure calculation [6.32] and (in the upper two panels) the energy position and the intensities of peaks *2* and *3* as a function of the energy of the incident radiation. It can be seen that in roughly the area of the gap (15–19 eV) the energies of the two peaks are constant and their intensities decrease. However, at no point does the intensity actually drop to zero. This finding can be interpreted as the observation of transitions into a gap. In the chosen geometry, namely normal emission, one is probably observing transitions near to or at the X-point of

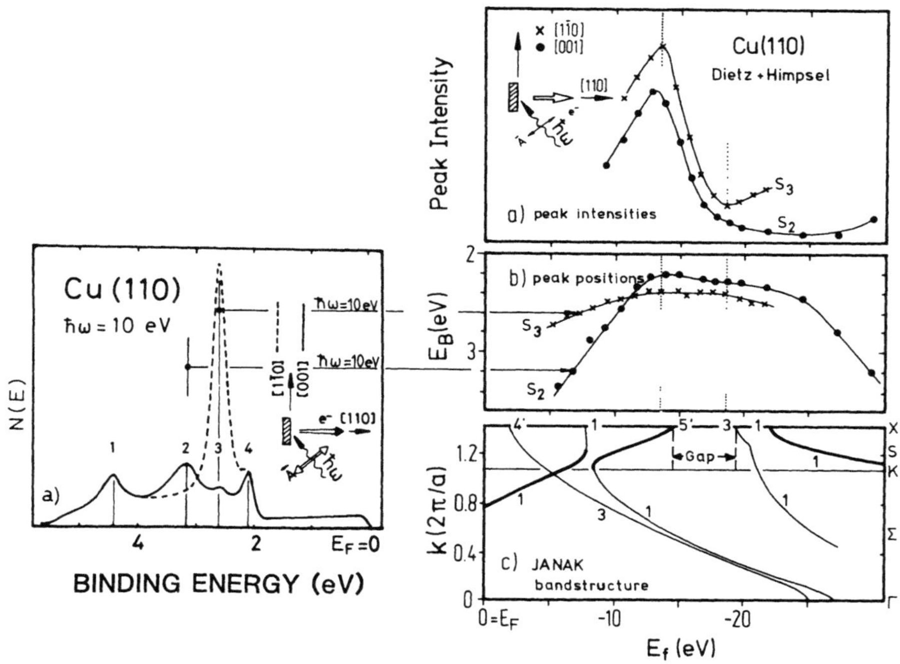

Fig. 6.19. *Left*: Cu(110) normal emission AREDCs obtained at $\hbar\omega = 10\,\mathrm{eV}$ with two different polarizations of radiation relative to the crystal azimuth φ. The change of polarization is achieved by rotating the crystal around the surface normal. The peaks are labelled with respect to the symmetry of the initial bands (Fig. 7.7). *Right*: (a) intensity of peaks 3 and 4 normalized to the incident photon flux. (b) Dispersion of peaks 3 and 4 as a function of the final state energy above E_F. (c) Energy-band structure for Cu(110) as calculated by Janak [6.32]. The full line gives the final state band into which transitions are expected under normal emission conditions. The dotted lines indicate the appearance and disappearance of the $X_5 - X_3$ gap [6.31]

the Brillouin zone into gap states between the X_5 and the X_3 point in the Brillouin zone (Sects. 7.3.2,7.3.3).

Emission via excitation into evanescent states in gaps can be used to measure occupied bands [6.30]. If the initial energies are the allowed bulk energies, one measures approximately a one-dimensionally $\boldsymbol{k}_{\|}$-resolved bulk density of states modulated by the appropriate transition matrix elements, if $k_{\perp}^{(2)}$ is comparable in size to the Brillouin zone diameter (Sect. 7.3.4).

It is not possible to make a clear distinction between damped band states and evanescent gap states because in both cases the damping of the final waves leads to a localization of the state near the surface and to a more or less undefined $k_{\perp}$.

For an extended discussion of surface effects the reader is referred to the work of Feuerbacher and Willis [6.17].

6.3.4 One-Step Calculations

The presentation so far has more or less followed the historical development, in which the three-step model of PES in solids has played an important part, because it allowed one, in an intuitively very acceptable way, to understand many features of a seemingly complicated process. On the other hand, the reasoning underlying the three-step model makes many simplifying assumptions. The most serious are probably the neglect of interference between bulk and surface emission, the neglect of interference between loss and no-loss transport from the place of photoexcitation to the surface, and the description of the escape of the electron through the surface by a simple transmission factor.

In principle, the "correct" treatment of the PE process has already been presented in (6.15, 6.16). The Golden Rule equation (6.15) with the proper functions for the initial and final states, and the dipole operator for the interaction of the electron and the incoming light, contains all the physics of the problem in a "correct" one-step formulation. In the form given in (6.15) the problem cannot be solved rigorously and various approximations have to be employed to make a one-step calculation feasible.

The best-founded approximation is certainly the use of the inverse LEED function for the final state as in (6.62,6.63). The use of the $\boldsymbol{p}_{\mathrm{op}} \cdot \boldsymbol{A}$ operator for the interaction is also an acceptable approximation. Here, however, one must question to what extent it is justified to use the vector potential of the vacuum also for the solid. The most difficult problems are how to treat the surface properly and how to introduce the electron damping.

Present-day one-step calculations for the valence band EDCs in solids are usually performed by starting out with a formulation presented by Pendry [6.25,6.33–6.37], who modifies the Golden Rule formula in the following way in order to make the problem tractable. As before, the damping of the electron states is taken into account by using a complex (optical) potential for the electrons in the crystal. The problem is then broken up into a surface

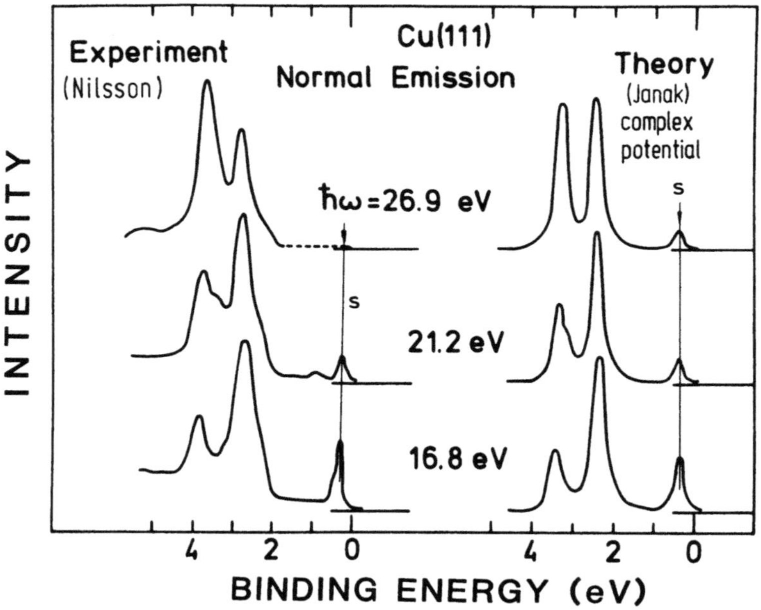

Fig. 6.20. Experimental and theoretical PE spectra for Cu(111) taken under normal emission conditions. Polarized light was incident with $\psi = 45°$ relative to the surface normal. The calculated EDCs were obtained with a complex potential [6.38]. S indicates a surface state

part (first layer) and an (undisturbed) bulk part, where the surface is approximated by a step function. In addition, to simplify matters, the surface potential is made real but the real part of the surface layer potential is allowed to differ from that of the bulk (which seems to be sufficient to bring out surface features). The total photocurrent is then written as a sum of the surface and the bulk photocurrent, which in a sense makes the whole procedure a two-step rather than a one-step calculation.

A number of calculations have been performed along these lines and the results have reproduced the measured valence-band EDCs surprisingly well, although not perfectly [6.30, 6.38–6.40]. As an example Fig. 6.20 shows the results of an investigation by Nilsson et al. [6.38] using the computer code of Pendry. The calculations were done using the crystal potential given by Janak et al. [6.32], which gives good agreement between the measured and calculated bulk-band structure of Cu (Chap. 7) and is therefore an appropriate potential.

The left-hand side of Fig. 6.20 shows the normal emission EDCs for Cu(111) taken at three energies and the right-hand side depicts the calculated EDCs. The features below $E_{\rm B} = 2\,{\rm eV}$ are due to direct transitions in the bulk-band structure of Cu. The peak near the Fermi energy (S) is a surface state and it is very gratifying to see that this feature is reproduced by the calculations.

In this section we have purposely given no equations, since they are quite complicated and only useful to those wishing to perform actual calculations. These readers are referred to the review of Braun [6.6] and the literature cited therein.

6.4 Thermal Effects

PES experiments are generally performed at room temperature. At such temperatures, both free molecules and solids will be vibrationally excited. Via the electron–phonon interaction these vibrations can couple to the electrons and, although the vibrational energies are small with respect to the electronic ones, their influence on the PES data can be seen. So far, not too much use has been made of this effect and thus it will not be dealt with at great length. It was already mentioned implicitly when we dealt with the effect of vibrations on the valence-band spectra of molecules in Chap. 5. There it was shown that the electron–phonon interaction produces, besides a pure electronic line, a number of vibrational side bands, whose intensity distribution can be accounted for by invoking the Frank–Condon principle for the transition (Figs 5.2 and 5.5).

The same effect is also found in solids and has been observed, e.g., in the core-level spectra of potassium halides. Using a diagram of the type shown in Fig. 5.2 the phonon contribution to the linewidth can be written as follows [6.41](using an insulating solid with an Einstein type phonon spectrum):

$$\Gamma_{\mathrm{Ph}} = 2.35\sqrt{\hbar\omega_{\mathrm{LO}} E_{\mathrm{R}}}\left[\coth\left(\frac{\hbar\omega_{\mathrm{LO}}}{2k_{\mathrm{B}}T}\right)\right]^{1/2} \tag{6.78}$$

(note that this approaches a finite value for $T \to 0$, which is due to the fact that the spontaneous emission of phonons is temperature independent).

Here $\hbar\omega_{\mathrm{LO}}$ is the energy of the optical phonon, assuming an Einstein-type spectrum, and E_{R} is the vibrational relaxation energy. In a Frank–Condon type of diagram E_{R} is the energy difference between the bottom of the excited-state phonon energy parabola and the energy of the phonon state into which the PE transition is most likely to occur. This energy can be calculated as [6.41]:

$$E_{\mathrm{R}} = e^2(6/\pi V_{\mathrm{m}})^{1/3}(1/\varepsilon_\infty - 1/\varepsilon_0) \tag{6.79}$$

where V_{m} is the volume of the primitive unit cell of the crystal, and ε_0 and ε_∞ are the low and high frequency limits of the dielectric constant. One obtains, e.g., $E_{\mathrm{R}} = 0.94\,\mathrm{eV}$ for KI [6.41]. (In the harmonic oscillator model E_{R} represents the energy of many phonons and thus its magnitude can far exceed the energy of a single phonon.)

Figure 6.21 shows the results of a study of the temperature dependence of the K 2p core lines in KI, KCl, and KF. The full circles are the raw measured

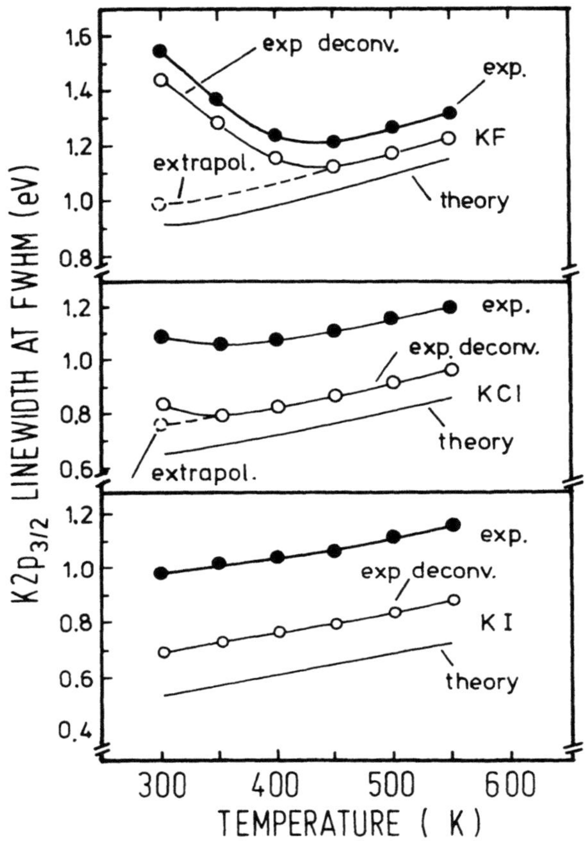

Fig. 6.21. Temperature dependence of the linewidth of the K $2p_{3/2}$ core line ($E_{\rm B} = 295\,{\rm eV}$) in KF, KCl and KI [6.41]. The full dots are the experimental points, and the open dots are deconvoluted for experimental broadening. The theory is based on (6.78,6.79). In KF (and to a smaller degree in KCl) the linewidth increases again at low temperatures presumably due to the charging which arises because the optical gaps in these materials are too large to provide the necessary low temperature conductivity. The phonon contribution to the linewidth does not extrapolate to zero for $T \rightarrow 0\,{\rm K}$ because of the spontaneous phonon emission channel

widths and the open circles are the widths obtained after subtracting out the lifetime and instrumental width. The lines with no symbols attached are the results calculated with the above equations. The same trend is observed in the experimental and calculated curves, indicating a realistic understanding of the effect. At low temperatures the experimental data for both KCl and KF show a further increase in linewidth. This is probably due to differential charging of the samples. The magnitude of this charging is proportional to the optical band gap $E_{\rm g}$ which increases in the sequence $E_{\rm g}({\rm KF}) > E_{\rm g}({\rm KCl}) > E_{\rm g}({\rm KI})$. The ionic mobility, which decreases the charging effect, goes in the reverse order to the band gap, and it thus requires lower temperatures to destroy the charging in KI than in KCl and KF as is actually observed.

A theory of the effects of thermal vibrations on EDCs of valence bands of metallic solids has been presented by Shevchik and coworkers [6.42–6.47], see also [6.48]. It has been used to interpret the temperature dependence of the intensities as observed in the valence-band spectra of metals. The essence of this approach is summarized in the following (see also [6.49]).

PES can, in principle, be thought of as a scattering process by the ions of a crystal. The detected electrons must thus reflect the effect of the lattice vibrations in a way similar to that found in X-ray diffraction or neutron scattering. In these techniques the coherently scattered intensities are multiplied by the Debye–Waller factor e^{-2W} (W contains the temperature), which leads to a decrease of intensity with increasing temperature. From this analogy one expects that the direct transition intensity, as seen, e.g., in the valence-band spectra of the metals, will also contain a Debye–Waller factor, which reduces the intensity with increasing temperature. Experiments have supported this view.

In developing a theory to describe this effect one encounters a slight pedagogical problem. The lattice vibrations are motions of the nuclei. The nuclei, however, have so far not been considered except in their role of providing potentials in which the electrons move. The valence electrons (for which thermal effects will be discussed here) are, however, described as "free" electrons, being only slightly disturbed by the crystal potential, which in turn would mean they are hardly effected by the lattice vibrations. In order to make the situation more transparent it is therefore useful to move to a description of the valence electrons in terms of *tight-binding functions* of the form

$$\psi(\boldsymbol{k}_i, \boldsymbol{r}) = \sum_m \exp(\mathrm{i}\boldsymbol{k}_i \cdot \boldsymbol{R}_m)\phi(\boldsymbol{r} - \boldsymbol{R}_m) \tag{6.80}$$

where $\boldsymbol{R}_m$ is the coordinate of atom m and $\phi(\boldsymbol{r} - \boldsymbol{R}_m)$ is an atomic orbital. In the atomic orbitals the electrons are "rigidly" connected to the vibrating ion cores and the effect of the lattice vibrations is thereby introduced. If one calculates the photocurrent (or energy distribution of photoemitted electrons) from the Golden Rule (6.15), using as the initial state the wave functions of (6.80) and as the final state a plane wave, one can split up the matrix element in the following way:

$$\langle f|H^{\mathrm{int}}|i\rangle \propto \sigma \cdot S \tag{6.81}$$

where σ is the photoabsorption cross section and S the scattering factor written as:

$$S = \sum_{\mu\nu} \exp[\mathrm{i}\Delta\boldsymbol{k} \cdot (\boldsymbol{R}_\mu - \boldsymbol{R}_\nu)]; \quad \Delta\boldsymbol{k} = \boldsymbol{k}_f - \boldsymbol{k}_i\ . \tag{6.82}$$

The indices μ, ν now run over the atomic sites (and not the various initial and final states). To take the lattice vibrations into account one writes

$$\boldsymbol{R}_\mu(t) = \boldsymbol{R}_\mu^0 + \boldsymbol{U}_\mu(t) \tag{6.83}$$

where $\boldsymbol{U}_\mu$ is the displacement of ion μ. One now has to calculate the thermal average of the scattering factor. This calculation yields two terms which correspond to two types of momentum conservation. The first represents the usual direct transitions and the second is made possible by the fact that at finite temperature many phonons are excited and can also provide momentum to make (indirect) transitions possible. For the photocurrent one obtains

$$N(E,\hbar\omega) \propto |\sigma|^2 \left\{ \underbrace{\exp(-\Delta k^2 U_0^2) \sum_{\boldsymbol{G}} \delta(\Delta \boldsymbol{k} - \boldsymbol{G})}_{\text{direct PE current}} + \underbrace{N[1 - \exp(-\Delta k^2 U_0^2)]}_{\text{indirect PE current}} \right\} . \tag{6.84}$$

Here N is the number of atoms that contribute to the photoemission signal and $\exp(-\Delta k^2 U_0^2)$ is the usual Debye–Waller factor e^{-2W} with $\Delta \boldsymbol{k} = \boldsymbol{k}_i - \boldsymbol{k}_f$. (Note that in the usual nomenclature [6.14,6.50] one has $W = \frac{1}{2}\overline{(\Delta k U_\mu(t))}^2 = (1/6)\Delta k^2 \bar{U}_\mu^2$; if one takes the three polarizations into account a factor 1/3 is removed. Thus one has $U_0^2 = \bar{U}_\mu^2$ for the average squared displacement of an atom).

In the Debye approximation one finds

$$U_0 = \frac{3\hbar^2}{M_{\mathrm{A}} k_{\mathrm{B}} \Theta_{\mathrm{D}}^2} T , \tag{6.85}$$

where M_{A} is the mass of the atoms, k_{B} is Boltzmann's constant and Θ_{D} is the bulk Debye temperature.

In this simplified model one sees that apart from a diffuse background produced by the indirect transitions, there is also a temperature dependence of the direct transition intensity.

A shortcoming of (6.84) is the fact that it averages out the electron–phonon coupling of different electronic states.

Larsson and Pendry [6.36] have incorporated the electron–phonon coupling into their one-step theory. In their approach the interaction operator and the scattering matrix contain $U(t)$, the displacement of an atom. The results can no longer be expressed in a simple form such as (6.84), and in fact these authors also conclude that the overall effect on the PE spectrum is akin to multiplying it by a Debye–Waller factor.

From the Debye–Waller factor formulation an interesting conclusion can be drawn. For XPS energies ($\hbar\omega \simeq 1500\,\mathrm{eV}$) one can neglect the initial-state momentum $\boldsymbol{k}_i$ and has $|\boldsymbol{K}_f| \simeq 2\pi/\lambda = 20\,\mathrm{\AA}^{-1}$. For a material like Cu the displacement U_0 at room temperature is about 0.1 Å, such that the Debye–Waller factor $\exp(-\Delta k^2 \cdot U_0^2)$ is e^{-4} since $\boldsymbol{k}_i - \boldsymbol{k}_f \simeq -\boldsymbol{k}_f$. For XPS energies this implies that about one percent of the transitions are direct transitions. Thus in the XPS regime, apart from an averaging produced by the finite acceptance angle of the electron analyzer and the finite size of the photon

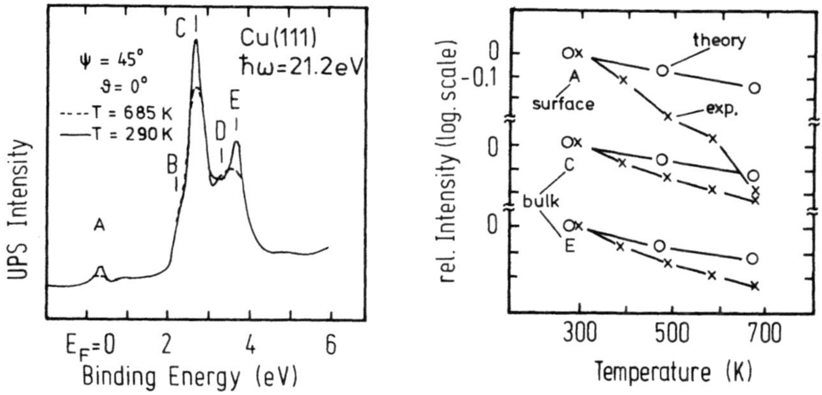

Fig. 6.22. *Left*: Experimental normal emission EDCs for Cu(111) ($\hbar\omega = 21.2\,\text{eV}$) at two different temperatures. Incident light at $\psi = 45°$. *Right*: Temperature dependence of structures A (*surface state*), C and E (*direct transitions*) as measured (*crosses*) and calculated (*open dots*) (6.86) [6.50]

wave vector, the temperature averaging also leads to spectra that are densities of states weighted by transition probabilities.

A detailed investigation of the temperature dependence of peak intensities in EDCs from Cu(111) with $\hbar\omega = 21.2\,\text{eV}$ was performed by Mårtensson et al. [6.50]. Figure 6.22 shows spectra at two different temperatures and the logarithms of the relative intensities of structures, $\ln(I/I_{\text{ref}})$ as a function of temperature $[I_{\text{ref}} = I(T = 273\,\text{K})]$. Note that according to (6.84 and 6.85) one has

$$\ln I = -bT + a; \quad b = \frac{3\hbar^2 \Delta k^2}{M_{\text{A}} k_{\text{B}} \Theta_{\text{D}}^2} . \tag{6.86}$$

The theoretical intensities for peaks A, C and E in Fig. 6.22 were calculated with the inverse LEED formalism [6.25] with a reference intensity at $T = 273\,\text{K}$. Peaks C and E are direct transitions and it can be seen that they have a similar temperature dependence, although generally the experimental slope of the temperature dependence is steeper than the theoretical one. Peak A is a Shockley surface state (Chap. 8), and its temperature dependence is stronger than that of the direct transitions C and E. The surface states are confined predominantly to the first atomic layer. In that layer the crystal binding is weaker than in the bulk crystal leading to a smaller Debye temperature. This results in a larger coefficient b in (6.86) for the surface as compared to the bulk meaning a stronger temperature dependence in agreement with experimental findings.

If the experimental data are fitted to straight lines one can extract the b coefficients; these are given in Table 6.1. If one uses b as given in (6.86) and sets $|\Delta k| = G(111)$ one can calculate from the b values effective Debye temperatures. These are also given in Table 6.1. Compared to the bulk Debye temperature obtained from X-ray scattering measurements ($\Theta_{\rm D} = 343\,{\rm K}$), those in the table tend to be lower. Again this must be attributed to surface effects. The differences in effective Debye temperatures point to the fact that the simple model with a Debye–Waller factor is an oversimplification and that the transition matrix elements may also be temperature-dependent. Other experiments along these lines have been reported in [6.48, 6.51–6.61] and will be discussed at the end of this section.

The electron–phonon interaction results also in temperature-dependent shifts and broadenings of direct transitions in PE spectra [6.62]. If a calculation of the electron self-energy is performed by perturbation theory to second order in the atomic displacements, two temperature-dependent terms are obtained. The Debye–Waller term represents the effect of the second-order electron phonon interaction taken to first-order perturbation theory. It results in a real self-energy. A second term (Fan term) arises if the first-order term in the electron–phonon interaction is taken to second-order perturbation theory. This term has a real (shift) and imaginary (broadening) part. The energy $E(\boldsymbol{k},T)$ of a valence-band state can be written as

$$E(\boldsymbol{k},T) = E^0(\boldsymbol{k}) + \Sigma(\boldsymbol{k},T) \tag{6.87}$$

where $E^0(\boldsymbol{k})$ is the valence-band energy for zero electron–phonon interaction, ${\rm Re}\Sigma(\boldsymbol{k},T)$ is the real part of the temperature-dependent self-energy (Debye–Waller and Fan term), and ${\rm Im}\Sigma(\boldsymbol{k},T)$ is the imaginary part of the self-energy. The temperature dependence in the shift and the width is given by a Bose–Einstein factor yielding [6.62]

$${\rm Re}\Sigma(\boldsymbol{k},T) = {\rm Re}\Sigma(\boldsymbol{k},0) + \frac{{\rm Re}\Sigma^0(\boldsymbol{k})}{{\rm e}^{(\Theta_{\rm D}/T)}-1} \tag{6.88}$$

$${\rm Im}\Sigma(\boldsymbol{k},T) = {\rm Im}\Sigma(\boldsymbol{k},0) + \frac{{\rm Im}\Sigma^0(\boldsymbol{k})}{{\rm e}^{(\Theta_{\rm D}/T)}-1} \tag{6.89}$$

Here ${\rm Re}\Sigma(\boldsymbol{k},0) = (1/2){\rm Re}\Sigma^0(\boldsymbol{k})$ is the renormalization of the valence-band energies due to the zero-point phonons. ${\rm Im}\Sigma^0(\boldsymbol{k}) \sim (N\bar{D}^2/\Theta_{\rm D})$ where N

Table 6.1. Experimental and theoretical temperature coefficients for the intensities of structures in the AREDCs taken from Cu(111)/ΓLUX with $\hbar\omega = 21.2\,{\rm eV}$, and experimental and theoretical effective Debye temperatures (Fig. 6.22) [6.51]

Structure	Temperature coefficient b		Effective Debye temperature [K]	
	experiment	theory	experiment	theory
A	15	3.6	–	–
C	6	3.1	186	260
E	6.3	3.1	181	260

is the electronic density of states at the Fermi energy and $\bar{D}^2$ the average electron-phonon deformation potential. $\mathrm{Im}\Sigma(\boldsymbol{k},0)$ includes, besides the contribution from the zero point phonons, the lifetime which is due to electron–electron interaction. The width of a direct transition $\Gamma(\boldsymbol{k},T)$ contains additional terms besides those given by the imaginary part of the self-energy. These are the energy resolution, the angular resolution, and the width induced by surface defects. If they are assumed to be temperature independent, one has for the total measured width of a direct transition [from(6.89)]

$$\Gamma(\boldsymbol{k},T) = \Gamma(\boldsymbol{k},0) + \frac{\Gamma^0(\boldsymbol{k})}{\mathrm{e}^{(\Theta_\mathrm{D}/T)} - 1} \tag{6.90}$$

where $\Gamma(\boldsymbol{k},0)$ contains all temperature independent contributions to the width of a direct transition.

These considerations describe the data on InSb well [6.62] as can be seen in Fig. 6.23. The spectrum in the bottom panel shows a low temperature ($T = 50\,\mathrm{K}$) spectrum ($\hbar\omega = 21.2\,\mathrm{eV}$) in the normal emission geometry ($\vartheta = 0°$). The x-axis was oriented along the [110] direction of the crystal. The top two panels of Fig. 6.23 give the difference in the temperature dependence of the energy shifts of line B_2 and line B_4 (absolute values are difficult to measure because the Fermi energy can show a temperature dependence in a semiconductor) and the temperature dependence of the width of structure B_4. The full lines are fits to a Bose–Einstein temperature dependence [6.62].

Finally it is worth mentioning a more or less trivial (small) temperature effect. It stems from the fact that solids have a thermal expansion, and thus the potential seen by a particular ion in the crystal will diminish upon raising the temperature [6.63]. This effect, which might be called a purely geometrical one, has been observed in core levels of insulating solids, where it is easy to understand via the change in the Madelung constant. A similar effect has been seen in the band structure of Cu as shown in Fig. 6.24 [6.64].[11]

As expected, the bands move nearer to E_F (are more weakly bound) as the temperature is raised. Note that in Fig. 6.24 the difference between the bands at 25° C and 499° C has been exaggerated by a factor of 2.5 so as to make the effect more clearly visible.

The temperature effects mentioned so far are intuitively appealing: A decrease in intensity with temperature due to a Debye–Waller factor, and a change of line position and line width given essentially by the phonon occupation number, namely the Bose–Einstein factor, in addition to a decrease in bandwidth with increasing lattice constant.

The experimental results, however, suggest that matters may be more complicated. One observes the following deviations from the simple appealing picture just outlined:

[11] A detailed analysis of these data in terms of the theory given in [6.62] is lacking. Therefore, the interpretation presented here has to be considered preliminary.

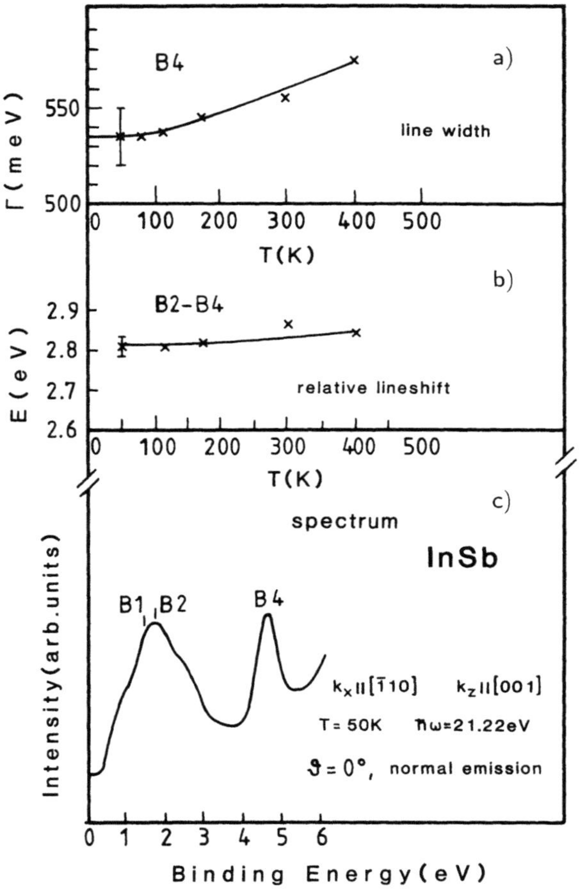

Fig. 6.23. Temperature dependence of the linewidth and the lineshift of direct transitions in InSb [6.62]. (**a**) Normal emission spectrum at $T = 50\,\mathrm{K}$ from a [001] crystal plane; (**b**) difference in energy shift of the structures B_2 and B_4 as a function of temperature; full curve is a Bose–Einstein temperature dependence; (**c**) temperature dependence of the linewidth of structure B_4; solid curve is a Bose–Einstein temperature dependence

1) The decrease in intensity of a particular line with temperature is usually faster than that predicted by the theory (see Fig. 6.22 and Table 6.1).
2) High resolution experiments on core lines, e.g., of the simple metals, show that at high temperature there can be a linear as well as a square root dependence on temperature.
3) The line shape of core lines is often a Gaussian and not a Lorentzian.

We shall now comment briefly on these problems, remarking in advance that the problem of temperature effects on PES data has not been solved completely.

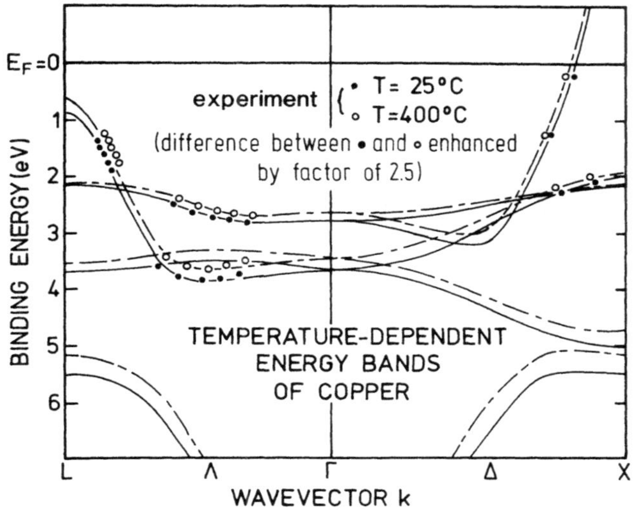

Fig. 6.24. Temperature-dependent energy bands for Cu (ΓX and ΓL directions) at $T = 25°$ C and $T = 400°$ C [6.56]. The shift in the energies at $T = 400°$ C relative to their position at 25° C has been amplified by a factor of 2.5 to make the effect clearly visible. Because the lattice constant a increases with increasing temperature and the crystal potential goes as $1/a$, the binding energy has to decrease with increasing temperature. The solid lines are the theoretical bands with the $T = 25°$ C lattice constant, and the dashed lines are the bands calculated with the lattice constant for $T = 400°$ C

Figure 6.25a shows part of a valence-band spectrum from a Cu(110) crystal [6.53] with the data taken in the ΓXWK plane (see Fig. 7.5 for the Brillouin zone of Cu). These spectra show a slight, although not an overwhelming, temperature dependence with respect to amplitude, area and width of the line. The data are summarized in Fig. 6.25b, where the open squares are the experimental data and the dots a simulation. If the simple model with a Debye–Waller factor is used, the decrease in peak intensity is smaller than that measured here. Therefore the authors have included two other possible mechanisms that can contribute to temperature effects, namely a scattering of the photoelectron out of the line by quasi-elastic interaction with the phonon and of course also the reverse process, where a photoelectron originally in a different peak gets scattered into the particular signal under consideration. By adding these two mechanisms, the authors were at least able to fit all the three pieces of data in a consistent, if not a priori way.

High resolution data on a Mo surface state in Mo(110) on the other hand seem to agree with the simple theory, provided the modifications introduced by the conduction electrons are taken into account [6.58].

In the most general form, the temperature-dependent width of a state in a metal comes from that in the electron–electron interaction term (generally small) and that in the electron–phonon term (generally dominant).

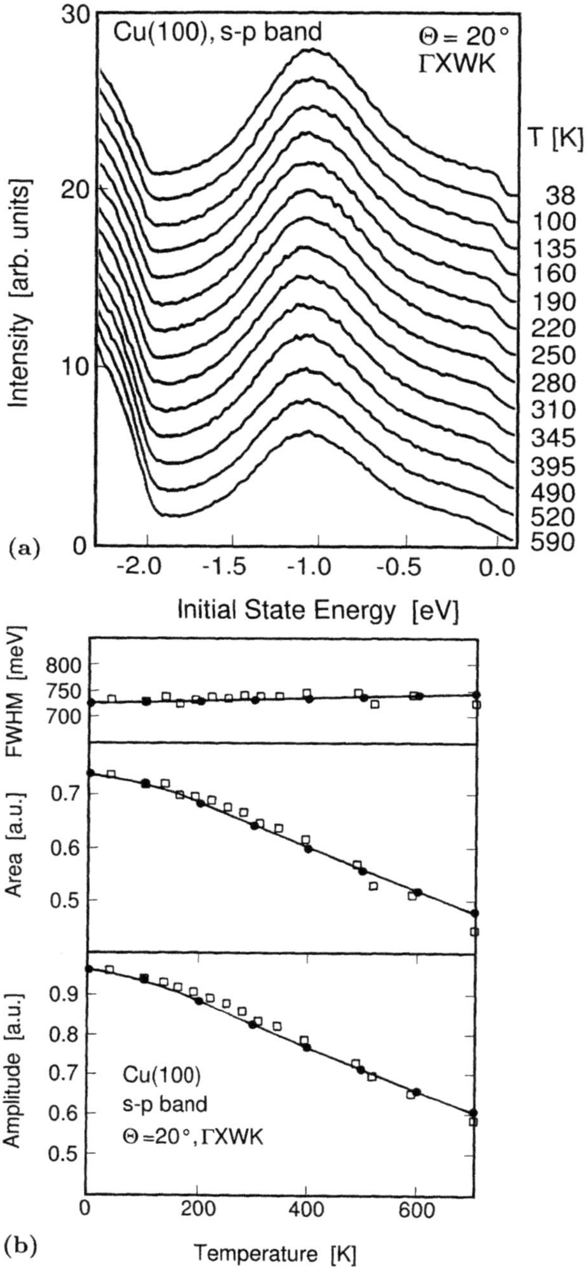

Fig. 6.25. (**a**) Temperature dependence of a transition out of the sp band in Cu from a Cu(100) crystal with He I radiation in the ΓXWK plane. (**b**) Amplitude area and full width half maximum of the line shown in (**a**) as a function of temperature. The full curves are the result of a simulation [6.53]

The width due to the electron–electron interaction is given by:

$$\Gamma_{\mathrm{el-el}}(E,T) = 2\beta[(\pi k_{\mathrm{B}}T)^2 + E^2] \tag{6.91}$$

where one has the usual Fermi liquid energy dependence but also a quadratic temperature dependence. In addition there is a width due to electron–phonon scattering, which is given in a metal by:

$$\Gamma_{\mathrm{el-ph}}(E,T) = 2\pi \int_0^\infty \mathrm{d}\omega \alpha^2 F(\omega)[2n(\omega,T) + f(\omega + E,T) - f(\omega - E,T)] \, , \tag{6.92}$$

where $\alpha^2 F(\omega)$ is the Eliashberg function, $n(\omega, T)$ the Bose–Einstein function, and $f(\omega \pm E, T)$ the Fermi function.

With present day high resolution experiments the two contributions can be separated, as was shown in the data of Fig. 1.29. The pure temperature dependence at a fixed energy is now determined almost entirely by the Bose–Einstein factor. The results for two energies ($E_{\mathrm{B}} = 0$ and $E_{\mathrm{B}} = 100\,\mathrm{meV}$) are shown in Fig. 6.26, where the agreement between theory (in essence reflecting the temperature dependence of the Bose-Einstein factor) and experiment is very good. The electron–phonon coupling constant deduced from the fits is different for the two energies but agrees with expectations.

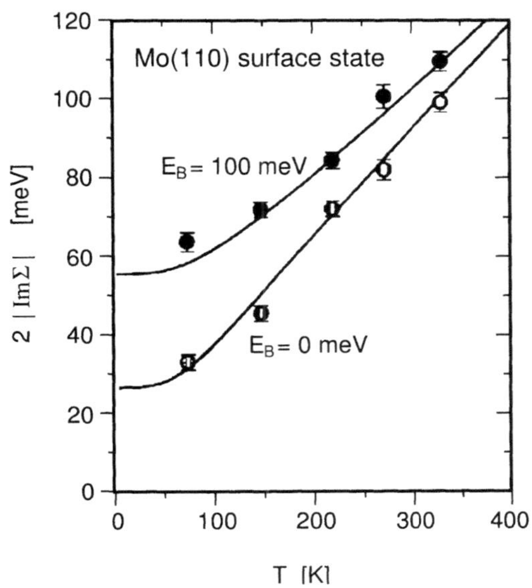

Fig. 6.26. Temperature dependence of the Mo(110) surface state shown in Fig. 1.29 at the Fermi energy and at 100 meV below the Fermi energy. The full curve is a fit with (6.92) [6.58] confirming the validity of the underlying model

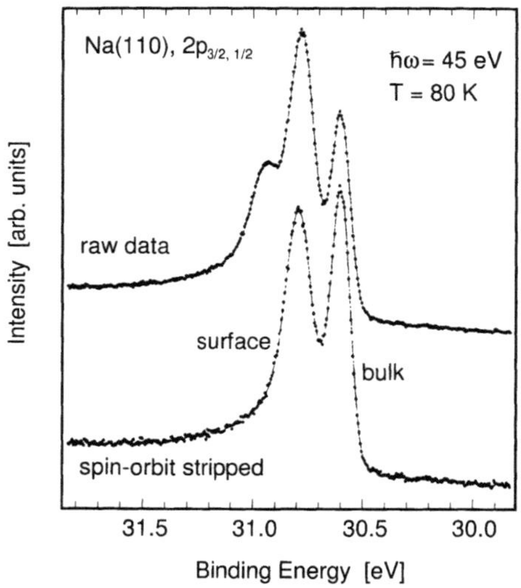

Fig. 6.27. *Top*: Raw data of the 2p core-level photoemission spectrum from Na(110) at $T = 80\,\mathrm{K}$. *Bottom*: Spectra stripped of the $2p_{1/2}$ component, showing only the $2p_{3/2}$ line. The solid curve is a smoothed representation of the data [6.65]

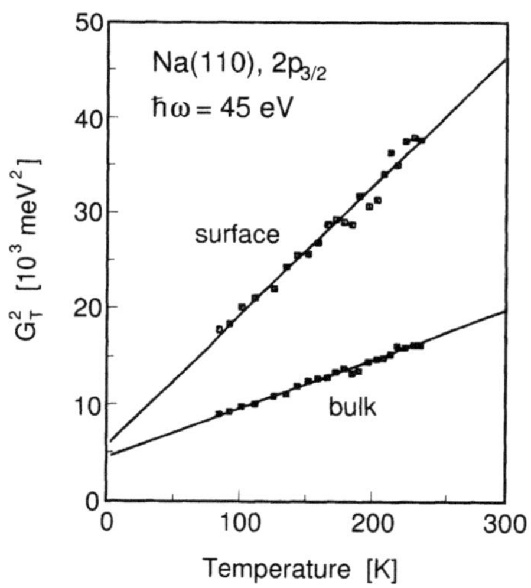

Fig. 6.28. Temperature dependence of the square of the total Gaussian width of the $2p_{3/2}$ bulk and surface line shown in Fig. 6.27. The temperature dependence of the surface and bulk line are different but both extrapolate to roughly the same number at zero temperature, namely to the square of the instrumental width [6.65]

We next present some data for temperature effects on core lines in simple metals. In Fig. 6.27 high-resolution low-temperature core-level spectra of Na(110) are shown, taken with a photon energy of 45 eV [6.65, 6.66]. This small photon energy guarantees that the escape depth of the photoemitted electron is small, and therefore the bulk and the surface line have about the same intensity. In addition the raw data are complicated by the spin–orbit interaction. If the $2p_{1/2}$ spectrum is removed (lower trace) one can distinguish the bulk and surface signal. An analysis of the lineshape of these two lines shows that they are almost completely Gaussian. If the square of the measured lineshape of the bulk and the surface component are plotted as a function of temperature, one obtains the results depicted in Fig. 6.28. These data are interesting in two respects. First the temperature dependence is obviously a function of the square root of the temperature (note that the simple expansion of the Bose–Einstein factor for a large temperature leads to a linear temperature dependence) and secondly that the widths of the surface and the bulk component have different temperature dependences.

These data can in principle be explained by a theory given by Rosengreen and Hedin [6.67]. These authors calculated the temperature dependence of the core lineshape by taking into account the change of the electronic potential produced by the lattice vibrations. After averaging over the phonon spectrum this leads (approximately) to a Gaussian lineshape; in contrast, an Einstein phonon spectrum (Fig. 6.21) leads to a Lorentzian PE spectrum. In addition one has to be careful in the calculation of the electron–phonon matrix element. Using a model like that depicted in Fig. 5.2 one finds that the energy is a parabolic function of the nuclear coordinates and depends on the equilibrium distances and the bonding forces. One can now consider two limiting cases. If only the equilibrium distance between the ion from which the photoelectron has been ejected and its nearest neighbors is changed, the energy change will depend linearly on the change of distance, yielding a $T^{1/2}$ dependence for the high temperature linewidth (linear case). But there is also another case, where the equilibrium distance remains the same after the photoionization. However, here the bonding forces change and in this case the energy change will depend quadratically on the nuclear coordinates (quadratic case) with a T^1 dependence of the linewidth in the high temperature limit.

Using a Debye model and expanding the Bose–Einstein factor for large T leads to the following temperature dependence for the square of the Gaussian linewidth in the linear case [6.65, 6.66]:

$$G^2_{\mathrm{el-ph}}(T) = G^2(0)\left[1 + 8\left(\frac{T}{\Theta_{\mathrm{D}}}\right)^4 \int\limits_0^{\Theta_{\mathrm{D}}/T} \pi \frac{x^3 \mathrm{d}x}{\mathrm{e}^x - 1}\right] \tag{6.93}$$

If the integral is now calculated for the approximation of large T, we finally obtain for the temperature dependence of the quadratic Gaussian width:

$$G^2_{\mathrm{el-ph}}(T) = \frac{c}{\Theta_{\mathrm{D}}}\left[1 + \left(\frac{8}{3}\frac{T}{\Theta_{\mathrm{D}}}\right)^2\right]^{1/2} . \tag{6.94}$$

In addition to the width produced by the electron–phonon interaction, we have a width produced by the resolution G^2_{res} and by some inhomogeneous contributions G^2_{inh}, and since we are dealing with Gaussian lineshapes they can all be summed up linearly leading to the following equation (also expanding the temperature dependence for large $(T/\Theta_{\mathrm{D}})^2$):

$$G^2(T) = G_{\text{res}}^2 + G_{\text{inh}}^2 + \frac{8CT}{3\Theta_{\text{D}}^2}\left[1 + \frac{1}{2}\left(\frac{3}{8}\frac{\Theta_{\text{D}}}{T}\right)^2 + \ldots\right] \qquad (6.95)$$

yielding approximately a linear temperature dependence for G^2, or a $T^{1/2}$ dependence for G, as found in the data of Fig. 6.28.

The straight lines in Fig. 6.28 end, for $T = 0$, at about the same value, which is the square of the instrumental width, thus providing an internal consistency check.

A fit of (6.93) to the bulk data of Fig. 6.28 is shown in Fig. 6.29, where also the linear fit to the high temperature data is given for comparison.

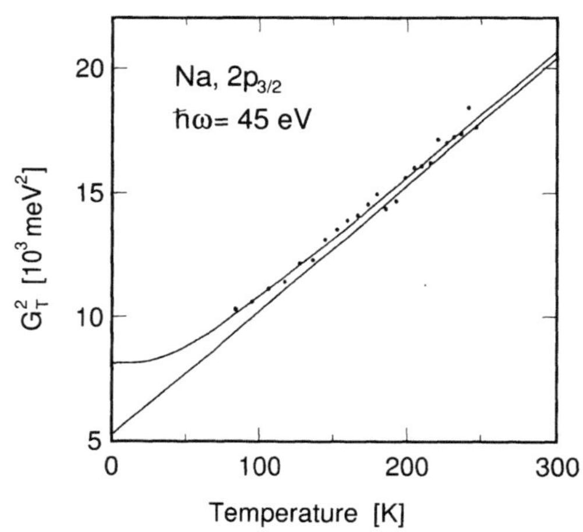

Fig. 6.29. Fit (6.93) to the bulk data from Fig. 6.28 [6.65]

A different type of behavior is observed for the Al 2p core line. Here it is found that the temperature dependence of the Gaussian width is roughly linear, which would correspond to the case where the equilibrium distances are not changed in the photoemission process and only the bonding forces change. Why aluminum is different from Na (but also K, Rb and Yb) is an open question at this point.

So far no effect of the temperature dependence of the electron–electron interaction on the core lines has been documented.

For valence bands the width behaves slightly differently to the above, because here one has an additional electron–electron interaction and, furthermore, the Fermi function has to be taken into account [6.58].

6.5 Dipole Selection Rules for Direct Optical Transitions

Since both the band states in the solid and the electric dipole operator responsible for the optical excitation process have definite symmetry, dipole selection rules are important and have to be taken into account when analyzing experimental PE data, especially if polarized radiation is employed. The selection rules can be used to determine the symmetry of the states that are involved in a direct transition, and in cases of complex spectra they can be employed to reduce the number of lines, making an analysis more tractable [6.68].

Table 6.2. Dipole Selection Rules (SR) for direct interband transitions observed in *normal emission*. For each direction the irreducible representations of the initial states and the final states for the fcc structure are listed in single group notation (NR: non-relativistic case) and double group notation (R: relativistic case). For each face the table gives the allowed direction of polarization of the vector potential along the conventional cubic axes shown in Fig. 6.25 [6.69]

Face/ symmetry	Irreducible representations		Final-state representation	Direction of polarization
(111)	NR:	Λ_1	Λ_1	$\hat{z}$
C_{3v}		Λ_3		$\hat{x}, \hat{y}$
	R:	$\Lambda_{6(1,3)}$	$\Lambda_{6(1)}$	$\hat{x}, \hat{y}, \hat{z}$
		$\Lambda_{4,5(3)}$		$\hat{x}, \hat{y}$
(110)	NR:	Σ_1	Σ_1	$\hat{z}$
C_{2v}		Σ_2		–
		Σ_3		$\hat{x}$
		Σ_4		$\hat{y}$
	R:	$\Sigma_{5(1,2,3,4)}$	$\Sigma_{5(1)}$	$\hat{x}, \hat{y}, \hat{z}$
(001)	NR:	Δ_1	Δ_1	$\hat{z}$
C_{4v}		Δ_2		–
		$\Delta_{2'}$		–
		Δ_5		$\hat{x}, \hat{y}$
	R:	$\Delta_{6(1,5)}$	Δ_6	$\hat{x}, \hat{y}, \hat{z}$
		$\Delta_{7(2,2',5)}$		$\hat{x}, \hat{y}$

Non-relativistic dipole selection rules for direct interband transitions in a fcc structure have been published for the special cases of normal emission [6.69] and emission in a mirror plane [6.70]. Relativistic dipole selection rules,

Table 6.3. Non-relativistic dipole selection rules for transitions on high symmetry lines of the fcc structure and emission in a mirror plane. (||) stands for $\boldsymbol{A}$ parallel to the symmetry line, and ($\perp$) stands for $\boldsymbol{A}$ normal to the symmetry line. (X) and (Y) stand for $\boldsymbol{A}$ along the x- and y-axes defined in Fig. 6.25 for emission from a (110) face [6.69]

Initial-state representation	Final-state representation				Representation of $\boldsymbol{A} \cdot \boldsymbol{p}$
	Λ_1	Λ_3			
Λ_1	$\|\|$	$\perp$			$\|\|: \Lambda_1$
Λ_3	$\perp$	$\|\|, \perp$			$\perp: \Lambda_3$
	Σ_1	Σ_2	Σ_3	Σ_4	
Σ_1	$\|\|$	...	X	Y	$\|\|: \Sigma_1$
Σ_2	...	$\|\|$	Y	X	$X: \Sigma_3$
Σ_3	X	Y	$\|\|$	...	$Y: \Sigma_4$
Σ_4	Y	X	...	$\|\|$	
	Δ_1	Δ_5	$\Delta_{2'}$		
Δ_1	$\|\|$	$\perp$	...		$\|\|: \Delta_1$
Δ_2	...	$\perp$	...		$\perp: \Delta_5$
$\Delta_{2'}$	...	$\perp$	$\|\|$		
Δ_5	$\perp$	$\|\|$	$\perp$		

which take into account the effects of spin-orbit coupling, exist only for the normal emission case [6.71, 6.72].

We now give a brief outline of non-relativistic selection rules for an fcc metal. For a comparison with ARPES data, one has to consider which final states can contribute to the emission process. In normal emission along the three principal directions [001] (Δ line), [110] (Σ line) and [111] (Λ line) (see Figs. 7.5 and 7.7 for the Brillouin zone and a band structure of the fcc structure) the final state in the optical transition has to be totally symmetric with respect to all symmetry operations along the surface normal. This requires that the excitation takes place into a Δ_1, Σ_1 or Λ_1 final state (non-relativistic notation). For the more general case of emission in a mirror plane, the final Bloch states are even with respect to the symmetry operation of that mirror plane. In the electric dipole approximation, a non-vanishing transition matrix element of the form $\langle f|\boldsymbol{A} \cdot \boldsymbol{p}|i\rangle$ implies that the initial state must have the symmetry of the dipole operator $\boldsymbol{A} \cdot \boldsymbol{p}$. Only then is the total matrix element symmetric with respect to the mirror plane as it must be if the optical transition is an allowed one. If the vector potential $\boldsymbol{A}$ lies in the mirror plane, then the initial state must be symmetric (or even), whereas if $\boldsymbol{A}$ is perpendicular to the mirror plane, then the initial state must be antisymmetric or odd.

These considerations and the application of group theory to the dipole matrix element lead to the selection rules given in Tables 6.2 and 6.3 for normal emission [6.69,6.71,6.72] and emission in a mirror plane [6.70] respectively. Figure 6.30 shows the (111), (110) and (001) planes in a fcc structure and the coordinate systems used for characterizing the polarization of the incident light [6.69]. Non-relativistic selection rules will never be obeyed rigorously because spin–orbit interaction is always present. However, they will give the dominant contribution to the spectra for a metal like Cu where one has only small relativistic effects. They thus serve as a useful guide for the interpretation of spectra taken, e.g., in the $(1\bar{1}0)$ mirror plane (a very common measuring geometry containing the Λ, Δ and Σ symmetry lines).

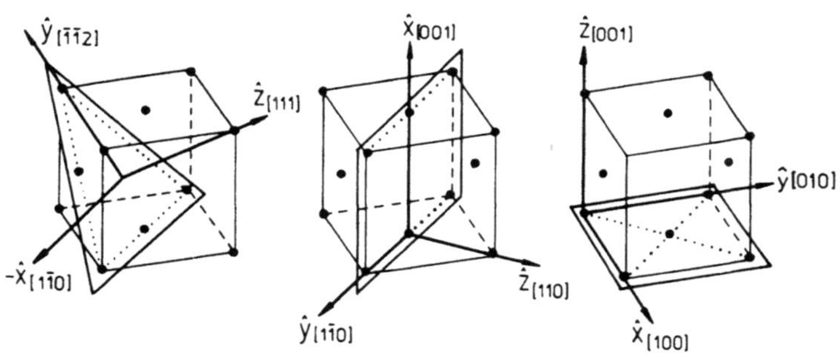

Fig. 6.30. Low index faces of the fcc structure with Cartesian coordinate axes $\hat{x}$, $\hat{y}$, $\hat{z}$ with respect to the conventional cubic axes [6.69]

The dipole selection rules as given in Tables 6.2 and 6.3 are not completely sufficient to describe the polarization dependence of a transition. For example, for a transition from a Λ_3 initial state to a Λ_1 final state, they do not predict where the allowed polarization direction occurs in the plane perpendicular to Λ. This can be determined only by a calculation of the transition momentum matrix elements $\langle f|\boldsymbol{p}|i\rangle$ [6.73].

A typical arrangement for measurements using polarized light is shown in Fig. 6.31. The figure also gives the definition of s and p polarization with respect to the chosen mirror plane. The light incidence angle ψ_i is not conserved inside the material, the transmitted angle ψ_t being given by the Fresnel equations, because of the strong absorption of light in metals.

The effects of selection rules will be seen in various experiments discussed in this volume. Here only a brief example will be given [6.68] (Fig. 6.32). The data are taken from a Cu(110) single crystal but under conditions ($\vartheta = 53°$) such that one is measuring spectra for the ΓL(Λ) direction (see Fig. 7.5 for definitions). The piece of the ΓL band structure relevant to the interpretation of the measured spectra is shown in the bottom part of Fig. 6.32. The spin–

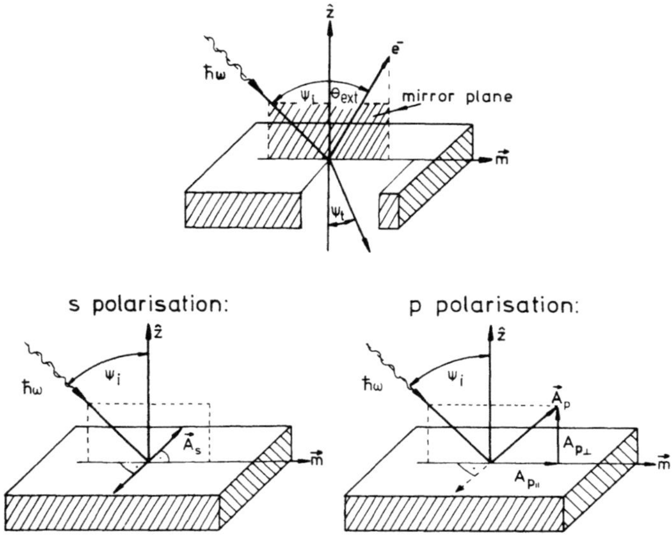

Fig. 6.31. Definition of directions used in a PE experiment with polarized light. The incident light and detected electrons are in a mirror plane

orbit interaction has been omitted in order that the non-relativistic selection rules can be applied. Also shown is the final state band (Λ_1-symmetry) to which transitions with $\hbar\omega = 21.2\,\mathrm{eV}$ radiation are possible. It is displaced by 21.2 eV so that one can identify the occupied states from which transitions are possible at this energy (Sect. 7.2). The spectra taken under these conditions, with the electron detection angle in the ΓKLUX plane [(110) plane] and the light polarized in that plane (p) or perpendicular to it (s) are also shown. According to the selection rules in Table 6.3, transitions of the type $\Lambda_3 \to \Lambda_1$ and $\Lambda_1 \to \Lambda_3$ are possible for s polarization. (The selection rules also allow a $\Lambda_3 \to \Lambda_3$ transition for light polarized perpendicular to the Λ line; for this case, however, the light has to be polarized within the symmetry plane). Since the final state is Λ_1, one should only see the $\Lambda_3 \to \Lambda_1$ transition, and the spectra show indeed only this transition (full curve). From the band-structure diagram of Fig. 6.32 one would expect for s polarization two transitions out of the two Λ_3 bands. The momentum matrix elements $\langle f|\boldsymbol{p}|i\rangle$ calculated by Smith [6.74] are very small for transitions from the lower Λ_3 band to the Λ_1 band that take place in the middle of the ΓL line. This is the situation in Fig. 6.32 (top) with $\hbar\omega = 21.2\,\mathrm{eV}$. Note that this is not an effect of the selection rules.

For p polarization (parallel to the symmetry plane), $\boldsymbol{A}$ can have components both parallel to the high symmetry line ($A_{\mathrm{p}||}$) and perpendicular to it ($A_{\mathrm{p}\perp}$). According to the selection rules in Table 6.3, $A_{\mathrm{p}||}$ excites a $\Lambda_1 \to \Lambda_1$ transition and $A_{\mathrm{p}\perp}$ a $\Lambda_3 \to \Lambda_1$ transition. Thus for A_{p} light (unless it is polarized parallel to the Λ line) both transitions are possible and this is indeed

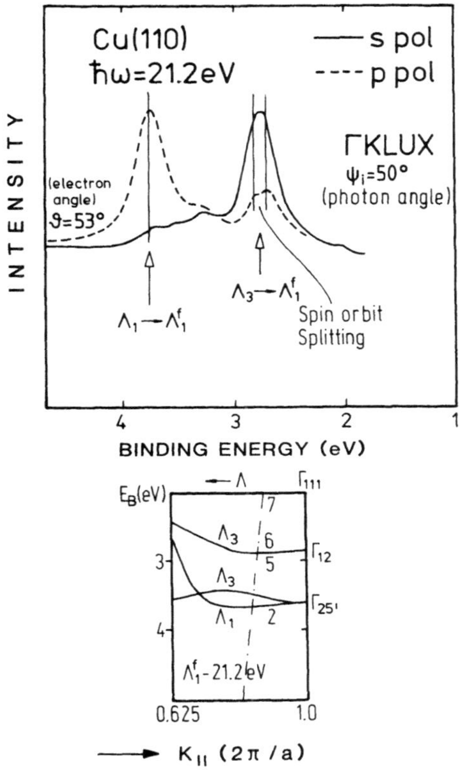

Fig. 6.32. AREDCs from Cu(110) taken with He I radiation ($\hbar\omega = 21.2$ eV) at $\vartheta = 53°$ with two different polarizations. The plane of measurement is a ΓKLUX plane (Fig. 7.5). The lower diagram gives the relevant part of the band structures (Fig. 7.7) where the final-state band (–·–·) has been displaced to lower energies by 21.2eV [6.68]

seen in the experimental data (Fig. 6.32). We remark in passing that the splitting in the p-polarized $\Lambda_3 \to \Lambda_1$ transition is a measure of the spin–orbit interaction, which is small as expected for a low-Z element such as Cu.

We now give a brief example to indicate the necessity of applying the Fresnel equations for the complete understanding of the intensity variations in PE spectra. Since we have found no good example for Cu metal we take data for another noble metal, namely Ag metal [6.73].

The p-polarized intensity in an EDC is strongly dependent on the angle of incidence of the light. In order to understand this completely, one has to realize that light is also refracted as it enters the crystal such that ψ_i becomes ψ_t. One has (Snell's law):

$$\sin\psi_i = n \sin\psi_t \tag{6.96}$$

where

$$n^2 = \varepsilon; \qquad \varepsilon = \varepsilon_1 + i\varepsilon_2 \, . \tag{6.97}$$

From this it follows that ψ_t is a complex angle. Using the known optical constants of the material under investigation, the intensity variation of structures as a function of ψ_i in p-polarization can in principle be accounted for.

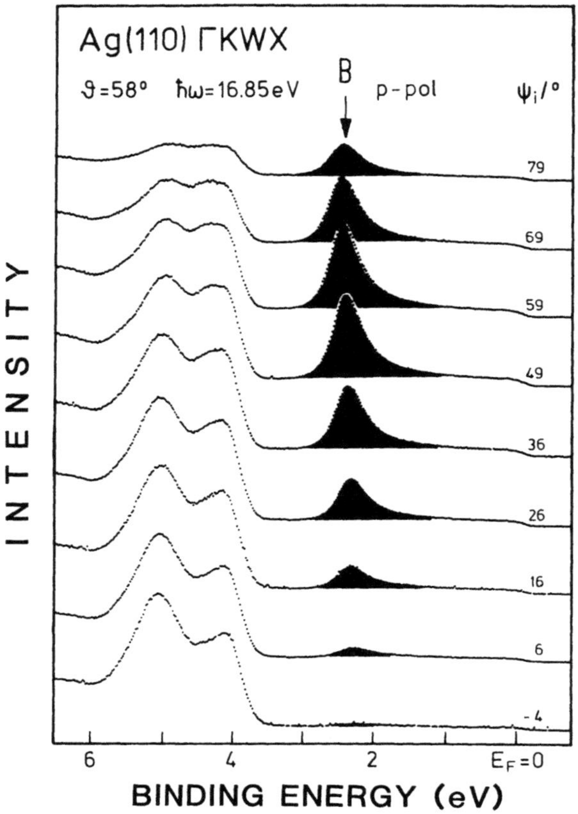

Fig. 6.33. EDCs from Ag(110) measured in the ΓKWX plane (for a definition of this plane, see Fig. 7.5). The electron detection angle is 58°, the radiation is p-polarized (in the ΓKWX plane, see Fig. 6.26). The variable parameter is the angle of incidence of the photon ψ_i (Fig. 6.26). The variation in the intensity of structure B is due to the change in the ratio of $A_{\mathrm{p}\|}$ to $A_{\mathrm{p}\perp}$, where $A_{\mathrm{p}\perp} = 0$ for $\psi_\mathrm{i} = 0$. The vanishing of the intensity in this situation indicates that the transition matrix element is perpendicular to the surface. For the band structure of Ag see Fig. 7.24

Since this involves quite complicated equations we give only a brief example (Fig. 6.33) to demonstrate the magnitude of the effect [6.73, 6.75].

In the experiment on a (110) surface of Ag with an electron detection angle ϑ of 58° (which is chosen so that structure B is well separated from the other lines) the angle of incidence of the radiation with p-polarization has been varied. One can detect a large effect on the intensity of structure B. This is caused by the variation in the relative magnitudes of $A_{\mathrm{p}\perp}$ and $A_{\mathrm{p}\|}$. As mentioned, the vector potential in the crystal is complex because of the complex dielectric constant.

Since the intensity is close to zero at $\psi_\mathrm{i} \sim 0$ (Fig. 6.33) one can conclude that the transition matrix element $\langle f|\boldsymbol{p}|i\rangle$ must be perpendicular to the surface. In a classical approximation the intensity of structure B would then be proportional to $\sim \sin^2 \psi_\mathrm{i}$ (this is obtained from the exact equations by setting $\varepsilon_1 = 1$ and $\varepsilon_2 = 0$). The decrease of the intensity at large ψ_i is an effect of the optical constants of Ag, and can be well described by them [6.73].

References

6.1 B. Dardel, D. Malterre, M. Grioni, P. Weibel, Y. Baer: Phys. Rev. Lett. **67**, 3144 (1991)
6.2 H. Raether: *Springer Tracts Mod. Phys.*, **38**, 84 (Springer, Berlin, Heidelberg 1965)
6.3 C.S. Fadley: In *Electron Spectroscopy, Theory, Techniques and Applications*, Vol.II, ed. by C.R. Brundle, A.D. Baker (Academic, New York 1978) p.1
6.4 C.N. Berglund, W.E. Spicer: Phys. Rev. A **136**, 1030 and 1044 (1964)
6.5 N.V. Smith: In *Photoemission in Solids I*, ed. by M. Cardona, L. Ley, Topics Appl. Phys., Vol.26 (Springer, Berlin, Heidelberg 1978) Chap.6
6.6 J. Braun: Rep. Prog. Phys. **59**, 1267 (1996)
6.7 G.D. Mahan: Phys. Rev. **163**, 612 (1967)
6.8 G.D. Mahan: Phys. Rev. B **2**, 4334 (1970)
6.9 C. Caroli, B. Roulet, D. Saint-James: Theory of Photoemission, in *Handbook of Surfaces and Interfaces*, Vol.1, ed. by L. Dobrzynski (Garland, New York 1978)
6.10 W. Eberhardt, E.W. Plummer: Phys. Rev. B **21**, 3245 (1978)
W. Eberhardt, E.W. Plummer, K. Horn, J. Erskine: Phys. Rev. Lett. **45**, 275 (1980)
6.11 H.J. Levinson, E.W. Plummer, P.J. Feibelman: Phys. Rev. Lett. **43**, 952 (1979)
6.12 H.J. Levinson, E.W. Plummer, P.J. Feibelman: J. Vac. Sci. Technol. **17**, 216 (1980)
6.13 A. Liebsch: Ferienkurs 1980, *Elektronenspektroskopische Methoden an Festkörpern und Oberflächen*, Vols.I and II, Kernforschungsanlage Jülich GmbH (1980)
6.14 C. Kittel: *Introduction to Solid State Physics*, 5th edn. (Wiley, New York, 1974)
6.15 R.Y. Koyoama, N.V. Smith: Phys. Rev. B **2**, 3049 (1970)
6.16 J.C. Phillips: In *Solid State Physics*, Vol.18, ed. by F. Seitz, D. Turnbull (Academic, New York 1966) p.55
6.17 B. Feuerbacher, R.F. Willis: J. Phys. C **9**, 169 (1976)
6.18 D.J. Spanjaard, D.V. Jepsen, P.M. Marcus: Phys. Rev. B **15**, 1728 (1977)
6.19 I. Adawi: Phys. Rev. **134**, A 788 (1964)
6.20 W.I. Schaich, N.W. Ashcroft: Solid State Commun. **8**, 1959 (1970)
6.21 W.I. Schaich, N.W. Ashcroft: Phys. Rev. B **3**, 2452 (1971)
6.22 C. Caroli, D. Lederer-Rozenblatt, B. Roulet, D. Saint-James: Phys. Rev. B **8**, 4552 (1973)
6.23 P.J. Feibelman, D.E. Eastman: Phys. Rev. B **10**, 4932 (1974)
6.24 A. Liebsch: Phys. Rev. B **13**, 544 (1976)
6.25 J.B. Pendry: Surf. Sci. **57**, 679 (1976); and Solid State Phys. **8**, 2413 (1975)
6.26 W. Bardyszewski, L. Hedin: Physica Scripta **32**, 439 (1985)
6.27 W.L. Schaich: In *Photoemission in Solids I*, ed. by M. Cardona, L. Ley, Topics Appl. Phys., Vol.26 (Springer, Berlin, Heidelberg 1978) Chap.2

6.28 N.V. Smith: Crit. Rev. Solid State Sci. **2**, 45 (1971); and Phys. Rev. B **3**, 1862 (1971)
6.29 M.J. Sayers, F.R. McFeely: Phys. Rev. B **17**, 3867 (1978)
6.30 L. Ley, M. Cardona, R.A. Pollak: In *Photoemission in Solids II*, ed. by L. Ley, M. Cardona, Topics Appl. Phys. Vol.27 (Springer, Berlin, Heidelberg 1979) Chap.1
T. Grandke, L. Ley, M. Cardona: Solid State Commun. **23**, 897 (1977)
T. Grandke, L. Ley, M. Cardona: Phys. Rev. Lett. **38**, 1033 (1977)
6.31 E. Dietz, F.J. Himpsel: Solid State Commun. **30**, 235 (1979)
6.32 J.F. Janak, A.R. Williams, V.L. Moruzzi: Phys. Rev. B **11**, 1522 (1975)
6.33 J.B. Pendry, D.T. Titterington: Commun. Phys. **2**, 31 (1977)
6.34 J.B. Pendry, J.F.L. Hopkinson: J. Phys. C **4**, 142 (1978)
6.35 J.F.L. Hopkinson, J.B. Pendry, D.J. Titterington: Comput. Phys. Commun. **19**, 69 (1980)
6.36 C.G. Larsson, J.B. Pendry: J. Phys. C **14**, 3089 (1981)
6.37 For recent calculations of photocurrents by a one-step theory see, e.g., M.A. Hoyland, R.G. Jordan: J. Phys. C **3**, 1337 (1991) and references therein
6.38 P.O. Nilsson, J. Kanski, C.G. Larsson: Solid State Commun. **36**, 111 (1980)
6.39 D. Westphal, D. Spanjaard, A. Goldmann: J. Phys. C **13**, 1361 (1980)
6.40 A. Goldmann, A. Rodriguez, R. Feder: Solid State Commun. **45**, 449 (1983)
6.41 P.H. Citrin, P. Eisenberger, D.R. Haman: Phys. Rev. Lett. 33, 965 (1974)
For an analysis of the temperature dependence of the linewidth of a core level in a metal (Na) see D.M. Riffe, G.K. Wertheim, P.H. Citrin: Phys. Rev. Lett. **67**, 116 (1991)
6.42 N.J. Shevchik, D. Liebowitz: Phys. Rev. B **18**, 1618 (1978)
6.43 M. Sagurton, N.J. Shevchik: J. Phys. C **11**, 2353 (1978)
6.44 M. Sagurton, N.J. Shevchik: J. Phys. C **11**, L353 (1978)
6.45 M. Sagurton, N.J. Shevchik: Phys. Rev. B **17**, 3859 (1978)
6.46 N.J. Shevchik: Phys. Rev. B **16**, 3428 (1977); and Phys. Rev. B **20**, 3029 (1979)
6.47 N.J. Shevchik, D. Liebowitz: Phys. Rev. B **18**, 1618 (1978)
6.48 R.C. White, C.S. Fadley, M. Sagurton, P. Roubin, D. Chandesris, J. Lecante, C. Guillot, Z. Hussani: Phys. Rev. B **35**, 1147 (1987)
6.49 J.M. Ziman: *Principles of the Theory of Solids* (Cambridge Univ., Cambridge 1972)
6.50 H. Mårtensson, P.O. Nilsson, J. Kanski: Appl. Surf. Sci. **11**, 652 (1982)
H. Mårtensson, C.G. Larsson, P.O. Nilsson: Surf. Sci. **126**, 214 (1983)
6.51 S.D. Kevan, D.A. Shirley: Phys. Rev. B **22**, 542 (1980)
6.52 R. Matzdorf, G. Meister, A. Goldmann: Surf. Sci. **286**, 56 (1993)
6.53 R. Matzdorf, G. Meister, A. Goldmann: Surf. Sci. **296**, 241 (1993)
6.54 A. Goldmann, R. Matzdorf: Prog. Surf. Sci. **42**, 331 (1993)
6.55 R. Matzdorf, A. Goldmann, J. Braun, G. Borstel: Solid State Commun. **91**, 163 (1994). Bringing to light the importance of the angular resolution, for measuring the width in valence bands of metals, in addition to the energy resolution
6.56 R. Matzdorf, R. Paniago, G. Meister, A. Goldmann: Solid State Commun. **92**, 839 (1994). Showing that for Cu(100) the lifetime width of a valence hole state has an energy dependence of $(E_\mathrm{B} - E_\mathrm{F})^2$, thus fulfilling the prediction of Fermi-liquid theory, see (Chap. 1)
6.57 R.S. Williams, P.S. Wehner, G. Apai, J. Stöhr, D.A. Shirley, S.P. Kowalczyk: J. Electron Spectrosc. Relat. Phenom. **12**, 477 (1977)
6.58 T. Valla, A.V. Fedorov, P.D. Johnson, S.L. Hulbert: Phys. Rev. Lett. **83**, 2085 (1999)

6.59 Reference [6.58] also gives data of the change of linewidth by disorder. This is a topic, which is important in the high resolution studies now possible (see K. Breuer, S. Messerli, D. Purdie, M. Garnier, M. Hengsberger, Y. Baer, M. Mihalik: Phys. Rev. B **56**, R7061 (1997)). For an approach of dealing with these problems see: F. Thulmann, R. Matzdorf, G. Meister, A. Goldmann: Phys. Rev. B **56**, 3632 (1997)

6.60 M.C. Desjonquères, D. Spanjaard: Concepts in Surface Physics, 2nd Edn. (Springer, Berlin, Heidelberg 1996)

6.61 Here the following considerations from Chap. 1 have to be kept in mind: The width of a valence-band state is the combination of the width of the hole state Γ_{h} and the electron state Γ_{e}: What one is interested in, is the hole-state width and this is not easy to separate from the total measured width Γ_{exp}. The relation between the measured width Γ_{exp}, the hole-state width Γ_{h} and the electron-state width Γ_{e} is given by

$$\Gamma_{\mathrm{exp}} = \frac{(\Gamma_{\mathrm{h}}/|v_{\mathrm{h}\perp}|) + (\Gamma_{\mathrm{e}}/|v_{\mathrm{e}\perp}|)}{|(1/v_{\mathrm{h}\perp})[1 - (m v_{\mathrm{h}||} \sin^2\vartheta/\hbar k_{||})] - (1/v_{\mathrm{e}\perp})[1 - (m v_{\mathrm{e}||} \sin^2\vartheta/\hbar k_{||})]|}$$

where $v_{\mathrm{h}||}$, $v_{\mathrm{h}\perp}$, $v_{\mathrm{e}||}$, $v_{\mathrm{e}\perp}$ are the group velocities given by $v_{\mathrm{h}||} = \hbar^{-1}(\partial \mathrm{E}_{\mathrm{h}}/\partial k_{||})$, etc., and ϑ is the angle between the electron-detection direction and the surface normal. In the case of a normal emission experiment one has

$$\Gamma_{\mathrm{exp}}(\vartheta = 0) = \frac{\Gamma_{\mathrm{h}} + \Gamma_{\mathrm{e}}|v_{\mathrm{h}\perp}/v_{\mathrm{e}\perp}|}{1 - |v_{\mathrm{h}\perp}/v_{\mathrm{e}\perp}|}\,.$$

In the case where the dispersion of the initial (hole) state is small with respect to that of the final (electron) state the equation can be further simplified:

$$\Gamma_{\mathrm{exp}}(\vartheta = 0) = \Gamma_{\mathrm{h}} + \Gamma_{\mathrm{e}}|v_{\mathrm{h}\perp}/v_{\mathrm{e}\perp}|\,.$$

Finally in the case where $v_{\mathrm{h}\perp}$ vanishes, i.e., where the dispersion of the band perpendicular to the surface is perfectly flat (which holds for surface states, or two- and one-dimensional electronic states) one can identify $\Gamma_{\mathrm{exp}})(\vartheta = 0)$ with Γ_{h} which is the imaginary part of the self energy; see N.V. Smith, P. Thiry, Y. Petroff: Phys. Rev. B **47**, 15476 (1993)

6.62 T. Grandke, L. Ley, M. Cardona: Phys. Rev. B **18**, 3847 (1978)
P.B. Allen, M. Cardona: Phys. Rev. B **27**, 4760 (1983)
J. Fraxedas, H.J. Trodahl, S. Gopalan, L. Ley, M. Cardona: Phys. Rev. B **41**, 10068 (1990)

6.63 M.A. Butler, G.K. Wertheim, D.L. Rousseau, S. Hüfner: Chem. Phys. Lett. **13**, 473 (1972)

6.64 J.A. Knapp, F.J. Himpsel, A.R. Williams, D.E. Eastman: Phys. Rev. B **19**, 2844 (1979)

6.65 G.K. Wertheim, D.M. Riffe, P.H. Citrin: Phys. Rev. B **49**, 2277 (1994)

6.66 D.M. Riffe, G.K. Wertheim, P.H. Citrin: Phys. Rev. Lett. **67**, 116 (1991)

6.67 L. Hedin, A. Rosengren: J. Phys. F **7**, 1339 (1977)

6.68 R. Courths, S. Hüfner: Phys. Rep. **112**, 53 (1984)

6.69 J. Hermanson: Solid State Commun. **22**, 9 (1977)

6.70 W. Eberhardt, F.J. Himpsel: Phys. Rev. B **21**, 5572 (1980)

6.71 G. Borstel, W. Braun, M. Neumann, G. Seitz: Phys. Stat. Sol. (b) **95**, 453 (1979)

6.72 G. Borstel, M. Neumann, G. Wöhlecke: Phys. Rev. B **23**, 3121 (1981)

6.73 H. Wern, R. Courths: Surf. Sci. **162**, 29 (1985)

6.74 N.V. Smith, R.L. Benbow, Z. Hurych: Phys. Rev. B **21**, 4331 (1980)

6.75 H. Wern: Winkelaufgelöste ultraviolett Photoelektronenspektroskopie an Silber. Thesis, Universität des Saarlandes (1985)

7. Band Structure and Angular-Resolved Photoelectron Spectra

The direct transitions which, as shown in Chap. 6, dominate the EDCs from UPS, can be used to determine the electronic band structure of a solid. Looking at standard textbooks on solid-state physics one notices that, while for phonons the band structures (i.e., dispersion relations) have been mapped out – sometimes many years ago – for a number of representative solids in considerable detail by neutron diffraction, such information has become available only more recently for the electronic case. Generally the Fermi surface, which is the surface of the energy $E = E_{\mathrm{F}}$ in $\boldsymbol{k}$-space has been measured. In addition, a few selected points in the $E(\boldsymbol{k})$ diagram are known from optical experiments. However, only recently, with the utilization of a wide body of material from PES (mostly employing synchrotron radiation), has one been able to determine the electronic dispersion curves for a number of solids. A few representative examples will be given in this chapter, but first the question of how band structure information can be extracted from measured EDCs will be dealt with.

It has been mentioned already, that the determination of a band structure from a measured EDC is not always a trivial matter. For the sake of completeness the problem will be sketched again.

In an EDC in the direct transition regime one observes a transition between an initial state E_i and a final state E_f, where both energies are measured with respect to the Fermi energy, which is an experimentally easily accessible reference. The measured quantity is the kinetic energy E_{kin} of photoelectrons detected at an angle ϑ relative to the sample surface normal (Fig. 6.12). Knowing the work function ϕ one obtains the final and initial state energies as ($E_i = -E_{\mathrm{B}}$)

$$E_f = E_{\mathrm{kin}} + \phi, \quad E_f = E_i + \hbar\omega \; . \tag{7.1}$$

The determination of the final state wave vectors[1] $\boldsymbol{k}_f$ and $\boldsymbol{K}_f$ is, however, more complex. For the wave vector parallel to the surface $K_{||}$, one has (Fig. 6.4)

---

[1] Note:
$\boldsymbol{k}$ is the wave vector of a Bloch state in the band structure;
$\boldsymbol{K}$ is the wave vector of a photoexcited electron in the crystal; we have $\boldsymbol{K} = \boldsymbol{k}+\boldsymbol{G}$, $\boldsymbol{G}$ being a wave vector in the reciprocal lattice;
$\boldsymbol{p}/\hbar$ is the wave vector of the photoexcited electron in the vacuum.

$$|\boldsymbol{p}_{||}|/\hbar = |\boldsymbol{K}_{||}| = \sqrt{(2m/\hbar^2)E_{\text{kin}}} \sin\vartheta \tag{7.2}$$

an equation that determines $\boldsymbol{K}_{||}$ exactly. The situation with respect to the perpendicular momentum $K_{\perp}$ is more complex, as has been discussed before, because there is no direct relation between the measured quantities $\boldsymbol{p}_{||}$ and $p_{\perp}$ and the quantity $\hbar K_{\perp}$ which one needs to obtain the band structure without any further assumption.

To overcome this problem two paths have been pursued. One can make an assumption about the final state using, e.g., the so-called "free-electron model" [7.1–7.6], or a calculated band structure. In a second approach one can try to devise experiments that allow $\boldsymbol{k}$ to be determined directly. In the next sections the different procedures to determine $\boldsymbol{k}$ and thereby $\boldsymbol{k}_f$ or $\boldsymbol{k}_i$ will be outlined.

The simplest, most frequently employed and most economical method is based on the assumption of a free-electron final state. Since the optical excitation takes place in the presence of a crystal potential, the free electron final state can only be an approximation. However, the higher the excitation energy, the better that approximation, because the effect of the crystal potential gets weaker with increasing kinetic energy of the electron. Unfortunately another effect works against this. In extending the UPS regime (in which one usually works for the determination of band structures) beyond about 30 eV there is a rapid decrease of the electron mean free path (Fig. 1.11) and thus an increase in the wave-vector broadening, which in turn smears out the direct transition features.

Next in simplicity to the free-electron model is the approach that uses final states obtained from a band structure calculation. The increasing accuracy of these calculations, due to the success of the local-density approximation [7.11–7.14], makes this quite a realistic method.

In an intermediate approach one approximates the theoretical bands for the excited states with a free-electron parabola; this is very convenient in the evaluation of the actual data.

One might at this point worry about the accuracy, even the usefulness, of data obtained by such procedures. A judgement can only be made after an inspection of a large body of experimental data and at this point one can state that the use of a free-electron type final state (possible adjusted with some theoretical insight) has generally given results that reproduced the relevant physics very well. Of course, this procedure must lead to erroneous results in particular cases, e.g., when one wants to determine gaps in the final-state band structure.

A circumstance that has helped to obtain rather accurate occupied bands from not very well-known final states is the fact that a large body of experimental material has been accumulated for transition metals and particularly for their d bands. These d bands are relatively flat. It is then apparent that a shift in the excited state parabola (which is relatively steep) by 1 or 2 eV will affect the dispersion of the initial state only slightly. This statement does not hold for free-electron metals. For these, however, the free electron final state should always be a good approximation.

Let us conclude this introduction by stating that a free-electron type final state is not only a convenient, but also a rather accurate approximation for the determination of $E(\boldsymbol{k})$ curves from measured EDCs. Much more elaborate methods (which by necessity are also more time consuming) have to be used, in order to find $E(\boldsymbol{k})$ curves without any a priori assumptions. These methods are interesting in their own right, and have their value in adjusting the free electron final states to a value which assures the most accurate evaluation of a band structure from measured spectra.

7.1 Free-Electron Final-State Model

In this model the dispersion relation of the final state is assumed to be that of a free electron:

$$E_f = (\hbar^2/2m^*)(\boldsymbol{k}+\boldsymbol{G})^2 - |E_0| \tag{7.3}$$

where m^* is the effective mass, and the energies are measured with respect to the Fermi energy. This leaves as the parameters to be determined the inner potential $V_0(V_0 = E_\mathrm{v} + |E_0| = |E_0| + \phi)$ and m^*. While for m^* the free-electron mass m is usually employed, for the determination of V_0 three methods have been used: either it is adjusted in such a way that the agreement between experimental and theoretical band structure for the occupied states is optimal, one uses the value of the theoretical muffin-tin[2] zero or one looks for symmetries in the experimental $E(k_\perp)$ curves. Having chosen V_0, the procedure is simple and is demonstrated for convenience for the normal emission case, $\boldsymbol{K}_{||} = 0$; see Fig. 7.1 (free-electron parabola means $m^* = m$).

Assume that a three-peak spectrum, as given in Fig. 7.1, has been observed in a normal-emission experiment. The energies of the three peaks give the initial energies $E_i(1)$, $E_i(2)$ and $E_i(3)$ with respect to the Fermi energy, which is always observable in the spectra. The final state free-electron parabola for this emission direction, calculated with a suitable energy E_0, is now shifted parallel to the energy axis to lower energies by $\hbar\omega$, which takes care of energy and momentum conservation. One then seeks points on the shifted parabola that have energies $E_i(1)$, $E_i(2)$ and $E_i(3)$, and the corresponding $k_{f\perp}$ values of these initial energies are thereby determined. In order to get the $k_{i\perp}$ value in the first Brillouin zone a shift by the corresponding $\boldsymbol{G}$ has to be performed. Using this procedure for a number of different photon

[2] The crystal potential is sometimes approximated for band structure calculations by a so-called muffin-tin potential. This potential is represented by that of an isolated ion within a sphere with a radius r_mt (of the order of half the nearest-neighbor distance) and centered around each lattice point. It is taken as zero elsewhere. This zero with respect to the vacuum level is often called the muffin tin zero.

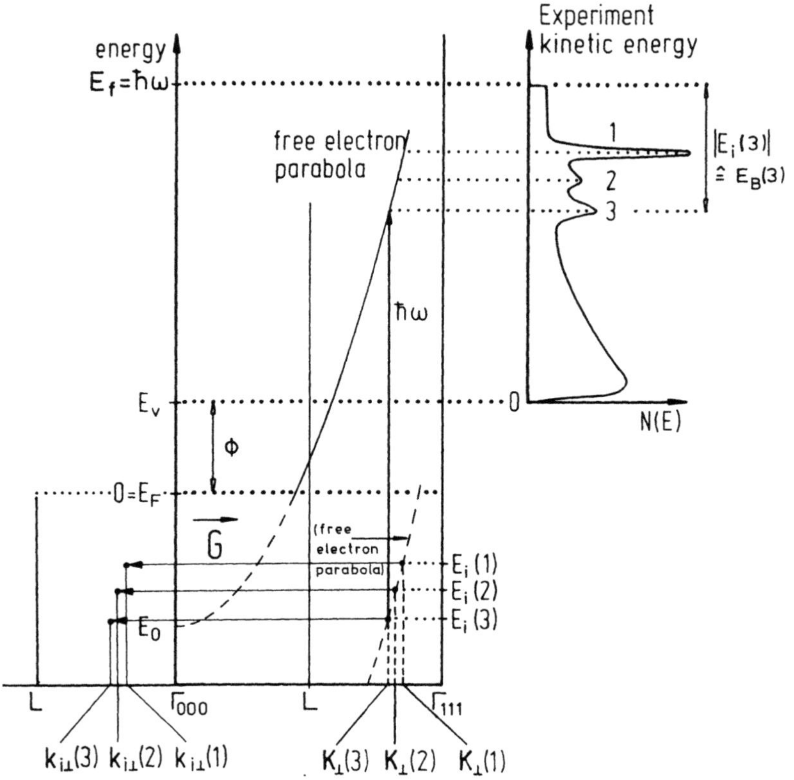

Fig. 7.1. Band-structure determination using a free electron final state parabola, assuming that a normal emission EDC has been measured from a (111) plane of a fcc crystal. The free electron parabola is "attached" to the bottom of the conduction band at E_0 and the measured spectrum with its low-kinetic energy cutoff at E_v. The intersection of the measured peak energies (1,2,3) with the free electron parabola starting at E_0 gives the wave vector $K_\perp$ of these states and the vertical lowering by $\hbar\omega$ yields the initial-state band-structure energies $E_\mathrm{i}(K_\perp)$. The $E_\mathrm{i}(K_\perp)$ are folded into the first Brillouin zone with G(111) to yield $E_\mathrm{i}(K_\perp)$

energies $\hbar\omega$, the $E_i(\boldsymbol{k})$ relation along the chosen direction (normal emission) can be determined experimentally, and by employing different crystal faces the band structure in the whole Brillouin zone can be mapped out.

To give an idea of the spread of the data we list in Table 7.1 a summary of some values of E_0 (and E_F) and effective masses m^*/m used or obtained in analyzing data for Cu. Although the values of E_0 differ by almost 2 eV the final band structures that emerge from the different data analyses are quite similar. The band structure of Cu along [110] determined in this way is given in Fig. 7.2 [7.3, 7.4]. It is in good agreement with that calculated by Burdick [7.19], which in itself is very similar to the most sophisticated later calculations [7.11].

Table 7.1. Inner potentials E_0 and effective masses m^* used for band-structure determination of Cu with free-electron final states (m is the free-electron mass). The muffin-tin zero of Burdick [7.19] is −7.55 eV

Experiment Reference	$E_F - \|E_0\|$ [eV]	m^*/m	Remarks
7.2	−8.9	1	Bottom of Burdick's sp band at Γ
7.3	−8.6	1	Bottom of experimental sp band at Γ
7.15	−7.0	1	Low energy fit (10–20 eV) to Burdick's band structure [7.19]
7.16	−8.9	1.15	Photoelectron refraction ($\hbar\omega = 45$ eV)
7.17	−7.25	1	Cu(001) and Cu(111) ΓKLUX plane
7.18	−6.9	–	LEED (10–50 eV) average value

7.2 Methods Employing Calculated Band Structures

Since the *free-electron approximation* is quite successful and one knows that, for metals at least (to a lesser degree for semiconductors), the current band-structure calculations give reasonable results (for "true" band electrons although not for f electrons), it is very tempting to try to improve on the free-electron model by using final states obtained in a theoretical calculation. An intermediate, but very convenient step consists in approximating the calculated final states by a free-electron parabola. This process can be iterated using the experimental data to adjust the zero of the final state parabola.

Until now the discussion has more or less implicitly assumed normal emission, and the availability of a large range of photon energies (as provided by synchrotron radiation). In a standard laboratory, however, this is not the case, and as far as energy goes one is restricted to a few resonance lines (noble gas lines, Sect. 1.5). Then the only way to change $\boldsymbol{k}$ in order to map the band structure is to change the electron detection angle ϑ with respect to the surface normal.

The analysis of the data is then mostly done by using so-called "structure plots". In its simplest form this means that one calculates the band structure for the energy and angles used in the experiment and compares an experimental $E(\vartheta)$ with the theoretical $E(\vartheta)$. In a more refined way the experimental structure plot is used to derive an experimental band structure via an inter-

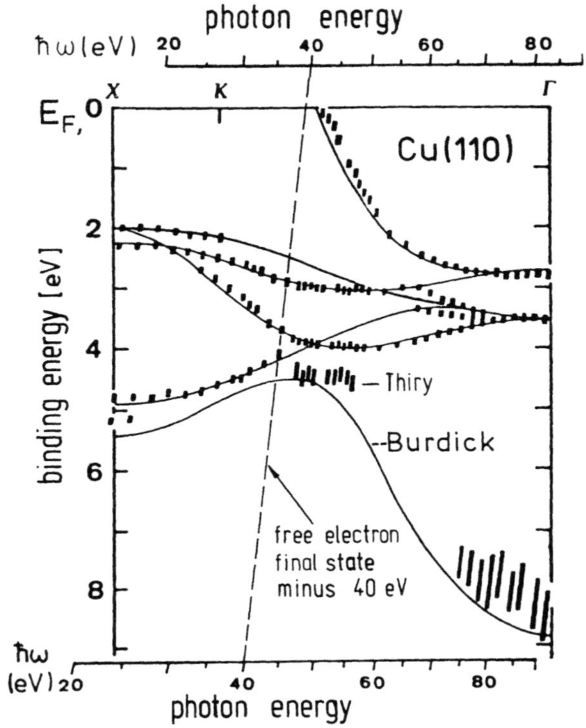

Fig. 7.2. Band structure of Cu(110). Data points have been obtained by analyzing spectra with a free-electron final state. An example for a photon energy of $\hbar\omega$ = 40 eV is shown [7.3]. Photon energies at the bottom and the top correspond to free electron parabolas displaced by the photon energies indicated. Solid curves give the band structure of Burdick [7.19]

polation scheme [7.20–7.22]. An example is given in Fig. 7.3, which shows the results of an experiment with Ne radiation ($\hbar\omega$ = 16.85 eV) where a Ni(110) crystal was used and the electron detection angle was tilted towards the [111] direction and the [100] direction [7.23]. The points in Fig. 7.3 are directly taken from the EDCs. The theoretical curves (cross-hatched "bands") were then obtained by using a band structure parameterized in an interpolation scheme for the occupied bands and a free electron final state. The parameters of the interpolation scheme and the E_0 for the free electron parabola were fitted to give the best agreement to the measured points. In order to take the lifetime broadening of the final state into account, it was given a width of 4 eV, which is reflected in the width of the calculated structures. Such plots were analyzed for two photon energies and two crystal faces of Ni. The final result of this procedure is presented in Fig. 7.4.

Figure 7.4 shows in the left panel (a) the best fit "experimental" band structure as obtained from the procedure just outlined and compares it to the

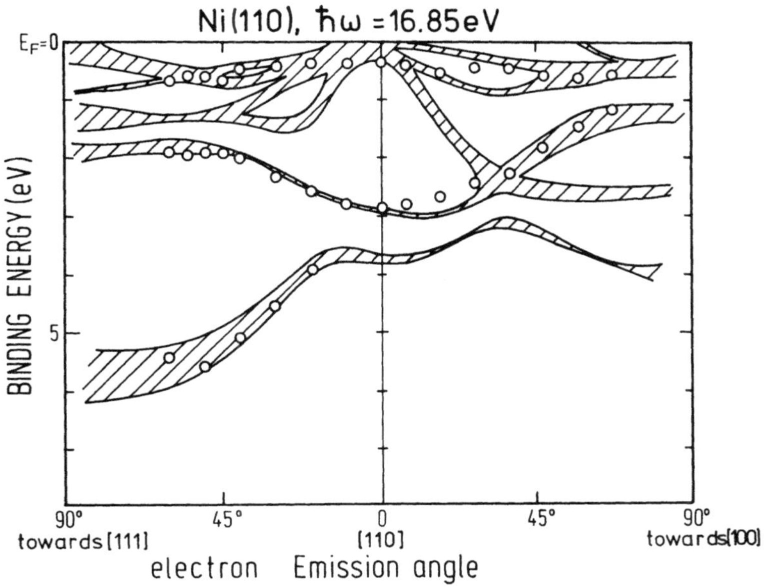

Fig. 7.3. "Structure plot" of the valence band of Ni obtained with $\hbar\omega = 16.85\,\mathrm{eV}$ from a Ni(110) crystal; the electron detection angle is tilted towards the [111] direction and the [100] direction (Fig. 7.5) [7.23]. The hatched bands are obtained using an interpolation scheme to fit a band structure of Ni to the data points (*open dots*); the width of the "bands" reflects a 4 eV broadening (lifetime) introduced into the final state

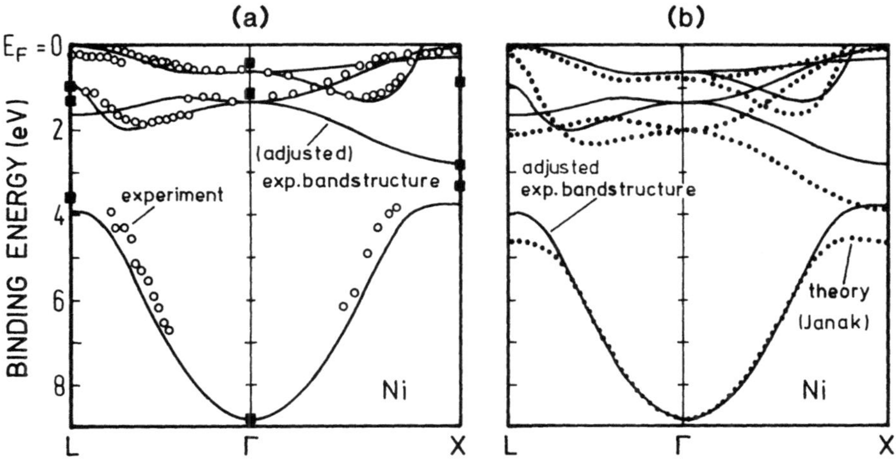

Fig. 7.4. "Experimental" band structure of Ni [7.23] derived with the procedure outlined in Fig. 7.3 (*full curves*). (**a**) Comparison with normal emission PES data taken from [7.24] (*open dots*) and [7.25] (*squares*). (**b**) (*dotted curves*) calculated band structure of [7.26]

available normal emission data obtained using synchrotron radiation (note that the exchange splitting has been omitted). In the right hand panel (b) the experimental band structure (full lines) is compared to a theoretical one (dotted line) [7.26]. As is well known, for Ni the deviation between theory and experiment, especially for the d band width, is quite large, and is actually the largest found so far in any system, except perhaps for Na [7.27]. In Ni it is believed to come from a considerable final state interaction for which other evidence will be given later [7.24, 7.28–7.38], and in Na it may have the same origin although the effects are less well documented (but see Chap. 4).

This section may give the reader the impression that one is dealing with some sort of alchemy rather than with physics. This impression is not completely unfounded. But the "accurate" determination of band structures from experimental data is so complex that it is usually worth trying approximate methods.

We add that the methods that will be described to determine band structures of occupied states can also be used to study band structures of unoccupied states.

7.3 Methods for the Absolute Determination of the Crystal Momentum

It is obvious from the two preceding sections that it is highly desirable to have methods that derive band structures from measured EDCs without involving any a priori assumptions, thus allowing an unbiased comparison with theoretical band structures. It will be seen that such methods do in fact exist. However, they are generally quite complicated and require many experimental spectra, very few of which are actually used in obtaining the final result. It has thus become customary to apply these methods only in order to check (or adjust) a few points in a calculated final-state band structure and use the adjusted final states for an evaluation of the occupied bands from measured EDCs. The examples will be taken from the noble metals in order to make this chapter as coherent as possible.

All the noble metals have the fcc crystal structure. The Brillouin zone for this structure is given in Fig. 7.5. Two slices through the Brillouin zone are also shown, namely a so-called ΓKLUX plane ($1\bar{1}0$ plane) and a ΓKWX plane (001-plane). If one works, e.g., with a (110) crystal surface, these two planes can be used as emission (mirror) planes.

For the further discussion it will also be useful to have sections through the Brillouin zone in the extended zone scheme (a scheme which repeats the Brillouin zone periodically in space) for the two planes that are shaded in Fig. 7.5. These sections are given in Fig. 7.6. Finally, Fig. 7.7 shows the band structure of the noble metal referred to most frequently in the following discussion, namely Cu. This band structure was obtained by adjusting an

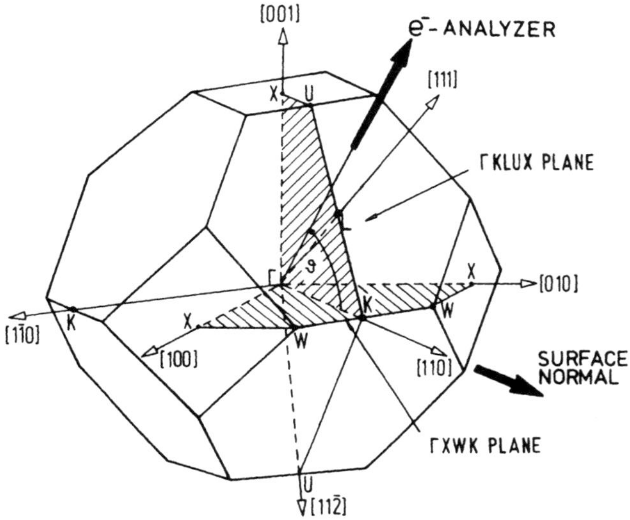

Fig. 7.5. Brillouin zone for a fcc cubic lattice. The high symmetry ΓKLUX and ΓKWX planes used in many PE experiments are indicated [7.39]. The arrow shows e^--emission at an angle ϑ with respect to the normal of a (110) surface

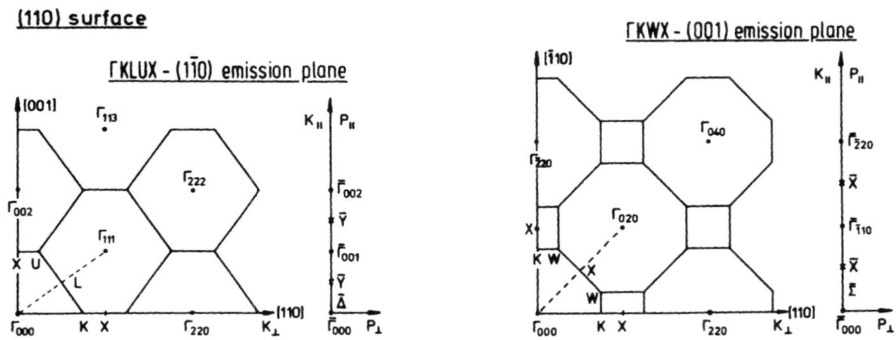

Fig. 7.6. Cuts through the extended zone scheme using the two planes hatched in Fig. 7.5. *Left*: ΓKLUX plane used to study PES from a (110) crystal face; the corresponding line in the surface Brillouin zone is also given (labels $\bar{\Gamma}$, $\bar{\Delta}$,$\bar{\mathrm{Y}}$). *Right* ΓKWX plane used to study PES from a (110) crystal face; the corresponding line in the surface Brillouin zone is also given (labels $\bar{\Gamma}$, $\bar{\Sigma}$, $\bar{\mathrm{X}}$) [7.39]

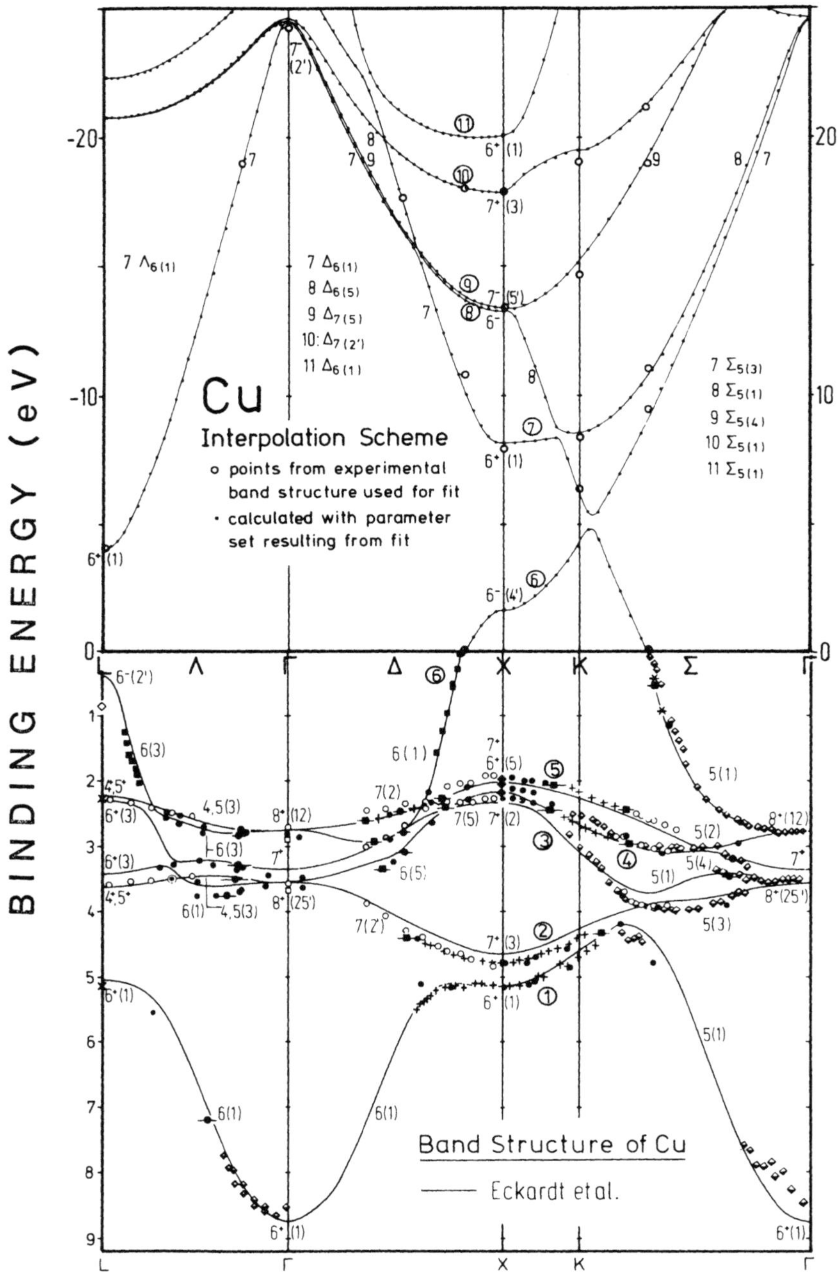

Fig. 7.7. "Experimental" band structure of Cu (*full lines*) [7.39, 7.40] obtained by fitting experimental points (*open dots*) with an interpolation scheme. The encircled numbers give the commonly used numbering of the bands

interpolation scheme to a number of experimental points (open circles) and it can therefore be termed an "experimental" band structure of Cu [7.3–7.6, 7.39, 7.40].

7.3.1 Triangulation or Energy Coincidence Method

This method was first proposed by Kane [7.41] and applied successfully by Neddermeyer and his group [7.42], with other work substantiating the usefulness of the procedure [7.15, 7.43–7.51]. In essence the method consists of observing a direct transition occurring at a well-defined point in the Brillouin zone from two crystal faces. This increases the amount of information available and in general allows $\boldsymbol{k}$ to be uniquely determined.

The principles of the procedures are given in Figs. 7.8–7.10. Assume one is interested in determining the band structure of a crystal (Cu for the present purposes) along the ΓL direction ([111] direction in real space). Then one measures a spectrum with 21.2 eV radiation at normal emission (surface 1, (111) surface). This spectrum is given as (a) in Fig. 7.8, and the various structures have been numbered 2–6. From this spectrum one can obtain the initial state energies, and the final state energies but, however, not the momenta k_Λ for the points between Γ and L in the Brillouin zone where transitions $2 \rightarrow 6$ have occurred.

For this determination a second measurement is performed. Now a (110) surface is used (surface 2, (110) surface). As can be seen from Fig. 7.5 the ΓKLUX emission plane for such a surface also contains the ΓL direction, which is now of course an off-normal direction. However, this ensures that the same point in the Brillouin zone that has been reached by the normal emission experiment from surface 1 can also be reached by a non-normal emission experiment from surface 2. Of course, due to refraction, one does not know the angle under which this point will be reached. Therefore a series of spectra have to be taken with varying electron detection angle in the ΓKLUX plane. They are displayed together with the Cu(111) normal-emission spectrum in Fig. 7.9. Here the structures are labelled (a)–(e); their energy as a function of detection angle is shown in the right panel of this figure. One sees that at $\approx 52.5°$ the energies of structures 2–6 (from the (111) crystal) and those of structures (a)–(e) (from the (110) crystal) coincide (hence the term *energy coincidence method*) signaling that under these two conditions one is sampling the same transitions in the Brillouin zone, which are located on the Λ line (ΓL). Figure 7.8 shows as spectrum (b) the one that agrees with respect to the energy positions with that taken at normal emission from the (111) crystal.

For comparison, Fig. 7.8 also shows a spectrum (c) taken from the (110) crystal in the ΓKLUX plane in the [111] direction, which is, of course, at an angle of 35.2° to the [110] direction. If there were negligible refraction at the surface, this spectrum should agree with spectrum (a). The obvious disagree-

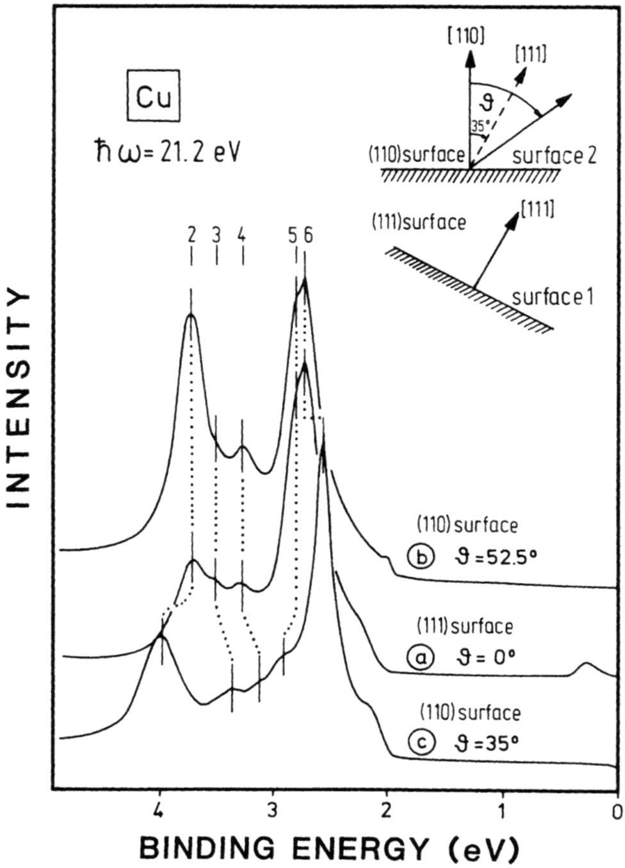

Fig. 7.8. Typical experimental data used in the application of the triangulation method for an absolute determination of $E(\boldsymbol{k})$ points (here along ΓL in $\boldsymbol{k}$-space). Spectra taken from a (110) crystal at $\vartheta = 35°$ and $\vartheta = 52.5°$ in the KLU-azimuth are compared with a normal emission spectrum from a (111) surface. Geometrically the [111] direction appears at an angle of 35° from the [110] direction. The fact that the (110) spectrum taken under this condition does not agree with the normal emission (111) spectrum shows the importance of refraction in PES (Figs. 6.4, 6.5) [7.39]

ment confirms the importance of refraction (via the inner crystal potential), which the electron experiences in crossing the solid–vacuum interface.

The spectra are analyzed according to Fig. 7.10. We know that in the off-normal emission experiment from the (110) plane the parallel component of the wave vector is conserved. Thus the wave vector of the electrons of the state E_i in the $\vartheta_{\text{ext}} = 52.5°$ spectrum is calculated from ($E_i = -E_{\text{B}}$)

$$|\boldsymbol{p}_{52.5°}| = \sqrt{2m(E_i + \hbar\omega - \phi)} \tag{7.4}$$

and the parallel component in the (110) plane is ($\boldsymbol{p}_{||} = \hbar\boldsymbol{K}_{||}$)

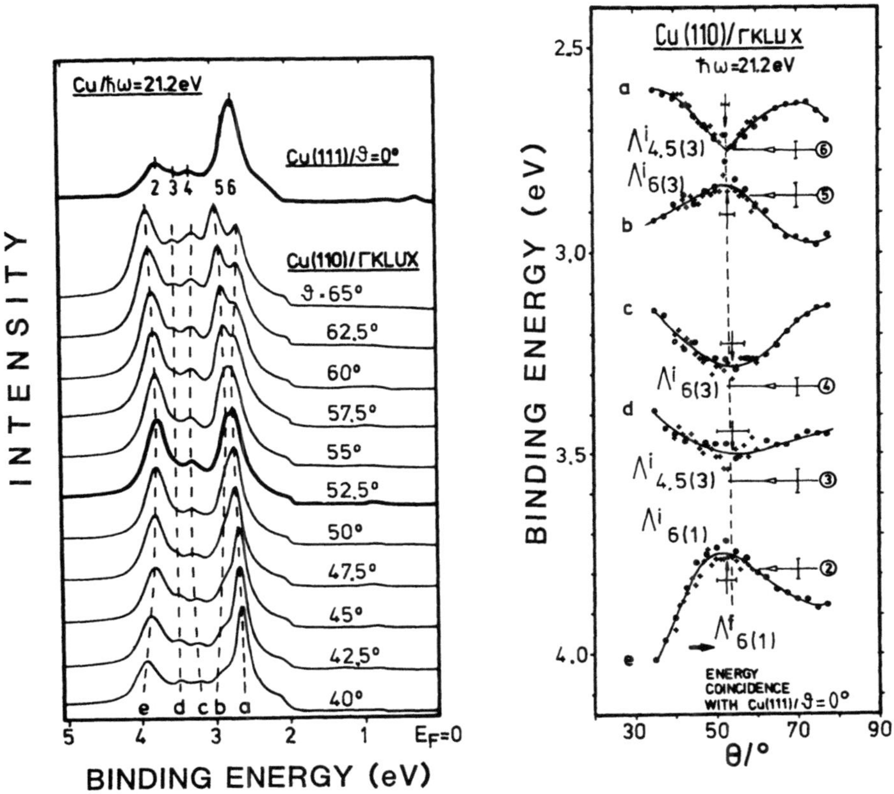

Fig. 7.9. *Left*: EDCs taken with $\hbar\omega = 21.2\,\text{eV}$ from a (110) surface in the ΓKLUX plane for varying electron detection angle ϑ. The top spectrum is a normal emission spectrum from a (111) surface. The $\vartheta = 52.5$ spectrum from the (110) surface (*thick line*) is the one for which the positions of the peaks correspond most closely in energy to those from the (111) normal emission spectrum (therefore also the term energy coincidence method). *Right*: peak positions taken from the spectra in the left panel, as a function of electron detection angle ϑ. Also given are the energies of the peaks (2), (3), (4), (5), (6) of the Cu(111) normal emission spectrum [7.39]. Note, that the "energy coincidence" generally occurs well within the error bars of typically only $\pm 0.07\,\text{eV}$

$$K_{||}(110) = \sqrt{(2m/\hbar^2)(E_i + \hbar\omega - \phi)}\sin 52.5^\circ \quad (\vartheta_{\text{ext}} = 52.5^\circ) \qquad (7.5)$$

where $K_{||}(110)$ is in the first Brillouin zone and therefore $K_{||}(110) = k_{f||}(110)$.

Since the spectrum measured at 52.5° in the ΓKLUX plane from the (110) crystal is the same as that obtained from the (111) crystal in normal emission, the total wave vector must be along ΓL, and must be the same in the two cases. In order to obtain it we therefore have to project $K_{||}(110)$ back onto the Λ line which yields

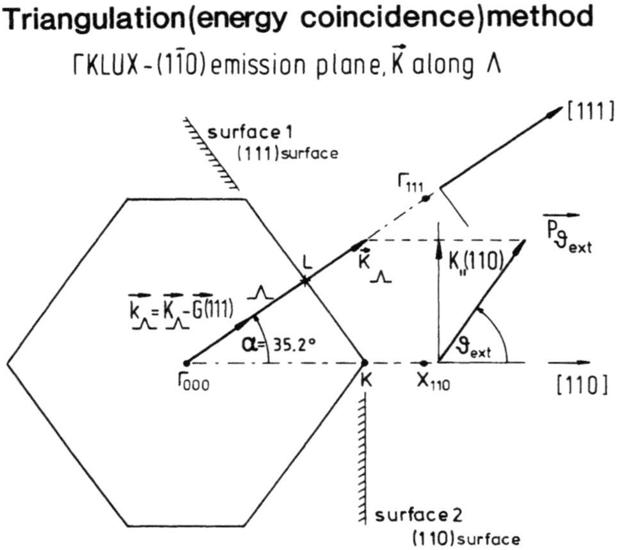

Fig. 7.10. Representation of the triangulation method (in a Brillouin zone cut) for the absolute determination of the final state wave vector $\boldsymbol{K}$ involved in a direct transition from a Cu(111) face in normal emission (Λ line, or ΓL direction). $\hbar^{-1}\boldsymbol{p}_{\vartheta_{\rm ext}}$ is the wave vector of the electrons emitted at 52.5° $= \vartheta_{\rm ext}$ from the (110) plane. From this $\boldsymbol{K}_{||}(110)$ is obtained by projecting $\hbar^{-1}\boldsymbol{p}_{\vartheta_{\rm ext}}$ onto the (110) surface. Since $\boldsymbol{K}_{||}$ is conserved in the PE experiment $\boldsymbol{k}_{f\Lambda}$ is obtained by projecting $\boldsymbol{K}_{||}(110)$ back onto the [111] direction. The whole procedure consists of drawing a line parallel to the [110] direction through the endpoint of $\hbar^{-1}\boldsymbol{p}_{\vartheta_{\rm ext}}$ and its intersection with the Λ line ([111] direction) gives $\boldsymbol{K}_\Lambda$ directly [7.39, 7.42]

$$K_\Lambda = K_{||}(110)/sin 35.2° \qquad (7.6)$$

In the case under consideration one finds

$$K_\Lambda > G_{111} \qquad (7.7)$$

and therefore, in order to obtain k_Λ in the first Brillouin zone one has to fold the measured K_Λ back into the first Brillouin zone via:

$$k_{i\Lambda} = |\boldsymbol{K}_\Lambda - \boldsymbol{G}_{111}| \,, \qquad (7.8)$$

where $\boldsymbol{G}_{111}$ is a reciprocal lattice vector in the [111] direction. The data points obtained in this way are shown together with a calculated band structure in Fig. 7.11 and one can see that there is quite good agreement between theory and experiment [7.50]. Similar data for Ag and Au are also displayed in this figure, showing for these two metals an equally good agreement between theory and experiment [7.50].

The procedure just outlined is evidently quite cumbersome. Having established the good agreement between the calculated and experimental final state band, which except at the zone boundaries does not deviate much from

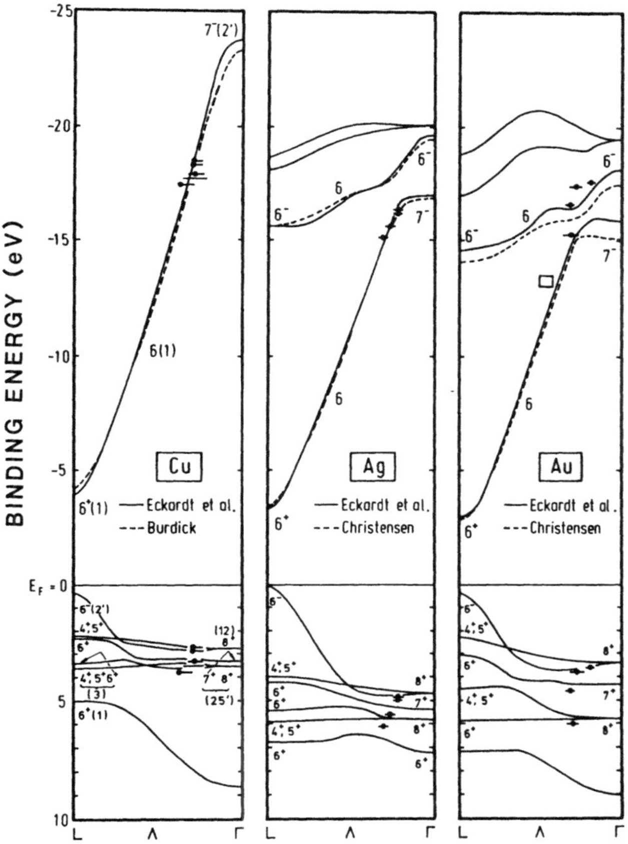

Fig. 7.11. Comparison of the band structures of Cu, Ag and Au along the ΓL direction [7.50]. The full curves are the calculations of Eckardt et al. [7.53]

a parabola, one may with good confidence use the now-established final state band to evaluate further normal emission spectra for a (111) crystal face, taken with different energies, to map the occupied band structure along ΓL.

7.3.2 Bragg Plane Method: Variation of External Emission Angle at Fixed Photon Frequency (Disappearance/Appearance Angle Method)

The next two methods to be described rest on a simple principle but are nevertheless quite complex to describe and to understand. Since the physical effects underlying them can, however, significantly disturb the EDCs from solids, and because they can sometimes reveal much about the photoemission process in solids, it seemed necessary to include them. Many of the principles to be described can be found in the work of Gerhardt's group [7.52].

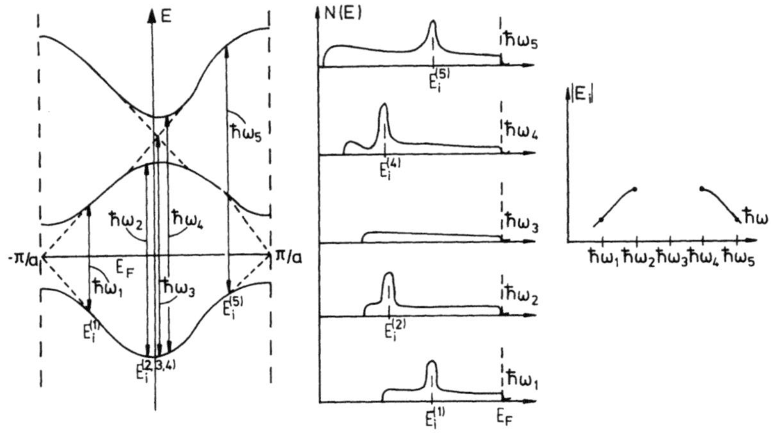

Fig. 7.12. The disappearance/appearance method for the determination of absolute points in a band structure, using a variable energy photon source. The dashed lines show the free electron parabolas. For $\hbar\omega_2 \leq \hbar\omega_3 \leq \hbar\omega_4$ no transitions are possible. Thus for $\hbar\omega_3 \geq \hbar\omega_2$ ($\hbar\omega_3 \leq \hbar\omega_4$) transitions start to disappear (appear)

The underlying physical principle is relatively straightforward and can be illustrated with reference to Fig. 7.12. Assume one has the simple band structure seen in that figure, where the free electron bands that have no gaps are shown as dashed lines. Now a PE experiment is performed with varying photon energy in normal emission. At $\hbar\omega_1$ the direct transition indicated is observed, and the corresponding spectrum is indicated in the middle panel of the figure. Further transitions are observed upon increasing the energy, until, at $\hbar\omega_2$, the last possibility for direct transitions is reached. If the photon energy is increased to $\hbar\omega_3$, the PE spectrum shows no direct transition peaks. They again start to appear, however, at $\hbar\omega_4$ and with increasing excitation energy the band structure is sampled further. Thus one sees (right panel) that at certain photon energies a transition can suddenly appear or disappear and this is why this method has also been termed the *disappearance/appearance angle method.* One also sees that the form of $E_i(\hbar\omega)$ is symmetrical about $\hbar\omega_3$, or, more exactly, around that point in the Brillouin zone where this transition should occur. This symmetry can be exploited in actual experiments. Note that the appearance or disappearance of the direct transitions is merely an effect of the gaps produced by the crystal potential. In Fig. 7.12 the free electron band structure is also indicated by dashed lines and one sees that in this case the sudden appearance or disappearance of the transitions would not occur. This is why the methods utilizing the effects mentioned, are also called Bragg plane methods. Note that the intensity variation of the optical transitions in the band-structure scheme can also be achieved by working with a fixed photon energy and varying the electron detection angle.

In an actual example one has to realize that the Brillouin zone will be three dimensional and that the situation will be more complicated although the physics remains the same.

PE experiments are performed in such a way that by changing the photon energy to excite the electrons, or by changing their detection angle, Brillouin zone boundaries are crossed, causing irregularities in the EDCs to occur. Since these irregularities are a consequence of band splitting at the Brillouin zone boundaries, one can determine when one is observing an EDC from a point on the Brillouin zone boundary and this additional information allows accurate determinations of the crystal momentum.

The disappearance/appearance angle method will now be illustrated in an experiment on Cu(110), where the plane defined by the electron detection and the incoming photon is the ΓKLUX plane (Fig. 7.5). The experiments will be performed with He I radiation ($\hbar\omega = 21.2\,\text{eV}$), and the electron detection angle ϑ will be varied away from the surface normal.

It is perhaps instructive, with the help of the band structure (Fig. 7.7), to first line out what the experimental result will look like. Using the equation $E_f - E_i = \hbar\omega$ and the band structure, one calculates the points where in the extended zone scheme the possible transitions from the uppermost d band (band 5) lie in the ΓKLUX plane. This curve is called a constant energy difference curve (CEDC). The result of this calculation is shown in Fig. 7.13 as the solid line. The noteworthy feature of this figure is that splittings in the full line occur (as expected) each time one crosses a Bragg plane in the Brillouin zone.

In the experiment, only one quadrant of the Brillouin zone needs to be studied, and this is shown in more detail in Fig. 7.14. The experiment is now performed ($\hbar\omega = 21.2\,\text{eV}$) first at normal emission ([110] direction) and then by tilting the angle of electron emission towards the [001] direction. For the Cu(110) crystal at normal emission, the transition out of band 5 takes place into band 10 (Fig. 7.14) because, as can be seen from Figs. 7.13 and 7.14, the momentum is along the XK line, meaning that the $\vartheta = 0$ transition is observed at A_1 (Fig. 7.14). If ϑ is increased in the ΓKLUX plane towards the [001] direction the Δ line ($\Gamma_{111} - X_{110}$) will be reached at the point A_2. Since Δ is a Bragg line, a gap occurs. If the angle is further increased there will thus be some angles at which no transitions out of band 5 should be observed until at the point B_1 transitions out of band 5 into band 7 become possible.

Between points A_2 and B_1 one thus observes a discontinuity in the angular variation of the PE signal as a function of the electron detection angle. In addition, a change of intensity and/or polarization behavior can occur, because the wave functions of band 7 and band 10 will in general be different. The wave vector at the point where the transition "changes" from A_2 to B_1 can be determined, because at this discontinuity $K_{||}$ equals K_Δ, the distance $A_2 - X_{110}$.

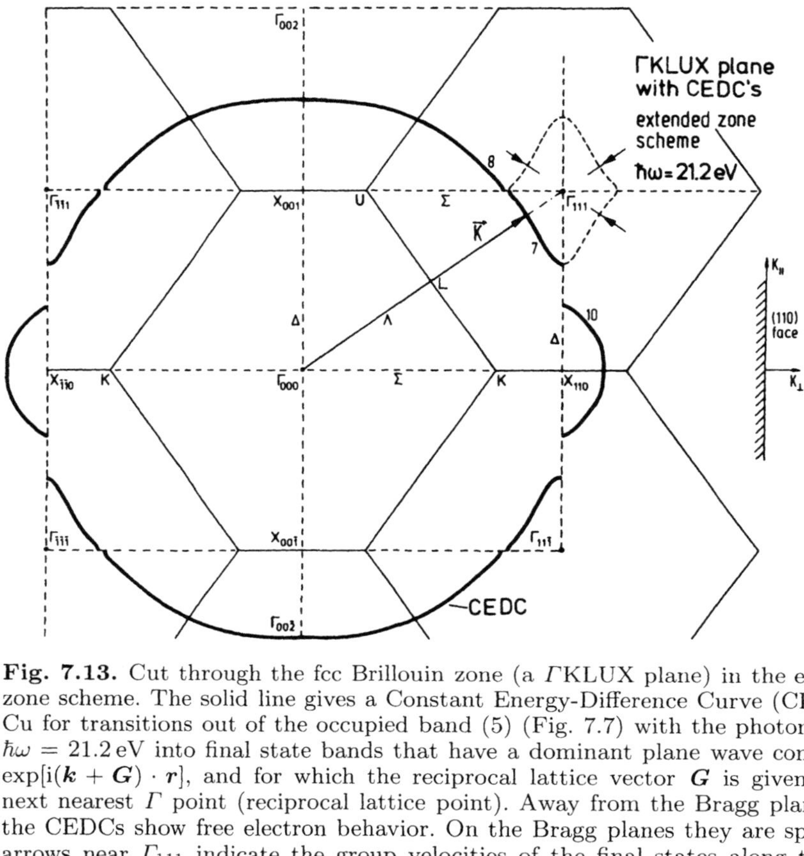

Fig. 7.13. Cut through the fcc Brillouin zone (a ΓKLUX plane) in the extended zone scheme. The solid line gives a Constant Energy-Difference Curve (CEDC) in Cu for transitions out of the occupied band (5) (Fig. 7.7) with the photon energy $\hbar\omega = 21.2\,\mathrm{eV}$ into final state bands that have a dominant plane wave component $\exp[\mathrm{i}(\boldsymbol{k} + \boldsymbol{G}) \cdot \boldsymbol{r}]$, and for which the reciprocal lattice vector $\boldsymbol{G}$ is given by the next nearest Γ point (reciprocal lattice point). Away from the Bragg planes (—) the CEDCs show free electron behavior. On the Bragg planes they are split. The arrows near Γ_{111} indicate the group velocities of the final states along the high symmetry lines [7.39]

An experiment which has been performed under the conditions just outlined [7.39] will now be discussed.

Figure 7.15 shows a series of AREDCs obtained from Cu(110) with $\hbar\omega = 21.2\,\mathrm{eV}$ sweeping the photoelectron detection angle ϑ in the ΓKLUX plane away from the surface normal towards the [111] direction; the data points obtained from these spectra are plotted in Fig. 7.16. Emission angle regions I, II and III are observed (marked by the heavy AREDCs in Fig. 7.15). The disappearance of region I and the appearance of region II are, respectively, characterized by the occurrence of new transitions (β) and the disappearance of transitions (A) at about $\vartheta = 20°$, and by changes in the slope of the energy dispersion (e.g., structure C) and/or intensity dispersion (e.g., structures C and α).

The border between regions II and III is manifested by the appearance of new transitions (structures c and d) and by a change of slope of energy

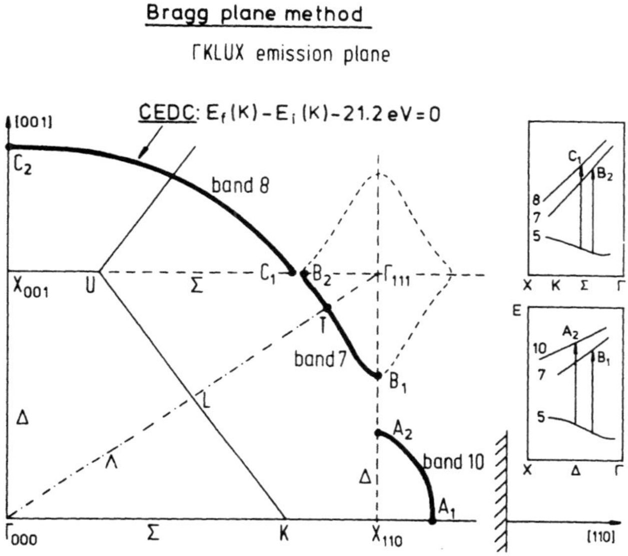

Fig. 7.14. Demonstration of the Bragg plane method (appearance/disappearance method) in one quadrant of the Brillouin zone of Fig. 7.13. The plot represents a PES measurement from a Cu(110) crystal in which the electron detection angle is varied. The strong lines give the loci of the CEDCs in Cu with $\hbar\omega = 21.2\,\mathrm{eV}$ for transitions out of band (5) (Fig. 7.7) into the final state bands indicated. One sees that in varying the electron detection angle, one sweeps through regions where no direct transitions are possible (between A_2 and B_1, and between B_2 and C_1). The corresponding energy-band schemes are shown as inserts

dispersion (structures β and b, c and D, and e) or of intensity dispersion (α, a and D, e). Each of the three regions is characterized by a number of structures with typical behavior in energy and intensity dispersion, the border between two regions being relatively sharp, but with some width as can be seen by the sometimes different critical angle setting in the energy structure plot and the intensity structure plot.

The occurrence of the three emission angle regions is interpreted as follows (compare with Fig. 7.14): in region I one observes direct transitions from initial states to band 10 which start to disappear at about $\vartheta = 13°$, where region II starts. The disappearance or appearance angles then correspond to transitions on the Δ line, $\mathrm{i} \to \Delta_7$ (point A_2 in Fig. 7.14). Region II represents gap emission up to a critical angle of about $\vartheta \simeq 30°$. The region III is dominated by direct transitions to band 7. One has $\mathrm{i} \to \Delta_6$ on the border between II and III (point B_1 in Fig. 7.14).

The set of data in Fig. 7.15 is of course similar to that obtained in the triangulation method (Fig. 7.9) and thus it comes as no surprise that at an angle of $\vartheta \approx 52°$ energy coincidence with the data from a (111) crystal (He I) is observed, meaning that we are crossing the Λ line (point T in Fig. 7.14).

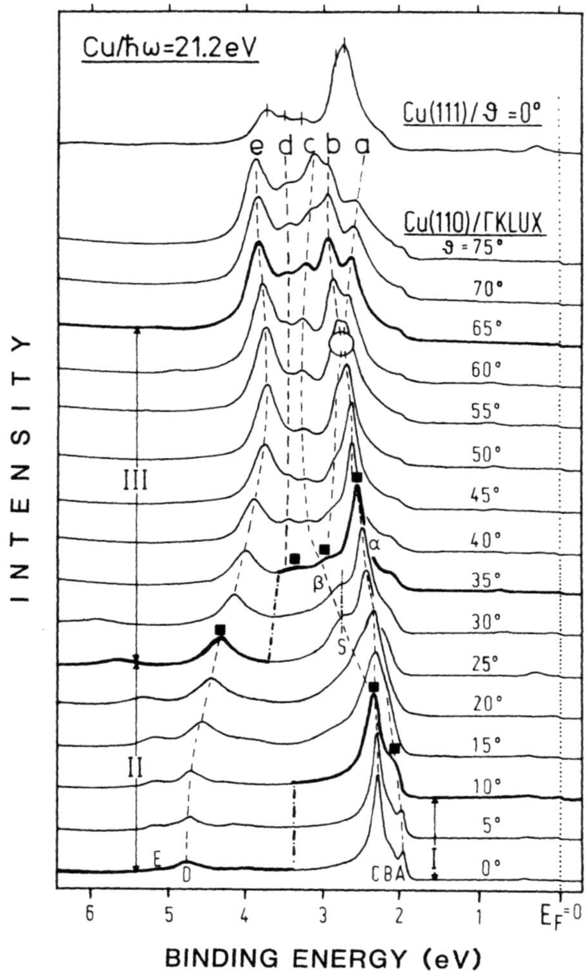

Fig. 7.15. Data taken in the geometry of Fig. 7.14 from Cu(110), with a ΓKLUX detection plane, $\hbar\omega = 21.2\,\mathrm{eV}$, and the electron detection angle as parameter. The variation in energy of the well-resolved peaks as a function of ϑ is indicated by dashed lines. The heavy lines separate three different regions, which are marked by Roman numbers

The excited bands $10(\Delta^{\mathrm{f}}_{7(2')})$ and $7(\Delta^{\mathrm{f}}_{6(5)})$ or $\Lambda^{\mathrm{f}}_{6(1)}$ (from the band structure in Fig. 7.7) into which transitions are observed at the critical angles (points A_2, B_2 and T) have been shifted by 21.2 eV and are also shown in Fig. 7.16.

The $E_i(k)$ points along the Δ line are now obtained as follows. The initial state energy E_i of a transition at a critical angle ϑ_Δ and its final state energy $E_f = \hbar\omega + E_i$ are taken from the spectra at the critical angle. The momentum

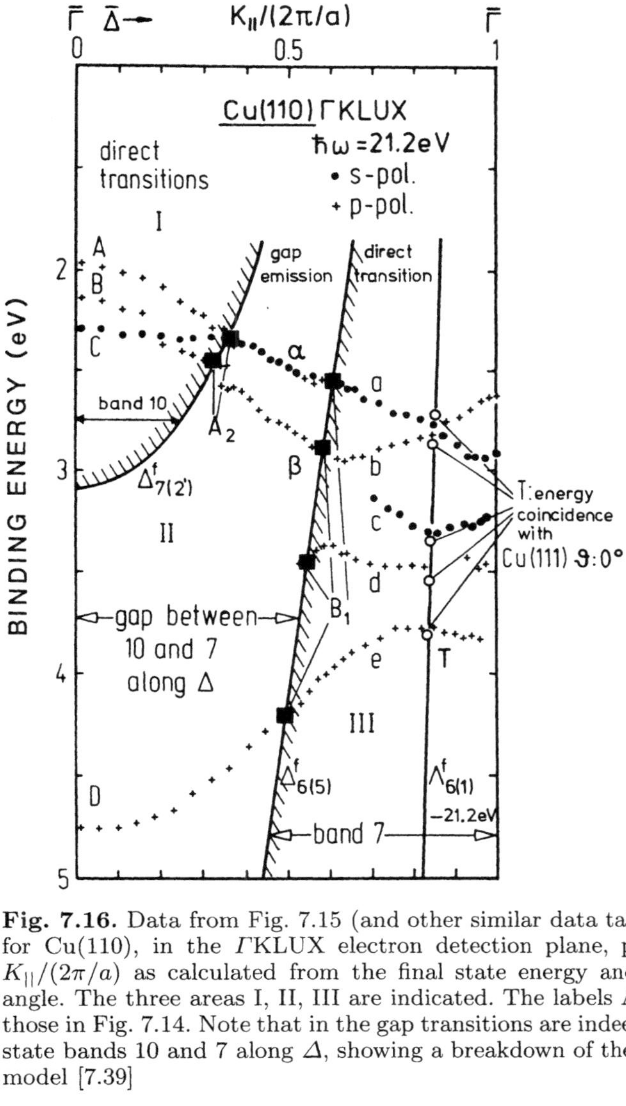

Fig. 7.16. Data from Fig. 7.15 (and other similar data taken with polarized light) for Cu(110), in the ΓKLUX electron detection plane, plotted as a function of $K_{||}/(2\pi/a)$ as calculated from the final state energy and the electron detection angle. The three areas I, II, III are indicated. The labels A_2 and B_1 correspond to those in Fig. 7.14. Note that in the gap transitions are indeed observed between final state bands 10 and 7 along Δ, showing a breakdown of the simple direct transition model [7.39]

is given by $K_\Delta = K_{||\Delta} = [(2m/\hbar^2)(E_i + \hbar\omega - \phi)]^{1/2} \sin\vartheta_\Delta$ where ϑ_Δ is the angle at which the crossing of the Δ line is observed.[3]

[3] K_Δ and $K_{||\Delta}$ are wave vectors of the photoexcited electron within the crystal; they are obtained from measured wave-vector components outside the crystal but note that $p_{||}/\hbar = K_{||}$. They are written as scalars because their directions are given by the subscript. Figure 7.16 indicates that they are so small that they fall into the first Brillouin zone and therefore they are also the wave vectors in the reduced zone scheme.

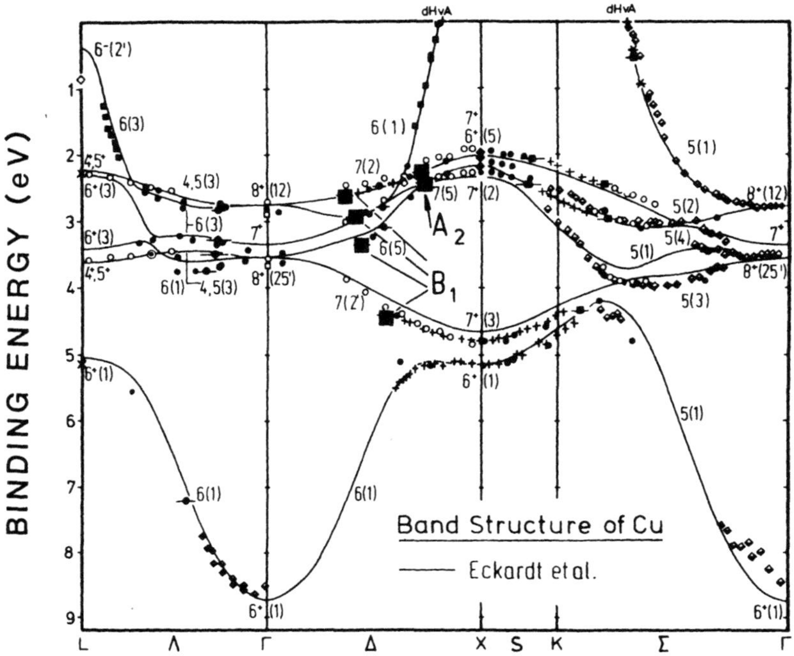

Fig. 7.17. Occupied part of the band structure of Cu [7.39] with data points from various sources and a theoretical result [7.53]. Also shown (*squares*) are the two A_2 points and the four B_1 points from Fig. 7.16

This is again demonstrated using the data of Fig. 7.16. At the first critical point, corresponding to point A_2 in Fig. 7.14, two transitions are observed at $E_i \approx 2.2\,\mathrm{eV}$ binding energy. Since one is on the Δ line, one has $K_{||} = K_{\Delta}$, which can be determined from the angle at which the transitions are observed. These two points are then put into a band-structure scheme (Fig. 7.17) and are marked there A_2. In Fig. 7.17 all the known experimental points on the occupied band structure are shown, and superimposed is a theoretical band structure from Eckardt et al. [7.53]. Moving vertically by 21.2 eV yields the points in the unoccupied band structures given as points A_2 in Fig. 7.18. In an analogous manner points B_1 are obtained in the occupied and empty bands.

Finally one should comment briefly on the spectra observed in region II of Fig. 7.15. In principle, since the final state is in a gap in this region, no spectra should be observable at all. The fact that one actually sees spectra supports the contention that the direct-transition model does not describe all possible transitions and, in particular, that the damping of the electron states due to the small escape depth leads to surface localized states that seem to fill the energy gaps.

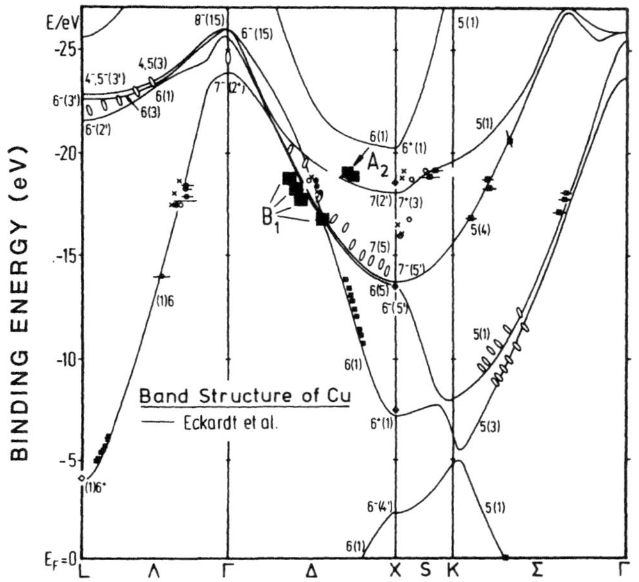

Fig. 7.18. Unoccupied part of the band structure of Cu [7.39] with data points from various sources, and the result of a band-structure calculation [7.53]. The two points A_2 and the four points B_1 from Fig. 7.16 are also shown

It is perhaps worth pointing out that in Fig. 7.16 the structure plot is roughly symmetric around the $\Lambda^{f}_{6(1)} - 21.2\,\mathrm{eV}$ line, a fact which, in addition to the energy coincidence on that line, can also be used to determine the momentum [7.54].

The initial-state energies as a function of the crystal momentum (or the electron detection angle, if a fixed photon energy is used as in Fig. 7.16) follow a curve that has zero slope at the line $\Lambda^{f}_{6(1)} - 21.2\,\mathrm{eV}$. Therefore, the method that uses this fact for the determination of the crystal momentum can be called the *zero-slope method* [7.54].

7.3.3 Bragg Plane Method: Variation of Photon Energy at Fixed Emission Angle (Symmetry Method)

This method, like the previous one, makes use of the splitting of the bands at the Brillouin zone boundaries. It is merely the detection method that is different: one now works with a fixed external electron detection angle, while the photon energy is varied.

The experiment to be described here was also performed on Cu in the ΓKLUX plane [7.55], but now using a (111) crystal surface. Again the phenomena to be expected are outlined first.

The experiment can be explained in the extended zone scheme shown in Fig. 7.19. It is helpful for this method if one has a certain knowledge of the band structure. The dotted lines are Constant Energy-Difference Curves (CEDCs) for three different photon energies (compare with Figs. 7.13 and 7.14), which show the familiar splitting on Bragg planes (the Σ line and Λ line in the case shown). The CEDCs in the second Brillouin zone (BZ2) represent the transition from an initial band i to the final band 7, i $\to$ 7, whereas the CEDCs in BZ3 contain transitions from an initial band to final band 8, i $\to$ 8. At a fixed external emission angle ϑ a transition i $\to$ 7 occurs for a photon energy $\hbar\omega_1$ which is smaller than a critical photon energy, $\hbar\omega_1 < \hbar\omega_c$, on a CEDC i $\to$ 7 with the wave vector $\boldsymbol{K}(\hbar\omega_1)$ as indicated in Fig. 7.19. For higher photon energies, $\boldsymbol{K}$ is located at the CEDCs i $\to$ 8. On passing through the critical photon energy $\hbar\omega_c$, $\boldsymbol{K}$ is originally located at the intersection of

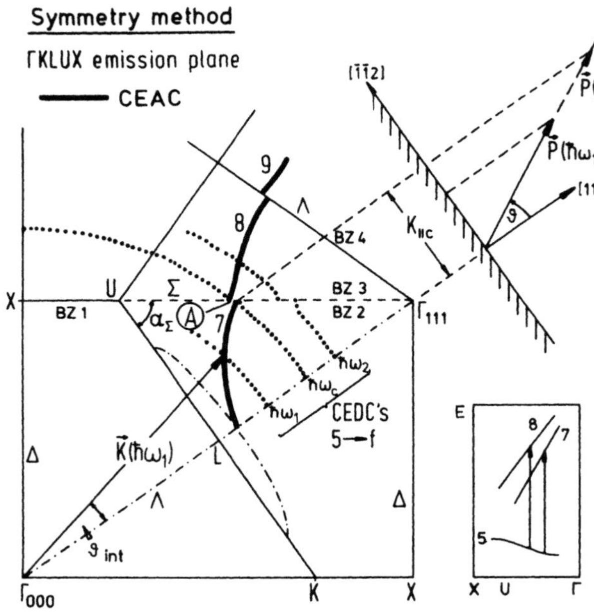

Fig. 7.19. Symmetry method for the determination of a band structure from PES data. ΓKLUX cut through the extended Brillouin zone in Cu. The geometry of a measurement from a (111) surface of Cu taking data under a constant electron acceptance angle but varying the photon energy is given. The positions in momentum space that are reached in this experimental mode are given by the heavy lines (Constant Emission Angle Curve – CEAC). The Constant Energy Difference Curves (CEDCs) for transitions out of band 5 into bands 7 and 8 are shown as dotted lines. The dash-dotted line is the constant final state energy curve for $E_f = E_{vac}$. In crossing the Bragg planes the CEACs show irregularities, which are reflected in the spectra and can thus serve to determine the momentum at the Bragg plane crossing. This is shown schematically for a crossing of the Σ line in the insert [7.55]

the Σ line and the CEDC (7, $\hbar\omega_c$) and then jumps discontinuously to the CEDC (8, $\hbar\omega_c$). This discontinuity can be used to locate the Bragg plane and, in conjunction with the other experimental parameters, the $\boldsymbol{k}$-vector of the transition. Since this method enables one to identify transitions on symmetry lines it is often called the *symmetry method.*

The Constant Emission Angle Curve (CEAC) in Fig. 7.19 (thick line) gives the $\boldsymbol{k}$-space points that are allowed for electron emission into the $\vartheta = 30°$ external direction (the internal emission angle ϑ_{int} is not constant). This curve is determined by energy and momentum conservation conditions. The exact form of the CEAC can only be obtained if the band structure is known. Only the critical points on the Σ line between BZ2 and BZ3 (or Λ line between BZ3 and BZ4) can be measured via disappearance/appearance effects. With increasing final-state energy (and increasing kinetic energy), the position of the transition in $\boldsymbol{k}$-space must wander away from the ΓL line towards the ΓU(Σ) line. Since $K_{||}$ is conserved during emission through the surface, the transition must be confined to the ΓKLUX plane. At the critical photon energy the Σ line is reached, and with increasing energy one enters the next Brillouin zone. If one wants to check a calculated band structure, e.g., along the Σ- , Λ- and Δ lines, one can make some approximate predictions for the photon energies and the corresponding external angles at which the critical behavior should be observed.

The actual measurements [7.55] were performed with s-polarized light from a synchrotron ($\boldsymbol{A}$ perpendicular to the ΓKLUX plane) and the incident light direction and the photoelectron detection direction had a constant angle of 53° with respect to one another. Due to the polarization selection rules, the spectra show fewer lines than the commonly measured unpolarized spectra.

The case in point can be demonstrated with the spectra in Fig. 7.20 [7.55], which were measured with varying photon energy at constant external (electron collection) angle of $\vartheta = 30°$. It can be seen that in two narrow photon energy intervals, between 14 eV and 15 eV and then again between 26 eV and 27 eV, the intensity of the structures in the EDCs changes and, in addition, new lines appear such as the line (A) (4 eV) in the $\hbar\omega = 16$ eV spectrum. This line must be due to the fact that one has reached the Σ line (Fig. 7.19) and is now observing transitions into band 8 instead of band 7.

The momentum of the initial and final states of the transition is determined as follows. The energy conservation conditions read:

$$E_{\text{kin}} = E_f - \phi, \quad E_f - E_i = \hbar\omega, \quad E_{\text{kin}} = (\hbar^2/2m)K_{||}^2(\sin^2\vartheta)^{-1} \tag{7.9}$$

and for wave-vector conservation one has

$$k_\Sigma(11\bar{2}) = K_{||} = [(2m/\hbar^2)E_{\text{kin}}]^{1/2}\sin\vartheta, \quad K_{||} = k_\Sigma\cos\alpha_\Sigma\ , \tag{7.10}$$

where k_Σ and $k(11\bar{2})$ are the reduced momenta along the Σ line and α_Σ is the angle between the Σ line and the surface (Fig. 7.19). Thus one has:

$$E_f = \phi + [(\hbar k_\Sigma)^2/2m]\cos^2\alpha_\Sigma/\sin^2\vartheta\ . \tag{7.11}$$

This is a parabola for E_f as a function of the $\hbar k_\Sigma$ starting out at $\hbar k_\Sigma = 0$ and $E_f = \phi$. At the point where this parabola intersects the final state energy for an experimental peak whose behavior indicates that it has been measured on the Σ line, one has k_Σ for this particular peak. For the case of the peak (A) in Fig. 7.20 this means the following: it has an initial energy of $E_\mathrm{B} = 4\,\mathrm{eV}$, the photon energy is $\hbar\omega = 16\,\mathrm{eV}$ and therefore $E_f = 12\,\mathrm{eV}$. Since the experiment has been performed with $\vartheta = 30°$, one has to look for the intersection of the parabola for $\vartheta = 30°$ with $E_f = 12\,\mathrm{eV}$. The final-state momentum, and hence also the initial-state momentum, obtained in this way is shown in Fig. 7.21. The initial state position is now obtained by going "down" vertically from the final-state point by $\hbar\omega = 16\,\mathrm{eV}$.

In a similar manner one can determine the momentum and energy of points (B) and (C).

The experiment has also been performed for other angles ϑ and all the data points are shown in Fig. 7.21; they have also been incorporated into Figs. 7.17 and 7.18.

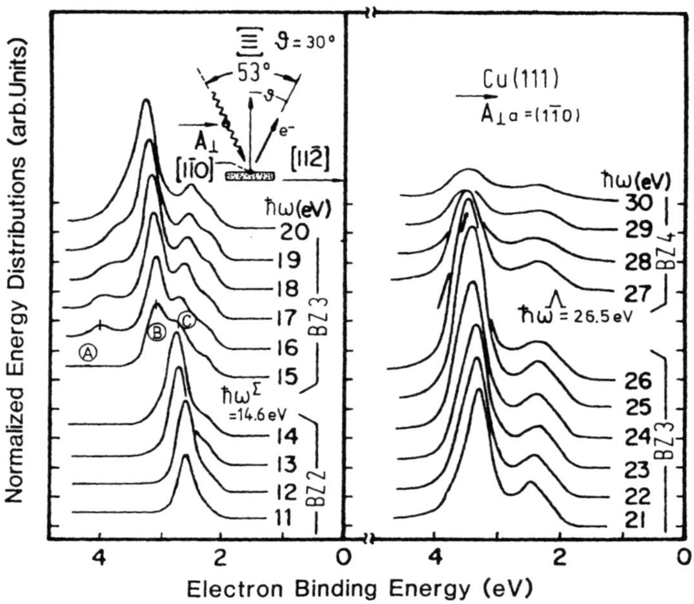

Fig. 7.20. PE spectra of Cu(111) taken with varying photon energy $\hbar\omega$ at constant electron detection angle ($\vartheta = 30°$) thus measuring a CEAC. In the $\hbar\omega = 16\,\mathrm{eV}$ spectrum a new feature (A) suddenly appears, indicating that one has crossed a Bragg plane in momentum space (between Brillouin zones 2 and 3, Σ line). This irregularity can now be used to determine the momentum. Features (B) and (C) are of similar origin. A less distinct irregularity is observed around $\hbar\omega = 26\,\mathrm{eV}$ as one crosses the Λ line in going from Brillouin zone 3 into Brillouin zone 4 [7.55]

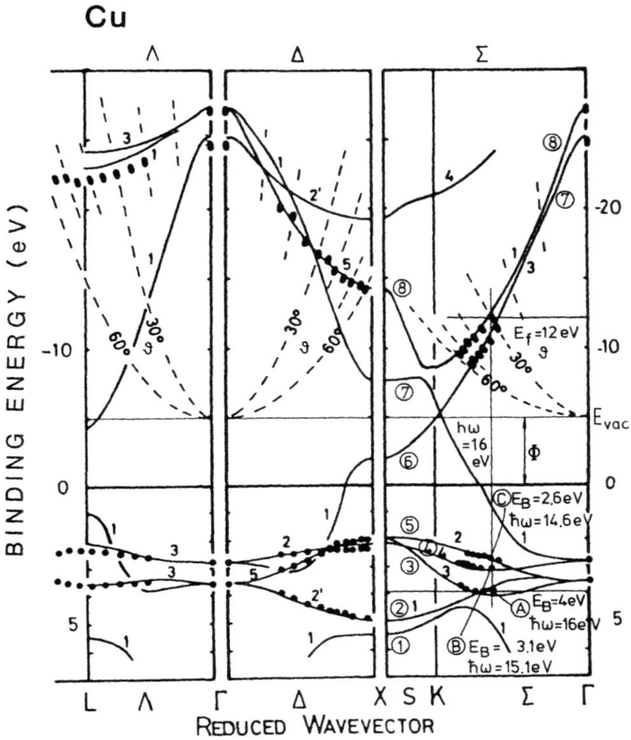

Fig. 7.21. Experimental band structure for Cu (*dots*) [7.55] determined with the symmetry method outlined in Figs. 7.19 and 7.20. The solid line is the band structure of Janak et al. [7.56]. The position of peaks (A), (B) and (C) is indicated; see Fig. 7.20

7.3.4 The Surface Emission Method and Electron Damping

It was shown in the discussion of the disappearance/appearance angle method (Sect. 7.3.2) that PE is also observed in cases where no direct bulk transitions are possible. It is assumed that in these cases the emission is due to transitions from bulk initial states into strongly damped final surface states (evanescent states) located in bulk-band gaps.

The *Surface Emission Method* (SEM) for the determination of the occupied band structure will again be discussed using emission measured in the ΓKLUX plane. The corresponding AREDCs are given in Fig. 7.15 and the corresponding structure plot is given in Fig. 7.16.

As was mentioned previously, under these conditions spectra are also observed in the band gap region namely in region II of Figs. 7.15 and 7.16, where, according to Fig. 7.14, with 21.2 eV radiation transitions out of occupied band states should not be observable at all. It is now assumed that in this area (angles between A_2 and B_1 in Fig. 7.14) transitions along the

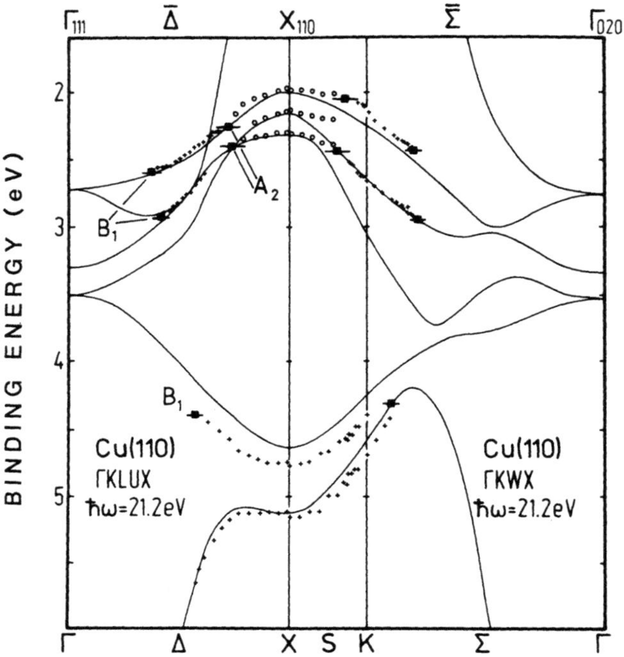

Fig. 7.22. Results obtained by the surface emission method for the band structure of Cu. Solid lines are the calculations of Eckardt et al. [7.53]. The crosses represent the energetic structures in the AREDCs from Cu(110) ΓKLUX (data along ΓX) and from Cu(110) ΓKWX (data along ΓKX), in the polar angle regime, where direct optical transitions are not possible. The full heavy squares represent appearance/disappearance structures [7.39]. The open circles are from direct transitions [7.39]

Δ line between A_2 and B_1 occur into surface states, where the initial state is restricted by the wave number of the outgoing electron parallel to the Δ line, a wave number that can be read off directly from Fig. 7.16. With these assumptions one obtains the crosses in Fig. 7.22 [7.39], where the data determined by the appearance angle/disappearance angle (full squares) and a theoretical band structure are also given. This figure strongly supports the validity of the procedure outlined. It also suggests a very simple method for actually measuring band structures by doing just the opposite of what is generally done, namely to measure, under appropriate conditions of course, transitions to final states in a gap. It has been found for Ag and Pt [7.57,7.58] that this can indeed be a very useful procedure.

Figure 7.22 also includes data from an experiment on Cu(110) but in the ΓKWX emission plane. They are obtained in the same way as those from the ΓKLUX plane and they give states along the XKΓ line.

It was pointed out in Sect. 6.3.3 that transitions into a band gap measure a one-dimensional density of states because the $k_\perp$ selection rule gets blurred by the electron damping. In this interpretation the peaks in the spectra originate from critical points in the one-dimensional density of states. The data on Cu presented here reveal that, in addition, there are direct $k_{||}$-conserving transitions present in the spectra that allow a band mapping.

7.3.5 The Very-Low-Energy Electron Diffraction Method

This method [7.58] combines the description of the photoemission final state as an inverse LEED wave function, with the possibility to determine the band structure by employing a non-normal emission transition into a gap, which leads to a $\boldsymbol{k}_{||}$-conserving transition (see previous section). In the first stage, a LEED experiment at very low energies (VLEED) is used to determine the unoccupied band structure. In the second stage, a constant final state (CFS) photoemission experiment (in off normal directions) is performed, where the final state energies are chosen to lie in the gaps that were determined in the VLEED experiment. This constant final state PES experiment conserves $\boldsymbol{k}_{||}$ and thereby directly gives the $E(\boldsymbol{k})$ curves as demonstrated in the previous section.

We now outline the details of the method (see also Chap. 6). The three-step approach to photoemission ignores the fact that photoemission has to be described as one coherent excitation process. In particular, in contrast to the three step model, in order to be more accurate the final photoelectron state, including its damping into the solid, has to be calculated as a solution of the Schrödinger equation for the semi-infinite crystal, with an asymptotic plane wave ($\exp(-\mathrm{i}\boldsymbol{k}\cdot\boldsymbol{r})$ at the detector. Therefore, it can be viewed as a time-reversed LEED state. Note that in LEED one directs an electron (plane) wave at the sample, and allows it to scatter off the (semi-infinite) crystal (potential). This relationship between LEED and PES can be utilized to determine the photoemission final states directly by very-low-energy electron diffraction (VLEED) as sketched in Fig. 7.23. For an electron beam impinging on a sample, the LEED reflectivity R (or, equivalently, the electron transmission $T = 1 - R$ from the vacuum into the crystal) will undergo rapid variations whenever, for a given vacuum wave vector $\boldsymbol{p}_{||}/\hbar$, the energy of the incident electrons passes through a critical point (CP) in the corresponding *perpendicular* dispersion $E(k_\perp)$, provided that the relevant band couples sufficiently strongly to the vacuum wave. CPs are defined as points where the band slope $\mathrm{d}E/\mathrm{d}k_\perp$ shows sharp changes or vanishes at a band edge. Because this occurs predominantly at the Brillouin zone edge, a CP corresponds to a well-defined value of $k_\perp$. The energy position of the CPs is derived by the extrema in the differential electron transmission $\mathrm{d}T/\mathrm{d}E$, measured by target current spectroscopy (TCS).

The VLEED constant final state PE method determines band dispersions along BZ directions in symmetry planes parallel to the surface. First, one

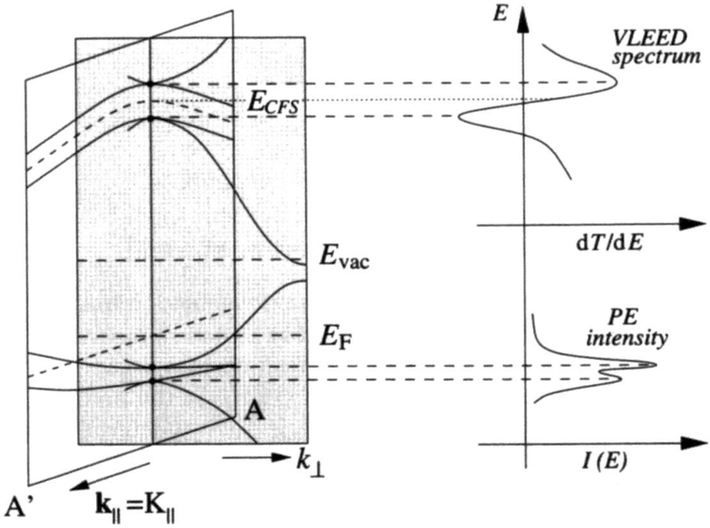

Fig. 7.23. Schematic illustration of the angle-dependent VLEED-CFS-PES method. AA' is a symmetry line parallel to the surface, $k_\perp$ denotes a wave vector component perpendicular to this line [7.58]

performs angle-dependent VLEED measurements varying the wave vector $\boldsymbol{k}_{||}$ of the incident electrons along the chosen BZ direction. The CPs in the symmetry plane (the lower and upper edge of a local band gap) show up as a characteristic minimum–maximum structures in $\mathrm{d}T/\mathrm{d}E$ (see Fig. 7.23). These CPs line up to a pair of unoccupied $E(\boldsymbol{k}_{||})$ dispersions in the symmetry plane. Second, with the final-state energy for each $\boldsymbol{k}_{||}$ chosen to lie between these bands (i.e., in the band gap) one performs angle-dependent CFS PE measurements (see Sect. 7.3.4). Thus $\boldsymbol{k}$ is pinned in the symmetry plane, and the peak positions in the PE spectra directly yield the valence-band dispersion along the chosen BZ line.

The method thus inherently employs photoemission from band-gap states, which for the semi-infinite crystal are Bloch waves with complex $\boldsymbol{k}_\perp$ that decay into the crystal due to elastic scattering off the crystal potential.

For an application of the method to an fcc metal such as Cu, the (110) surface is most convenient. It gives access to the entire symmetry plane parallel to the surface ΓKLUX (see Fig. 7.5 for the Brillouin zone of Cu).

The two stages of the method will now be demonstrated by an experiment using the (110) surface of Cu.

Unoccupied states studied by VLEED

The VLEED experiment is performed with a standard four-grid LEED unit operated in the retarding field mode. The resulting data, i.e., the extremal points of $\mathrm{d}T/\mathrm{d}E$, yield the $E(\boldsymbol{k}_{||})$ map of Fig. 7.24. The points are grey-shaded proportional to $-\frac{\mathrm{d}^2}{\mathrm{d}E^2}(\mathrm{d}T/\mathrm{d}E)$. The sign distinguishes the minima and maxima in $\mathrm{d}T/\mathrm{d}E$. The magnitude characterizes their location. The experimental $E(\boldsymbol{k}_{||})$ map is basically a direct image of the edges of unoccupied upper bands, except for those which couple too weakly to the vacuum.

For the interpretation of the VLEED data, it is helpful to have a general idea of the upper band structure, achieved, e.g., by a qualitative comparison to an empirical pseudopotential band calculation.

Proceeding in this manner we have obtained the PE final state energies pertinent to the ΓKLUX plane for the entire $\boldsymbol{k}_{||}$ range of interest. They are displayed in Fig. 7.24 as bold dashed curves. These energies have discontinuities in regions of extended multiple-band hybridization where the principal coupling band disperses almost vertically.

From Figs. 7.5 and 7.6 and Appendix A2 one realizes that in order to get the ΓX and the ΓKX band structure two different azimuthal scans have to be performed: one from the [110] normal towards the [001] direction, which is the $\bar{\Gamma}\bar{Y}\bar{\Gamma}$ line in the surface Brillouin zone, and one from the [110] normal towards the $[\bar{1}10]$ direction, which is the $\bar{\Gamma}\bar{X}\bar{\Gamma}$ line in the surface Brillouin zone (see Fig. 7.6).

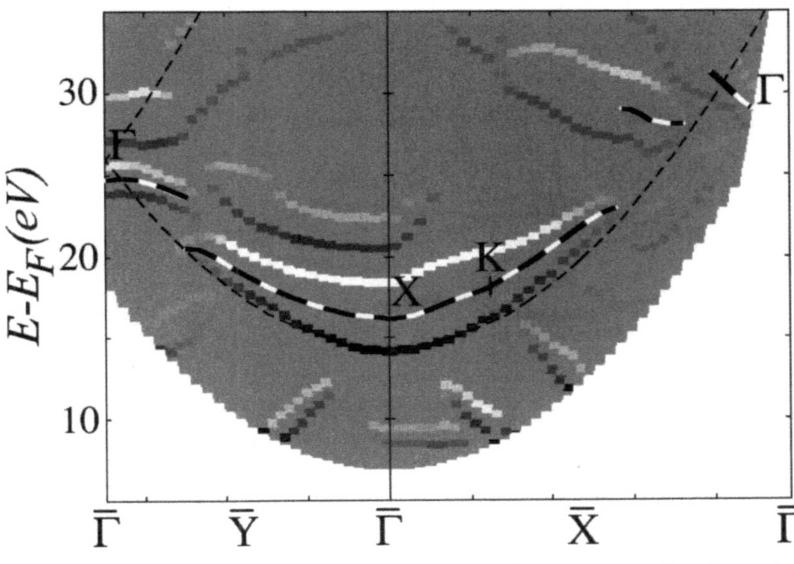

Fig. 7.24. Experimental upper (unoccupied) bands for Cu derived from the $\mathrm{d}T/\mathrm{d}E$-VLEED data on a Cu(110) surface, plotted as a function of the wave vector component parallel to the surface $\boldsymbol{K}_{||}$. Dark and light shading corresponds to $\mathrm{d}T/\mathrm{d}E$ minima and maxima, respectively. The bold dashed curves show the final state dispersion, chosen for the CFS measurements, with the corresponding high symmetry points in the ΓKLUX plane of the bulk Brillouin zone. The dashed curve represents the free electron approximation [7.58]

The bold black-white line in Fig. 7.24 connects the energies chosen as the final states for the constant final state PES experiment. Γ, X, K, Γ indicate the locations in that diagram where the transitions to these $\boldsymbol{k}$-points are observed.

Occupied states observed by constant final state photoemission

The results of the constant final state photoemission using the energies characterized by the dashed line in Fig. 7.24 are given in Fig. 7.25. Note that along ΓX the whole Brillouin zone was mapped, while for ΓKX, due to instrumental limitations a small part of the Brillouin zone near Γ could not be mapped by VLEED and is therefore also missing in Fig. 7.25. For an unbiased representation of the PE data, Fig. 7.25 displays a map of the negative second derivative $-\frac{\mathrm{d}^2 I}{\mathrm{d}E^2}$ of the photocurrent, cut off below a threshold to suppress negative values and noise, in the $(E, \boldsymbol{k}_{||})$ plane. The peaks in this map have their maxima located at the position of the PE peaks, with the width reflecting the width of the PE peaks and shoulders (the error bars are in fact much smaller). This map is in principle a direct image of the occupied valence bands (except for those whose matrix element is too small). The experimental bands are easily identified from a comparison with a band calculation also shown in Fig. 7.25 (thin lines).

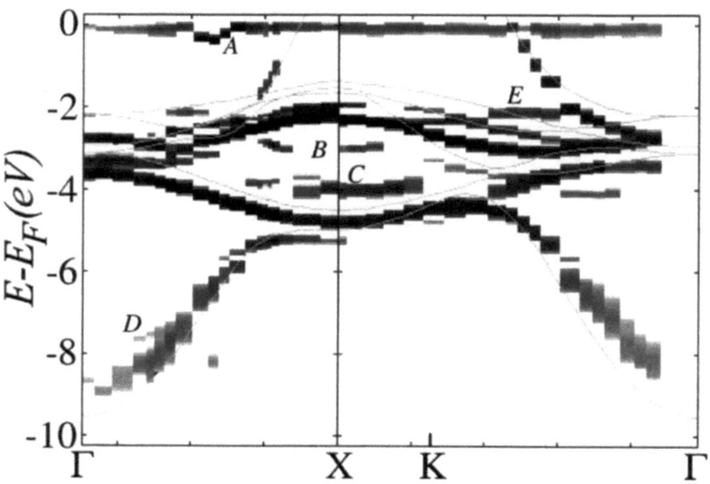

Fig. 7.25. Experimental valence bands for Cu obtained by ARPES in the CFS-mode using the final state energies of Fig. 7.24. The experimental data are obtained from the negative second energy derivative of the measured intensity and displayed in a logarithmic grey scale. The non-dispersive structure at E_f results from the Fermi edge. Also shown is the result of a density-functional band-structure calculation [7.58]

As a first test of the method one can check the internal consistency of the PE data. Both the measured d bands as well as the highly dispersive sp bands show a smooth dispersion, even where the final-state energies undergo discontinuities. As a critical test, constant-$\boldsymbol{k}_{||}$ final state spectra were taken in such regions, scanning the final-state energy through the discontinuity. Although the intensity varied considerably, the peak binding energies remained stable to within $\pm 50\,\mathrm{meV}$.

The data in Fig. 7.25 are in broad agreement with those collected from many different experiments and displayed in Fig. 7.7. Note that the results of Fig. 7.25 were essentially obtained in a "one shot" experiment, indicating the power of this method. The structure at zero binding energy reflects the Fermi step.

In Fig. 7.25 there are obvious discrepancies between the experimental and the theoretical band structure. A recent band structure calculation which includes self-energy corrections [7.189] leads to much better agreement between theory and experiment. Marini et al. [7.189] show that the exchange-correlation contributions to the self-energy arising from the 3s and 3p core levels are very important.

7.3.6 The Fermi Surface Method

The Fermi surface method has so far been employed only rarely [7.59,7.60]. As in the two Bragg plane methods described before, a reasonable knowledge of the band structure is advantageous, and may be essential if one wants to apply this method. The method involves finding AREDCs that show structures due to transitions from initial states residing at the Fermi level, and using these to determine $E(\boldsymbol{k})$ of the final states of these transitions. This is possible since the Fermi surface, which is a surface of constant energy $E_{\mathrm{F}}(\boldsymbol{k}_{\mathrm{F}}) = 0$ in $\boldsymbol{k}$-space, is well known. From the experimentally determined $\boldsymbol{k}_{||}^{\mathrm{exp}}$ of such a Fermi transition and the Fermi surface, $\boldsymbol{k}$ is fixed via the relation $\boldsymbol{k}_{||}^{\mathrm{exp}} = \boldsymbol{k}_{\mathrm{F}||}$. The indeterminacy in the $k_{\perp}$ of the transition is removed. However, PE data alone are not sufficient when one wants to apply this technique. It relies on an exact knowledge of the Fermi surface geometry, and this information must come from other sources.

The Fermi surface method for the absolute determination of the final-state band structure can be applied in two ways: firstly, one may fix the external electron emission angle ϑ and tune the photon energy. The intersection of the Constant Emission Angle Curve (CEAC, Sect. 7.3.3) with the Fermi surface then gives the location of the Fermi surface transition. Alternatively, one may fix the photon energy and tune ϑ; the transition is then located at the intersection of the corresponding Constant Energy-Difference Curve (CEDC, Sect. 7.3.2) with the Fermi surface.

The second approach is demonstrated for Cu(110)/ΓKWX ($\hbar\omega = 21.2\,\mathrm{eV}$) in Fig. 7.26 [7.40]. Emission due to transitions out of the sp band (band 6,

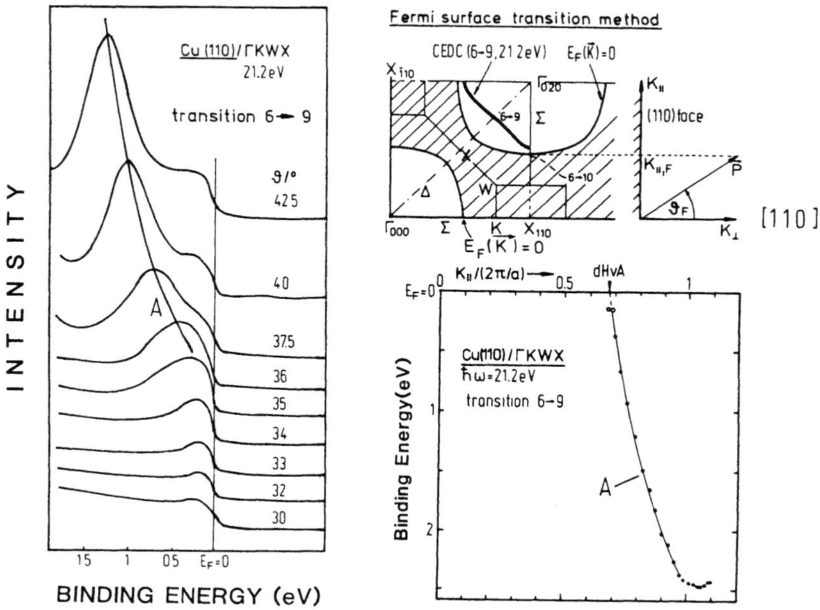

Fig. 7.26. Fermi surface method for determination of a band structure. The left panel gives spectra with $\hbar\omega = 21.2\,\mathrm{eV}$ taken from Cu(110) ΓKWX with varying polar angle for the region near the Fermi energy. The band-structure plot (top right) is a ΓKWX section through the Cu Brillouin zone. The solid line is a CEDC for transitions from band 6 (which intersects the Fermi energy) to band 9 for $\hbar\omega = 21.2\,\mathrm{eV}$. The shaded area is that in which band 6 is below E_F. The plot at the bottom right shows the energy of structure A as a function of $K_{||}/(2\pi/a)$ calculated from ϑ and the electron kinetic energy. It intersects the Fermi energy at the position anticipated from de Haas–van Alphen experiments

Fig. 7.7) appears at polar angles at about 32° with slowly increasing intensity and width, but no peak dispersion is observed up to $\vartheta \simeq 34.5°$. The emission shows a rapid intensity increase and strong dispersion for $\vartheta \geq 35°$. In Fig. 7.26 the energy of the peak A as a function of $K_{||}$, where $K_{||}$ is measured parallel to the (110) plane, is also given. One notices that in this plot the energy of peak A extrapolates well to the Fermi surface radius [7.61] measured by the de Haas–van Alphen effect along the Σ line. One is thus led to the conclusion that at the angle where peak A starts to appear, transitions out of band 6 into an unoccupied band (band 9) become possible, because band 6 crosses the Fermi energy. This of course immediately fixes the momentum of the initial state.

This assumption is supported by calculations performed for the conditions in the experiment [7.40]. Figure 7.26 also gives a $\boldsymbol{k}$-space plot for a cut along ΓKWX with the transitions $6 \to 9$ and $6 \to 10$ as calculated from the band structure in Fig. 7.7. Also shown are the Fermi surface dimensions $[E_\mathrm{F}(\boldsymbol{k}) = 0]$

and the shaded areas are those in which band 6 is *not* occupied. This shows that transitions $6 \rightarrow 10$ are not expected under the conditions of the present experiment and that indeed the CEDC for transitions $6 \rightarrow 9$ coincide with the point where the Fermi surface intersects the Σ line, as expected from the plot given below the Brillouin zone cut, where it can be seen that the Fermi surface dimensions as measured by PES and de Haas–van Alphen effect agree.

7.3.7 Intensities and Their Use in Band-Structure Determinations

The intensity of a direct transition also shows a variation because the transition matrix element and the transmission factor through the surface will vary over the Brillouin zone. In principle, this effect can be used for a priori determination of $\boldsymbol{k}$ in direct transitions as will be demonstrated in the following. Unfortunately no data are available in the literature for Cu, so we shall deal instead with Ag. In this method one compares theoretical intensity variations with experiment in order to find intensity extrema, and to relate these to special points in the band structure. This means that to apply the method successfully one has to have a rough knowledge of the band structure, as is also the case for other methods. For pedagogical reasons, Fig. 7.27 first presents the result one is interested in, namely the band structure of Ag. The full lines are the result of a calculation by Eckardt et al. [7.53] and the symbols represent experimental results from various sources [7.57]. The intensity method will be demonstrated for the ΓL part of the band structure in Fig. 7.27. The essence of the argument will be to compare calculated intensities with measured ones, and to adjust extrema found in theory and experiment so as to refine a calculated band structure [7.53] using the experimental data.

Thus the experimental intensity variation of the structures in the EDCs as a function of photon energy will be compared with calculated intensities, which are the product of bulk *Momentum Matrix Elements* (MMEs) and transmission factors along the Λ line (ΓL direction). We use relativistic MMEs for direct transitions (M_{fi}) from the occupied bands to band 7 and band 8 along Λ (first two empty bands above E_{F}, see Fig. 7.27) as calculated by Benbow and Smith [7.62] using the combined interpolation scheme. In calculating emission intensities I_{fi} one also has to take into account the transmission factors $T(\boldsymbol{k}_\Lambda)$ of the final states, according to the three-step formula of ARUPS (6.14a)

$$N(\boldsymbol{k}_\Lambda) \propto I_{fi}(\boldsymbol{k}_\Lambda) \propto \left|\tilde{M}^1_{fi}(\boldsymbol{k}_\Lambda)\right|^2 |T(E_f, \boldsymbol{K}_{||}, \boldsymbol{k}_\Lambda)|^2 \,. \tag{7.12}$$

(The parts of (6.14a, 6.14b) that are not necessary in the following discussion have been omitted.) The transmission factors $T_f(\boldsymbol{k}_\Lambda) = T(E_f, \boldsymbol{K}_{||}, \boldsymbol{k}_\Lambda)$ can be approximated by the squared amplitude of the relevant plane wave with reciprocal lattice vector $\boldsymbol{G} = (2\pi/a)(111)$ in the final state Bloch wave, see

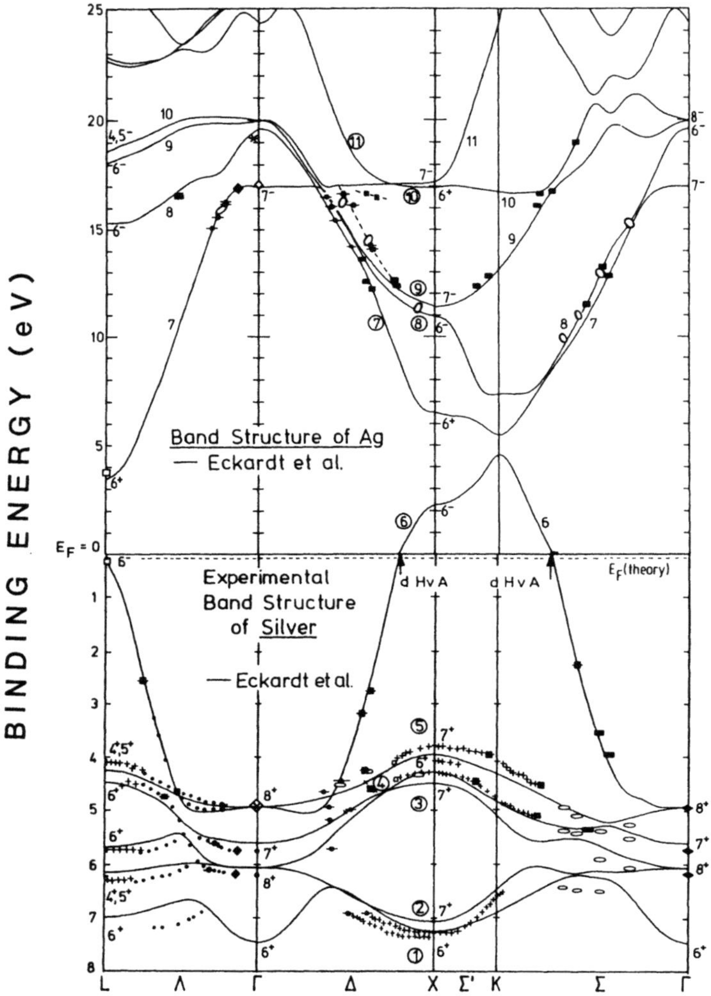

Fig. 7.27. Band structure of Ag giving experimental data from various sources [7.40] and a calculation by Eckardt et al. [7.53]

(6.11) [7.63, 7.65]. If

$$\psi_f(\boldsymbol{k}_\Lambda) = \sum_{\boldsymbol{G}} u_{\boldsymbol{G}}(\boldsymbol{k}_\Lambda) \exp[\mathrm{i}(\boldsymbol{k}_\Lambda + \boldsymbol{G}) \cdot \boldsymbol{r}] \; , \tag{7.13a}$$

then one has

$$T_f(\boldsymbol{k}_\Lambda) = |u_{(111)}|^2 \; . \tag{7.13b}$$

Using the parameters of Benbow and Smith [7.62], transmission factors $|u_{(111)}|^2$ were calculated for the final bands 7 and 8 (Fig. 7.28). This figure

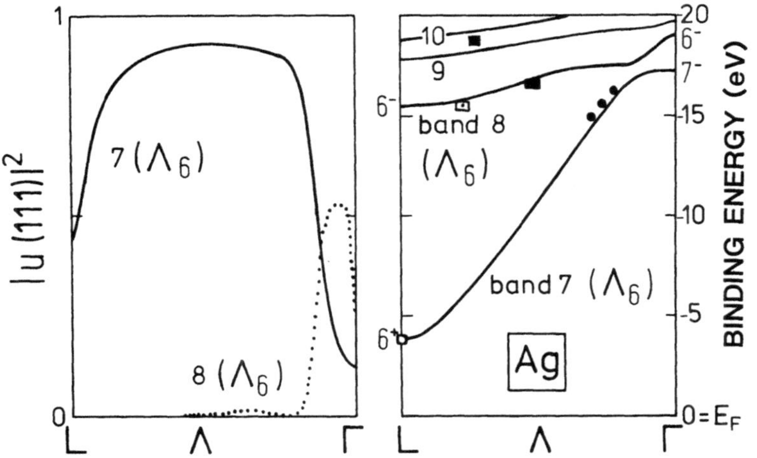

Fig. 7.28. *Right*: ΓL direction of the unoccupied band structure of Ag (Fig. 7.27). *Left*: $|u_{(111)}|^2$ as a function of momentum for bands 7 and 8 along ΓL

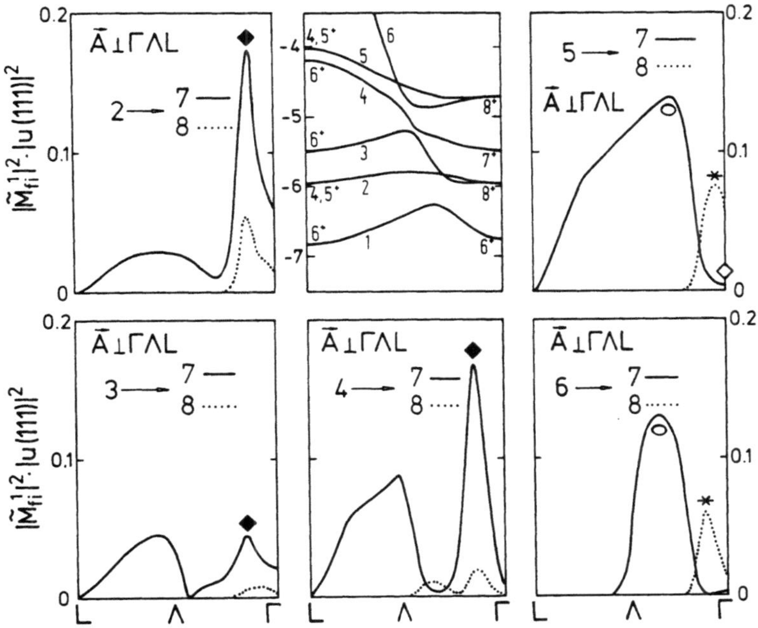

Fig. 7.29. The emission intensities for the transitions in Ag(111) from the occupied bands (*middle upper panel*) to the final state bands 7 and 8, with polarization of the light perpendicular to the ΓL direction. The emission intensities are given by $|\tilde{M}_{fi}^1|^2 \times |u_{(111)}|^2$

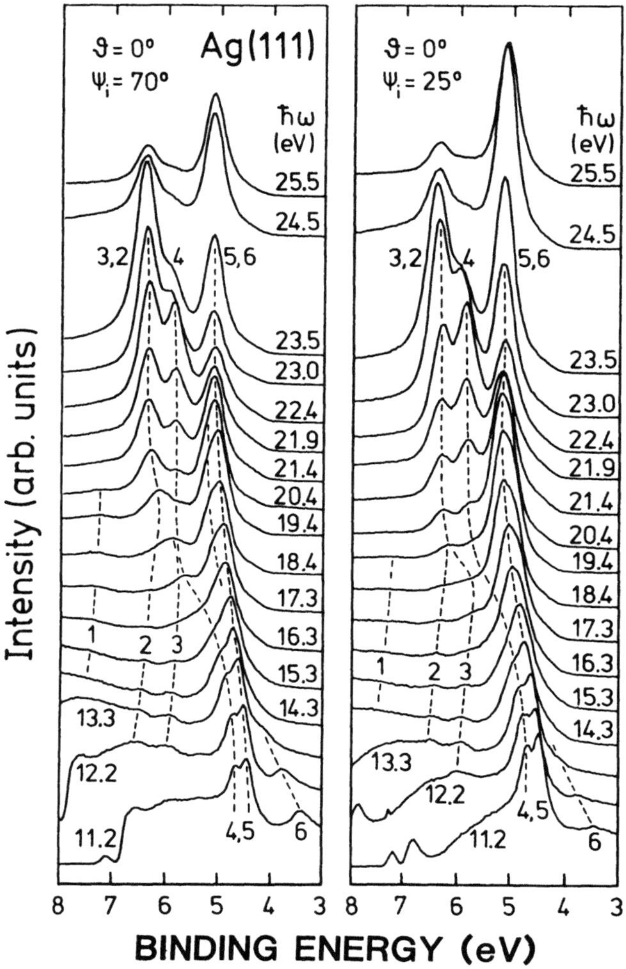

Fig. 7.30. Spectra for Ag(111) taken with $\vartheta = 0$ and varying the energy $\hbar\omega$. The angle of incidence for the light $\psi_{\mathrm{i}} = 25°$ enhances $\mathrm{A} \perp [111]$, and $\psi_{\mathrm{i}} = 70°$ enhances $\mathrm{A} \| [111]$ [7.40]

also shows the ΓL band structure above E_{F} for Ag. The emission intensities for the direct transitions along Λ, $\left|\tilde{M}_{fi}^{1}\right|^{2} |u_{(111)}|^{2}$, for light polarized perpendicular to Λ are given in Fig. 7.29.

The transitions out of band 2 ($\Lambda_{4,5}$ symmetry) into band 7 (Λ_6) and out of band 4 (Λ_6) into band 7 will peak strongly at $k_\Lambda \simeq 0.12$ $(2\pi/a)$ near the Γ-point, where the final band 7 is essentially flat (Figs. 7.28 and 7.29), whereas the transitions from the same occupied bands into band 8 (Λ_6) also peak in intensity near Γ but remain weak compared to the transitions into

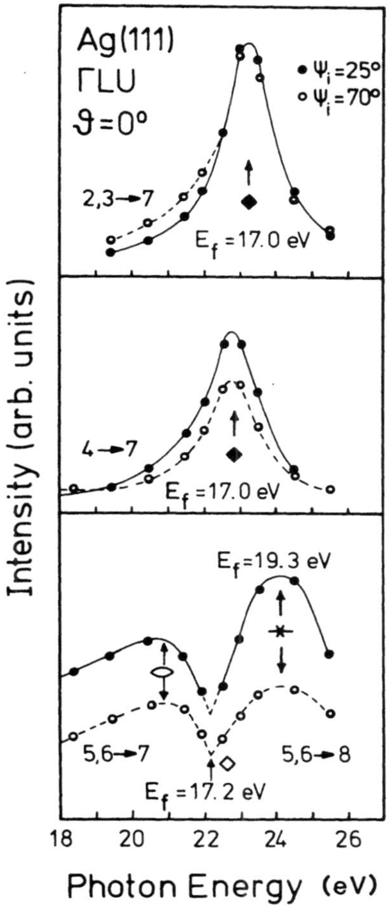

Fig. 7.31. Experimental intensities for transitions 2, 3 → 7, 4 → 7, and 5, 6 → 7, 5, 6 → 8 as a function of photon energy. The symbols marking the extrema can be correlated with similar extrema in Fig. 7.29, and in this way these data can be used to determine points in the band structure [7.40]. The corresponding points are marked with the same symbols in Fig. 7.27

band 7. The intensity of the transitions out of band 5 ($\Lambda_{4,5}$) into band 7 drops sharply to near zero at Γ, but the intensity of the transition from band 5 into band 8 shows a maximum very close to Γ. The behavior of the latter intensity is also seen for the transitions out of band 6 (Λ_6) into bands 7 and 8.

We now discuss some synchrotron radiation experiments which can be explained with this simple model. Spectra taken at normal emission in the [111]-direction with variable photon energy are shown in Fig. 7.30. At the energies chosen the data correspond to the case in point, i.e. they are produced by transitions from the occupied bands into the empty bands 7 and 8. Besides an energy dispersion one also observes a strong intensity variation showing pronounced maxima and minima, as illustrated in Fig. 7.31. Using the band-structure diagram in Fig. 7.27 it is now easy to correlate the extrema in the intensities with special points in the band structure. The full diamond (Fig. 7.27) marks the onset of the flat dispersion of band 7 along Λ due to the hybridization gap between bands 7 and 8. The open diamond marks the intensity minimum for transition out of bands 5 and 6 into band 7 and the star marks the maximum in the intensity of transitions out of bands 5 and 6 into band 8. In view of good agreement between theory and experiment found in these cases, one can also use the maxima for transitions out of band 5 and 6 into band 7 marked with an open oval to determine a further point in the band-structure diagram. Intensity resonances (strong maxima) in direct transitions have also been observed in Pd(111) [7.66] and Au(111) and Pt(111) [7.67] normal emission EDCs, and are interpreted as being due to excitation into a flat f-like Λ_6 final band near Γ.

This is an additional method for obtaining special points in the band structure in a direct and independent way. It is especially noteworthy that the simple model employed seems to describe the intensities in a photoemission process remarkably well.

7.3.8 Summary

Of necessity, this chapter has so far been somewhat detailed and the material was at times complicated. But in order to be able to judge the reliability of the many electronic structure determinations presented in the literature, we felt that a comprehensive account of the underlying procedures is necessary.

The main conclusion thus far is that one can, without any further assumptions, use PES alone to determine the band structure in a solid. This generally involves a considerable amount of effort. The task can be greatly simplified without much sacrifice of accuracy by working with a free-electron parabola for the final state, with the bottom of the conduction band taken from theoretical results for the material under consideration, or by adjusting it iteratively to the experimental data. Alternatively, one can take the final states directly from a calculation. Matters can be greatly improved if one determines a single point of the free electron or theoretical final state by one of the absolute methods mentioned previously, and thus adjusts the relevant final state band. This usually gives very accurate final states. Finally, there are a variety of iterative procedures that may be employed.

The relevant question, of course, is how accurate are the results of these procedures? This can be judged by comparing the results of studies of the same material using different methods. In order to choose a moderately complicated case, results will be given for Ag, for which a number of PES investigations exist [7.48,7.50–7.52,7.68–7.98]. A summary is presented in Table 7.2, which includes the results of a number of theoretical investigations and those of the two most detailed PES studies [7.57, 7.58, 7.97].

The first three calculations in Table 7.2 [7.53, 7.99, 7.100] are state of the art band-structure calculations, while that of Benbow and Smith [7.62] employs an interpolation scheme and therefore reflects a condensation of experimental data rather than a purely theoretical result. The four sets of the-

Table 7.2. Theoretical and experimental energy levels for silver at Γ, L and X. Energies are given in eV relative to the Fermi energy as initial-state energies ($E_i = E_B$)

Symmetry	Theory: references				Experiment: reference/method		
	[7.53]	[7.99]	[7.100]	[7.62]	[7.58] ARUPS	[7.97] ARUPS	[7.74–7.79] other work
Γ_{6+}	−7.21	−7.50	−7.92	−6.63	–	–	
Γ_{8+}	−8.52	−5.90	4.69 (spans Γ_{8+}, Γ_{7+})	−5.97	−6.19	−6.23	
Γ_{7+}	−5.37	−5.46		−5.49	−5.76	−5.80	
Γ_{8+}	−4.69	−4.75	−3.69	−4.73	−4.95	−4.95	
Γ_{7-}	+16.98	+16.84	–	+17.17	+17.10	+17.00	
Γ_{6-}	+19.61	+19.49	–	+18.99	+19.20	+23.00	
L_{6+}	−6.74	−6.94	−6.34	−6.85	–	−7.13	
L_{4+}, L_{5+}	−5.91	−5.99	4.76 (spans L_{4+}, L_{5+} and L_{6+})	−5.97	−6.27	−6.28	
L_{6+}	−5.44	−5.53		−5.50	−5.74	−5.74	
L_{6+}	−4.23	−4.20	2.87 (spans L_{6+} and L_{4+}, L_{5+})	−4.20	–	−4.31	
L_{4+}, L_{5+}	−4.01	−3.97		−4.04	−4.11	−4.06	
L_{6-}	−0.03	−0.16	−0.59	−0.62	−0.20	–	L_{6-} 0.31[a], −0.30[b]
L_{6+}	+3.44	+3.33	+3.79	+3.82	–	–	L_{6+} +3.86[a], +3.85[c]; +3.8(0.1)[d,e]; +3.77(0.09)[f]
X_{6+}	−7.00	−7.13	−6.54	−7.12	−7.38	−7.35	
L_{7+}	−6.82	−6.99	−6.22	−7.04	−7.32	−7.00	
L_{7+}	−4.25	−4.21	−2.93	−4.21	−4.35	−4.20	
L_{6+}	−4.05	−4.03	−2.64	−4.11	−4.12	–	
L_{7+}	−3.71	−3.73	−2.64	−3.79	−3.82	−3.90	
L_{6-}	+2.27	+2.02	+1.70	+1.88	–	–	+2.1(0.1)[f]

[a] Photoemission [7.74]
[b] Electron tunneling [7.75]
[c] Thermoreflection and thermotransmission [7.76]
[d] Reflectance [7.77]
[e] Inverse photoemission, $\hbar\omega = 1{,}486.7\,\text{eV}$ [7.78]
[f] Inverse photoemission, $\hbar\omega = 9.7\,\text{eV}$ [7.79]

oretical results, however, show remarkable agreement considering that the spin–orbit interaction in Ag is large ($\approx 0.5\,\text{eV}$). On the experimental side Nelson et al. [7.97] have extracted the dispersion relation by using for the final state a plane wave fitted to the band structure of Christensen [7.99], while Wern et al. [7.57] used absolute methods. Again quite good agreement between the two sets of data is found to within about $\pm 0.10\,\text{eV}$ with the notable exception of the Γ_6 energy which is given as 19.2 eV by Wern et al. [7.57] and 23.0 eV by Nelson et al. [7.97]. The reason for this discrepancy remains unclear at present and we simply note that the determination of this energy is not without problems from the experimental point of view. There is also quite close agreement between the experimental and theoretical numbers.

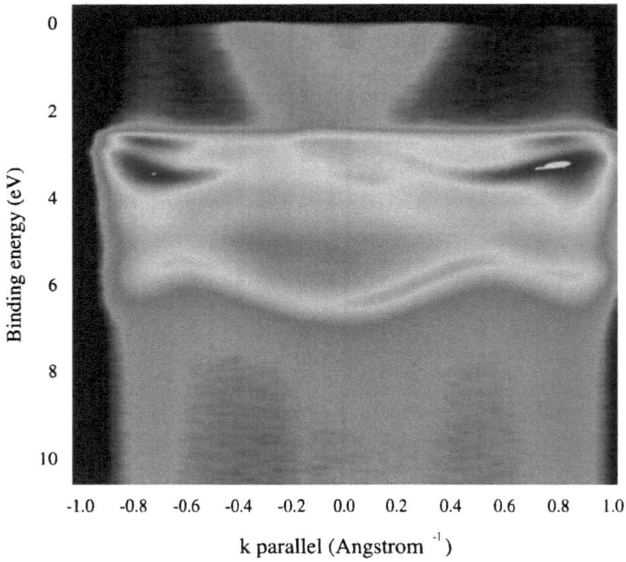

Fig. 7.32. Photoemission spectra taken from the Cu(100) surface. The temperature is $T = 140\,\mathrm{K}$, the photon energy $\hbar\omega = 580\,\mathrm{eV}$. The center of the diagram corresponds to a point along the ΓX direction (see Fig. 7.6). The angle was changed in the ΓKLUX plane which means that a path in the Brillouin zone away from the (001) direction in Fig. 7.6 towards the ΓL direction is scanned. The exact momentum values have not been worked out at this time [7.190].

A very interesting new perspective in the determination of band structures has now been found by the combination of soft X-ray synchrotron radiation and very high energy and momentum resolution [7.190] . The first result in this respect is given in Fig. 7.32, which shows data taken from a (100) surface of Cu. The temperature was $T = 140\,\mathrm{K}$ and the photon energy $\hbar\omega = 580\,\mathrm{eV}$. The center of the figure (0.0) corresponds to a point along the ΓX direction about half way between Γ and X. The parallel momentum was obtained by changing the angle in the ΓKLUX plane by ± 6 degrees. With respect to Fig. 7.6 this means that the center of the data corresponds to the (001) direction and that the momentum changes from this direction in the ΓKLUX plane towards L. The exact momentum values for these data have not been worked out at this moment.

The most interesting aspect of this set of data is the fact that even at these high photon energies direct transitions are still responsible for the majority of the photoemission intensity. Thus, if one only takes care of the momentum resolution by working with small angular acceptance a band structure is easily observed in this mode of operation. For an exact evaluation of the data the momentum of the photon has to be taken into account. This method of determining band structures seems to be quite promising and will probably be used in the future.

7.4 Experimental Band Structures

It was the intention in previous sections of this chapter to indicate how electronic bulk structures can (in principle) be determined by PES: This was done by using explicit examples, but in view of the importance of the band structure for the understanding of many solid state properties, a few additional examples of actual results will be given. We will not detail all the assumptions and shortcomings associated with the collection of the experimental data. The objective of this section is to present some examples to illustrate the quality and range of available data. For a complete tabulation of such data see [7.96].

7.4.1 One- and Two-Dimensional Systems

Two-dimensional systems [7.64] have played an important part in band-structure determinations because in these cases an experimental band structure can be obtained from EDCs with relative ease [7.63], without recourse to assumptions or to any of the complicated procedures described in preceding sections. We first summarize the reasons behind this assertion. If $E_i(\boldsymbol{k})$ and $E_f(\boldsymbol{k})$ are the initial and final state energies of an electron in a band structure, the energy conservation relation for a direction transition reads:

$$E_f(\boldsymbol{k}) - E_i(\boldsymbol{k}) = \hbar\omega \; . \tag{7.14a}$$

Now in the photoemission process the wave vector parallel to the crystal surface is conserved, so that

$$\boldsymbol{K}_{||} = \boldsymbol{k}_{i||} + \boldsymbol{G}_{||} \tag{7.14b}$$

with

$\boldsymbol{K}$: wave vector of the photoexcited electron

$\boldsymbol{k}_i$: wave vector of the initial state

$\boldsymbol{G}$: wave vector of a reciprocal lattice vector.

If E is the energy of the outgoing electron with respect to the vacuum level (kinetic energy) and ϑ the angle between the surface normal and the direction of thc outgoing electron, one has

$$K_{||} = \sqrt{2mE/\hbar^2}\sin\vartheta \; . \tag{7.15}$$

Equation (7.15) yields the magnitude of $\boldsymbol{K}_{||}$. Its direction can be obtained from the azimuthal orientation of the electron detector relative to the crystal. If the experimental conditions are chosen such that one is working in the first

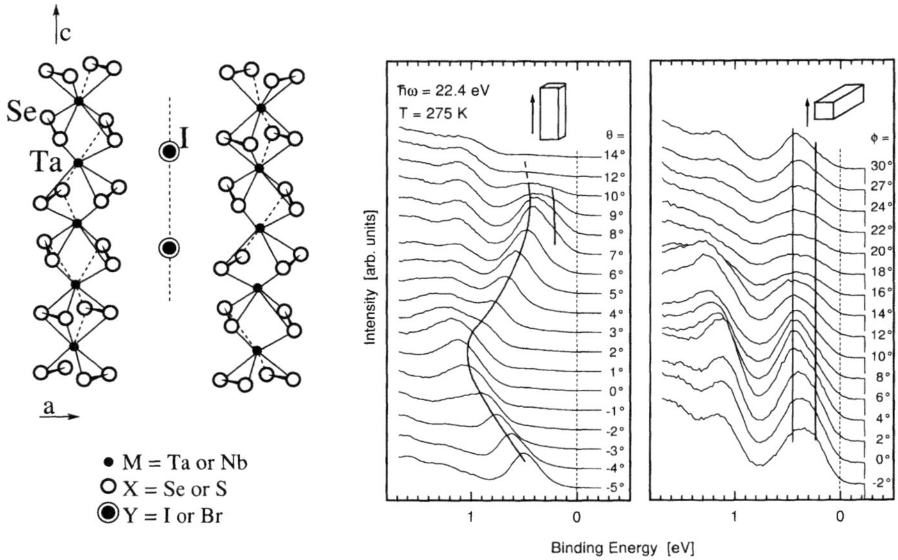

Fig. 7.33. *Left part*: Sketch of the crystal structure of $(TaSe_4)_2I$ showing the one-dimensional nature of the structure, which results in very anisotropic electronic properties. The *center* and *right-hand* panels show dispersion curves parallel (*center*) to the chains and in one direction perpendicular to them (*right*) [7.64]

Brillouin zone and uses the first Mahan cone, one has $\boldsymbol{G} = 0$ and thus $E_i(\boldsymbol{k}_{||})$ can be determined directly from the experimental data for a two-dimensional layered compound.[4]

All these considerations are naturally also true for one-dimensional compounds, except that dispersion will (in principle) only be visible in the one-dimensional direction. One-dimensional metals are of interest, because it has been suggested that they are not Fermi liquids (as two- and three-dimensional metals are) but Luttinger liquids (see Chap. 5 for a discussion of this point). In the context of the present chapter it will simply be demonstrated to what degree a three-dimensional solid can indeed show one-dimensional dispersion relations, because of one-dimensional crystallographic elements. A case in point is $(TaSe_4)_2I$, for which dispersion relations measured with $\hbar\omega = 22.4\,\mathrm{eV}$ radiation in all three crystallographic directions are shown in Fig. 7.33. Fig. 7.33 also shows a sketch of the crystallographic structure, from which it is apparent, that there are rod like $TaSe_4$ pyramids, which are responsible for the one-dimensional electronic behavior. These one-dimensional structures are indicated in the two sets of spectra and the arrows indicate the direction in which the dispersion has been measured (by tilting the electron detection

[4] If one works in a higher Brillouin zone or uses higher Mahan cones one has to fold the data back with the appropriate reciprocal lattice vectors into the first Brillouin zone in order to make a comparison with the theoretical band structure.

angle into this direction, see (7.15)). It is evident that, as anticipated, there is a large dispersion along the rod (c-axis) but none in the perpendicular directions (the spectra in the second perpendicular direction agree with those shown in the right hand panel of Fig. 7.33 and are therefore not given explicitly). The data can be taken as a measure of the degree of one-dimensionality that can be observed in nominally three-dimensional systems. The spectra in Fig. 7.33 have all been taken at 275 K, above the Peierls temperature of $T_{\mathrm{p}} = 263\,\mathrm{K}$, which marks the transition from the insulating into the metallic state. Nevertheless, as predicted for a one-dimensional metal, in no case is there a Fermi step visible (see Chap. 5).

There are two types of *"layered" compounds* that are of interest. Firstly, a number of three-dimensional solids actually display a nearly two-dimensional structure [7.101], the prototype material being graphite. Secondly, ordered overlayers of atoms or molecules on a substrate surface can often be approximately described as two-dimensional structures, although here the overlayer–substrate interaction will clearly tend to introduce deviations from two-dimensionality.

Graphite. Data will be presented for graphite because it has become the prototype compound in this field.

The crystal structure and the corresponding Brillouin zone of graphite are shown in Fig. 7.34 [7.102]. Note that the Brillouin zone is of course three-dimensional, but for the purpose of this section one is only interested in the k_x-k_y plane of $\boldsymbol{k}$-space because this corresponds to the two-dimensional electron states in the x-y plane.

Atomic carbon has the electronic configuration $1\mathrm{s}^2 2\mathrm{s}^2 2\mathrm{p}^2$ and since the 1s state is at 284 eV below the Fermi energy, only 2s and 2p electrons contribute to the valence band. In the x-y plane the electronic structure is composed of 2s, $2\mathrm{p}_x$ and $2\mathrm{p}_y$ atomic orbitals. There is weak interlayer π-bonding from the $2\mathrm{p}_z$ orbitals, but this can be ignored for the present purposes.

Figure 7.35a shows an XPS spectrum of crystalline graphite [7.103]. This should give an approximate representation of the DOS of this material (apart from differences in cross-sections). By comparison with X-ray emission data and calculated band structures, the spectrum can be interpreted as follows: peak I is due to the 2s band, peak III represents the 2p band and peak II is of mixed s- and p-character [7.102]. These XPS data provide only a crude description of the band structure of graphite.

The results of a band-mapping experiment with synchrotron radiation [7.104] are given in Fig. 7.35b. The points shown in this figure are determined without any inherent assumption because of the two-dimensionality of the system. One can see that there is good agreement between theory [7.102] and experiment, and that a real band-structure mapping gives information superior to simple DOS data. Additional experimental data on the band structure of graphite is presented in [7.105–7.114].

The unoccupied bands of graphite have been determined by IPES measurements (Fig. 7.35b) [7.115]. Here again, the dispersion (in the ΓM direction) has been obtained directly because of the two-dimensional nature of graphite. In this case comparison has been made with the band structure of Holzwarth et al. [7.116] mainly because this calculation gives the best agreement with the measured data for the empty bands.

Of special interest in Fig. 7.35b is the dispersion obtained in the ΓA direction, which is the direction perpendicular to the layer (Fig. 7.34) and therefore measures the three-dimensionality of graphite. The experiments for this direction can only be evaluated with the assumption of a free electron initial state (note that one is dealing with an IPES experiment); however, the general trend obtained by this procedure should be correct. This band can be thought of as a 3s unoccupied band of graphite and it is important for the understanding of the intercalation compounds produced on graphite bases [7.117]. In this direction a surface state is also observed [7.118].

There exists a number of additional investigations of the occupied [7.105, 7.106] and unoccupied band structure [7.119,7.120] of graphite. A comparison of Figs. 7.35b and 7.36 indicates the progress that has been achieved in this field in a relatively short time. In total one may state that the occupied bands

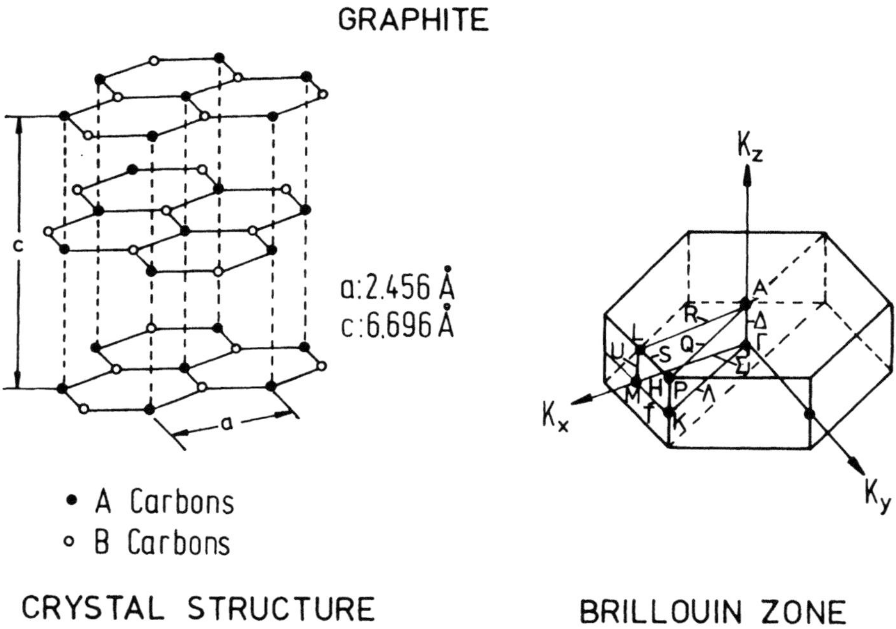

Fig. 7.34. Crystal structure and Brillouin zone of graphite. The crystal structure shows the two-dimensional sheets separated by $c/2$. Note that any dispersion of the bands along ΓA will indicate a deviation from a purely two-dimensional behavior [7.102]

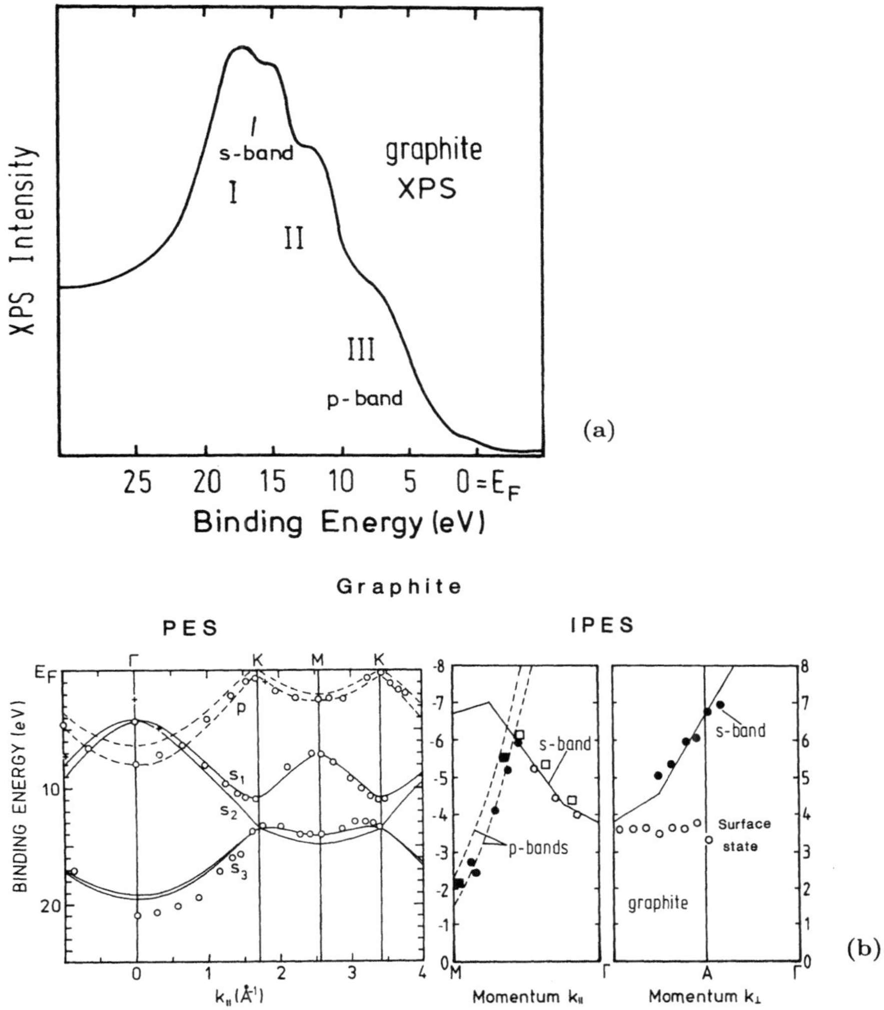

Fig. 7.35. (**a**) XPS spectrum of graphite [7.103]. This gives a first-order representation of the density of states of this material. An approximate interpretation of this spectrum is I: 2s of C; III: 2p of C; II: s and p contributions of C. (**b**)(*Left*) Band structure of graphite measured with synchrotron radiation [7.104] along ΓKM. Full (σ-symmetry) and dashed (π symmetry) lines are results from a band-structure calculation [7.102]. The data points have been read off directly from the spectra by assuming $k_\perp = 0$. (*Right*) Unoccupied band structure of graphite measured by IPES [7.115]. The full curve is the band structure calculated by [7.116]. Note that ΓA is the direction along the c axis in real space and therefore the dispersion relations seen in this direction are a measure of the three-dimensionality of graphite. The s band can be viewed as the empty 3s band (which is important in the formation of intercalation compounds). The surface state at $E_\mathrm{B} = -3.6\,\mathrm{eV}$ is assigned via a calculation [7.118]

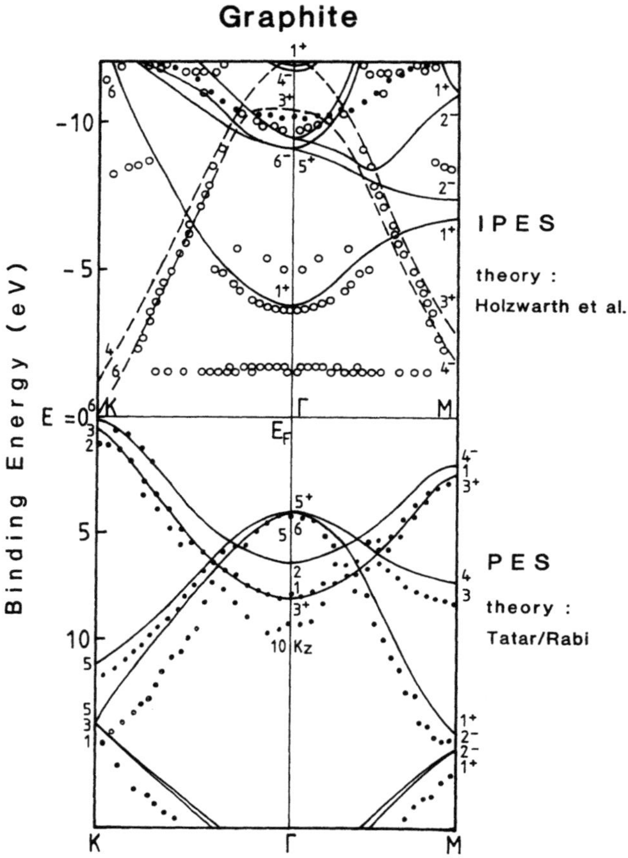

Fig. 7.36. Band structure of graphite [7.119, 7.120]. The difference to the data in Fig. 7.35a demonstrates the improvement possible in the technique of PES/IPES over a short time; the IPES data have been obtained from highly oriented pyrolytic graphite. Therefore, the ΓM and ΓK directions are indistinguishable

seem to be fairly well understood. The empty bands, however, are so far not so well established, especially since they seem to be better described by a theory that does not do equally well for the occupied bands.

Ordered Overlayers. As mentioned previously, a particularly simple form of quasi-two-dimensional systems is the ordered overlayer of atoms or molecules on a substrate. Such overlayers have a strictly two-dimensional structure only if the substrate–overlayer interaction is zero, which, by necessity, it never can be.

Here data on a controversial system, namely O on Cu(110) will be presented [7.121–7.125].

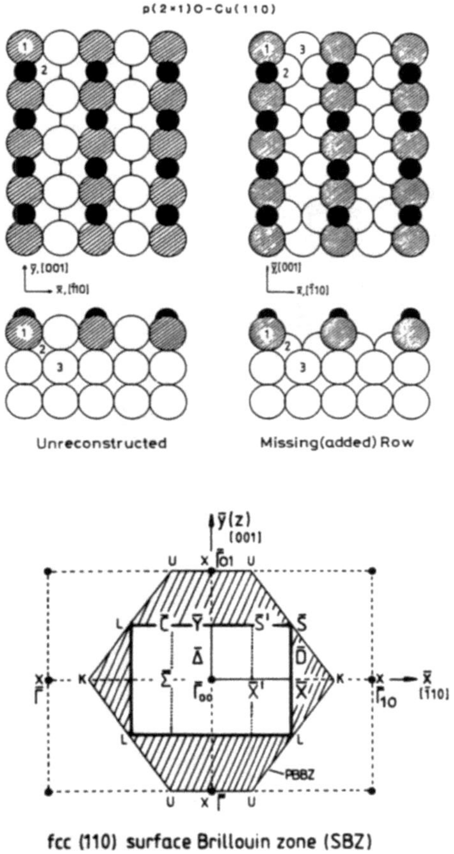

Fig. 7.37. (*Above*): Top and side views of the geometrical arrangement of the p(2 × 1)O/Cu(110)structure in the missing row model. The oxygen atoms are shown by filled circles. 1,2 and 3 denotes the first, second and third Cu planes. (*Below*): Surface Brillouin zone for the clean (110) surface and the 2(2×n)O-surface (dashed lines and symmetry labels with bars). The dimension of the p (2 × 1) surface Brillouin zone is halved in the X-direction because of the doubling of the unit cell in real space. The hatched area gives the Projected Bulk Brillouin Zone (PBBZ). The bulk symmetry points are also indicated

The structure of O/Cu(110) has been investigated by various methods. However, the results do not show complete agreement [7.126–7.147]. The basic structure, agreed upon by all researchers, is p(2×1), where p(n×1) structures may be precursors to the formation of the final p(2 × 1) structure (Fig. 7.37). It is believed that the atomic structure of the Cu(110) surface reconstructs under oxygen adsorption with three possible configurations: Missing row, saw tooth, buckled row. The reconstruction, which seems to be established beyond any doubt [7.144], the missing row model (or rather added row model, if the growth mode is taken to label the structure), is also shown in Fig. 7.37. In all the reconstruction models the oxygen atoms sit on long bridge sites along the [001] rows, as indicated in Fig. 7.37. The surface Brillouin zone of the system is also depicted in Fig. 7.37. Whereas details about the reconstruction cannot be deduced from PES data, the strong anisotropy of the bonding of the oxygen atom to the substrate should be discernible in the spectra. Indeed PES does not find any dispersion of the oxygen induced bands for the $\bar{\Gamma}\bar{X}$ direction in

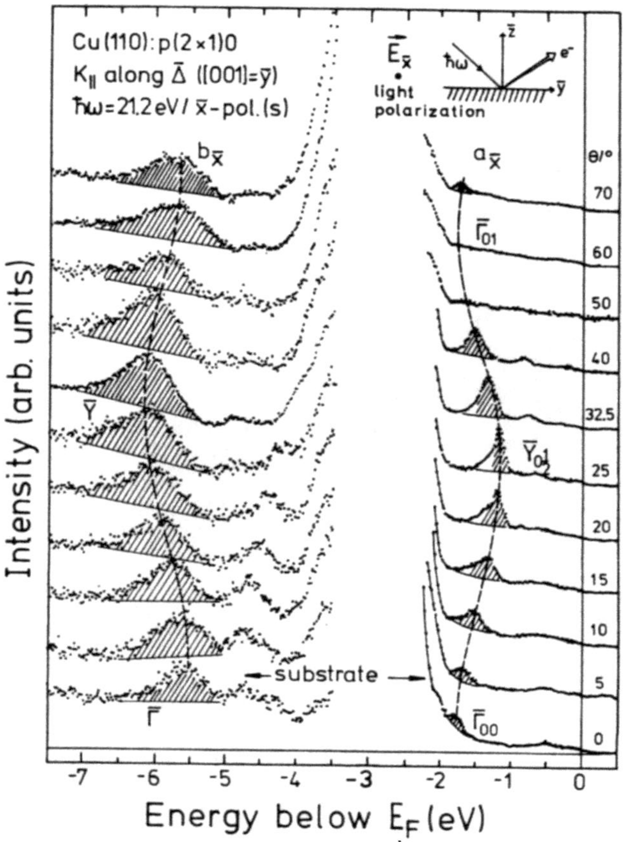

Fig. 7.38. ARPE spectra for p(2×1)O/Cu(110) taken in the $\bar{y}\bar{z}$-plane at a photon energy of 21.2 eV. The light has $\bar{x}$ polarization and therefore one measures only the dispersion along $\bar{y} = [001]$ of the odd states

the surface Brillouin zone (Fig. 7.37), emphasizing that the bonding is mainly one-dimensional in the $\bar{y}$ direction along the (001) rows, where the oxygen atoms are located [7.144].

Spectra of the O/Cu(110) system, with the p(2×1) structure as identified by LEED, are shown in Figs. 7.38–7.42.

Spectra taken with $\boldsymbol{K}_{\|}$ perpendicular to the [001] direction (which is the direction of the oxygen–copper rows) are not depicted here, because they show, as anticipated, negligible dispersion for the structures b (bonding bands, below the Cu 3d band) and a (antibonding bands, above the Cu 3d band), which are both oxygen induced, because they are lacking in the spectra taken from a clean (110) Cu surface.

The spectra taken with light polarized along $\bar{x}$ are easy to interpret and show in essence the expected $b_{\bar{x}}$ and $a_{\bar{x}}$ surface bands (Fig. 7.38). The data

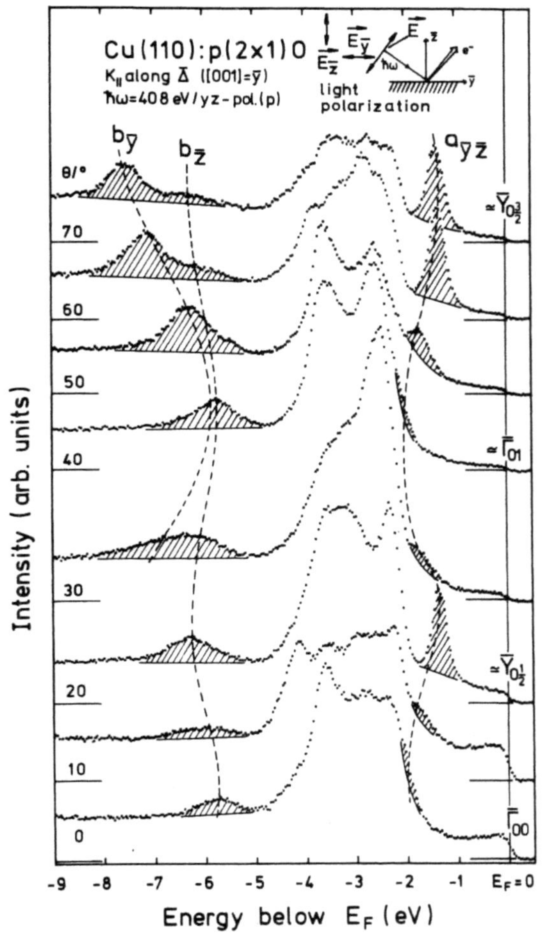

Fig. 7.39. Same as Fig. 7.38 for the even states along $\bar{y}$, $\hbar\omega = 40.8\,\mathrm{eV}$ and $\bar{y}\bar{z}$ polarization of the light. The label $a_{\bar{y}\bar{z}}$ indicates that at present it is not possible to determine whether the state has $\bar{y}$ or $\bar{z}$ character.

taken with the $\bar{y}\bar{z}$ polarization (polarization in the measuring plane) are more difficult to analyze (the $\bar{z}$ direction is perpendicular to the surface). The spectra in Fig. 7.39 indicate two bonding and one antibonding band. At larger detection angle with respect to the surface normal (small take-off angle of the photoelectrons) the bond along the $\bar{y}$ direction (direction of the O-Cu rows) is most likely to be detected and therefore it is appealing to suggest that the strong bonding peak in the high Θ spectra belongs to the $b_{\bar{y}}$ state. This is in agreement with the reasoning that the bond along the O-Cu row is the strongest, yielding the largest dispersion. The relative dispersion of the $b_{\bar{y}}$ and $b_{\bar{z}}$ bands cannot be obtained from the spectra in Fig. 7.39. A second set

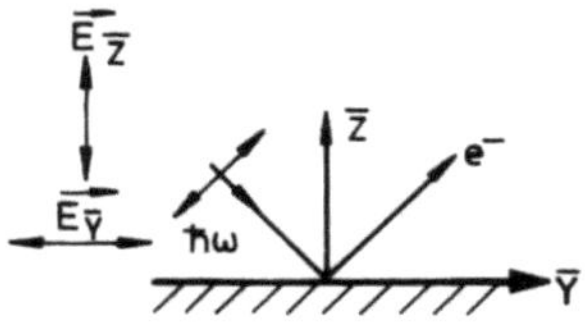

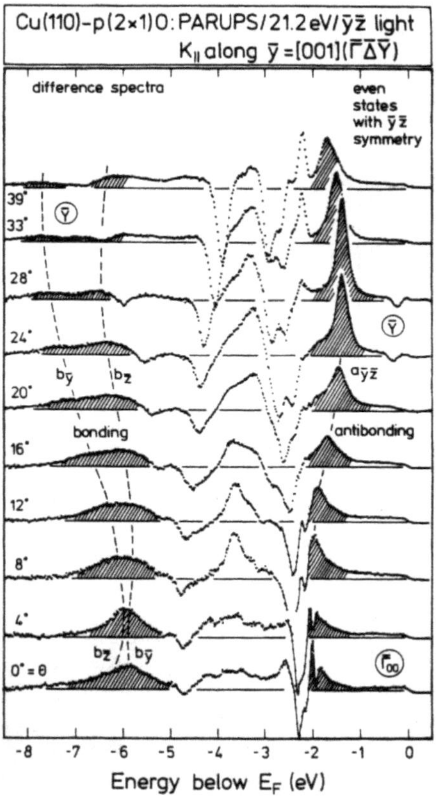

Fig. 7.40. Angle dependent PARUP spectra from p(2 × 1)O/Cu(110), taken along the added rows with $\bar{y}\bar{z}$-polarized He I light. Shown are the differences between spectra from the oxygen covered and the clean surfaces. Normalization of the original spectra has been performed at about 8 eV binding energy. The spectra give the dispersion of the oxygen induced states with even reflection symmetry (*hatched areas*) along $\bar{y}$

of data using the more intense $\hbar\omega = 21.2\,\text{eV}$ line is given in Fig. 7.40, but again it is difficult to extract from these data the dispersion of the $b_{\bar{y}}$ and the $b_{\bar{z}}$ bands through the whole Brillouin zone. In this dataset the spectra from the oxygen covered surface were subtracted from those of the clean surface to gain better contrast. In order to reveal the ordering of the bands at the points $\bar{\Gamma}$ and $\bar{Y}$ of the surface Brillouin zone, data at these points in the Brillouin zone were taken.

The data at $\bar{Y}$ (Fig. 7.41) are simple to interpret and show the position of the $b_{\bar{x}}$, $b_{\bar{y}}$, $b_{\bar{z}}$, $a_{\bar{x}}$ and $a_{\bar{y}\bar{z}}$ bands. The antibonding band observed with $\bar{y}\bar{z}$ polarized light was labelled $a_{\bar{y}\bar{z}}$ because the exact orbital character of this state could not be determined even by using polarized light.

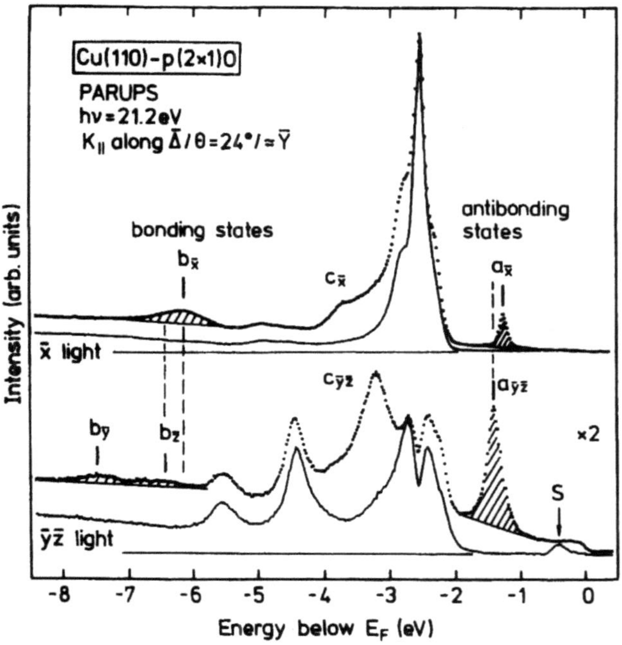

Fig. 7.41. PARUP spectra of clean Cu(110) (*solid lines*) and of p(2 × 1)O/Cu(110) (*data points*) taken in the $\bar{y}\bar{z}$ plane parallel to the added rows and excited with polarized He I radiation. The electron polar angle $\vartheta = 24°$ corresponds approximately to the $\bar{\mathrm{Y}}$ symmetry point of the surface Brillouin zone. The oxygen induced features are indicated by the hatched area. The spectra have been normalized to the maximum intensity

Figure 7.42 gives the bonding bands at $\bar{\Gamma}_{00}$ measured with $\bar{x}$-, $\bar{x}\bar{z}$-, $\bar{y}\bar{z}$- and $\bar{y}$-polarized light. These data show that the ordering of the levels at $\bar{\Gamma}_{00}$ is (starting with that with the largest binding energy) $b_{\bar{z}}$, $b_{\bar{y}}$, $b_{\bar{x}}$, while at $\bar{Y}$ it was $b_{\bar{y}}$, $b_{\bar{z}}$, $b_{\bar{x}}$. Thus, between $\bar{Y}$ and $\bar{\Gamma}_{00}$ the two lowest bands exchange their wave function, meaning that there is an anticrossing in the dispersion probably near $\bar{\Gamma}_{00}$ (see $\Theta = 4°$ spectrum in Fig. 7.40). In Fig. 7.43 the experimental dispersion for the Cu(110)-p(2 × 1) O overlayer structure is shown and compared to a theoretical band structure [7.147]. The latter has been shifted in order to bring it in the best possible agreement with the experimental data. While for the bonding bands the agreement between theory and experiment is good, for the antibonding bands the agreement is poor. There is one particularly troublesome observation. Cu in the Cu(110)-(2 × 1) O p structure is believed to be in a $3d^9$ configuration, which means that there should be two completely filled antibonding bands and one half-filled antibonding band (as is obvious in the theoretical band structure). Unfortunately it has not so far been possible to detect the half-filled antibonding band.

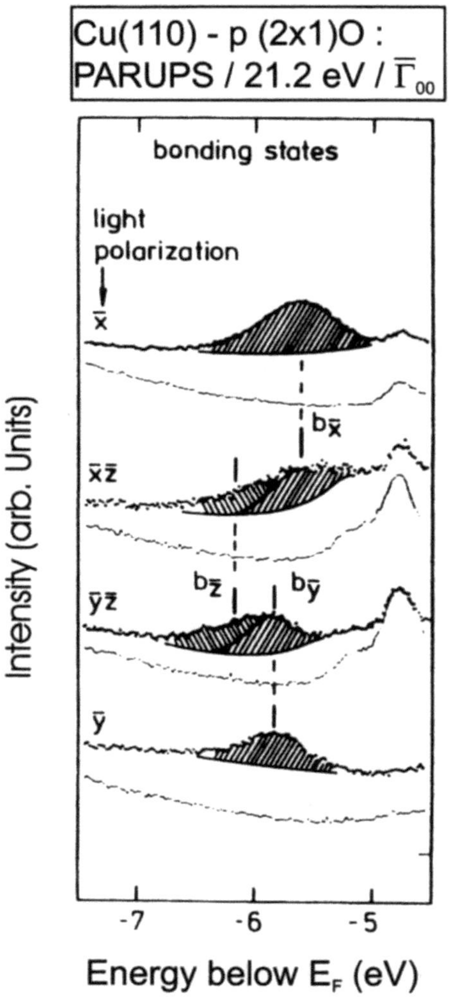

Fig. 7.42. Enlarged view of the oxygen derived states in the normal emission PARUP spectra taken with He radiation of p(2×1)O/Cu(110) corresponding to the center of the surface Brillouin zone. Only bonding states are shown

Finally we show in Fig. 7.44 experimental results for all the bands induced by the oxygen–copper interaction. The unoccupied bands have been measured using inverse photoemission spectroscopy (ARIPES) by Jacob et al. [7.125, 7.148]. The state centered at 4 cV above the Fermi energy (Fig. 7.45) may represent a $Cu3d^{10}$ state in keeping with the $3d^9$ interpretation for the ground state given above, because with the ARIPES technique an additional electron is added to the copper–oxygen system.

The obviously symmetric dispersion of this state with respect to the occupied oxygen O $2p_{\bar{y}}$ state is in agreement with the assumption of a two-level approximation for the (O, Cu, σ)-bond [7.147]. Justification for this interpretation comes from the Electron Energy Loss (EEL) data of Spitzer and

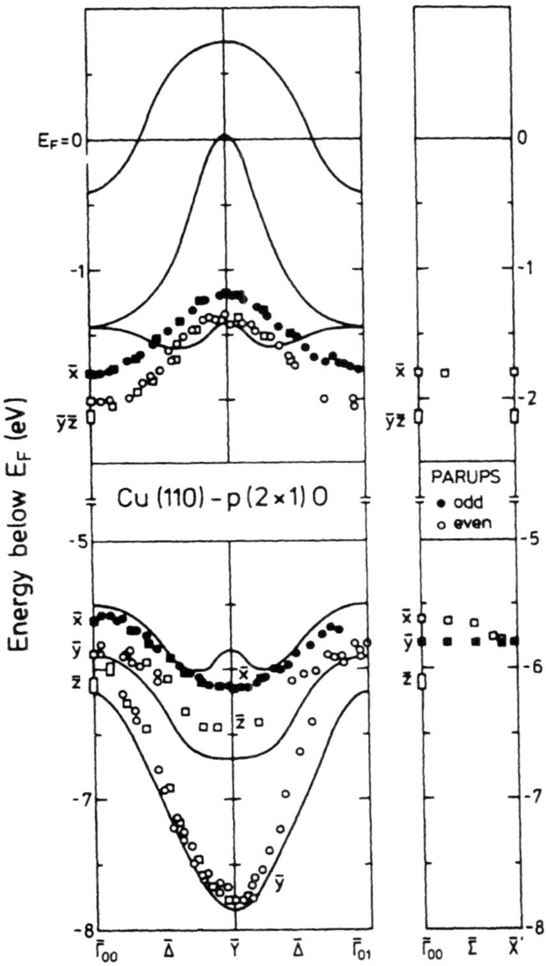

Fig. 7.43. Experimental dispersion of the oxygen-induced surface bonds along (*left*) and perpendicular (*right*) to the (added) O-Cu-O rows. The symmetry character $\bar{x}, \bar{y}$ and $\bar{z}$ is indicated. In the left panel ($\bar{\Gamma}\bar{Y}$ direction) the results of the calculation by Weimert et al. [7.147] are also given (antisymmetric states). The calculated curves have been shifted in order to give the best agreement with experiment

Lüth [7.123]. They found that the EEL spectra of Cu(110) show an additional signal at $\Delta E = 9.3\,\text{eV}$ when a (2×1) oxygen layer is present. The loss energy agrees well with the combined PES and IPES data. Since the strongest signal in such spectra is from charge transfer transitions, which are between the oxygen 2p electrons and the empty Cu states, the signal observed in the loss measurements should give the energy separation of the oxygen 2p and the empty Cu states. This interpretation of the state above E_F is, however, not

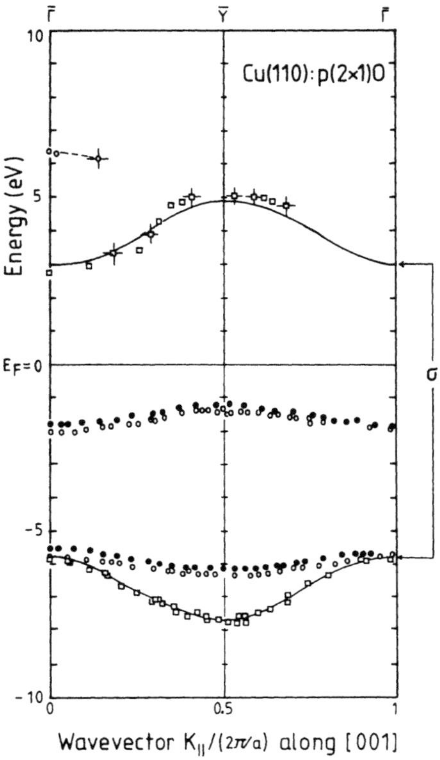

Fig. 7.44. Dispersion relations along $\bar{y}$ = [001] for p(2 × 1)O/Cu(110) as measured with photoemission below E_F and with inverse photoemission [7.125, 7.145] above E_F. The full lines come from a best fit to the $b\bar{y}/b\bar{z}$ band which was mirrored and adjusted to the unoccupied band

the only one possible. In [7.121] it has been suggested that the band above E_F is a surface state which means that the agreement of its dispersion with that of the O $2p_{\bar{y}}$ state below E_F is accidental. At this point in time there seems to be no definite answer to the problem [7.149].

One might, however, take a very pragmatic view to that problem. Any state in a monoatomic surface layer can be viewed as a surface state. In this sense the two different interpretations given for the unoccupied state of O on Cu(110) may not be all that different.

We now discuss briefly the case of CO on a transition metal surface. This was already mentioned in Sect. 5.11.2, where the gross electronic structure with respect to the chemisorption bond was described. Here we present data for the low temperature phase ($T \leq 180$ K) of CO on Ni(110), in which the CO molecules order in the two-dimensional (2 × 1)p2mg structure [7.150–7.152]. Figure 7.46 gives a geometrical picture of this overlayer structure along with the two-dimensional surface Brillouin zone. The CO molecules occupy bridge positions along the rows of Ni atoms in the [1$\bar{1}$0] direction. There is one CO molecule per surface Ni atom. In order to avoid the strong lateral repulsion between adjacent CO molecules, a structure develops in which the

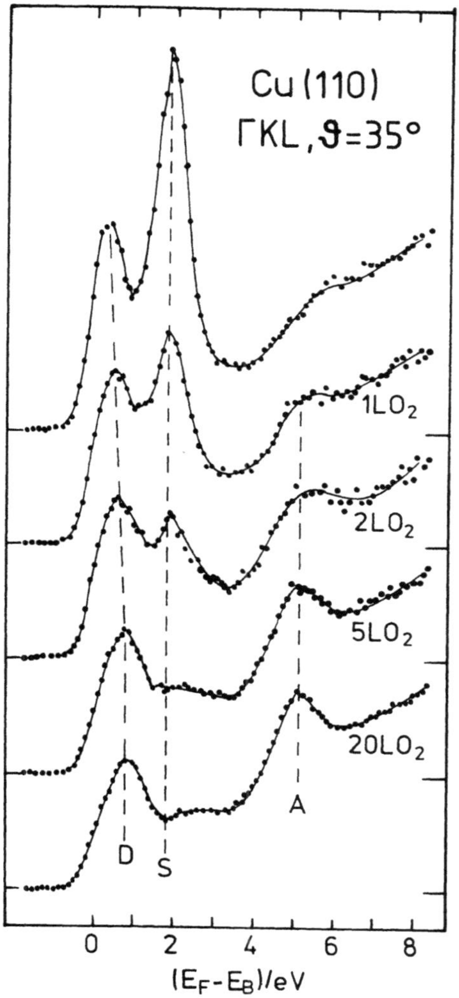

Fig. 7.45. IPES spectra of Cu(110)/O ($\hbar\omega$ = 9.7 eV) [7.145]. The feature D is a direct transition in the bulk band structure, while S is a surface state. Feature A is an oxygen induced orbital

CO molecules in adjacent rows along [001] tilt away from each other. Thus in each row along [001] the CO molecules are tilted parallel to each other, away from the surface normal. There are two CO molecules in the surface unit cell, and they are tilted with respect to each other (Fig. 7.46). The direction of the tilt is indicated by arrows in the unit cell diagram.

The tilting of the CO molecules can be observed directly, e.g., by photoelectron diffraction (Chap. 11) and the results of a corresponding experiment are given in Fig. 7.47 [7.152]. If in a PE experiment electrons are emitted from the C1s level of an adsorbed layer of CO molecules on a substrate, these electrons can be scattered by the O atoms (note that the CO molecules are attached to the Ni surface by the C ion), which may lead to an interfer-

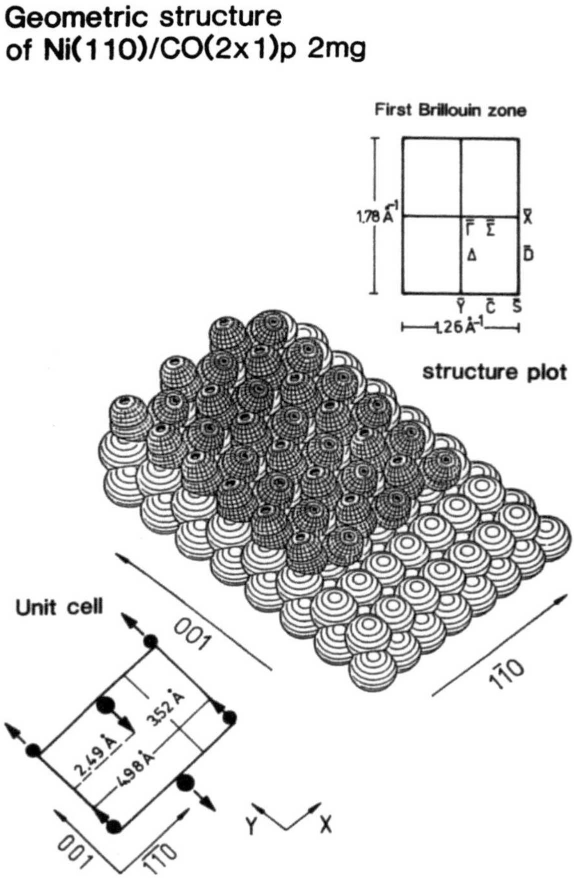

Fig. 7.46. Geometric structure of Ni(110)/CO(2 × 1)p2mg [7.150] at low temperatures ($T \leq 180\,\text{K}$). The CO molecules occupy bridge positions along the rows of Ni atoms in the [1$\bar{1}$0] direction. There is one CO molecule per surface Ni atom. To reduce the strong lateral repulsion between adjacent CO molecules, the two CO molecules in each unit cell bend in opposite directions along the [001] azimuth. A diagram of the surface unit cell is shown in the lower left of the figure. The tilt of the molecules is indicated by arrows. The surface Brillouin zone is sketched in the upper right corner

ence between these scattered electrons and the primary PE electrons. This interference pattern is stationary with respect to the molecule–surface geometry. Therefore, if we change the electron detection angle with respect to the surface normal, the interference pattern is monitored and its shape allows one to deduce the direction of the CO axis with respect to the surface normal. Note that like in a simple optical single-slit experiment, the intensity of the

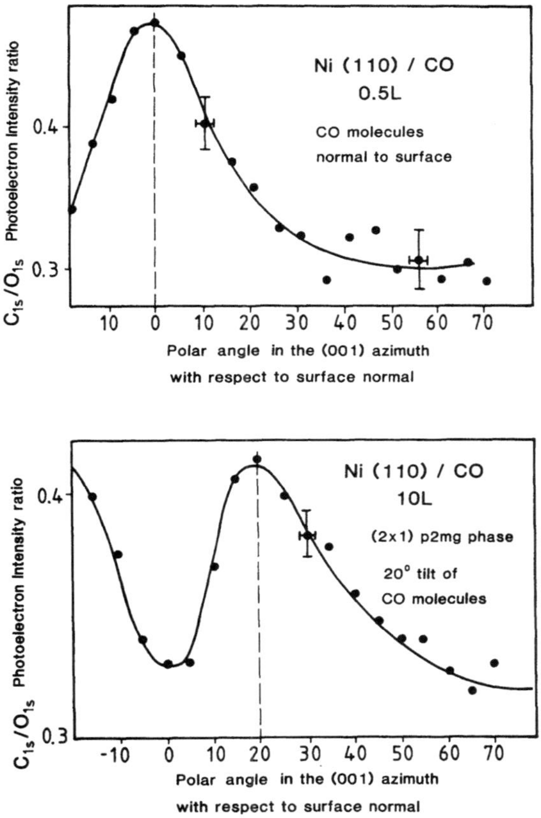

Fig. 7.47. Photoelectron diffraction from a Ni(110) surface covered with CO [7.152]. Plotted is the C 1s to O 1s photoelectron intensity ratio as a function of the angle in the [001] azimuth as measured with respect to the surface normal. The top panel presents data from a disordered layer of CO on Ni(110). The constructive interference in the zero degree direction shows that the CO molecules are oriented parallel to the surface normal. In the lower panel, data from the ordered (2×1)p2mg surface are shown. The constructive interference now occurs at $\pm20°$ indicating a tilt of the CO molecules by that angle away from the surface normal along the [001] azimuth

PE current has a maximum in the forward direction, i.e., when the electron detector is oriented parallel to the molecular axis.

With these points in mind the interpretation of the data in Fig. 7.47 is straightforward. In the top panel the coverage of CO on Ni(110) is low and therefore the CO molecules are oriented with their axes perpendicular to the substrate surface. The PE intensity measured with respect to the surface normal in the [001] azimuth shows, as expected, a maximum at zero degrees. At higher coverages, when the (2×1)p2mg phase has developed, the PE current develops a minimum at zero degrees, whereas maxima of equal

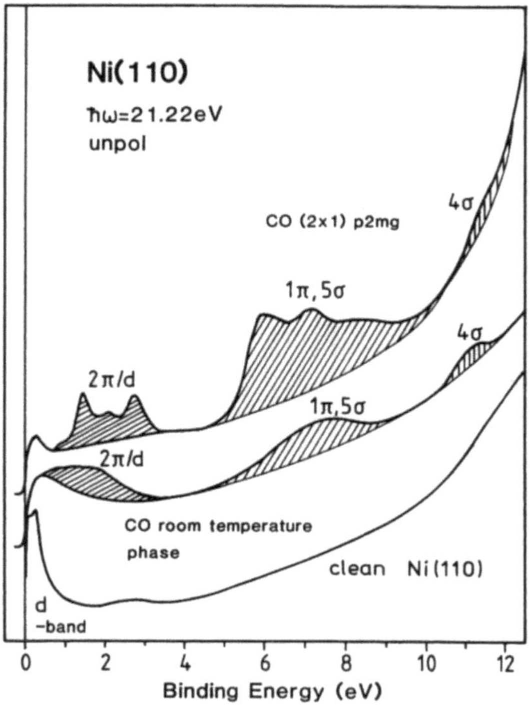

Fig. 7.48. PE spectra taken with He I radiation (21.2 eV) from a clean Ni(110) surface, a Ni(110) surface covered with CO molecules at room temperature, and a Ni(110) surface on which the CO(2×1)p2mg ordered structure has developed. In the spectra from the CO-covered surfaces, CO-derived features are visible, namely the 2π/d orbitals, the 1π, 5σ orbitals and the 4σ orbitals. In the data from the ordered adsorbate, the splittings caused by the dispersion are visible [7.150–7.152]

height show up at $\pm 20°$. This is evidence that there are an equal number of CO molecules tilted at $\pm 20°$ with respect to the surface normal in the [001] azimuth, in agreement with the structure given in Fig. 7.46.

PE spectra from a clean Ni(110) surface, a CO covered surface at room temperature (CO molecules oriented perpendicular to the surface normal), and a Ni(110) surface, on which at low temperature a (2×1)p2mg structure has developed, are compared with one another in Fig. 7.48 [7.152]. The data from the room temperature phase are similar to those shown in Figs. 5.95 and 5.96, displaying the 4σ orbital and the almost degenerate 1π and 5σ orbitals. In addition, this spectrum shows the orbital that develops upon interaction of the $2\pi^*$ (unoccupied) CO orbital and the 3d (occupied) Ni band, see Fig. 5.93.

In the ordered, low-temperature (2×1)p2mg structure, a two-dimensional band structure develops. This is reflected in the distinct splitting of the peaks, which, in the room-temperature disordered phase, showed only a considerable

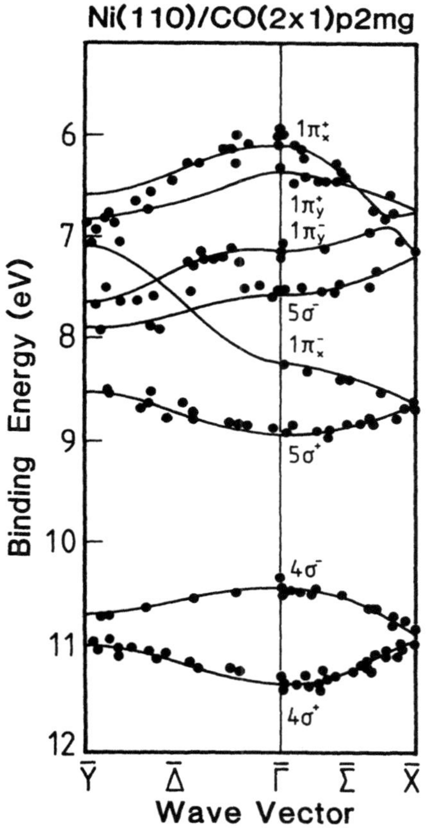

Fig. 7.49. Two-dimensional band structure of CO (2×1)p2mg on Ni(110) along two principle directions of the surface Brillouin zone (Fig. 7.40). The splittings observed at $\bar{\Gamma}$ between $4\sigma^+$ and $4\sigma^-$, $5\sigma^+$ and $5\sigma^-$, etc. are a result of the tilting of the two CO molecules in the unit cell in opposite directions along the [001] azimuth [7.150–7.152]

width. A mapping of this band structure in the two principal directions (see Fig. 7.46 for the surface Brillouin zone), namely along $\bar{\Sigma}$ (along $[1\bar{1}0]$) and $\bar{\Delta}$ (along [001]), is given in Fig. 7.49 [7.152]. The full lines are the result of a simple tight-binding calculation, which represents the data quite well (similar data have been obtained for the 2π/d-orbitals near to the Fermi energy).

The data in Fig. 7.49 allow a qualitative discussion. Each band that is derived from a singly degenerate molecular orbital must show a two-fold splitting at $\bar{\Gamma}$ because there are two inequivalent CO molecules per unit cell: this is indeed observed for the 4σ and 5σ bands where the additional labels + and − indicate the symmetry with respect to the glide plane along the $[1\bar{1}0]$ azimuth of the Ni(110) surface. Using the same reasoning, the 1π orbital shows a splitting into a $1\pi_x$ and $1\pi_y$ orbital, where each of these is further split because of the different symmetry with respect to the glide plane, resulting in four 1π bands.

Thus instead of two signals as visible in Fig. 5.96 or the middle panel of Fig. 7.48 (because of the near coincidence of the 1π and 5σ orbital) one observes eight in the ordered overlayer with the (2×1)p2mg structure.

We point out one additional detail. The splitting at $\bar{\Gamma}$ is 2 eV for the $1\pi_x$ bands and 0.6 eV for the $1\pi_y$ bands. This can be understood from the fact that along x the distance between adjacent molecules is 2.5 Å, whereas it is 3.5 Å along y, and the strength of the overlap integrals that determine the dispersion decreases strongly with intermolecular distance.

The PES data on CO/Ni(110) show that the intermolecular interactions in ordered overlayers can be considerable and that the interpretation of spectra from CO covered surfaces in terms of simple molecular orbitals can only be a first approximation. We mention that dispersion-related splittings for this system have also been observed in the BIS spectra of the $2\pi^*$ derived level (Chap. 9).

7.4.2 Three-Dimensional Solids: Metals and Semiconductors

To date, a considerable number of three-dimensional band structures have been explored by PES and have added much to our understanding of the electronic states in solids [7.96]. The most important conclusion that can be drawn from the material accumulated so far is that for metals and semiconductors there is surprisingly good general agreement between calculated band structures (using a single particle model) and experimental ones derived from measured EDCs, the most noteworthy exceptions so far being Ni and Na. What is surprising about this is that PES, as stated many times before, measures a final state with one electron missing, whereas the calculated (one-electron) band structures are those for the (initial) ground state.

In interpreting these findings we have to touch upon a subject which was already mentioned several times in this volume. Within the simple screening argument this good agreement between the calculated one-electron bands and the experimental EDCs means that the measuring process (with only few exceptions) involves a very effective screening, which results in a final state that is quite similar to the (initial) ground state.

Within the Green's function formalism (1.16) a good agreement between a measured EDC and a one-electron band structure means a dominant intensity in the coherent part of the spectral function and small electron–electron correlations. There rests the problem of the number of electrons: a neutral final state in the screening approach and a one hole state (one positive charge) in the Green's function formalism. Upon closer inspection one realizes that this difference is small. In a one-electron system, the ejection of an electron from a valence state does not change the energy of the other electrons. Therefore, with respect to the energy of the final state, the screening picture and the Green's function approach lead to the same result. One can also argue that the hole in a wide band one-electron system is very delocalized, which

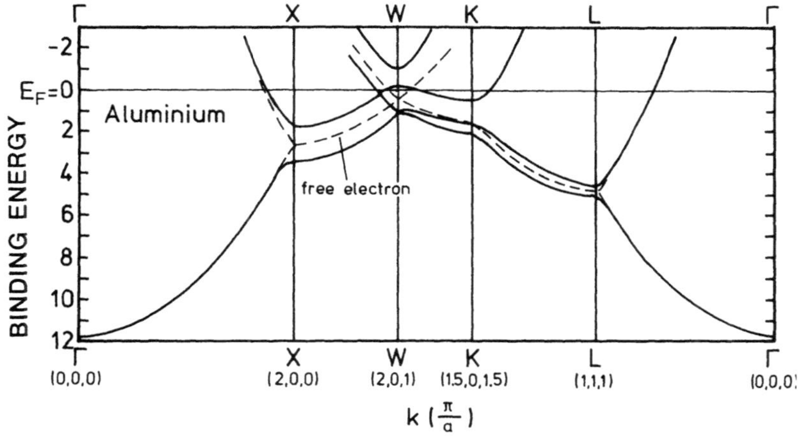

Fig. 7.50. Band structure of Al (fcc) [7.155]. The dashed line gives the free-electron approximation. The splitting of these dashed bands is an effect of the crystal potential

makes a system with all electrons and one with one electron removed almost identical.

Upon closer inspection of the experimental data it will be seen that there are differences between theory and experiment. It is surprising and somewhat disturbing that these disagreements are rather large in the simple metals (Table 4.5) [7.27, 7.153] where one would have expected that the understanding of the electronic structures is most advanced. In contrast, the agreement is observed to be very good for the filled d shell metals Cu, Ag and Au. For semiconductors, e.g., GaAs [7.154], the energy-band dispersion as observed by PES correlates quite well with calculated bands. However, the fact that measured *band gaps* are not easy to reproduce in band-structure calculations has been known for a long time.

In the following, a few selected examples are presented; these reflect to a large extent the preference of the author rather than a particular system.

Simple Metals. Although these materials constitute the backbone of solid state theory, their band structures have only recently been elucidated in detail [7.153]. We begin our discussion with Al [7.155] since the data for this material are the most detailed and complete.

(i) Aluminium. The occupied band structure of Al (fcc) is shown in Fig. 7.50, where the dashed line gives the free electron ("empty lattice") approximation ($m^* = m$ and $V_G \equiv 0$). The splitting in the "real bands" between XWKL as compared to the free electron bands is a manifestation of the crystal potential $V_G \neq 0$, and thus a deviation from free electron behavior [7.155].

The PES data for Al demonstrate particularly clearly the effects of the Brillouin zone boundary and are therefore presented here. The experimental

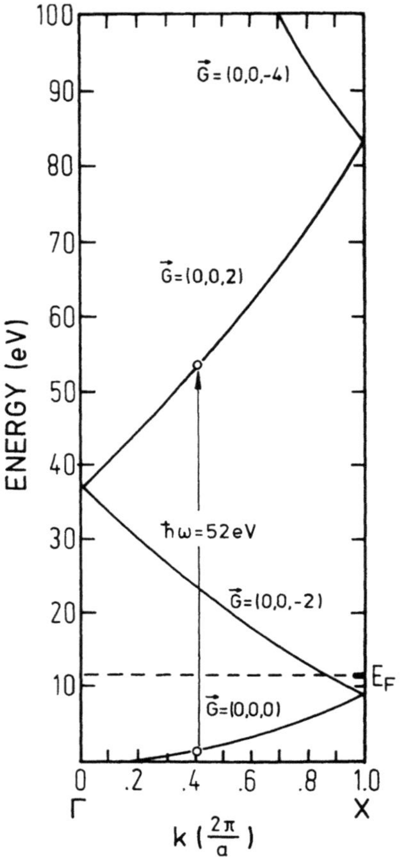

Fig. 7.51. Free-electron band structure of Al along the ΓX direction in the reduced zone scheme [7.155]. The $\boldsymbol{G}$ values are given. A radiative transition with $\hbar\omega = 52\,\mathrm{eV}$ is indicated

method applied is also a good example of what was called the Bragg plane method in Sects. 7.3.2 and 7.3.3.

The Brillouin zone of an fcc crystal was shown in Fig. 7.5. The schematic free electron band structure of Al along the ΓX direction in the reduced zone scheme is given in Fig. 7.51 [7.155–7.157]. The gaps have been omitted because in the present discussion they are of no importance. Transitions along ΓX occur in the experimental situation where one has a (001) crystal plane with the detector parallel to the surface normal ($\boldsymbol{K}_{||} = 0$ for every detected electron!). In the PES experiment energy and wave vector are conserved: $E_f - E_i = \hbar\omega$ and $\boldsymbol{K}_f = \boldsymbol{k}_i + \boldsymbol{G}$. The momentum conservation equation corresponds to the situation in an extended zone scheme. In the reduced zone scheme, as shown in Fig. 7.51, the wave vector of the final state is folded back into the first Brillouin zone via $\boldsymbol{k}_f = \boldsymbol{K}_f - \boldsymbol{G} = \boldsymbol{k}_i$ (in the reduced zone scheme we are observing a vertical transition!). By this construction, however, each band of the reduced zone scheme corresponds to a different $\boldsymbol{G}$. Under the

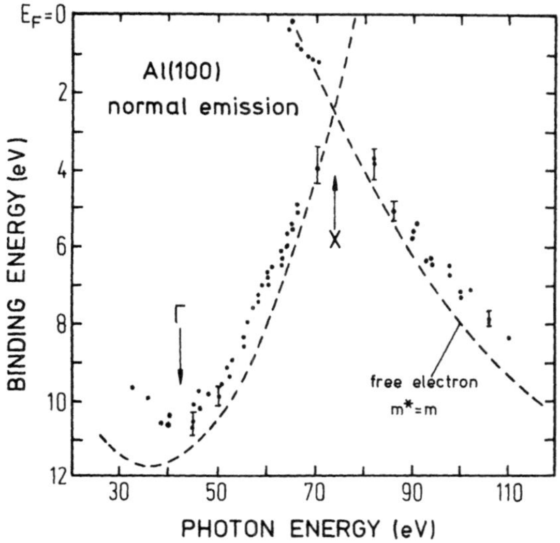

Fig. 7.52. Normal emission peak energies of a PE measurement on Al(100) [7.155] as a function of the photon energy $\hbar\omega$. The dashed curve indicates the initial state energies expected for a free-electron gas ($m = m^*$), with the experimental electron density of Al used to adjust the position of E_{F}

assumed conditions of electron detection normal to a (001)-surface ($\boldsymbol{k}||[001]$), the first band has $(2\pi/a)\ (0,0,-1) \leq \boldsymbol{k} \leq 2\pi/a\ (0,0,+1)$. The second band has $\boldsymbol{K} = \boldsymbol{k} - (2\pi/a)\ (0,0,2)$, the third band has $\boldsymbol{K} = \boldsymbol{k} + (2\pi/a)\ (0,0,2)$, etc. and these are the bands shown in Fig. 7.51. They are selected such that they always satisfy the condition $k_z \neq 0$; k_x, $k_y = 0$. There are, of course, also other $\boldsymbol{G}$ values that can be used to fold a band from the extended zone scheme back into the first zone along ΓX. However, these represent electrons traveling off-normal and therefore cannot be detected under the chosen experimental conditions.

Figure 7.51 also shows what happens if one exposes the crystal to photons of $\hbar\omega = 52\,\mathrm{eV}$. The requirement that $\boldsymbol{k}_i = \boldsymbol{k}_f$ (in the reduced zone scheme) means that one is performing a vertical transition, and since the initial state has to be occupied, the only transition possible is the one indicated in the figure.

Assume now that the photon energy is reduced. The possible transitions then move nearer to Γ and the initial-state energy, as measured in the PES experiment will decrease. At $\hbar\omega \simeq 37\,\mathrm{eV}$ the center of the Brillouin zone, Γ, is reached, and if the photon energy is reduced further, the initial states energy will *increase* again. A similar behavior will occur around $\hbar\omega = 84\,\mathrm{eV}$, where the X-point is reached. The actual data for the initial states along the ΓX direction are shown in Fig. 7.52. Unfortunately, measurements for $70\,\mathrm{eV} \leq \hbar\omega \leq 89\,\mathrm{eV}$ are hard to evaluate because the 2p binding energy of Al is 72 eV, and the Auger electrons filling the 2p holes interfere with the genuine valence-band structures produced by direct PE. Nevertheless, the observed behavior of the initial state energy as a function of the photon energy is in agreement with that anticipated from Fig. 7.51 and also roughly in agreement with the free-electron energy bands.

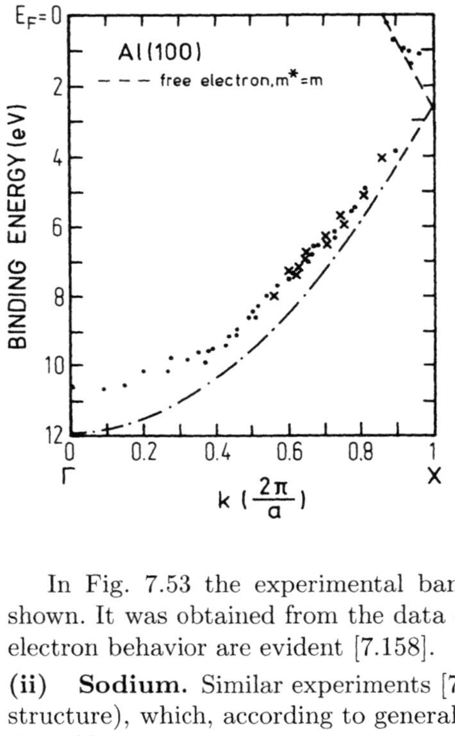

Fig. 7.53. Experimental band structure of Al along ΓX obtained from the data of Fig. 7.52 [7.155]. The free-electron band is also indicated

In Fig. 7.53 the experimental band structure for the ΓX direction is shown. It was obtained from the data of Fig. 7.52. The deviations from free electron behavior are evident [7.158].

(ii) Sodium. Similar experiments [7.27] have been performed for Na (bcc structure), which, according to general belief, is even more free-electron-like than Al.

Figure 7.54 shows the experimental band structure in the ΓN direction as derived from the data and compares it with the free electron result. A free electron-type calculation with $m^* = 1.28\,m$ gives a very good agreement with the experimental data, which again shows the considerable deviation of the experimental data from the free-electron model.

The experiments have been repeated with better accuracy leading to essentially the same result [7.159], namely an effective mass of $m^* = 1.23\,m$ or a bandwidth of 2.65 ± 0.05 eV. These authors also interpret the observed band narrowing via a calculation of the self-energy that goes beyond the random phase approximation. Another approach tries to explain the anomalous PE data of Na (and other free electron metals) by taking surface and lifetime effects into account [7.160]. One might also conjecture that the electron–hole coupling, which is especially large in Na, could be at least partially responsible for the observed discrepancy (Chap. 4).

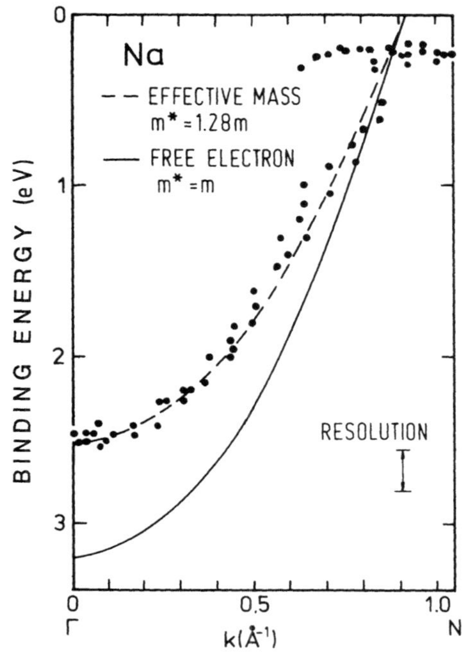

Fig. 7.54. Experimental band structure derived from PES data of Na along the ΓN direction (Na is bcc; ΓN corresponds to the [110] direction in real space). Also given are the free electron bands for $m^* = m$ (*full curve*) and $m^* = 1.28m$ (*dashed curve*) [7.159]

The Magnetic Transition Metal Nickel

The energy dispersion curves measured so far for Ni are given in Fig. 7.55 [7.161]. The full lines in Fig. 7.55 are drawn to connect the experimental points.

For Ni it now seems established that the measured width of the d bands is about 30 % smaller than the calculated one. The same statement holds for the exchange splitting. This exchange splitting can and has been obtained either by PE experiments with polarized light or by detecting the spin of the photoemitted electrons. Both methods arrive at the same results (Chap. 10).

The electronic structure of nickel metal has been investigated by the Fermi surface mapping method [7.162], described in detail in Chap. 11. In short this method consists of scanning the photoemission intensity over the 2π geometry above a surface plane, using an energy window closely confined around the Fermi energy. In a three-dimensional solid, this in principle produces a cut through the Fermi surface and a number of such cuts can be used to construct the complete Fermi surface of a particular material. The experiment described here has been performed on a (100) surface of Ni metal. The transitions possible in the Brillouin zone with He I radiation are indicated in Fig. 7.56, where the final state for excitation with He I radiation is approximated by a free electron parabola resulting here in a circle around the Γ-point. A hemispherical electron distribution for this geometry with the energy

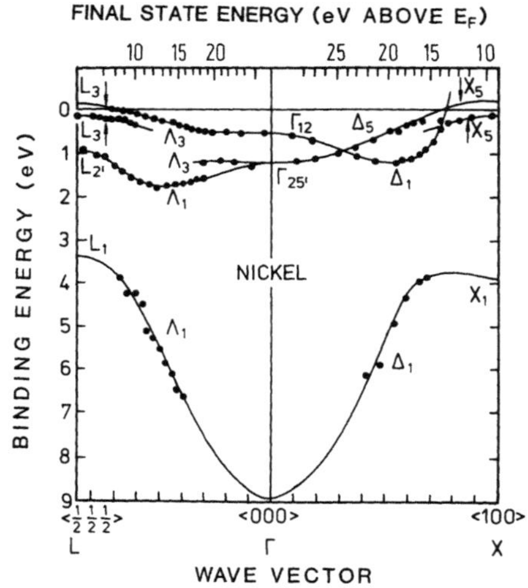

Fig. 7.55. Experimental band structure of Ni along ΓL and ΓX. The analysis of the data was performed with a semiempirical free electron final state, for which the energies below E_{F} are shown. The exchange splitting is resolved near the L point. Full curves are drawn to connect the data points [7.24, 7.161]

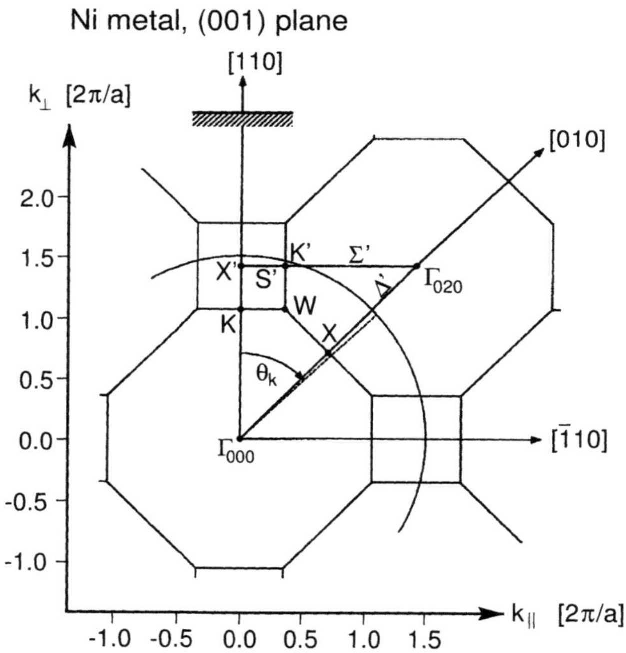

Fig. 7.56. View of the (001) plane in reciprocal space indicating the Brillouin zone boundaries for nickel in the extended zone scheme. High symmetry points and lines are indicated, as is the (110) surface normal. The circle represents the free electron final-state wave vector for emission from the Fermi edge at a photon energy at 21.2 eV. The full range of polar emission angle Θ_K is indicated by the curved arrow [7.162]

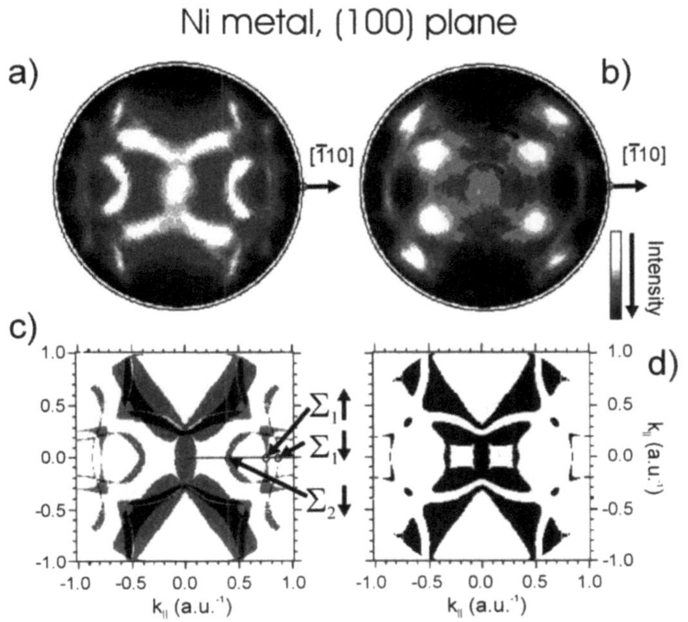

Fig. 7.57. (**a**) Room temperature ($T/T_c \approx 0.48$) mapping of the intensity of He I excited photoelectrons collected within an energy window of about 30 meV centered at E_F. The plot is linear in $k_{||}$. The outer circle indicates an emission angle of 90°. The center represents normal emission or $k_{||} = 0$. Intensities are given in a linear scale as indicated. (**b**) Same as (**a**) but for $T/T_c \approx 1.1$ meaning above the Curie temperature. (**c**) Calculated cut through the bulk Fermi surface using the spin polarized layer-Korringa-Kohn-Rostoker (LKKR) scheme for the initial state and a free electron final state. (**d**) LKKR calculation, unpolarized [7.162]

window around the Fermi energy is shown in Fig. 7.57 for room temperature (below T_c) and for a temperature slightly above T_c. While Fig. 7.57a,b gives the experimental results, Fig. 7.57c,d shows corresponding theoretical Fermi surface electron distributions. The agreement between theory and experiment is by no means perfect, but it can at least be seen that passing through the transition temperature has a pronounced effect on the energy distribution, as expected, although at this point it is not clear how the various bands behave in detail as a function of temperature. Looking at a cut along (110), the resulting spectrum is a function of angle (momentum) at the Fermi energy. Such data at and below T_c are shown in Fig. 7.58. It can be seen that the sp band collapses upon reaching the transition temperature. Finally, Fig. 7.59 displays a cut through the Fermi surface accessible with the present geometry (see Fig. 7.5) with the Fermi surface contours as measured by the de Haas–van Alphen effect, and with ellipsoids displaying the photoemission results. It can be seen that there is basic agreement between the two datasets. However,

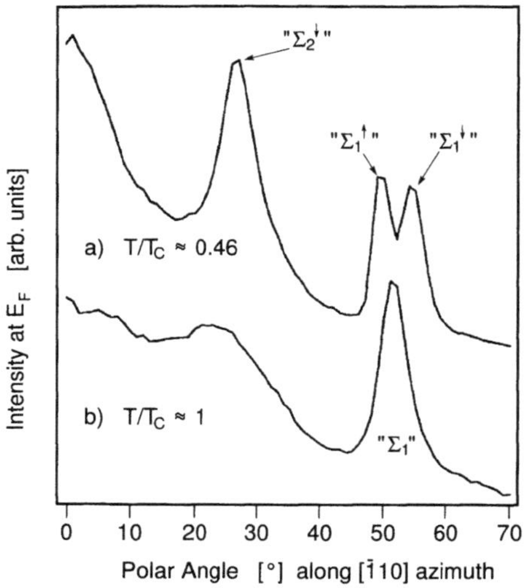

Fig. 7.58. (a) Polar scan through the room temperature intensity map of Fig. 7.56 along the (110) azimuth. (b) Same as in (a) but for $T/T_c \approx 1$, i.e. above the Curie temperature [7.162]

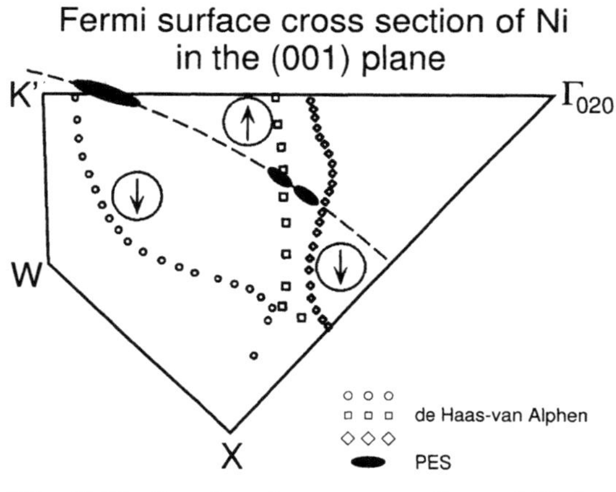

Fig. 7.59. Fermi surface cross sections of Ni in the (100) plane obtained from de Haas–van Alphen experiments (*open symbols*) compared to the data extracted from Fig. 7.57 (*full ellipses*) [7.162]

the accuracy in the de Haas–van Alphen experiment is better than that of the photoemission experiment. Further improvements in PES will probably lead to its achieving an accuracy similar to that of the de Haas–van Alphen experiment (see Chap. 11 for further details).

The Semiconductor Gallium Arsenide

Of the three best-known semiconductors Si, Ge, GaAs (for complete electronic-structure data of these three semiconductors, see [7.96]), we show results for

GaAs [7.154]; they are compared with a theoretical band structure [7.163] in Fig. 7.60. In this case the data were analyzed by assuming a free electron final state. The experimental points above the valence-band maximum at Γ_{15} (energy zero) were obtained in an IPES experiment [7.164].

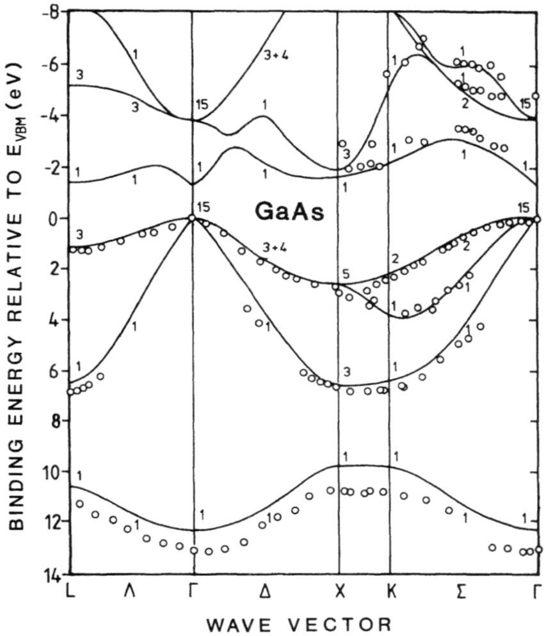

Fig. 7.60. Band structure of GaAs derived from PES (or IPES above E_F) data with the use of a free electron final state [7.154]. The solid curves are the result of a calculation [7.163]. Energy zero is referred to the top of the valence band at Γ_{15} (E_{VBM}). We use binding energies, yielding negative numbers above E_{VBM}

Note that the experimental valence-band and conduction-band edges were adjusted to the theoretical ones in order to circumvent the problem that the theoretical gap deviates from the experimental one.

7.4.3 UPS Band Structures and XPS Density of States

The measurement of valence bands using XPS will not be dealt with at great length here because this topic is implicitly contained in the material presented on the band-structure determination by ARUPS. It seems worthwhile, however, to compare the data obtained by ARUPS from single crystals with the XPS data from polycrystalline samples. Of course one can also perform XPS experiments on single crystals. However, in our judgement, the additional information obtained from a single crystalline sample does not justify the extra effort in crystal preparation and measuring time (except for the case of photoelectron diffraction, see Chap. 11).

A UPS experiment, when suitably performed, measures the $E(\boldsymbol{k})$ relation of the one-hole state. The amazing feature of the data presented in the preceding sections is the remarkably good agreement between these one-hole state data and the calculations performed for the effective single-particle energies in the ground state. (This statement is not intended to play down the deviations found, e.g., in Ni and Na!) As "good" agreement we define deviations of $\leq 1\,\mathrm{eV}$. This may seem a generous limit, and upon inspecting the data in the preceding sections we see that, over wide ranges of the Brillouin zone, the agreement is significantly better. This good agreement can only mean that the electron–electron interactions are small, a result that one might not have anticipated.

An exact statement of what an XPS experiment on a polycrystalline sample actually measures is hard to give. In principle, the EDC from this type of experiment provides a density of states of the material, since the $\boldsymbol{k}$ selection rules are averaged over a 4π solid angle and are effectively "lifted". However, a density of states usually contains contributions from electrons with various angular momenta ℓ, and these electrons have different photoelectron cross-sections. Thus the XPS spectrum is actually a sum of the ℓ-projected densities of states weighted by their respective transition probabilities into a plane-wave final state. This makes it difficult to analyze an XPS spectrum using the approach applied to EDCs from ARUPS data; for an exact deconvolution with respect to the different ℓ contributions the data do not contain enough information. Progress can only be made by using calculated densities of states, in which the various ℓ components are weighted properly and by comparing these with a measured XPS spectrum. This procedure leads, as shown below, to quite satisfying results.

d Band Metals and Alloys

In the case of d band metals or alloys, the density of states (and the XPS spectrum) is dominated by the d electron contribution ($\ell = 2$), which is also the contribution of greatest interest. It is then sufficient to use the total calculated density of states for the comparison with an experimental XPS spectrum. In this restricted sense, one can say that XPS experiments measure densities of states. The following examples are intended to demonstrate the degree of validity of this approximation [7.165, 7.166].

Figure 7.61 shows an XPS spectrum of Cu obtained using monochromatized Al-K_α radiation (resolution $0.6\,\mathrm{eV}$) with background subtraction [7.30]. One can recognize the features anticipated from the band structure (Fig. 7.7). At the Fermi energy, a weak emission is seen, which is due to the 4s band. At about $1.5\,\mathrm{eV}$ the d-emission starts to show up, with maximum intensity being reached at $\sim 2\,\mathrm{eV}$ below E_F. Further structure follows and the strong d-intensity terminates at $\sim 5.5\,\mathrm{eV}$ below E_F. Thus one sees the d electron den-

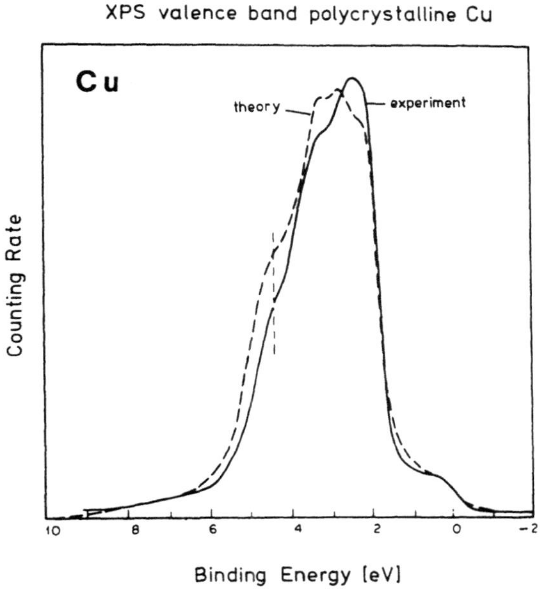

Fig. 7.61. XPS spectrum of Cu (undeconvoluted data, after only a background subtraction) and calculated density of states [7.19], adjusted in height to match the measured spectrum [7.30]. The calculated density of states was convoluted with a 0.5 eV wide instrumental function

sity of states exactly where it is expected from the band structure (Fig. 7.7). The bottom of the s band cannot be detected in these data.

Such a spectrum naturally contains much less information than a series of EDCs taken at various photon energies and angles with UV radiation; it can, however, be obtained with much greater ease.

The measured XPS spectrum is compared in Fig. 7.61 with a density of states calculated from Burdick's [7.19] band structure and broadened with a 0.5 eV Lorentzian to take lifetime effects and resolution into account. The calculated band structure is in good agreement with the experimental band structure (Fig. 7.17) and deviates from that of Eckart et al. [7.53], which was also given in Fig. 7.17, by only a few tenths of an eV. There is good agreement between theory and experiment in Fig. 7.61, and the only discrepancy is the too low intensity at the top of the d band in the calculated spectrum; this will be commented upon shortly.

One can, of course, reverse the procedure and use XPS densities of states to check band structure calculations. The good agreement found for example in the case of Cu, long before any dispersion relation was measured, gave a good indication that the band structure of Burdick [7.19] was essentially "correct". This was later confirmed by actual dispersion relation measurements. It was also XPS density of states that led to the discovery of band narrowing in Ni [7.167], although for this material UPS measurements on polycrystalline samples [7.28], which measure a EDJDOS, see (6.45, 6.49), had already suggested this band narrowing.

One can now go a step further and try to understand the intensity differences between the measured and calculated density of states. The most likely source of these differences is the influence of the surface. To investigate this Steiner et al. [7.168] calculated the density of states of a five layer slab of Cu(001). From this they derived the individual layer densities of states, in order to see the effect of the surface (Fig. 7.62). One sees that the surface-layer density of states (S) is, as expected, narrowed and has its intensity maximum near the top of the d band. The density of states in the first layer below the

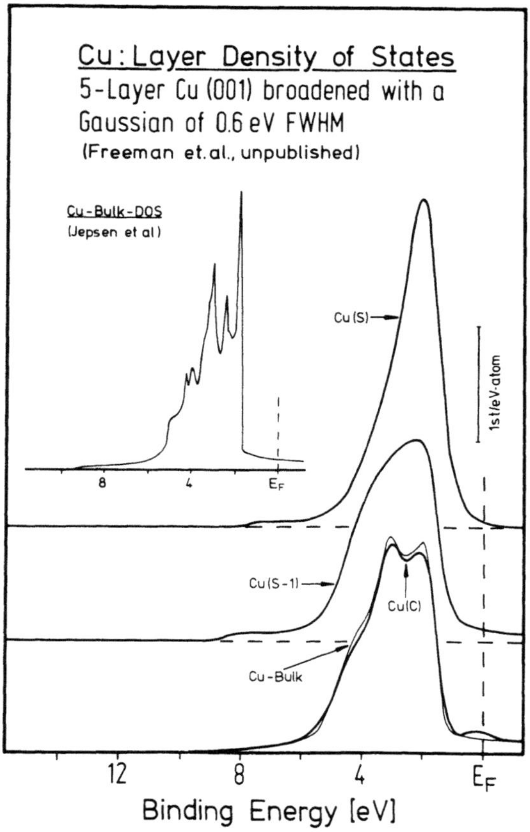

Fig. 7.62. Layer densities of states for Cu [7.168]. A 5-layer sample was used: S stands for surface, S-1 for the first layer beyond the surface layer and C for the center layer of the 5-layer sample. The center-layer density of states and a calculated bulk density of states [7.100] agree quite well with each other, indicating the short penetration depth of surface effects in metals

surface (S-1) is already close to the bulk density of states, and the center layer (C) has a density of states very close to the bulk one, as can be seen by a comparison to the broadened density of states calculated by Jepsen et al. [7.100]. Since an XPS experiment samples only a few layers below the surface, the band narrowing in the surface layer will clearly show up in the measured spectrum.

The sampling depth in an XPS experiment can be varied by varying the electron escape angle and the results of such an experiment [7.169] are presented in Fig. 7.63. The calculated curves in this figure have been obtained by adding up surface layer intensity (S), subsurface layer intensity (S-1) and center (C) layer intensity weighted with a mean escape depth $\lambda = 15\,\text{Å}$. The effective escape depth is $\lambda_{\text{eff}} = \lambda \cos\vartheta$, where ϑ is the angle between the electron detection direction and the surface normal (yielding $\lambda_{\text{eff}} = \lambda$ at $\vartheta = 90°$ and $\lambda_{\text{eff}} = \lambda$ at $\vartheta = 0°$). The appropriately weighted intensities are then calculated as

$$\begin{aligned} I_{0°} &= 0.13 I_{\text{S}} + 0.12 I_{\text{S}-1} + I_{\text{C}} \\ I_{80°} &= 1.44 I_{\text{S}} + 0.80 I_{\text{S}-1} + I_{\text{C}} \,. \end{aligned}$$

The agreement between theory and experiment [7.169] as shown in Fig. 7.63 is good, and demonstrates that even the details of XPS spectra are understood. However, to judge the quality of a calculated density of states, the investigation of such details is seldom necessary.

The value of XPS data is greatly increased in cases where single crystal specimens are not available (or are difficult to prepare) and where the band

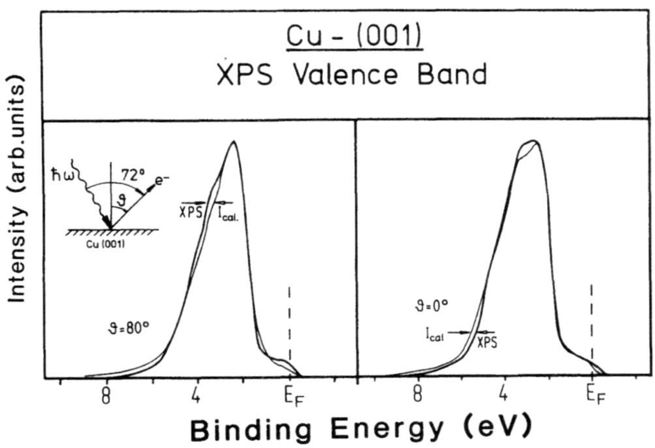

Fig. 7.63. Calculated density of states of Cu(001) (*thin curve*) and measured XPS spectrum (*solid curve*) for grazing electron emission angle ($\vartheta = 80°$) and for normal electron emission ($\vartheta = 0°$) [7.168,7.169]. The calculated curve was obtained by using an electron escape depth and calculating a superposition of the layer densities of states as given in Fig. 7.62 with the appropriate weighting factors

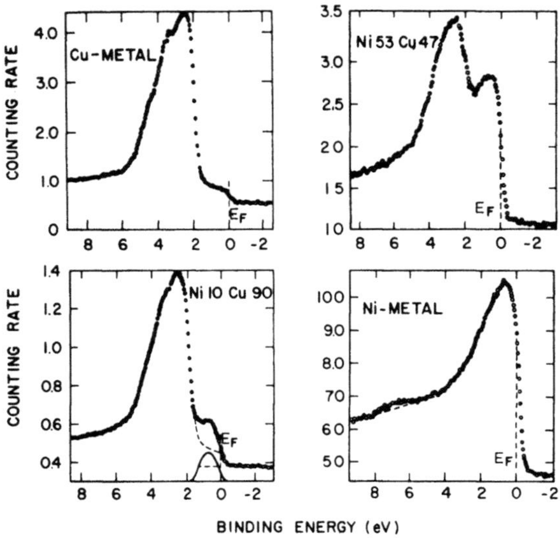

Fig. 7.64. XPS spectra (raw data) of polycrystalline samples of Cu, $Cu_{47}Ni_{53}$, $Cu_{90}Ni_{10}$ and Ni [7.165]. The data clearly show that the d bands of the constituents survive the alloying, although increasing dilution leads to a narrowing, which finally, at very small concentration, produces the virtual bound state

structures are complicated or not yet calculated. This is true for ordered alloys, intermetallic compounds and disordered alloys. Here the XPS technique has proven very valuable. We therefore demonstrate with a few examples, the kind of information that can be obtained. Figure 7.64 shows two XPS spectra (raw data) of CuNi alloys [7.30, 7.165] together with the spectra of the pure metals. The data indicate that even in the alloy the d-bands of the constituents remain roughly intact and there is no formation of a common d band that is gradually filled up or emptied as one or the other constituent is added. Thus a rigid band model for the electronic structures of such alloys or intermetallic compounds can definitely be ruled out on the basis of these data (and many others). Of particular interest is the spectrum of the $Cu_{90}Ni_{10}$ alloy.[5] In this system the Ni 3d electrons are located in the area of the flat sp band of the Cu host and form a "virtual bound state" with a Lorentzian-shaped distribution of binding energy [7.171–7.174]. The expression "virtual bound state" derives from the fact that the d level interacts with the sp band of the host, such that an electron in the d level can be scattered in and out of this level, making it only virtually bound. A well-developed theory exists to describe this situation and thus further experimental data are of great interest.

The theoretical model considers a d metal impurity dissolved in a host with an sp band to form a virtual bound state. An experimental case which

[5] The situation in Pd_5Ag_{95} is very similar [7.30, 7.170].

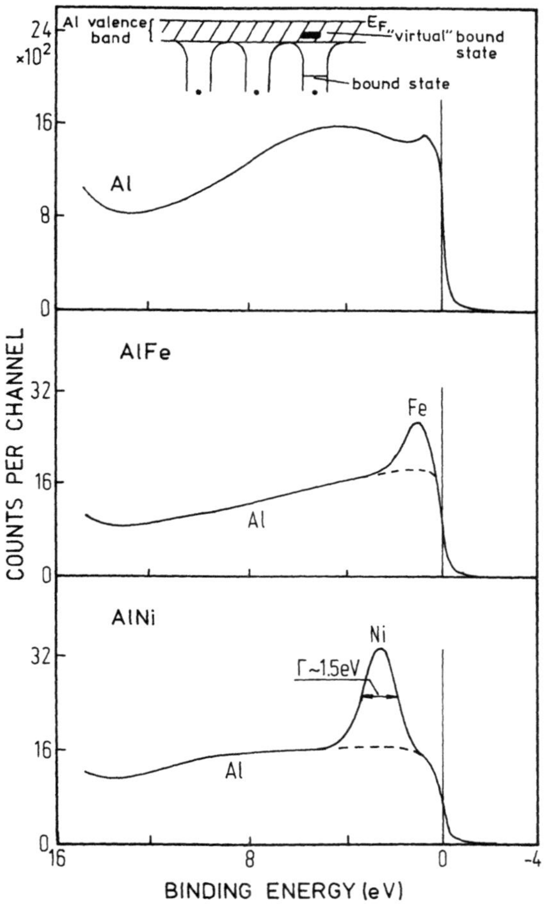

Fig. 7.65. XPS spectra of Al metal (*top*) and of dilute alloys of Fe and Ni in Al metal [7.176]. The d metal impurities lead to the formation of a narrow resonance (virtual bound state)

comes close to this theoretical picture arises for 3d metal atoms dissolved in Al. XPS data for these systems are shown in Fig. 7.65 and Fig. 7.66 [7.175–7.177]. Figure 7.65 shows raw data for the host metal Al and for two typical alloys, namely NiAl and FeAl. In the alloys one can clearly discern the additional intensity produced by dissolving the d metal in the host at very low concentrations. Figure 7.66 shows a compilation of difference spectra (the alloy spectrum minus the host metal spectrum), giving a reasonable representation of the virtual bound state density of states. One sees, as expected, that the virtual bound state shifts away from the Fermi energy as the number of d electrons increases.

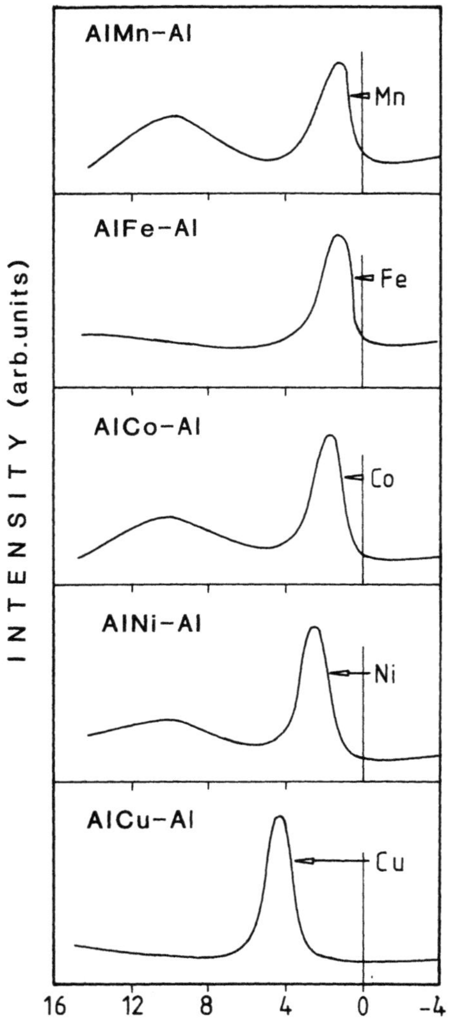

Fig. 7.66. XPS difference spectra (alloy spectrum minus pure Al spectrum) for Mn, Fe, Co, Ni, Cu as (dilute) impurities in Al

Finally one should address the problem of data analysis in this area [7.178]. In order to extract the virtual bound state shapes from the dilute aluminium alloys the pure Al spectrum had to be subtracted from the alloy spectrum. This produced quite satisfactory results. However, such a subtraction procedure becomes uncertain in the case of a d metal host. Upon alloying, the host band structure may now also change appreciably making a simple subtraction meaningless. An example is given in Fig. 7.67, where the XPS spectrum of pure Au is compared with that of $Au_{85}Zn_{15}$. Zn has a wide sp band and the d band is at a binding energy of 10 eV, such that the Au and the Zn d band structures do not interfere. Nevertheless, there are changes in the

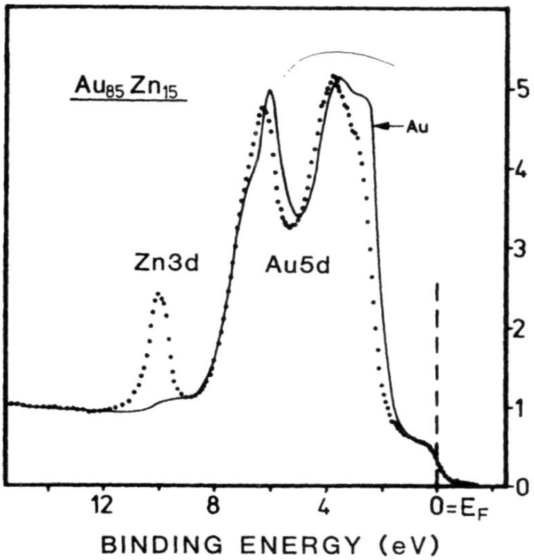

Fig. 7.67. XPS valence-band spectrum of polycrystalline $Zn_{15}Au_{85}$ (*dotted line*) and of polycrystalline Au metal (*full line*). The Zn 3d band is at $E_B = 10\,eV$, and well resolved from the Au 5d band which is significantly changed by the alloying process [7.176]

Au d band part of the spectrum on alloying, which are due to the distortion of the lattice periodicity brought about by the replacement of 15 % of the Au atoms by Zn atoms.

It will now be shown how, under such circumstances, useful experiments can nonetheless be performed in order to recover the impurity contribution to a spectrum. As an example, we choose data on Ni dissolved in Au. The task is to obtain the Ni contribution from the spectra of the NiAu alloys.

If one forms, e.g., the alloy of $Au_{90}Ni_{10}$ and wants to recover the Ni d band structure (virtual bound state) one cannot simply subtract the XPS spectrum of pure Au from that of $Au_{90}Ni_{10}$. Rather one has to perform the subtraction with a system that has the same lattice distortion as $Au_{90}Ni_{10}$ but no d electrons that cause added "unwanted" structures. An example is the alloy of $Au_{90}Zn_{10}$. Zn and Ni have similar radii and the d states of Zn are at an energy, where they do not interfere either with those of Au or those of Ni. The result of such an experiment is given in Fig. 7.68. One can now clearly see the contribution of the Ni 3d electrons to the measured spectrum and subtract it appropriately.

The final example concerns the investigation of a dilute alloy which exhibits a magnetic splitting [7.174]. In an alloy like dilute Cu in Al (Fig. 7.66) the Cu ions behave non-magnetically (having a filled d^{10} configuration). For a dilute alloy in which the transition metal ion has a non-filled d band and the host is an sp band metal or a noble metal, the impurity can be non-

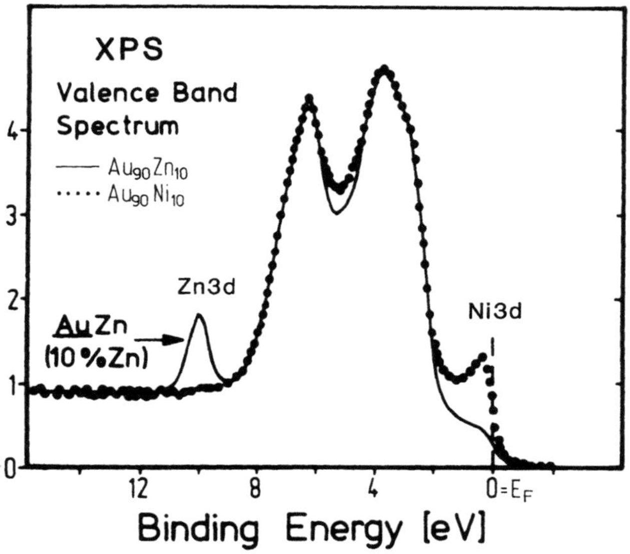

Fig. 7.68. XPS valence-band spectra of polycrystalline samples of $Zn_{10}Au_{90}$ and $Ni_{10}Au_{90}$. The 3d bands of Ni and Zn are well separated from each other. In contrast to the situation in Fig. 7.67 the Au 5d bands for the two alloys are now quite similar allowing a subtraction procedure [7.178]

magnetic or magnetic, where "magnetic" is defined by the observation of a Curie–Weiss law at (not too) low temperatures. Classical examples of magnetic alloys are FeCu and MnAg. For such systems Anderson [7.174] derived a simple model which explains the magnetic behavior. In essence it considers an exchange interaction between the spin-up and spin-down electrons in the virtual bound state formed by a dilute d impurity in an sp band or noble metal. This exchange energy pushes the spin-down state above the spin-up state, such that it comes near to or above the Fermi energy. The spin-down state is then less populated (minority state) than the spin-up state (majority state), resulting in a net magnetic moment of the d impurity state.

A combined PES/BIS spectrum of the $Ag_{95}Mn_5$ alloy system, which from other data is known to exhibit a magnetic moment due to the Mn impurities, is given in Fig. 7.69 [7.179]; see also [7.180]. The spectra of Ag metal and the $Ag_{95}Mn_5$ alloy are not very different; however, the difference spectrum of the alloy and the metal brings out clearly the Mn states at $E_B \simeq +3\,eV$ (below E_F) for the spin-up state and $E_B \simeq -2\,eV$ (above E_F) for the spin-down state. (The "overshoots" are common features in difference spectra like the one shown; they arise because the Mn ions in the alloy change the d band structure of the host as compared to the pure metal. For a possibility to avoid this effect see Fig. 7.68).

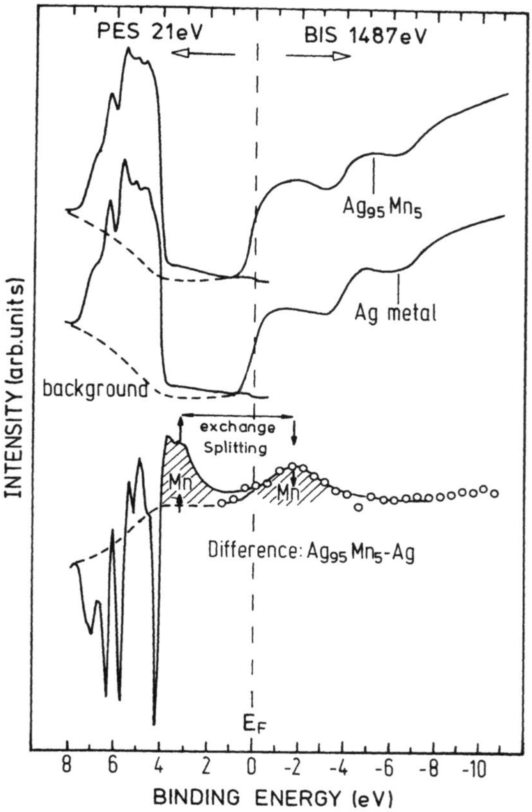

Fig. 7.69. PES (21 eV) and BIS (1487 eV) spectra of a $Ag_{95}Mn_5$ alloy and Ag metal. The difference spectrum gives the contribution from the Mn impurities, showing the spin splitting of the virtual bound state of Mn in Ag. The additional structure for binding energies larger than 4 eV comes from the fact that the alloying also slightly changes the host d band in $Ag_{95}Mn_5$ [7.179]

The separation of the two peaks gives the exchange splitting of the dilute Mn impurity in Ag. The magnetic moment calculated from these data is 4.7 μ_B, in good agreement with the results from magnetic measurements. These results verify an older experiment [7.181,7.182] where optical measurements, employed to probe the spin-down state above the Fermi energy, were combined with PES investigations.

Semiconductors

As was the case for metals, XPS and UPS from polycrystalline semiconductors measure essentially the density of states. Such work has helped to extend our knowledge of their electronic structure [7.183–7.186]. The "classic" semiconductors are those with 4 valence electrons per atom in an sp^3

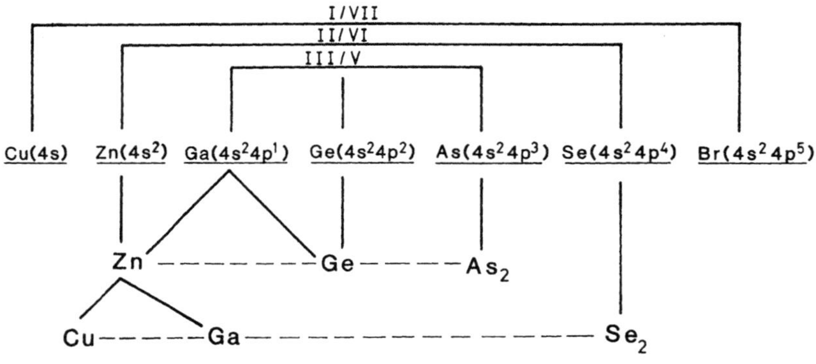

Fig. 7.70. Atomic electronic structure of the elements Cu to Br indicating the generation of IV/IV, III/V, II/VI and I/VII semiconductors

configuration. The construction of the valence electronic structure of these semiconductors can be seen from the diagram in Fig. 7.70 which gives the atomic electronic structure of the elements Cu through Br.

Two Ge atoms ($4s^24p^2$) can combine their electrons to form a closed-shell s^2p^6 configuration, i.e., an sp^3 configuration for each spin direction. Equally Ga ($4s^24p$) and As ($4s^24p^3$) can combine their valence electrons

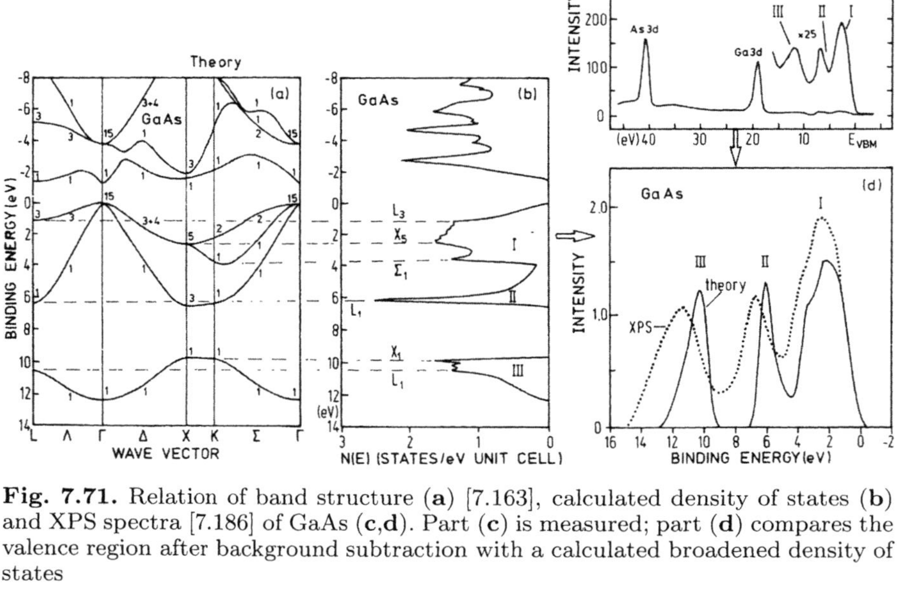

Fig. 7.71. Relation of band structure (**a**) [7.163], calculated density of states (**b**) and XPS spectra [7.186] of GaAs (**c**,**d**). Part (**c**) is measured; part (**d**) compares the valence region after background subtraction with a calculated broadened density of states

to give an s^2p^6 configuration, resulting again in an sp^3 configuration for each spin direction. In the same way in $CuGaSe_2$ four sp^3 configurations are formed.

The band structure of a typical semiconductor of this class (GaAs), without spin–orbit interaction, is shown in Fig. 7.71a [7.163]. This band structure in its general form is the same for all semiconductors that have four valence electrons per atom and that crystallize in the zinc-blende lattice. Only in cases where the d electrons also contribute to the valence bands (e.g., CuBr) or in cases of different crystal lattices (e.g., in wurtzite structures such as CdS) or in the ternary semiconductors, are there additional bands, these, however, will be ignored in what follows. The density of states of GaAs, which is shown in Fig. 7.71b, has a typical three-peaked structure. It is easy to attribute the three peaks I, II and III to distinct parts of the band structure: I represents mostly p states, II sp-hybridized states, and III mostly s states. Figure 7.71c gives an XPS spectrum of GaAs, displaying the three-peak structure anticipated from theory. In Fig. 7.71d a comparison of theory and experiment is presented [7.186].

7.5 A Comment

At the end of this Chapter it is reiterated that, with respect to the determination of the electronic structure, PES has produced important and unique information. This point is discussed with the help of Fig. 7.72. This figure shows for Cu metal the crystal structure (a), the Brillouin zone (b), the Fermi surface (c), the phonon dispersion curves (d) and the electron dispersion curves (e). The crystal-structure determination (by X-rays or neutrons) gives the position of the atoms in the crystal (and the force constants between them, because of the known atomic potentials). The Brillouin zone (b) is calculated from the crystallographic unit cell (a). Now we have two sorts of excitations that must reflect the symmetry of the crystal: the elastic excitations (phonon-dispersion curves (d)) and electron excitations (electron-dispersion curves (e)). The phonon dispersion curves of solids have been measured by neutron scattering, and data on a wide range of materials are available. Until about 1980, with respect to the electron dispersion curves for solids, they were mostly known at one particular energy, namely the Fermi energy. This energy distribution is called the Fermi surface (c). The complete electron wave vector dispersion curves (e) for a number of materials have been measured since 1980 [7.96]. These data have filled a gap in our understanding of solids, and in this area PES has yielded important new information which obviously could not have been obtained by any other experimental method.

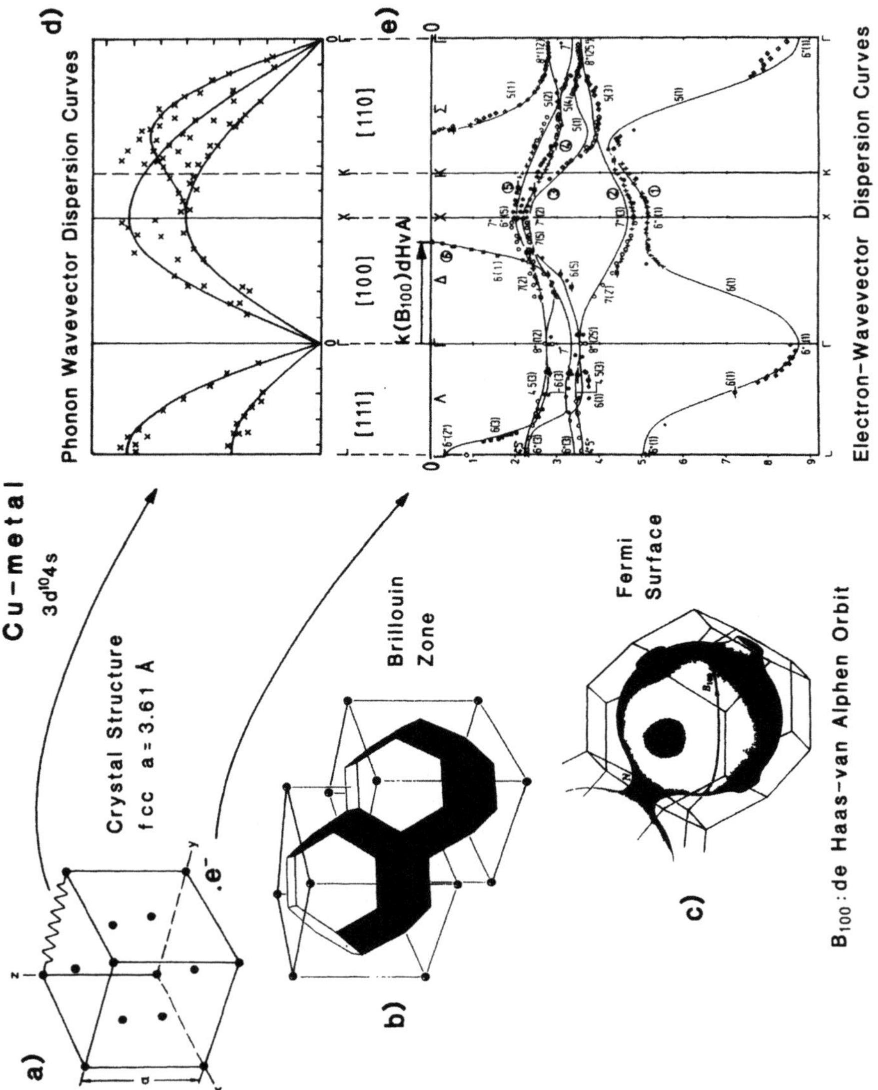

Fig. 7.72. Electrons and phonons in Cu metal: (**a**) fcc crystal structure, (**b**) Brillouin zone of the fcc crystal structure, (**c**) Fermi surface of Cu metal, (**d**) phonon dispersion curves for Cu metal [7.188], (**e**) electron-dispersion curves for Cu metal [7.187]

References

7.1 For a recent comprehensive treatment of the field of angle-resolved PES see S.D. Kevan (ed.): *Angle-Resolved Photoemission, Theory and Current Applications*, Studies in Surf. Sci. Catal., Vol.74 (Elsevier, Amsterdam 1992); a discussion of the relation between ground-state and excited-state band structures via PES can be found in R.W. Godby, R.J. Needs: Physics Scripta T **31**, 227 (1990)

7.2 J. Stöhr, P.S. Wehner, R.S. Williams, G. Apai, D.A. Shirley: Phys. Rev. B **17**, 587 (1978)

7.3 P. Thiry, D. Chandesris, J. Le Cante, G. Guillot, P. Pinchaux, Y. Petroff: Phys. Rev. Lett. **43**, 82 (1979)
Y. Petroff, P. Thiry: Appl. Optics **19**, 3957 (1980)

7.4 P. Thiry: La photoémission angulaire dans les solides. Thêse d'État, Université Paris VI, 1979

7.5 R. Courths, H. Schulz, S. Hüfner: Solid State Commun. **29**, 667 (1979)

7.6 R. Courths, S. Hüfner, H. Schulz: Z. Phys. B **35**, 107 (1979)

7.7 D.E. Eastman, J.A. Knapp, F.J. Himpsel: Phys. Rev. Lett. **41**, 825 (1978)

7.8 J.A. Knapp, F.J. Himpsel, D.E. Eastman: Phys. Rev. B **19**, 4952 (1979)

7.9 E. Dietz, F.J. Himpsel: Solid State Commun. **30**, 235 (1979)

7.10 D.T. Ling, J.N. Miller, D.L. Weiszmann, P. Pianetta, L.I. Johansson, I. Lindau, W.L. Spicer: Surf. Sci. **92**, 350 (1980)

7.11 A.R. Mackintosh, O.K. Andersen: *In Electrons at the Fermi Surface*, ed. by M. Springford (Cambridge Univ. Press, Cambridge 1980)

7.12 P. Hohenberg, W. Kohn: Phys. Rev. **136**, B864 (1964)

7.13 W. Kohn, L.J. Sham: Phys. Rev. **140**, A1133 (1965)

7.14 L. Hedin, B.L. Lundquist: J. Phys. C **4**, 2064 (1971)

7.15 R. Courths, B. Cord, H. Wern, S. Hüfner: Physica Scripta T **4**, 144 (1983)

7.16 R.S. Williams, P.S. Wehner, J. Stöhr, D.A. Shirley: Surf. Sci. **75**, 216 (1978)

7.17 D. Westphal, A. Goldmann: Surf. Sci. **131**, 92 (1983)

7.18 P.J. Jennings, S.M. Thurgate: Surf. Sci. **104**, L210 (1981)

7.19 G.A. Burdick: Phys. Rev. **129**, 138 (1963)

7.20 L. Hodges, H. Ehrenreich, N.D. Lang: Phys. Rev. **152**, 505 (1966)

7.21 F.M. Müller: Phys. Rev. **153**, 659 (1967)

7.22 L. Ehrenreich, L. Hodges: In *Methods in Computational Physics*, Vol.8 (Academic, New York 1968)

7.23 H. Mårtensson, P.O. Nilsson: Phys. Rev. B **30**, 3047 (1984)

7.24 F.J. Himpsel, J.A. Knapp, D.E. Eastman: Phys. Rev. B **19**, 2919 (1979)

7.25 W. Eberhardt, E.W. Plummer: Phys. Rev. B **21**, 3245 (1980)

7.26 V.L. Moruzzi, J.F. Janak, A.R. Williams: *Calculated Electron Properties of Metals* (Pergamon, New York 1978)

7.27 E. Jensen, E.W. Plummer: Phys. Rev. Lett. **55**, 1912 (1985)

7.28 D.E. Eastman: In *Electron Spectroscopy*, ed. by D.A. Shirley (North-Holland, Amsterdam 1972) p.487

7.29 C.S. Fadley, D.A. Shirley: In *Electronic Density of States* ed. by L.H. Bennet (Nat'l Bur. Stand. (U.S.), Spec. Publ. 232) (U.S.GPO, Washington, D.C. 1971) p.163

7.30 S. Hüfner, G.K. Wertheim, N.V. Smith, M.M. Traum: Solid State Commun. **11**, 323 (1972)

7.31 R.J. Smith, J. Anderson, J. Hermanson, G.J. Lapeyre: Solid State Commun. **21**, 459 (1977)

7.32 C. Guillot, Y. Ballu, J. Paigne, J. Lecante, K.P. Jain, P. Thiry, R. Pinchaux, Y. Petroff, L.M. Falicov: Phys. Rev. Lett. **39**, 1632 (1977)

7.33 D.E. Eastman, F.J. Himpsel, J.A. Knapp: Phys. Rev. Lett. **40**, 1514 (1978)
7.34 M.M. Traum, N.V. Smith, H.H. Farrel, D.P. Woodruff, D. Norman: Phys. Rev. B **20**, 4008 (1979)
7.35 W. Eberhardt, E.W. Plummer, K. Horn, J. Erskine: Phys. Rev. Lett. **45**, 273 (1980)
7.36 J. Barth, G. Kalkoffen, C. Kunz: Phys. Lett. A **74**, 360 (1979)
7.37 R. Clauberg, W. Gudat, E. Kisker, G.M. Rothberg: Phys. Rev. Lett. **47**, 1314 (1981)
7.38 D.R. Penn: Phys. Rev. Lett. **42**, 921 (1979)
7.39 R. Courths, S. Hüfner: Phys. Rep. **112**, 53 (1984)
7.40 H. Wern: Winkelaufgelöste Ultraviolett Photoelektronenspektroskopie an Silber. Thesis, Universität des Saarlandes (1985)
7.41 E.D. Kane: Phys. Rev. Lett. **12**, 97 (1964)
7.42 P. Heimann, H. Miosga, H. Neddermeyer: Solid State Commun. **29**, 463 (1979)
7.43 P.O. Nilsson, N. Dahlbaeck: Solid State Commun. **29**, 303 (1979)
7.44 H. Asonen, M. Lindroos, M. Pessa, N. Dahlbaeck: Solid State Commun. **35**, 69 (1980)
7.45 R. Courths, V. Bachelier, S. Hüfner: Solid State Commun. **38**, 887 (1981)
7.46 M. Lindroos, H. Asonen, M. Pessa, N.V. Smith: Solid State Commun. **39**, 285 (1981)
7.47 R. Courths: Solid State Commun. **40**, 529 (1981)
7.48 R. Courths, V. Bachelier, B. Cord, S. Hüfner: Solid State Commun. **40**, 1059 (1981)
7.49 M. Pessa, M. Lindroos, H. Asonen, N.V. Smith: Phys. Rev. B **25**, 738 (1982)
7.50 R. Courths, H. Wern, U. Hau, B. Cord, V. Bachelier, S. Hüfner: J. Phys. F., Metal Phys. **14**, 1559 (1984)
7.51 R. Courths, H. Wern, U. Hau, B. Cord, V. Bachelier, S. Hüfner: Solid State Commun. **49**, 989 (1984)
7.52 H. Becker, E. Dietz, U. Gerhardt, A. Angermueller: Phys. Rev. B **12**, 2084 (1975)
7.53 H. Eckardt, L. Fritsche, J. Noffke: J. Phys. F **14**, 97 (1984)
7.54 M. Wöhlecke, A. Baalmann, M. Neumann: Solid State Commun. **49**, 217 (1984)
A. Baalmann, M. Neumann: BESSY Annual Report 1984
7.55 E. Dietz, D.E. Eastman: Phys. Rev. Lett. **41**, 1674 (1978)
7.56 J.F. Janak, A.R. Williams, V.L. Moruzzi: Phys. Rev. B **11**, 1522 (1975)
7.57 H. Wern, R. Courths, G. Leschik, S. Hüfner: Z. Physik B **60**, 293 (1985)
7.58 V.N. Strocov, R. Claessen, G. Nicolay, S. Hüfner, A. Kimura, A. Harasawa, S. Shin, A. Kakizaki, P.O. Nilsson, H.I. Starnberg, P. Blaha: Phys. Rev. Lett. **81**, 4943 (1998)
7.59 R. Rosei, R. Lässer, N.V. Smith, R.L. Benbow: Solid State Commun. **35**, 979 (1980)
7.60 J.A. Knapp, F.J. Himpsel, D.E. Eastman: Phys. Rev. B **19**, 4952 (1979)
7.61 P.T. Coleridge, I.M. Templeton: Phys. Rev. B **25**, 7818 (1982)
7.62 R.L. Benbow, N.V. Smith: Phys. Rev. B **27**, 3144 (1983)
7.63 N.V. Smith: In Photoemission in Solids I, ed. by M. Cardona, L. Ley, Topics Appl. Phys., Vol.26 (Springer, Berlin, Heidelberg, 1978) Chap.6
7.64 B. Dardel, D. Malterre, M. Grioni, P. Weber, Y. Baer: Phys. Rev. Lett. **67**, 3144 (1991)
R. Claessen, G.H. Gweon, F. Reinert, J.W. Allen, W.P. Ellis, Z.X. Shen, C.G. Olson, L.F. Schneemeyer, F. Lévy: Proc. 6th Int'l Conf. on Electron Spectrosc (Rome 1995), Journ. Electr. Spectrosc. **76**, 121 (1995)

These communications, among others, show the special properties of one-dimensional metals, as observed by PES, most notably the absence of a sharp Fermi cut-off, an energy renormalization of the bands calculated for non-interacting electrons which is larger than one, and possibly a splitting of the one-electron band into a spinon and a holon branch. A system exhibiting these properties is sometimes called a *Luttinger liquid* (see Chap. 5)
J.M. Luttinger: J. Math. Phys. **4**, 1154 (1963)
V. Meden, K. Schönhammer: Phys. Rev. B **46**, 15763 (1992); ibid. B **47**, 16205 (1993)
J. Voit: Phys. Rev. B **47**, 6740 (1993)
7.65 H. Przbylski, A. Baalmann, G. Borstel, M. Neumann: Phys. Rev. B **27**, 6669 (1983)
7.66 F.J. Himpsel, D.E. Eastman: Phys. Rev. Lett. **41**, 507 (1978)
7.67 K.A. Mills, R.F. Davis, S.D. Kevan, G. Thornton, D.A. Shirley: Phys. Rev. B **22**, 581 (1980)
7.68 D.P. Woodruff, N.V. Smith, P.D. Johnson, W.A. Royer: Phys. Rev. B **26**, 2943 (1982)
7.69 W. Altmann, V. Dose, A. Goldmann, U. Kolac, J. Rogozik: Phys. Rev. B **29**, 3015 (1984)
7.70 C.N. Berglund, W.E. Spicer: Phys. Rev. **136**, A1030 (1964)
7.71 C.N. Berglund, W.E. Spicer: Phys. Rev. **136**, A1044 (1964)
7.72 N.V. Smith: Crit. Rev. Solid State Sci. **2**, 45 (1971)
7.73 T. Gustafsson, P.O. Nilsson, L. Wallden: Phys. Lett. A **37**, 121 (1971)
7.74 L. Wallden, T.G. Gustafsson: Phys. Scr. **6**, 73 (1972)
7.75 R.C. Jaklevic, J. Lambe: Phys. Rev. B **12**, 4146 (1975)
7.76 R. Rosei, C.H. Culp, J.H. Weaver: Phys. Rev. B **10**, 484 (1974)
7.77 P.O. Nilsson, B. Sandell: Solid State Commun. **8**, 721 (1970)
7.78 D.v.d. Marel, G.A. Sawatzky, R. Zeher, F.U. Hillebrecht, J.C. Fuggle: Solid State Commun. **50**, 47 (1984)
7.79 B. Reihl, R.R. Schlittler: Phys. Rev. B **29**, 2267 (1984)
7.80 N.V. Smith, M.M. Traum: Phys. Rev. Lett. **29**, 1243 (1972)
7.81 P.O. Nilsson, D.E. Eastman: Phys. Scr. **8**, 113 (1975)
7.82 H.F. Roloff, H. Neddermeyer: Solid State Commun. **21**, 561 (1977)
7.83 P. Heimann, H. Neddermeyer, H.F. Roloff: Phys. Rev. Lett. **37**, 775 (1976)
7.84 F.L. Battye, A. Goldmann, L. Kasper, S. Hüfner: Z. Phys. B **27**, 209 (1977)
7.85 D. Liebowitz, N.J. Shevchik: Phys. Rev. B **17**, 3825 (1978)
7.86 D. Liebowitz, N.J. Shevchik: Phys. Rev. B **16**, 2395 (1977)
7.87 G.V. Hansson, S.A. Flodström: Phys. Rev. B **18**, 1562 (1978)
7.88 S.P. Weeks, J.E. Rowe: Solid State Commun. **27**, 885 (1978)
7.89 P.S. Wehner, R.S. Williams, S.D. Kevan, D. Denley, D.A. Shirley: Phys. Rev. B **19**, 6164 (1979)
7.90 D.P. Spears, R. Melander, L.G. Petersson, S.B.M. Hagström: Phys. Rev. B **21**, 1462 (1980)
7.91 G. Borstel, M. Neumann, G. Wöhlecke: Phys. Rev. B **23**, 3121 (1981)
7.92 A. Goldmann, D. Westphal, R. Courths: Phys. Rev. B **25**, 2000 (1982)
7.93 H.A. Padmore, C. Norris, G.C. Smith, C.G. Larsson, D. Norman: J. Phys. C, Solid State Phys. **15**, L155 (1982)
7.94 K.C. Prince, G. Paolucci, B. Hayden, A.M. Bradshaw: Bessy Report 1983
7.95 H. Wern, G. Leschik, U. Hau, R. Courths: Solid State Commun. **50**, 581 (1984)
7.96 For a compilation of band structures determined by photoemission spectroscopy see *Landolt-Börnstein*, New Series, Group III, Vols. 23A (1989), 23B (1994), and 23C2 (1999) ed. by A. Goldmann, E.E. Koch (Springer,

Berlin, Heidelberg)
7.97 J.G. Nelson, S. Kim. W.J. Gignac, R.S. Williams, J.G. Tobin, S.W. Robey, D.A. Shirley: Phys. Rev. B **32**, 3465 (1985)
7.98 H. Wern, R. Courths: Surf. Sci. **162**, 29 (1985)
7.99 N.E. Christensen: Phys. Status Solidi (b) **54**, 551 (1972)
7.100 O. Jepsen, D. Gloetzel, A.R. Mackintosh: Phys. Rev. B **23**, 2683 (1981)
7.101 J.A. Wilson, A.D. Yoffe: Adv. Phys. **18**, 193 (1969)
7.102 R.C. Tatar, S. Rabii: Phys. Rev. B **25**, 4126 (1982)
7.103 F.R. McFeely, S.P. Kowalczyk, L. Ley, D.A. Shirley: Phys. Lett. A **49**, 301 (1974)
F.R. McFeely, S.P. Kowalczyk, L. Ley, R.G. Cavell, R.A. Pollak, D.A. Shirley: Phys. Rev. B **9**, 5268 (1974)
7.104 A.R. Law, M.T. Johnson, H.P. Hughes: Phys. Rev. B **34**, 4289 (1986)
7.105 D. Marchan, C. Fretigny, M. Lagues, F. Batallan, Ch. Simon, I. Rosenman, R. Pinchaux: Phys. Rev. B **30**, 4788 (1984)
7.106 T. Takahashi, H. Tokailin, T. Sagawa: Solid State Commun. **52**, 765 (1984)
7.107 A.R. Law, J.J. Barry, H.P. Hughes: Phys. Rev. B **28**, 5332 (1983)
7.108 I.T. McGovern, W. Eberhardt, E.W. Plummer, J.E. Fischer: Physica B **99B**, 415 (1980)
7.109 P.M. Williams, D. Latham, J. Wood: J. Electron Spectrosc. Relat. Phenom. **7**, 281 (1975)
7.110 R.F. Willis, B. Feuerbacher, B. Fitton: Phys. Rev. B **4**, 2441 (1971)
7.111 V. Dose, G. Reusing, H. Scheidt: Phys. Rev. B **26**, 984 (1982)
7.112 Th. Fauster, F.J. Himpsel, J.J. Donelson, A. Marx: Rev. Sci. Instrum. **54**, 68 (1983)
7.113 R.F. Willis, B. Fitton, G.S. Painter: Phys. Rev. B **9**, 1926 (1974)
7.114 L.S. Caputi, G. Chiarello, E. Colavita, A. Santaniello, L. Papagno: Surface Sci. **152/153**, 278 (1985)
7.115 Th. Fauster, F.J. Himpsel, J.E. Fischer, E.W. Plummer: Phys. Rev. Lett. **51**, 470 (1983)
7.116 N.A.W. Holzwarth, S.G. Louie, S. Rabii: Phys. Rev. B **26**, 5382 (1982)
7.117 M. Posternak, A. Baldereschi, A.J. Freeman, E. Wimmer, M. Weinert: Phys. Rev. Lett. **50**, 761 (1983)
7.118 M. Posternak, A. Baldereschi, A.J. Freeman, E. Wimmer: Phys. Rev. Lett. **52**, 863 (1984)
7.119 I. Schäfer, M. Schlitzer, M. Skibowski: Phys. Rev. B **35**, 7663 (1987)
7.120 F. Maeda, T. Takahashi, H. Ohsawa, S. Suzuki: Phys. Rev. B **37**, 4482 (1988)
The complete occupied band structure of graphite has been measured by resonant soft x-ray fluorescence: J.A. Carlisle, E.L. Shirley, E.A. Hudson, L.J. Terminello, T.A. Callcott, J.J. Jia, D.L. Lederer, R.C.C. Perera, F.J. Himpsel: Phys. Rev. Lett. **74**, 1234 (1995)
7.121 C.T. Chen, N.V. Smith: Phys. Rev. B **40**, 7487 (1989); and Surf. Sci. **247**, 133 (1991)
7.122 R. Courths, B. Cord, H. Wern, H. Saalfeld, S. Hüfner: Solid State Commun. **63**, 619 (1987)
7.123 A. Spitzer, H. Lüth: Surf. Sci. **118**, 121 (1982)
7.124 R.A. Didio, D.M. Zehner, E.W. Plummer: J. Vac. Sci. Techn. A **62**, 852 (1984)
7.125 W. Jacob, V. Dose, A. Goldmann: Appl. Phys. A **41**, 145 (1986)
7.126 R.P.N. Bronckers, A.G.J. De Wit: Surf. Sci. **112**, 133 (1981)
7.127 J. Lapujoulade, Y. LeCruer, M. Lefort, Y. Lejay, E. Maurel: Surf. Sci. **118**, 103 (1982)

7.128 V. Ponthier, C. Ramseyer, C. Girardet, P. Zeppenfeld, V. Dierks, R. Halmer: Phys. Rev. B **58**, 998 (1998)
7.129 R. Feidenhans'l, I. Stensgaard: Surf. Sci. **133**, 453 (1983)
7.130 J.F. Wendelken: Surf. Sci. **108**, 605 (1981)
7.131 U. Doebler, K. Baberschke, J. Haase, A. Puschmann: Phys. Rev. Lett. **52**, 1437 (1984)
7.132 G. Ertl: Surf. Sci. **6**, 208 (1967)
7.133 G. Ertl, J. Kueppers: Surf. Sci. **24**, 104 (1971)
7.134 A.J. de Wit, R.P.N. Bronckers, J.M. Fluit: Surf. Sci. **82**, 177 (1979)
7.135 J. Lapujoulade, Y. Le Cruer, M. Lefort, Y. Lejay, E. Maurel: Phys. Rev. B **22**, 5740 (1980)
7.136 M. Bader, A. Puschmann, C. Ocal, J. Haase: Phys. Rev. Lett. **57**, 3273 (1986)
7.137 U. Doebler, K. Baberschke, D.D. Vvedensky, J.B. Pendry: Surf. Sci. **178**, 679 (1986)
7.138 P. Hofman, R. Unwin, W. Wyrobisch, A.M. Bradshaw: Surf. Sci. **72**, 635 (1978)
7.139 H. Niehus, G. Comsa: Surf. Sci. **140**, 18 (1984)
7.140 F.M. Chua, Y. Kuk, P.J. Silverman: Phys. Rev. Lett. **63**, 386 (1989)
7.141 J.A. Stroscio, M. Persson, W. Ho: Phys. Rev. B **33**, 6758 (1986)
7.142 D.T. Ling, J.N. Miller, D.L. Weissman, P. Pianetta, P.M. Stefan, I. Lindau, W.E. Spicer: Surf. Sci. **95**, 89 (1980)
7.143 D.T. Ling, J.N. Miller, D.L. Weissman, P. Pianetta, L.I. Johansson, I. Lindau, W.E. Spicer: Surf. Sci. **92**, 350 (1980)
7.144 D.J. Coulman, J. Wintterlin, R.J. Behm, G. Ertl: Phys. Rev. Lett. **64**, 1761 (1990)
7.145 R. Ozawa, A. Yamane, K. Morikawa, M. Okwada, K. Suzuki, H. Fukutani: Surf. Sci. **346**, 237 (1996)
7.146 R. Courths, S. Hüfner, P. Kemkes, G. Wiesen: Surf. Sci. **376**, 43 (1997)
7.147 B. Weimert, N. Noffke, L. Fritsche: Surf. Sci. **264**, 365 (1992)
7.148 V. Dose: Surf. Sci. Rep. **5**, 337 (1986)
7.149 F. Bertel: Appl. Phys. A **53**, 356 (1991)
7.150 H.J. Freund, M. Neumann: Appl. Phys. A **47**, 3 (1988)
7.151 H. Kuhlenbeck, M. Neumann, H.J. Freund: Surf. Sci. **173**, 194 (1986)
7.152 H. Kuhlenbeck, H.B. Saalfeld, U. Buskotte, M. Neumann, H.J. Freund, E.W. Plummer: Phys. Rev. B **39**, 3475 (1989)
7.153 E.W. Plummer: Surf. Sci. **152/153**, 162 (1985)
7.154 T.-C. Chiang, J.A. Knapp, M. Aono, D.E. Eastman: Phys. Rev. B **21**, 3513 (1980)
D. Straub, M. Skibowski, F.J. Himpsel: Phys. Rev. B **32**, 5239 (1985)
For data on Si see L.S.O. Johansson, P.E.S. Persson, U.O. Karlsson, R.I.G. Uhrberg: Phys. Rev. B **42**, 8991 (1990)
Those for Ge are found in X.H. Chen, W. Ranke, E. Schröder-Bergen: Phys. Rev. B **42**, 7429 (1990)
7.155 H.J. Levinson, F. Greuter, E.W. Plummer: Phys. Rev. B **27**, 727 (1983)
7.156 N.W. Ashcroft: Phil. Mag. **8**, 2055 (1963)
7.157 J.R. Anderson, S.S. Lane: Phys. Rev. B 2, 298 (1970)
7.158 S.-K. Ma, K.W.-K. Shung: Phys. Rev. **49**, 10617 (1994). Calculating the angle-resolved PE spectra of Al(001), taking into account surface, many-body and band-structure effects. The conclusion is that the conventional one-electron picture works for simple metals, provided that these effects are treated properly
7.159 I.W. Lyo, E.W. Plummer: Phys. Rev. Lett. **60**, 1558 (1988)

7.160 K.W.K. Shung, G.D. Mahan: Phys. Rev. B **38**, 3856 (1988)
S.K. Ma, K.W.K. Shung: Phys. Rev. B **49**, 10617 (1994)
7.161 F.J. Himpsel: Adv. Phys. **32**, 1 (1983)
7.162 P. Aebi, T.J. Krantz, J. Osterwalder, R. Fosch, P. Schwaller, L. Schlapbach: Phys. Rev. Lett. **76**, 1150 (1996)
7.163 C.S. Wang, B.M. Klein: Phys. Rev. B **24**, 3393 (1982)
7.164 D. Straub, M. Skibowski, F.J. Himpsel: Phys. Rev. B **32**, 5239 (1985)
7.165 S. Hüfner, G.K. Wertheim, J.H. Wernick: Phys. Rev. B **8**, 4511 (1973)
7.166 S. Hüfner: In *Photoemission in Solids II*, ed. by L. Ley, M. Cardona, Topics Appl. Phys., Vol.27 (Springer, Berlin, Heidelberg 1979) Chap.3
7.167 H. Höchst, S. Hüfner, A. Goldmann: Z. Physik B **26**, 133 (1977)
7.168 P. Steiner, S. Hüfner, A.J. Freeman, D. Wang: Solid State Commun. **44**, 619 (1982)
7.169 L.F. Wagner, Z. Hussain, C.S. Fadley, R.J. Baird: Solid State Commun. **21**, 453 (1977)
7.170 S. Hüfner, G.K. Wertheim, J.H. Wernick: Solid State Commun. **17**, 1585 (1975)
7.171 J. Friedel: Can. J. Phys. **34**, 1190 (1956)
7.172 J. Friedel: J. Phys. Radium **19**, 573 (1958)
7.173 J. Friedel: Suppl. Nuovo Cimento **VII**, 287 (1958)
7.174 P.W. Anderson: Phys. Rev. **124**, 41 (1961)
7.175 H. Höchst, P. Steiner, S. Hüfner: J. Magn. Magn. Mater. **6**, 159 (1977)
7.176 P. Steiner, H. Höchst, W. Steffen, S. Hüfner: Z. Physik B **38**, 191 (1980)
7.177 P. Steiner, H. Höchst, S. Hüfner: J. Phys. F **7**, L105 (1977)
7.178 H. Höchst, P. Steiner, S. Hüfner: Z. Phys. B **38**, 201 (1980)
7.179 D.v.d. Marel, G.A. Sawatzky, F.U. Hillebrecht: Phys. Rev. Lett. **53**, 206 (1984)
7.180 R.G. Jordan, W. Drube, D. Straub, F.J. Himpsel: Phys. Rev. B **33**, 5280 (1986)
7.181 L. Wallden: Phil. Mag. **21**, 571 (1970)
7.182 H.P. Meyers, L. Wallden, A. Karlson: Phil. Mag. **18**, 725 (1968)
7.183 L. Ley, M. Cardona, R.A. Pollack: In Topics in Appl. Phys. **27**, 11 (1979)
7.184 A. Goldmann: Phys. Status Solidi (b) **81**, 9 (1977)
7.185 N.J. Shevchik, J. Tejeda, M. Cardona: Phys. Rev. B **9**, 2627 (1974)
7.186 L. Ley, R.A. Pollak, F.R. McFeely, S.P. Kowalczyk, D.A. Shirley: Phys. Rev. B **9**, 600 (1974)
7.187 A.B. Pippard: Rep. Prog. Phys. **23**, 176 (1960)
7.188 S.K. Shinha: Phys. Rev. **143**, 422 (1966)
7.189 A. Marini, G. Onida, R. Del Sole: Phys. Rev. Lett. **88**, 016403 (2002)
7.190 O. Tjernberg, C. Dallera, M. Finazzi, L.H. Tjeng, M. Mansson, Th. Claesson, C. de Nada, F. Venturini, N.B. Brookes: private communication (2002)

8. Surface States, Surface Effects

It has been stated already that PES, because of the small escape depth of the electrons in a solid, is in principle a surface sensitive technique. However, the agreement found in the preceding chapter between calculated and experimental bulk-band structures was remarkable. We take this as evidence that most of the information obtained in a PES experiment is representative of the bulk solid. This is due to the fact that the electronic and lattice relaxation at the surface of a solid are generally not very large. This means, for example, that for most materials the interlattice spacing perpendicular to the surface is changed measurably only for the topmost layer, typically only by a few percent of the bulk lattice constant. A similar statement holds with respect to the perturbation of the electron states perpendicular to a surface: the unperturbed bulk electronic states are generally found to within one lattice constant from the surface. Therefore in conventional PES experiments, which probe several atomic layers perpendicular to the surface, the genuine surface contribution is not a large fraction of the total measured spectrum and is often even hard to recover (but see Sect. 7.3.4 for notable exceptions).

These general statements have to be qualified for cases where one is working with high resolution (100 meV or smaller) and is looking at electrons with a kinetic energy near the minimum escape depth (Fig. 1.11). Under these circumstances the volume contained in the first surface layer can be responsible for roughly half the PE signal and even if the shift between the bulk and surface is small it can be detected. As a case in point we show in Fig. 8.1 a He II spectrum of Cs metal [8.1] for the region of the $5p_{3/2}$ emission. The surface layer and the bulk of the sample give rise to two distinct signals, with a surface-atom core-level shift of (228 ± 5) meV. The sign and magnitude of the shift can be rationalized in a way which will be described later in this chapter. The escape depth calculated from the spectrum in Fig. 8.1 is 5.5 Å(29 eV kinetic energy of the electrons) which has to be compared with the lattice constant of Cs (6.07 Å) or the lattice spacing in the [110] direction, which is 4.29 Å.

In a covalently bonded crystal (e.g., Si) and in those in which the rather localized and highly anisotropic d-orbitals contribute significantly to the bonding, the electron states produced by the creation of the surface in a hitherto infinite crystal can be viewed as so-called *dangling bond states*. These states

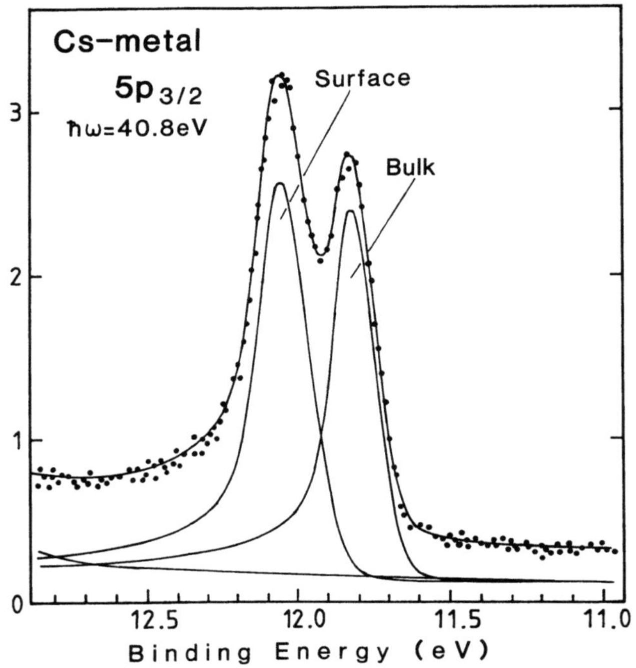

Fig. 8.1. Cs $5p_{3/2}$ PE-spectrum taken with He II radiation. The full line gives a fit to the measured spectrum. The sample was prepared by evaporating Cs on a Cu(111) surface at 90 K [8.1]

are produced by cutting the bonds between atoms and are directed as partly filled orbitals out of the surface. It is intuitively obvious that these bonds can be especially chemically "aggressive" and will tend to saturate by binding atoms and molecules available in the finite vacuum. This fact leads to a decrease in the intensity of surface state features in PES with time. On the other hand it also provides a useful method to distinguish surface from bulk PES data. Those features in PES or IPES that are sensitive to a thin overlayer on the sample, produced by admitting reactive gases to the vacuum in a controlled way, can be expected to be surface features, although one should be wary of using this as the only criterion.

For the most part this chapter treats two types of surface feature. The first is the so-called *surface core-level shift* (Fig. 8.1): the quasi-atomic energy levels in a solid, e.g., the core levels, often have a binding energy that differs between bulk and surface atoms because of the changed potential at the surface. This effect will also change the position and width of bands derived from pseudo-localized orbitals like the d bands (Sect. 7.4.3) in the transition metals or the f bands in the rare earth metals [8.2].

A second electronic peculiarity of the surface are the so-called surface states. These are (additional) states which may occur in the conduction bands, valence bands, or in gaps, as a result of the changes introduced into the wave functions of the infinite crystal by making it semi-infinite at the surface. Such a surface state is spatially confined to the surface region. This means that its wave function decays exponentially not only from the surface into the vacuum, as do all crystal states below the vacuum level, but also into the solid: the real part of the wave function inside the crystal will consist of an exponential modulated by a sin or cos function.

For good reviews of surface effects in PES see [8.3, 8.4].

8.1 Theoretical Considerations

In order to understand the basic origin of the surface states, a simple model as given in Fig. 8.2 is helpful. In Fig. 8.2a one assumes that, inside the crystal, there is a constant potential and the surface can be modelled by a step potential. Such a model yields wave functions that decay exponentially outside the crystal, whereas inside the crystal they are traveling waves with a real $\boldsymbol{k}$-vector. This means that in the free-electron-gas model with a step-like surface potential surface states do not exist, because every energy inside the crystal corresponds to a traveling wave. In other words, the matching conditions at the surface under no circumstances allow a wave decaying into the crystal.

Consider next the situation in which the crystal has a periodic but weak potential (nearly free-electron model), with the termination at the surface again achieved by a step potential (Fig. 8.2b).

In this case the crystal potential may generally be written in one dimension as:

$$V(z) = V_0 + \sum_g V_g e^{igz}; \quad g = \frac{2\pi}{a} n, \quad n = \pm 1, \pm 2, \ldots . \tag{8.1a}$$

For simplicity, the constant potential is chosen such that $V_0 = 0$. In addition it is assumed that the potential is real (no damping) and only the leading terms ($n = \pm 1$) in g are kept [8.4, 8.6–8.11]:

$$V(z) = 2V_g \cos \frac{2\pi}{a} z . \tag{8.1b}$$

From now on we also drop the index g in V; z is the coordinate perpendicular to the surface (Fig. 8.2a).

In Fig. 8.2b the potential $V(z)$ is depicted for the case in which the maxima of $V(z)$ are at the sites of the ion cores while the minima are between the ions. From Fig. 8.2b one then sees that the sign of V in this case depends on the position of the origin: if the origin $z = 0$ is between the cores of the ions, one has $V < 0$ and if the origin is at the site of the ion cores,

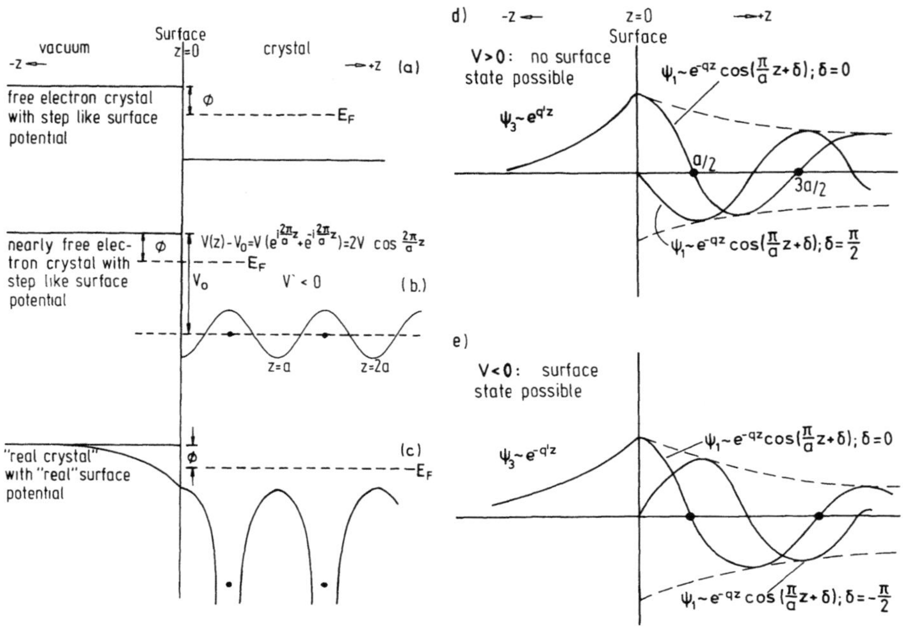

Fig. 8.2. Representation of the surface of a crystal. (**a**) Free-electron metal with a step-like surface potential. (**b**) Nearly-free-electron solid with a potential: $V(z) = V_0 + 2V\cos(2\pi/a)z$ $(V < 0)$. The origin is taken between the ion cores. The usual pseudopotential is repulsive at the ion cores and attractive between ion cores; $2V\cos(2\pi/a)z$ is its leading term in a Fourier series. (**c**) Approximate representation of the potential in a "real" crystal with a "real" surface potential. (**d**,**e**) Matching condition for wave functions at the surface. (**d**): $V > 0$ [with origin as in (**b**)]. The wave function at the top of the gap $(\delta = \pi/2)$ and at the bottom of the gap $(\delta = 0)$ cannot match the wave function outside the crystal: No surface state. (**e**) $V < 0$ [with origin as in (**b**)]. The wave function at the bottom of the gap $\delta = -\pi/2$ can obviously be made to approximately match the vacuum wave function. This shows that a surface state will occur somewhere near the bottom of the gap

one has $V > 0$. In our case the surface is put between the ions such that the situation is as in Fig. 8.2b, a convention first introduced by Maue [8.6]. Other workers have put the origin through the ion cores and then one has $V > 0$, see, e.g., [8.7]. The potential $V(z)$ modifies the free-electron band structure such that it opens up gaps at the Brillouin zone boundaries at $k_z = \pm\pi/a$, $\pm 2\pi/a$, ... in which no extended electronic states can exist.

The one-dimensional Schrödinger equation for this situation reads:

$$\frac{\hbar^2}{2m}\frac{\mathrm{d}^2\psi(z)}{\mathrm{d}z^2} + [E - V(z)]\psi(z) = 0\,. \tag{8.2}$$

One is interested in the solutions near the Brillouin-zone boundary, i.e. for $k_z \simeq \pm\pi/a$. These have the form:

$$\psi(z) = \exp \mathrm{i}(k_z z)[a + b\exp(-\mathrm{i}g_z z)] = a\exp(\mathrm{i}k_z z) + b\exp(ik_z - \mathrm{i}g_z)z \ , \quad (8.3)$$

where $g_z = 2\pi/a$ and $k_z \simeq \pm\pi/a$ and, for convenience, only the point $k_z \simeq \pi/a$ is regarded here.

For the wave functions at the top and bottom of the gap one finds the following situation (for the origin given in Fig. 8.2b one can take $V > 0$ – not shown – or alternatively $V < 0$, as depicted there, depending upon the actual form of the potential (8.1a)):

$V < 0$:	top	$\psi \sim \cos\left(\frac{\pi}{a}z\right)$,	s type
	bottom	$\psi \sim \cos\left(\frac{\pi}{a}z - \frac{\pi}{2}\right)$, $\sim \sin\left(\frac{\pi}{a}z\right)$	p type
$V > 0$:	top	$\psi \sim \cos\left(\frac{\pi}{a}z + \frac{\pi}{2}\right)$, $\sim \sin\left(\frac{\pi}{a}z\right)$	p type
	bottom	$\psi \sim \cos\left(\frac{\pi}{a}z\right)$,	s type

In atomic systems the s level is usually found below the p level. Thus, the resulting gap for $V < 0$ (origin between atomic planes) is sometimes called a "(Shockley) inverted gap".

Note that if the origin in Fig. 8.2b is chosen as the plane through the ion cores, then the signs are reversed and one has the "(Shockley) inverted gap" for $V > 0$.

One can extend the solutions of the Schrödinger equation into the gap by using a complex $\boldsymbol{k}$-vector:

$$k_z = p \pm \mathrm{i}q$$

with $p = \pi/a$ for a solution near the Brillouin zone boundary. In addition we now drop the index z, remembering, however, that all wave vectors are normal to the surface.

Our wave function now reads:

$$\psi(z) = \mathrm{e}^{\mathrm{i}(p \pm \mathrm{i}q)z}(a + b\mathrm{e}^{-\mathrm{i}gz}) \quad (8.4a)$$

$$= \mathrm{e}^{\mp qz}\,\mathrm{e}^{\mathrm{i}pz}(a + b\mathrm{e}^{-\mathrm{i}gz})\,. \quad (8.4b)$$

This wave function (8.4b) reveals the physics involved. In the infinite crystal both of these solutions are forbidden because the first exponential diverges either for $z \to -\infty$ or for $z \to +\infty$.

This does not hold, however, for the semi-infinite crystal. For the situation chosen, namely a crystal extending to $+\infty$, the following wave function is a possible solution

$$\psi(z) = \mathrm{e}^{-qz}\mathrm{e}^{\mathrm{i}pz}(a + b\mathrm{e}^{-\mathrm{i}gz})\,, \quad (8.5)$$

since it decays into the crystal.

Setting

$$a = \mathrm{e}^{\mathrm{i}\delta} \quad \text{and} \quad b = \mathrm{e}^{-\mathrm{i}\delta} \quad (8.6)$$

leads, after some calculation, to the following wave function (decaying inside the crystal as $z \to \infty$):

$$\psi = A\mathrm{e}^{-qz} \cos\left(\frac{\pi}{a}z + \delta\right) , \tag{8.7}$$

where (for $q > 0$)[1] one has for δ:

$$\begin{aligned} V < 0 \ :\ & \text{top} && 0 = \delta \\ & \text{bottom} && -\pi/2 = \delta \ , \\ V > 0 \ :\ & \text{top} && \pi/2 = \delta \\ & \text{bottom} && 0 = \delta \ . \end{aligned}$$

For the wave function outside the crystal one has to assume that it decays exponentially, meaning

$$\psi = B\mathrm{e}^{+q'z} \qquad (z < 0) \ . \tag{8.8}$$

The coefficients A and B are obtained by matching ψ and $\mathrm{d}\psi/\mathrm{d}z$ at the surface.

Figure 8.2d attempts to represent the wave functions for the potential of Fig. 8.2b. One will see that for the case $V > 0$ a matching of the wave functions for $z = 0$ at the surface is impossible. The wave function decaying into the vacuum has

$$\left.\frac{\mathrm{d}\psi}{\mathrm{d}z}\right|_{z=0} = Bq'\mathrm{e}^{q'z}\Big|_{z=0} = Bq' > 0 \ . \tag{8.9}$$

On the other hand, the wave function decaying into the crystal has

$$\begin{aligned} \left.\frac{\mathrm{d}\psi}{\mathrm{d}z}\right|_{z=0} &= A\{p\mathrm{e}^{-qz}[-\sin(pz+\delta)] - q\mathrm{e}^{-qz}\cos(pz+\delta)\}\big|_{z=0} \\ &= A[p(-\sin\delta) - q\cos\delta] < 0 \end{aligned} \tag{8.10}$$

for $p > 0$ and $0 < \delta < \pi/2$, making a matching impossible for V positive, see Fig. 8.2d.

For the second possibility, namely $V < 0$ the situation is shown in Fig. 8.2e. In this case one has, e.g., for the wave function decaying into the crystal at the bottom edge of the band gap ($\delta = -\pi/2$):

$$\left.\frac{\mathrm{d}\psi}{\mathrm{d}z}\right|_{z=0} = Ap[-\sin(-\pi/2)] > 0 \ . \tag{8.11}$$

Therefore, for the case $V < 0$ one can in principle match the slopes of the wave functions inside and outside the crystal. Figure 8.2e shows only the two limiting cases $\delta = 0$ and $\delta = -\pi/2$. Matching is not possible for either of these limiting cases because in the $\delta = -\pi/2$ case the wave function itself

[1] It is emphasized again that the phases given here apply to the case where the origin is between the ions [8.4, 8.6]. Other workers [8.9, 8.11] have chosen the origin through the ion cores and then find different phases.

is zero inside the crystal, and in the $\delta = 0$ case the derivative of the wave functions cannot be matched. However, for $0 > \delta > -\pi/2$ there is a region where matching of the wave function and its slope is possible and a surface state can exist.

In the literature use is often made of the logarithmic derivative $[(\mathrm{d}\psi/\mathrm{d}z)/\psi]$. This yields the matching condition:

$$\left.\frac{\mathrm{d}\psi/\mathrm{d}z}{\psi}\right|_{\text{outside}} = \left.\frac{\mathrm{d}\psi/\mathrm{d}z}{\psi}\right|_{\text{inside}}$$

and therefore

$$q' = -q - p\tan\delta \ . \tag{8.12}$$

Let us summarize the physics of the problem. In an infinite crystal the "propagating" electronic states in a real potential have a real wave vector $\boldsymbol{k}$. The crystal potential modifies the free-electron states and this modification is most easily seen at the Brillouin zones ($k = \pm\pi/a,\ \pm 2\pi/a,\ \ldots$). States with energies lying in the gaps are forbidden in the infinite crystal.

Electron states in the gap can exist if one accepts a complex wave vector $k = p + \mathrm{i}q$. The resulting wave functions contain exponentials with real exponents $[\exp(\pm qz)]$ for a one-dimensional crystal in the $\pm z$ direction. These states are forbidden in the infinite crystal because they diverge as $z \to \pm\infty$. However, for a semi-infinite crystal there are exponentially decaying eigenfunctions outside the crystal (into the vacuum) but also such wave functions inside the crystal, which lead to states confined to the surface (surface states).

Another elegant way of looking at surface states has been introduced by Echenique and Pendry [8.12]. They view these states as electrons trapped between the surface of the crystal (via a band gap) and the surface barrier potential, which prevents them from escaping into the vacuum (Fig. 8.3). Alternatively one may say that an electron outside the crystal surface can still be bound to it by its own image potential, which is why these states are often called "image potential states".

Echenique and Pendry [8.11–8.18], by taking this approach at face value, devised a model in which an electron outside the crystal travels to (ψ_+) and away from the surface (ψ_-). It is multiply reflected at the crystal surface and at the barrier potential (which is the image potential at large distances). If ψ_+ travels towards the crystal (index C) the reflected beam may be written

$$\psi_- = r_\mathrm{C}\mathrm{e}^{\mathrm{i}\phi_\mathrm{C}}\psi_+ \ . \tag{8.13}$$

Reflection of that beam at the surface barrier (index B) leads to

$$\psi_+ = r_\mathrm{B}\mathrm{e}^{\mathrm{i}\phi_\mathrm{B}} r_\mathrm{C}\mathrm{e}^{i\phi_\mathrm{C}}\psi_+ \ , \tag{8.14}$$

where r_B and r_C are the real parts of the reflection coefficients and ϕ_B and ϕ_C are the phase changes occurring upon reflection ($r\mathrm{e}^{\mathrm{i}\phi}$ is thus the total reflection coefficient).

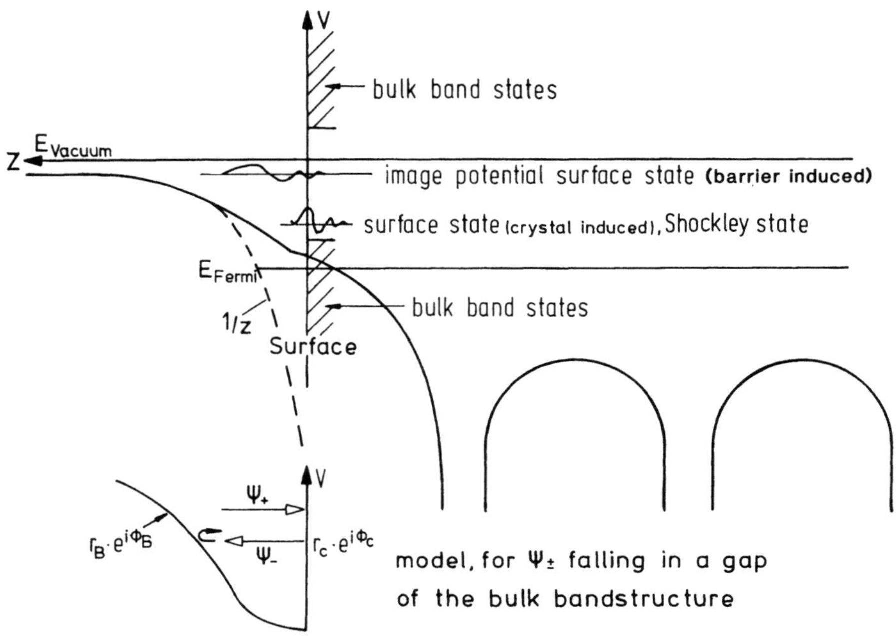

Fig. 8.3. The Echenique–Pendry view of a surface state [8.12]. It is assumed that an electron is trapped between the crystal surface and the image potential and travels back and forth between the two barriers. The lowest order solution ($n = 0$) in this model gives the "normal" surface state while higher order solutions ($n \geq 1$) give the so-called image potential states, which extend out into the vacuum. The Shockley state ($n = 0$) can be situated above E_{F} (as in this figure) or below E_{F}, as shown in Fig. 8.4

If one sums up an infinite number of reflections, one finds that the total amplitude of ψ_+ is proportional to

$$\{1 - r_{\mathrm{B}} r_{\mathrm{C}} \exp[\mathrm{i}(\phi_{\mathrm{B}} + \phi_{\mathrm{C}})]\}^{-1} . \tag{8.15}$$

If all the intensity of ψ_+ is piled up at the surface, then, in the one-dimensional model discussed here, the total intensity must be infinite [pole in (8.15)]. The existence of a surface state thus requires

$$r_{\mathrm{B}} r_{\mathrm{C}} = 1 \quad \text{and} \quad \phi_{\mathrm{C}} + \phi_{\mathrm{B}} = n 2\pi \qquad (n \text{ integer}) . \tag{8.16}$$

Since by definition

$$r_{\mathrm{B}}, r_{\mathrm{C}} \leq 1$$

the first condition may also be read as

$$r_{\mathrm{B}} = r_{\mathrm{C}} = 1 . \tag{8.17}$$

In physical terms this means: since one is dealing with states that have $E < E_{\mathrm{V}}$ one has $r_{\mathrm{B}} = 1$ and because the states are positioned energetically in a gap of the crystal-band structures, one has $r_{\mathrm{C}} = 1$. In order to make further predictions, one has to make assumptions about the crystal wave functions in the gap and the form of the barrier potential.

The crystal wave functions in the gap have been estimated before and we shall make use of these results.

According to (8.3) the wave function in the vacuum is

$$\psi_{\text{outside}} = \mathrm{e}^{ikz} + \mathrm{e}^{-ikz}\mathrm{e}^{\mathrm{i}\phi_{\mathrm{C}}} \quad (r_{\mathrm{C}} = 1) \, . \tag{8.18}$$

The matching condition $(\psi'/\psi)_{\text{outside}} = (\psi'/\psi)_{\text{inside}}$ gives on the right-hand side the result from (8.12) while the left-hand side is calculated from (8.18).

$$(\psi'/\psi)_{\text{outside}} = -q - p\tan\delta \tag{8.19}$$

$$k\tan(\phi_{\mathrm{C}}/2) = -q - p\tan\delta \, ; \quad k = \sqrt{E/f^2} \, , \tag{8.20}$$

where E is the energy of the electron, and $f^2 = \hbar^2/2m$.

In a rough approximation one can set $q \simeq 0$ whence $k \simeq p$ and one has $\phi_{\mathrm{C}}/2 = -\delta$.

It is now necessary to check whether this matching condition has a correspondence to the one previously calculated with a different model. Let us consider the lowest energy state, see (8.16), in the Echenique–Pendry model ($n = 0$), i.e.,

$$\phi_{\mathrm{C}} + \phi_{\mathrm{B}} = 0 \, . \tag{8.21}$$

In the "standard" model the surface was represented by a potential step, instead of the Coulomb-like image potential. Thus one has a wave function inside the crystal (Fig. 8.2d) of the type $\mathrm{e}^{-ikz} + \mathrm{e}^{ikz}\exp(\mathrm{i}\phi_{\mathrm{B}})$ and another one decaying as $\exp(q'z)$ into the vacuum. Applying the matching conditions at the surface leads to

$$\tan\left(\frac{1}{2}\phi_{\mathrm{B}}\right) = -\frac{q'}{k} \tag{8.22}$$

with $\phi_{\mathrm{C}} = -\phi_{\mathrm{B}}$, this is put into the above matching condition (8.20) yielding:

$$+k(q'/k) = -q - p\tan\delta \qquad \text{and} \qquad q' = -q - p\tan\delta \tag{8.23}$$

which is exactly the equation found before (8.12).

From this it is obvious that the $n = 0$ state in the Echenique–Pendry model and the surface state in the "standard" model are the same. The Echenique–Pendry model, however, gives additional states for $n = 1, 2, \ldots$. Since these states have their wave function amplitude centered closer to the surface barrier potential, they are also called the image potential states. They have indeed been detected, at least the $n = 1$ member of the series, in IPES experiments (Chap. 9). Further members of the series have been resolved in high-resolution two-photon PES measurements [8.15].

The exact form of the barrier (image) potential (Fig. 8.3) is difficult to estimate accurately. If one makes the admittedly crude approximation of a purely "Coulombic" potential for the barrier one has

$$V_B \approx \frac{1}{4z}; \quad z < 0 \tag{8.24}$$

and with this potential one finds the approximate condition [8.11,8.13,8.14]:

$$\frac{\phi_B}{\pi} = \sqrt{\frac{3.4\,\mathrm{eV}}{E_v - E}} - 1 .$$

Here E_v is the vacuum energy and E the energy of the bound states.

The equation for ϕ_B/π can be rationalized in the following way: in a "Coulombic" potential $V(z) \sim 1/(4z)$ which is terminated midway between atomic planes by an infinitely high potential barrier, the energies for an electron are calculated as (E_v: vacuum level) [8.11–8.13,8.20,8.21]:

$$E_v - E_m = \frac{0.85}{m^2}; \quad m = 1, 2, \ldots; \quad 0.85\,\mathrm{eV} = \frac{1}{16}\mathrm{Ry} . \tag{8.25}$$

This equation offers the simplest way to calculate the energy of an electron trapped between the surface of a solid and its image potential. Thus, in this simple model, one would expect that a hydrogenic series of image potential states exists on any material, and that the energetics does not depend on the nature of the material. This point will be raised again in the discussion of actual experiments.

In the Echenique–Pendry treatment, the nature of the barriers at both sides of the "box" in which the electron is being scattered back and forth are taken into account by considering explicitly the phase changes in the scattering process. In this approach the phase change at the crystal face (ϕ_C) and at the barrier (ϕ_B) are related by

$$\phi_B + \phi_C = 2\pi n .$$

For an infinitely high barrier at the crystal surface one has $\phi_C = -\pi$. ϕ_B must be a function of the energy and the simplest ansatz is $\phi_B \sim a(E_v - E)^b$. In order to start out at $-\pi$, such that $\phi_B + \phi_C = 2\pi n$, it is convenient to write

$$\phi_B = -\pi + a(E_v - E)^b \tag{8.26}$$

yielding

$$\begin{aligned} -2\pi + a(E_v - E)^b &= 2\pi n , \\ a(E_v - E)^b &= 2\pi(n+1) . \end{aligned} \tag{8.27}$$

If the image barrier is assumed to be $\sim 1/4z$, one has $E \sim 1/n^2$ meaning that $b = -1/2$, and therefore

$$\frac{1}{E_v - E} = \left(\frac{2\pi}{a}\right)^2 (n+1)^2 \quad \text{or} \quad E_v - E = \frac{(a/2\pi)^2}{(n+1)^2} . \tag{8.28}$$

Setting this equal to the energy in the Coulomb potential implies

$$\frac{(a/2\pi)^2}{(n+1)^2} = +\frac{0.85}{m^2} . \tag{8.29}$$

(Note that this series starts with $m = 1$, while the Echenique–Pendry model gives the lowest state with $n = 0$. This is reflected in the above equation). Thus

$$a = \sqrt{2\pi(0.85)} \tag{8.30}$$

yielding

$$\frac{\phi_B}{\pi} = -1 + \sqrt{\frac{4 \cdot 0.85}{E_v - E}} = \sqrt{\frac{3.4\,\text{eV}}{E_v - E}} - 1\,. \tag{8.31}$$

For the (Shockley) inverted gap, in which a surface state exists, we had (8.20) $\phi_C = -2\delta(-\pi/2 \le \delta \le 0)$. (This should not be confused with ϕ_C for an infinitely high potential barrier at the crystal face). Therefore one can write

$$\phi_C/\pi = +\varepsilon \quad \text{with} \quad 0 < \varepsilon < 1 \tag{8.32}$$

and from $\phi_C + \phi_B = 2\pi n$ we obtain

$$\phi_B/\pi + \varepsilon = 2n \quad \text{or} \quad \phi_B/\pi + 1 = 2n + 1 - \varepsilon\,. \tag{8.33}$$

If ϕ_B varies more rapidly with E than ϕ_C, which is the case for $n \ge 1$, then from the equation for ϕ_B we obtain

$$E_v - E = \frac{3.4\,\text{eV}}{(\phi_B/\pi + 1)^2} \tag{8.34}$$

and substituting $\phi_B/\pi + 1$ one finally arrives at

$$E_v - E_n = \frac{0.85\,\text{eV}}{(n + 1/2 - \varepsilon/2)^2}\,; \quad n = 0, 1, 2, \ldots\,. \tag{8.35}$$

From experience one knows that the "usual" surface states are near the band edge. At the lower gap edge one has $\varepsilon \simeq 0$ and

$$E_v - E_0 \simeq 3.4\,\text{eV}$$

for $n = 0$. This is typical for what may be called the normal surface state, which has already been dealt with. On the other hand for $n = 1$ one finds[2]

$$E_v - E_1 = 0.55\,\text{eV}\,, \quad \varepsilon = 1/2$$
$$E_v - E_1 = 0.85\,\text{eV}\,, \quad \varepsilon = 1\,.$$

This state is only loosely bound, and will therefore have its wave function mostly outside the crystal. It is called an image potential state. Between E_v and E_1 an infinite number of states with energies $\approx n^{-2}$ will occur; these are called a "Rydberg-like" series. Because of the proximity of the levels

[2] Here we are only interested in the basic properties of the model and therefore we try out two limiting values for ε. The most accurate experimental results fall between the energies calculated for $\varepsilon = 1/2$ and $\varepsilon = 1$ [8.15]. For a more sophisticated analysis see [8.16].

they will not be resolved by IPES in detail, but have been clearly detected in two-photon spectroscopy [8.15]. (For a more detailed discussion of the Echenique–Pendry model see [8.11]).

The dependence of a barrier state on $k_{||}$ will be the following:

$$E(\boldsymbol{k}) = E_{\mathrm{v}} - E_n + \hbar^2 k_{||}^2/2m^* \; ; \quad n = 1, 2 \ldots \, , \tag{8.36}$$

where m^*/m is an effective mass.

It has become customary to distinguish between Shockley [8.8] and Tamm [8.22] surface states. The states just discussed are usually called Shockley states because they are produced by adding a weak potential to the free-electron-gas model of a solid, and thus emerge from a formalism that is generally applied to the description of the sp bands in metals.

Tamm [8.22], who was the first to actually consider the problem of the surface electronic states, used the Kronig–Penny model in the version with

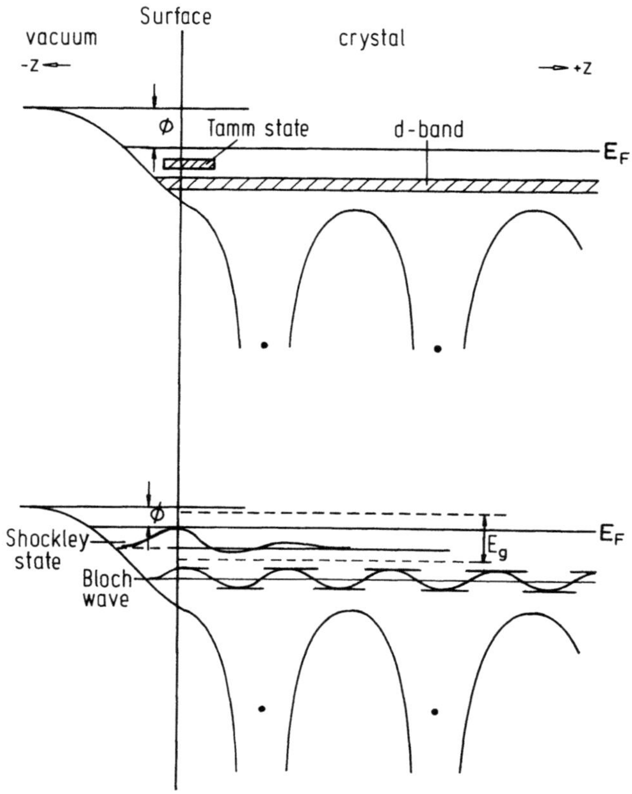

Fig. 8.4. Tamm and Shockley states. The Tamm state (*top*) is split-off from a band (e.g., a d band) of the crystal by the change in potential at the surface. A Shockley state (*bottom*) is a state created in a gap of the bulk-band structure due to the termination of the crystal by a surface (semi-infinite crystal)

infinitely high potential barriers between the lattice cells. This model is approximately valid for tight binding states (d- and f-valence band states). It leads to surface states split off from the unperturbed band, which are often called Tamm states. Note that in a rigorous sense the electron states in a crystal are always Bloch waves and thus the distinction between Shockley states and Tamm states is somewhat artificial.

The difference between Tamm states and Shockley states is shown schematically in Fig. 8.4. It is customary to distinguish in the Shockley picture between surface states (located in energy gaps and decaying rapidly into the crystal) and surface resonances (whose decay into the crystal is slow, and which may overlap energetically with bulk-band states).

The Tamm state in this simple picture derives from the fact that the d band states at the surface experience a weaker crystal potential than the bulk d states. This leads to a split-off state. The Shockley state originates in the gap at the surface due to new boundary conditions not present inside the bulk crystal.

8.2 Experimental Results on Surface States

PES is a surface sensitive technique and can thus measure the electronic properties of the topmost surface layers. In the preceding section it was shown that the very geometry of the solid, a crystal terminated by a surface, leads to a modification of the bulk-band structure reflected by the occurrence of surface states (Shockley [8.8] and Tamm [8.22] states).

For semiconductor crystals, the existence of surface states has been commonly assumed for many years. Indeed many of the properties of interfaces containing semiconductors are intimately related to the existence of these surface states (both empty and occupied). It is therefore astonishing that the PES results available seem more abundant and probably better understood for metals than for semiconductors.

By definition, a surface state has to appear in a band gap of the bulk-band structure. If for simplicity one assumes that it is confined just to the topmost surface layer, its $\boldsymbol{k}$-dependence will be described in the two-dimensional surface Brillouin zone (SBZ). The spatial arrangement of the atoms of a fcc metal is given in Fig. 8.5. A bulk state is characterized by $E_{\mathrm{B}}(\boldsymbol{k})$ where $\boldsymbol{k} = (k_{\perp}, \boldsymbol{k}_{||})$. A surface state is characterized by $E_{\mathrm{S}}(\boldsymbol{k}_{||})$. If the surface state $E_{\mathrm{S}}(\boldsymbol{k}_{||})$ is to be in a gap, then there can be no $E_{\mathrm{B}}(k_{\perp}, \boldsymbol{k}_{||})$ with the same $\boldsymbol{k}_{||}$, irrespective of $k_{\perp}$. From this it follows that for a particular surface one will find surface states in those areas of the SBZ that represent gaps in the projected three-dimensional density of states.

Let us take the (111) surface of Cu as an example. The three-dimensional band structure is shown in Fig. 7.7 and the corresponding (111) SBZ is shown in Fig. 8.9. In order to identify gaps available for surface states at $\bar{\Gamma}$ in the SBZ one has to project the band structure along [111] (ΓL direction) onto

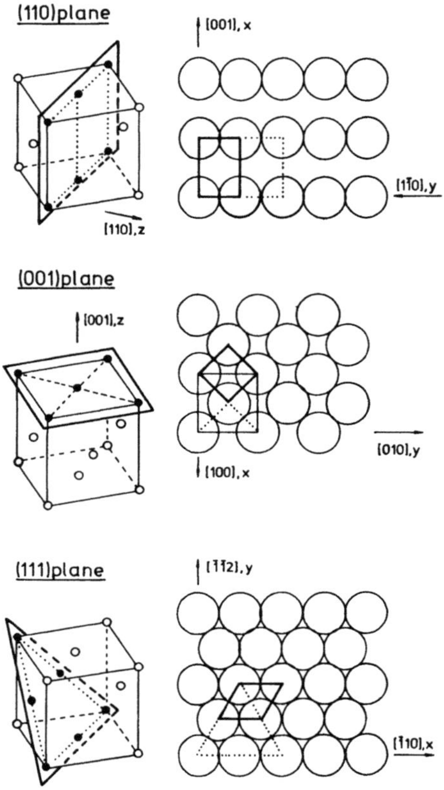

Fig. 8.5. (100), (001) and (111) surfaces of a fcc crystal. *Left*: (110), (001) and (111) cuts through the crystal unit cell and atomic arrangement in the plane of the section. *Right*: Surface Brillouin zones corresponding to the cuts on the left-hand side [8.23]

this surface. From the inspection of the three-dimensional band structure (Fig. 7.7) it is evident that a gap will exist around $\bar{\Gamma}$ for binding energies $1.0\,\mathrm{eV} \geq E_1 \geq E_\mathrm{F}$ and $5.1\,\mathrm{eV} \geq E_2 \geq 3.7\,\mathrm{eV}$ because along ΓL there are evidently no states in these energy intervals. Thus if one performs a normal emission experiment on a Cu(111) crystal, states observed in the two energy regimes E_1 and E_2, can only be surface states. The (111)-projected density of states of Cu is shown in Fig. 8.6 and at $\bar{\Gamma}$ (which is the projection of the ΓL line onto the SBZ) the two predicted gaps are observed [8.23]. A similar result is found for the two other noble metals Ag and Au.

The experimental verification of the surface state near the Fermi energy is given in Fig. 8.7 [8.24], where normal emission spectra for Cu(111), Ag(111), Au(111) are displayed, and indeed for each material a surface state (S) is observed, situated between the top of the d band (around 2 eV in Cu and Au and around 4 eV in Ag) and the Fermi energy. (Note that the spectra have been taken with an Ar discharge lamp which induces the satellite "sat". This satellite originates from the Ar II radiation which is excited, although weakly, in the lamp).

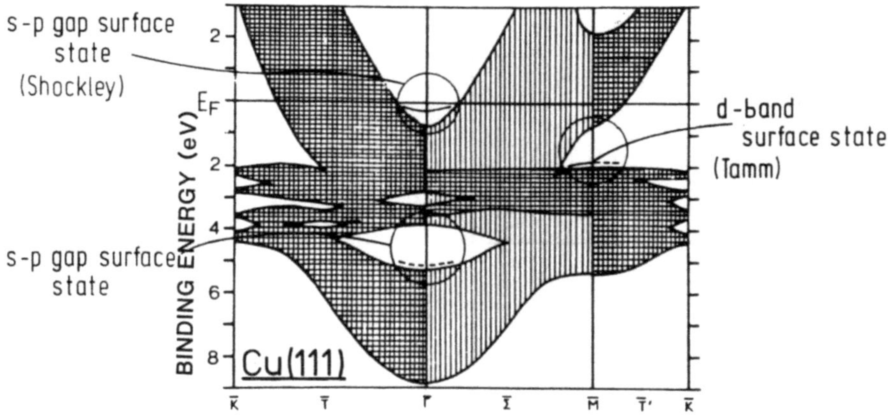

Fig. 8.6. Density of states of Cu projected onto the (111) plane [8.23]. This provides the gaps in which a surface state can be expected to show up: near $\bar{\Gamma}$ with binding energies $E_\mathrm{F} \le E_\mathrm{S} \le 1$ eV, near $\bar{\Gamma}$ with $3.7\,\mathrm{eV} \le E_\mathrm{S} \le 5.1\,\mathrm{eV}$ and near $\bar{\mathrm{M}}$ with $1\,\mathrm{eV} \le E_\mathrm{S} \le 2\,\mathrm{eV}$

Since a surface state has a definite $\boldsymbol{k}_{||}$ whereas its $k_\perp$ is undefined, one has an obvious prerequisite that a structure in an EDC must fulfill in order to be classified as a surface state. Namely its energy must be independent of the exciting photon energy for a particular fixed $\boldsymbol{k}_{||}$.

A measurement of the dispersion $E(\boldsymbol{k}_{||})$ [$= E_\mathrm{S}(\boldsymbol{k}_{||})$] of the surface state in the L gap of Cu(111) is shown in Fig. 8.8 [8.25]. Here the electron detection angle was progressively tilted away from the normal towards the $[11\bar{2}]$ direction, which allows one to measure the dispersion along the $\bar{\Gamma}\bar{\mathrm{M}}$ direction of the SBZ. The spectra taken with $\hbar\omega = 16.85\,\mathrm{eV}$ are shown in Fig. 8.8 and a plot of the dispersion is given in Fig. 8.9. This latter figure also shows the appropriate SBZ and the part of the projected band structure relevant to the experiment. The gap closes at the wave vector of the "neck" of the Fermi surface which is approximately $0.25\,\mathrm{\AA}^{-1}$. Note that the Fermi surface of the noble metals consists of spheres which are connected by so-called "necks" along the (111) direction. The neck area is the shaded insert in the surface Brillouin zone in Fig. 8.9 or the wave vector at which the boundary of the projected occupied bands intersects the Fermi energy. The dispersion curves for the surface state taken with two different photon energies roughly agree, as indeed they should if the state is to be a true surface state because its energy should not depend on $k_\perp$. The surface state dispersion can be approximated by $E(\mathrm{eV}) = +0.36 + 0.03k_{||} + 5.29k_{||}^2$ ($k_{||}$ in units of $\mathrm{\AA}^{-1}$), showing no deviation of the dispersion from the parabolic free-electron type.[3]

[3] The linear term ($0.03k_{||}$) is a consequence of the accuracy of the determination of $k_{||}$ in the experiment and therefore indicates an experimental problem. Any

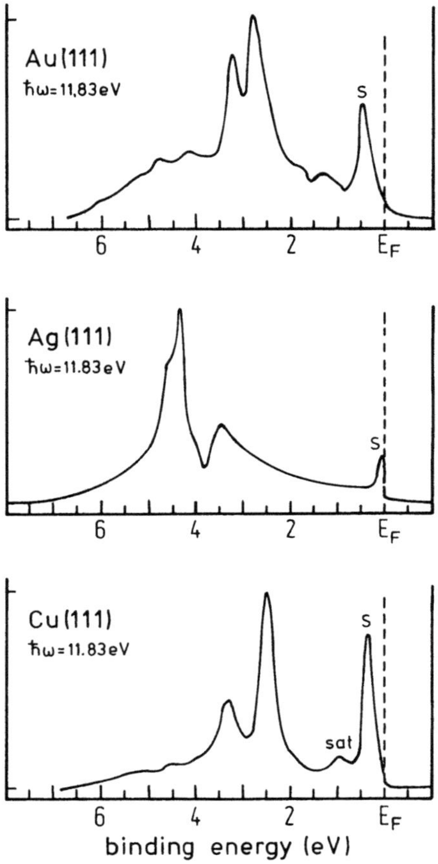

Fig. 8.7. UPS spectra of the (111) surfaces of Cu, Ag and Au with $\hbar\omega = 11.83\,\mathrm{eV}$. For all three metals a surface state (label S) is observed near E_F in the so-called L-gap at $E_F \leq E_S \leq 1\,\mathrm{eV}$, see Fig. 8.6 [8.24]

A summary of the dispersion curves for the L-gap surface states in the three noble metals is given in Fig. 8.10 [8.26]. In each case the dispersion can be approximated by a parabola yielding effective masses m^*/m_e of 0.46, 0.53 and 0.28 for Cu(111), Ag(111) and Au(111), respectively [8.27]. High resolution spectra of the L-gap surface states on the (111) surfaces of the noble metals have been presented in Figs. 1.26, 1.27 and 1.28, where also the most recent literature is given.

According to Fig. 8.6 the Cu(111) surface should also contain a d-band split-off surface state at the $\bar{\mathrm{M}}$ point of the SBZ. This surface state should show up in experiments performed on a (111) crystal with the electron detection angle scanned in the $\bar{\Gamma}\bar{\mathrm{M}}$ azimuth. The corresponding data are reproduced in Fig. 8.11 (where the plane of the electron detection is the ΓKLUX

deviation from the free-electron-type behavior would show up in terms even in $k_{||}$.

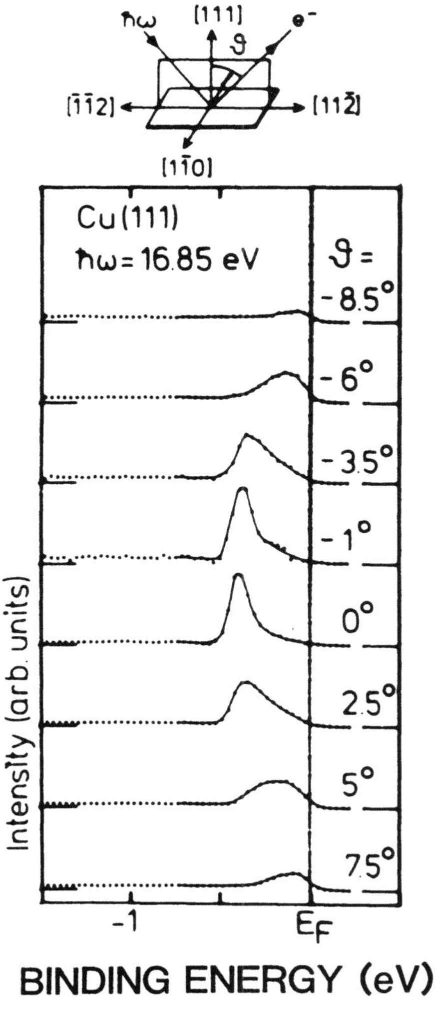

Fig. 8.8. Measurement of the dispersion of the L-gap surface state for Cu(111) (near normal emission, $\hbar\omega = 16.85\,\mathrm{eV}$) The emission direction is in a $(1\bar{1}0)$ plane and it is tilted towards $[\bar{1}\bar{1}2]$ and $[11\bar{2}]$. In going away from the normal the surface state disperses towards the Fermi energy. The $\vartheta = 0$ energy is the same as in Fig. 8.7, ($\hbar\omega = 11.83\,\mathrm{eV}$), indicating that one is indeed observing a surface state. Note that the contribution of the satellite of the Ne I radiation to the spectra has been subtracted from the measured experimental data [8.25]

plane). Indeed a very sharp structure [8.28] marked with S shows up in an angular range of $\approx 50°$. This structure cannot be accounted for by direct transitions, thus indicating that it is the expected surface state.

We so far have the information that the state S occurs at the anticipated position in a gap, and that it cannot easily be ascribed to a bulk direct transition. But is this the only evidence for assigning this state to a surface state? Actually, it is not: one can also exploit the fact that the surface state should show a $\boldsymbol{k}_{\|}$ dispersion that is independent of the incident photon energy. Data from such an experiment are shown in Fig. 8.12, where it is seen that the measured dispersions for $\hbar\omega = 21.2\,\mathrm{eV}$ and $\hbar\omega = 16.8\,\mathrm{eV}$ agree. The shaded

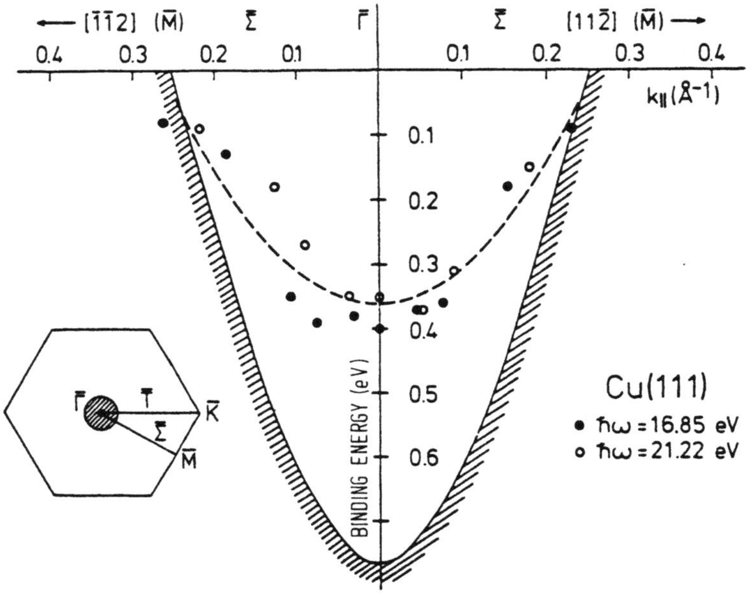

Fig. 8.9. Energy dispersion of the L-gap surface state in Cu(111) along the $\bar{\Sigma}$ line ($\bar{\Gamma}\bar{M}$). Data points (obtained as in Fig. 8.8) are given for $\hbar\omega = 16.85\,\mathrm{eV}$ and $\hbar\omega = 21.22\,\mathrm{eV}$, and the coincidence of these data shows that one is observing a surface state. The dashed line is a parabola: $E(\mathrm{eV}) = +0.36 + 0.03k_{||} + 5.29k_{||}^2$ ($k_{||}$ in units of $Å^{-1}$). The full line is the boundary of the projected three-dimensional density of states (Fig. 8.6) [8.25]

area is the projected density of states and again one sees that the surface state is observed only in the gap of the projected bulk-band structure [8.29].

Further evidence that the structure S in Fig. 8.11 is emission from a surface state comes from the fact that the same point in $\boldsymbol{k}$-space can be reached by using a (100)-crystal with detection in the ΓKLUX plane. Figure 8.13 gives a section through the three-dimensional Brillouin zone with the SBZs for the [111] direction and that for the [100] direction. The area in momentum space covered by the spectra in Fig. 8.11 is shaded around the [111] direction [ΓL direction; a $\bar{\Gamma}\bar{\mathrm{M}}$ scan in the (111) SBZ]. An experiment with a (100) crystal and ΓKLUX electron detection plane, will cover the momentum space also shaded in Fig. 8.13 [$\bar{\Gamma}\bar{\mathrm{X}}$ scan in the (100) SBZ]. This figure indicates that a limited area in $\boldsymbol{k}$-space is accessible using both crystal faces, a trivial fact which underlies the triangulation procedure (Sect. 7.3.1).

The spectra observed in the ΓKLUX electron detection plane taken with $\hbar\omega = 21.2\,\mathrm{eV}$ radiation and using Cu(111) and Cu(100) crystals are shown in Fig. 8.14. From the arguments given in Sect. 7.3.1 a measurement at the same

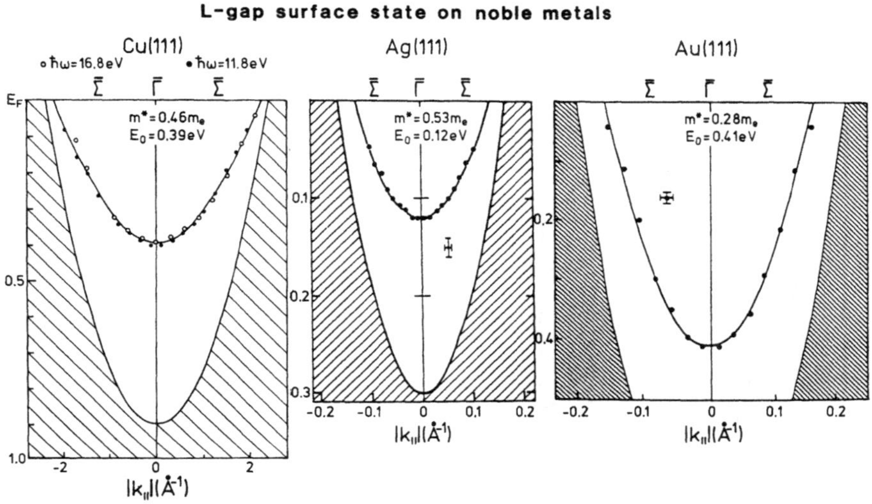

Fig. 8.10. Energy dispersion relations for the L-gap surface states on Cu(111), Ag(111) and Au(111) [8.26]. The solid line is a parabolic fit to the data yielding effective masses m^*/m_e of 0.46, 0.53 and 0.28 for Cu, Ag and Au, respectively. The shaded regions are the projected bulk continua of states. The surface states enter the bulk continua above the Fermi energy

point in $\boldsymbol{k}$-space from the two surfaces is achieved when the energies of the bulk transitions observed from the (100) crystal agree with the corresponding ones from the (111) crystal. We see that spectra at $\approx 50°$ from the (111) crystal and those at $\approx 20°$ of the (100) crystal show this energy coincidence, however, the feature S is only present in the (111) crystal data, as it should be if it is a surface state in that plane and not a bulk direct transition [8.29].

There is another (very popular) test to determine whether a structure observed in an EDC is a surface state. Since surface states are confined to the topmost layer(s) of a crystal, they will be significantly affected if the surface is altered somehow, for example by exposing the sample to small amounts of an adsorbate gas (H, O, CO). One can then watch the surface state disappear, or at least alter its intensity as compared to that of a bulk state.

Such an experiment is shown for the (111) surface of Cu in Fig. 8.15 [8.23, 8.30, 8.31]. It compares an EDC of the clean surface with that of the same surface exposed to 1200 L (1 L $= 10^{-6}$ Torr · s) oxygen (the spectra are normalized at point D). Structures ascribed to surface states (S_1 and S_3; according to Fig. 8.6 state S_1 is the surface state in the L gap, see also Fig. 8.8; S_3 is less well documented and is perhaps a state expected in the gap between ~ 4 eV and 5.5 eV binding energy) are significantly reduced in intensity after the gas exposure. However, there are, in addition, other intensity changes in the spectrum. These are also likely to be attributable to the coverage of the

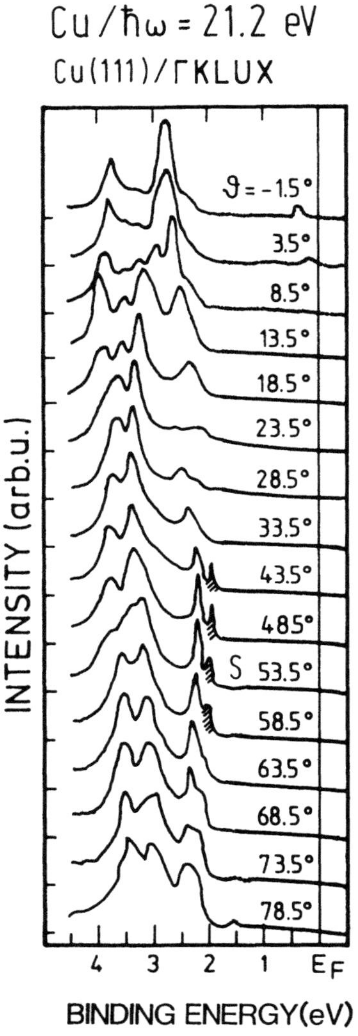

Fig. 8.11. Measurement of the surface state around $\bar{\mathrm{M}}$ on a Cu(111) crystal with $\hbar\omega = 21.2\,\mathrm{eV}$ (see Fig. 8.6 for identification of the area in the two-dimensional Brillouin zone) [8.28]. The sharp structure S showing up at the top of the d band around $\vartheta = 53.5°$ is the anticipated Tamm state

surface with oxygen; the adsorption disrupts the periodicity of the top layer, and thus results in a relaxation of the $\boldsymbol{k}$-selection rule making new transitions possible. This leads to density-of-states contributions in the spectrum. It is known that the region at the top of the d band is especially sensitive to surface modifications and therefore the large decrease in intensity in this area is no surprise (Sect. 7.4.3; Fig. 7.52). The adsorbate test, however, may give misleading information if used in isolation.

So far we have dealt with the surface states below the Fermi energy and their detection and characterization by PES. There are also surface states between the Fermi energy and the vacuum level. These may be viewed as

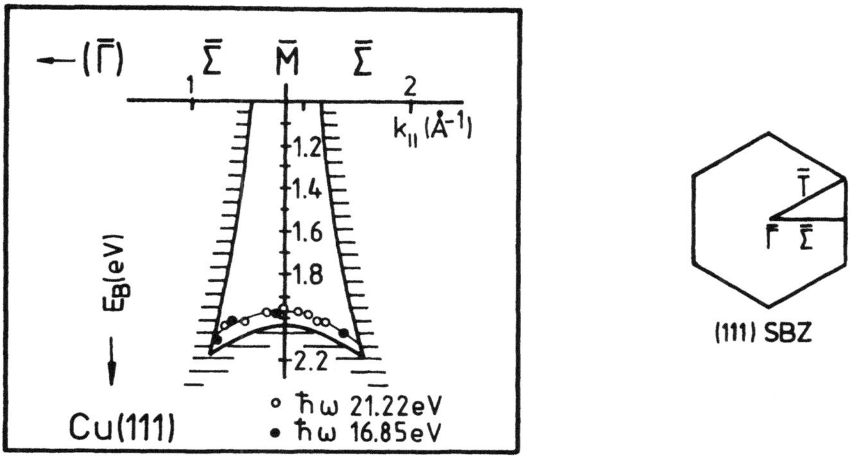

Fig. 8.12. Dispersion of the surface state (Tamm state) around $\bar{M}$ in the (111) surface Brillouin zone of Cu obtained from the spectra in Fig. 8.11. The fact that the $\hbar\omega = 16.85$ eV data and the $\hbar\omega = 21.22$ eV data agree, signals a surface state [8.29]

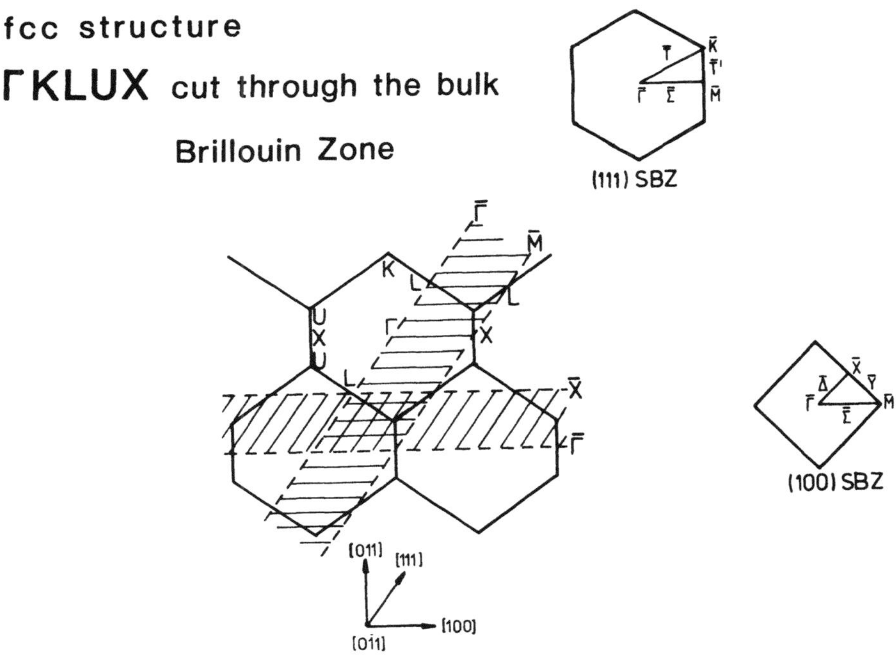

Fig. 8.13. ΓKLUX cut through the fcc Brillouin zone with the (111) and (110) surface Brillouin zone respectively. The data in Fig. 8.11 have been taken in the area shaded between the $\bar{\Gamma}$ and $\bar{M}$ rods in the cut, corresponding to a scan along $\bar{\Gamma}\bar{M}$ in the (111) surface Brillouin zone [8.29]

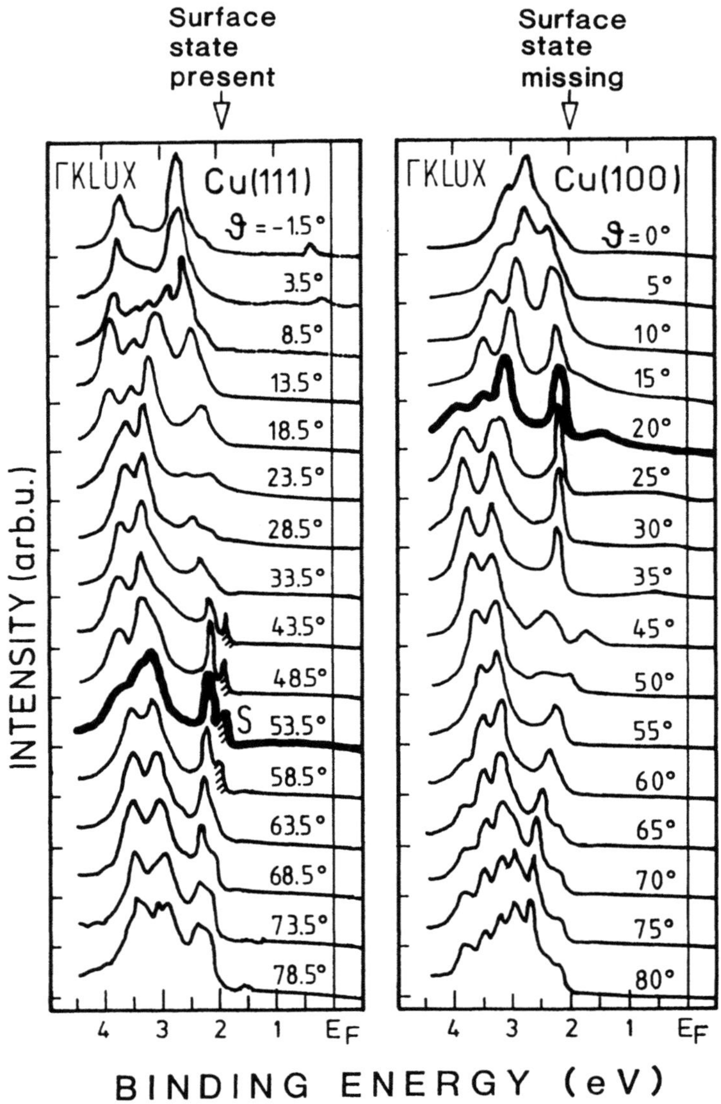

Fig. 8.14. Triangulation measurement to confirm the existence of a surface state around $\bar{\mathrm{M}}$ (Fig. 8.12). Data from a Cu(111) and a Cu(100) crystal both taken in the ΓKLUX electron emission plane (Fig. 8.13) are compared. The measurement from the (100) plane corresponds to a $\bar{\Gamma}\bar{\mathrm{X}}$ scan in the surface Brillouin zone; it is shaded in Fig. 8.13. Figure 8.13 also shows that there is an area in $\boldsymbol{k}$-space that is reached from both crystal faces. The spectra from the (111) crystal around $\vartheta \simeq 50°$ coincide in energy with the spectra from the (100) crystal around $\vartheta \simeq 20°$, indicating that one is observing transitions at the same position in the bulk Brillouin zone. However, the surface state is only visible in the (111) crystal data because it only occurs on the (111) and not on the (100) surface [8.29]

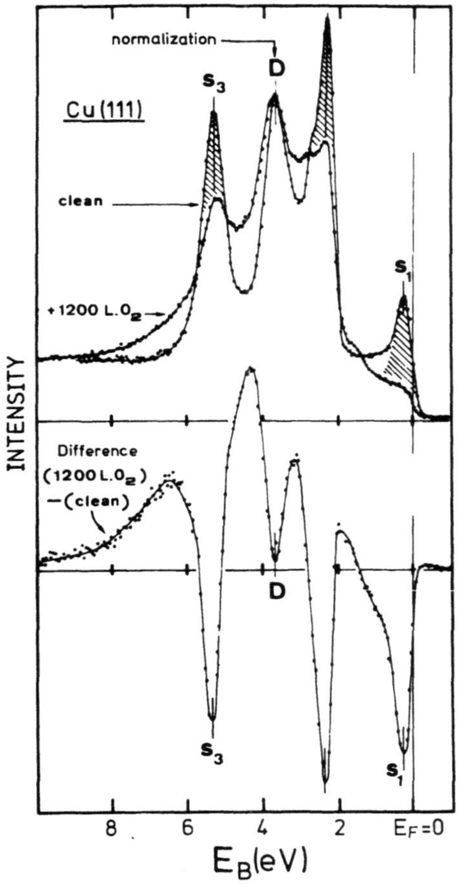

Fig. 8.15. Adsorption test for the presence of a surface state: Cu(111) [8.23, 8.30]. Upon adsorption of 1200 L O_2, structures at $\approx 0\,\text{eV}$, $\approx 2\,\text{eV}$ and $5.5\,\text{eV}$ binding energy disappear (normalization of the spectra at structure D) as shown clearly by the difference spectra of the clean and the oxygen covered crystal. S_1 and S_3 are probably the surface states anticipated around $\bar{\Gamma}$ (Fig. 8.6). The weakening upon adsorption of the emission from the top of the d band can indicate that density of states transitions also contribute to the spectrum and it is known that the surface density of states peaks near the top of the d band (Fig. 7.52)

electrons trapped between the crystal surface and potential barrier which terminates the crystal [Fig. 8.3; (8.35)]. Such states, which cannot be detected by ordinary PES because they are not occupied, have been extensively investigated by IPES (Chap. 9). This technique, in which an electron is stopped on the surface, is clearly well suited to measure states above the Fermi energy. There is, however, a very clever mode of PES, namely two-photon PES, which can also be used with good success to investigate the surface states between the Fermi energy and vacuum level. The technique consists of using a laser with a fixed frequency for the PES experiment where a first photon populates an intermediate "long-lived" state between the Fermi energy and the vacuum level and a second photon subsequently excites electrons out of this intermediate state to above the vacuum level such that they can be observed as in an ordinary PES experiment. The energy of the intermediate state can be calculated from the photon energy and the kinetic energy of

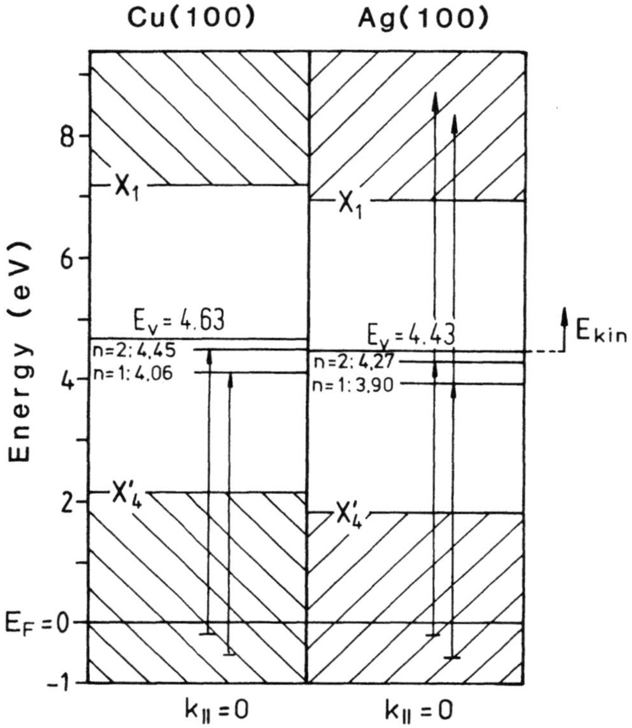

Fig. 8.16. Energy diagram for the states on the Cu(100) and the Ag (100) surface. Energies are referenced to the Fermi energy. The hatched areas give occupied states of the projected two-dimensional band structure. The gaps between ≈ 2 eV and ≈ 7 eV are clearly seen. Also indicated are the positions of the $n = 1$ and $n = 2$ image potential states. A laser photon energy of 4.55 eV for Cu and 4.35 eV in Ag can populate the image potential states on the surfaces. Out of the populated image potential states, electrons can be photoemitted into the vacuum by a second laser photon [8.32]

the liberated electron [8.32–8.35]. We note that empty bulk states situated between the vacuum level and the Fermi energy can also be determined by two-photon PES [8.35].

The measurement of unoccupied surface states with two-photon PES will be exemplified by data from the Cu(100) and the Ag(100) surface. It can be seen from Figs. 7.7 and 7.24 that for both surfaces there is a gap in the bulk-band structure ranging from about 2 eV to 7 eV above the Fermi energy. The projected bulk-band structure for Cu(100) and Ag(100) is given in Fig. 8.16, which also indicates the position of the $n = 1$ and $n = 2$ image potential (barrier-induced) surface states. If laser light with an energy slightly smaller than the work function [e.g., 4.55 eV for Cu(100) and 4.35 eV for Ag(100)] is used, all states between the Fermi energy and the vacuum level can be

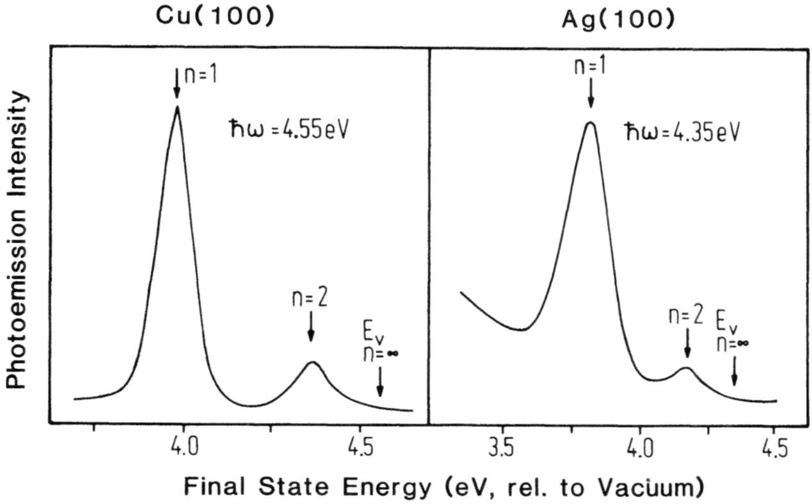

Fig. 8.17. Two-photon PES spectra from the Cu(100) and Ag(100) surface for the situation sketched in Fig. 8.16. The $n = 1$ and $n = 2$ image potential states are visible and their position relative to E_v can be calculated by subtracting the photon energy from the measured final state energy [8.32]

populated, but no electrons can be liberated into the vacuum from the bulk states below the Fermi energy. Thus no photocurrent from the states below the Fermi energy is observed. However, the high intensity of the laser light enables the $n = 1$ and $n = 2$ image potential states, populated by a first photon, to emit electrons via a second photon. The corresponding PES intensity is shown in Fig. 8.17 giving the positions of the $n = 1$ and $n = 2$ states for Cu(100) and Ag(100). The energy relative to E_F of the $n = 1$ state in Cu(100) is calculated as $\phi + E_{\text{kin}} - \hbar\omega = 4.63\,\text{eV} + 3.98\,\text{eV} - 4.55\,\text{eV} = 4.06\,\text{eV}$ above E_F (or 0.57 eV below E_v).

In two-photon photoemission spectroscopy with time resolution, the lifetime of some image potential states has been measured [8.36]. In these experiments the pump-probe technique was used. This means that the image potential state is populated out of the bulk conduction electron states by a short pulse of laser radiation (95 fs duration). After a delay time the magnitude of the population in the image potential state is measured by a probe pulse (70 fs duration). After the population pulse has been switched off, the image potential state decays exponentially with a characteristic image potential lifetime. Therefore for positive delay times between the pump and the probe pulse the two photon photoemission signal should show an exponential decay (straight line on a semi-logarithmic plot). Data for this effect for the $n = 1, 2, 3$ image potential states on Cu(100) and Ag(100) (see Fig. 8.17) are shown in Figs. 8.18 and 8.19. The full lines in these figures are the expo-

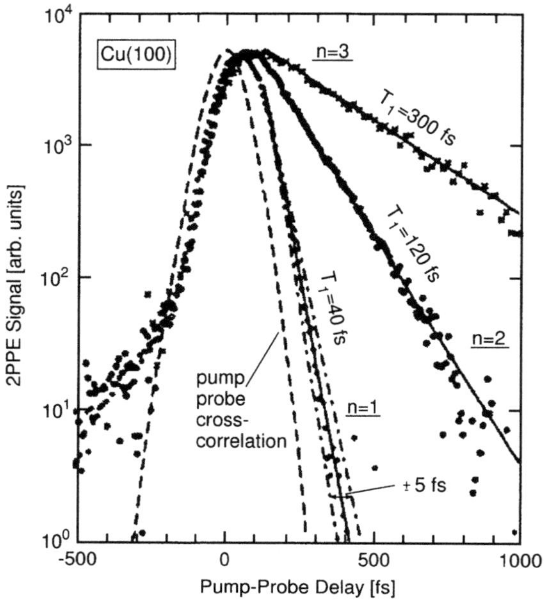

Fig. 8.18. Lifetime of the image potential states on Cu(100) as measured by a pump-probe experiment. The pump-probe cross correlation curve is shown by a dashed line [8.36]

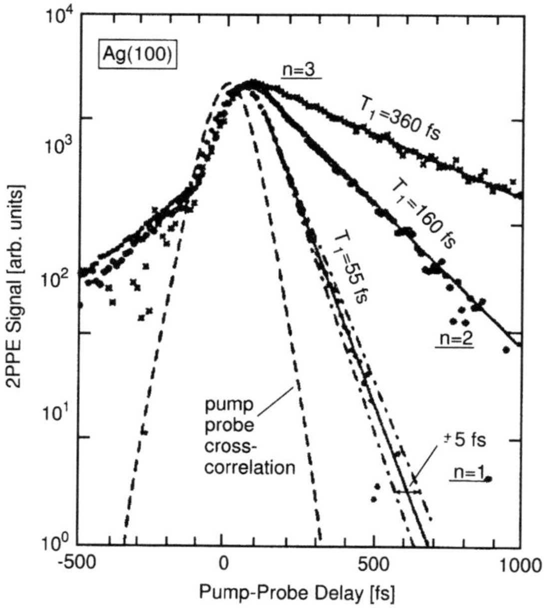

Fig. 8.19. Same as Fig. 8.18 for Ag(100) [8.36]

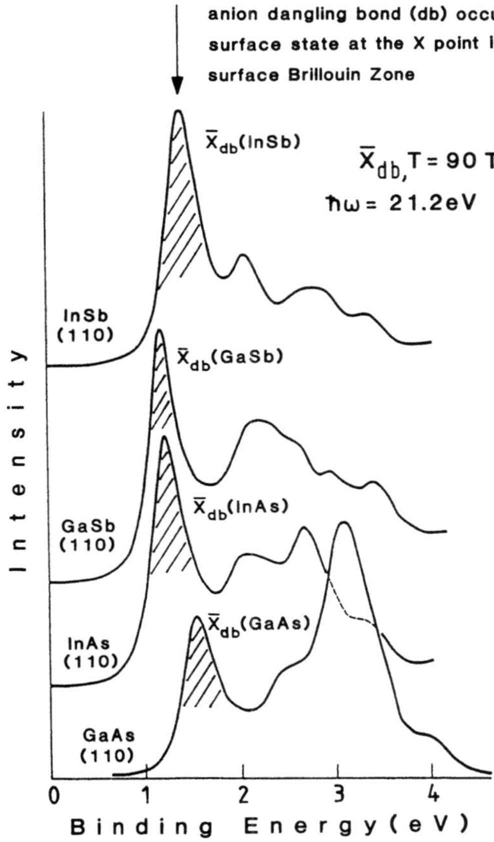

Fig. 8.20. Anion dangling bond surface states on the (110) surface of InSb, GaSb, InAs and GaAs. Experimental conditions are chosen such that the $\bar{X}$ point is sampled in the surface Brillouin zone [8.38]

nentials yielding the lifetimes indicated on the decay curves. For the $n = 1$ state the spread of a ± 5 fs lifetime is indicated by the dash-dotted line. The dashed line is the pump-probe cross correlation reference (zero delay time). According to the Echenique–Pendry model, the lifetimes should scale with n^3 (n is the quantum number of the image potential state (8.35)), a prediction which is roughly fulfilled [8.12].

Surface states have also been observed on semiconductor surfaces [8.5, 8.37, 8.38]. A typical example is given in Fig. 8.20 [8.38], which shows spectra for the III-V semiconductors InSb, GaSb, InAs and GaAs measured from the (110) surface under conditions where the anion dangling bond occupied surface state is clearly visible (hatched structure).

The analysis and understanding of data as in Fig. 8.20 is not yet as complete as that for, e.g., the surface states on the surfaces of the noble metals.

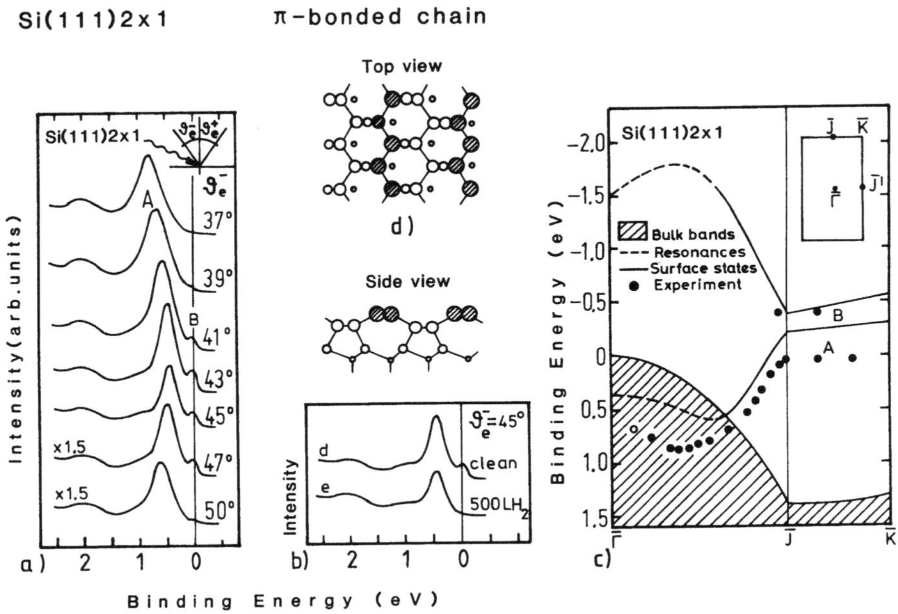

Fig. 8.21. Surface state on the Si(111) 2 × 1 surface [8.5]. (**a**) PE spectra from a heavily n-doped crystal along the $\bar{\Gamma}\bar{J}$ direction of the surface Brillouin zone. A and B correspond to emission from the bonding and antibonding surface state respectively. The $\bar{J}$ point is reached at $\vartheta_e^- = 43°$. (**b**) Sensitivity of the surface states A and B to hydrogen exposure ($\vartheta_e^- = 45°$). (**c**) Comparison between the experimental surface states (A and B) and a theoretical prediction. The hatched area is the projected bulk density of states. (**d**) Geometrical view (π-bonded chain model) for the Si(111)-2 × 1 surface

There are many reasons for this situation, among them the complicated reconstructions that take place on the semiconductor surfaces and the derivation of the surface states on the semiconductor surfaces from dangling bond states. The surface states on semiconductor surfaces are important because of their role in the properties of interfaces of semiconductors. An in-depth treatment of this matter would, however, need much space and therefore we refer the reader to the excellent review of Hansson and Uhrberg [8.5] and only give one (and almost the only one) well understood example, namely that of the surface state on the Si(111)2 × 1 surface (Fig. 8.21). [The situation for Ge(111) 2 × 1 is of course equally well understood.] In Fig. 8.21a a few spectra are shown taken with $\hbar\omega = 10.2\,\mathrm{eV}$ radiation at different electron detection angles in order to measure the dispersion. Figure 8.21b gives the comparison of one spectrum taken from a clean surface with one taken after the clean surface has been exposed to 500 L of H_2. The dangling bonds get saturated by hydrogen making the surface states disappear. Indeed, the surface state B disappears almost completely while the state A is only attenuated by the hydrogen exposure. Figure 8.21c depicts the dispersion of the surface states A and B from Fig. 8.21a in the surface Brillouin zone and compares it with a theoretical prediction. The hatched area indicates the projection of the bulk Brillouin zone. The energy is referenced in the diagram relative to the top of the valence band E_{VB}. The data have been taken on a heavily n-doped sample, which allowed the observation of states above the valence-band maximum. In Fig. 8.21d the geometrical structure of the Si(111) 2×1 surface is indicated. The hatched atoms in the top surface layer carry the dangling bonds [8.5].

8.3 Quantum-Well States

A particularly interesting group of surface states are the so-called quantum-well states, which occur in thin overlayers on a substrate. The basic explanation for the occurrence of these states is simple. It can be thought of as originating from an electron traveling perpendicular to the surface within the overlayer. If the thickness of the layer corresponds to a full wavelength, it can be considered as a standing electron wave or what is generally called a quantum-well state. Thus the overlayer–vacuum interface and the overlayer–substrate interface are regarded as potential barriers and in the simplest approximation one views the overlayer system as a (square) potential well. The quantum-well states can thus be calculated by simple quantum mechanics.

The first model for an analysis of these quantum-well states is the Echenique–Pendry model [8.12] as already shown in Fig. 8.3. The extension of the model for the image potential states to the quantum-well states is shown in Fig. 8.22. For the image potential states, the phase relation for their occurrence is $\phi_{\mathrm{B}} + \phi_{\mathrm{C}} = 2\pi n$ where ϕ_{C} is the phase shift at the vacuum–crystal interface and ϕ_{B} the phase shift at the vacuum barrier. In order to take into account the overlayer of thickness $d = ma$, which can be considered as producing a phase shift mka, where a is the lattice constant perpendicular to the surface; the same reasoning yields the phase relation $\phi_{\mathrm{B}} + \phi_{\mathrm{C}} + 2mka = 2\pi n$.

In order to calculate the energy dependence with respect to the two quantum numbers m and n, one uses the relation given below (8.24) [8.39] for the phase shift ϕ_{B}

$$\frac{\phi_{\mathrm{B}}}{\pi} = \left[\frac{3.4\,\mathrm{eV}}{E_{\mathrm{V}} - E}\right]^{1/2} - 1 \tag{8.37}$$

and for ϕ_{C} the following empirical formula:

$$\phi_{\mathrm{C}} = 2 \arcsin\left[\frac{E - E_{\mathrm{L}}}{E_{\mathrm{U}} - E_{\mathrm{L}}}\right]^{1/2} - \pi \tag{8.38}$$

with

E_{V} = vacuum level

E_L = energy at the lower band edge

E_U = energy at the upper band edge;

k is obtained from a two-band model. The graphical solutions for the phase accumulation relation ($\phi_C + \phi_B + 2mka = 2\pi n$) are obtained as demonstrated in Fig. 8.23; these data apply to the system Ag on Fe. Here the thick full lines connect the results for the phase $2\pi n - \phi_C - \phi_B$ and the dashed curve represents the results for $2mka$. In the phase $2\pi n - \phi_C - \phi_B$ the energy dependence comes from (8.37) and (8.38). In the phase $2mka$ the energy dependence is contained in k, which is calculated from the two band model with a lower band edge at -5.5 eV [8.39]. The crossing points for these two sets of curves gives a solution for the phase relation in the case of the quantum-well states. It can be seen that they come out as a function of $m-n$, which can be used as a new quantum number $\nu = m-n$. This simple model describes the energetics of the quantum-well state surprisingly well to a first approximation.

A typical example of experiments in this field is given in Fig. 8.24, which has been obtained for the system of Ag on Fe(100) [8.40]. The coverages are

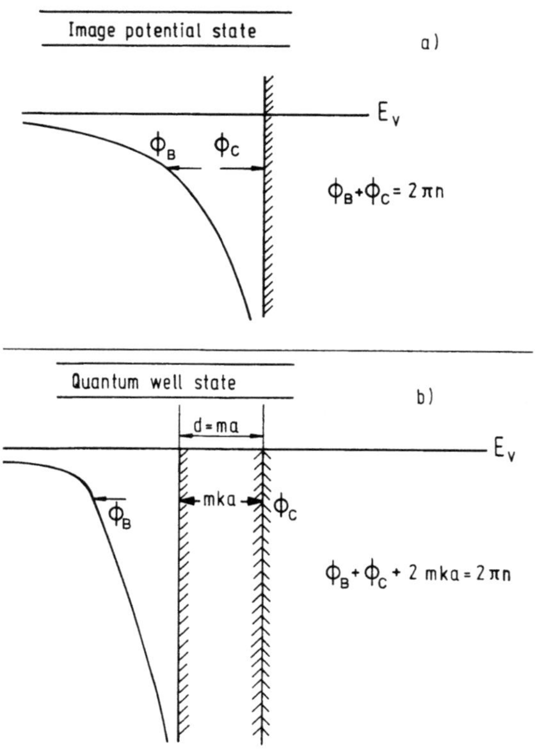

Fig. 8.22. Schematic diagram indicating the origin of the phase accumulation equation for an image potential state (**a**), and a quantum-well state (**b**)

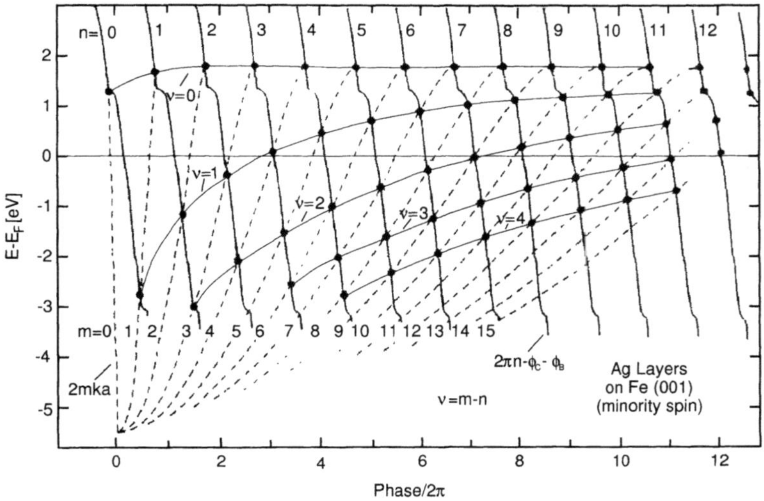

Fig. 8.23. Solution of the phase accumulation model for the quantum well state for the special case of Ag layers on Fe(001) [8.39]

given in monolayers and, except for the cases of 27.5 and 42.5 monolayers, one sees a series of sharp lines as anticipated from the solution depicted in Fig. 8.23. With increasing thickness of the sample, m is increased and, as can be seen from Fig. 8.23, this leads to a closer spacing of the solution for the phase accumulation equation.

The data for 27.5 and 42.5 monolayers consist of spectra from two quantum-well states and will not be considered any further.

One can improve the analysis of these data (Fig. 8.24) by taking the picture of the square well potential for the overlayer on the substrate at face value and using Fabry–Perot solutions for the analysis. This is surprisingly successful as will be briefly outlined now [8.40]. Note that the full lines in Fig. 8.24 are a fit to a Fabry–Perot-type analysis which describes the data perfectly. The background, which has also been fitted in this procedure, is indicated by a full line. The reasoning behind the Fabry–Perot analysis will now be indicated. The starting point, as before, is the model of a standing wave between two potential barriers.

The initial state for an electron in a quantum-well can be thought of a wave being reflected back and forth between the interface barrier and the surface barrier. This results in a modulation of the electron wave of the form (where the phase accumulation relation has been condensed into $2k(E)Nt + \phi(E) = 2n\pi$, meaning $\phi_B + \phi_C = \phi(E)$; Nt is the film thickness, where t is the monolayer thickness and N the number of monolayers in the overlayer:

$$\frac{1}{1 - R_{\exp}[\mathrm{i}(2kNt + \phi)] \exp(-Nt/\lambda)}, \tag{8.39}$$

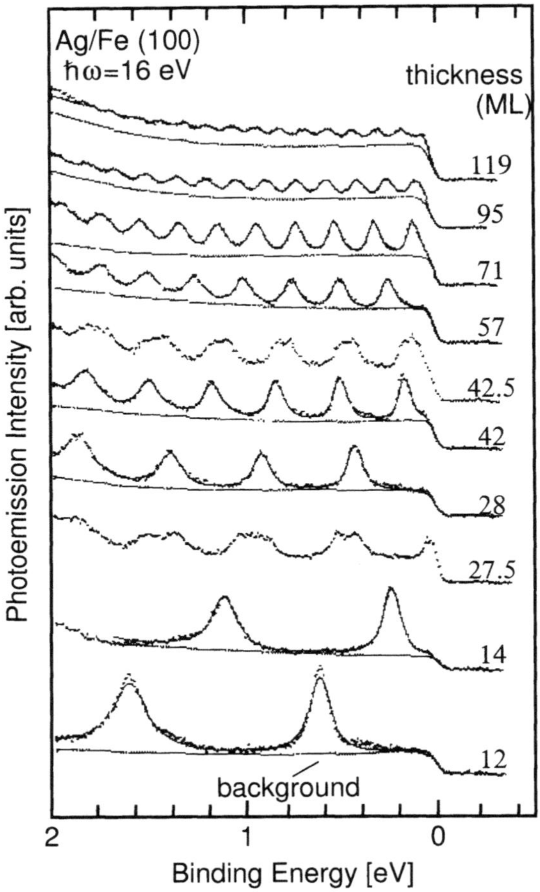

Fig. 8.24. Experimental quantum-well states for Ag on Fe(100) taken with $\hbar\omega = 16\,\mathrm{eV}$ radiation. The thickness of the Ag layer is given in monolayers [8.40]

where R is the product of the reflectivities at the surface and the interfaces, and λ is the quasi-particle mean free path. The mean free path gives rise to a damping and is related to the quasi-particle inverse lifetime Γ, via $\Gamma = v/\lambda$, where v is the group velocity.

The photoemission from a clean bulk (100)-Ag crystal below E_F down to the beginning of the d bands at 4 eV below E_F is a featureless, rising signal cut-off at the Fermi energy, produced by surface photoemission. In the case of a quantum well system, the initial state is not a simple Bloch wave but is modulated by the interference factor (8.39). Thus, into the transition matrix elements between the initial and the final state for the calculation of the photocurrent (1.19) the interference factor above enters in the initial state wave function. The photocurrent, which is proportional to the square of the matrix element, also contains the square of the interference factor, and thus for the photocurrent I we have

$$I \propto \frac{1}{1+(4f^2/\pi^2)\sin^2(kNt+\phi/2)} A(E) + B(E) \ . \qquad (8.40)$$

Here A is a smoothly varying function of energy that describes the strength of the matrix element and $B(E)$ is the smoothly varying background due to inelastic scattering. The factor f is given by:

$$f = \frac{\pi R^{1/2}\exp(-Nt/2\lambda)}{1 - R\exp(-Nt/\lambda)} \ . \qquad (8.41)$$

The equation for the intensity obviously has maxima where the sin function is zero, meaning that one has

$$\begin{aligned} \sin(kNt) + \frac{\phi}{2} &= 0 \\ kNt + \frac{\phi}{2} &= n\pi \\ 2kNt + \phi &= n \cdot 2\pi \end{aligned}$$

which is the phase accumulation relation, also derived in the beginning of this section from Fig. 8.22 following the reasoning of Echenique and Pendry. Thus, as expected, the classical Fabry–Perot treatment coincides with the quantum mechanical results.

While the quantization condition gives the position of the quantum-well states, their width is given by

$$\mathit{\Delta} E = \mathit{\Gamma}\xi \frac{1 - R\exp(-1/\xi)}{R^{1/2}\exp[(-1/2\xi)]} \ , \qquad (8.42)$$

where

$$\xi = \frac{\lambda}{Nt} .$$

When the mean free path is large and the reflectivity close to one, the width of the peak corresponds approximately to the quasi-particle lifetime.

Note that $\boldsymbol{k}$ and ϕ are related to the peak positions, and $\mathit{\Gamma}$ and R are related to the peak width. For fitting the above intensity relation to the actual spectra, it was assumed that the band structure can be described by a two-band model and that the reflectivity, the inverse lifetime and the phase shift are functions of the energy up to second order. The same was assumed for the functions A and B. The values determined for the parameters of the fits are given in Fig. 8.25. It is particularly gratifying that, e.g., the width of the state seems to be in very close agreement with the prediction of Fermi liquid theory, namely behaving as the square of the energy (relative to the Fermi energy).

Nice examples of quantum-well states are found in thin overlayers of xenon on a metallic substrate, in this case Pt(100) [8.41, 8.42]. Here the first layer

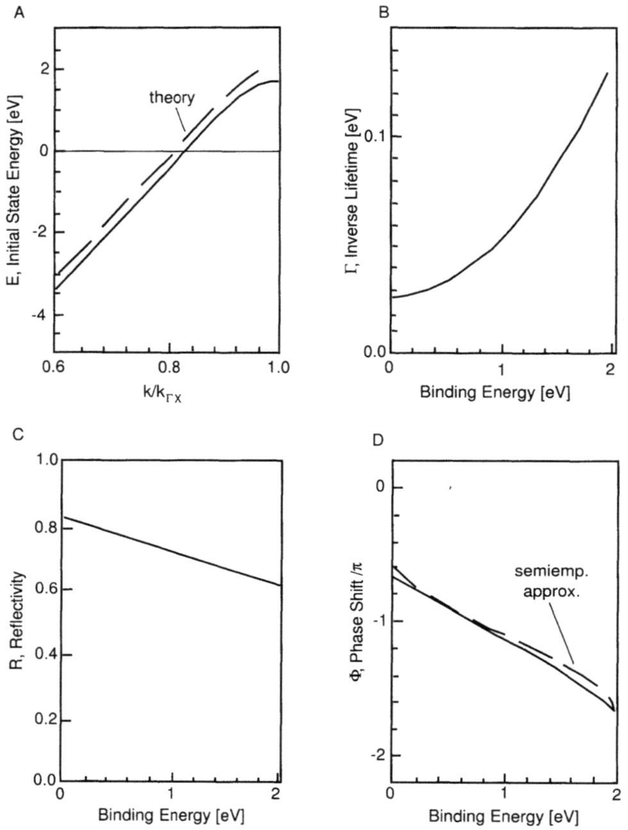

Fig. 8.25a–d. Values for the parameters used to fit the data in Fig. 8.24 using the Fabry–Perot model (parameters are defined in (8.40)–(8.42)) [8.40]

provides a kind of an insulation between the subsequent, undisturbed layers and the metallic substrate and can be regarded as a boundary layer. Beginning with the second layer one observes the quantum-well state. Examples for one additional layer (note that the layers are labelled with $N+1$, where the 1 stands for the first insulation layer) and two additional layers, leading to one and two quantum-well states, are given in Fig. 8.26, where the small shoulder is due to indirect transitions. Figure 8.27 displays the quantum-well state derived from the 5p level of adsorbed xenon and we see that the number of quantum-well states, as expected, rises linearly with the number of xenon layers, seen especially clearly for the $5p_{1/2}$ line.

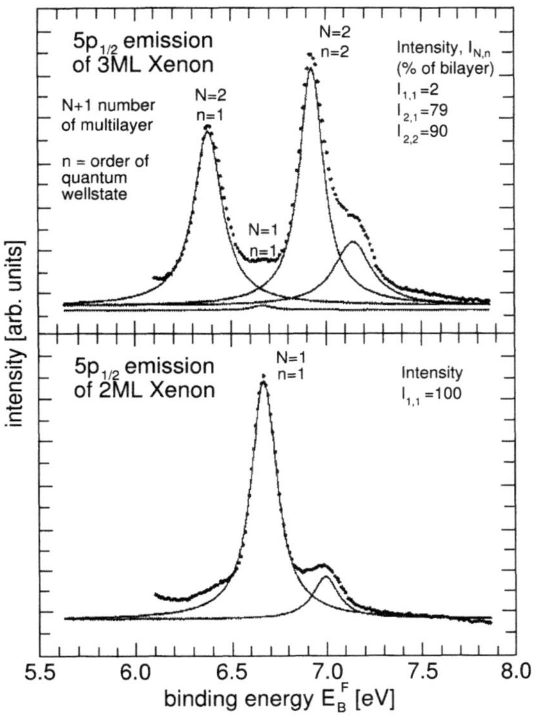

Fig. 8.26. Photoemission spectrum from the $5p_{1/2}$ state of xenon for two and three monolayers of xenon on Pt(100). $N+1$ is the number of monolayers, where the first layer is a spacer layer. n is the order of the quantum-well state [8.41]

8.4 Surface Core-Level Shifts

It was shown in Sect. 7.4.3 that the bulk valence DOS of a material can be different from the surface DOS of that same material. This is due to the fact that the coordination number for atoms in the surface is different from that of atoms in the bulk, leading to a smaller valence-electron density in the surface region. Thus one expects the energy of the core levels of surface atoms to be different from that in the bulk, since the chemical shift of the core levels is largely determined by the valence-electron distribution [8.1,8.3,8.4,8.39–8.75]

Such core-level shifts have indeed been observed (Fig. 8.1) and analyzed, and examples for Ir metal and Yb metal are shown in Figs. 8.28, 8.29 [8.61,8.62]. In Fig. 8.28 the shift is clearly visible in the spectrum from an Ir(100)-(1×1) crystal face, and the surface lines are shifted to smaller binding energies. This can be verified by H_2 adsorption, which lowers the intensity of the surface-core levels while leaving those of the bulk largely unaltered. In addition the figure shows the 4f-core levels of an Ir(100)-(5 × 1) surface. This structure is more densely packed than the unreconstructed (1 × 1) surface

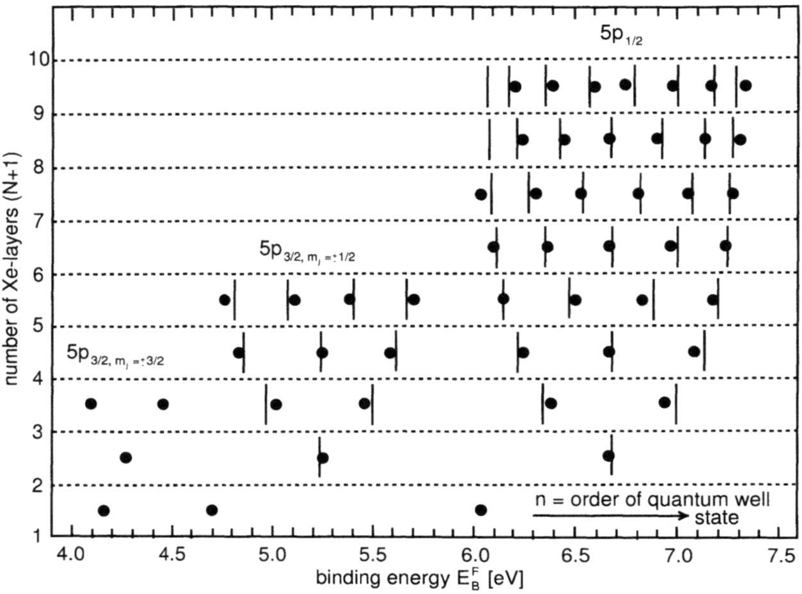

Fig. 8.27. Quantum-well states of xenon on Pt(100) as measured for a number of xenon layers [8.41] They are especially clearly seen for the $5p_{1/2}$ line

and thus, as intuitively anticipated, the surface core-level shift is smaller for this face.

A comparison of the core levels of an Ir(100)-(1 × 1) and an Ir(111) face again shows a difference in the magnitude of the core-level shifts as anticipated from the different coordination numbers for the two faces: 9 for the (111) and 8 for the (100) surface.

A second example of this type is shown in Fig. 8.29 with data from Yb metal [8.62].

The spectra show the $f_{7/2}/f_{5/2}$ spin–orbit split doublet, where the surface line is now shifted to larger binding energies (contrary to the findings for Ir metal). The surface contribution can be recognized by comparing data taken with two different photon energies. At the higher photon energy, the escape depth is increased and the surface contribution is correspondingly reduced [8.63].

Surface core-level shifts are, as can be seen from the two examples, rather small, $\approx 1\,\mathrm{eV}$ or less. For their detection one thus needs high resolution, which is most easily achieved with synchrotron radiation. Secondly one requires a short electron escape depth in order to give a high surface-to-bulk intensity ratio; this argues for electron kinetic energies in the region of 50 eV, a range at which synchrotron radiation is very effective.

Some first principle calculations of the surface core-level binding energy shifts have been made, but here only a qualitative analysis will be given.

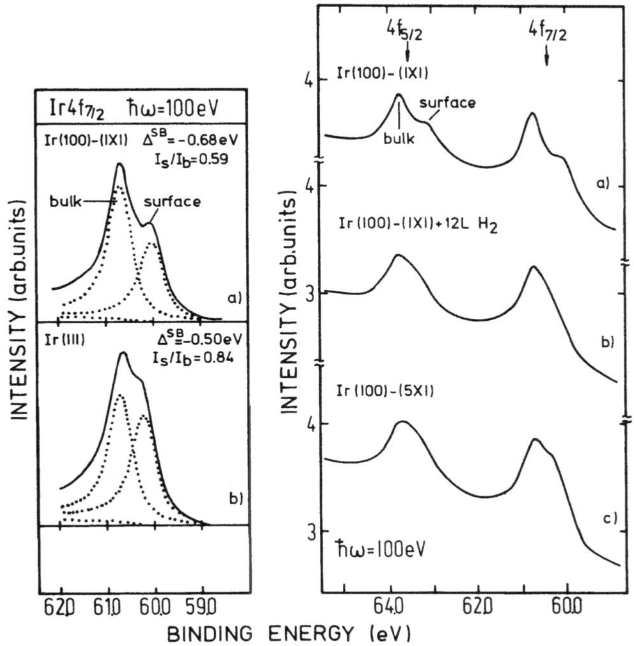

Fig. 8.28. Surface core-level shift in Ir. *Left panel*: $4f_{7/2}$ state from spectra measured with $\hbar\omega = 100\,\text{eV}$ for the Ir(100)-(1 × 1) surface and the Ir(111) surface. The shifts between the bulk and the surface line are $\Delta^{\text{SB}} = 0.68\,\text{eV}$ and $0.50\,\text{eV}$ respectively. *Right panel*: $4f_{7/2,5/2}$ spectrum of Ir(100) taken with $\hbar\omega = 100\,\text{eV}$. The top spectrum is from an Ir(100)-(1 × 1) surface, the bottom one from a more densely packed Ir(100)-(5 × 1) surface, which gives a smaller core-level shift. The adsorption of hydrogen (middle panel) on the (100)-(1 × 1) surface lowers the intensity of the surface peak, which is the one nearer to the Fermi energy [8.61]

In addition we discuss the model of Johansson and Mårtensson [8.49, 8.64], which uses empirical data, yet reproduces the measured data reasonably well.

It is an experimental finding that metals whose valence band consists essentially of sp bands, e.g., Al or Ag (where the d levels are 4 eV below E_{F}), have only small surface core-level shifts (Fig. 8.1), whereas metals that have d electrons at or near the Fermi energy have large shifts (Figs. 8.28, 8.29). This suggests that band narrowing is an important factor in this effect because the former is large for d band metals (in this sense the rare earths are d band metals) and small for sp band metals.

For the purposes of this discussion we take a simple triangular DOS for the d states. One has to distinguish between the cases $n_{\text{d}} > 5$ and $n_{\text{d}} < 5$ because it has been found that the shift changes sign approximately in the middle of a d-series (n_{d} is the number of electrons in the d band) [8.2]. For $n_{\text{d}} < 5$ (Fig. 8.30) one sees that $E_{\text{F}}^{\text{S}} > E_{\text{F}}^{\text{B}}$ if one assumes charge neutrality at the surface, and, in addition, that the energy level scheme stays rigid and

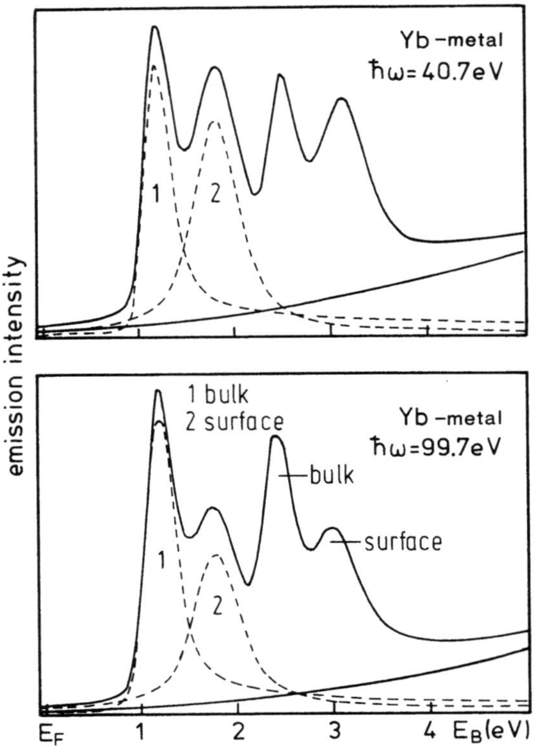

Fig. 8.29. Surface core-level shift on polycrystalline Yb. Each line of the $4f_{7/2,5/2}$ doublet is split because of the surface shift. At $\hbar\omega = 40.7\,\text{eV}$ the surface sensitivity is larger than at $\hbar\omega = 99.7\,\text{eV}$ and therefore at the lower photon energy the surface peak has an enhanced intensity [8.62]

the surface DOS becomes narrowed. If the surface and bulk Fermi energies are brought into coincidence, the surface core levels shift to larger binding energies, as observed.

The opposite effect is seen for $n_{\mathrm{d}} > 5$ (Fig. 8.31). Thus in Ir (with $5d^8$) one has $\Delta^{\mathrm{SB}} = E_{\mathrm{F}}^{\mathrm{S}} - E_{\mathrm{F}}^{\mathrm{B}} < 0$ and in Yb (with $4f^{14}5d^26s$) one has $\Delta^{\mathrm{SB}} > 0$, as observed experimentally.

A quantitative analysis of the *surface core-level shift* can be performed with the thermodynamic model of Johansson and Mårtensson [8.64], which was discussed previously for the calculation of binding energies in metals (Chap. 2). The chemical shift between a free atom and an atom in the metallic state is

$$E^{\mathrm{B}}_{\mathrm{c},E_{\mathrm{F}}}(Z)^{\mathrm{atom-metal}} = I(Z+1) + E^{\mathrm{B}}_{\mathrm{coh}}(Z+1) \\ \times - E^{\mathrm{B}}_{\mathrm{coh}}(Z) + E^{\mathrm{B}}(Z+1, Z);, \tag{8.43}$$

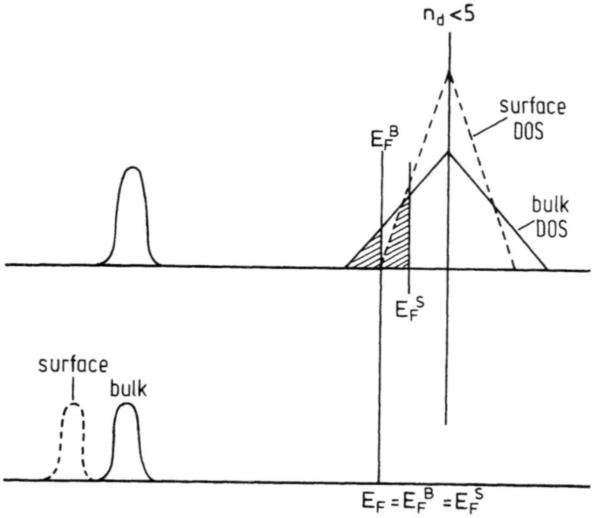

Fig. 8.30. Explanation of the surface core-level shift for $n_{\mathrm{d}} < 5$ and a triangular density of states. The surface density of states is narrowed with respect to that of the bulk. Assuming that the d electron count is the same in the bulk and the surface (alignment of the Fermi energies) and that there is a rigid shift in the energy levels, the surface shifted level is below the bulk state [8.3]

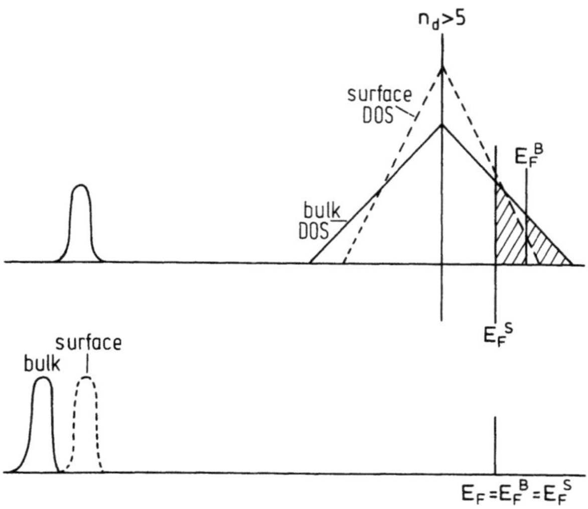

Fig. 8.31. Explanation of the surface core-level shift for $n_{\mathrm{d}} > 5$ and a triangular density of states (see also Fig. 8.24). The surface density of states is narrowed with respect to that of the bulk. Assuming that the d-count is the same for the surface and the bulk species (alignment at the Fermi energies), the surface core level is nearer to the Fermi energy than the bulk one [8.3]

where $E^{\mathrm{B}}_{\mathrm{coh}}$ is the cohesive energy, $E^{\mathrm{B}}(Z+1, Z)$ is the solution energy of metal $Z+1$ in metal Z, the upper index B stands for bulk, and $I(Z+1)$ is the ionization energy of the $Z+1$ ion.

One can derive an analogous equation for the binding energy of an atom on a metal surface

$$E^{\mathrm{S}}_{\mathrm{c},E_{\mathrm{F}}}(Z)^{\mathrm{atom-metal}} = I(Z+1) + E^{\mathrm{S}}_{\mathrm{coh}}(Z+1) \\ \times - E^{\mathrm{S}}_{\mathrm{coh}}(Z) + E^{\mathrm{S}}(Z+1, Z)\,, \tag{8.44}$$

where the upper index S stands for surface. Johansson and Mårtensson [8.64] assumed that

$$E^{\mathrm{S}}_{\mathrm{coh}} \simeq 0.8 E^{\mathrm{B}}_{\mathrm{coh}}\,,$$

which yields

$$\Delta^{\mathrm{SB}} \simeq 0.2[E^{\mathrm{B}}_{\mathrm{coh}}(Z+1) - E^{\mathrm{B}}_{\mathrm{coh}}(Z)] + [E^{\mathrm{S}}(Z+1, Z) - E^{\mathrm{B}}(Z+1, Z)]\,. \tag{8.45}$$

The second term in this equation is frequently neglected because solution energies $E(Z+1, Z)$ are only small fractions of the cohesive energies $E_{\mathrm{coh}}(Z)$ or $E_{\mathrm{coh}}(Z+1)$.

One thus sees that the difference between the surface and bulk binding energy is given essentially by the difference of two cohesive energies. With this simple approach, surface core-level shifts (Δ^{SB}) have been calculated [8.49]

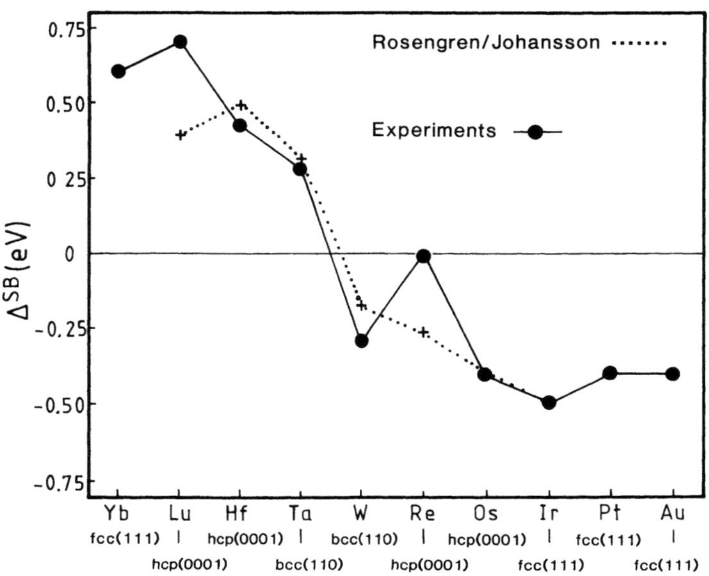

Fig. 8.32. Surface core-level shifts for 5d metals [8.73]. The trend of the thermodynamic calculation reproduces the experimental data quite well

and results for the 5d transition metals are shown in Fig. 8.32 together with some experimental data [8.73]. As one can see, this thermodynamic model predicts a trend in the surface core-level shifts which is in agreement with the qualitative discussion given at the beginning of this section. Since binding energies are positive, a positive Δ^{SB} means that the surface component has a larger binding energy than the bulk component, which is the case for $n_d < 5$ [8.3].

Mårtensson et al. [8.73] and Andersen et al. [8.75] have analyzed the Z-dependence of the core-level shifts in 5d elements in greater detail, as indicated here.

In some cases (Ta, W) in addition to the shift from atoms in the surface layer, shifts from atoms in the first subsurface layer have also been detected (Fig. 8.33). The detection of shifts from the second subsurface layer in W(111) is not certain at this point [8.72]. The data and their analysis in Fig. 8.33 for the $4f_{7/2}$ line of W from a (111) surface yield so far only two surface components. The hard sphere model for a cut through the (111) surface indicates that in principle it should be possible to observe a second sub-surface component because the bulk is reached only at the fourth layer.

We also mention that it has been shown [8.76] that α-type Ce-intermetallics form a surface layer of γ-type material (Sects. 3.3,3.4). This has implications for a quantitative application of the Gunnarsson–Schönhammer theory (Sect. 3.3).

There has been some discussion as to whether the surface core-level shift is an initial- or a final-state effect. The model given in Figs. 8.30 and 8.31 is

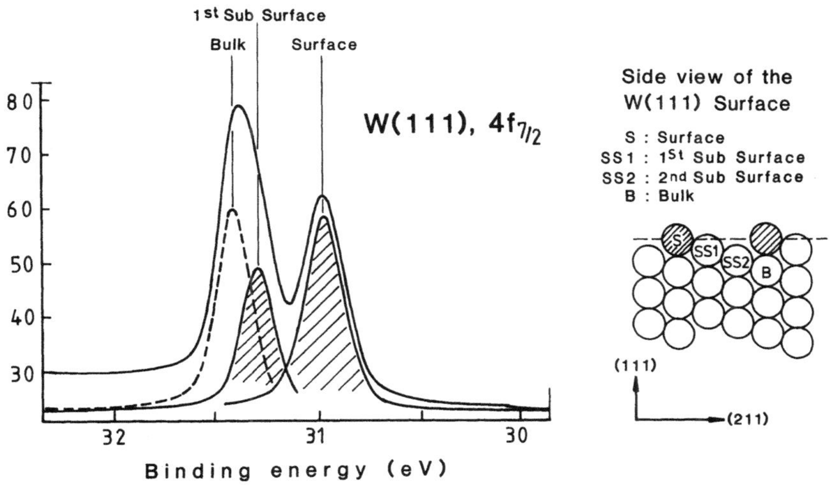

Fig. 8.33. W(111) $4f_{7/2}$ core-level shift [8.72]. The surface component is clearly visible, while the component originating from the first sub-surface layer can only be detected from a least-squares analysis

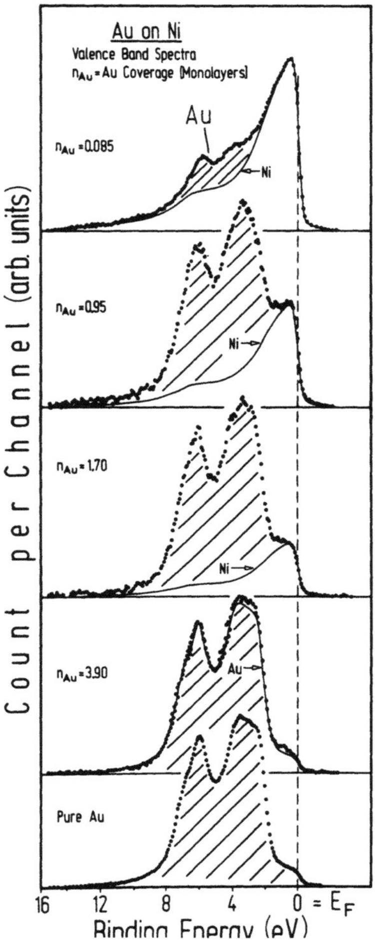

Fig. 8.34. XPS valence-band spectra of Au overlayers on Ni. The overlayer thickness n_{Au} has been measured with a quartz crystal which changes its frequency upon deposition of the Au

based solely on an initial-state picture. Since it works so well, one can assume that the surface core-level shift is indeed predominantly an initial-state effect, a finding that is also confirmed by a more complex analysis [8.3, 8.77].

Finally, some data are presented for a trivial effect in the valence band of the d metal Au, namely the decrease in bandwidth as a function of coordination number (Sect. 7.4.3). Valence-band spectra for Au overlayers on Ni are shown in Fig. 8.34 and one observes that the bandwidth decreases with decreasing coverage. In Fig. 8.35 the data are plotted as a function of the square root of the mean Au-Au coordination number, calculated under the assumption that the Au layers deposit to form a (111) surface. One finds a linear relation between the square-root of the Au-Au coordination number and the Au bandwidth, which is in agreement with the assumption

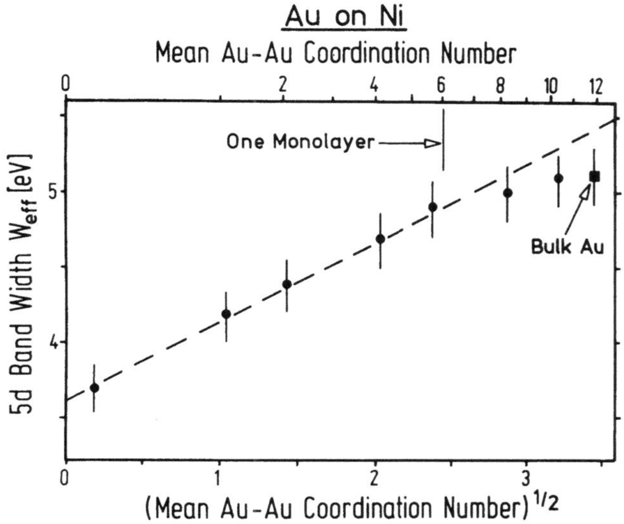

Fig. 8.35. Width of the Au 5d-valence band (from Fig. 8.28) as a function of the square root of the mean Au–Au coordination number under the assumption that the Au deposits as a (111) surface on the substrate (one monolayer corresponds to an Au–Au coordination number of 6 in the fcc structure) and also assuming layer-by-layer growth

that the bandwidth is determined mainly by the nearest-neighbor overlap integrals [8.43].

To end this chapter, an interesting application of a surface core-level shift is presented in Fig. 8.36 where the $4f_{7/2}$ line from a clean W(110) surface is shown and compared to data taken from the same surface covered with one monolayer of O, Cs, K and Na [8.78]. In the spectrum from the clean surface, the signal from the bulk and the surface layer are well separated. Upon covering the surface with one layer of an alkali metal very little change in the W spectrum is observed.

Only if oxygen is adsorbed on the W surface does a large shift in the surface peak occur, while the bulk peak again rests at the position which it had in the clean condition. The interpretation given by Riffe et al. [8.78] for these data concludes that for the alkali metals in the absorption process, very little charge transfer occurs from the adsorbate (alkali metal) to the substrate (W metal) while the adsorption of oxygen on tungsten occurs via a charge transfer.

However, more recent investigations [8.79–8.81] show that these considerations are too simple. One has to take into account the change of the coordination of the surface layer of W relative to that in the bulk in order to analyze the core-level shifts induced by the adsorbates correctly. If this is done, one arrives at the conclusion that the commonly employed charge

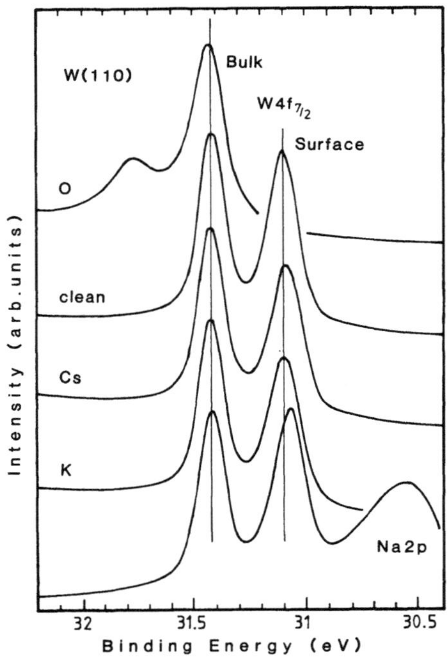

Fig. 8.36. PE spectra of the $4f_{7/2}$ electrons from a clean and adsorbate covered W(110) surface. For O the spectrum is for 1/2 monolayer. For Cs, K and Na the spectra correspond to the densest single layer coverage, i.e. (5.2, 5.9 and 8.0)$\times 10^{14}$ atoms/cm^2 respectively. For Na the emission from the Na 2p levels is also visible [8.78]

transfer picture for the binding of alkali atoms on a metal substrate is at low coverages a useful approximation. On the other hand, further experiments by Wertheim et al. [8.82] seem to support the picture of a covalent bond between the alkali metal and the transition metal substrate.

A particularly nice experiment with respect to the effect of quantum confinement on the nature of an electronic state was performed by Ortega et al [8.83]. These authors measured the photoemission spectra from stepped Au(111) surfaces, namely an Au(322) and an Au(788) surface. These two surfaces represent superlattices with terraces, where the widths of the terraces are $d = 12.8$ Å and $d = 38.7$ Å respectively. A scanning tunneling microscope picture of these two terraces is shown in Fig. 8.37. Noting the very different scales on the diagrams one realizes that the terrace widths are indeed different by about a factor of three. The photoemission spectra from these two superlattices are different as can also be seen from Fig. 8.37. The spectra of Au(322), if analyzed, yield two different parabola as dispersion curves, whereas the spectra of Au(788) result in just two straight lines. This very different behavior can be understood in terms of the energy of the electronic states on terraces of different widths. For one-dimensional states the energy depends dramatically on the width of the terrace and decreases with increasing terrace width. Therefore for the wide terraces of Au(788) one-dimensional states can develop. Their energy is given by

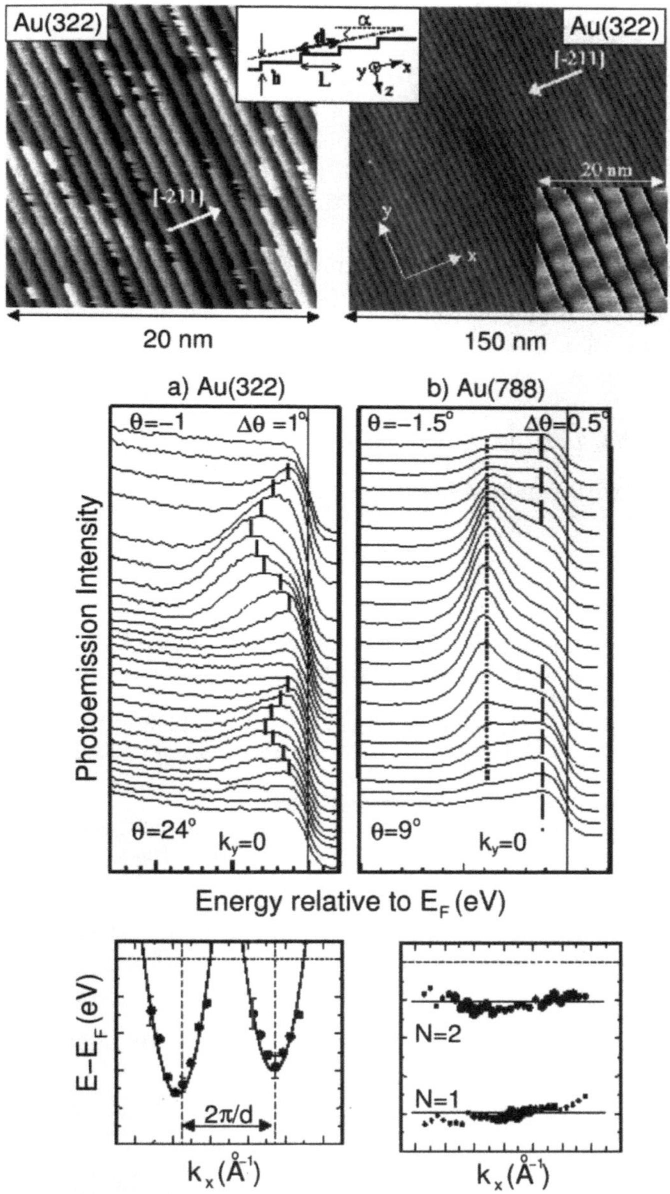

Fig. 8.37. Quantum confinement on terraces of Au(322) and Au(788) crystals. The top panel shows an STM picture of the crystal used, demonstrating the perfection of the terraces. Note their different length scales. The middle panel shows photoemission spectra as a function of angle taken with high resolution. The lowest panels show the dispersions derived from the spectra; they display a markedly different behavior for the two crystal faces investigated [8.83]

$$E_N = E_0 + \frac{\hbar^2\pi^2}{2m^*L^2}N^2$$

which yields energies for $N = 1$ and $N = 2$ of $-0.4\,\text{eV}$ and $-0.1\,\text{eV}$ for $L = 38.7\,\text{Å}$, $E_0 = -0.5\,\text{eV}$ and $m^* = 0.26m_e$. If the terrace width is decreased the energy of the one-dimensional states obviously increases dramatically (note that L appears in the denominator of the formula) and therefore for a small terrace width the system adopts a two-dimensional behavior, which decreases the energy of the electronic states.

The superlattice wave vector of the two-dimensional state is $g = \frac{2\pi}{L}$, which, for the wave vector at the bottom of the parabola, yields

$$k_x(g/2) = \pi/d = 0.25\,\text{Å}^-1$$

and

$$k_x(3g/2) = 3\pi/d = 0.74\,\text{Å}^-1$$

in very good agreement with experiment.

References

8.1 G.K. Wertheim, D.N.E. Buchanan: Phys. Rev. B **43**, 13815 (1991)

8.2 E. Weschke, C. Laubschat, C.T. Simmons, M. Domke, O. Strebel, G. Kaindl: Phys. Rev. B **44**, 8304 (1991)

8.3 P.H. Citrin, G.K. Wertheim, M. Schlüter: Phys. Rev. B **20**, 4343 (1979)

8.4 D. Spanjaard, C. Guillot, M.C. Desjonqueres, G. Treglia, J. Lecante: Surf. Sci. Reports **5**, 1 (1985);
W.F. Egelhoff, Jr.: Surf. Science Reports **6**, 253 (1987)
J.W. Holland, D.E. Ingelsfield: In *The Chemical Physics of Solid Surfaces and Heterogeneous Catalysis I*, ed. by D.A. King, D.P. Woodruff (Elsevier, Amsterdam 1981)
G.V. Hansson, R.I.G. Uhrberg: Surf. Sci. Rep. **9**, 197 (1988)
Landoldt-Börnstein, New Series III, Vol.24-D (Springer, Berlin, Heidelberg 1994) Sect. 8.2 by A.M. Bradshaw, R. Hammen, Th. Schedel-Niedrig, D. Richen: Photoemission and inverse photoemission
E. Bertel: Surf. Sci. **331-333**, 1136 (1995)

8.5 PES investigations of quantum well states in metallic multi-layers can be found:
W.E. McMahon, M.A. Mueller, T. Miller, T.C. Chiang: Phys. Rev. B **49**, 10426 (1994) N.V. Smith, N.B. Brookes, Y. Chang, P.D. Johnson: Phys. Rev. B **49**, 332 (1994)
N.B. Brookes, Y. Chang, P.D. Johnson: Phys. Rev. B **50**, 15330 (1994)
T. Miller, A. Samesavar, T.-C. Chiang: Phys. Rev. B **50**, 17686 (1994)
F. Patthey, W.D. Schneider: Phys. Rev. B **50**, 17560 (1994)
U. Alkemper, C. Carbone, E. Vescovo, W. Eberhardt, O. Rader, W. Gudat: Phys. Rev. B **50**, 17495 (1994)

8.6 A.W. Maue: Z. Phys. **94**, 717 (1935)

8.7 E.T. Goodwin: Proc. Cambridge Philos. Soc. **35**, 205 (1935)

8.8 W. Shockley: Phys. Rev. **56**, 317 (1939)

8.9 F. Forstmann: Z. Phys. **235**, 69 (1970)

8.10 D.S. Boudreaux, V. Heine: Surf. Sci. **8**, 426 (1967)
8.11 N.V. Smith: Phys. Rev. B **32**, 3549 (1985)
8.12 P.M. Echenique, J.B. Pendry: J. Phys. C **11**, 2065 (1978)
For a review of the results obtained with the Echenique–Pendry method see N.V. Smith, C.T. Chen: Surf. Sci. **247**, 133 (1991)
8.13 E.G. McRae: Rev. Mod. Phys. **51**, 541 (1979)
8.14 E.G. McRae: J. Vac. Sci. Technol. **16**, 654 (1979)
8.15 W. Steinmann: Appl. Phys. A **49**, 365 (1989)
8.16 N.V. Smith, C.T. Chen, M. Weinert: Phys. Rev. B **40**, 7565 (1989)
8.17 N.V. Smith: Rep. Prog. Phys. **51**, 1227 (1988)
8.18 N.V. Smith: Phys. Rev. B 35, 975 (1987)
8.19 K. Giesen, F. Hage, F.J. Himpsel, J.H. Riess, W. Steinmann: Phys. Rev. Lett. **55**, 300 (1985)
8.20 M.W. Cole, M.H. Cohen: Phys. Rev. Lett. **23**, 1238 (1969)
8.21 C.C. Grimes, G. Adams: Phys. Rev. Lett. **42**, 795 (1979)
8.22 I. Tamm: Z. Phys. **76**, 848 (1932); see also Physikalische Zeitschrift der Sowjet Union **1**, 733 (1932)
8.23 P. Thiry: La photoémission angulaire dans les solides. Thèse d'état, Université Paris, 1979
8.24 P. Heimann, H. Neddermeyer, H.F. Roloff: J. Phys. C **10**, L17 (1977)
8.25 P. Heimann, J. Hermanson, H. Miosga, H. Neddermeyer: Surf. Sci. **85**, 263 (1979)
8.26 S.D. Kevan: Phys. Rev. Lett. **50**, 526 (1983); Phys. Rev. B **33**, 4364 (1986)
S.D. Kevan, R.H. Gaylord: Phys. Rev. Lett. **36**, 5809 (1987)
8.27 B.A. McDougall, T. Balasubramanian, E. Jensen: Phys. Rev. B **51**, 13891 (1995). In this communication a careful analysis of the temperature and energy dependences of a Cu(111) surface state has been reported. The temperature dependence is linear, however, no variation of the width with the binding energy was observed
8.28 P. Heimann, J. Hermanson, M. Miosga, H. Neddermeyer: Phys. Rev. Lett. **42**, 1782 (1979)
8.29 P. Heimann, J. Hermanson, M. Miosga, H. Neddermeyer: Phys. Rev. B **20**, 3059 (1979)
8.30 Y. Petroff, P. Thiry: Appl. Optics **19**, 3957 (1980)
8.31 S. Louie, P. Thiry, R. Pinchaux, Y. Petroff, D. Chandesris, J. Lecante: Phys. Rev. Lett. **44**, 549 (1980)
8.32 K. Giesen, F. Hage, F.J. Himpsel, J.H. Ries, W. Steinmann: Phys. Rev. B **35**, 971 (1987)
8.33 K. Giesen, F. Hage, F.J. Himpsel, J.H. Ries, W. Steinmann, N.V. Smith: Phys. Rev. B **35**, 975 (1987)
8.34 T. Wegehaupt, D. Rieger, W. Steinmann: Phys. Rev. B **37**, 10086 (1988)
8.35 Th. Fauster, W. Steinmann: In *Photonic Probes of Surfaces*, ed. by P. Halevi, Electromagnetic Waves: Recent Developments in Researvh, Vol.2 (North-Holland, Amsterdam, 1995) Chap. 8, p. 347
An investigation of image potential states including the spin splitting for those on the Fe(110) surface has been reported by U. Thomann, Ch. Reuß, Th. Fauster, F. Passek, M. Donath: Phys. Rev. B **61**, 16163 (2000)
8.36 R.W. Schoenlein, J.G. Fujimoto, G.L. Eesley, T.W. Capehart: Phys. Rev. Lett. **61**, 2596 (1988),,
R.W. Schoenlein, J.G. Fujimoto, G.L. Eesley, T.W. Capehart: Phys. Rev. B **41**, 5436 (1990);
R.W. Schoenlein, J.G. Fujimoto, G.L. Eesley, T.W. Capehart: Phys. Rev. B **43**, 4866 (1991);

T. Hertel, E. Knoesel, M. Wolf, G. Ertl: Phys. Rev. Lett. **76**, 535 (1996); E. Knoesel, T. Hertel, M. Wolf, G. Ertl: Chem. Phys. Lett. **240**, 409 (1995); M. Wolf, E. Knoesel, T. Hertel: Phys. Rev. B **54**, 5295 (1996); U. Höfer, I.L. Shumay, Ch. Reuss, U. Thomann, W. Wallauer, Th. Fauster: Science **277**, 1480 (1997)
8.37 R. Manzke, M. Skibowski: Physica Scripta **31**, 2213 (1988)
8.38 J. Fraxedas, H.J. Trodahl, S. Gopalan, L. Ley, M. Cardona: Phys. Rev. B **41**, 10068 (1990)
J. Fraxedas, M.K. Kelly, M. Cardona: preprint (1991)
H. Carstensen, R. Claessen, R. Manzke, M. Skibowski: Phys. Rev. B **41**, 9880 (1990)
8.39 N.V. Smith, N.B. Brookes, Y. Chang, P.D. Johnson: Phys. Rev. B **49**, 332 (1994)
8.40 J.J. Paggel, T. Miller, T.C. Chiang: Science **283**, 1709 (1999);
J.J. Paggel, T. Miller, T.C. Chiang: Phys. Rev. Lett. **83**, 1415 (1999)
8.41 T. Schmitz-Hübsch, K. Oster, J. Radnik, K. Wandelt: Phys. Rev. Lett. **74**, 2595 (1995)
8.42 R. Paniago, R. Matzdorf, G. Meister, A. Goldmann: Surf. Sci. **325**, 336 (1995)
8.43 M.C. Desjonqueres, F. Cyrot-Lackmann: J. Phys. F **5**, 1368 (1975)
8.44 J.A. Appelbaum, D.R. Hamann: Solid State Commun. **27**, 881 (1978)
8.45 R.C. Baetzold, G. Apai, E. Shustorovitch, R. Jaeger: Phys. Rev. B **26**, 4022 (1982)
8.46 P.J. Feiblman, J.A. Appelbaum, D.R. Hamann: Phys. Rev. B **20**, 1433 (1979)
8.47 P.H. Citrin, G.K. Wertheim, Y. Baer: Phys. Rev. B **27**, 3160 (1983)
8.48 M.C. Desjonqueres, D. Spanjaard, Y. Lassailly, C. Guillot: Solid State Commun. **34**, 807 (1980)
8.49 A. Rosengren, B. Johansson: Phys. Rev. B **22**, 3706 (1980)
8.50 G. Treglia, M.C. Desjonqueres, D. Spanjaard, Y. Lassailly, C. Guillot, Y. Jugnet, Tran Minh Duc, J. Lecante: J. Phys. C **14**, 3464 (1981)
8.51 A. Rosengren: Phys. Rev. B **24**, 7393 (1981)
8.52 D. Tomanek, V. Kumar, S. Holloway, K.H. Bennemann: Solid State Commun. **41**, 273 (1982)
8.53 D. Tomanek, P.A. Dowben, M. Grunze: Surf. Sci. **126**, 112 (1983)
8.54 G.A. Benesh, R. Haydock: J. Phys. C **17**, L 83 (1984)
8.55 E. Wimmer, A.J. Freeman, J.R. Hiskes, A.M. Karo: Phys. Rev. B 28, 3074 (1980)
8.56 P.H. Citrin, G.K. Wertheim, Y. Baer: Phys. Rev. B **6**, 425 (1977)
8.57 P.H. Citrin, G.K. Wertheim, Y. Baer: Phys. Rev. Lett. **41**, 1425 (1978)
8.58 G. Apai, R.C. Bätzold, E. Shustorovitch, R. Jaeger: Surf. Sci. 116, L 191 (1982)
8.59 K. Düchers, H.P. Bonzel, D.A. Wesner: Surf. Sci. **166**, 141 (1986)
8.60 C. Laubschat, G. Kaindl, W.D. Schneider, B. Reihl, N. Mårtensson: Phys. Rev. B **33**, 6675 (1986)
8.61 J.F. van der Veen, F.J. Himpsel, D.E. Eastman: Phys. Rev. Lett. **44**, 189 (1980)
8.62 S.F. Alvarado, M. Campagna, W. Gudat: J. Electron Spectrosc. Relat. Phenom. **18**, 43 (1980)
8.63 M. Alden, H.L. Skriver, I.A. Abrikosov, B. Johansson: Phys. Rev. B **51**, 1981 (1995). Analyzing the surface core-level shifts for rare earth metals
8.64 B. Johansson, N. Mårtensson: Phys. Rev. B **21**, 4427 (1980)
For recent reviews on surface core-level shifts see
A. Flodström, R. Nyholm, B. Johansson: In *Synchrotron Radiation Research, Advances in Surface and Interface Science*, ed. by R.Z. Bachrach (Plenum,

New York 1992) Vol.1 and
N. Mårtensson, A. Nilsson: J. Electron. Spec. Related Phenom. **75**, 209 (1995)
8.65 J.F. van der Veen, F.J. Himpsel, D.E. Eastman: Solid State Commun. **37**, 555 (1981)
8.66 J.F. van der Veen, F.J. Himpsel, D.E. Eastman: Solid State Commun. 40, 57 (1981)
8.67 C. Guillot, C. Thuault, Y. Jugnet, D. Chauveau, R. Hoogewijs, J. Lecante, Tran Minh Duc, G. Treglia, M.C. Desjonqueres, D. Spanjaard: J. Phys. C **15**, 4023 (1982)
8.68 C. Guillot, M.C. Desjonqueres, D. Chauveau, G. Treglia, J. Lecante, D. Spanjaard, Tran Minh Duc: Solid State Commun. **50**, 393 (1984)
8.69 C. Guillot, P. Roubin, J. Lecante, M.C. Desjonqueres, G. Treglia, D. Spanjaard, Y. Jugnet: Phys. Rev. B **30**, 5487 (1984)
8.70 R. Nyholm, J. Schmidt-May: J. Phys. C **17**, L113 (1984)
8.71 M.C. Desjonqueres, F. Cyrot-Lackmann: Solid State Commun. **18**, 1127 (1976)
8.72 G.K. Wertheim, P.H. Citrin: Phys. Rev. B **38**, 7820 (1988)
8.73 N. Mårtensson, H.B. Saalfeld, H. Kuhlenbeck, M. Neumann: Phys. Rev. B **39**, 8181 (1989)
8.74 G. Le Lay, M. Gothelid, T.M. Grelik, M. Björkquist, U.O. Karlson, V.Yu. Aristov: Phys. Rev. B **50**, 14277 (1994). Reviewing the situation with respect to an understanding of the surface core-level shifts observed for the Si(111) 7×7 surface
8.75 J.N. Andersen, D. Henning, E. Lundgren, M. Methfessel, R. Nyholm, M. Scheffler: Phys. Rev. B **50**, 17525 (1994). Discussing surface core-level shifts in 4d transition metals
8.76 C. Laubschat, E. Weschke, C. Holtz, M. Domke, O. Strebel, G. Kaindl: Phys. Rev. Lett. **65**, 1639 (1990)
8.77 A more detailed analysis indicates, however, that initial as well as final state effects have to be considered for a complete understanding of surface core-level shifts;
for Gd see
J.E. Ortega, I.J. Himpsel, Dongqi Li, P.A. Dowlen: Solid State Commun. **91**, 807 (1994)
A.V. Federov, E. Arenholz, K. Starke, E. Navas, L. Baumgarten, C. Laubschat, G. Kaindl: Phys. Rev. Lett. **73**, 601 (1994)
and for Be(0001) see
P.J. Feibelman, R. Stumpf: Phys. Rev. B **50**, 17480 (1994)
8.78 D.M. Riffe, G.K. Wertheim, P.H. Citrin: Phys. Rev. Lett. **64**, 571 (1990)
8.79 M. Scheffler, Ch. Droste, A. Fleszar, M. Maca, G. Wachuta, G. Barzel: Physica B **172**, 143 (1991)
8.80 J. Neugebauer, M. Scheffler: Phys. Rev. B **46**, Dez. (1992)
8.81 G.A. Benesh, D.A. King: Chem. Phys. Lett. **191**, 315 (1992)
8.82 G.K. Wertheim, D.M. Riffe, P.H. Citrin: Phys. Rev. B **49**, 4824 (1994)
A.B. Andrews, D.M. Riffe, G.K. Wertheim: Phys. Rev. B **49**, 8396 (1994)
8.83 J.E. Ortega, A. Mugarza, V. Pérez-Dieste, V. Repain, S. Rousset, A. Mascaraque: Phys. Rev. B **65**, 165413 (2002)

9. Inverse Photoelectron Spectroscopy

It has already been pointed out in Chap. 1 that photoemission experiments can also be performed in a reversed mode, namely by sending electrons of varying energy onto a sample and detecting the photons that are thereby produced by them (Bremsstrahlung) [9.1]. If only photons of one particular energy are detected, which is a common measuring mode, the technique is called Bremsstrahlung Isochromate Spectroscopy (BIS) and, a little misleadingly, this term is sometimes used to characterize the IPES technique in general. For the sake of accuracy, the technique as such – sending electrons onto a sample and detecting the photons produced in the Bremsstrahlung process – will henceforth be called Inverse PhotoElectron Spectroscopy (IPES) and only the mode in which the photon detection energy is kept constant while varying the energy of the incoming electrons will be called BIS.

The techniques of PES and IPES are complementary to one another (Fig. 9.1) in that PES connects the energy levels below the Fermi energy and above the vacuum level and IPES connects the energy levels above the vacuum level and between the vacuum level and the Fermi level. For each of the techniques there is thus an energy range which is inaccessible.

Figure 9.1 can be used to demonstrate the complementary nature of PES and IPES. If one uses the convention that energies are referenced to the Fermi energy E_F, one has

PES:

$$\hbar\omega = E_f - E_i \,, \tag{9.1}$$

$$-E_i = \hbar\omega - E_\mathrm{kin} - \phi \,, \tag{9.2}$$

$$E_f = E_\mathrm{kin} + \phi \,, \tag{9.3}$$

and since $-E_i = E_\mathrm{B}$,

$$E_\mathrm{B} = \hbar\omega - E_\mathrm{kin} - \phi \,. \tag{9.4}$$

IPES:

$$\hbar\omega = E_i - E_f \,, \tag{9.5}$$

$$E_f = -\hbar\omega + E_\mathrm{kin} + \phi \,, \tag{9.6}$$

$$E_i = E_\mathrm{kin} + \phi \,, \tag{9.7}$$

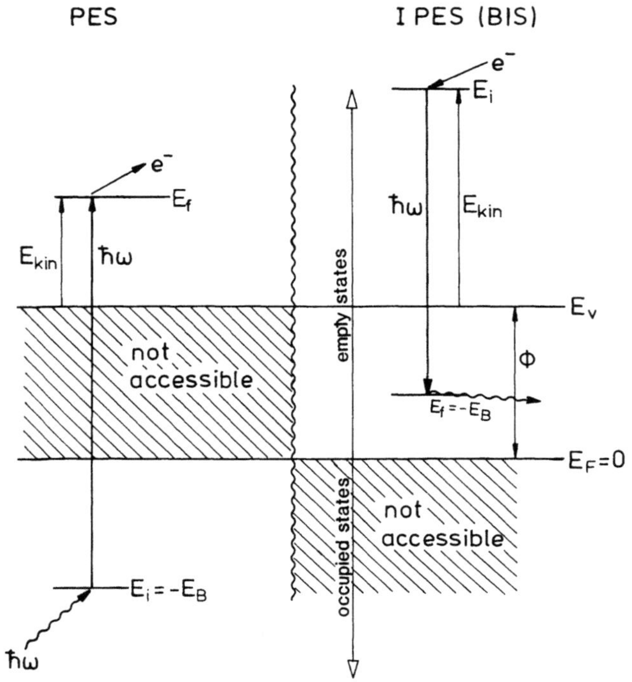

Fig. 9.1. Schematic diagram of PhotoEmission Spectroscopy (PES) and Inverse PhotoEmission Spectroscopy (IPES). In PES the energy range between the Fermi energy and the vacuum level is not accessible, while in IPES the unaccessible range is that below E_F. Thus the two techniques complement each other

and since $-E_f = E_\mathrm{B}$

$$-E_\mathrm{B} = -\hbar\omega + E_\mathrm{kin} + \phi \,. \tag{9.8}$$

The comparison of (9.1–9.4) with (9.5–9.8) gives a nice picture of the complementary nature of the two processes. The equations also illustrate why it is sometimes claimed that PES measures an initial state energy while IPES measures a final state energy. This statement, however, is somewhat misleading because in both spectroscopies one measures a final state energy.

The problem of screening in the final states for PES has been discussed at length in this volume and that of screening in IPES has not yet been explored sufficiently for us to make definite statements. On a formal basis, IPES measures the electron addition spectrum or the spectral function (above the Fermi energy) multiplied by a transition matrix element.

Given their similarity, it might seem desirable to describe the results of PES and IPES in the same context. This separate chapter on IPES might thus seem unnecessary at first sight. However, IPES does have certain distinct features which may be usefully presented in a separate chapter on this technique. Broadly speaking, we use the same pattern as will be used for treating the technique of spin polarized photoemission (Chap. 10) and photoelectron diffraction (Chap. 11).

A large effort in IPES began in the 1960s by the group of Ulmer in Karlsruhe [9.2–9.7] who chose X-rays as the photons to be detected with a crystal monochromator to make the bandpass sufficiently narrow. This, however, had the disadvantage that quite high fluxes of electrons had to be employed, which in turn put some restrictions on the samples that one could investigate. This approach was refined by Lang and Baer [9.8] by literally "reversing" an XPS instrument, and very interesting data, for example, on rare earth metals and their compounds, were obtained by this group (Chap. 2). This whole field received, however, a new dimension by virtue of Dose's discovery [9.9] that a gas proportional counter could be used with good success for the high efficiency detection of photons of 9.7 eV energy, thereby making ultraviolet IPES, or rather BIS, investigations quite easy.

The Dose detector [9.10] is a Geiger–Müller counter consisting of a stainless steel tube of 20 mm in diameter, with a stainless steel electrode, filled with about 100 mbar of Ar as the multiplication gas and a few iodine crystals to produce iodine vapor in the cell. This serves as the detecting agent via the reaction

$$I_2 + \hbar\omega \rightarrow I_2^+ + e^- \ , \tag{9.9}$$

which has a threshold at 9.23 eV. Its cross section is shown in Fig. 9.2. The entrance window of the proportional counter is a 2 mm thick CaF_2 crystal, whose transparency cut-off is at around 10.2 eV (see also Fig. 9.2). The combination of the photoionization cross section of I_2 and the transmission coefficient of CaF_2 gives a detector for photons of an energy $(9.7 \pm 0.40\,\text{eV})$, extremely well suited for BIS experiments in the UV energy range. The sensitivity of the method can be further increased by using a mirror to collect the radiation [9.11] as shown in Fig. 9.3, a system which is, of course, especially suited for making $\boldsymbol{k}$-resolved measurements. For angle-integrated measurements, a much simpler arrangement, relying simply on a hairpin electron emitter and a Geiger–Müller counter, is sufficient.

Another photon counter with a fixed photon energy consists of an open multiplier, with KBr being evaporated on the first dynode in combination with a CaF_2 window. This yields a photon energy of $(9.9 \pm 0.3)\,\text{eV}$ [9.12].

In the meantime, grating monochromators have also been used to increase the photon resolution [9.12, 9.13]. On the other hand, the use of SrF_2 for the entrance window of the Geiger–Müller counter shifts the photon energy to 9.5 eV and increases the resolution by about a factor of two (Fig. 9.4) [9.14]. For other experimental details see, e.g., [9.15–9.21].

The bulk of the IPES data produced so far have been in the UV energy range and since the interesting data of the group of Baer [9.22] have already been dealt with, the discussion will be restricted to results from UV IPES

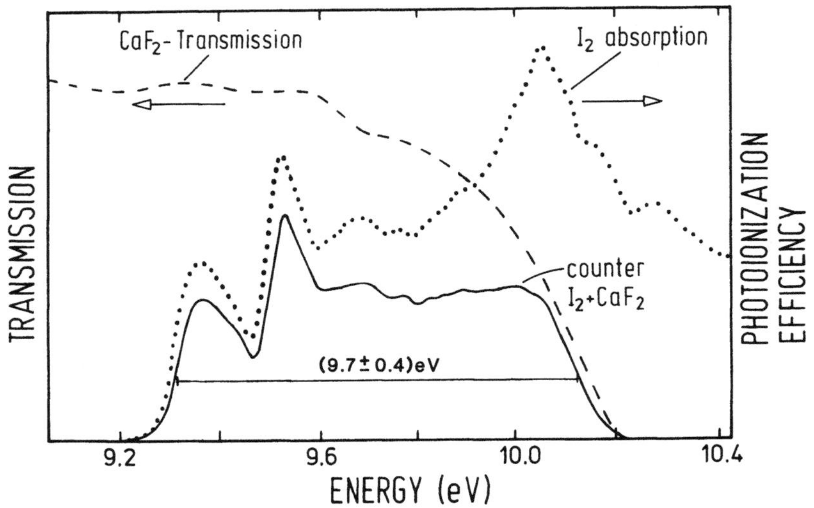

Fig. 9.2. Transmission spectrum of CaF_2 (window of the Dose counter [9.10]), absorption spectrum of I_2 vapor (the counting gas) and the convolution of the two curves giving the "acceptance window" of (9.7 ± 0.4) eV for the Dose counter

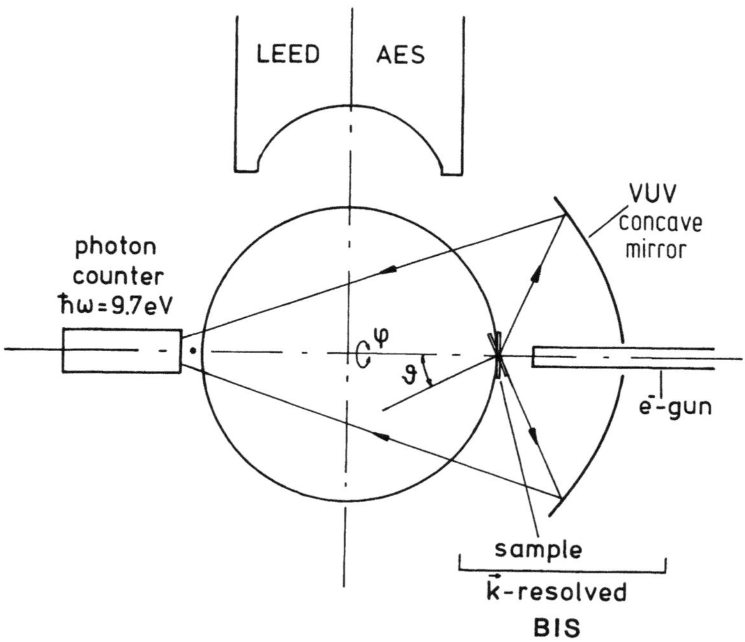

Fig. 9.3. Arrangement for an experiment in the $\boldsymbol{k}$-resolved IPES mode. A mirror serves to enhance the counting efficiency of the arrangement [9.11]

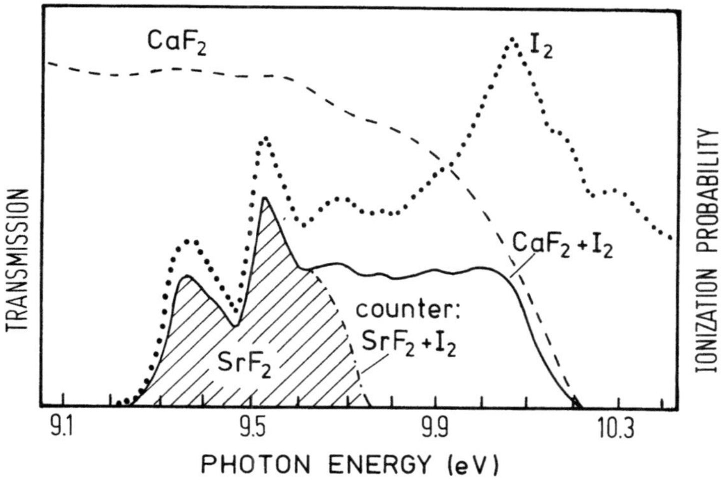

Fig. 9.4. Comparison of the resolution of the Dose counter using a CaF_2 and a SrF_2 window [9.11]. The SrF_2 window increases the resolution by about a factor of two and lowers the mean bandpass-photon energy from 9.7 to 9.5 eV

experiments[1] [9.6, 9.14, 9.17, 9.24–9.32]. In principle, IPES produces data in the same areas as PES. However, it is obvious from Fig. 9.1, that this technique will be of special value for detecting surface (and bulk) states in the region between the Fermi energy and the vacuum level, because this area is not accessible to PES. For investigating adsorbate energy levels above the Fermi energy IPES is also highly successful and, therefore, we shall discuss experiments of these types to demonstrate the power of IPES.

9.1 Surface States

In addition to the surface states already discussed at length [9.33, 9.34] in the energy region below the Fermi level [9.35–9.41], there is a type of surface state detectable by IPES which can be viewed as an electron trapped (in a band gap) between the surface and the surface barrier potential (image potential), which prevents it from escaping into the vacuum (see Chap. 8 and, in particular, Fig. 8.3).

This state, which was predicted in the elegant work of Echenique and Pendry [9.42], is generally called a barrier state or *image potential state* and since its energy lies between the Fermi level and the vacuum level it cannot be detected by PES but is seen by IPES. The physics of this state was discussed

---

[1] See, e.g., [9.23] for BIS data with XPS photon energies, and a general account of IPES.

in Chap. 8 (see, e.g., [9.43,9.44]) and here it will be treated with reference to the actual results measured by IPES.

The situation for such a barrier state is depicted in Fig. 9.5, which shows the band structure of the (100) surface of Cu [9.45]. Also given is the $X_{4'} - X_1$ gap above the Fermi energy and the shaded areas indicate projected band states of the two-dimensional density of states. For the surface barrier potential a Coulomb potential $1/4z$ is indicated, as are the surface states derived in the Echenique–Pendry model for $n = 0$ (Shockley type – this is a surface resonance rather than a surface state because it falls into the occupied part of the projected two-dimensional density of states), and $n = 1$ (surface barrier state).

A BIS spectrum for Cu(100) from a clean and a CO covered sample ($\hbar\omega = 9.7\,\mathrm{eV}$) is shown in Fig. 9.6 [9.46]. The clean spectrum is dominated

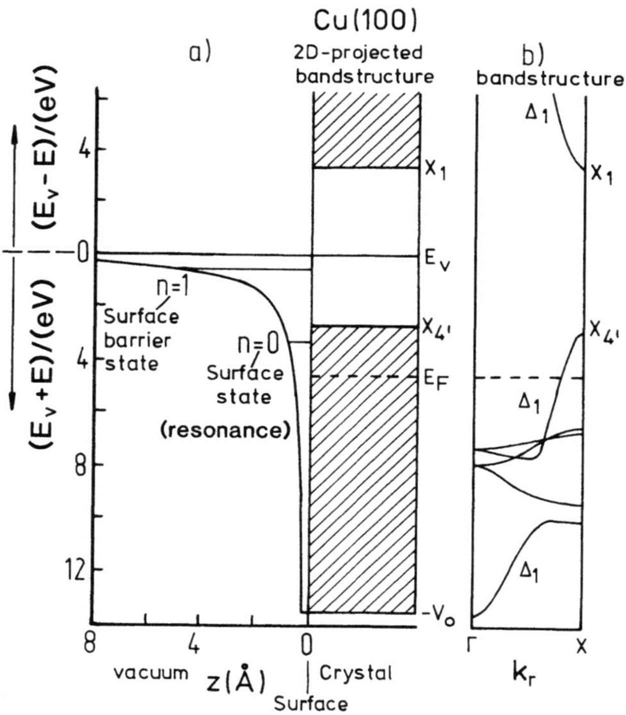

Fig. 9.5a,b. Surface states on the Cu(100) surface [9.45]. The right panel gives the band structure in the ΓX direction; the left panel is a two-dimensional projection of this band structure with the projected band areas shaded; this representation shows the $X_{4'} - X_1$ gap in which surface states are expected; a "Coulombic" barrier potential ($\sim 1/4z$) is also indicated in which the ($n = 0$) surface state and the ($n = 1$) barrier state are indicated

by an asymmetric peak near the Fermi energy, which can be resolved (as shown) into two peaks. The main peak has been interpreted [9.47] as a direct transition in the bulk-band structure, as indicated in Fig. 9.7, while the peak in the shoulder is due to the surface state calculated in (8.35), at 3.4 eV below the vacuum level ($E_{\mathrm{vac}} = 4.5\,\mathrm{eV}$ above E_{F}, yields $E_{n=0} = 1.1\,\mathrm{eV}$ above E_{F} as found in the experiment). Covering the surface with CO causes the surface state to disappear while the direct transition is hardly changed at all.

A more extended spectrum [9.47] from the Cu(100) surface is shown in Fig. 9.8, where a second spectrum for higher temperature ($T = 900\,\mathrm{K}$) is also given. In these spectra, an additional structure at a binding energy of $E_{\mathrm{B}} = -4\,\mathrm{eV}$ is seen. It was interpreted by the authors as the image potential state expected at 0.55 eV below the vacuum level [see (8.35)]. Confirmation

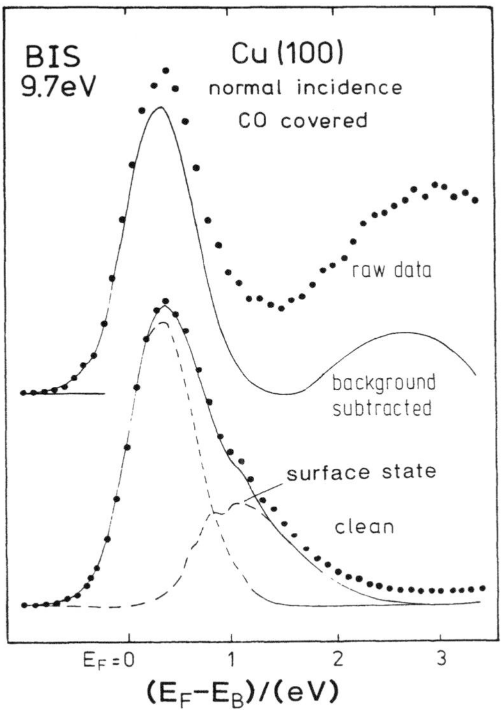

Fig. 9.6. BIS spectra of Cu(100). The clean surface spectra display a peak near the Fermi energy and a shoulder at around 1 eV above the Fermi energy [9.46]. The main peak is a direct transition in the bulk-band structure (Fig. 9.7). The shoulder is a surface state calculated at 3.4 eV below E_{vac}, which with a work function of 4.5 eV puts it at 1.1 eV above E_{F}. For a CO covered surface, the surface state, but not the main peak (direct transition), disappears

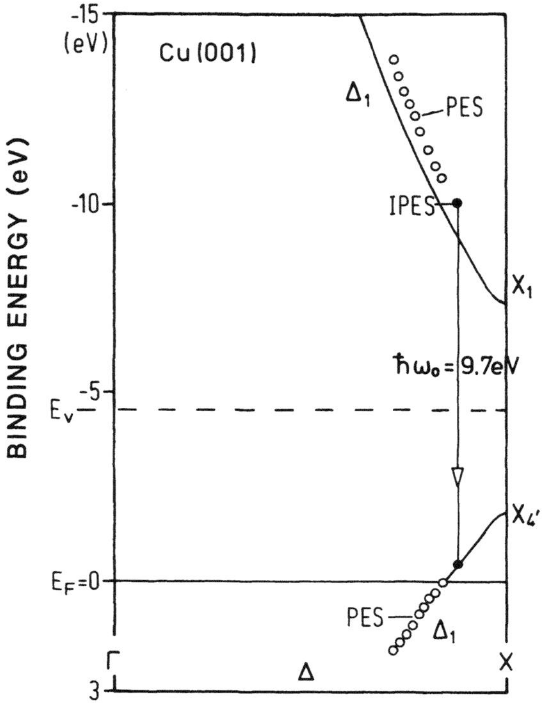

Fig. 9.7. Part of the band structures for Cu(100) obtained with PES and BIS data [9.47]. The origin of the main peak (near E_F) in Fig. 9.6 is given by the 9.7 eV transition (CaF_2 window)

of this assignment comes from the temperature independence of the intensity of this state as shown in Fig. 9.8. In contrast, the direct transition shows a considerable decrease in intensity, which can be ascribed to a Debye–Waller effect (Sect. 6.4) [9.48].

Another strong indication that one is indeed observing an image-potential or barrier-induced surface state comes from the fact that the state is pinned to the vacuum level as predicted by (8.35). Figure 9.9 shows the relevant data. The spectrum of the clean Cu(100) surface is compared with that of a Cu(100) surface covered with a c(2 × 2) overlayer of Cl [9.49] for which the work function is increased by 1.1 eV. The surface barrier state increases in energy, with respect to the Fermi energy (!), by 1.1 eV, as indeed it should if it is pinned to the vacuum level. Conversely, for Pt(111) the adsorption of potassium lowers the work function and thus the image potential state is observed to shift to a correspondingly lower energy.

A compilation of image potential states [9.43–9.67] is given in Table 9.1. It shows that the ($n = 1$) binding energies fall in a narrow range between 0.85 eV and 0.4 eV (below E_v) with hardly any exception from this rule. It

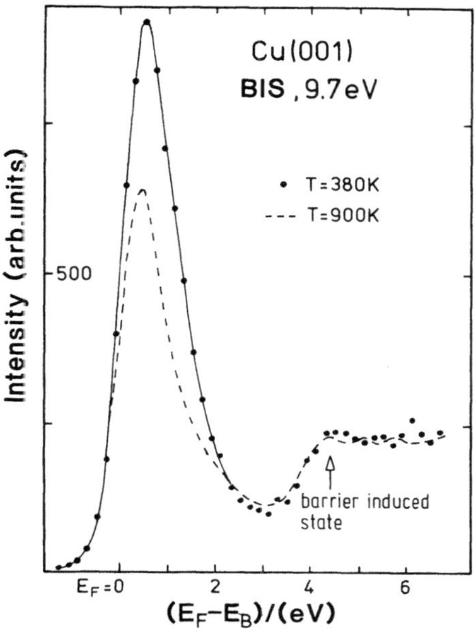

Fig. 9.8. BIS spectrum of Cu(100) at two different temperatures, 380 K and 900 K [9.47]. Because of the Debye–Waller factor, the direct transition becomes weaker at the higher temperature whereas the surface state (barrier state) at $\sim 4.0\,\mathrm{eV}$ does not change in intensity

has been shown that if one uses a purely "Coulombic" potential ($\sim 1/4z$) for the image potential,[2] the energy E_1 of the lowest image potential state is $E_1 = E_\mathrm{v} - 0.85\,\mathrm{eV}$, assuming that the crystal provides an infinitely high barrier (8.25). This equation is completely independent of the material under investigation. The near constancy of $E_\mathrm{v} - E_1$, as given in Table 9.1 [Sb(100): $E_\mathrm{v} - E_1 = 0.76\,\mathrm{eV}$, Ag(111): $E_\mathrm{v} - E_1 = 0.77\,\mathrm{eV}$], shows that this very simple model describes the relevant physics quite well.

The theory can be improved by using a finite crystal potential as in the Echenique–Pendry approach, which treats the electron as being scattered back and forth between the now finite crystal potential and the barrier potential. This leads to $E_1 = E_\mathrm{v} - 0.55\,\mathrm{eV}$ [(8.35) for the middle of the gap ($\varepsilon = \frac{1}{2}$)] which gives more or less the lower limit of the experimentally observed trend. Even in this calculation the explicit properties of the material enter only marginally.

[2] The force on an electron at distance z from a metal surface is given by its interaction with its image charge and calculated as $F \propto 1/(2z)^2$ ($2z$ being the distance between the electron and its image charge). From this force, the potential (image potential) is calculated by integration yielding $V_\mathrm{image} \propto 1/4z$.

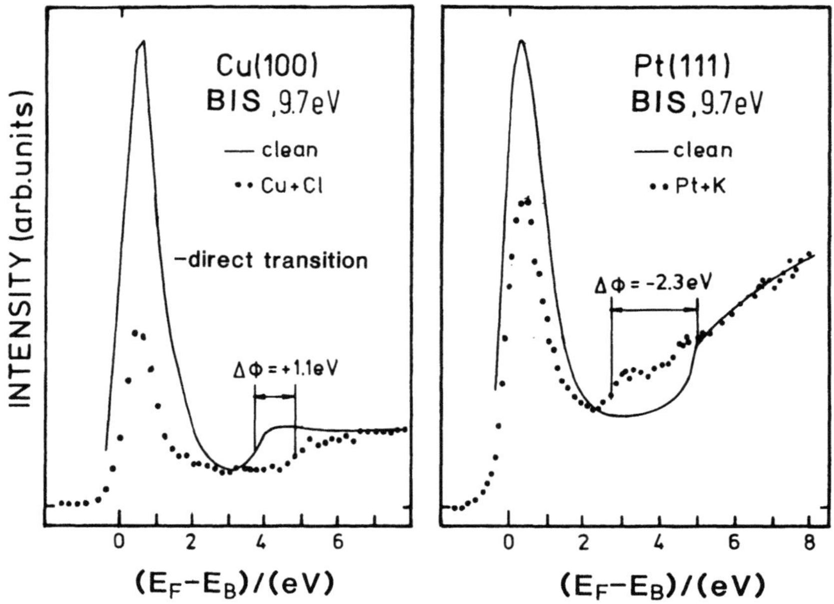

Fig. 9.9. Effect of work function change on the energy of the barrier-induced surface state [9.49]. An overlayer of Cl on Cu [c(2×2)] changes the work function by +1.1 eV and indeed the barrier-induced state moves by that amount to higher energies. As a check, a corresponding experiment was performed by covering a Pt(111) surface with K, which lowers the work function by 2.3 eV. A shift of the barrier-induced state by this amount is seen in the data

The investigation of barrier states by IPES (and two photon spectroscopy) has reached a considerable sophistication [9.50–9.54]. These data can, in turn, be used to model the image potential [9.51].

9.2 Bulk Band Structures

In the same way that PES can be used to measure transitions between the occupied and empty states via direct optical transitions [9.68–9.70], IPES can be employed to measure transitions between empty bands [9.11, 9.47, 9.58, 9.71–9.73]. IPES, however, has the disadvantage that in the BIS mode, only measurements at around $\hbar\omega = 10\,\mathrm{eV}$ can be done in a reasonable time, so that the information obtainable is somewhat restricted. This restriction is not inherent to the technique, but is rather a consequence of the fact that measurements not employing the Dose counter still take a long time.

An example of the determination of an empty energy band via IPES has already been given for the cases of Cu (Figs. 9.6, 9.7) and graphite (Fig. 7.33); here two more examples will be added.

Table 9.1. Binding energies (relative to the vacuum level) and effective masses of a number of image potential states

Sample	Inverse photoemission			Two-photon photoemission		
	E_B ($n = 1$) [eV]	m^*/m_e	Ref.	E_B [eV]	m^*/m_e	Ref.
Au(100)	0.63		[9.55]			
Au(111)	0.42		[9.56]			
Ag(100)	0.5	1.2	[9.58]	0.53($n = 1$)	1.15	[9.50]
			[9.45]	0.16($n = 2$)		
Ag(111)	0.65	1.3	[9.56]	0.77($n = 1$)	1.3	[9.50]
				0.23($n = 2$)		
Cu(100)	0.6	1.2	[9.47]	0.57($n = 1$)	0.9	[9.50]
				0.18($n = 2$)		
Cu(111)	0.94	1.0	[9.63]	0.83($n = 1$)	0.9	[9.50]
				0.26($n = 2$)		
Ni(100)	0.4	1.2	[9.45]			
			[9.58]			
Ni(111)	0.6	1.6	[9.45]	0.81($n = 1$)	1.1	[9.50]
			[9.58]	0.27($n = 2$)		
				0.10($n = 3$)		
Pd(111)	0.85		[9.45]			
			[9.58]			
Sb(100)	0.76		[9.56]			
$1T-TiS_2$	0.60		[9.56]			

For Cu the unoccupied bands are well known from PES studies and therefore IPES measurements do not give much more information on the bulk-band structure. There are, however, cases where the available empty bands were measured, for the first time, by the IPES technique. In principle, one must tackle the same problem as with PES, i.e., since one is measuring a transition between two bands, it is necessary to use one of the procedures outlined in Chap. 7, in order to derive the bands from the measured spectra. The IPES technique has thus been used mostly to check the accuracy of the assumed position of unoccupied bands, although in a few cases the energy coincidence technique has been employed [9.58] to make an absolute determination [9.26] of the wave vector of unoccupied bands.

In Fig. 9.10 an example is given [9.30], where band mapping was performed in germanium by a combination of PES and IPES. It can be realized that for Ge the data from the two techniques complement each other quite nicely. In addition it can be seen that the theory is in good agreement with the experiment.

Figure 9.11 shows the results of an interesting experiment in which a combination of PES and IPES was used to measure the exchange splitting of Co [9.25]. Both spectra were taken at normal emission and the conditions were such that the PES experiment measured the occupied part and the IPES experiment the unoccupied part of an exchange-split band near the Fermi energy, which, however, intersects the Fermi energy over a small region in momentum space. In the region where the upper band is below E_F, its position can of course be determined by PES. By varying the photon energy

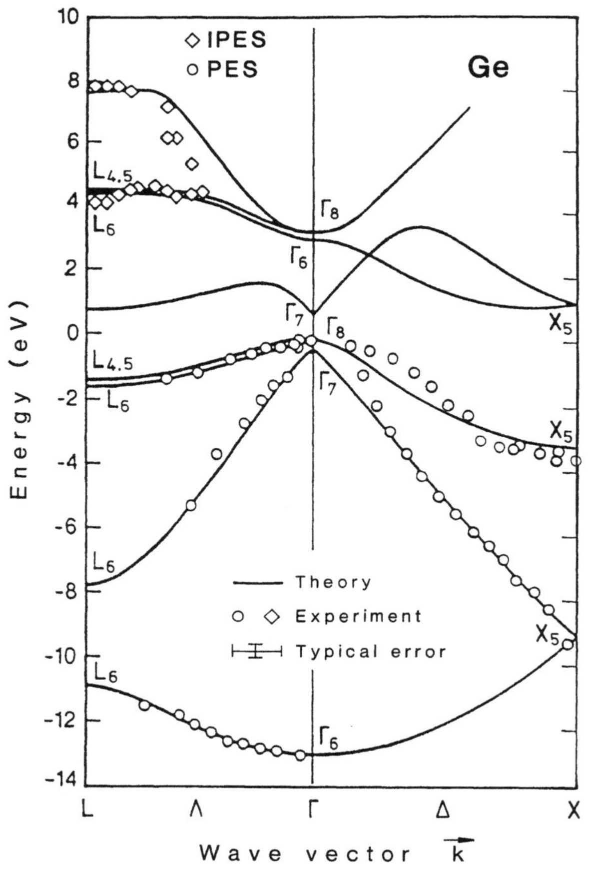

Fig. 9.10. Band structure in the LΓX part of the Brillouin zone for Ge [9.30]. Data from PES and IPES are combined to check the theoretical band structure

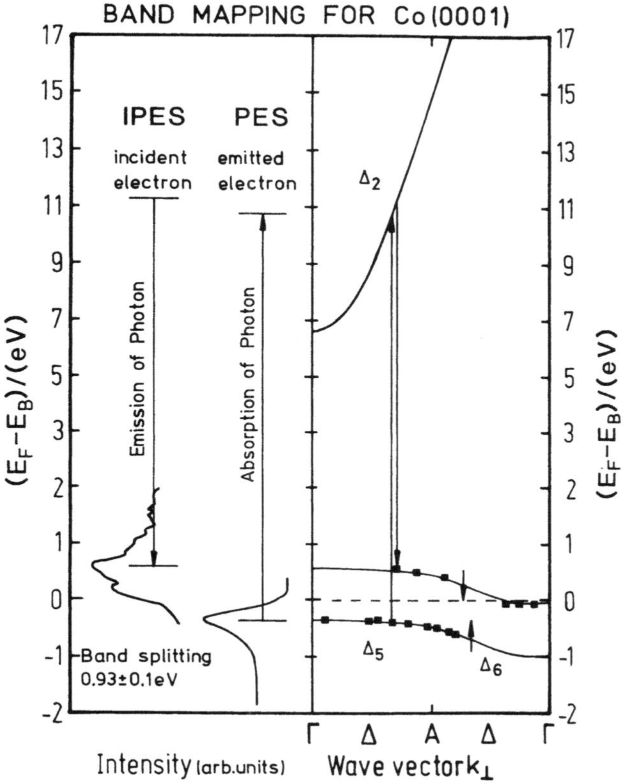

Fig. 9.11. PES and IPES on Co(0001) [9.25]. The exchange splitting of the Δ_5, Δ_6 band leads to a band below, and another one (partly) above E_F. In a combined PES/IPES experiment with the same photon energy this exchange splitting can be measured directly as indicated in the figure

in both (!) experiments, the dispersion of the two bands could be swept as shown in the figure. For the upper band, the data obtained with PES and with IPES join smoothly together. This experiment indicates that the screening is the same in PES and IPES. Since in the PES experiment the final state is probably fully screened (only coherent intensity), the same must hold for the IPES experiment.

9.3 Adsorbed Molecules

By its very nature IPES is suited to study energy levels between the Fermi energy and the vacuum level (Fig. 9.1). A case in point is the antibonding orbital of molecules adsorbed on substrate surfaces, and as a simple example of the power of the method we discuss comparatively IPES data for CO and

NO adsorbed on Pd [9.74]. Palladium is chosen as a substrate because it induces very little NO decomposition [9.75–9.77].

CO (Sect. 5.11.2) has (besides the 1s shells of C and O) the ground-state orbitals $(\sigma_g\,2s)^2$ $(\sigma_u\,2s)^2$ $(\sigma_g\,2p)^2$ $(\pi_u\,2p)^4$ [9.78, 9.79] and in NO one has the same molecular orbitals with an additional 2p electron in the $(\pi_g\,2p)^1$ orbital (singly occupied) [9.78, 9.79]. This latter orbital is antibonding [9.78, 9.79]. The smallest ionization energy of gaseous CO is the $(\pi_u\,2p)^4$ photoionization at 14 eV, whereas the smallest ionization energy of gaseous NO is the $(\pi_g\,2p)^1$ photoionization at 9.3 eV [9.78, 9.79]. (In CO this orbital is not occupied and therefore cannot be detected).

The chemisorption process is described in the Blyholder model [9.80, 9.81] by backdonation from the metal to the adsorbed species [9.80, 9.81] (For a discussion of the CO metal bonding see [9.82–9.89]; it was already pointed out in Chap. 5 that the Blyholder model [9.80, 9.81] is not universally accepted; however, it is beyond the scope of this book to discuss this point in any detail). This leads to a lowering of the energy levels and sometimes even to an occupation of levels that are not occupied in the gas phase. The $2\pi^*$ antibonding orbital of adsorbed CO can be observed (Fig. 9.12) by IPES. It

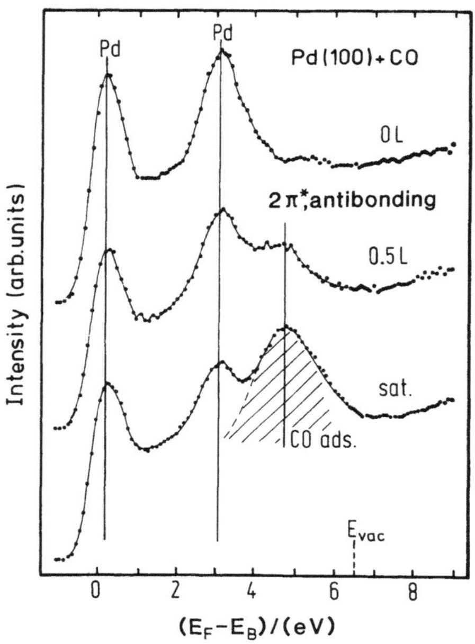

Fig. 9.12. BIS of CO on Pd [9.90]. The top trace gives the spectrum of the clean Pd(100) surface which shows a two-peaked structure. After CO exposure a new peak corresponding to the CO antibonding orbital appears, which can be identified from its exposure dependence

is, however, less tightly bound than the corresponding level in NO, as can be seen by a comparison with Fig. 9.13 [9.90, 9.91].

The clean Pd(001) surface shows a two-peaked IPES spectrum (Fig. 9.13). Adsorption of CO or NO adds a further peak, but leaves the metal signals largely unaltered. The NO signal is nearer to the Fermi energy than that of CO, as is to be expected from the above discussion. These findings have implications for the stability of the intra-atomic bond of the molecule in the chemisorptive state. In NO the $2\pi^*$ antibonding orbital can easily be filled by backdonation from the metal because it is at E_F; thus NO will exhibit a greater tendency towards dissociative adsorption (since an antibonding orbital is filled) than CO, as is actually the case [9.92, 9.93].

The observation of the $2\pi^*$ level for Pd(100)/NO (or on other metal substrates) [9.94] by IPES may seem puzzling at first sight because this level has also been observed in PES experiments on Pd(111) [9.77] and Ni(001) [9.95]. It can, however, be understood along the lines used to analyze the PES/IPES data of transition metal compounds (Sect. 5.8). Figure 9.14 gives a combined PES/BIS diagram for the Pd/NO system. It shows the $2\pi^*$ level at $E_B \simeq 2.5$ eV (PES) and at $E_B \simeq -1.5$ eV (BIS). A possible reason for this

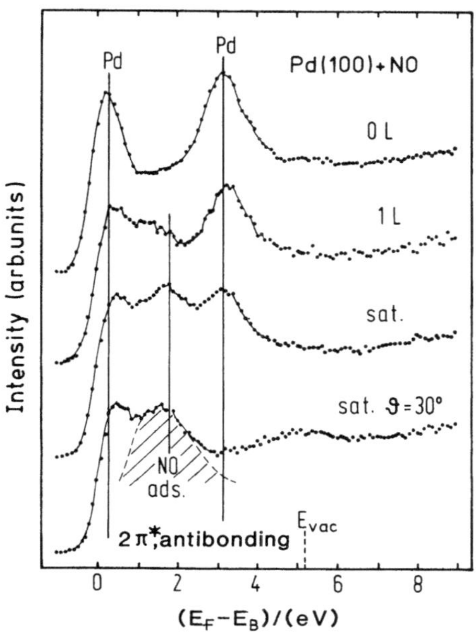

Fig. 9.13. BIS of NO on Pd [9.90]. The top trace gives the spectrum of the clean Pd(100) surface while the next two traces are measured for increasing NO adsorption, showing the emergence of the NO antibonding level. Upon tilting the crystal (bottom trace) the Pd peak at 3 eV in normal geometry disperses while the NO orbital stays at the same energy, as expected if the former comes from a direct transition and the latter from an adsorbate level

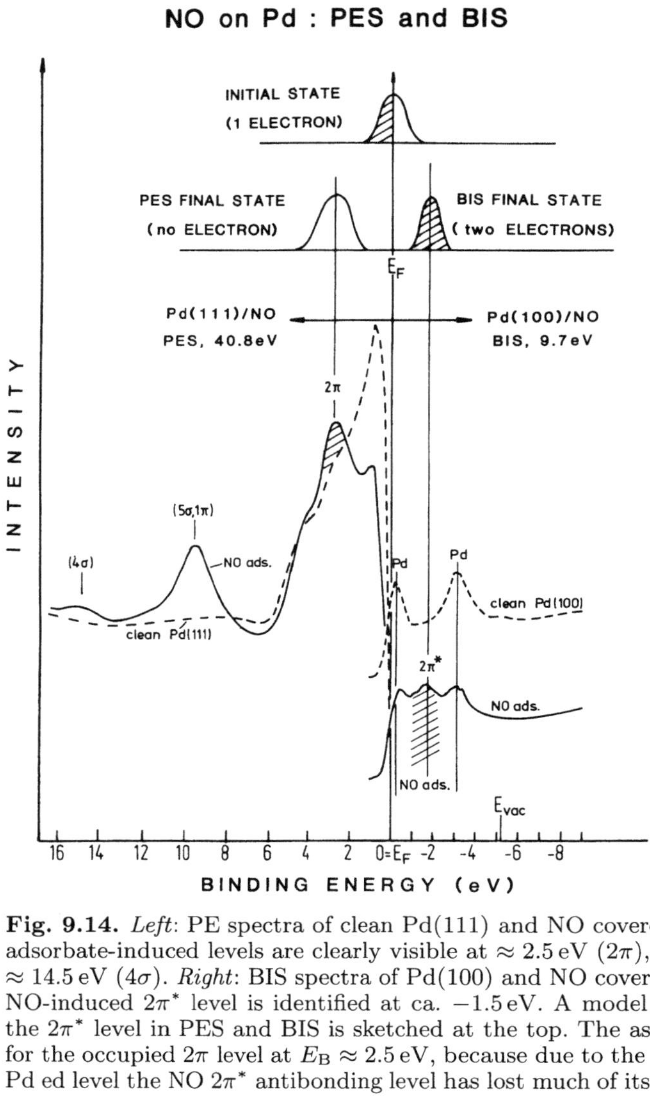

Fig. 9.14. *Left*: PE spectra of clean Pd(111) and NO covered Pd(111) [9.76]. The adsorbate-induced levels are clearly visible at $\approx 2.5\,\text{eV}$ (2π), $\approx 9,5\,\text{eV}$ ($5\sigma, 1\pi$) and $\approx 14.5\,\text{eV}$ (4σ). *Right*: BIS spectra of Pd(100) and NO covered Pd(100) [9.90]. The NO-induced $2\pi^*$ level is identified at ca. $-1.5\,\text{eV}$. A model for the observation of the $2\pi^*$ level in PES and BIS is sketched at the top. The asterisk has been left off for the occupied 2π level at $E_\text{B} \approx 2.5\,\text{eV}$, because due to the hybridization with the Pd ed level the NO $2\pi^*$ antibonding level has lost much of its antibonding character

"double" observation is sketched at the top of Fig. 9.14. In the initial state, the level is half occupied (1 electron) and is thus pinned to be Fermi energy. In a PES experiment this electron is removed and one observes the 2π hole state at $E_\text{B} \simeq 2.5\,\text{eV}$ below E_F. Conversely, in the BIS experiment, a two-electron final state is observed, which is located above the Fermi energy. Note that this experiment nicely allows one to determine the 2p correlation energy in the $2\pi^*$ state which turns out to be $U \simeq 4\,\text{eV}$ in this case, as compared to the gas phase value of 9.5 eV [9.90].

As another example, Fig. 9.15 shows IPES data for N_2, CO and NO adsorbed on Ni(001) [9.94]. N_2 and CO are isoelectronic with an empty $2\pi^*$ level in the free molecule and thus the position of this level in the adsorbed case is very similar. For NO, on the other hand, this level contains one electron in the free molecule and thus its empty part shifts towards the Fermi energy upon adsorption.

For further IPES data from adsorbed molecules, see, e.g., [9.96, 9.97].

IPES experiments have also been performed in an attempt to analyze the screening mechanism in rare-gas overlayers on metal substrates, however, so far, the results have been inconclusive [9.98, 9.99].

The examples of CO and NO on Pd have already demonstrated the usefulness of IPES as a tool to investigate the bonding properties of adsorbate systems. The unoccupied or partly occupied orbitals in such systems are, almost by definition, sensitive to small changes of the geometry and/or electronic structure. Thus by doing IPES measurements, changes in the spectra can be seen originating from these small alterations. Such investigations are virtually impossible with PES, since this method senses the much more tightly bound occupied orbitals. An important example is the promotion of catalysts by alkali metals and we choose the Pt(111)+K+CO system as an

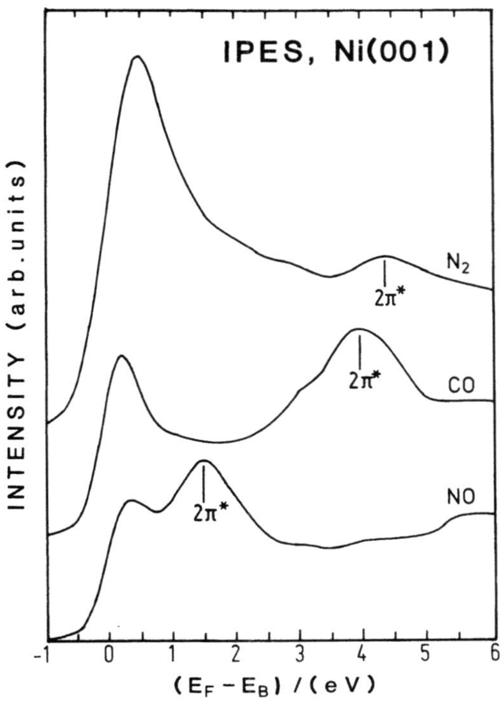

Fig. 9.15. $2\pi^*$ orbital observed via IPES for N_2, CO and NO on Ni(001) [9.94]

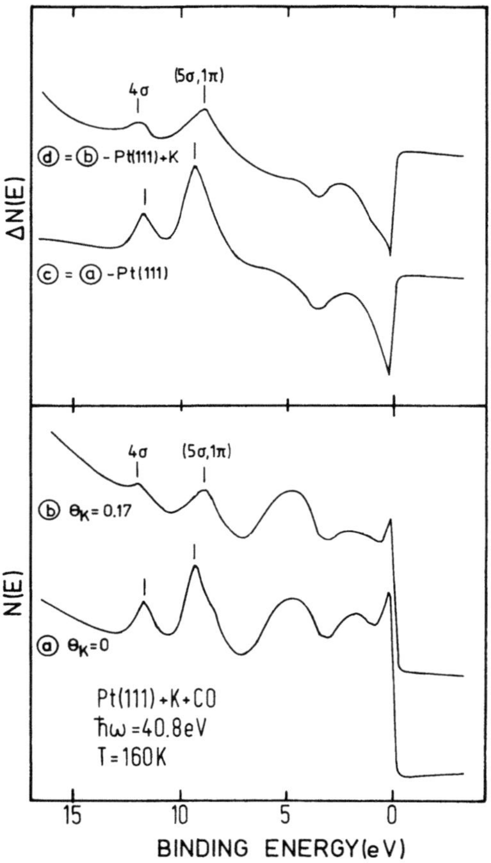

Fig. 9.16. UPS spectra ($\hbar\omega = 40.8\,\mathrm{eV}$) of Pt(111)+K+CO. (a) CO on Pt(111); (b) CO on Pt(111), with $\theta = 0.17$ monolayer K coadsorbed. Curve (c) is spectrum (a) minus the spectrum from the clean Pt (111) surface, giving approximately the orbitals of CO adsorbed on Pt(111). Curve (d): Spectrum (b) minus the spectrum from the K-covered Pt(111) surface. Even the difference spectra indicate little change in the CO orbital position due to K coadsorption [9.101]

illustration. Figure 9.16 shows He II spectra of this system. Curve a shows a He II spectrum of CO on Pt(111). Below the Pt d-band, one sees the two-peaked structure created by the adsorption of CO, representing photoionization from the 4σ and (5σ, 1π) orbitals [9.100, 9.101] (see Fig. 5.77 for comparison with CO on Ni). The spectrum with an additional small coverage of K ($\theta_K = 0.17$ monolayers) is shown as spectrum b and reveals hardly any change at all. The changes become slightly more evident if the platinum contributions to the spectra are subtracted out as is done in spectra c and d. It can now be seen more clearly that, upon covering the Pt surface with the small amounts of the promotor K, the 4σ to (5σ, 1π) splitting is increased.

However, since the PES spectrum contains initial and final state effects, the interpretation of the spectra is not straightforward.

For comparison, Fig. 9.17 displays the BIS spectra of CO on Pt, NO on Pt and CO on Pt+K. The hatched areas in the figure are the adsorbate induced contributions to the spectra [9.102]. It is apparent that the adsorption of the promotor on the Pt surface lowers the energy of the CO antibonding $2\pi^*$ orbital considerably. This implies a significant backdonation of charge into the CO $2\pi^*$ orbital by the K coadsorption, which strengthens the CO-substrate bond. This, on the other hand, means a weakening of the C-O bond, which has indeed been detected by infrared spectroscopy [9.103]. Thus in comparing Figs. 9.16 and 9.17 one realizes immediately the considerable advantages of IPES over PES for studying the chemical bond.

Finally we mention a point which was already raised in Chap. 5 during the discussion of the PE data of absorbed molecules. Here we are interested in the interaction of one isolated atom or molecule with a substrate. Such a system does not exist and all the data presented reflected experimental conditions which corresponded to coverages in the monolayer range. Under these conditions the adsorbate-adsorbate interaction is not negligible. It is then a more adequate approach to choose a system, which is an ordered overlayer on a metal substrate. The electronic structure of such a system is

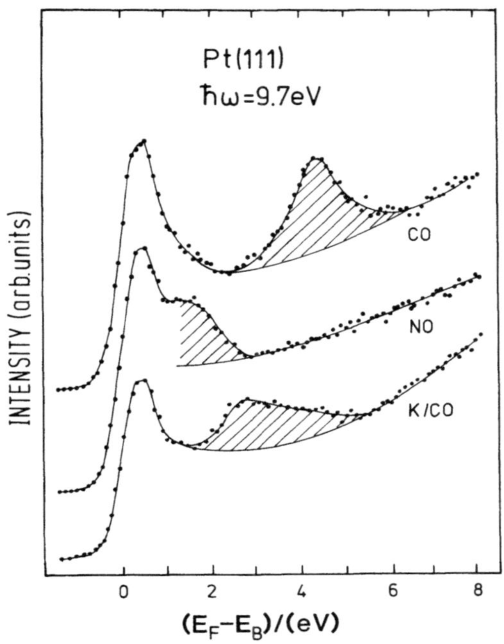

Fig. 9.17. BIS spectra of CO, NO and K/CO on Pt(111) [9.102] (see also Fig. 9.12). The coadsorption of K and CO lowers the CO orbital drastically

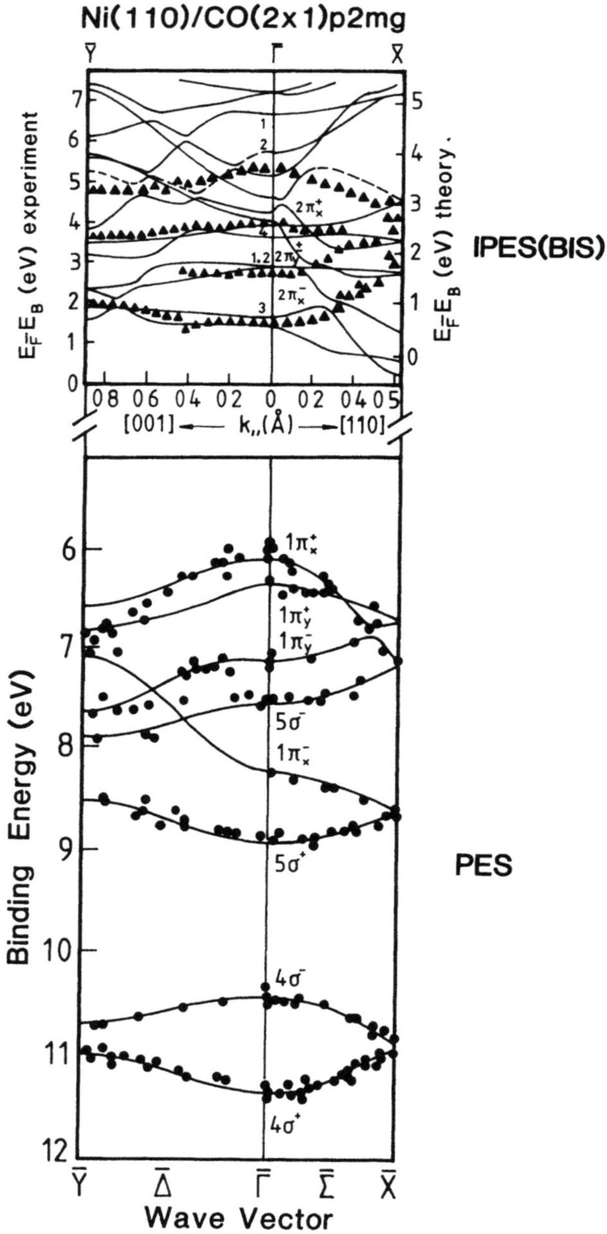

Fig. 9.18. Combined PES/IPES (BIS) data of the CO-derived bands for the Ni(110)/CO(2 × 1)p2mg structure. The full lines are the result of a tight binding calculation for the occupied bands and of a slab calculation for the unoccupied bands. The theoretical energy scale for the data above E_F has been expanded by 25 and shifted by 0.8 eV [9.104]

given by the overlayer-substrate and the intra-overlayer interaction making the analysis of the data more complicated, however, also more correct. As an example we show data on the system Ni(110)/CO(2 × 1)p2mg, representing an ordered CO overlayer structure of CO on Ni(110), which was already used in Chap. 5.

In Fig. 9.18 the band structure of the unoccupied bands as determined in a BIS experiment [9.104] has been added to that of the occupied CO-derived bands. It is immediately apparent that the bands above E_F make a description in terms of a single level reflecting the substrate–adsorbate chemical bond not very useful. In analyzing the data of Fig. 9.18 in terms of a band structure, Memmel et al. [9.104] come to the conclusion that the basic features of the Blyholder model are correct [9.80, 9.81].

References

9.1 W. Duane, F.L. Hunt: Phys. Rev. **6**, 166 (1915)
9.2 H. Claus, K. Ulmer: Z. Phys. **173**, 462 (1963)
9.3 H. Merz, K. Ulmer: Z. Phys. **197**, 409 (1966)
9.4 H. Rempp: Z. Phys. **267**, 181 (1974)
9.5 K. Ulmer: In *Band Structure Spectroscopy of Metals and Alloys*, ed. by D.J. Fabian, L.M. Watson (Academic, New York 1973)
9.6 H. Scheidt: Fortsch. Phys. **31**, 357 (1983)
9.7 G. Böhm, K. Ulmer: Z. Phys. **228**, 473 (1969)
9.8 J.K. Lang, Y. Baer: Rev. Sci. Instrum. **50**, 221 (1979)
9.9 V. Dose: Appl. Phys. **14**, 117 (1977)
9.10 G. Denninger, V. Dose, H. Scheidt: Appl. Phys. **18**, 375 (1979)
9.11 K. Desinger, V. Dose, M. Glöbl, H. Scheidt: Solid State Commun. **49**, 479 (1984)
9.12 N. Bahke, W. Drube, I. Schäfer, M. Skibowski: J. Phys. E **18**, 158 (1985)
9.13 G. Chauvet, R. Baptist: J. Electron Spectrosc. Relat. Phenom. **24**, 255 (1981)
Th. Fauster, F.J. Himpsel, J.J. Donelson, A. Marx: Rev. Sci. Instrum. **54**, 68 (1983)
9.14 V. Dose: J. Phys. Chem. **88**, 1683 (1984)
9.15 G. Denninger, V. Dose, H.P. Bonzel: Phys. Rev. Lett. **48**, 279 (1982)
9.16 A. Kovacs, P.O. Nilsson, J. Kanski: Phys. Scr. **25**, 791 (1982)
9.17 D.P. Woodruff, P.D. Johnson, N.V. Smith: J. Vac. Sci. Technol. A 1, 1104 (1983)
9.18 M. Conrad, V. Dose, Th. Fauster, H. Scheidt: Appl. Phys. **20**, 37 (1979)
9.19 V. Dose, H.J. Grossman, D. Straub: Phys. Rev. Lett. **47**, 608 (1981)
9.20 V. Dose, M. Glöbl, H. Scheidt: J. Vac. Sci. Technol. A 1, 1115 (1983)
9.21 P. Weibel, M. Grioni, D. Malterre, B. Dardel, Y. Baer: Phys. Rev. Lett. **72**, 1252 (1994). Showing that the analogue to resonance PES (Sect. 3.2.1) can be found in IPES if the incoming electrons energy is able to scatter a core electron into an empty valence-band state
9.22 J.K. Lang, Y. Baer, P.A. Cox: J. Phys. Metal. Phys. **11**, 121 (1981)
9.23 J.C. Fuggle, J.E. Inglesfield (eds.): *Unoccupied Electronic States*, Topics Appl. Phys., Vol.69 (Springer, Berlin, Heidelberg 1992)
D.v.d. Marel, G.A. Sawatzky, R. Zeller, F.U. Hillebrecht, J.C. Fuggle: Solid State Commun. **50**, 47 (1984)
D. Malterre, M. Grioni, P. Weibel, Y. Baer: Phys. Rev. Lett. **68**, 2656 (1992)

9.24 V. Dose: Progr. Surf. Sci. **13**, 225 (1983); J. Chem. Phys. **88**, 1681 (1984); Surf. Sci. Rep. **5**, 337 (1985); J. Vac. Sci. Technol. A **5**, 2032 (1987)
9.25 F.J. Himpsel, Th. Fauster: J. Vac. Sci. Technol. A **2**, 815 (1984)
9.26 V. Dose: Appl. Surf. Sci. **22/23**, 338 (1985)
9.27 N.V. Smith: Vacuum **33**, 803 (1983); Rep. Prog. Phys. **51**, 1227 (1988)
9.28 B. Reihl: Surf. Sci. **162**, 1 (1985)
9.29 N.V. Smith, D.P. Woodruff: Prog. Surf. Sci. **21**, 295 (1986)
9.30 F.J. Himpsel: Comments Cond. Mat. Phys. **12**, 199 (1986); Surf. Sci. Reports **12**, 1 (1990)
9.31 Th. Fauster, V. Dose: In: *Chemistry and Physics of Solid Surfaces, VI*, ed. by R. Vanselow, R. Howe (Springer, Berlin, Heidelberg 1986)
9.32 G. Borstel, G. Thörner: Surf. Sci. Rep. **8**, 1 (1988)
9.33 I. Tamm: Z. Phys. **76**, 848 (1932), see also Physikalische Zeitschrift der früheren Sowjetunion **1**, 733 (1932)
9.34 W. Shockley: Phys. Rev. **56**, 317 (1939)
9.35 P. Heimann, J. Hermanson, H. Miosga, H. Neddermeyer: Phys. Rev. B **20**, 3059 (1979)
9.36 P. Heimann, J. Hermanson, H. Miosga, H. Neddermeyer: Phys. Rev. Lett. **43**, 1757 (1979)
9.37 R.G. Jordan, G.S. Sohal: J. Phys. C **15**, L663 (1982)
9.38 H. Asonen, M. Lindroos, M. Pessa, R. Prasad, R.S. Rao, A. Bansil: Phys. Rev. B **25**, 7075 (1982)
9.39 D. Westphal, A. Goldmann: Surf. Sci. **95**, L249 (1980)
9.40 A. Goldmann, E. Bartels: Surf. Sci. **122**, L629 (1982)
9.41 F.J. Arlinghaus, J.G. Gay, J.R. Smith: Phys. Rev. B **23**, 5152 (1981), and references therein
9.42 P.M. Echenique, J.B. Pendry: J. Phys. C **11**, 2065 (1978)
9.43 N.V. Smith: Phys. Rev. B **32**, 3549 (1985)
9.44 D.P. Woodruff, S.L. Hubert, P.D. Johnson, N.V. Smith: Phys. Rev. B **31**, 4046 (1985)
9.45 A. Goldmann, V. Dose, G. Borstel: Phys. Rev. B **32**, 1971 (1985)
For a review of the experimental results on image potential states in Cu, see M. Grass, J. Braun, G. Borstel, R. Schneider, H. Dürr, Th. Fauster, V. Dose: J. Phys. Condens. Matter **5**, 599 (1993)
9.46 G. Thörner, G. Borstel, V. Dose, J. Rogozik: Surf. Sci. **157**, L379 (1985)
9.47 W. Altmann, V. Dose, A. Goldmann, U. Kolac, J. Rogozik: Phys. Rev. B **29**, 3015 (1984)
9.48 R.S. Williams, P.S. Wehner, G. Apai, J. Stoehr, D.A. Shirley, S.P. Kowalczyk: J. Electron Spectrosc. Relat. Phenom. **12**, 477 (1977)
9.49 V. Dose, W. Altmann, A. Goldmann, U. Kolac, J. Rogozik: Phys. Rev. Lett. **52**, 1919 (1984)
9.50 W. Steinmann: Appl. Phys. A **49**, 365 (1989)
W. Steinmann, Th. Fauster: In *Laser Spectroscopy and Photochemistry on Metal Surfaces*, ed. by H.L. Dai, W. Ho (World Scientific, Singapore 1993)
9.51 N.V. Smith, C.T. Chen, M. Weinert: Phys. Rev. B **40**, 7565 (1989)
9.52 S. Yang, K. Garrison, R.A. Bartynski: Phys. Rev. B **43**, 2025 (1991)
9.53 S. Schuppler, N. Fischer, W. Steinmann, R. Schneider, E. Bertel: Phys. Rev. B **42**, 9403 (1990)
9.54 N. Fischer, S. Schuppler, Th. Fauster, W. Steinmann: Phys. Rev. B **42**, 9717 (1990)
9.55 D. Straub, F.J. Himpsel: Phys. Rev. Lett. **52**, 1922 (1984)
9.56 D. Straub, F.J. Himpsel: Phys. Rev. B **33**, 2256 (1986)

9.57 G. Binnig, K.H. Frank, H. Fuchs, N. Garcia, B. Reihl, H. Rohrer, F. Salvan, A.R. Williams: Phys. Rev. Lett. **55**, 991 (1985)
9.58 A. Goldmann, M. Donath, W. Altmann, V. Dose: Phys. Rev. B **32**, 837 (1985)
9.59 K. Giesen, F. Hage, F.J. Himpsel, H.J. Riess, W. Steinmann: Phys. Rev. Lett. **55**, 300 (1985)
9.60 P.D. Johnson, N.V. Smith: Phys. Rev. B **27**, 2527 (1983)
9.61 B. Reihl, K.H. Frank: Phys. Rev. B **32**, 8282 (1985)
9.62 P.O. Gartland: Phys. Norv. **6**, 201 (1972)
9.63 S.L. Hulbert, P.D. Johnson, N.G. Stoffel, W.A. Royer, N.V. Smith: Phys. Rev. B **31**, 6815 (1985)
9.64 G.B. Baker, B.B. Johnson, G.L.C. Maire: Surf. Sci. **24**, 572 (1971)
9.65 W. Altmann, M. Donath, V. Dose, A. Goldmann: Solid State Commun. 53, 209 (1985)
9.66 D.A. Wesner, P.D. Johnson, N.V. Smith: Phys. Rev. B **30**, 503 (1984)
9.67 H. Scheidt, M. Glöbl, V. Dose, J. Kirschner: Phys. Rev. Lett. **51**, 1688 (1983)
9.68 F.J. Himpsel: Adv. Phys. **32**, 1 (1983)
9.69 E.W. Plummer, W. Eberhardt: Advances in Chemical Physics **49**, 533 (1982)
9.70 R. Courths, S. Hüfner: Phys. Rep. **112**, 53 (1984)
9.71 D.P. Woodruff, N.V. Smith, P.D. Johnson, W.A. Royer: Phys. Rev. B **26**, 2943 (1982)
9.72 G. Thörner, G. Borstel: Solid State Commun. **47**, 329 (1983)
9.73 G. Thörner, G. Borstel: Solid State Commun. **49**, 997 (1984)
9.74 S. Ishi, I. Ohno, B. Viswanathan: Surf. Sci. **161**, 349 (1985)
9.75 J.C. Tracy, P.W. Palmberg: J. Chem. Phys. **51**, 4852 (1969)
9.76 H. Conrad, G. Ertl, J. Koch, E.E. Latta: Surface Sci. **43**, 462 (1974)
9.77 H. Conrad, G. Ertl, J. Küppers, E.E. Latta: Surf. Sci. **65**, 235 (1977)
9.78 D.W. Turner, A.D. Baker, C. Baker, C.R. Brundle: *Molecular Photoelectron Spectroscopy* (Wiley Interscience, New York 1970)
9.79 I.P. Batra, C.R. Brundle: Surf. Sci. **57**, 12 (1976)
9.80 G. Blyholder: J. Phys. Chem. **68**, 2772 (1964)
9.81 G. Blyholder: J. Vac. Sci. Technol. **11**, 865 (1975)
9.82 B. Gumhalter, K. Wandelt, Ph. Avouris: Phys. Rev. B **37**, 8048 (1988)
9.83 J.P. Muncat, D.M. Newns: Prog. Surf. Sci. **9**, 1 (1978)
9.84 Ph. Avouris, P.S. Bagus, C.J. Nelin: J. Electron Spectrosc. Relat. Phenom. **38**, 269 (1986)
9.85 R. Hoffmann: Rev. Mod. Phys. **60**, 601 (1988)
9.86 P.S. Bagus, C.J. Nelin, S.W. Bauschlicher: J. Vac. Sci. Technol. A **2**, 905 (1984)
9.87 J.K. Norskov, S. Holloway, N.D. Lang: Surf. Sci. **137**, 65 (1984)
9.88 P.S. Bagus, K. Hermann, Ph. Avouris, A.R. Rossi, K.C. Prince: Chem. Phys. Lett. **118**, 311 (1985)
9.89 J. Rogozik, V. Dose, K.C. Prince, A.M. Bradshaw, P.S. Bagus, K. Hermann, Ph. Avouris: Phys. Rev. B **32**, 4296 (1985)
9.90 J. Rogozik, J. Küppers, V. Dose: Surf. Sci. **148**, L653 (1984)
9.91 H. Conrad, G. Ertl, J. Küppers, W. Sesselmann, H. Haberland: Surf. Sci. **121**, 161 (1982)
9.92 T.N. Rhodin, J.W. Gadzuk: In *The Nature of the Surface Chemical Bond*, ed. by T.N. Rhodin, G. Ertl (North-Holland, Amsterdam 1979)
9.93 W.F. Egelhoff: In *The Chemical Physics of Solid Surfaces and Heterogeneous Catalysis*, ed. by D.A. King, D.P. Woodruff (Elsevier, Amsterdam 1982)
9.94 P.D. Johnson, S.L. Hulbert: Phys. Rev. B **35**, 9427 (1987)
9.95 D.E. Peebles, E.L. Hardegree, J.M. White: Surf. Sci. **148**, 635 (1984)

9.96 F.J. Himpsel, Th. Fauster: Phys. Rev. Lett. **49**, 1583 (1982)
9.97 J. Rogozik, H. Scheidt, V. Dose, K.C. Prince, A.M. Bradshaw: Surf. Sci. **145**, L481 (1984)
9.98 K. Horn, K.H. Frank, J.A. Wilder, B. Reihl: Phys. Rev. Let. **57**, 1064 (1986)
9.99 K. Wandelt, W. Jakob, N. Memmel, V. Dose: Phys. Rev. Lett. **57**, 1643 (1986)
9.100 M. Kishinova, G. Pirug, H.P. Bonzel: Surf. Sci. **133**, 321 (1983)
9.101 H.P. Bonzel: J. Vac. Sci. Technol. A **2**, 866 (1984)
9.102 V. Dose: Surf. Sci. Rep. **5**, 337 (1986)
9.103 E.L. Garfunkel, J.E. Crowell, G.A. Somorjai: J. Phys. Chem. **86**, 310 (1982)
9.104 N. Memmel, G. Rangelov, E. Bertel, V. Dose, K. Kometer, N. Rösch: Phys. Rev. Lett. **63**, 1884 (1989)

10. Spin-Polarized Photoelectron Spectroscopy

10.1 General Description

Besides mass and charge, electrons also possess a spin. By detecting the direction of the spin of photoemitted electrons one can gain important information about the photoexcitation process and the properties of the sample under investigation. There are two modes in which a PE experiment can produce a spin polarization of the photoemitted electron. One can use unpolarized light to excite polarized electrons in a sample, which is a good way to investigate, e.g., magnetic materials, and one can employ circularly polarized light to excite transitions between states that are split by spin-orbit interactions, thereby obtaining spin-polarized electrons in the final state. This second method is useful for producing beams of polarized electrons, which can then be used for other experiments. Unfortunately, the measurement of spin polarization is a non-trivial problem and there are relatively few laboratories that can employ this very elegant method. Therefore, we shall simply outline the principles of such experiments whilst referring the reader to the excellent reviews [10.1–10.6] for further details.

The quantity that is measured is the polarization of a beam of electrons with respect to a direction in space, where the polarization P is defined as

$$P = \frac{n\uparrow - n\downarrow}{n\uparrow + n\downarrow} \tag{10.1}$$

with $n\uparrow$ and $n\downarrow$ being the number of up- and down-polarized electrons. The measurement of the polarization is often performed by using the spin-dependent scattering of electrons at high energies (Mott detector) [10.7]. The spin-dependent diffraction of low energy electrons by a solid can also be used (SPLEED detector) [10.8]. The most promising detector seems to be one which uses the spin-dependent absorption of electrons in a solid [10.9, 10.10].

10.2 Examples of Spin-Polarized Photoelectron Spectroscopy

The application of spin-polarized photoemission is of particular interest for magnetic materials. Figure 10.1 shows the schematic density of states of ferromagnetic Ni. The spin-up and spin-down bands are shifted with respect to one another by the exchange splitting ΔE_{ex}, which has a theoretical value of $\Delta E_{ex} = 0.6\,\mathrm{eV}$. Experimentally, however, one observes $\Delta E_{ex} = 0.3\,\mathrm{eV}$ as determined by photoemission. Note that Fig. 10.1 is an oversimplification (rigid band splitting); the exchange splitting is actually energy and wave-vector dependent, but the essence of the band magnetism is contained in the figure. A critical test of this picture can be made by measuring the spin polarization of electrons photoemitted from Ni near the photothreshold. The results of such a measurement are shown in Fig. 10.2. As expected from Fig. 10.1, the electrons indeed show a large negative spin polarization at the photothreshold, which rapidly changes sign with increasing photon energy, as likewise expected from the schematic density of states, since with increasing photon energy the number of photoexcited majority-spin electrons rises [10.11].

Another very instructive example for the use of polarization analysis is the case of the 6 eV satellite in Ni metal [10.12]. It was demonstrated earlier (Sect. 3.2), that this feature is caused by a two-hole state at the site of one Ni atom, or in other words, it is a state that is screened by sp electrons, whereas the main line is due to a hole screened by a d electron. In the valence band this satellite shows a resonance enhancement at the 3p photothreshold because direct photoemission and the $M_{23}M_{45}M_{45}$ super-Koster–Kronig Auger decay can then overlap coherently (Sect. 3.2.1). But in ferromagnetic Ni the Auger electrons should be polarized, as can be understood with reference to Fig. 10.3.

At threshold photoabsorption ($\hbar\omega = 67.7\,\mathrm{eV}$) only spin-down (minority) electrons are photoexcited, since only the spin-down band has empty states

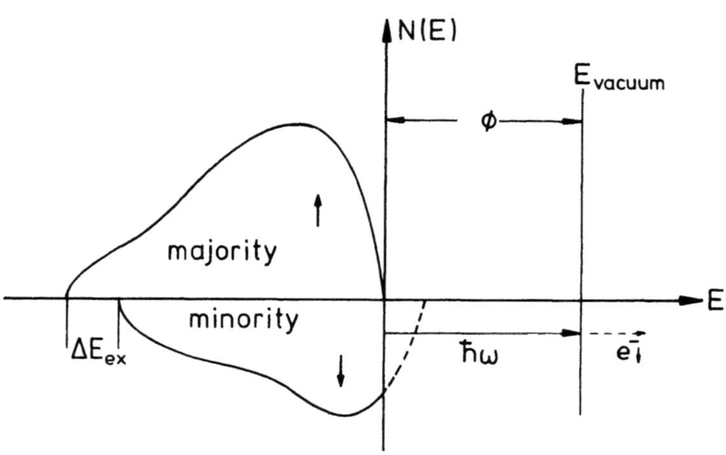

Fig. 10.1. Photoemission just above the threshold ($\hbar\omega \simeq \phi$) for Ni. Under these conditions only minority spin (spin down) photoelectrons, $e_{\downarrow}^{-}$, can be observed [10.11]

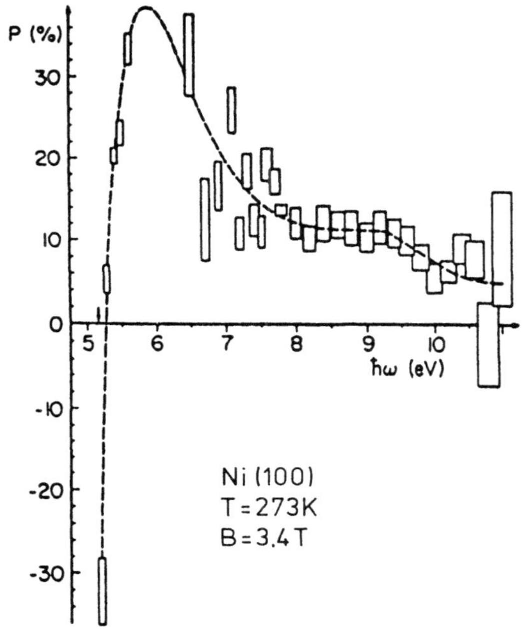

Fig. 10.2. Measurement of the spin polarized PE from Ni(100) at $T = 273\,\mathrm{K}$ [10.11]. Near threshold a negative spin polarization is observed as expected (Fig. 10.1). On increasing the photon energy one also probes the states further below E_F and thus the resulting spin polarization becomes positive

at the Fermi energy. This results in an intermediate "spin-up" polarized state with a "spin-down" hole in the 3p shell. The deexcitation process which leads to the final $3d^8$ state in the valence band, takes place via the emission of two electrons out of the valence band. Since the Auger electrons lead to a singlet ($S = 0$) final state and only spin-down electrons can go into the $3p^5$ state, spin-up electrons must be ejected. As the data in Fig. 10.4 demonstrate, a positive spin polarization ($P \approx 60\,\%$) is indeed observed experimentally. This provides further evidence to support our picture of the nature of the 6 eV satellite in Ni metal [10.13, 10.14].

We now give a brief outline of the production of polarized electrons by circularly polarized light from non-magnetic solids. This is illustrated for the case of GaAs [10.15, 10.16] which has been studied extensively [10.17, 10.18]. The energy bands for GaAs along ΓL are shown in Fig. 10.5 together with a diagram giving the possible optical transitions at the Γ point (direct gap) and their weight [10.19].

At Γ the conduction band has symmetry Γ_6 while the states at the top of the valence band have symmetry Γ_8 from which a Γ_7 state is split off, via the spin–orbit interaction, by an energy of $\Delta = 0.34\,\mathrm{eV}$. The Γ_6 state has the symmetry of an $s_{1/2}$ state, while the Γ_8 and Γ_7 state have the symmetries

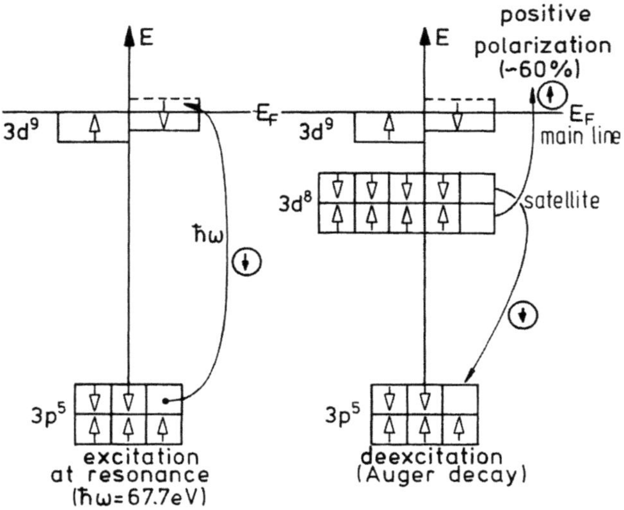

Fig. 10.3. Spin polarization for the 6 eV satellite in Ni metal [10.13,10.14] measured at threshold for the photoexcitation of Ni 3p electrons ($\hbar\omega = 67.7$ eV). At threshold there are only empty minority states in Ni metal, and thus only minority holes are produced in the 3p shell:
$3p^6 3d_\uparrow^5 3d_\downarrow^4 + \hbar\omega \rightarrow 3p_\uparrow^3 3p_\downarrow^2 3d_\uparrow^5 3d_\downarrow^5$
which leads to the decay $\rightarrow 3p_\uparrow^3 3p_\downarrow^3 3d_\uparrow^4 3d_\downarrow^4 + e_\uparrow^- (E_{\rm kin})$.
One should thus measure a positive (spin-up) spin polarization

of $p_{3/2}$ and $p_{1/2}$ states, respectively. In that approximation, one can use atomic selection rules to derive the possible transitions (and their intensities) for right and left circularly polarized light (σ^+ and σ^-). At the gap energy (threshold), one sees that, e.g., with σ^+ radiation electrons with positive polarization ($-3/2 \rightarrow -1/2$) are produced with a relative intensity of 3, whereas electrons with a negative polarization ($-1/2 \rightarrow +1/2$) are produced with an intensity of 1, thus one has

$$P = \frac{3-1}{3+1} = 1/2 \tag{10.2}$$

meaning that one expects a 50 % polarization for electrons produced at the threshold from a GaAs crystal by σ^+-polarized light. The experimental results are in agreement with this prediction (Fig. 10.6) giving a polarization from a GaAs crystal close to threshold of almost 50 % [10.19].

This method can of course be turned around to study the nature of transitions between two bands via the measured spin polarization of the photoemitted electrons. Such a study has been performed, e.g., for Pt metal [10.20]. However, for this purpose one may also study the intensity variation of the photoelectrons produced by circularly polarized light. This yields very similar information, but without the effort necessary to measure the electron spin polarization. In addition, from the perspective of band structures, the sym-

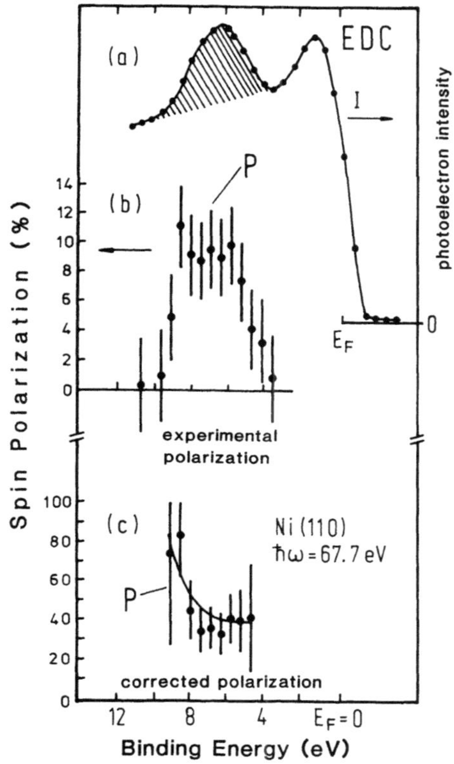

Fig. 10.4. Measurement of the spin polarization of the 6 eV satellite at threshold ($\hbar\omega = 67.7\,\mathrm{eV}$) [10.13, 10.14]. (a) EDC at resonance; (b) polarization (positive) as measured; (c) polarization corrected for background

metry properties of bands are known with sufficient certainty from theory. Thus the study of a band structure by spin-polarized PES is usually only of interest when one has very little prior knowledge about it.

The technique of detecting the spin direction of the photoemitted electrons is of special value in cases where the spin degeneracy of the solid state bands has been lifted (ferromagnets), and one can thus measure spin- and energy-resolved PES. Figure 7.4 showed the band structure of Ni, but omitted the exchange splitting of the bands. The fact that electrons photoemitted from polycrystalline Ni (Fig. 10.2) show spin polarization, indicates that these electrons are polarized in the solid and, according to our present understanding of magnetic d band metals (Fe, Co, Ni), this means that the d bands are exchange split. A spin- and angle-resolved photoemission experiment should therefore be able to detect this splitting and this is indeed the case [10.21].

Figure 10.7 shows the result of such an experiment on Ni(110), where the conditions have been chosen such that only electrons out of the X_2 band are excited within 0.5 eV below the Fermi energy [10.22, 10.23]. The data show convincingly that the $X_2\uparrow$ and the $X_2\downarrow$ bands have different energies. The splitting of the bands at the X point is also indicated in the figure.

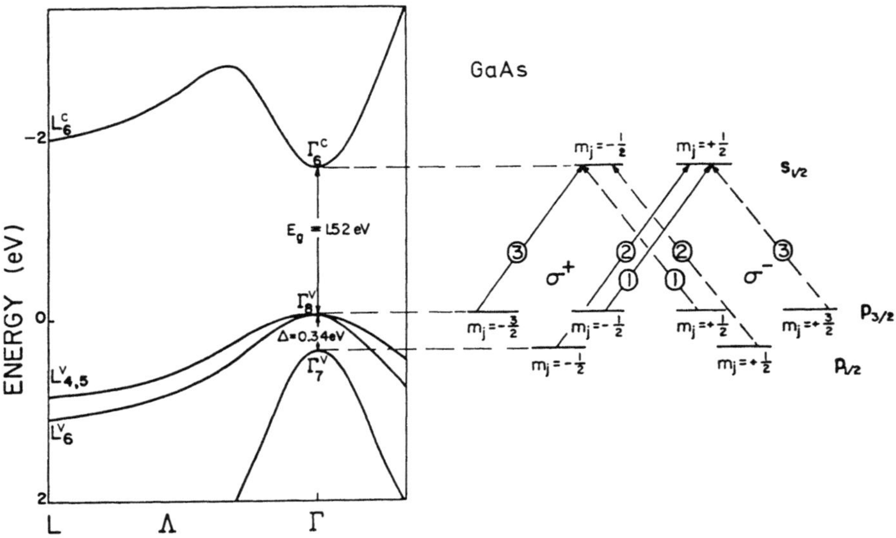

Fig. 10.5. Energy bands of GaAs near Γ and L and the transitions possible in this band structure using right (σ^+) and left (σ^-) circularly polarized light. The numbers in circles are the relative transition probabilities, and E_g is the energy gap [10.19]

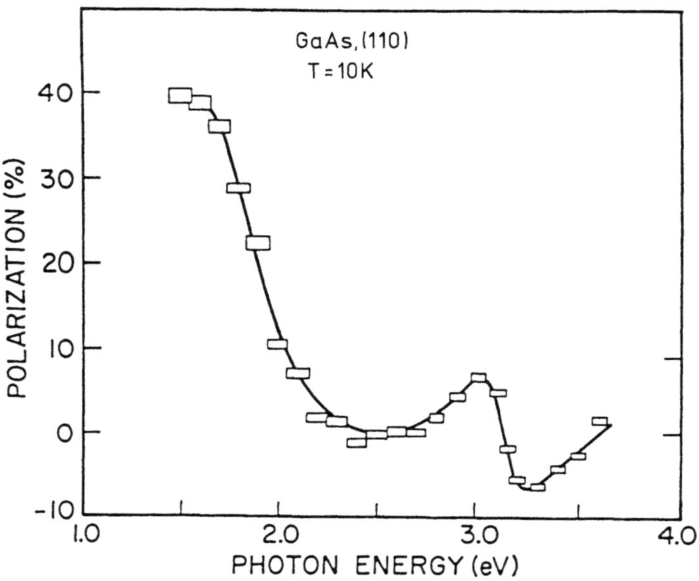

Fig. 10.6. Polarization of electrons emitted from a (110) surface of GaAs at 10 K. The polarization is largest at the threshold and is quite close to the expected 50 %. [10.19]

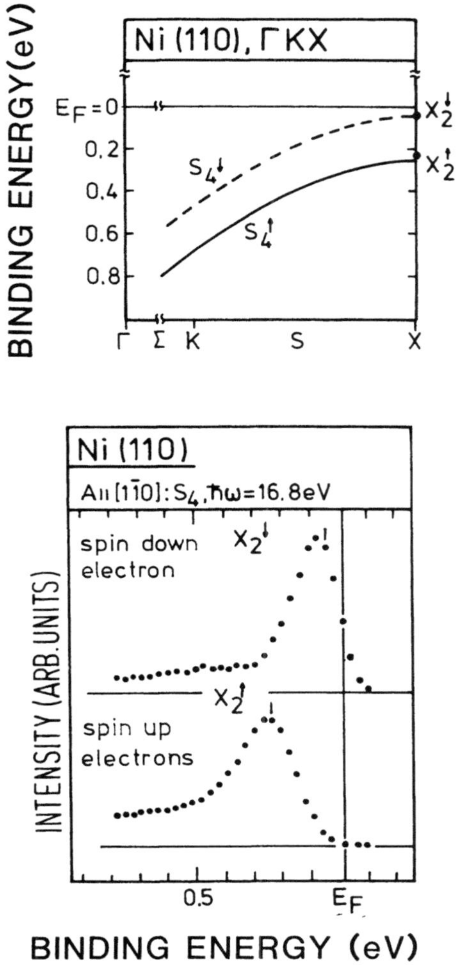

Fig. 10.7. Spin-, momentum- and energy-resolved PES from a Ni(110) surface. Near the X point the splitting into the $X_2^{\uparrow}$ and the $X_2^{\downarrow}$ state can be measured directly by monitoring the polarization of the photoemitted electrons [10.21]. The spectra for spin-up and spin-down electrons are given in the lower panel, while the upper panel shows the spin-resolved band structure [10.22]

The temperature dependence of the exchange splitting in Ni has also been measured [10.24].

One has to admit, however, that in this case the same kind of information has been obtained earlier, although less directly, from a normal (non-spin-polarized) PE experiment [10.25]. The data are shown together with the corresponding band structure in Fig. 10.8, and two special spectra are given in Eig. 10.9, together with their analysis. In this case one must assume from the outset that the small splitting observed is due to the exchange splitting of the bands [10.26], since the experiment itself cannot discriminate between spin-up and spin-down electrons. The analysis also shows that the band further below the Fermi energy has a larger width. This is a manifestation of the fact that the Auger decay rate increases in going away from the Fermi energy.

The spin-split band structure of Ni in the vicinity of the X point is given in Fig. 10.10 [10.25]. One realizes that the exchange splitting is not a rigid shift of the paramagnetic bands by an exchange energy. Rather the splitting is $\boldsymbol{k}$-dependent as can be seen from Fig. 10.10 (the full curves are scaled theoretical bands by Zornberg [10.27]), and also by a comparison of the two spectra in Fig. 10.9 which correspond to PES at different points in the Brillouin zone.

A similar investigation on iron [10.21, 10.28, 10.29] will now be discussed, in which the exchange effects are more noticeable. Figure 10.11 provides a nice demonstration of the additional spectral features that are observed with spin-polarized detection. The top curve is an EDC of Fe(001) without spin detection. The next curve shows the net spin polarization for iron in the valence-band states. The final plot includes spin and energy discrimination, giving the intensity of the spin-up and the spin-down electrons separately. The experimental conditions are chosen such that one is observing transitions near the Γ point of the Brillouin zone. A comparison with the band structure (Fig. 10.12) shows that one is indeed observing the exchange splitting of the

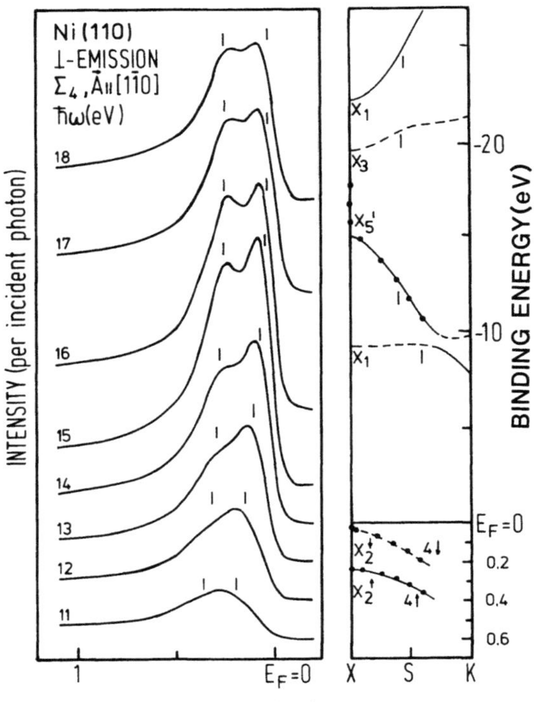

Fig. 10.8. Exchange splitting of Ni near the X point of the Brillouin zone measured by unpolarized PES. The left panel shows the spectra and the right one the interpretation in terms of the spin-split band structure [10.25]. For $\hbar\omega \geq 15\,\mathrm{eV}$ the observed splitting stays constant, indicating a transition into a gap as shown

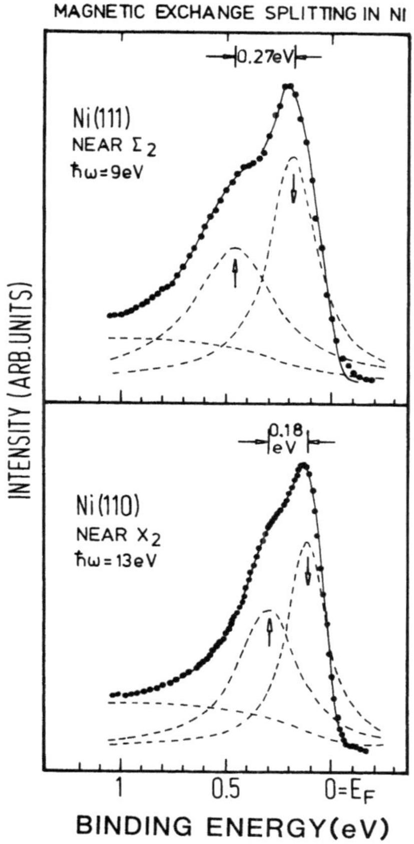

Fig. 10.9. Analysis of high resolution spectra of Ni [10.25]. The larger binding energy component shows a greater width due to the larger Auger decay rate

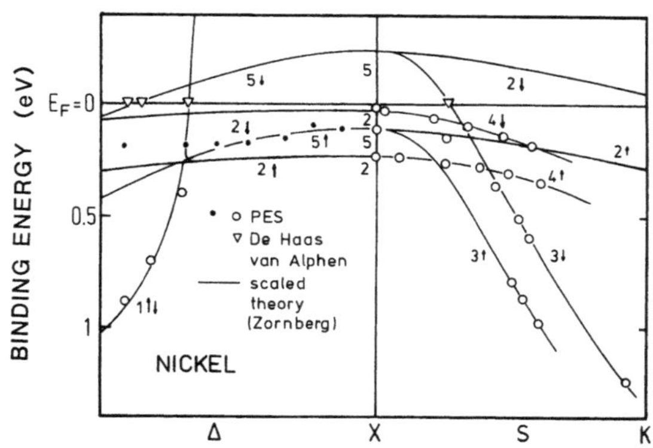

Fig. 10.10. Detailed band structure investigation of Ni around the X point [10.25]. The comparison with the de Haas–van Alphen data is good. The full line is the result of a band structure calculation of Zornberg [10.27] which has been scaled to give the best agreement with the experimental data

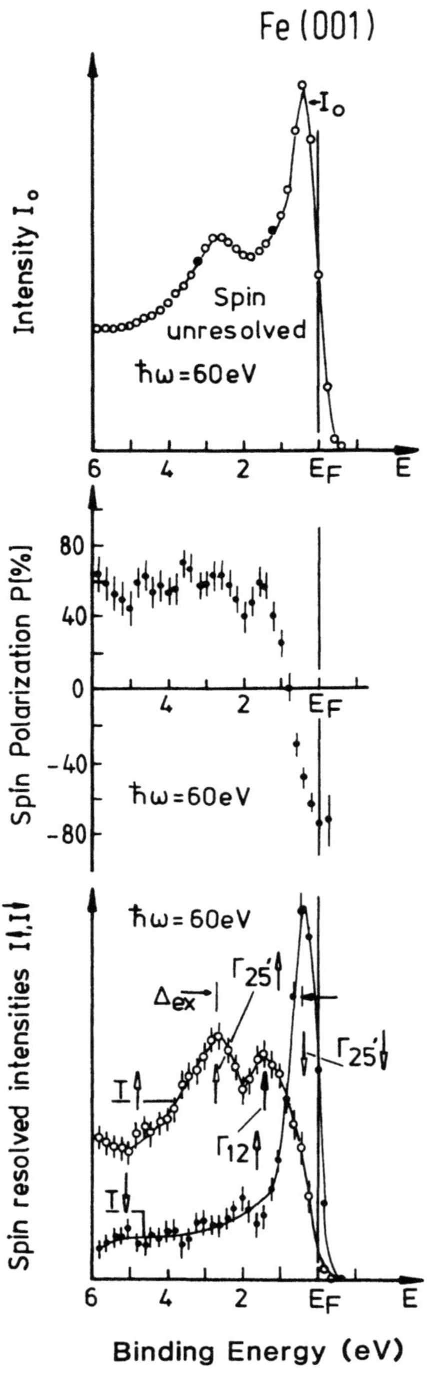

Fig. 10.11. Spin polarization in Fe(001) [10.21]. *Top*: EDC (unpolarized) taken with photons of $\hbar\omega = 60\,\mathrm{eV}$. *Middle*: Total spin polarization as a function of initial state energy measured with $\hbar\omega = 60\,\mathrm{eV}$. *Bottom*: Spin- and energy-resolved PE from Fe(001) measured with $\hbar\omega = 60\,\mathrm{eV}$

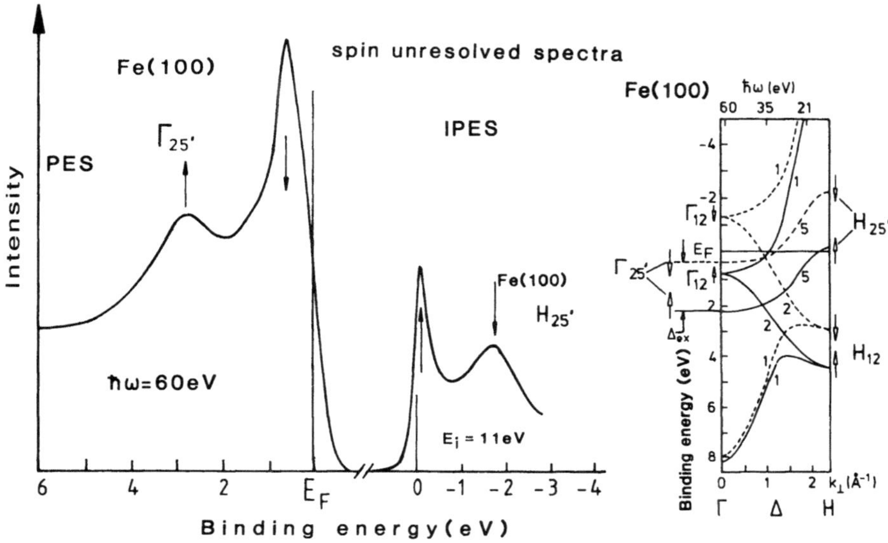

Fig. 10.12. Combination of spin unresolved PES/IPES data for Fe metal, compared to a band structure in the ΓH direction. From a comparison with spin resolved data it can be concluded that the PES data give the $\Gamma_{25'\uparrow} - \Gamma_{25'\downarrow}$ splitting and the IPES data the $H_{25'\uparrow} - H_{25'\downarrow}$ splitting [10.21, 10.32]

$\Gamma'_{25}\uparrow$ and $\Gamma'_{25}\downarrow$ bands and that its magnitude is $\approx 2\,\mathrm{eV}$. The splitting of the $H'_{25}\uparrow$ and $H'_{25}\downarrow$ bands can be estimated from spin-resolved IPES data and has also been measured [10.30].

Polarization effects have also been observed in the PES from the 3p core levels of Fe metal [10.31].

The band-structure diagram in Fig. 10.12 shows that if a spin unresolved PE spectrum is measured at the Γ point, and a spin unresolved IPE spectrum is measured at the H point the exchange splittings at the $\Gamma_{25'}$ point and the $H_{25'}$ point dominate the respective spectra [10.21, 10.32]. These data are also given in Fig. 10.12 in order to demonstrate the magnitude of the exchange splittings of Fe metal.

Exchange splittings have also been seen in adsorbate derived bands and a typical example for the Ni(110)/p(2 × 1)O structure is shown in Fig. 10.13. The splitting is small although measurable in spin resolved experiments [10.33].

We mention that experiments resolving the spin give an additional selection rule and can therefore be favorably employed to determine the character of states between which a PE transition takes place also in non-magnetic materials [10.34].

Good recent reviews on spin-polarized photoelectron spectroscopy can be found in [10.35, 10.36].

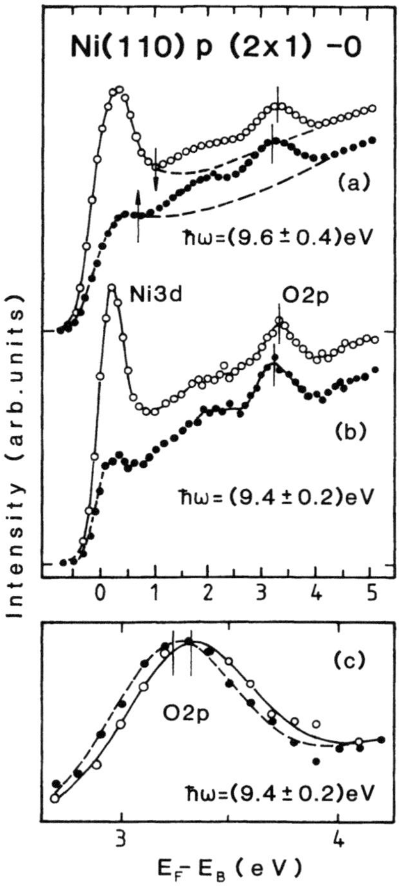

Fig. 10.13. Spin resolved IPES data for the Ni(110)/p(2 × 1)O system [10.33]: (**a**) CaF_2 window for the iodine counter, (**b**) SrF_2 window for the iodine counter, and (**c**) O 2p part of the spectrum in panel (**b**) on an extended scale

10.3 Magnetic Dichroism

High resolution photoemission experiments with subsequent spin analysis are very elegant. They are hampered, however, by the fact that generally the spin analyzer does not have a high efficiency, which makes such experiments time consuming. In looking for alternatives it was realized that, in principle, one can get information similar to that of a spin analysis by using dichroic effects [10.36]. This means that the intensity of a photoemission signal varies with the polarization of the incoming radiation (works for linear polarization and circular polarization) and/or, in a magnetic material, with the direction of the magnetization of the sample (in this case one can even use unpolarized light). One can summarize these facts by stating that, in principle, in order to observe dichroic effects, one just needs "handedness" in the experimental geometry [10.36].

Although the concern in this chapter is with magnetic materials, dichroic effects can also be observed from non-magnetic samples provided the condition of handedness in the measurement arrangements is fulfilled.

The simplest way of introducing handedness is to use a geometry in which the directions of the light incidence, the surface normal and the electron emission are non-coplanar. This geometry gives rise to a dichroic effect termed circular dichroism in the angular distribution of photoelectrons (CDAD) and is observed for clean surfaces but also for surfaces covered with molecular absorbates. In essence CDAD is an interference effect between symmetric and antisymmetric contributions in the final state of the photoexcitation step and does not involve any spin-dependent interaction. In this sense it does not really belong in this chapter, but is included for the sake of completeness.

This effect is very different from the case of magnetic CDAD (MCDAD), which requires the combination of spin–orbit coupling and/or exchange interaction in the initial and/or final state. This latter effect is a spin-dependent effect and here, in the simplest case, the handedness enters the system through the polarization vector characterizing the sample magnetization. The changes in the intensity of the spectra due to changes in the polarization or the magnetization reversal can be used to obtain information on the spin state of the initial state. This then means that such experiments are more or less equivalent to those using direct spin analysis. In principle there are three different ways of performing magnetic dichroic angular distribution experiments. One can use circularly polarized light (MCDAD), one can use linear polarized light (MLDAD), and one can even work with unpolarized light (MUDAD). We shall give a brief example of each of these three cases.

If one designates the photocurrent produced by right (left) circular polarized light by $I(\sigma^+)$ ($I(\sigma^-)$) and the spin polarization in the photocurrent I by $P(\sigma^+)$ ($P(\sigma^-)$), the dichroic experiments can be approximately classified as follows:

1. $I(\sigma^+) = I(\sigma^-); \quad P(\sigma^+) \neq P(\sigma^-)$

This process describes the optical orientation from non-magnetic substances, for example the production of polarized electrons from non-magnetic samples (Fig. 10.5).

2. $I(\sigma^+) \neq I(\sigma^-); \quad P(\sigma^+) \neq P(\sigma^-)$

This condition holds for the experiments showing magnetic circular dichroism in the angular distribution of photoelectrons (MCDAD).

3. $I(\sigma^+) \neq I(\sigma^-); \quad P \equiv 0$

This describes the circular dichroism in the angular distribution of photoelectrons (CDAD)

The principle of a magnetic dichroic experiment is outlined in Fig. 10.14 [10.36]. There it is assumed that the system under investigation has an occupied twofold degenerate state (like a $p_{1/2}$ state), which is split by the magnetic interaction assumed to be present in the system into two states, designated here as spin-up and spin-down. The final state in the photoemission process, far above the vacuum level, has no polarization. Right or left polarized light (σ^+, σ^-) excites electrons only out of one of these two states, and therefore the signals originating from a photoemission experiment with left or right circularly polarized light lead to lines with slightly different binding energy. Thus by analyzing the photoemission spectra taken with light of the two different polarizations, one can deduce from the difference in the spectra the splitting in the initial state. Instead of working with right- or left-polarized light, one can also just change the magnetization direction of the sample and thereby change the directions of the spins in the initial state relative to the polarization of light and also obtain a dichroic effect. Since circularly polarized light can be considered to consist of two beams of linearly polarized light, linearly polarized light can likewise be thought of as originating from two oppositely polarized circularly polarized beams. Therefore it is intuitively apparent that a dichroic effect can also be obtained with linearly polarized light. It was stated above that, in order to observe a dichroic effect, one needs a handedness in the experiment. In the magnetic case this is already provided by the magnetization direction of the sample, and therefore one can anticipate that even unpolarized light can yield a dichroic signal if the geometrical conditions are chosen appropriately.

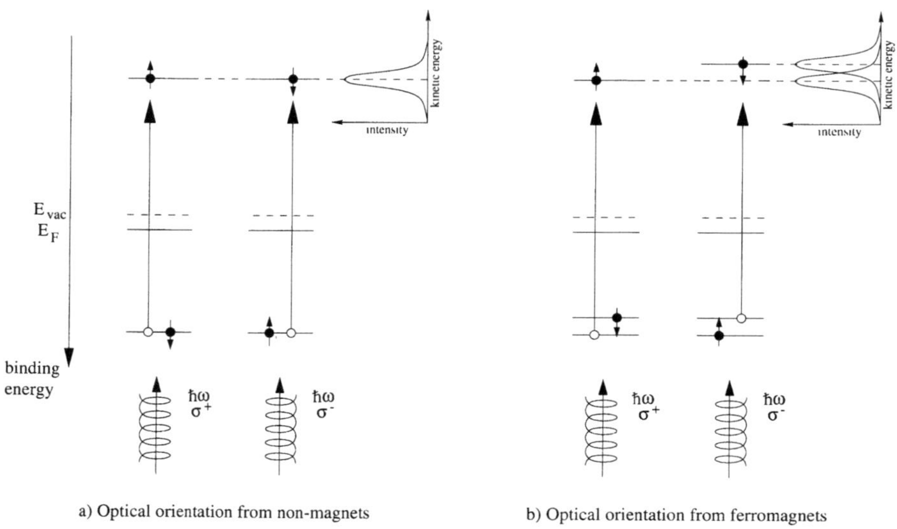

Fig. 10.14a,b. Principle of optical orientation experiments with right and left circularly polarized light from a non-magnetic sample and from a ferromagnetic sample. In the ferromagnetic sample the photoemission line occurs at slightly different energies for the two light polarizations, while in the non-magnetic sample it always occurs at the same energy [10.36]

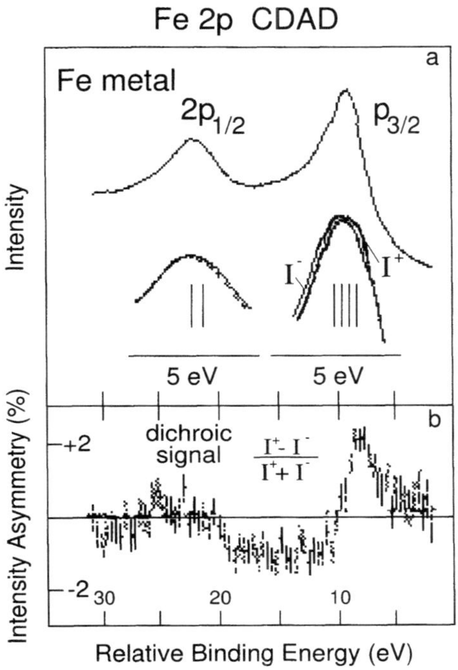

Fig. 10.15a,b. Circular dichroism in the angular distribution of photoelectrons (CDAD) measured with the 2p electron of iron metal [10.37]. The top trace shows a spectrum with unpolarized light and also traces with left and right circularly polarized light. The lower signal shows the dichroitic signal, namely the intensity asymmetry calculated as $A = (I^+ - I^-)/(I^+ + I^-)$

These general considerations have been verified by many experiments and here we only show the classic example of the 2p photoemission spectra of metallic iron [10.37]. Dichroism in photoemission spectra was observed for the first time in this system and these data are shown in Fig. 10.15. The top trace shows the spin–orbit split $2p_{1/2}$, $2p_{3/2}$ spectrum and below are the spectra observed with the two different circular polarizations of the light. The bar diagrams are meant to indicate the splitting of the initial state. The size of the dichroism is usually given by an equation similar to (10.1), where now instead of the number of electrons the intensity of the photocurrent produced by right or left circularly polarized light is used, leading to an asymmetry in the photoelectron distribution ($A = (I^+ - I^-)/(I^+ + I^-)$). This quantity is also depicted in Fig. 10.15 and shows non-zero values.

An experiment using linearly polarized light [10.38] is shown in Fig. 10.16 and again a considerable asymmetry is visible in the observed intensity. Finally Figs. 10.17a, 10.17b shows an experiment with unpolarized light [10.39]. The geometry is given in (a) and the result in (b). Here too, as expected, an asymmetry in the intensity is observed. In the latter two experiments the dichroic effect is produced by changing the magnetization direction in the sample.

Modulations of the spectra and, in particular, of the observed asymmetries by assuming a simple magnetic splitting of the initial states leads to a good

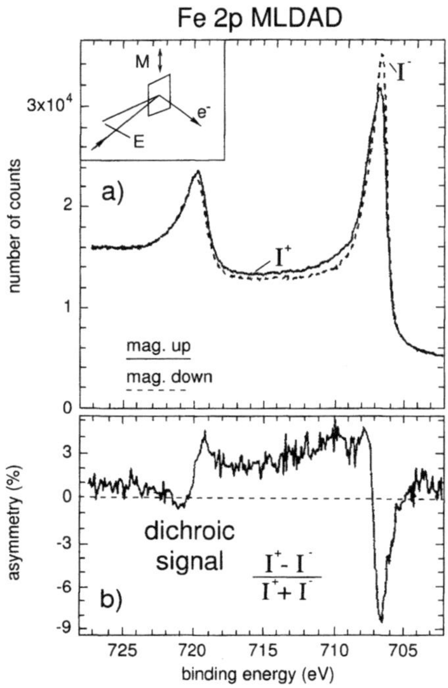

Fig. 10.16a,b. Magnetic linear dichroism angular distribution (MLDAD) for the iron 2p lines in iron metal. The experiment was performed by switching the direction of the magnetization relative to the orientation of the linearly polarized light. The lower trace shows again the dichroic signal calculated as $A = (I^+ - I^-)/(I^+ + I^-)$ [10.38]

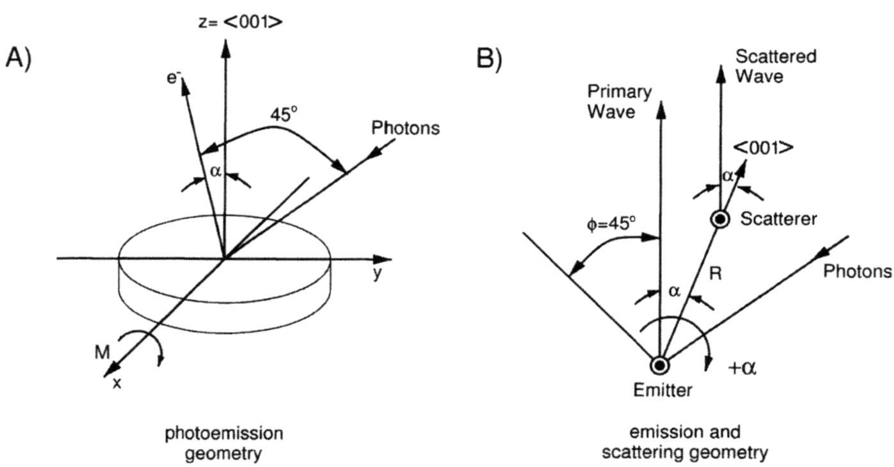

Fig. 10.17a. Geometry of the experiment to measure a magnetic unpolarized dichroitic angular distribution (MUDAD) from a sample of iron [10.39]. A) Mg K_α unpolarized radiation impinges on the sample in the $y-z$ plane. The sample magnetization is switched between the $+x$ and $-x$ direction in order to produce the dichroic signal. B) The electron emission and scattering geometry for the environment A

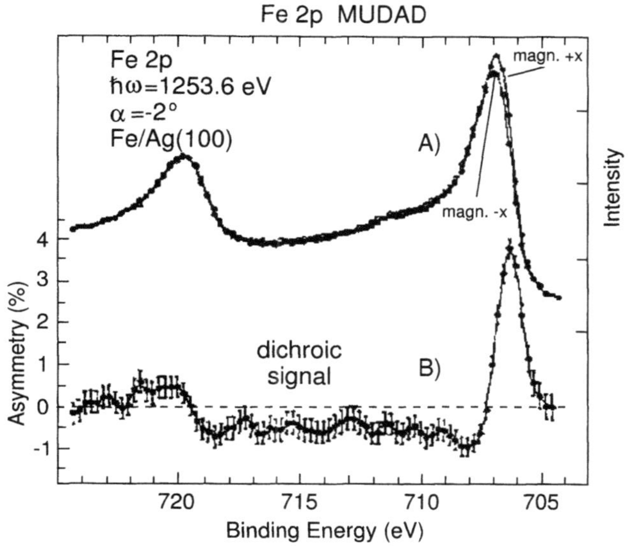

Fig. 10.17b. A) Results for the experiment using unpolarized light (MUDAD) to measure the 2p spectra of iron metal [10.39]; the two different spectra are for magnetization in the $+x$ and $-x$ direction (Fig. 10.17) B) A dichroic signal ($A = (I^+ - I^-)/(I^+ + I^-)$ is observed

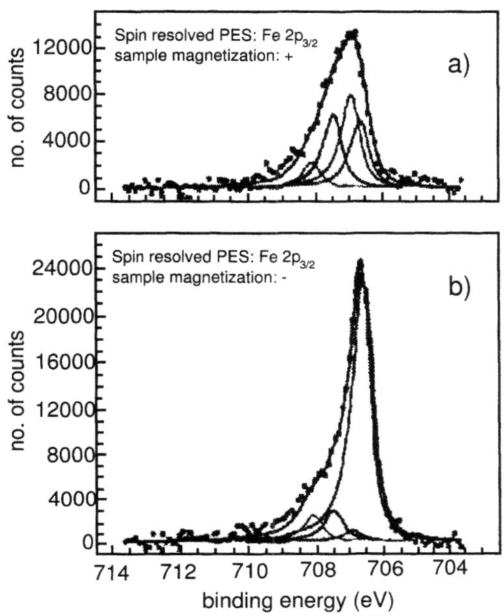

Fig. 10.18a,b. Spin-resolved photoemission spectra from the Fe2p$_{3/2}$ level with the magnetization in two different directions (+ and -). The spectra have been fitted by assuming a magnetic splitting of the 2p$_{3/2}$ level into four lines [10.40]

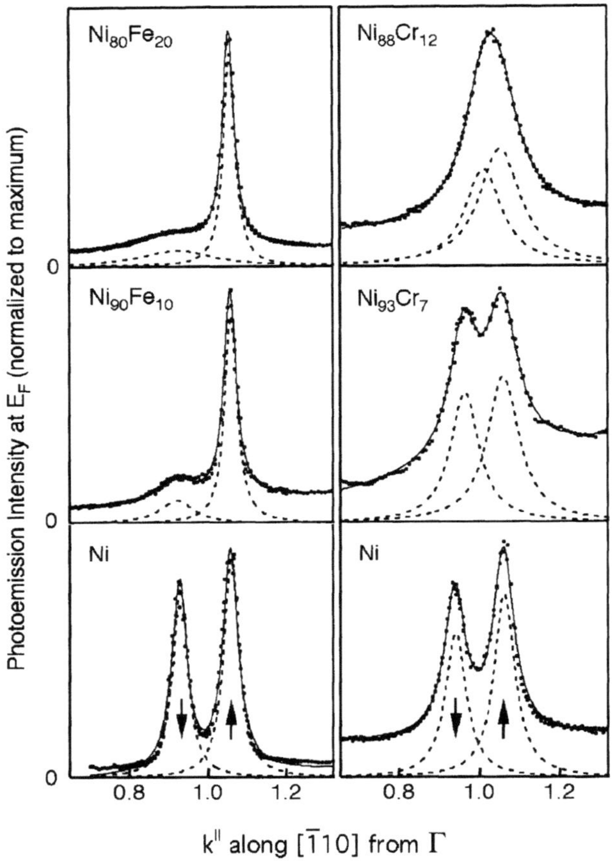

Fig. 10.19. Valence-band photoemission from pure Ni and Ni doped with Fe or Cr. The data indicate that Fe impurities scatter predominantly spin down electrons while Cr impurities seem to scatter electrons of both spin directions equally [10.42]

agreement between theory and experiment. A high resolution experiment has tried to observe the different Zeeman components in the $2p_{3/2}$ state of iron metal directly and was partly successful as the analysis on the data show in Fig. 10.18. Although the spectra can be fitted by four lines, it is obvious that these fits are suggested by the input of the magnetically split 3/2 initial state [10.40].

Larger dichroic effects in magnetic materials have been observed in a number of impressive investigations of rare earths [10.41].

The field of magnetoelectronics is a rapidly developing area of electronics, in which spin currents replace charge currents. As in semiconductor electronics, one may dope the host material. The analogues of acceptors and donors are spin-polarized impurities, which change the spin balance towards spin-up or spin-down carriers. In addition to manipulating the carrier density n, one now also has the option of changing the mean free path ℓ by spin-selective scattering, introduced by the spin doping.

An experiment demonstrating that different spin impurities have different effects on the spin scattering has been performed by introducing iron or chromium impurities into a Ni host [10.42]. The changes in the band structure caused by this spin doping have been measured by high-resolution photoemission spectroscopy, and the results are shown in Fig. 10.19. This diagram shows the emission out of a particular band, where in the undoped material, pure nickel, the spin-up and spin-down peaks show roughly the same intensity (Fig. 10.19). The exact position of that band is of no interest for the effect described here. Iron doping leads to a strong decrease in the intensity of the spin-down peak, which manifests itself by a broadening. This means that the doping causes the mean free path for spin-down electrons to decrease, while that for spin-up electrons stays approximately the same. This indicates that the iron impurities scatter only spin-down electrons.

The experiment with a chromium-doped sample shown on the right-hand side of Fig. 10.19 yields a very different behavior. Now the intensity of the two peaks hardly changes at all with doping, while both peaks, namely those of the spin-up and the spin-down electrons, become broader with increasing doping. This shows that chromium scatters electrons of both spin directions, in turn demonstrating that there is a strong effect of magnetic impurities on the electronic states near the Fermi energy. The simple anticipation that chromium and iron show opposite behavior because of their opposite spin directions is actually not fulfilled. This means that spin scattering is a much more complex process than suggested by simple arguments.

References

10.1 F. Meier, B.P. Zakharchenya (eds.): *Modern Problems in Condensed Matter Science, Optical Orientation* (North-Holland, Amsterdam 1984)

10.2 H.C. Siegman: Phys. Rep. C **17**, 38 (1975)

10.3 S.F. Alvarado, W. Eib, F. Meier, H.C. Siegmann, D. Zürcher: In *Photoemission and the Electronic Properties of Surfaces*, ed. by B. Feuerbacher, B. Fitton, R.F. Willis (Wiley, New York 1978)

10.4 J. Kessler: *Polarized Electrons*, 2nd edn., Springer Ser. Atoms Plasmas, Vol.1 (Springer, Berlin, Heidelberg 1985)

10.5 E. Feder (ed.): *Polarized Electrons in Surface Physics* (World Scientific, Singapore 1985)

10.6 J. Kirschner: *Polarized Electrons at Surfaces*. Springer Tracts Mod. Phys., Vol.106 (Springer, Berlin, Heidelberg 1976)

10.7 N.F. Mott: Proc. Roy. Soc., London A **135**, 429 (1932)

10.8 J. Kirschner, R. Feder: Phys. Rev. Lett. **42**, 1008 (1979)

10.9 M. Erbudak, N. Müller: Appl. Phys. Lett. **38**, 575 (1981)

10.10 R.J. Celotta, D.T. Pierce, H.C. Siegmann, J. Unguris: Appl. Phys. Lett. **38**, 577 (1981)

10.11 W. Eib, S.F. Alvarado: Phys. Rev. Lett **37**, 444 (1976)

10.12 S. Hüfner, G.K. Wertheim: Phys. Lett. A **51**, 299 (1975)
10.13 L.A. Feldkamp, L.C. Davis: Phys. Rev. Lett. **43**, 151 (1979)
10.14 R. Clauberg, W. Gudat, E. Kisker, G.M. Rothberg: Phys. Rev. Lett. **47**, 1314 (1981)
10.15 E. Garwin, D.T. Pierce, H.C. Siegmann: Helv. Phys. Acta **47**, 343 (1974)
10.16 D.T. Pierce, F. Meier, P. Zürchner: Appl. Phys. Lett. **26**, 670 (1975)
10.17 D.T. Pierce, F. Meier, P. Zürchner: Phys. Lett. A **51**, 465 (1975)
10.18 D.T. Pierce, F. Meier: Phys. Rev. B **13**, 5484 (1976)
10.19 F. Meier, B.P. Zakharchenya: In [10.1]
10.20 A. Eyers, E. Schäfers, G. Schönhense, U. Heinzmann, H.P. Öppen, K. Heinlich, J. Kirschner, G. Borstel: Phys. Rev. Lett. **52**, 1559 (1984)
10.21 M. Campagna: J. Vac. Sci. Technol. A **3**, 1491 (1985)
10.22 R. Raue, H. Hopster, R. Clauberg: Phys. Rev. Lett. **50**, 1623 (1983); a discussion of the temperature dependence of the exchange splitting in Ni metal can be found in A. Kakizaki, J. Fujii, K. Shimada, A. Kamata, K. Ono, K.H. Park, T. Kinoshita, T. Ishii, H. Fukutani: Phys. Rev. Lett. **72**, 2781 (1994) and references therein
10.23 H. Hopster, R. Raue, E. Kisker, G. Güntherodt, M. Campagna: Phys. Rev. Lett. **50**, 70 (1983)
10.24 H. Hopster, R. Raue, G. Güntherodt, E. Kisker, R. Clauberg, M. Campagna: Phys. Rev. Lett. **51**, 829 (1983)
10.25 P. Heimann, F.J. Himpsel, D.E. Eastman: Solid State Commun. **39**, 219 (1981)
10.26 C.S. Wang, J. Callaway: Phys. Rev. B **15**, 298 (1977)
10.27 E.J. Zornberg: Phys. Rev. B **1**, 244 (1970)
10.28 E. Kisker, K. Schröder, M. Campagna, W. Gudat: Phys. Rev. Lett. **52**, 2285 (1984)
10.29 E. Kisker, K. Schröder, W. Gudat, M. Campagna: Phys. Rev. B **31**, 329 (1985)
10.30 J. Kirschner, M. Glöbl, V. Dose, H. Scheidt: Phys. Rev. Lett. **53**, 612 (1984)
10.31 C. Carbone, E. Kisker: Solid State Commun. **65**, 1107 (1988); a discussion of spin-dependent effects in PES of core levels can be found in G. Rossi, F. Sirotti, N.A. Cherepkov, F. Combet, F.G. Panaccione: Solid State Commun. **90**, 557 (1994)
D.G. Van Campen, L.E. Klebanoff: Phys. Rev. B **49**, 2040 (1994)
M. Getzlaff, Ch. Ostertag, G.H. Fecher, N.A. Cherepkov, G. Schönhense: Phys. Rev. Lett. **73**, 3030 (1994). Reporting on the observation of magnetic dichroism in photoemission with unpolarized light
F.U. Hillebrecht, T. Kinoshita, D. Spanke, J. Dresselhaus, Ch. Roth, N.B. Rose, E. Kisker: Phys. Rev. Lett. **75**, 2224 (1995). Observing a linear magnetic dichroism in the total photoelectron yield from the Fe 3p and Co 3p levels, and using the effect for magnetic-domain imaging
10.32 A. Santoni, F.J. Himpsel: Phys. Rev. B **43**, 1305 (1991)
10.33 M. Donath: Appl. Phys. A **49**, 351 (1989)
10.34 B. Vogt, B. Kessler, N. Müller, G. Schönhense, B. Schmiedeskamp, U. Heinzmann: Phys. Rev. Lett. **67**, 1318 (1991)
See also B. Schmiedeskamp, N. Irmer, R. David, U. Heinzmann: Appl. Phys. A **53**, 418 (1991)
10.35 R. Kappert: Spectroscopic studies of local magnetic properties in metals. Thesis, University of Nijmegen (1992); see also J. Magn. Magn. Mater. **100**, 363 (1991)
10.36 C.M. Schneider, J. Kirschner: Crit. Rev. Solid State and Mater. Sci. **20**, 179 (1995)

10.37 L. Baumgarten, C.M. Schneider, H. Petersen, F. Schäfers, J. Kirschner: Phys. Rev. Lett. **65**, 492 (1990)
10.38 F.U. Hillebrecht, Ch. Roth, H.B. Rose, W.G. Park, E. Kisker, N.A. Cherepkov: Phys. Rev. B **53**, 12182 (1996)
10.39 A. Fanelsa, R. Schellenberg, F.U. Hillebrecht, E. Kisker, J.G. Menchero, A.P. Kaduwela, C.S. Fadley, M.A. Van Hove: Phys.Rev. B **54**, 17962 (1996)
10.40 C. Bethke, N. Weber, F.U. Hillebrecht: ESRF Newsletter, October (1999), p.20
10.41 K. Starke, E. Navas, E. Arenholz, L. Baumgarten, G. Kaindl: Appl.Phys. A **60**, 179 (1995)
10.42 K.N. Altmann, N. Gilman, J. Hayoz, R.F. Willis, F.J. Himpsel: Phys. Rev. Lett. **87**, 137201 (2001)

11. Photoelectron Diffraction

So far in this volume, we have neglected a seemingly obvious effect in PES, namely the scattering of the photoexcited electron on its way through the crystal to the surface, by the crystal potential. This is a straightforward scattering problem leading to intensity modulations as a function of electron wavelength and/or crystal orientation. Such intensity modulations have indeed been observed and the method of measuring PES in order to bring out most clearly the scattering features of the final-state electron intensity is called *PhotoElectron Diffraction* (PED).

In the conventional description of PED (and EXAFS, a short hand for Extended X-ray Absorption Fine Structure) [11.1–11.6] one treats these phenomena as a scattering problem, an approach which brings out nicely the interference effects (Fig. 11.1). Thus one views the electron beam in the detector, as being composed of a primary beam, originating from the site of the photoemission process, plus a number of secondary beams created in the elastic scattering of the primary beam from the other ions in the crystal.[1] In order to make the problem tractable one considers only a finite cluster around the primary ion (site of the photoemission process). Furthermore, one usually considers only single scattering events [11.2], a procedure which does not, however, seem to give full agreement with experiment [11.3]. The elastically scattered beams can interfere with each other and with the primary beam thus giving interference patterns. These patterns are sensed in PED and EXAFS. In PED the sensing can be achieved by working in the mode with a fixed photon energy and thus a fixed final state kinetic energy. The interference pattern is then measured by varying the azimuth and/or polar detection angle of the photoelectrons [11.2,11.4–11.7]. An alternative mode of detection leaves the geometry of the experiment constant, while varying the photon energy and thereby the electron kinetic energy. The accompanying change of electron wavelength also produces interference oscillations which are then measured [11.3].

We now consider briefly the question of why photoelectron diffraction effects have not been mentioned explicitly up to now. Actually this is not

[1] EXAFS is not always measured by detecting the electrons. The direct absorption of the photon beam can also be monitored. But since electron emission is determined by the photon absorption cross section, the analysis remains the same in both cases.

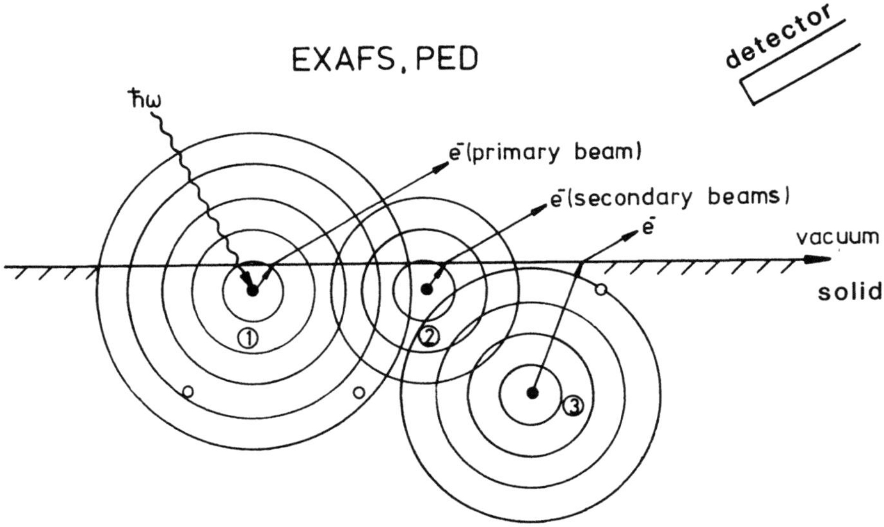

Fig. 11.1. Real-space diagram of a PED (or EXAFS) experiment. A photon excites an electron from an atom (1), which can be represented as a (primary) spherical wave. The part of this wave that escapes directly into the vacuum is the primary detected beam. Inside the solid the primary wave can be scattered by other ions (2) and (3) creating secondary beams that can also penetrate into the vacuum and which have to be added (taking phases properly into account) to the primary beam in order to get the total electron intensity distribution due to photon absorption

completely true. The electronic band structure of a material consists of the electronic states (as a function of crystal momentum) that are created by the scattering of the electrons in the material. Thus the band-to-band transition in a PES experiment actually takes into account the scattering of the electrons in the crystal. In this sense, PED was dealt with implicitly in the chapter describing the investigation of band structures in solids. The technique referred to as PED is thus only a particular mode of performing PES, namely one in which the structural parameters are extracted directly from the experimental data, in contrast to the ordinary PES experiment in which energy states are determined as a function of crystal momentum.[2]

---

[2] There seems to be a difference between a band-structure state which is formed by the scattering of an electron by a very large number of atoms in the crystal and a photoemitted electron, which seemingly gets scattered only by a few atoms on its way to the surface. This difference is, however, not real. The photoabsorption process excites an electron out of a band state into another (unoccupied) band state. This is the concept of the direct transitions.

As a matter of convenience the evaluation of EXAFS and PED data have been performed in single scattering calculations employing small clusters of atoms. The equivalence of the band-structure approach and the scattering approach for the evaluation of EXAFS data has been given by Schaich [11.8]. He shows that already for relatively small clusters of atoms of a particular element the valence-

In short, PED measures the direction dependence of the transition matrix element $M^1_{fi}(\boldsymbol{k}_f, \boldsymbol{k}_i)$, as given in (6.1). The above statements can be illustrated with reference to Fig. 7.45. It was demonstrated in this figure that if one measures the EDCs from a single crystal [(001) direction in the present case] in normal emission as a function of the photon energy, changes in the slopes of the $E_i(\hbar\omega)$ curves occur when the photon energy reaches the center or the boundary of the Brillouin zone. The actual data for Al(100), corresponding to the case shown schematically in Fig. 7.45, are given in Fig. 7.46, and the positions of the Γ and the X points are also indicated.

The separation between Γ and X in Fig. 7.46 gives $k_x = 2\pi/a_x$, thus allowing one to measure k_x or a_x. Usually, of course, in the systems on which $E(\boldsymbol{k})$ experiments are performed, the lattice parameters are known with good accuracy and one is mainly interested in determining the energy at a specific point in $\boldsymbol{k}$-space.

In the example given, $\boldsymbol{k}$ was not determined absolutely from the PES curves and thus, in order to analyze the data further, an additional assumption has to be made, such as that of a free-electron final state in Al, which is not a bad approximation. It was shown, however, how PES experiments can be used to determine $\boldsymbol{k}$ absolutely (Chap. 7) and therefore also lattice constants.

Although it has been demonstrated that PED data can be analyzed by taking the band-structure wave function as the final state [11.8], this is rarely done. Instead a scattering approach is employed.

We stress that lattice constants can, in principle, be inferred from PES experiments because the data implicitly contain this information. If then one is interested in such a determination it is clearly sensible to perform the measurement in such a way that $|\boldsymbol{k}| = k$ (or the lattice parameter a) can easily be extracted from the data. This is the main point of PED. No new physics is involved and the measuring mode is simply chosen such that $\boldsymbol{k}$ rather than E is easily accessible.

To put these statements into perspective Fig. 11.2 shows four different types of measurement, which all yield approximately the same final state: PES, X-ray absorption (EXAFS), PED and BIS, where it is assumed that the final state energy is the same in each case and that the initial state in PES, EXAFS and PED is a sharp core level. The structure in the measured spectrum in each technique will always reflect that of the final state. As is obvious from Fig. 11.2, the techniques PES, EXAFS and PED are essentially similar. The difference is mainly in the measuring procedure. BIS is slightly different from the three other techniques in that it lacks the core hole.

PES is performed such that it measures the energy difference between the initial and final states. PED, EXAFS and BIS, in principle, do the same;

band states are similar to those of a crystal made up of the same atomic species. Therefore, if the electron damping is not too strong (escape depth not too small) the band structure and the scattering formalism will give quite similar results for the analysis of EXAFS or PED data.

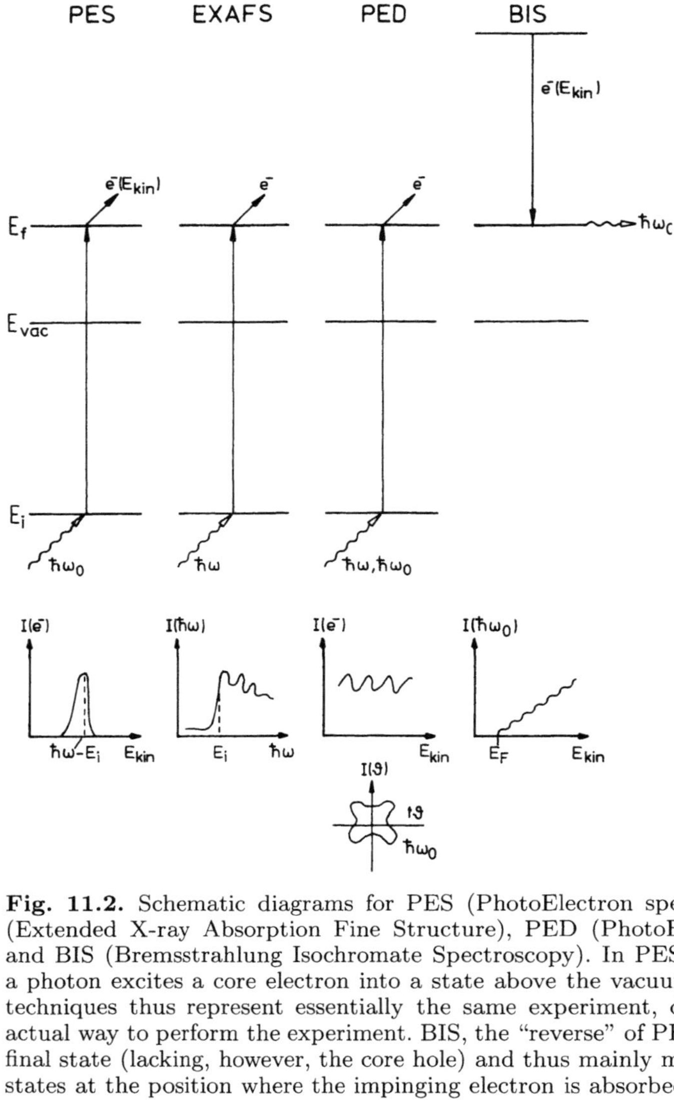

Fig. 11.2. Schematic diagrams for PES (PhotoElectron spectroscopy), EXAFS (Extended X-ray Absorption Fine Structure), PED (PhotoElectron Diffraction) and BIS (Bremsstrahlung Isochromate Spectroscopy). In PES, EXAFS and PED a photon excites a core electron into a state above the vacuum level. These three techniques thus represent essentially the same experiment, differing only in the actual way to perform the experiment. BIS, the "reverse" of PES, leads to a similar final state (lacking, however, the core hole) and thus mainly measures the electron states at the position where the impinging electron is absorbed

however, PED and EXAFS are performed in such a way that these experiments "filter out" the final-state properties. This is achieved by sampling the phase shifts of the photoexcited electron beam that are created when the final-state electron scatters from the ions surrounding the ion from which it was photoemitted (Fig. 11.1). BIS, by its very nature, measures the final state of the sample produced by the incident electron.

All the techniques mentioned here involve a process that can be linked to a LEED process with an internal source, insofar as the final state wave function is a LEED-type wave function. This shows the great similarity of the various measurements. Equation (6.64) with the wave functions of (6.65,6.66) can thus also be used to describe PED and has indeed been applied for that purpose. The LEED wave function can be evaluated either by summing up the secondary beams to infinity, which leads eventually to a band-structure description, or by considering the scattering of the photoemitted wave by a discrete number of neighbors, which leads to various types of scattering approaches. Since the photoelectrons are heavily damped (small mean free path λ for the relevant kinetic energies), the cluster sizes in a scattering calculation can be relatively small (≈ 100 atoms).

11.1 Examples

The general features described above will now be briefly exemplified with data for Cu. Fig. 11.3 shows the absorption coefficient of Cu at the K-edge, measured as a function of photon energy [11.9]. One sees the typical EXAFS structure "behind" the K-edge, where it is apparent that the structure contains more than one frequency, implying that scattering from more than one shell of neighbors has to be taken into account. While the EXAFS data are usually measured by the absorption coefficient μ as a function of the energy of the photons $\hbar\omega$, the data are more conveniently represented as a function of the electron wave vector $|\boldsymbol{k}| = k$

$$k = \sqrt{(2m/\hbar^2)(\hbar\omega - E_0)} \tag{11.1}$$

where $\hbar\omega$ is the photon energy and E_0 the energy at which the kinetic energy of the electrons is zero (the vacuum level).

Equation (11.1) is correct for a free electron with kinetic energy $(\hbar\omega - E_0)$. However, in EXAFS the transitions by which electrons are excited above the vacuum level (such that they can be detected) are also within the band structure of the solid. Therefore the description of those electrons as free electrons is only an approximation near the absorption edge ($E \leq 50\,\mathrm{eV}$) but becomes progressively better the further one moves away from the edge. These considerations are important for EXAFS experiments on "solid" samples, but for "isolated" impurities such as adsorbates they are of less concern.

The function measured in EXAFS is [11.10]:

$$\chi(\hbar\omega) = (\mu - \mu_0)/\mu(\hbar\omega_0) \tag{11.2}$$

with

μ : measured absorption coefficient at $\hbar\omega$,

μ_0 : smooth average absorption coefficient around $\hbar\omega$ obtained by a polynomial fit through the oscillatory structure above the edge,

$\mu(\hbar\omega_0)$: jump in absorption coefficient at the edge given by the count rate difference above and below the edge.

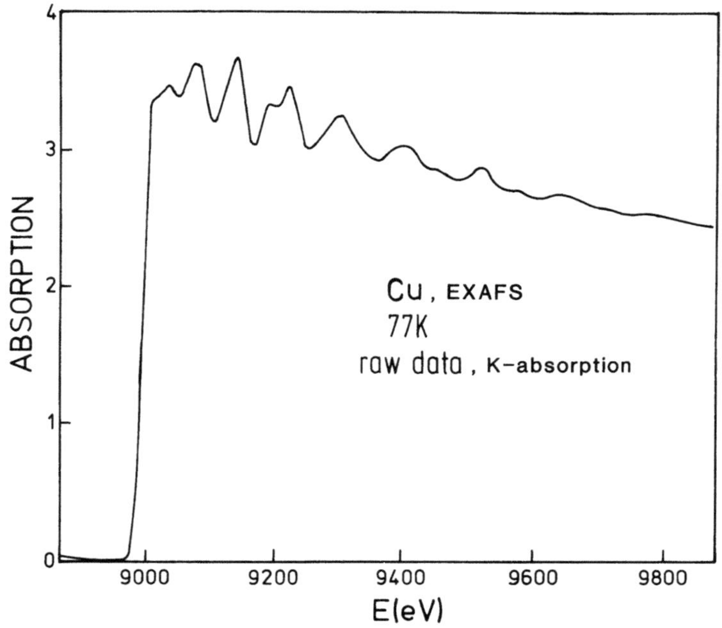

Fig. 11.3. K_α X-ray absorption spectrum for Cu at 77 K [11.9]. The raw data already show the intensity oscillations behind the edge, caused by the interference between the primary and secondary beams

(Note that some researchers denote $\mu(\hbar\omega_0)$ in (11.2) by μ_0!) The functions $\chi(\hbar\omega)$ can be transformed into $\chi(k)$ using (11.1) which yields for a K-edge (for higher angular-momentum edges the formula is more complex)

$$\chi(k) = (-1)^\ell \sum_i A_i(k) \sin[2kr_i + \beta_i^\ell(k)] \,. \qquad (11.3)$$

Here ℓ is the dominant partial wave component of the final state wave function determined by the dipole selection rule. (For an s initial state – K-edge – one has $\ell = 1$). The summation extends over all neighboring shells i, separated from the absorbing atom by a distance r_i; $\beta_i^\ell(k)$ is the total phase shift, which the photoelectrons have experienced from the absorbing and backscattering atoms.

Figure 11.3 presents the absorption of Cu at 77 K around the K-edge; the structures beyond the edge are clearly visible. From these $\chi(\hbar\omega)$ data, $\chi(k)$ can be determined and is plotted as the full line in Fig. 11.4. These data can now be Fourier analyzed, to obtain the various neighbor distances, r_i, contributing to the EXAFS signal and the result of this procedure is given in Fig. 11.5 (note that (11.3) is a function periodic in $2kr_i$, where r_i denotes the various neighbor distances). Using only the first neighbor shell, one calculates the EXAFS function given as the dashed line in Fig. 11.4, which is in good

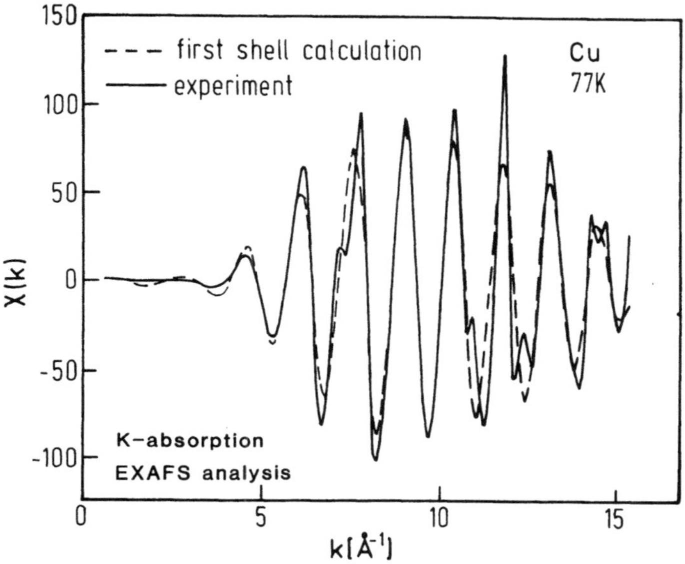

Fig. 11.4. EXAFS function $\chi(k)$ for Cu, derived from the $\chi(\hbar\omega)$ data of Fig. 11.3. Also shown is a theoretical $\chi(k)$ curve obtained by summing over first neighbors only [11.9]

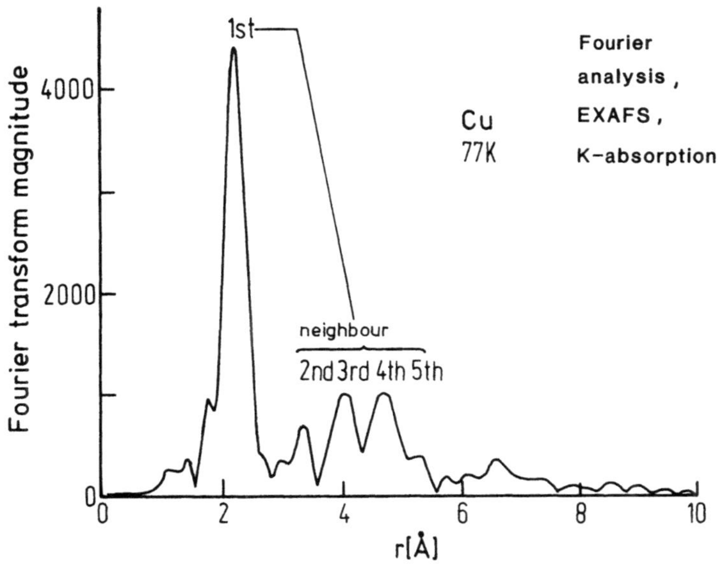

Fig. 11.5. Fourier analysis of the experimental $\chi(k)$ curve for Cu at 77 K as given in Fig. 11.4. The dominance of the first neighbor shell over the next four shells is visible and explains the success of the first shell calculation in Fig. 11.4 [11.9]

agreement with the measured function. An EXAFS spectrum of Cu calculated by the band-structure approach also agrees quite well with the experimental data, indicating the equivalence between this approach and the scattering analyzes.

Figures 11.6 and 11.7 enable one to compare EXAFS with BIS [11.11, 11.12]. Figure 11.6 shows a measured BIS spectrum, in which the BIS intensity is plotted versus $(E - E_1)^{1/2}$, where E_1 corresponds to the bottom of the conduction band. This reveals the free-electron-type behavior of this curve [$N(E) \sim E^{1/2}$ in the free-electron approximation]. One again observes a modulation in the intensity, which, because the electron has the same final state as in the EXAFS experiment (apart from the core hole in the latter experiment), should equal that seen in the EXAFS data. Figure 11.7 shows the detailed comparison [11.11]. The small discrepancy between the two curves is most probably due to the core hole in the final state of the EXAFS experiment, which leads to an additional phase shift.

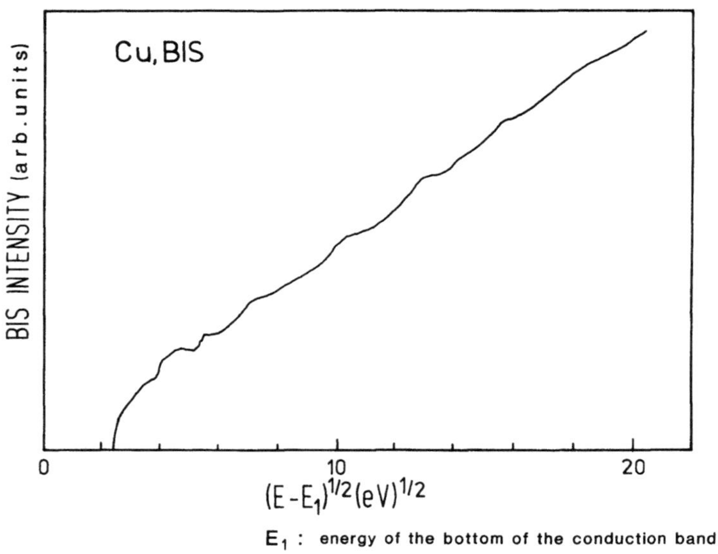

Fig. 11.6. BIS spectrum of Cu [11.11,11.12], plotted against $(E-E_1)^{1/2}$, where E_1 is the bottom of the conduction band of Cu. This way of plotting the data brings out the free electron behavior, for which $N(E) \sim \sqrt{E}$

The data in Figs. 11.3–11.7 serve to show how EXAFS and BIS are used to obtain structural information. To summarize these considerations we give in Fig. 11.8 the analysis of a LEED experiment on Cu(001) in terms of the band structure [11.13] and using a dynamical calculation [11.14]. LEED is generally a probe for determining crystal structures (usually at the surface) and the

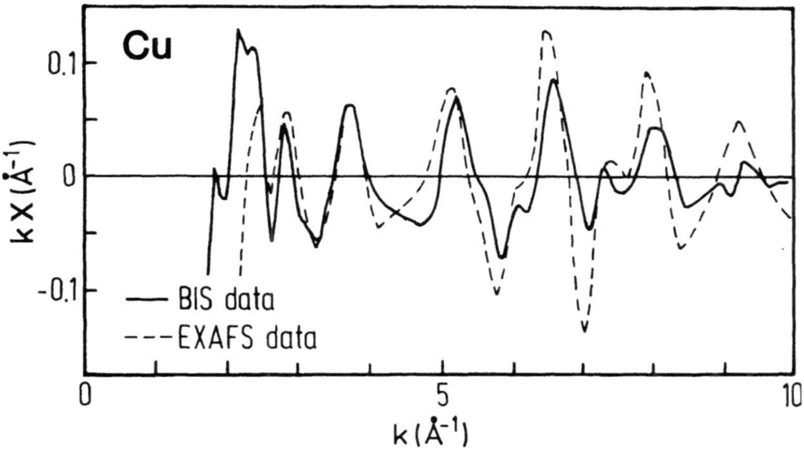

Fig. 11.7. Comparison of the $k\chi(k)$ function derived from the BIS data in Fig. 11.6 and EXAFS data showing good agreement of the results from the two different types of experiment [11.11, 11.12]

data are evaluated in this framework. An example is shown in the lower part of Fig. 11.8 where the measured I-V curve of the 00-beam (the specular beam) is compared with a calculated curve. Such a comparison can be used in a fitting procedure to derive the structural parameters of the sample. In the top part of Fig. 11.8 the same experimental data [I-V curve of the 00-beam, Cu(001)] are compared with the free-electron-like band structure of Cu(001) and the critical points that give rise to the maxima in the I-V curve are indicated. Thus the scattering approach and the band-structure approach are equivalent for the description of the movement of electrons in a solid. The type of analysis depends only on the final result one wants to obtain.

PED measurements on Cu single crystals in the fixed photon energy mode with varying electron detection angle have been performed extensively by Fadley and his group [11.2]. Figure 11.9 gives an example of data from a Cu(001) face, where azimuthal scans are plotted for the Cu $2p_{3/2}$ PE intensity excited with $\hbar\omega = 1487\,\text{eV}$ radiation; the parameter for the various curves is the polar angle ϑ, here measured relative to the surface normal. As can be expected for the fcc structure, the plots are symmetric about the [100] and [010] directions, corresponding to $\varphi = 0°$ and $\varphi = 90°$ and the [110] direction corresponding to $\varphi = 45°$. A comparison of some of the data with an analysis based on single scattering is shown in Fig. 11.10.

The experimental data presented thus far were intended merely to demonstrate the method of PED and to put it into perspective with respect to other similar methods. Obviously the crystal structure of Cu is well known and need not be explored further by PED.

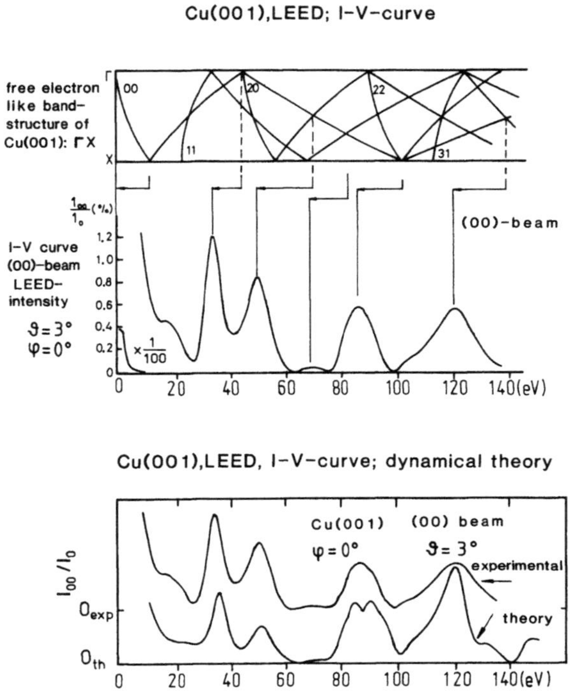

Fig. 11.8. Analysis of a LEED *I-V* curve from Cu(001) in terms of a band structure and a dynamical structure calculation. In the top two panels the correspondence of the maxima in the $I - V$ curve with critical points in the free-electron band structure for Cu(001) is indicated [11.13]. In the lower panel a comparison between the $I - V$ curve and a dynamical structure calculation [11.14] is given. These data emphasize that electron scattering in a solid can be analyzed in terms of a band structure, or by a scattering calculation

The situation is different for systems consisting of (ordered) overlayers on ordered substrates. Here the method chosen for the determination of the structure has traditionally been LEED. But there are certain problems associated with this method and alternative tools in this field are thus welcome.

Advantages of PED over LEED include:

- The core-level energies used in PED are element specific, which allows one to explore the bonding of a specific atom in more detail.
- In PED one is dealing with the interference of a primary beam (produced "in" the sample) with a number of diffracted beams, whereas in LEED one has only the diffracted beams. This makes the PED theory simpler, at least in the energy intervals studied so far.
- The cross section for photon stimulated desorption is smaller than that for electron-stimulated desorption.

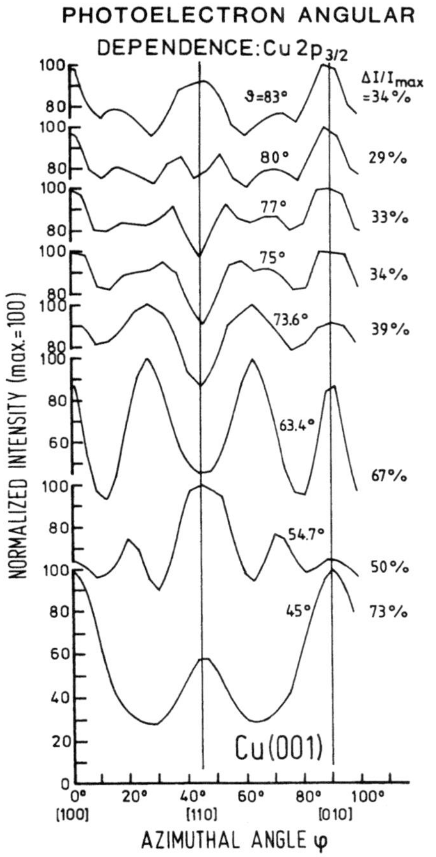

Fig. 11.9. PED data for Cu(001) employing the Cu 2p(3/2 photoemission line. Shown are azimuthal plots for various polar angles [11.2]. As expected from the crystal symmetry, the data are symmetric around $\varphi = 0°$, $45°$ and $90°$

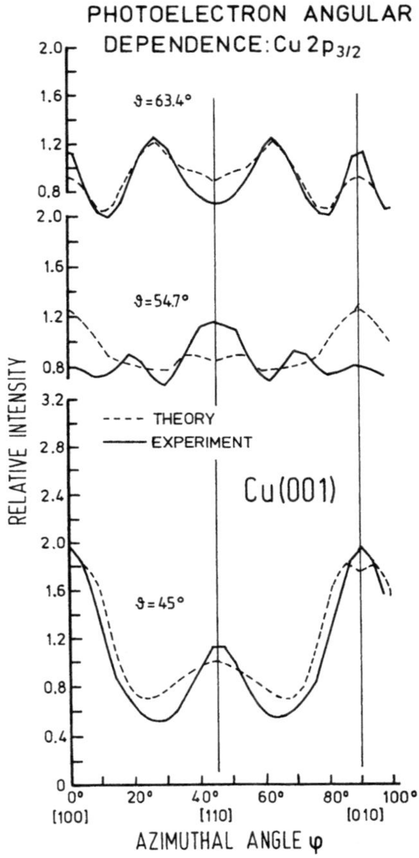

Fig. 11.10. Analysis of PED data (*full curves*) from Fig. 11.9 using a single-scattering approximation (*dashed curves*) for the $2p_{3/2}$ line of Cu(001) [11.2]

11.2 Substrate Photoelectron Diffraction

There is a very simple and intuitively appealing interpretation of PED patterns [11.2, 11.15, 11.16], which comes from the fact that high energy photoelectrons emitted from a particular atom are scattered by neighboring atoms predominantly into the forward directions, as is evident from the scattering amplitudes in Fig. 11.11. This means that one can expect an enhancement of the photoelectron intensity in directions which include both the emitting atom and one (or more) neighbors.

Atomic electron scattering amplitude for Ni

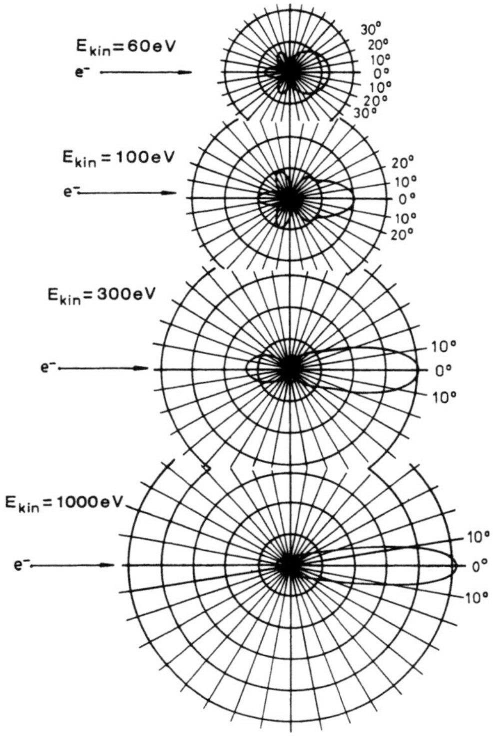

Fig. 11.11. Atomic scattering amplitudes of Ni calculated for electrons with different kinetic energies. With increasing kinetic energy the scattering cross section becomes peaked in the forward direction [11.16]

As an example, let us consider the (111) face of an fcc crystal, as shown in Fig. 11.12. It is assumed that the emitting atom is situated in the second subsurface layer (a total of 3 layers are considered). Trajectories of electrons emitted from this atom into the vacuum, which also contain a neighbor of this atom, are

1st neighbors: $\vartheta = 35.3°$; $\varphi = 0°,\ 120°,\ 240°$
2nd neighbors: $\vartheta = 54.7°$; $\varphi = 60°,\ 180°,\ -60°$
3rd neighbors: $\vartheta = 20°$; $\varphi = 60°,\ 180°,\ -60°$
4th neighbors: $\vartheta = 35.3°$; $\varphi = 0°,\ 120°,\ 240°$.

The fourth nearest neighbors (not shown in Fig. 11.12) are thus situated at the same angles as the first nearest neighbors. A similar condition holds for higher order neighbors and therefore only the angular conditions for the first three neighbors need to be considered.

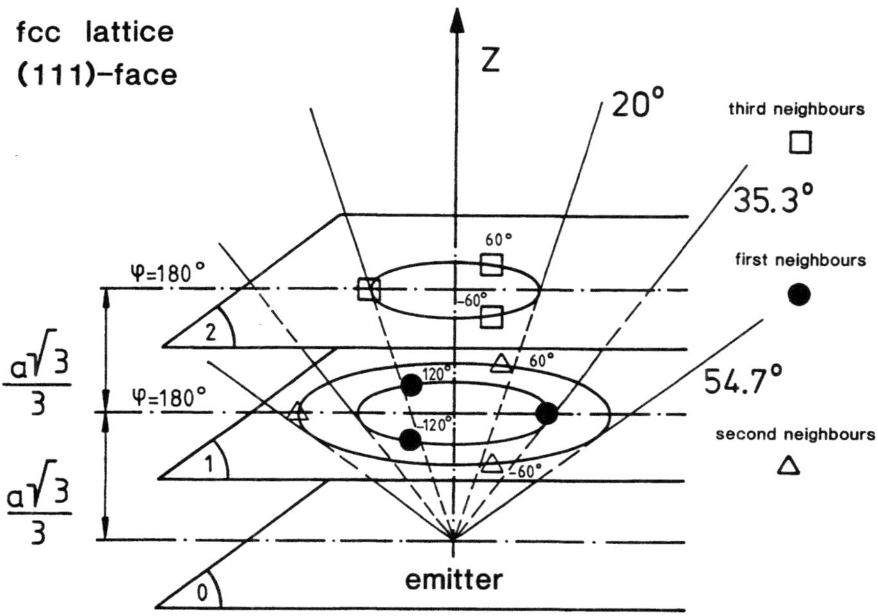

Fig. 11.12. Three surface layers of the (111) face of an fcc crystal (Ni). The photoemitting atom is in the third layer from the surface. The first and second nearest neighbors are in the layer above the emitting atom, and the third nearest neighbors in the surface layer [11.15]

Remaining within the simple forward scattering argument given above, we expect a peak in the intensity of the photoelectrons at the angles just given for the three sets of neighbors.

Plots of the photoelectron intensity as a function of the polar angle for $\varphi = 60°$ (2nd and 3rd neighbors) and for $\varphi = 120°$ (1st neighbors) are given in Fig. 11.13 where one can see the expected intensity enhancements for the Ni 2p photoelectron intensity (excitation with Al K_α radiation) for $\vartheta = 20°$ (3rd neighbors), 55° (2nd neighbors) and 35° (1st neighbors). This provides confirmation of our simple picture of the PED process, where the small effect of refraction has been neglected.

The diffraction patterns can be compared with calculations. In order to investigate the effects of the thickness of the surface for the development of the diffraction pattern, theoretical intensity distributions have been calculated for layers containing one to six atomic planes (Fig. 11.14).

The photoelectron intensity emitted from an atom in the topmost layer shows practically no structure as a function of the emission direction, which provides proof of the predominantly *forward scattering nature* of the photoemitted electrons (little backscattering) at this high energy. If the diffraction pattern for two layers is calculated, a strong modulation of the photo-

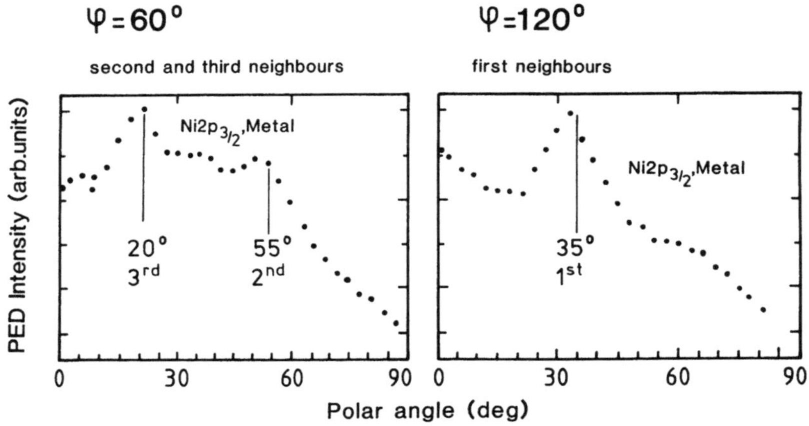

Fig. 11.13. Photoelectron intensity in the Ni 2p lines from a (111) face of nickel as a function of polar angle ϑ for two azimuthal angles $\varphi = 60°$ and $\varphi = 120°$ [11.15]

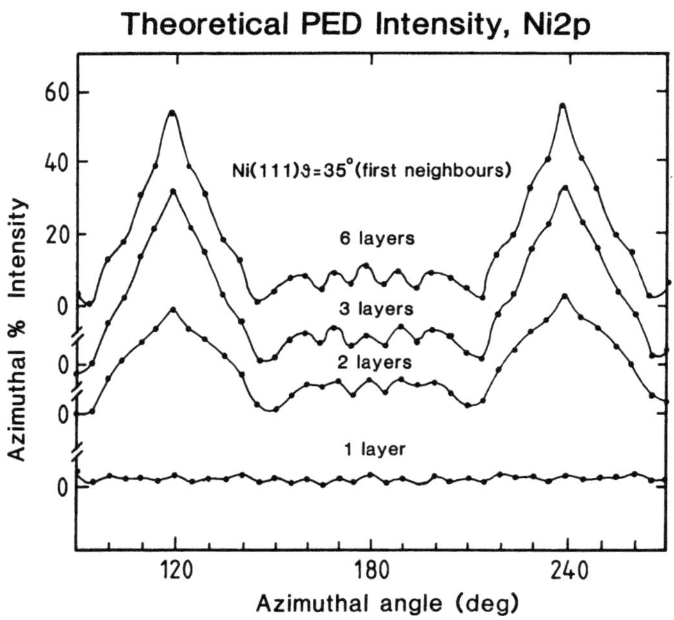

Fig. 11.14. Calculated photoelectron intensity from a Ni(111) surface (2p lines) at $\vartheta = 35°$ as a function of azimuthal angle for increasing number of layers. For one layer the photoemitting atom is in the surface, for 2 layers the photoemitting atom is in the layer below the surface, etc. [11.15]

electron intensity is observed. The addition of further layers adds only little to the pattern, demonstrating that the scattering from higher order neighbors (e.g., fourth neighbors in the case $\vartheta = 35°$) is not very important.

A comparison of the experimental and theoretical diffraction patterns for Ni 2p photoelectrons excited with Al K_α radiation at polar angles $\vartheta = 20°$, $35°$ and $55°$ as a function of the azimuthal angle is given in Fig. 11.15. There is good qualitative agreement between theory and experiment.

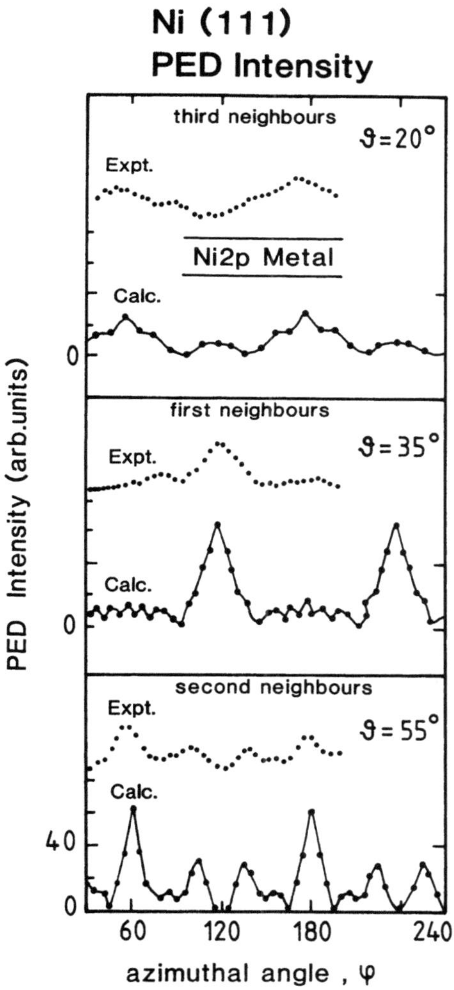

Fig. 11.15. Comparison of measured and calculated photoelectron intensity in the Ni 2p lines from a Ni(111) surface. Data are given for three polar angles, $\vartheta = 20°$, $35°$ and $55°$ as a function of the azimuthal angle φ [11.15]

Things are, however, slightly more complicated, as is indicated by the model calculations [11.16] in Fig. 11.16. These calculations have been performed for a simple configuration, namely the (001) surface of an fcc metal

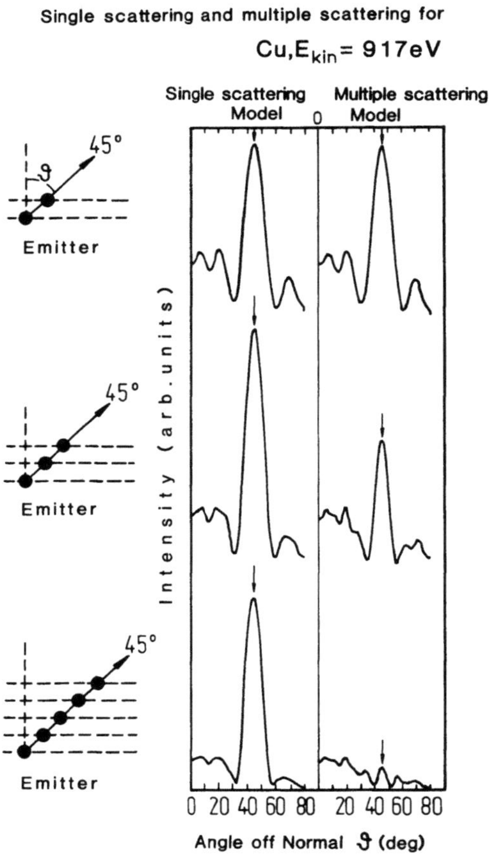

Fig. 11.16. Polar angle intensity distribution for a Cu(001) single crystal from a point source (emitter) in the direction of a chain of Cu atoms along the [101] direction [(100) azimuth]; kinetic energy of the electrons $E_{\text{kin}} = 917\,\text{eV}$. The left-hand diagrams illustrate the chain geometries for each of the calculations. The right-hand panels give the intensity distributions obtained for a single-scattering and for a multiple-scattering calculation. The forward scattering peak is indicated by an arrow [11.16]

with electron detection at 45° in the (100) azimuth corresponding to a (101) direction. The scattering configuration consisted of an emitter and 1, 2 and 4 scatterers along the (101) direction. For a single scattering model the scattering by one atom produces a large forward scattering peak, which is fully developed with 2 scatterers and shows only a very small attenuation with 4 scatterers. The situation is quite different for a multiple scattering calculation. Here, even 2 scatterers lead to a considerable attenuation of the forward scattering intensity and 4 scatterers almost completely wipe out the forward scattering intensity. This effect was termed "defocussing" by Xu et al. [11.16] in analogy with the optical properties of multiple lenses.

These calculations explain the existing data [11.17] displayed in Fig. 11.17 quite well. Here a (001) single crystal of Ni was covered with one monolayer of Cu, and these Cu ions were photoionized in the subsequent PED experiment. As scattering ions, successive layers of Ni were deposited on the Cu monolayer such that the evolution of the PED signal as a function of the number of scatterers could be observed.

The top panel in Fig. 11.17 shows the PED patterns observed by scanning the intensity of the Cu $2p_{3/2}$ line in the (010) plane as a function of the polar angle ϑ. With no Ni on the Cu monolayer the polar intensity distribution shows only very small modulation. With two monolayers of Ni a strong maximum is observed at $\vartheta = 45°$, which is the direction of the rows of closely packed ions. This illustrates the strong forward scattering characteristics of the electron scattering. For four monolayers the maximum at $\vartheta = 45°$ is attenuated considerably giving evidence of the defocussing effects. Thus

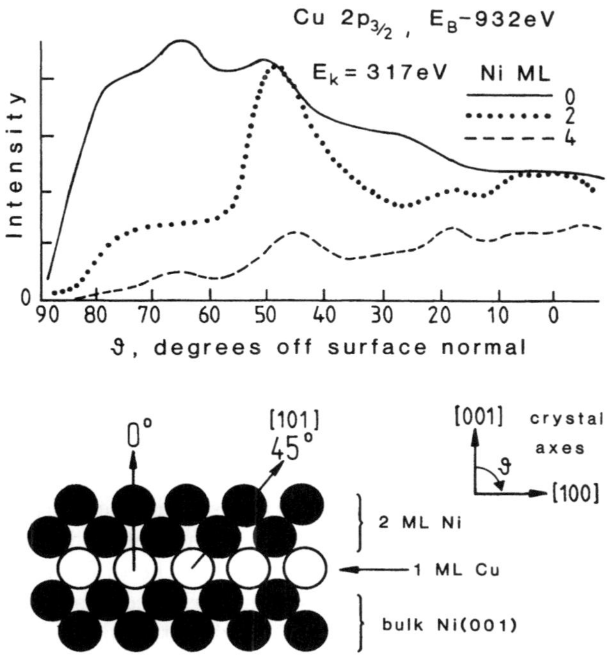

Fig. 11.17. Development of a forward-scattering feature in a PED experiment. A monolayer of Cu is deposited on a Ni(001) single crystal. This provides the atoms which produce the photoelectrons. On the Cu monolayer layers of Ni are then deposited, providing the scattering atoms. It can be seen that two Ni monolayers produce a large forward scattering peak, which, however, is attenuated by four layers [11.17]

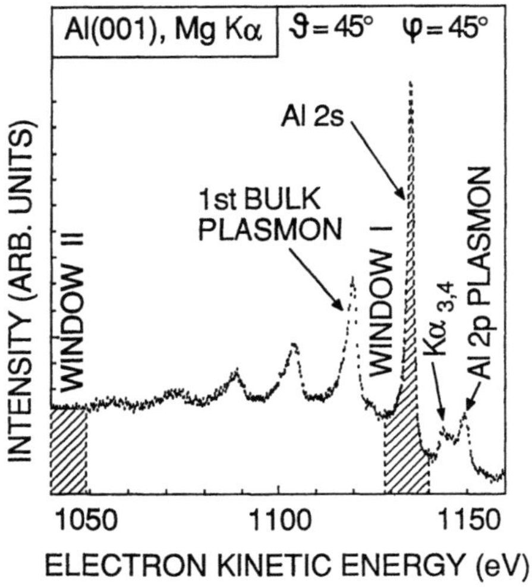

Fig. 11.18. Al 2s energy spectrum from Al(001) surface showing the main line and five plasmon loss lines, measured along the low symmetry direction $\Delta\varphi = 45°$ off [011] using Mg K_α radiation. Shaded areas indicate the energy windows used for obtaining the data in Fig. 11.19, with no background subtracted [11.18, 11.19]

the experimental results in Fig. 11.17 are in qualitative agreement with the results of the calculations in Fig. 11.16.

The possibilities of the PED technique will be further elaborated with data from aluminium [11.18, 11.19]. Figure 11.18 gives a 2s PE spectrum from a A1(001) single crystal. This spectrum is similar to one shown in Fig. 4.4 for Mg, in displaying besides the 2s "elastic" line also a number of bulk plasmons, created by intrinsic and extrinsic losses. In addition the background is visible. The PED curves measured for the electrons detected in the energy windows I and II are given in Fig. 11.19 together with two graphs of crystallographic data for an fcc crystal, which can help to understand the symmetry of the diffraction patterns. An fcc crystal with a [001] surface normal displays a fourfold symmetry which can be discerned in the diffraction data. For an electron detection angle of $\vartheta = 45°$ there are four [110] directions which contain rows of densely packed atoms, and therefore produce a high diffraction intensity because of the large forward scattering cross section (Fig. 11.16). These directions are clearly visible in the data, taken with window I. Interestingly, the electrons recorded with window II also show a fourfold symmetric diffraction pattern, however, with an intensity distribution that is notably different from that obtained with the elastic line. The development of the diffraction pattern in going from the elastic line to the various plasmons accompanying it can be seen in Fig. 11.20. The intensity along the [110] direction decreases with increasing order of the excited plasmon. In a qualitative way this can be understood by realizing that the average escape depth of the detected electrons increases with the number of plasmons that are excited. Because of the short defocussing length as evident from Fig. 11.16, the data of Fig. 11.20 indicate that this defocussing is a main mechanism in the destruction of the PED patterns with increasing plasmon numbers.

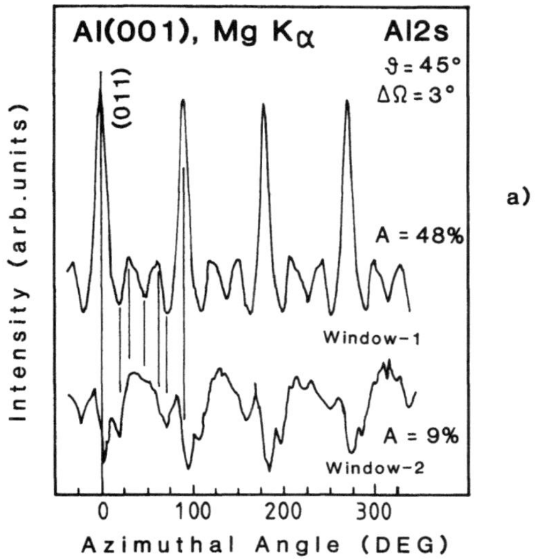

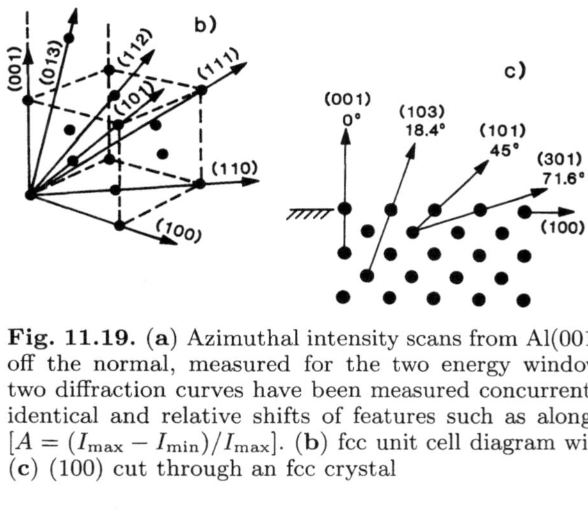

Fig. 11.19. (**a**) Azimuthal intensity scans from Al(001) at a polar angle of $\vartheta = 45°$ off the normal, measured for the two energy windows given in Fig. 11.18. The two diffraction curves have been measured concurrently, i.e. the angle settings are identical and relative shifts of features such as along [011] are real [11.18, 11.19] [$A = (I_{\max} - I_{\min})/I_{\max}$]. (**b**) fcc unit cell diagram with a few principle directions, (**c**) (100) cut through an fcc crystal

In Fig. 11.21 we compare the PED diffraction patterns of Figs. 11.19, 11.20 with those produced in a direct electron scattering experiment with electrons of 1200 eV, impinging on the (001) surface of Al. One realizes that the diffraction of the specular beam at the energy of the zero loss line and at the energy of the first surface and first bulk plasmon accompanying the zero loss line is almost identical to the PED patterns in Figs. 11.19, 11.20. This is, of course, not unexpected, it only emphasizes the similarities of the various electron spectroscopies as indicated in Fig. 11.2. Note that the technique employed to record the data in Fig. 11.21 is sometimes called MEED (Medium Energy Electron Diffraction). Figure 11.22 compares, again to demonstrate similarities in the various electron spectroscopies, diffraction patterns from an Al(001) crystal at $\vartheta = 45°$, for Mg K_α excitation of valence-band electrons, for 1200 eV MEED electrons, for 2p and 2s electrons excited with Mg K_α radiation, for 1s electrons excited with Si K_α radiation, and for LVV Auger electrons. The first five curves are rather similar, indicating that the simple forward-scattering picture describes the principal features of the data well. The Auger electron diffraction pattern is noticeably different, showing that at very low energies the simple forward scattering picture is no longer valid, as can be inferred from Fig. 11.11. For these low kinetic energy electrons a full diffraction analysis has to be performed, a fact which is well known in the field of LEED.

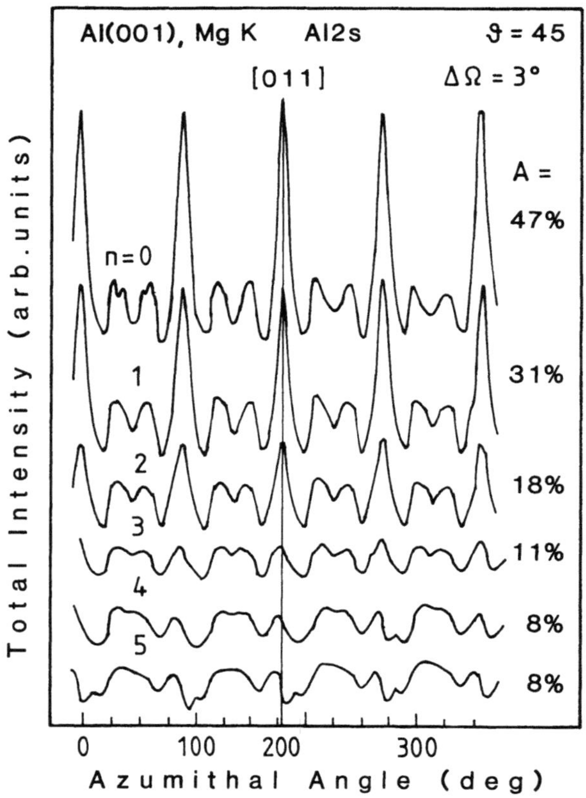

Fig. 11.20. Azimuthal intensity scans at $\vartheta = 45°$ from an Al(001) crystal measured in the Al 2s no-loss line ($n = 0$) and the consecutive plasmon loss lines ($n = 1$ through 5) using energy windows 15 eV apart and no background subtraction. Again, different curves have been measured concurrently and relative peak shifts are real [11.18, 11.19]; $A = (I_{\max} - I_{\min})/I_{\max}$

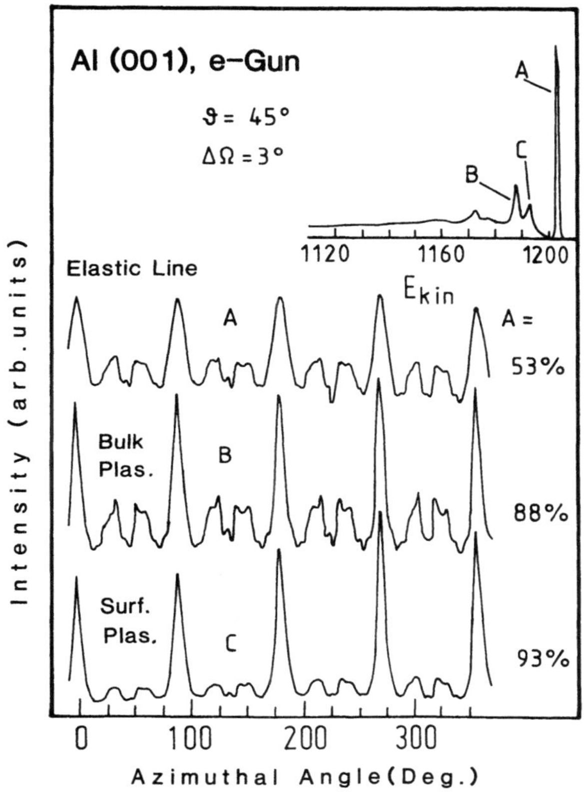

Fig. 11.21. Azimuthal MEED intensity curves measured at a polar angle of 45° using a primary energy E_p of 1200 eV from an Al(001) crystal. Shown are curves measured for the elastic line (A), which in the scattering geometry employed corresponds to the specular beam, with the first bulk plasmon (B) and with the first surface plasmon (C). The insert gives an energy scan [11.18, 11.19]

The technique of PED has obtained a new dimension by recording, instead of the line scans shown so far in this chapter, the electron intensity distribution over most of the hemisphere above the crystal plane [11.20]. A typical example, again for an Al(001) crystal, is given in Fig. 11.23. The fourfold symmetry of the diffraction pattern is easily recognized and also the four (110) "points" with the very high scattering intensity are apparent. In this kind of recording the data, also the diffractions from the crystal planes, are visible – a feature which is usually termed a Kikuchi band.

Figure 11.23 also gives the results from two single scattering simulations, one with a 6-atom cluster and another one with a 172-atom cluster. The calculation employing the small cluster reproduces only the main features of the measured diffraction pattern, namely the large intensity into the [110] directions. The simulation employing the larger cluster reproduces the measured

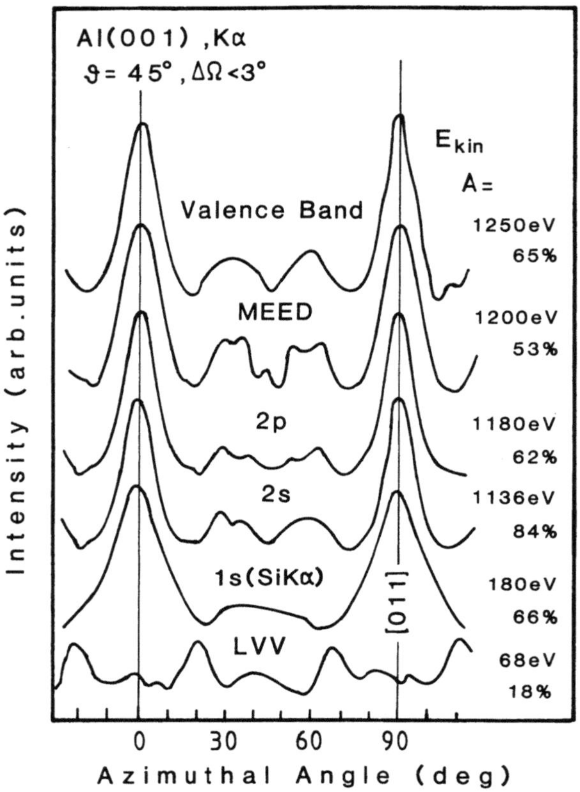

Fig. 11.22. Azimuthal electron intensity distributions at a polar angle of $\vartheta = 45°$ as a function of the azimuthal angle φ for (*from top to bottom*) valence electrons excited with Mg K_α radiation, MEED electrons with 1200 eV, 2p electrons and 2s electrons excited with Mg K_α radiation, 1s electrons excited with Si K_α radiation and LVV Auger electrons [11.18, 11.19]

spectra, to a surprising degree of accuracy in view of the fact that multiple scattering events have been neglected.

The data presented in this section have shown the importance of forward scattering of the photoelectrons. While this effect has, as demonstrated, some advantages in the determination of structures it has the disadvantage that the weaker scattering features in the diffraction patterns are hard to discover. This makes a full three-dimensional structure determination difficult by photoelectron diffraction. Omori et al. [11.31] have given an algorithm that allows to suppress the strong forward scattering intensity and thereby came to an easy three-dimensional real-space image for the crystal structure from a 2Π photoelectron hologram. A test experiment on Cu[001] demonstrated the power of this method.

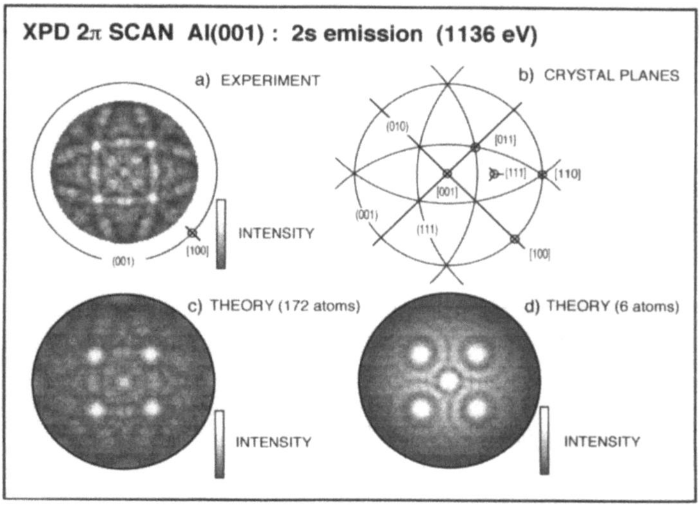

Fig. 11.23. Photoelectron intensity distribution of Mg K_α excited 2s electrons over the hemisphere of an Al(001) single crystal. (**a**) measured intensity distribution. (**b**) Low index crystal planes. (**c**) Single scattering calculation with a 172-atom cluster. (**d**) Single scattering calculation with a 6-atom cluster [11.20]

11.3 Adsorbate Photoelectron Diffraction

There is another advantage of PED over LEED: it does not need an ordered long-range structure for the determination of a particular short-range geometric structure. In this respect it resembles the EXAFS technique. This advantage is particularly important for the determination of adsorbate geometries for which we shall now give two examples [11.21].

In Fig. 11.24, we show data [11.22] for the so-called tilted α_3 state of CO on Fe(001). From simple chemical arguments it follows that in CO adsorption on a metal, it is always the C atom that binds to the substrate. The geometrical configuration, however, cannot be predicted. Fig. 11.24a shows polar intensity scans of the C 1s/O 1s intensity for two different azimuths, [001] and $[1\bar{1}0]$. Since there are no atoms on top of the oxygen atom, the O 1s intensity will not have strong modulations. On the other hand, the C 1s intensity will show an enhancement whenever the electron detection is along the CO axis, because one has the *forward scattering enhancement* by the O ions. Plotting the C 1s/O 1s ratio instead of the C 1s intensity takes out the instrumental asymmetries. Figure 11.24a shows that while in the $[1\bar{1}0]$ azimuth the C 1s/O 1s intensity is featureless, it shows a distinct maximum at $\vartheta = 55°$ in the [100] azimuth. This indicates that the CO molecule is tilted at $\vartheta = 55°$ (from the surface normal) in the [100] azimuth, as shown in Fig. 11.24c. The azimuthal orientation of the CO molecule can finally be con-

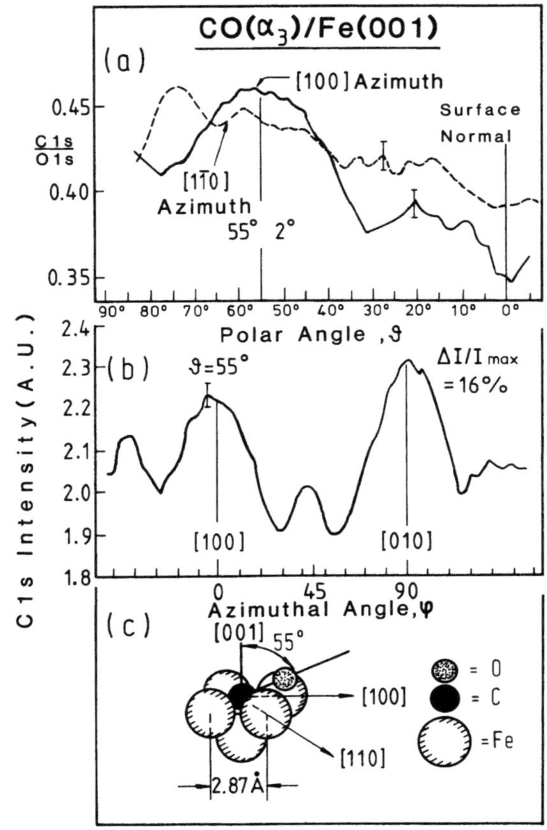

Fig. 11.24. Determination of the geometry of CO(α_3) on Fe(001). (**a**) Polar C 1s/O ls intensity ratio measured in the [100] and the [1$\bar{1}$0] azimuth; the intensity in the former case shows a maximum at $\vartheta = 55°$ which immediately allows one to deduce the structure of CO(α_3) on Fe(001) as shown in (**c**). In panel (**b**), as a check, the C 1s azimuthal intensity for $\vartheta = 55°$ is shown. The expected forward scattering peaks at 0° and 90° are clearly visible [11.22]

firmed by performing an azimuthal scan with the C 1s line at a polar angle of $\vartheta = 55°$. The result is shown in Fig. 11.24b. It displays pronounced maxima in the [100] and [010] direction, in agreement with the structure depicted in Fig. 11.24c. Note that for the orientation of the CO molecule the [100], [$\bar{1}$00], [0$\bar{1}$0] and [010] directions are equivalent, and domains will form with the tilt in any of these four directions. The small maxima displaced by 45° from the strong maxima in Fig. 11.24b indicate that, in addition to the simple forward scattering, there are other more complicated scattering processes giving rise to additional constructive interference.

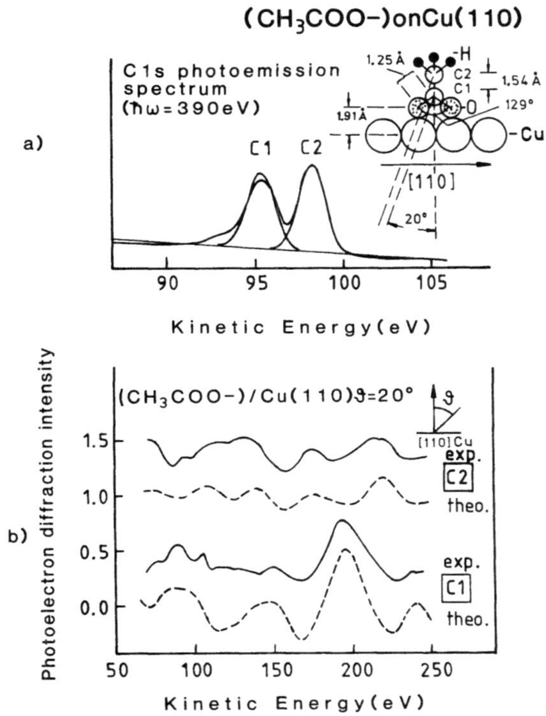

Fig. 11.25. Photoelectron diffraction data of (CH_3COO-) on Cu(110) [11.23]. (**a**) Chemical shift of the C 1s lines, the insert gives the structure obtained from the photoelectron diffraction experiment. (**b**) Photoelectron diffraction intensity of the C 1s lines (C1 and C2) for a detection angle 20° off normal for the photoelectrons. The theoretical curves have been calculated in a multiple scattering approach using the structural data given in the insert of the spectrum in panel (**a**)

In another example we shall also demonstrate the second method of taking data in the PED technique, namely varying the photon energy (and thereby the kinetic energy of the escaping electrons) at a fixed detection geometry.[3] Figure 11.25 gives the results of a PED experiment of (CH_3COO-) formed on a Cu(110) surface by decomposition of acetic acid [11.23]. The C 1s photoelectron spectrum gives two lines (C1 and C2) which can be identified with the two differently bonded carbon atoms of the (CH_3COO-) molecule as indicated in the insert of the spectrum [11.23].

The chemical shift observed between the C1 and C2 carbon species is in agreement with the systematics given in Chap. 2. The carboxyl carbon atom (C1) has a higher binding energy (smaller kinetic energy) than the

[3] This method of taking data allows one to utilize the smaller (compared to the forward scattering) back-scattering amplitude and is therefore particularly suited for investigating adsorbate geometries.

methyl (C2) carbon atom. On a qualitative basis this can be understood by realizing that in the carboxyl carbon atom more charge is removed from the carbon atom than in the methyl carbon atom. The photoelectron diffraction intensity of the C 1s line as a function of the kinetic energy of the escaping electrons for a detection angle of $\vartheta = 20°$ is given for the C1 line and C2 line in Fig. 11.25. The photoelectron detection angle of 20° was chosen to detect the intensity produced by the backscattering of the primary 1s intensity from the first Cu layer. As expected, the diffraction patterns are markedly different for the electrons excited from the C1 and C2 atoms. The theoretical intensity patterns were calculated by a multiple scattering procedure, and the curves shown in Fig. 11.25 which give the best agreement with the experimental results yield the following structural parameters (assuming that the aligned bridge site is the correct position); bond length: [Cu–O] = 1.91 Å;[C1–O] = 1.25 Å; [C1–C2] = 1.54 Å; O–C1–O bond angle: 129°.

Surface electronic structures have also been determined by PED, using surface core-level shifts [11.24]. A surface structure determination employing a holographic reconstruction of the PED pattern has been reported in [11.25].

The paper by Hofmann and Schindler [11.21] describes a new method for transforming scanned-energy-mode PED spectra from adsorbates into real-space images. The idea is based on the suggestion of Barton [11.21] that PED data can be viewed as a hologram of the surface structure. Adsorbate PED spectra taken in the backscattering (180°) geometry will exhibit large oscillations (if the experiment is performed in the energy dispersion, fixed geometry mode) which are mainly caused by one scatterer. This fact can be utilized to determine the position of the nearest-neighbor substrate atom

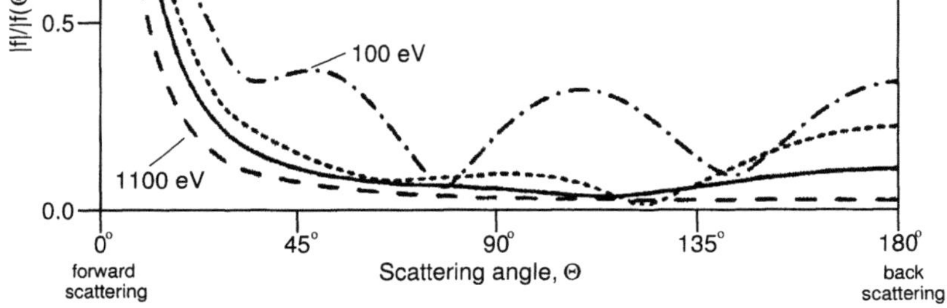

Fig. 11.26. Normalized electron scattering factor for copper for various kinetic energies. 0° is forward scattering and 180° is backscattering. Backscattering is important only for small kinetic energies [11.21]

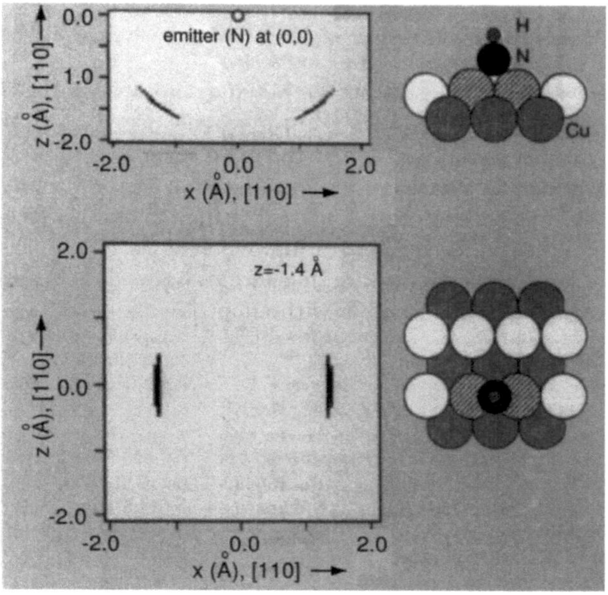

Fig. 11.27. Result of a photoelectron diffraction experiment on the NH species. *Top trace*: Emitter (N) over the top Cu layer. The position of the nearest Cu atoms (*shaded*) is indicated. *Lower trace*: Position of the nearest Cu atoms within the plane (*shaded*) [11.21]

as shown by Fritsche and Woodruff [11.21]. In an actual experiment the modulation function $\chi_{\text{exp}}(\boldsymbol{k})$ is measured at various emission angles and then Fourier transformed to yield

$$u(\boldsymbol{r}) = \int \chi_{\text{exp}}(\boldsymbol{k}) \mathrm{e}^{\mathrm{i}\boldsymbol{k}\cdot\boldsymbol{r}} \mathrm{d}\boldsymbol{k} \; . \tag{11.4}$$

The object of the method is to identify the directions of the main (substrate) backscattering nearest-neighbor atoms relative to the emitter. This method, however, leads to inaccuracies because the scattering phase shifts are neglected. Hofmann and Schindler [11.21] have improved this procedure by replacing the Fourier transform with the projection of the experimentally determined modulation function $\chi_{\text{exp}}(\boldsymbol{k})$ onto the calculated modulation function $\chi_{\text{th}}(\boldsymbol{k}, \boldsymbol{r})$ expected for all possible different substrate atom locations relative to the emitter. This is equivalent to replacing the pure harmonic phase function of the Fourier transform by $\chi_{\text{th}}(\boldsymbol{k}, \boldsymbol{r})$, leading to

$$c(\boldsymbol{r}) = \int \chi_{\text{exp}}(\boldsymbol{k}) \chi_{\text{th}}(\boldsymbol{k}, \boldsymbol{r}) \mathrm{d}\boldsymbol{k} \; . \tag{11.5}$$

The $c(\boldsymbol{r})$ functions for several experimental modulation functions taken in different emission directions are then combined to yield a coefficient $C(\boldsymbol{r})$ derived from the whole data set given by

$$C(\boldsymbol{r}) = \sum_i \exp[c_i(\boldsymbol{r})\boldsymbol{r}] \,. \tag{11.6}$$

The usefulness of backscattering experiments employing the scanned energy mode is revealed by Fig. 11.26. This diagram gives the magnitude of the scattering factor for copper as a function of the scattering angle. Zero means forward scattering, which is taken as the normalization of the various scattering factors, and 180° means complete backscattering. It can be seen that at 1100 eV there is practically no backscattering, whereas at 100 eV the backscattering is appreciable. Therefore, for kinetic energies between 100 and 500 eV, backscattering experiments are feasible without problems.

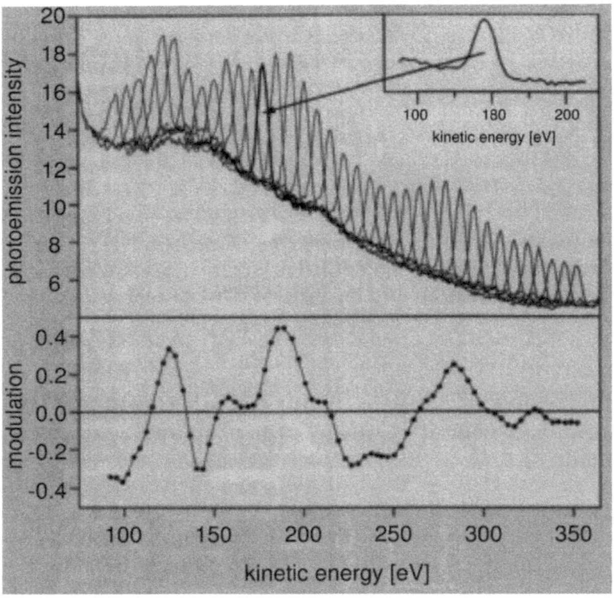

Fig. 11.28. Data set used to determine the position of the NH fragment (Fig. 11.27). *Top trace*: Set of spectra at various kinetic energies; a typical individual spectrum is shown in the inset. *Lower trace*: The modulation of the intensity of the spectra in the top trace. A series of about 30 such spectra have to be measured in order to obtain reasonable accuracy in the final result [11.21]

We shall now briefly indicate the result of an experiment using the projection method developed by Hofmann and Schindler [11.21]. This method, as described above, consists of finding those directions in which the detector, the adsorbate surface atom whose position is to be determined, and a nearest neighbor substrate atom form one line. In this direction the backscattering is particularly intense. As pointed out above, the adsorbed atom is now considered to be the origin of the coordinate system and the distance of the nearest neighbor atoms is determined relative to this zero. A particularly simple example described by Schaft and Bradshaw [11.21] is shown in Fig. 11.27, which gives the position of the NH molecular fragment as absorbed on a Cu(110) surface. This molecule absorbs on a bridge site. The positions, as determined by the experiment shown in Fig. 11.28, are also indicated in that diagram, as is an indication of the accuracy of the method. Figure 11.28 shows, from top to bottom, a typical 1s nitrogen photoemission spectrum, measured with moderate resolution; a number of photoemission spectra with different photon-excitation energies, which result in different kinetic energies; and, finally, the modulation factor as defined in (11.2), but with the absorption coefficients replaced by the photoemission intensity. In order to get a structure determination as in Fig. 11.27 one typically has to accumulate 30 sets of spectra such as those shown in Fig. 11.28 for different directions in space. Figure 11.29 compares experimental and theoretical modulation functions using the structure parameters depicted in Fig. 11.27. It provides a good estimate of the kind of agreement between experiment and simulation that can be achieved with this technique.

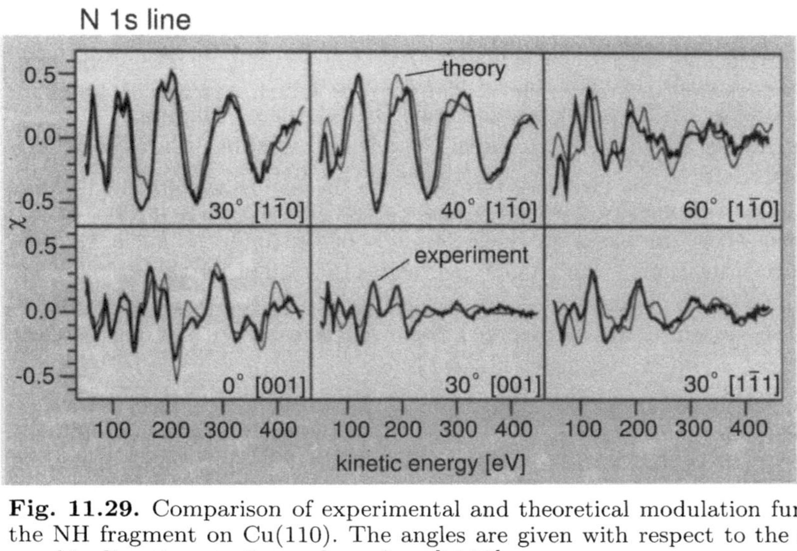

Fig. 11.29. Comparison of experimental and theoretical modulation function for the NH fragment on Cu(110). The angles are given with respect to the crystallographic directions in the surface plane [11.21]

Adsorbate photoelectron diffraction can also be detected in a 2π geometry in the way as presented in Fig. 11.23 for the core levels of Al metal. The power of such an experiment was demonstrated by applying it to photoelectron diffraction from a SiO_2/Si(111) interface with the Si 2p line. The 2p photoelectron spectrum from an oxidized Si(111) surface shows a splitting into 5 lines from the following species: Si^0 (unoxidized Si substrate), Si^{1+},

Si^{2+}, Si^{3+} and Si^{4+}. The diffraction patterns from these 5 lines differ and by comparing them to those obtained from multiple scattering calculations the structures of the SiO_2/Si(111) interface can be determined [11.32].

11.4 Fermi Surface Scans

Finally, to end this chapter and this book, we present what we personally think is a very appealing application of PES in the mode of accumulating the data over the hemisphere of the crystal [11.26]. It consists of setting the energy window for the photoelectrons to a small interval around the Fermi energy. In this way the Fermi surface of a one- or two-dimensional metal can be measured directly and, for three-dimensional metals, sections through the Fermi surface can be obtained.

The Fermi surface of the noble metals Cu, Ag and Au consists of spheres. In the periodic Brillouin zone scheme these spheres are connected by the so-called *necks* to the spheres in the next Brillouin zone. The necks point in the [111] directions. A representation of the Fermi surface of Cu metal is given in Fig. 11.30. Also shown is a PE spectrum near the Fermi energy, which samples the flat 4s band. In looking on the Fermi surface along the [001]-direction, one realizes that one observes only four small areas, where the Γ-L directions push through the Brillouin zone which contain no states at the Fermi energy. At all other places of the Brillouin zone there are states at the Fermi energy allowing direct transitions. If one samples in a PES experiment, which detects all electrons emitted into the hemisphere over the (001) surface, only those electrons which are photoemitted from the Fermi energy, one observes those positions in the Brillouin zone of the Fermi surface which allow direct transitions – and this expectation is verified by the actual data in Fig. 11.30 [11.27]. Thus the measured picture in Fig. 11.30 can be regarded as a representation of the (projected) Fermi surface of Cu. Analogous data are shown for a Cu(111) surface [11.27, 11.28].

While the data presented in Fig. 11.30 give an intuitive understanding of the determination of a Fermi surface with photoemission spectroscopy, the details of the method are more complex. For the benefit of readers intending to work in this field, we elaborate upon the actual application of this method [11.27, 11.29].

If one is working with low dimensional systems, where the relevant one- or two-dimensional directions lie in the surface, the problem is easy. In this case all the states are determined by $\boldsymbol{k}_{||}$, and since $\boldsymbol{k}_{||}$ is exactly determined by the kinetic energy of the photoelectrons (which is measured in the experiment) and the geometry of the experiment, the interpretation of the 2π electron energy distributions in terms of the Fermi surface is straightforward. In the case of Cu, a three-dimensional system, we recall that here one faces the problem of the indeterminacy of $k_{\perp}$, which makes the interpretation of the data more difficult.

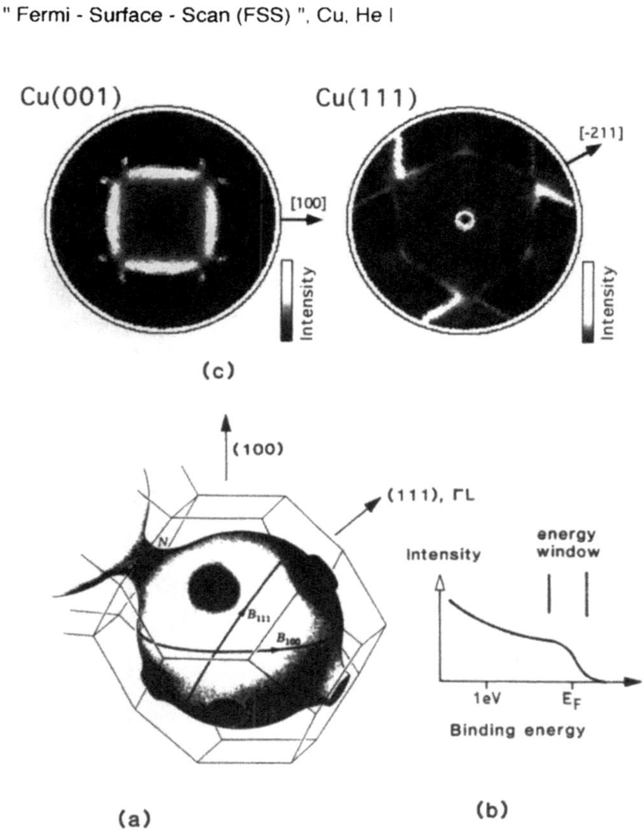

Fig. 11.30. (**a**) Fermi surface of Cu metal in the fcc Brillouin zone. B_{111}, B_{100} and N give electron orbits used for its determination. (**b**) PES spectrum of Cu metal near the Fermi energy. The energy window used in the experiment is indicated. (**c**) Intensity distribution of the photoelectrons in the energy window in (**b**) as measured over the hemisphere of the (001) and (111) Cu single crystal with He I radiation

In what follows we shall briefly explain how the Fermi surface scans for Cu are generated and how in principle one can determine from them an accurate three-dimensional Fermi surface. Figure 11.31 shows the band structure of Cu (given in more detail in Fig. 7.7). This representation of the electronic structure displays, along certain principle directions of the Brillouin zone (see Fig. 7.5), the band energy as a function of the momentum. Another common way of representing the electronic structure is to display all momentum vectors for one particular energy, namely the Fermi energy (Fig. 11.31). This way of representing the electronic structure is useful if one wants to compare thermodynamic data, which generally sample the electronic structure

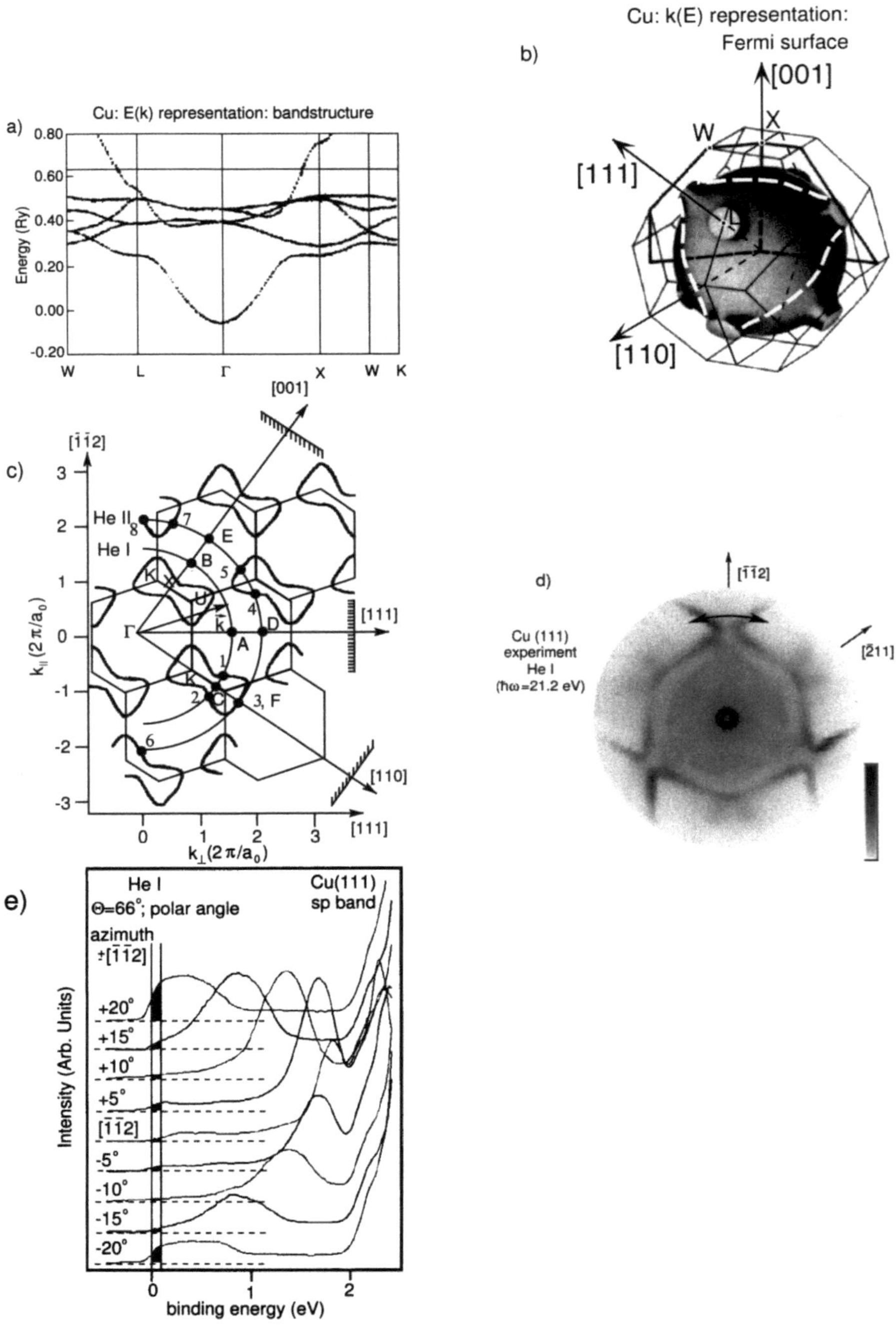

Fig. 11.31a–e. (Caption see opposite page)

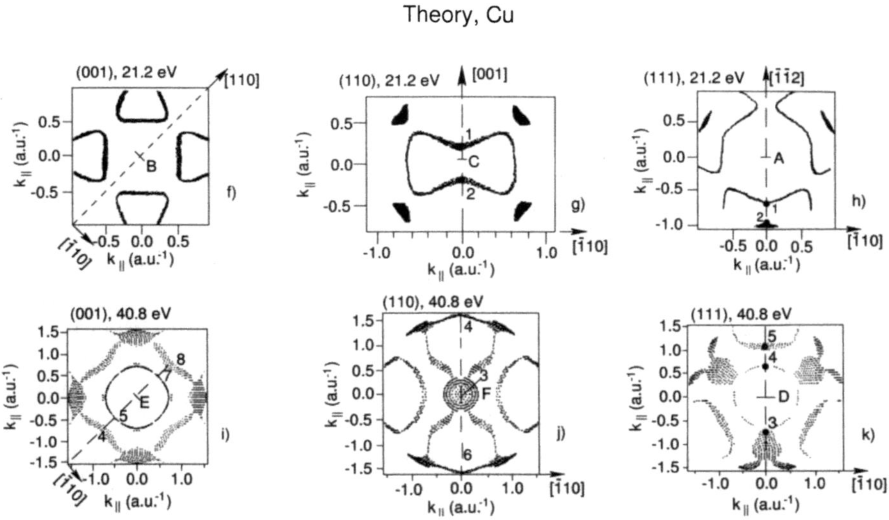

Fig. 11.31a–k. Series of diagrams to indicate how a Fermi surface is determined from a series of 2p photoelectron distributions for the case of Cu metal [11.29]. (**a**) Band structure of Cu for the principle directions. (**b**) Fermi surface of Cu, where the white dashed line shows the cut made when the 2π energy distribution is measured from a (111) surface with He I radiation. (**c**) ΓKLUX plane (plane normal to ($\bar{1}$10) for Cu. The parts of the Fermi surface that intersect this cut are shown along with the free electron parabola for He I and He II radiation and also specific points in $\boldsymbol{k}$-space, which can be used to explain the following experimental results. (**d**) 2π electron distribution from a Cu(111) crystal with He I radiation. (**e**) Azimuthal energy scans from a Cu(111) crystal with a polar angle of 66° relative to the [111] direction and the azimuthal angle measured around the [$\bar{1}\bar{1}$2] direction. (**f**) Calculated Fermi surface scan for Cu(001) and He I radiation. (**g**) Same as (**f**) for Cu(110). (**h**) Same as (**f**) for Cu(111). (**i**) Calculated Fermi surface scan for Cu(001) and He II radiation. (**j**) Same as (**i**) for Cu(110). (**k**) Same as i) for Cu(111)

at 300 K (or 30 meV), around the Fermi energy [the white dashed line indicates the points on the Fermi surface from which electrons are sampled if a (111) surface and He I radiation are used]. One can make some correlation between these two representations of the electronic structure. In going from Γ to L, one does not touch the Fermi surface and therefore no crossing of a band with the Fermi edge is observed in going in this direction. In going from L to W in the band structure (Fig. 11.31), a band crosses the Fermi energy shortly after one has left L. In the Fermi surface representation, this means one goes from the center of one of the necks at L, up to its radius and the radius is met where the band crosses the Fermi edge between L and W. A similar observation is made by going from Γ to X, where in the band structure one hits a Fermi level crossing of a band near X and this, in terms of the Fermi surface, means that one meets a large radius of the Fermi surface. Thus, along ΓX one determines the radius of the large belly, whereas along

LW one determines the radius of the so-called neck. Equally, in going from Γ to K (see Fig. 7.7) one again hits the Fermi surface roughly at the radius of the large belly.

A Fermi surface scan is now performed as shown in Fig. 11.30. An energy window is set around the Fermi energy and then the 2π electron distribution photoemitted above the surface is scanned. We know from Chaps. 6 and 7, that the photoemitted electrons come from regions of $\boldsymbol{k}$-space, where direct transitions can occur. This can be demonstrated for one particular cut, namely the one perpendicular to the $(\bar{1}10)$ direction, which is shown in Fig. 11.31c (ΓKLUX plane in $\boldsymbol{k}$-space). The calculated band structure of Cu for this cut in the extended zone scheme is also shown in Fig. 11.31c, together with the Brillouin zone boundaries. In order to readily identify the points in $\boldsymbol{k}$-space where transitions occur, one assumes (for the moment) that the final states are of free-electron type. Then the circles around Γ with the momentum of the free-electron final states for excitation with He I radiation and He II can be drawn as shown. They touch the (ground state) band structure at various places and can be correlated to the Fermi surface scans as described below.

The experimental Fermi surface scan for the (111) plane with He I radiation is depicted in Fig. 11.31d, and the comparison with the calculation as represented in Fig. 11.31h is good. The dashed line in Fig. 11.31h is the plane depicted in Fig. 11.31c. This is a plane normal to Fig. 11.31d, which represents a (111) plane, while Fig. 11.31c represents a plane containing the [111] direction. A detailed comparison of Figs. 11.31c, 11.31d and 11.31h will be presented below. We show in addition in Fig. 11.31e an azimuthal scan around the $[\bar{1}\bar{1}2]$ direction with a polar angle of 66° with respect to the [111] direction . It can be seen by comparing Fig. 11.31e with Fig. 11.31d, how the intensity variation in Fig. 11.31d is correlated to the photoemission spectra in Fig. 11.31e. The region scanned by the spectra in Fig. 11.31e is indicated by a curved arrow in Fig. 11.31d.

A summary of calculated Fermi surface scans of Cu(001), Cu(111) and Cu(110) for He I radiation is given in Fig. 11.31f–h. The calculated spectra, which agree with experimental ones, are used because they have more contrast than the experimental ones. We will briefly deal with the data for Cu(111), where the results for the other surfaces can be explained in a similar way. Note that the dashed line in Fig. 11.31h is the measuring plane given in 11.31c. This also applies to the other spectra in the panel. For He I radiation and Cu(111), the center of the diagram is designated by A (in 11.31c and 11.31h). In moving around the He I circle towards the $[\bar{1}\bar{1}2]$ direction one encounters no intersection with the Fermi surface, and therefore no intensity occurs in going in this direction in Fig. 11.31h. Moving, however, towards the (110) direction, we find that the free electron parabola cuts the Fermi surface in two places, numbered 1 and 2. These two points are visible in the measurements and more clearly still in the theoretical data.

Figures 11.31i–k represent the same Fermi surface scans, now, however, with He II radiation. They are obviously very different but can again be rationalized with the help of Fig. 11.31c. Again the data for Cu(111) (Fig. 11.31k) will be commented upon. The dashed line gives the intersection of the plane of Fig. 11.31c with the calculated diagrams. For He II, looking at the data along the [111] direction, we see that the center of the diagram is given by D. Moving towards the $[\bar{1}\bar{1}2]$ direction, one intersects the Fermi surface twice at the points labelled 4 and 5. Moving towards [110] one meets a rather broad intersection numbered 3, which is also identified in the actual data and the calculations.

The data for Cu(001) and Cu(110) for He I and He II radiation can be interpreted analogously.

One sees from the above explanation that, for three-dimensional systems, it is far from trivial to correlate measured electron distributions with the Fermi surface. However, by acquiring a large set of data, it is indeed possible to construct the Fermi surface of a material from the PES data.

The situation is simpler in two-dimensional systems because, here, taking one 2π scan above the two-dimensional crystal plane immediately yields the completely two-dimensional Fermi surface of the material. As an example we show data for a classical two-dimensional system, namely $TiTe_2$, for which the layer structure and the atomic structure are shown schematically in Fig. 11.32. It is seen that this material essentially consists of $TiTe_2$ layers, with the Ti layer sandwiched between two Te layers, these "sandwiches" themselves separated by van der Waals gaps from the next such layer. This

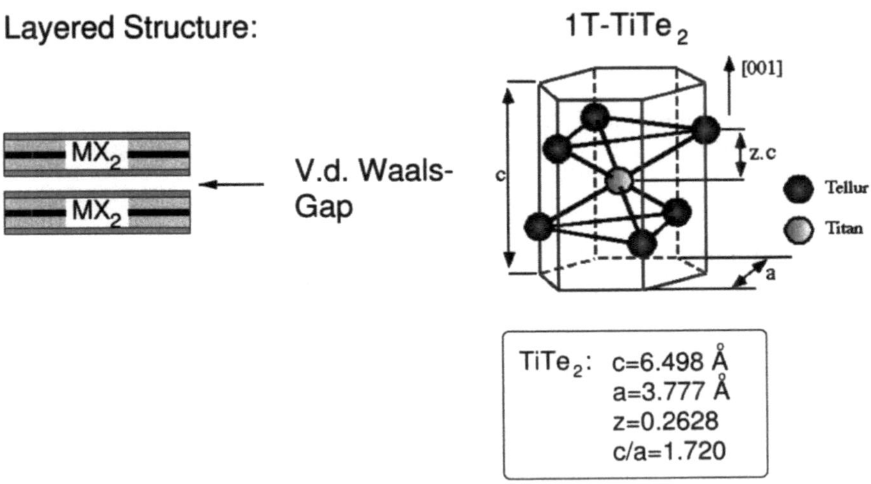

Fig. 11.32. Structural elements of 1T-$TiTe_2$. *Left*: Principle building blocks of the $TiTe_2$ planes, which are separated by van der Waals gaps. *Right*: Unit cell of 1T-$TiTe_2$ with position of the Te and the Ti ions [11.30]

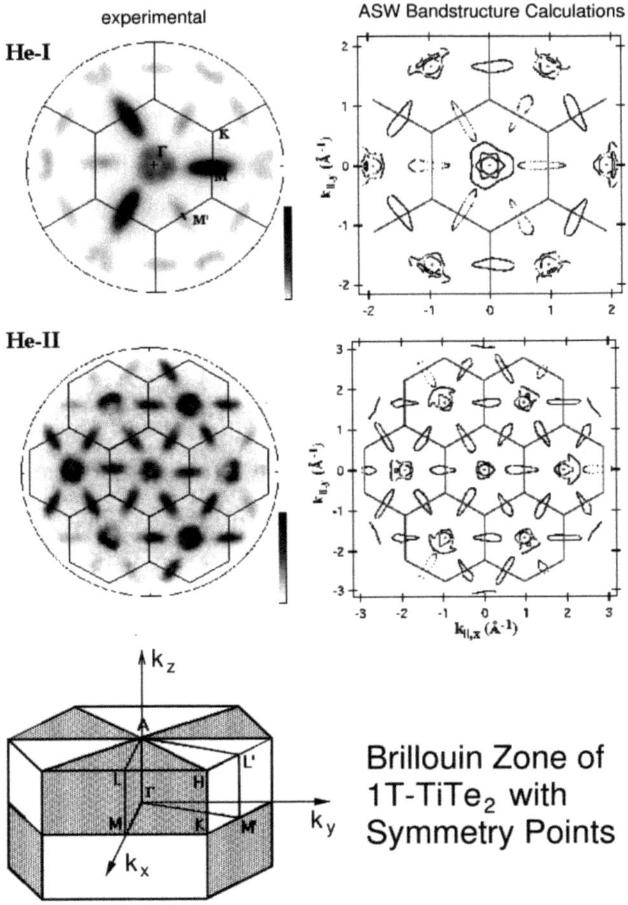

Fig. 11.33. Brillouin zone of 1T-$TiTe_2$ with the symmetry points. Also shown are the He I and He II 2π photoelectron distributions together with the result of an ASW band structure calculation [11.30]

system, and in particular the Ti layer, is a surprisingly good two-dimensional system. Figure 11.33 shows the Brillouin zone, He I and He II Fermi surface scans, and a comparison with the calculated Fermi surface, which shows quite good agreement [11.30].

These data (and those shown in Sect. 5.9.2 for the high temperature superconductors) show that photoemission spectroscopy yields complementary information to "conventional" methods for studying Fermi surfaces.

References

11.1 T.M. Hayers, J.B. Boyce: *Solid State Physics* **37**, 173 (Academic, New York 1982)
D.C. Koningsberger, R. Prins (eds.): *Principles, Applications, Techniques of EXAFS, SEXAFS and XANES* (Wiley, New York 1988)
For a recent comprehensive treatment of NEXAFS spectroscopy see J. Stöhr: *NEXAFS Spectroscopy*, Springer Ser. Surf. Sci., Vol.25 (Springer, Berlin, Heidelberg 1992); NEXAFS means that the EXAFS experiment is performed only near the absorption threshold
J.J. Rehr, J. Mustre de Leon, S.I. Zabinsky, R.C. Albers: J. Am. Chem. Soc. **113**, 5135 (1991)
L. Tröger, T. Yokoyama, D. Arvanitis, T. Lederer, M. Tischer, K. Baberschke: Phys. Rev. B **49**, 888 (1994)
11.2 C.S. Fadley: Progress in Surf. Sci. **16**, 275 (1984)
11.3 J.J. Barton, S.W. Rohey, D.A. Shirley: Phys. Rev. B **34**, 778 (1986)
11.4 C.S. Fadley: Phys. Scr. T **17**, 39 (1987)
11.5 C.S. Fadley: In *Synchrotron Radiation Research: Advances in Surface Science*, ed. by R.C. Bachrach (Plenum, New York 1989)
11.6 W.F. Egelhoff, Jr.: Crit. Rev. Solid State Mat. Sci. **16**, 213 (1990); Solid State Materials Sciences **16**, 213 (1990)
L. Fonda: Phys. Stat. Sol. (b) **182**, 9 (1994)
11.7 S.A. Chambers: In *Advances in Physics* ed. by S. Doniach (Taylor and Francis, London 1991)
11.8 W.L. Schaich: Phys. Rev. B **8**, 4028 (1973)
11.9 B. Lengeler: In *Elektronenspektroskopische Methoden an Festkörpern und Oberflächen*, Vols.**I** and **II** (Kernforschungsanlage, Jülich 1980)
11.10 J. Stöhr, R. Jäger, S. Brennan: Surf. Sci. **117**, 503 (1982)
11.11 D. v.d.Marel, G.A. Sawatzky, R. Zeller, F.U. Hillebrecht, J.C. Fuggle: Solid State Commun. **50**, 47 (1984)
11.12 W. Speier, T.M. Hayes, J.W. Allen, J.B. Boyce, J.C. Fuggle, M. Campagna: Phys. Rev. Lett. **55**, 1693 (1985)
11.13 S. Anderson: Surf. Sci. **15**, 231 (1969)
11.14 D.W. Jepson, P.M. Marcus, F. Jona: Phys. Rev. B **5**, 3933 (1972)
11.15 Y. Jugnet, G. Grenet, N.S. Prakash, Tran Minh Due, H.C. Poon: Phys. Rev. B **38**, 5281 (1988)
11.16 M.L. Xu, J.J. Barton, M.A. van Hove: Phys. Rev. B **39**, 8275 (1989)
11.17 W.F. Egelhoff, Jr.: Phys. Rev. Lett. **59**, 559 (1987)
11.18 J. Osterwalder, T. Greber, S. Hüfner, L. Schlapbach: Phys. Rev. B **41**, 12495 (1990)
11.19 S. Hüfner, J. Osterwalder, T. Greber, L. Schlapbach: Phys. Rev. B **42**, 7350 (1990)
11.20 J. Osterwalder, A. Stuck, Th. Greber, L. Schlapbach, S. Hüfner: Unpublished (Université de Fribourg)
11.21 D.P. Woodruff, A.M. Bradshaw: Rep. Prog. Phys. **57**, 1029 (1994)
M. Zharnikov, M. Weinelt, P. Zelisch, M. Stichler, H.P. Steinrück: Phys. Rev. Lett. **73**, 3548 (1994)
R. Davis, D.P. Woodruff, O. Schaff, V. Fernandez, K.M. Schindler, Ph. Hofmann, K.U. Weiss, R. Dippel, V. Fritzsche, A.M. Bradshaw: Phys. Rev. Lett. **74**, 1621 (1995)
Ph. Hofmann, K.M. Schindler: Phys. Rev. B **47**, 13941 (1993)
All these publications deal with the structure determination of adsorbates by photoelectron diffraction

O. Schaff, A.M. Bradshaw: Phys. Bl. **52**, 997 (1996)
J.J. Barton: Phys. Rev. Lett. **61**, 1356 (1988)
V .Fritsche, P.D. Woodruff: Phys. Rev. B **46**, 16128 (1992)
P.M. Leu, C.S. Fadley, G. Materlik: In *X-Ray and Inner-Shell Processes: 17th Int. Conf.*, R.L. Johnson, H. Schmidt-Boekering, B.F. Sonntag (eds.), AIP Conf. Proc. No. 389, p. 295 (1997)
11.22 R.S. Saiki, G.S. Herman, M. Jamada, J. Osterwalder, C.S. Fadley: Phys. Rev. Lett. **63**, 283 (1989)
11.23 K.U. Weiss, R. Dippel, K.M. Schindler, P. Gardner, V. Fritzsche, A.M. Bradshaw, A.L.D. Kilcoyne, D.P. Woodruff: Phys. Rev. Lett. **69**, 3196 (1992)
11.24 A. Locatelli, B. Brena, S. Lizzit, G. Comelli, G. Cantero, G. Paolucci, R. Rosei: Phys. Rev. Lett. **73**, 90 (1994)
E.L. Bullock, R. Gunnella, L. Pathey, T. Abukawa, S. Kono, C.R. Natoli, L.S.O. Johansson: Phys. Rev. Lett. **74**, 2756 (1995)
11.25 M. Zkarnikov, D. Mehl, M. Weinel, D. Zebisch, H.P. Steinrück: Surf. Sci. **312**, 82 (1994). Reporting on the holographic reconstruction of the Pt(110) surface by using multiple wave-number photoelectron diffraction patterns
11.26 A. Santoni, L.J. Terminello, F.J. Himpsel, T. Takahashi: Applied Physics A **52**, 299 (1991)
11.27 J. Osterwalder, A. Stuck, T. Greber, P. Aebi, L. Schlapbach, S. Hüfner: *Proc. 10th VUV Conf.*, ed. by F.J. Wullenmier, Y. Petroff, N. Nenner (World Scientific, Singapore 1993) p.475
P. Aebi, J. Osterwalder, R. Fasel, D. Naumovic, L. Schlapbach: Surf. Sci. **307–309**, 917 (1994); the Fermi surface of $Bi_2Sr_2Ca_1Cu_2O_8$ has been determined by this method by P. Aebi, J. Osterwalder, P. Schwaller, L. Schlapbach, M. Shimoda, T. Mochiku, K. Kadowaki: Phys. Rev. Lett. **72**, 2757 (1994)
11.28 T.J. Kreutz, P. Aebi, J. Osterwalder: Solid State Commun. **96**, 339 (1995). J. Osterwalder, P. Aebi, D. Schwaller, L. Schlapbach, M. Shimoda, T. Mochiku, K. Kadowaki: Appl. Phys. A **60**, 247 (1995). Giving examples for the determination of Fermi surfaces with the photoelectron-diffraction technique
11.29 P. Aebi, R. Fasel, D. Nanmovic, J. Hayoz, Th. Pillo, M. Bovet, R.G. Agostino, L. Patthey, L. Schlapbach, F.P. Gil, H. Berger, T.J. Krentz, J. Osterwalder: Surface Science **402–404**, 614 (1998)
11.30 Th. Straub, Thesis, Saarbrücken (1998)
11.31 S. Omori, Y. Nihei, E. Rotenberg, J.D. Denlinger, S. Marchesini, S.D. Kevan, B.P. Tonner, M.A. Van Hove, C.S. Fadley: Phys. Rev. Lett. **88**, 055504 (2002)
11.32 S. Dreiner, M. Schürmann, C. Westphal, H. Zacharias: Phys. Rev. Lett. **86**, 4068 (2001)

Appendix

In order to make this book as self-consistent as possible, it was decided to have some often used information in the index. The obvious information needed most frequently in dealing with photoelectron spectroscopy are the binding energies because this is the property that photoemission measures. Therefore a table of binding energies was compiled from various sources. As was mentioned before the notorious question with the binding energies is that of referencing. The natural (obvious) referencing is with respect to the vacuum level because in essence this is the natural zero of energy, meaning that an electron at the vacuum level has zero energy. If one wants to reference binding energies of solids with respect to the vacuum level, a problem occurs. In the instrument binding energies are always measured with respect to the Fermi energy, which is a very easily accessible zero and with some care can be determined with an accuracy of $\pm 0.5\,\text{eV}$ or even better. In order to obtain the binding energy with respect to the vacuum level the work function has to be added. Work functions are usually known with an accuracy of $\pm 0.1\,\text{eV}$. In addition work functions can depend on the crystal surface from which they are determined, as shown in the Appendix A3, where some work functions are given, and the two extremal numbers for silver are Ag(111), $\phi = 4.74\,\text{eV}$ and Ag (polycrystalline), $\phi = 4.26\,\text{eV}$. This shows that referencing binding energies with respect to the vacuum level produces some uncertainty.

Therefore it is common practice and very convenient for the experiments to reference binding energies in solids with respect to the Fermi energy. Core line positions can usually be obtained with an accuracy of about 0.1,eV, and generally the Fermi energy can be determined with an accuracy of less than that, namely $\pm 0.05\,\text{eV}$. This would make it possible to determine binding energies with an accuracy of about $\pm 0.2\,\text{eV}$. However, in looking at data from different groups it is obvious that the errors are larger, and the reason for these discrepancies are not always understood. Therefore it is save to assume that the accuracy of binding energies is usually of the order of $\pm 1\,\text{eV}$.

Another piece of information which carries, however, some personal judgement are the Brillouin zones of the fcc and the bcc cubic crystal structure. One of important areas of research to which photoelectron spectroscopy has contributed considerably to our knowledge is the determination of electronic band structures. For this kind of work the Brillouin zones are indispensable. Of course it is not possible to show here all the Brillouin zones or, at least, the most important ones. But, as in many other cases, it is important to understand the principles of a field, and the principles of band-structure determination have been established by investigating systems with the fcc or bcc structure. Therefore in understanding the principles of band-structure determination it was thought that the availability of the fcc and bcc Brillouin structures (and surface Brillouin structures) would be quite helpful.

Finally a number of work functions is given here because they are important if one wants to reference binding energies with respect to the vacuum level. This is also done to demonstrate that a work function is not a unique number for a particular material but depends on the crystal phase of the material. In a material where also the polycrystalline material has been measured the work function of this polycrystalline state is not the average of the work function determined for the various crystal faces, as is seen already in the case of silver.

A.1 Table of Binding Energies

The table of binding energies has been compiled by using published tabulated energies [A.1-5] as a starting point for searching additional data in the literature. In choosing numbers and averaging them (at times) a high degree of personal judgement had been invoked. It is our impression that researchers tend to overestimate the accuracy of their data. Generally speaking, the numbers given in this table should be accurate to about $\pm 1\,\mathrm{eV}$, in the valence-energy region ($E_B \leq 30\,\mathrm{eV}$) the variations are larger, of course. Good calibration points are the Au $4f_{7/2}$ line at $84.00\,\mathrm{eV}$ (error about $\pm 0.15\,\mathrm{eV}$), the carbon 1s line (of carbon contaminants in the spectrometer) at $284.0\,\mathrm{eV}$ (error $\pm 0.25\,\mathrm{eV}$) and, to a lesser degree, the O 1s line as a contaminant on metallic samples with $531.9\,\mathrm{eV}$ (error $\pm 1.0\,\mathrm{eV}$).

	1s	2s	$2p_{1/2}$	$2p_{3/2}$	3s	$3p_{1/2}$	$3p_{3/2}$	3d	4s
^{1}H gas	16.0								
^{2}He gas	24.6								
^{3}Li	54.7	VB							
^{4}Be	111.5	VB							
^{5}B	188.0	VB	VB						
^{6}C	284.7 (graphite) 283.5 (diamond)	VB	VB						
^{7}N gas	409.9	37.3	VB						
^{7}N solid	399.0	12.0	4.0						
^{8}O gas	543.1	41.6	VB						
^{8}O solid	531.0	22.0	6.0						
^{9}F gas	696.7		VB						
^{9}F solid	686.0	31.0	11.0						
^{10}Ne	870.2	48.4	21.7	21.6					
^{11}Na	1070.8	63.7	30.7	30.5	VB				
^{12}Mg	1303.0	88.6	49.6	49.2	VB				
^{13}Al	1559.0	118.1	72.9	72.5	VB	VB			
^{14}Si		149.8	99.8	99.2	VB	VB			
^{15}P		189.0	136.0	135.0	VE	VB			
^{16}S		230.9	163.6	162.5	16.0	VB			
^{17}Cl solid		270.0	202.0	200.0	18.0	VB			
^{18}Ar		326.3	250.6	248.5	29.3	15.94	15.76		
^{19}K		378.6	297.3	294.6	34.8	18.3			
^{20}Ca		438.7	350.0	346.4	44.2	25.5	25.1		

	2s	$2p_{1/2}$	$2p_{3/2}$	3s	$3p_{1/2}$	$3p_{3/2}$	$3d_{3/2}$	$3d_{5/2}$	4s
^{21}Sc	498.0	403.6	398.7	51.1	28.3		VB		VB
^{22}Ti	560.9	461.2	453.7	58.6	32.6		VB		VB
^{23}V	626.7	519.8	512.1	66.3	37.2		VB		VB
^{24}Cr	695.7	583.8	574.1	74.1	42.1		VB		VB
^{25}Mn	769.0	650.0	638.7	82.3	47.2		VB		VB
^{26}Fe	844.6	719.9	706.8	91.3	52.7		VB		VB
^{27}Co	925.1	793.4	778.4	101.0	59.0		VB		VB
^{28}Ni	1008.6	870.5	853.0	111.0	67.5	66.0	VB		VB
^{29}Cu	1096.7	952.5	932.5	122.5	77.2	75.2	VB		VB
^{30}Zn	1200.7	1044.9	1021.6	139.7	91.4	66.5	10.1		VB

	2s	$2p_{1/2}$	$2p_{3/2}$	3s	$3p_{1/2}$	$3p_{3/2}$	$3d_{3/2}$	$3d_{5/2}$	4s	4p
^{31}Ga	1299.0	1143.4	1116.6	159.4	106.5	103.5	18.7	18.3	VB	VB
^{32}Ge	1414.6	1248.1	1217.0	180.1	124.9	120.8	29.9	29.2	VB	VB
^{33}As		1359.1	1323.6	204.7	146.2	141.2	41.7		VB	VB
^{34}Se		1474.3	1433.9	229.6	166.5	160.7	55.5	54.6	VB	VB
^{35}Br				257.0	189.0	182.0	70.0	69.0	VB	VB
^{36}Kr				292.8	222.2	214.4	95.0	93.8	27.5	14.0

	3s	$3p_{1/2}$	$3p_{3/2}$	$3d_{3/2}$	$3d_{5/2}$	4s	$4p_{1/2}$	$4p_{3/2}$	5s	4d
^{37}Rb	326.7	248.7	239.1	113.0	112.0	30.5	16.1	15.2	VB	
^{38}Sr	358.4	279.7	269.8	135.9	134.1	38.8	21.4	20.2	VB	
^{39}Y	392.0	310.6	298.8	157.7	155.8	50.6	28.5	27.7	VB	VB
^{40}Zr	430.3	343.4	329.7	181.2	178.8	50.6	28.5	27.7	VB	VB

	3s	$3p_{1/2}$	$3p_{3/2}$	$3d_{3/2}$	$3d_{5/2}$	4s	$4p_{1/2}$	$4p_{3/2}$	$4d_{3/2}$	$4d_{5/2}$	5s	5p
^{41}Nb	466.6	376.1	360.6	205.5	202.3	56.4	32.6	30.8	VB	VB	VB	
^{42}Mo	506.3	410.6	394.0	231.1	227.9	63.2	37.6	35.5	VB	VB	VB	
^{43}Tc	544.0	447.6	417.6	257.3	253.5	69.6	42.3	39.9	VB	VB	VB	
^{44}Ru	586.1	483.5.1	461.3	284.1	280.0	75.0	46.5	43.2	VB	VB	VB	
^{45}Rh	628.1	521.3	496.5	311.9	307.2	81.4	50.65	47.3	VB	VB	VB	
^{46}Pd	671.7	560.0	532.3	340.5	335.2	87.6	55.7	50.9	VB	VB	VB	
^{47}Ag	719.1	603.8	573.0	374.0	368.1	97.0	63.7	58.3	VB	VB	VB	
^{48}Cd	771.7	652.3	618.2	411.8	405.1	109.4	64.0		11.5	10.6	VB	
^{49}In	827.1	703.1	665.2	451.4	443.9	122.8	73.5		17.4	16.6	VB	VB
^{50}Sn	884.6	756.5	714.6	493.1	484.9	137.1	84.0		24.8	23.8	VB	VB

	3s	$3p_{1/2}$	$3p_{3/2}$	$3d_{3/2}$	$3d_{5/2}$	4s	$4p_{1/2}$	$4p_{3/2}$	$4d_{3/2}$	$4d_{5/2}$	5s	$5p_{1/2}$	$5p_{3/2}$	$4f_{5/2}$	$4f_{7/2}$	6s	5d
^{51}Sb	946.0	812.6	766.3	537.7	528.2	153.0	95.6		33.4	32.1	VB	VB					
^{52}Te	1006.0	870.1	820.8	582.8	572.7	169.6	103.3		41.8	40.3	VB	VB					
^{53}I	1072.0	931.0	875.0	631.0	620.0	186.0	123.0		50.0		VB	VB					
^{54}Xe	1148.7	1002.1	940.6	689.0	676.4	213.2	146.7	145.5	69.5	67.5	23.3	13.4	12.1				
^{55}Cs	1211.0	1071.0	1003.0	740.5	726.6	232.3	172.4	161.3	79.8	77.5	22.7	14.2	12.1			VB	
^{56}Ba	1293.0	1137.0	1063.0	795.7	780.5	253.0	192.0	178.7	92.6	90.0	30.1	16.8	14.6			VB	
^{57}La	1362.0	1209.0	1128.0	853.0	836.0	274.7	210.0	196.0	105.3	102.5	34.3	19.3	16.8			VB	VB
^{58}Ce	1436.0	1274.0	1187.0	902.7	884.2	291.0	222.0	206.5	109.0		37.8	19.8	17.0	2.0(0.1)		VB	VB
^{59}Pr				948.3	928.8	305.0	237.0	218.0	115.1		37.4	22.3	22.3	3.5(1.5)		VB	VB
^{60}Nd				1003.3	980.4	319.2	244.0	225.0	120.5		37.5	21.1	21.1	4.5		VB	VB
^{61}Pm		1472.0	1357.0	1052.0	1027.0	331.0	255.0	237.0	121.0							VB	VB
^{62}Sm				1110.9	1083.4	347.2	267.0	249.0	129.0		37.4	21.3	21.3	6.0		VB	VB

	$3d_{3/2}$	$3d_{5/2}$	4s	$4p_{1/2}$	$4p_{3/2}$	$4d_{3/2}$	$4d_{5/2}$	5s	$5p_{1/2}$	$5p_{3/2}$	$4f_{5/2}$	$4f_{7/2}$	6s	5d
^{63}Eu	1158.6	1127.5	360.0	284.0	257.0		127.7	32.0	22.0	22.0	2.0		VB	VB
^{64}Gd	1221.9	1189.6	378.6	289.0	271.0		142.6	43.5	20.0	20.0	8.6		VB	VB
^{65}Tb	1278.8	1243.2	369.0	322.4	284.1	150.5		45.6	28.7	22.6	2.0		VB	VB
^{66}Dy		1292.6	414.2	333.5	293.2	153.6		49.9	29.5	23.1	4.0		VB	VB
^{67}Ho			432.4	345.0	308.2	160.0		49.3	30.8	24.1	5.2		VB	VB
^{68}Er			449.8	366.0	320.2	167.6		50.6	31.4	24.7	4.7		VB	VB
^{69}Tm			470.9	381.0	332.6	175.5		54.7	31.8	25.0	4.6		VB	VB
^{70}Yb			480.0	388.4	339.5	190.8	182.0	51.3	30.0	23.7	2.5	1.3	VB	VB

	4s	$4p_{1/2}$	$4p_{3/2}$	$4d_{3/2}$	$4d_{5/2}$	5s	$5p_{1/2}$	$5p_{3/2}$	$4f_{5/2}$	$4f_{7/2}$	$5d_{3/2}$	$5d_{5/2}$	6s
^{71}Lu	506.8	412.4	359.2	206.0	196.2	57.3	33.7	26.8	8.8	7.3	VB	VB	VB
^{72}Hf	538.0	438.2	380.7	220.0	211.5	64.2	38.0	29.9	15.9	14.2	VB	VB	VB
^{73}Ta	563.4	463.4	400.9	237.9	226.4	69.7	42.2	32.7	23.5	21.6	VB	VB	VB
^{74}W	594.1	490.4	432.7	256.0	243.4	75.6	45.3	36.8	33.6	31.5	VB	VB	VB
^{75}Re	625.4	518.7	446.8	273.9	260.5	82.8	45.6	34.6	42.9	40.5	VB	VB	VB
^{76}Os	658.2	549.1	470.7	293.1	278.5	88.0	58.0	44.5	52.4	50.7	VB	VB	VB
^{77}Ir	691.1	577.8	495.8	311.9	296.3	95.2	63.0	48.0	63.8	60.8	VB	VB	VB
^{78}Pt	725.4	609.1	519.3	331.6	314.6	101.7	65.3	51.7	74.7	71.1	VB	VB	VB
^{79}Au	762.1	642.7	546.3	353.3	335.1	107.2	74.2	57.2	87.7	84.0	VB	VB	VB
^{80}Hg	802.2	680.2	576.6	378.2	358.8	127.0	83.1	64.5	104.0	99.9	VB	VB	VB

	4s	$4p_{1/2}$	$4p_{3/2}$	$4d_{3/2}$	$4d_{5/2}$	$4f_{5/2}$	$4f_{7/2}$	5s	$5p_{1/2}$	$5p_{3/2}$	$5d_{3/2}$	$5d_{5/2}$	6s	$6p_{1/2}$	$6p_{3/2}$	7s	6d	5f
^{81}Tl	846.2	720.5	609.5	405.7	385.0	122.2	117.8	136.0	94.6	73.4	14.7	12.5	VB	VB				
^{82}Pb	891.8	761.9	643.5	434.3	412.2	141.7	136.9	147.0	106.6	83.3	20.2	17.6	VB	VB				
^{83}Bi	939.0	805.2	678.8	464.0	440.1	162.3	157.0	159.3	119.1	93.6	26.9	23.8	VB	VB				
^{84}Po	995.0	851.0	705.0	500.0	473.0	184.0		177.0	132.0	104.0	31.0		VB	VB				
^{85}At	1042.0	886.0	740.0	533.0	507.0	210.0		195.0	148.0	115.0	40.0		VB	VB				
^{86}Rn	1097.0	929.0	768.0	567.0	541.0	238.0		214.0	164.0	127.0	48.0		26.0	11.0				
^{87}Fr	1153.0	980.0	810.0	603.0	577.0	268.0		234.0	182.0	140.0	58.0		34.0	15.0		VB		
^{88}Ra	1208.0	1058.0	879.0	636.0	603.0	299.0		254.0	200.0	153.0	68.0		44.0	19.0		VB		
^{89}Ac	1269.0	1080.0	890.0	675.0	639.0	319.0		272.0	215.0	167.0	80.0					VB	VB	
^{90}Th	1330.0	1168.0	966.4	712.1	675.2	342.4	333.1	290.0	229.0	182.0	92.5	85.4	41.4	24.5	16.6	VB	VB	
^{91}Pa	1387.0	1224.0	1007.0	743.0	708.0	371.0	360.0	310.0	232.0		94.0					VB	VB	VB
^{92}U	1439.0	1271.0	1043.0	778.3	736.2	388.2	377.3	321.0	257.0	192.0	102.7	93.7	44.5	26.4	16.9	VB	VB	VB

A.2 Surface and Bulk Brillouin Zones of the Three Low-Index Faces of a Face-Centered Cubic (fcc) Crystal Face

In order to make the material [A.6,7] more comprehensible the information is given in a detailed way; the bulk lattice and its Brillouin zone are presented. For each of the three principal planes the positions of the planes in the crystal structure, the surface structure, the surface Brillouin zone and the combination of the bulk and the surface Brillouin zones are given.

FCC LATTICE WITH BASIS VECTORS

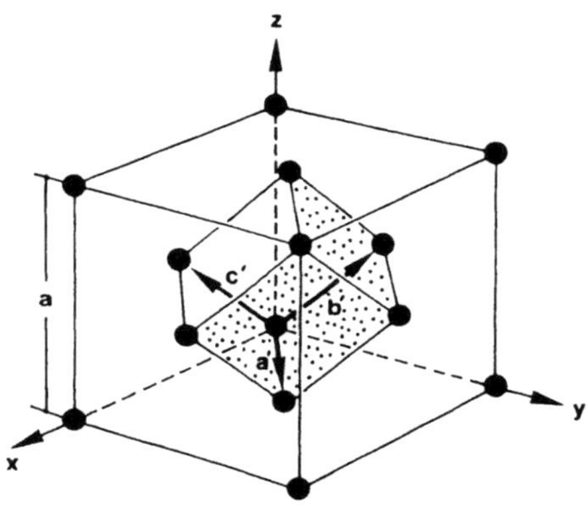

RECIPROCAL LATTICE OF THE FCC LATTICE
(IT IS AN BCC LATTICE) WITH THE
CORRESPENDING BRILLOUIN ZONES

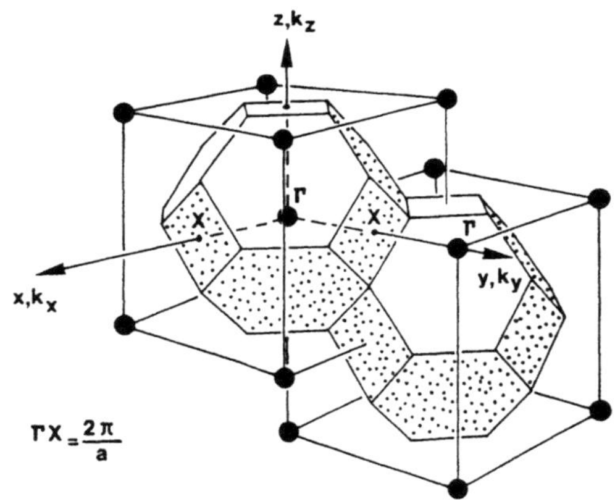

BRILLOUIN ZONE OF THE FCC LATTICE

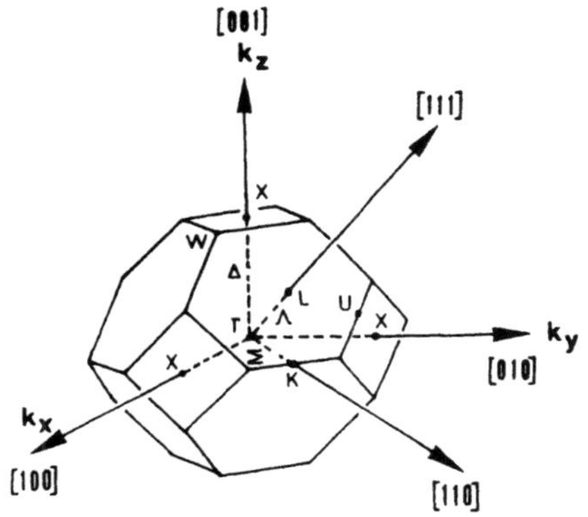

FCC LATTICE : (100) FACE

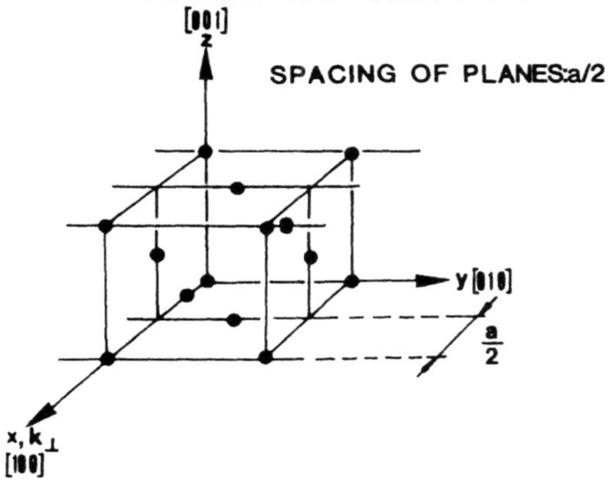

UNIT CELL OF THE (100) FACE

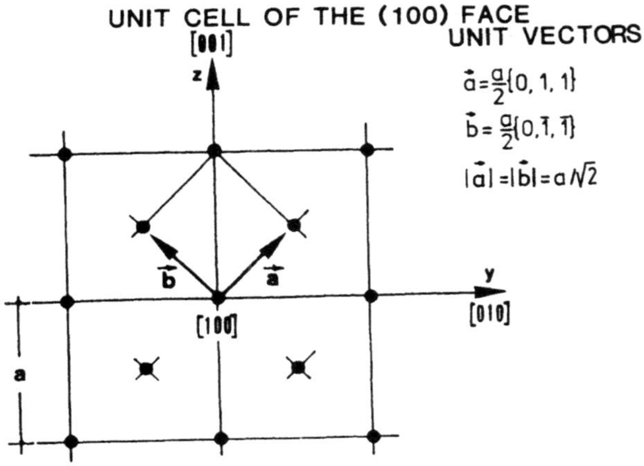

FCC (100) SURFACE : RECIPROCAL LATTICE WITH VECTORS

$A=\frac{2\pi}{a}\{0\ 1\ 1\}$ and $B=\frac{2\pi}{a}\{0\ \bar{1}\ 1\}$; $|A|=|B|\equiv\frac{2\pi}{a}\sqrt{2}$

— · — · — · — FIRST SURFACE BRILLOUIN ZONE

– – – – – – INTERSECTION OF THE FIRST BULK BRILLOUIN ZONE WITH THE (100) PLANE

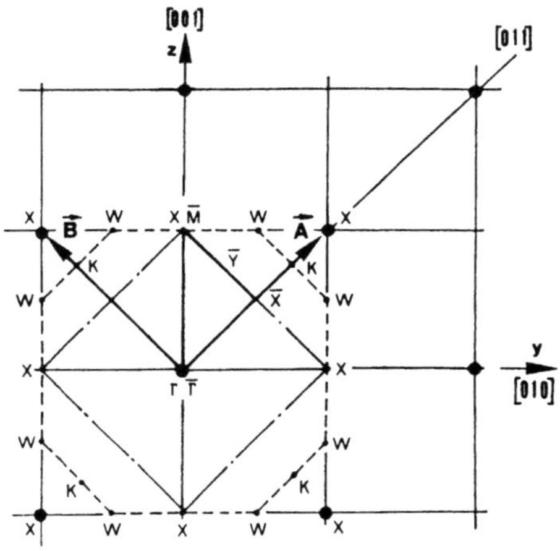

FCC LATTICE : BULK BRILLOUIN ZONE

AND (100) SURFACE BRILLOUIN ZONE

SURFACE NORMAL

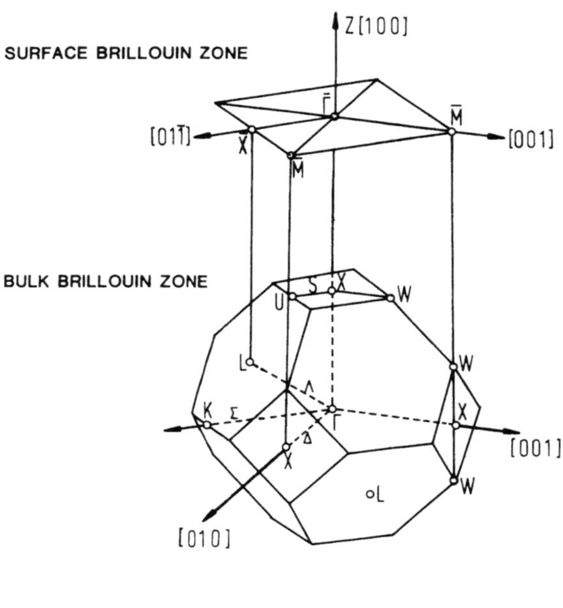

FCC LATTICE : (110) FACE

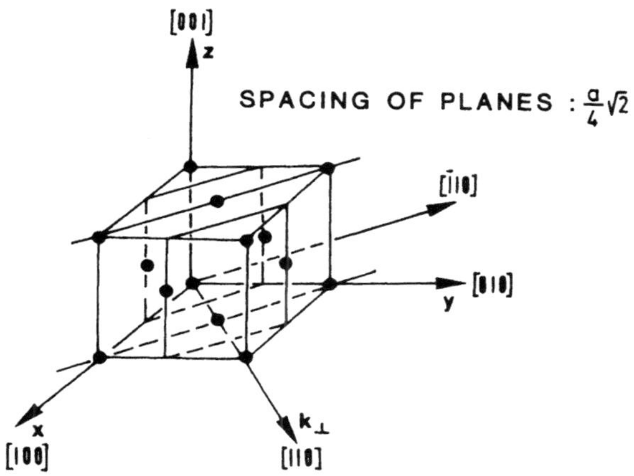

UNIT CELL OF THE (110)FACE

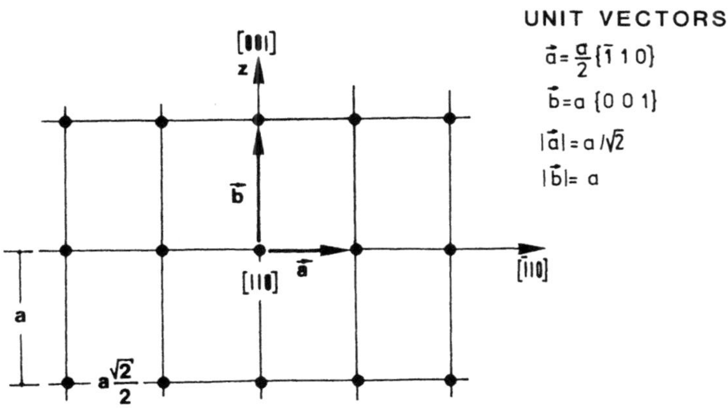

FCC (110) SURFACE: RECIPROCAL LATTICE WITH VECTORS

$\vec{A} = \frac{2\pi}{a}\{\bar{1},1,0\}$ and $\vec{B} = \frac{2\pi}{a}\{0,0,1\}$; $|\vec{A}| = \frac{2\pi}{a}\sqrt{2}$ $|\vec{B}| = \frac{2\pi}{a}$

— · — · — · — FIRST SURFACE BRILLOUIN ZONE

— — — — — — INTERSECTION OF THE FIRST BULK BRILLOUIN ZONE WITH THE (110) PLANE

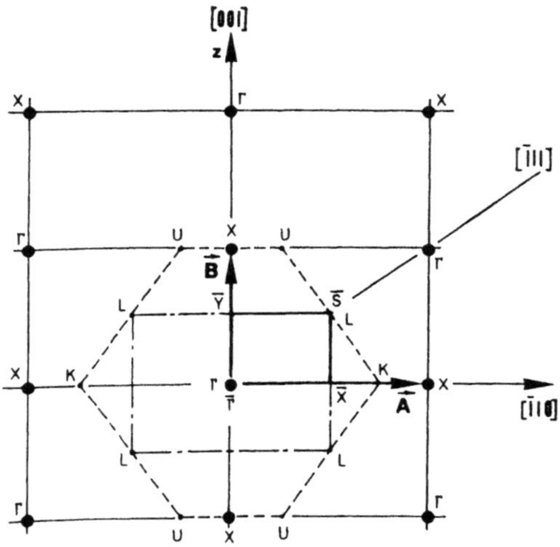

FCC LATTICE : BULK BRILLOUIN ZONE

AND (110) SURFACE BRILLOUN ZONE

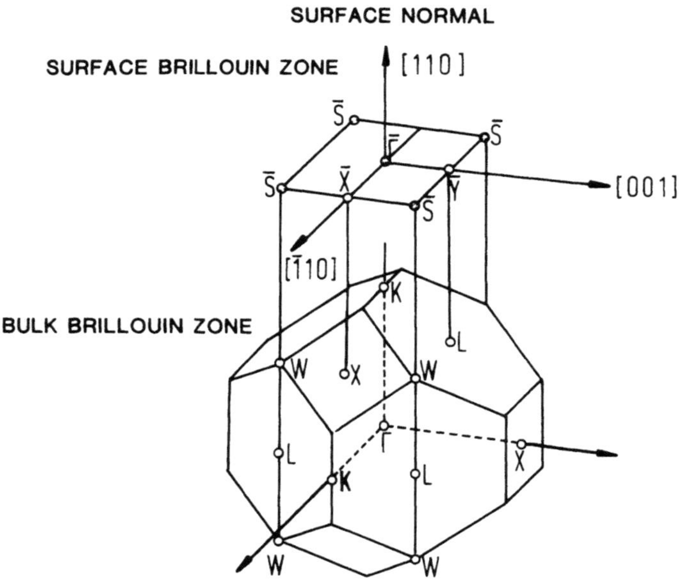

FCC LATTICE (111) PLANE

SPACING OF PLANES : $a \cdot \frac{\sqrt{3}}{3}$

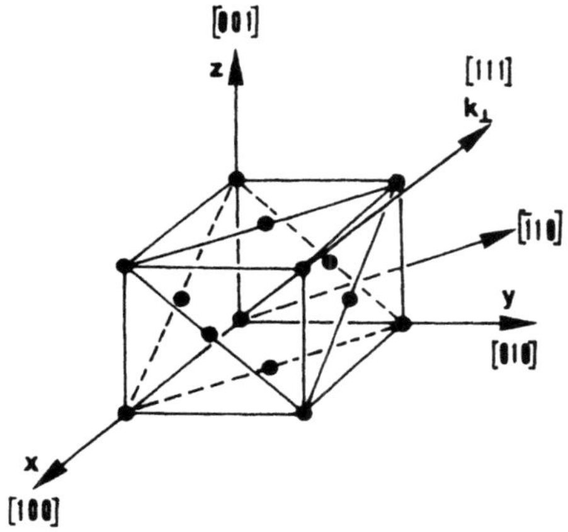

UNIT CELL OF THE (111) FACE

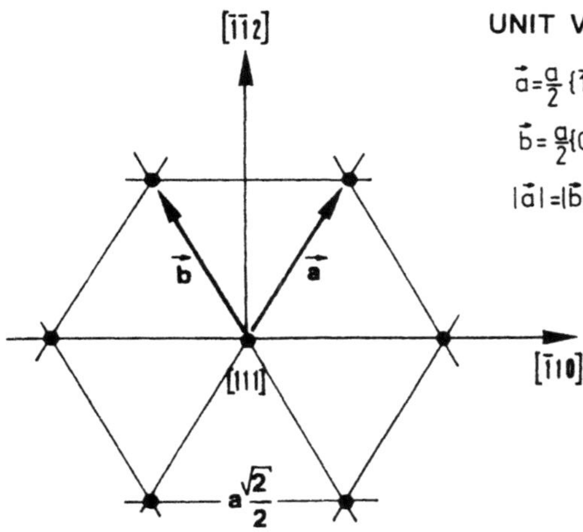

UNIT VECTORS

$\vec{a} = \frac{a}{2}\{\bar{1}\ 0\ 1\}$

$\vec{b} = \frac{a}{2}\{0\ \bar{1}\ 1\}$

$|\vec{a}| = |\vec{b}| = \frac{a}{\sqrt{2}}$

FCC (111) SURFACE : RECIPROCAL LATTICE WITH VECTORS

$$\vec{A} = \frac{2\pi}{a} \times \frac{2}{3} \{\bar{2}, 1, 1\} \text{ and } \vec{B} = \frac{2\pi}{a} \times \frac{2}{3} \{1, \bar{2}, 1\} ; |\vec{A}| = |\vec{B}| = \frac{2\pi}{a} \sqrt{\frac{8}{3}}$$

—·—·—·— FIRST SURFACE BRILLOUIN ZONE

— — — — — INTERSECTION OF THE FIRST BULK BRILLOUIN ZONE WITH THE (111) PLANE

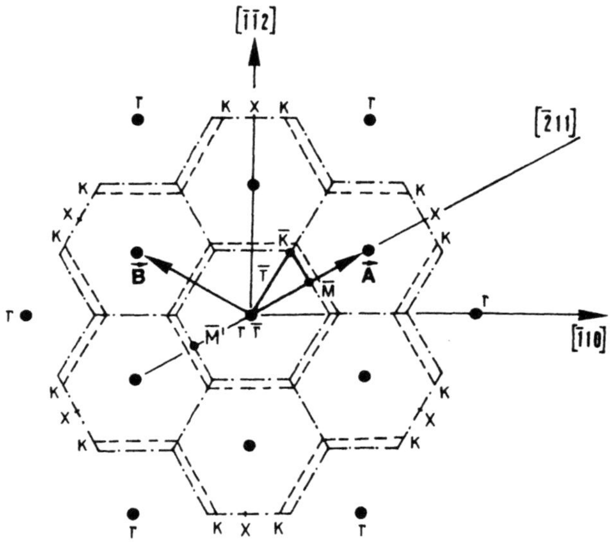

FCC LATTICE : BULK BRILLOUIN ZONE

AND (111) SURFACE BRILLOUIN ZONE

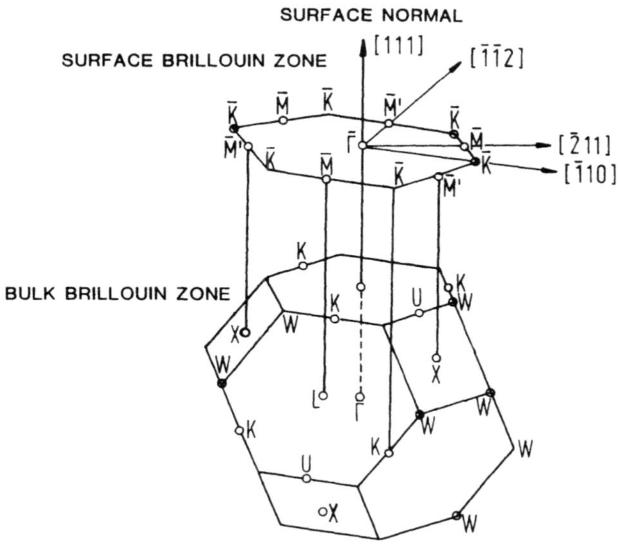

A.3 Compilation of Work Functions

From the data given in [A.8] a realistic general error estimate seems to be ±0.3 eV.

Element	Work function (eV)
Ag(poly)	4.26
Ag(100)	4.64
Ag(110)	4.52
Ag(111)	4.74
Al(poly)	4.28
Al(100)	4.41
Al(110)	4.28
Al(111)	4.23
Au(poly)	5.1
Au(100)	5.47
Au(110)	5.37
Au(111)	5.31
Ba(poly)	2.52
Be(poly)	4.98
Bi(poly)	4.34
Ca(poly)	2.87
Cd(poly)	4.08
Ce(poly)	2.9
Co(poly)	5.0
Cr(poly)	4.5
Cs(poly)	1.95
Cu(poly)	4.65
Cu(100)	4.59
Cu(110)	4.48
Cu(111)	4.94
Er(poly)	2.97
Eu(poly)	2.5
Fe(poly)	4.5
Fe(100)	4.67
Fe(111)	4.81
Ga(poly) at $T = 0°$ C	4.32
Gd (poly)	3.1
Hf(poly)	3.9
Hg (poly)$T = -100°$ C	4.43
In(poly)	4.09
Ir(110)	5.42
Ir(111)	5.76
K(poly)	2.28
La(poly)	3.5
Li(poly)	2.93

Element	Work function (eV)
Mg(poly)	3.66
Mo(poly)	4.6
Mo(100)	4.53
Mo(110)	4.95
Mo(111)	4.55
Mn(poly)	4.1 ±0.2
Na(poly)	2.36
Nb(poly)	4.3
Nb(100)	4.53
Nb(001)	4.02
Nb(110)	4.87
Nb(111)	4.36
Nd(poly)	3.2
Ni(poly)	5.15
Ni(100)	4.89
Ni(110)	4.64
Ni(111)	5.22
Os(poly)	5.93
Pb(poly)	4.25
Pd(poly)	5.22
Pt(poly)	5.5
Pt(100)	5.84
Pt(111)	5.93
Rb(poly)	2.16
Re(poly)	4.72
Ru(poly)	4..71
Sb(amorph)	4.55
Sb(100)	4.7
Sc(poly)	3.5
Sm(poly)	2.7
Sn(poly)	4.42
Ta(100)	4.15
Ta(110)	4.8
Ta(111)	4.0
Tc(poly)	4.88
Ti(poly)	4.33
Nb(001)	4.02
V(poly)	4.3
W(poly)	4.6
W(100)	4.63
W(110)	5.25
W(111)	4.47
Y(poly)	3.1
Yb(poly)	2.6
Zn(poly)	3.63
Zr(poly)	4.05

References

A.1 *Photoemission in Solids I, ed. by M. Cardona, L. Ley, Topics Appl. Phys.*, Vol. 26 (Springer, Berlin, Heidelberg 1978)

A.2 *Photoemission in Solids II, ed. by M. Cardona, L. Ley, Topics Appl.* Phys., Vol. 27 (Springer, Berlin, Heidelberg 1979)

A.3 J.C. Fuggle, N. Mårtensson: J. Electr. Spectrosc. Relat. Phenom. **21**, 275 (1980)

A.4 G.P. Williams: X-ray Data Booklet: Lawrence Berkeley Laboratory, Berkeley, California (1986)

A.5 Unoccupied Electronic States, ed. by J.C. Fuggle, J.E. Inglesfield, Topics Appl. Phys., Vol. 69 (Springer, Berlin, Heidelberg 1992)

A.6 E.W. Plummer, W. Eberhardt: Adv. Chem. Phys. **49**, 533 (1982)

A.7 P. Thiry: La photoémission angulaire dans les solides. Thêse d'État, Université Paris VI (1979)

A.8 J. Hölzl, F.K. Schulte, H. Wagner: Work function of metals, in *Solid Surface Physics*, Springer Tracts Mod. Phys., Vol. 85 (Springer, Berlin, Heidelberg 1979)

Index

MIX
Papier aus verantwortungsvollen Quellen
Paper from responsible sources
FSC® C105338

FSC
www.fsc.org

If you have any concerns about our products,
you can contact us on
ProductSafety@springernature.com

In case Publisher is established outside the EU,
the EU authorized representative is:
Springer Nature Customer Service Center GmbH
Europaplatz 3, 69115 Heidelberg, Germany

Printed by Libri Plureos GmbH
in Hamburg, Germany